Modern
Lens Design

Modern Lens Design

Warren J. Smith
Chief Scientist
Kaiser Electro-Optics, Inc., Carlsbad, California
and Consultant in Optics and Design

Second Edition

McGraw-Hill

New York   Chicago   San Francisco   Lisbon   London   Madrid
Mexico City   Milan   New Delhi   San Juan   Seoul
Singapore   Sydney   Toronto

The **McGraw·Hill** Companies

Cataloging-in-Publication Data is on file with the Library of Congress.

Copyright © 2005, 1992 by The McGraw-Hill Companies, Inc. All rights reserved. Printed in the United States of America. Except as permitted under the United States Copyright Act of 1976, no part of this publication may be reproduced or distributed in any form or by any means, or stored in a data base or retrieval system, without the prior written permission of the publisher.

1 2 3 4 5 6 7 8 9 0 DOC/DOC 0 1 0 9 8 7 6 5 4

ISBN 0-07-143830-0

The sponsoring editor for this book was Stephen S. Chapman and the production supervisor was Pamela A. Pelton. It was set in Century Schoolbook by International Typesetting and Composition. The art director for the cover was Anthony Landi.

Printed and bound by RR Donnelley.

 This book was printed on recycled, acid-free paper containing a minimum of 50% recycled, de-inked fiber.

McGraw-Hill books are available at special quantity discounts to use as premiums and sales promotions, or for use in corporate training programs. For more information, please write to the Director of Special Sales, McGraw-Hill Professional, Two Penn Plaza, New York, NY 10121-2298. Or contact your local bookstore.

Information contained in this work has been obtained by The McGraw-Hill Companies, Inc. ("McGraw-Hill") from sources believed to be reliable. However, neither McGraw-Hill nor its authors guarantee the accuracy or completeness of any information published herein, and neither McGraw-Hill nor its authors shall be responsible for any errors, omissions, or damages arising out of use of this information. This work is published with the understanding that McGraw-Hill and its authors are supplying information but are not attempting to render engineering or other professional services. If such services are required, the assistance of an appropriate professional should be sought.

To Rudolf Kingslake (1903–2003)—professor, mentor, colleague, gentleman, and friend

Contents

Preface xiii

Chapter 1. Introduction 1

    1.1 Lens Design Books 1
    1.2 Reference Material 2
    1.3 Specifications 2
    1.4 Lens Design 4
    1.5 Lens Design Program Features 7
    1.6 About This Book 28

Chapter 2. Automatic Lens Design: Managing the Lens Design Program 11

    2.1 Optimization 11
    2.2 The Merit Function 13
    2.3 Local Minima 19
    2.4 The Landscape Lens 21
    2.5 Types of Merit Functions 28
    2.6 Stagnation 29
    2.7 Generalized Simulated Annealing 30
    2.8 Considerations about Variables for Optimization 31
    2.9 How to Increase the Speed or Field of a System and Avoid Ray Failure Problems 36
    2.10 Test Plate Fits, Melt Fits, Thickness Fits, and Reverse Aberration Fits 37
    2.11 Spectral Weighting 40
    2.12 How to Get Started 41

Chapter 3. Improving a Design 47

    3.1 Lens Design Tip Sheet: Standard Improvement Techniques 47
    3.2 Glass Changes: Index and V-value 51
    3.3 Splitting Elements 52
    3.4 Separating a Cemented Doublet 55

3.5	Compounding an Element	55
3.6	Vignetting and Its Uses	58
3.7	Eliminating a Weak Element—the Concentric Problem	60
3.8	Balancing Aberrations	60
3.9	The Symmetrical Principle	67
3.10	Aspheric Surfaces	68

Chapter 4. Evaluation: How Good Is This Design? 71

4.1	The Uses of a Preliminary Evaluation	71
4.2	OPD versus Measures of Performance	71
4.3	Geometric Blur Spot Size versus Certain Aberrations	80
4.4	Interpreting MTF—The Modulation Transfer Function	82
4.5	Fabrication Considerations	83

Chapter 5. Lens Design Data 85

5.1	About the Sample Lens Designs	85
5.2	Lens Prescriptions, Drawings, and Aberration Plots	87
5.3	Estimating the Potential of a Redesign	92
5.4	Scaling a Design, Its Aberrations, and Its Modulation Transfer Function	96
5.5	Notes on the Interpretation of Ray Intercept Plots	98
5.6	Various Evaluation Plots	103

Chapter 6. Telescope Objectives 109

6.1	The Thin Airspaced Doublet	109
6.2	Merit Function for a Telescope Objective	110
6.3	The Design of an f/7 Cemented Doublet Telescope Objective	115
6.4	Spherochromatism	118
6.5	Zonal Spherical Aberration	123
6.6	Induced Aberrations	124
6.7	Three-Element Objectives	125
6.8	Secondary Spectrum (Apochromatic Systems)	125
6.9	The Design of an f/7 Apochromatic Triplet	133
6.10	The Diffractive Surface in Lens Design	145
6.11	A Final Note	150

Chapter 7. Eyepieces and Magnifiers 151

7.1	Eyepieces	151
7.2	A Pair of Magnifier Designs	155
7.3	The Simple, Classical Eyepieces	155
7.4	Design Story of an Eyepiece for a 6×30 Binocular	160
7.5	Four-Element Eyepieces	176
7.6	Five-Element Eyepieces	187
7.7	Very High Index Eyepiece/Magnifier	187
7.8	Six- and Seven-Element Eyepieces	200

Contents   ix

Chapter 8. Cooke Triplet Anastigmats   201

    8.1   Airspaced Triplet Anastigmats   201
    8.2   Glass Choice   205
    8.3   Vertex Length and Residual Aberrations   206
    8.4   Other Design Considerations   209
    8.5   A Plastic, Aspheric Triplet Camera Lens   215
    8.6   Camera Lens Anastigmat Design "from Scratch"—The Cooke Triplet   223
    8.7   Possible Improvements to Our "Basic" Triplet   234
    8.8   The Rare Earth (Lanthanum) Glasses   236
    8.9   Aspherizing the Surfaces   237
    8.10 Increasing the Element Thickness   246

Chapter 9. Split Triplets   247

Chapter 10. The Tessar, Heliar, and Other Compounded Triplets   259

    10.1 The Classic Tessar   259
    10.2 The Heliar/Pentac   266
    10.3 The Portrait Lens and the Enlarger Lens   266
    10.4 Other Compounded Triplets   272
    10.5 Camera Lens Anastigmat Design "from Scratch"—
The Tessar and Heliar   272

Chapter 11. Double-Meniscus Anastigmats   297

    11.1 Meniscus Components   297
    11.2 The Hypergon, Topogon, and Metrogon   297
    11.3 A Two Element Aspheric Thick Meniscus Camera Lens   299
    11.4 Protar, Dagor, and Convertible Lenses   302
    11.5 The Split Dagor   305
    11.6 The Dogmar   305
    11.7 Camera Lens Anastigmat Design "from Scratch"—The Dogmar Lens   305

Chapter 12. The Biotar or Double-Gauss Lens   319

    12.1 The Basic Six-Element Version   319
    12.2 Twenty-Eight Things That Every Lens Designer Should
Know About the Double-Gauss/Biotar Lens   329
    12.3 The Seven-Element Biotar—Split-Rear Crown   334
    12.4 The Seven-Element Biotar—Broken Contact Front Doublet   340
    12.5 The Seven-Element Biotar—One Compounded Outer Element   340
    12.6 The Eight-Element Biotar   340
    12.7 A "Doubled Double-Gauss" Relay   350

Chapter 13. Telephoto Lenses   355

    13.1 The Basic Telephoto   355
    13.2 Close-up or Macro Lenses   356

x   Contents

    13.3 Telephoto Designs 358
    13.4 Design of a 200-mm f/4 Telephoto for a 35-mm Camera "from Scratch" 367

Chapter 14. Reversed Telephoto (Retrofocus and Fish-Eye) Lenses 395

    14.1 The Reversed Telephoto Principle 395
    14.2 The Basic Retrofocus Lens 397
    14.3 Fish-Eye, or Extreme Wide-Angle Reversed Telephoto, Lenses 402

Chapter 15. Wide-Angle Lenses with Negative Outer Elements 415

Chapter 16. The Petzval Lens; Head-up Display Lenses 423

    16.1 The Petzval Portrait Lens 423
    16.2 The Petzval Projection Lens 423
    16.3 The Petzval with a Field Flattener 426
    16.4 Very High Speed Petzval Lenses 429
    16.5 Head-up Display (HUD) Lenses, Biocular Lenses, and Head/Helmet Mounted Display (HMD) Systems 437

Chapter 17. Microscope Objectives 441

    17.1 General Considerations 441
    17.2 Classical Objective Design Forms: The Aplanatic Front 442
    17.3 Flat-Field Objectives 446
    17.4 Reflecting Objectives 446
    17.5 The Microscope Objective Designs 447

Chapter 18. Mirror and Catadioptric Systems 455

    18.1 The Good and the Bad Points of Mirrors 455
    18.2 The Classical Two-Mirror Systems 456
    18.3 Catadioptric Systems 469
    18.4 Aspheric Correctors and Schmidt Systems 473
    18.5 Confocal Paraboloids 476
    18.6 Unobscured Systems 476
    18.7 Design of a Schmidt-Cassegrain "from Scratch" 482

Chapter 19. Infrared and Ultraviolet Systems 503

    19.1 Infrared Optics 503
    19.2 IR Objective Lenses 504
    19.3 IR Telescopes 507
    19.4 Laser Beam Expanders 511
    19.5 Ultraviolet Systems 514
    19.6 Microlithographic Lenses 514

Chapter 20. Zoom Lenses 521

    20.1 Zoom Lenses 521
    20.2 Zoom Lenses for Point and Shoot Cameras 526

20.3	A 20x Video Zoom Lens	539
20.4	A Zoom Scanner Lens	541
20.5	A Possible Zoom Lens Design Procedure	542

Chapter 21. Projection TV Lenses and Macro Lenses   551

21.1	Projection TV Lenses	551
21.2	Macro Lenses	553

Chapter 22. Scanner/f-θ, Laser Disk and Collimator Lenses   561

22.1	Monochromatic Systems	561
22.2	Scanner Lenses	561
22.3	Laser Disk, Focussing, and Collimator Lenses	571

Chapter 23. Tolerance Budgeting   573

23.1	The Tolerance Budget	573
23.2	Additive Tolerances	578
23.3	Establishing the Tolerance Budget	583

Chapter 24. Formulary   587

24.1	Sign Conventions, Symbols, and Definitions	587
24.2	The Cardinal Points	588
24.3	Image Equations	590
24.4	Paraxial Ray Tracing (Surface by Surface)	592
24.5	Invariants	594
24.6	Paraxial Ray Tracing (Component by Component)	594
24.7	Two-Component Relationships	595
24.8	Third-Order Aberrations—Surface Contributions	596
24.9	Third-Order Aberrations—Thin Lens Contributions: The G-Sum Equations	598
24.10	Stop Shift Equations	600
24.11	Third-Order Aberrations—Contributions from Aspheric Surfaces	601
24.12	Conversion of Aberrations to Wavefront Deformation (Optical Path Difference)	601

Glossary   605
References   621
Index   623

Preface

My personal optical design experience has spanned more than five decades. They have been exciting, fascinating, and delightful decades; I have enjoyed each one. During that half century, lens design has changed radically. In the mid-twentieth century, lens design was still a semi-intuitive art, practiced by a few dedicated individuals of great perseverance, knowledge, and skill. And by mid-century most of the classic lens design forms had already been created. To this day, these designs are still the basis of many excellent modern optical systems.

Of course the practice of lens design today is radically different from what it was in the 1940s and 50s. Then, most optical design was done with an electromechanical desk calculator (e.g., Marchant, Frieden, and Monroe), and the raytracing rate, measured in terms of the number of surfaces through which one could trace the path of a ray in a given amount of time, was to the order of one ray surface in about 250s (if one were to work at it continuously through the day). Thus, using the current dimensions for raytracing speed, one did about 0.004 ray-surfaces per second. And these were only meridional two-dimensional rays, not the three-dimensional general rays ordinarily traced today. A great deal of ingenuity (and elegant theory) went into finding ways to avoid tracing any more rays than were absolutely necessary.

Thanks to the modern personal computer or PC, the computing rate has increased almost unbelievably. Today a run-of-the-mill PC is capable of calculating several million ray-surfaces per second; this is about nine or ten orders of magnitude faster. Needless to say the techniques of lens design today differ mightily from those of fifty or sixty years ago. Then, the designer might calculate the derivatives of a few aberrations with respect to a limited number of constructional parameters and solve a small set of simultaneous, linear equations in the course of correcting his lens. These limited calculations were all carefully selected on the basis of theory, experience, and intuition. (Interestingly, one of the very real problems facing designers today is that the computer spews out numbers so rapidly that it takes strong self-discipline just to make oneself stop and think.)

In modern lens design work, a computer program almost instantaneously calculates and solves equations which are far more than an order of magnitude more complex and extensive than those cited above. It is not atypical for the computer program to control about 50 lens performance characteristics by adjusting the values of some 20 or 30 construction parameters of the optical system. These latter numbers imply a design space with 20 or 30 dimensions, a complex space indeed.

There are, however, some real limitations on the power of a so-called *automatic lens design* program. The typical program proceeds from a given starting design and drives the design to the nearest *local optimum*, a form at which any small structural changes will degrade the system performance. System performance is judged by a set of calculated characteristics defined in a *merit function*, which would be better termed a *defect* or *error* function, since the characteristics in it represent departures from desired values.

Obviously then, the final automatic design solution is completely and uniquely determined by (a) the merit function, (b) the starting design form, and (c) the algorithm by which the computer solves the problem of locating an optimum design form with the minimum value of the merit function.

When the first edition of *Modern Lens Design* (MLD) was published, there was a great need for a collection of suitable design forms at which to start the design process, and MLD provided almost 300 lens designs for this purpose. These designs were selected not only as starting points, but also as illustrations of important design principles. At the present time the need for sample designs, while still real, is significantly less, largely because most optical design programs now include libraries of lens designs. (These programs also include random search design capabilities which permit large changes in lens forms.) For example, all of the lens designs in the first edition of MLD (plus many others) are included in the lens libraries of the optical design program OSLO (a product of Lambda Research Corp.). Another program, LensVIEW by Brian Caldwell, is a compilation of over 30,000 lens designs and patents.

That said, it is (at least it is for me) far more easy and convenient to scan and compare a series of printed design pages than it is to do the same thing on a computer screen (even with the multiwindow capabilities of many programs). For this and other reasons this second edition of MLD has retained about half of the original designs and has added some new ones. The reader may also find some additional designs in the works referenced at the end of the book.

The practice of lens design is now essentially an engineering discipline. While this book is intended to be self-contained, we deliberately do not include a lot of derivations, or even the mechanics of exact ray tracing. And as valuable and cherished as they may be in academia, we happily omit any derivations from first principles, Maxwell's equations, or Fermat's principle. These are simply not necessary for a book on lens design. We make one exception to the no "ray tracing" rule, namely for the tracing of paraxial rays, which a lens designer often carries out by hand, or with a programmed pocket calculator. This topic is covered in the Formulary of Chapter 24, along with other valuable and frequently used geometrical optics relationships.

However, there is currently a growing need for a more detailed exposition of basic lens design and theory in a single volume. The first edition of MLD was a "companion" volume to the author's *Modern Optical Engineering*. Several very basic lens design books have recently appeared; some are almost extended user manuals written for a specific design program. This edition of MLD is definitely not intended as a user manual, or as a guide to any specific program. It is an attempt to go well beyond this level by presenting both the basics of, and a more advanced approach to, lens design. The intent is to advise the reader how to get the most from any computer lens design program. To this end, about half of the lens designs in the first edition of MLD have been eliminated to make room for quite a bit of new material.

The text is, as far as possible, completely program neutral. I have tried to make the material regarding design programs as generic as I could, discussing features that are available in almost all commercial software. I have used OSLO for the design work demonstrated in the text, and for preparing the new figures. (The lens analysis figures in the first edition of MLD were prepared with a customized version of the program GENII, using a new and unique presentation style which is now widely available; for an example, see the OSLO aberration plots herein.)

Most neophyte lens designers very quickly get past the basics and learn to use their computer programs with a high level of proficiency. At this point, what they need most is an answer to the question, "What do I do *now*?" Much of the new material in this edition is designed to this end and takes the form of actual design projects carried out from scratch, warts and all. (In other words, I have not papered over the blunders I made in the design process.) These designs include a cemented doublet, a triplet anastigmat, a Tessar, a Heliar, a Dogmar, a telephoto, a Schmidt cassegrain, a binocular eyepiece, an apochromatic triplet, and a landscape lens. Many of these design stories are carried out to some length to illustrate all of the possible steps that can be taken to improve a design. Every initial assumption is explained and justified. These design descriptions not only show the basic design approach, but continue on with advanced steps and the rationale for them.

I have surveyed the literature at some length for any design techniques which might have a general applicability (as well as the reported specific use for the writer's specific problem). Some were found in the references listed at the end of the book. For the most part, the design techniques described here are those which I have found to be useful in working with an optimization program. Many of the techniques have been developed or refined during more than two decades of teaching courses in lens design; indeed some of these ideas were suggested or inspired by my students. Other valuable sources were the many informal discussions that I have been fortunate to have with my colleagues.

For better or worse, one can never seem to squeeze all the material that you want into a book. At the manuscript deadline date there is always at least one more feature that you wished there was enough time to develop, write, and include. But I suppose that if there were time, no book would ever be finished.

Surprisingly, there are only a modest number of well-understood and widely utilized principles of optical design. If one can master a thorough understanding of these principles, their effects, and their mechanisms, it is easy to recognize them in existing designs and also easy to apply them to one's own design work. It is the intent here to promote such understanding by presenting both expositions and annotated design examples of these principles.

Readers are free to use the designs in this book as starting points for their own design efforts, or in any other way they see fit. The reader must accept full responsibility for meeting whatever limitations are imposed on the use of these designs by any patent, copyright, or other (whether indicated herein or not).

And finally, I want to express my gratitude to Kaiser Electro-Optics and Jerry Carollo for accomodating me, and to Lambda Research Corp. and Leo Gardner for providing the lens design program OSLO and for helping me use it to its full potential. PBGT.

WARREN J. SMITH
Vista, California

Modern
Lens Design

Chapter 1

Introduction

1.1 Lens Design Books

Modern Lens Design has several primary aims. The text contains extensive discussions of the principles and design techniques, which are appropriate to design work with a modern automatic lens design computer program. It is effectively a compendium of the many design techniques available to lens designers today. It is also a prescription source book for a variety of already designed lens types; these serve the dual purpose of providing starting points for the lens designer's efforts and also providing a basis for discussions of the application of many of the classic lens design principles. In the formulary of Chap. 24 we have provided a compact collection of most of the equations and relationships, which find frequent day-to-day use in lens design.

Modern Lens Design is intended as an aid to lens designers who work with the commercially available lens design computer programs. The approach used here assumes that the reader understands basic optical principles and may, in fact, have a command of the fundamentals of classical optical design methods. For those who want or need information in these areas, the following books should prove helpful. This author's *Modern Optical Engineering: The Design of Optical Systems*, 3rd ed., McGraw-Hill, 2000, is a comprehensive coverage of optical system design; it includes several chapters dealing specifically with lens design in considerable detail. Rudolf Kingslake's *Optical System Design* (1983), *Fundamentals of Lens Design* (1978), and *A History of the Photographic Objective* (1989), all by Academic Press, are complete, authoritative, very well written, and a pleasure to read.

1.2 Reference Material

There have been many books written about the subject of lens design. Some go deeply into the mathematical aspects of the subject; some are effectively expanded "user manuals" for a particular lens design program. (And then there are the program manuals themselves, many of which could benefit from an expanded and cross-referenced index, at the least.) Without further editorial comment I refer the reader to the list of references that follows Chap. 24. These are books, chapters, and articles whose subject is primarily *lens design, not optical engineering* or *optical system* design. Many of these primary references contain additional references that may be of interest to the reader who wishes to explore the edges and corners of the field of lens design. Many of the references are worthy of a thorough reading and study, in and of themselves. In the text we have cited these publications by their reference numbers as superscripts.

Adjacent to the reference list is a very complete glossary of the terms common to the practice of lens design. It is as complete, current, and authoritative as I could make it without becoming encyclopedic. I strongly recommend it to the reader. For the experienced designer it should serve to establish a common language. For the less experienced, a perusal of the glossary can serve almost as an introductory text.

1.3 Specifications

Before beginning the actual lens design work on a project, it is extremely wise to firmly nail down as many specifications as possible. Note that by "nail down" we do not mean cast in concrete; we mean that they should be at least considered and preliminarily agreed upon. An important part of defining a specification is determining which specifications are minimums, which are really just goals on someone's wish list, and which have some wiggle room, and if so, how much. It *always* pays to question the specifications. Are they theoretically possible and practically feasible in light of the diffraction limit and the conservation of radiance? Can they be met within the constraints of materials, spatial requirements, and the expected state of the fabrication technology? Which ones are those that define what the customer really wants and needs?

Note well that specifications are often prepared by someone who is not knowledgible in optics or optical design. The designer must convert the customer specifications into terms meaningful in optics, and should also be sure that the reality of fabrication tolerances is taken into account.

The following is a check list of possible specifications, which ought to prevent overlooking some requirement that might bite you later on.

1.3.1 Possible specification list

A checklist of specifications and considerations, which should be established before the lens design begins. Note that these items may be redundant or contradictory.

Effective focal length
Field angle or object size
F-number (infinity versus working)

Numerical aperture
Clear aperture
Size, weight, and packaging constraints
Wavelength and bandwidth
Spectral distribution and weighting
Transmission
Object and image distance
Object to image distance (*track length*)
Internal image required
Erect or inverted image
Magnification (and range)
Flat or curved field
Back focal length
Working distance (clearance)
Front focal length
Pupil location(s) and size; eye relief
Stops: Aperture, field, and cold
Space for iris diaphragm
Telecentric
Vignetting/illumination uniformity
Distortion
Zoom range: Continuous or stepped
Focus range
Focusing: Shift part or all of system
Anamorphic ratio

Performance
> versus diffraction limit
> versus field
> versus F-stop
> Resolution
> *Modulation transfer function* (MTF)
> Encircled/ensquared energy
> *Root mean square* (rms) spot or blur size
> Point spread function
> Wavefront aberration, *optical path difference* (OPD) (rms, P-V)
> Specific aberrations

Depth of focus/field

Lens type

Refractive versus mirror versus catadioptric

Plastic elements

Aspheric surfaces

Diffractive surfaces

Cost, quantity, delivery schedule

Materials

Environment, ambient conditions

Athermalization

Sensor, spectral response

Visual, photographic, projection

Patent considerations

Ergonomics

1.4 Lens Design

There are many commonly asked questions about lens design. Many of them are variations of the following:

1. Where do the aberrations come from?
2. Do we get spherical aberration because the surfaces are spherical?
3. Are aspheric surfaces free of aberration?
4. Are the aberrations related to each other?

The *complete* answer to no. 1 is that aberrations arise from Snell's law, $n \sin I = n' \sin I'$, which governs the refraction of rays at a surface. Because this is a trigonometric expression, it is nonlinear and the effect on a ray's direction varies across the aperture of the lens surface. And Snell's law is especially nonlinear as the angles I or I' approach 90°.

The answer to no. 2 is that spherical surfaces usually do have spherical aberration, but when used in a certain way (the aplanatic case) they are free of spherical aberration, coma, and astigmatism.

For question no. 3 the answer is similar to that for no. 2. The conic aspherics (e.g., paraboloid, ellipsoid, hyperboloid, and the sphere) have no spherical aberration if the object point is at one focal point and the image is at the other. But if we depart from this condition we definitely do get spherical aberration. A paraboloid, noted for having no spherical aberration when the object is at infinity, has spherical aberration when the object is at any other distance.

Question no. 4: The aberrations are in fact related, as is illustrated by Fig. 1.1, where a single spherical refracting surface is shown with an aperture stop to its left. Consider an axial object point A that has its paraxial image at point A'. If we pivot the axis about the center of curvature of the surface, we generate a curved object surface, A-C, and a curved image surface, A'-C'. The departure of these surfaces from the ideal flat object and image surfaces, A-B and A'-B', is the source of the *Petzval field curvature*. If we were to move object point C back to point B, putting it on the flat object surface, then the image point C' will also shift to the left, increasing the field curvature at the image. Note that this is a basic field curvature and does not depend on astigmatism.

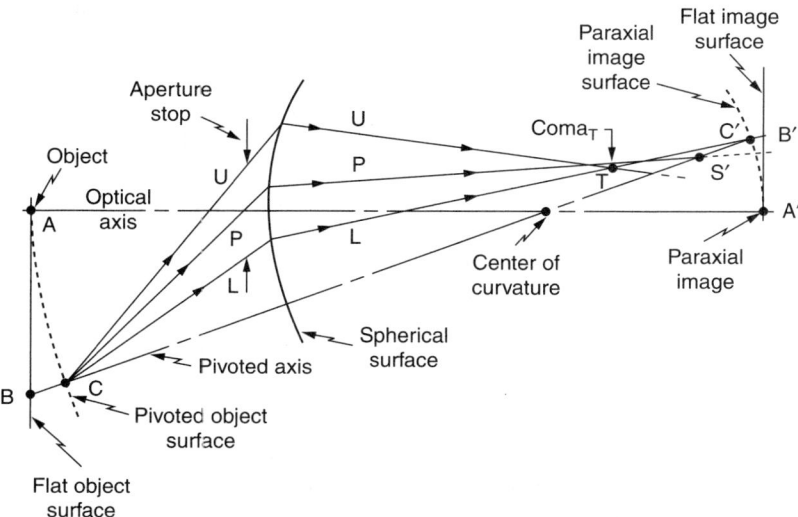

Figure 1.1 A single spherical refracting surface with an aperture stop to its left

Three rays are shown traced from object point C through the aperture stop: an upper rim ray, U; a lower rim ray, L; and a principal ray, P. Spherical aberration causes them to intersect the pivoted axis (C-C′) at distances from the paraxial focus (C′), which increase as the square of the height at which they strike the surface. The intersection of rays U and L at point T is vertically displaced from the ray P; this is *tangential coma*. The focus of the upper and lower rays is at point T; this is the tangential (or meridional) focus of these marginal rays. The corresponding focus of the sagittal rays can be generated approximately by rotating the diagram of the principal ray about the pivoted axis (C-C′)—this focus is at point S. Points T and S are indicative of the *tangential* and *sagittal field curvature*, and the distance between them is the *astigmatism*. Note that T is three times as far from the curved image surface as is S. Object point B in the ideal flat object plane has an ideal image point at B′. The distance from B′ to the intersection of the principal ray with the ideal image plane is the *distortion*.

Until fairly recently the term "lens design" implied the design of telescope and camera lenses almost exclusively. Currently the applications of optics cast a much broader net. Nonetheless, we strongly urge a careful study of Kingslake's excellent *A History of the Photographic Lens*. Although you may not be designing a photographic lens, the techniques and principles that have been used to improve camera lens designs through the years are the same techniques and principles that may well allow you to improve whatever optical system on which you may be working, and Kingslake explains them very well (and interestingly).

Lens design is both a science and an art, possibly best learned at the feet of a master (lens designer). Although lens design today is carried out almost entirely on a personal computer, often with only a very minimal amount of preliminary manual calculation, a knowledge and understanding of the basic principles of geometrical optics is essential to success in this field. Learning how and why a successful lens design works and why a less successful design doesn't work is assuredly worth the effort. Many, foolishly and blindly, hope and believe that the optical software will miraculously find and deliver a great and practical solution to almost any lens problem. If only it were so! The best designs begin solidly grounded in a physically correct concept, and only then are they optimized and polished by the power of the computer programs.

There are many interesting and elegant analytical studies of optical system design problems, usually based on the third-order aberrations (see, for example, the equations in the formulary of Chap. 24). These *third-order aberration* (TOA) results can also be found using the optimization routines of a computer program without the labor of algebraic derivations and the like. The equations of the formulary can be useful in getting to a good starting point for further computer work, but note that the equations also can yield a survey of the entire field (and not simply find a single local optimum as the optimization routines do). They can find a *good* solution, or multiple solutions, or show the impossibilities, and give a better understanding of the problem at hand. This is true for first-order imagery as well as for TOA, and the Chap. 24 formulary equations or the computer can equally well investigate the paraxial possibilities.

The basics should always be established before the actual design begins. As Bob Hopkins was fond of saying, "Keep your hands off the keyboard until you thoroughly understand the first- and third-order problem." Very wise advice indeed.

1.5 Lens Design Program Features

Most full-scale optical design programs have the following features or their equivalents. There are many specialized facilities in the major programs. We list here only those most widely utilized in ordinary practice.

Variable parameters. Any construction parameter, including:

Surface curvatures (radii)

Surface asphericities

Surface spacings (thickness and airspaces)

Materials (glass indices and dispersions)

Decentration or tilts

Vignetting

Variables may be subject to maximum and minimum bounds.

Solves

Angle solve for a curvature to get a desired ray slope after the surface. Applied to the last surface, it can establish the magnification or focal length.

Height solve to get a desired ray height at the next surface. If the axial ray height is zero it locates the paraxial focus.

Pick up of a previous data item

Targets (operands)

Any construction value (e.g., r, t, n)

Any ray data (height, slope, exact, or paraxial)

First-, third-, and fifth-order aberration contributions or sums

Field curvature (X_s and X_t)

Distortion

Offence against the sine condition (OSC)

S2T—aperture coefficient of coma

Optical path difference (OPD)

$(D - d)\delta n$

Derivatives of ray height or slope

Ray intersection differences

Averages, rms

Mathemetical combinations of targets (+, −, ×, ÷, **, >, <)

Graphic representations (plots)

Ray intersections

Distortion versus field

Lateral chromatic versus field

Axial chromatic versus ray height

Spherical aberration versus ray height

OPD versus aperture or field

Spot size versus field

MTF versus frequency, focus, or field

1.6 About This Book

The book proper begins with Chap. 2 with a discussion of automatic lens design programs and how to use them effectively. Here the merit function, optimization, variables, and the various and several techniques that are useful in connection with a software program are covered. In Chap. 3 we discuss many specific improvement strategies that can be applied to an existing design to improve its performance. Chapter 4 deals with the evaluation of a design from the standpoint of both ray and wave aberrations, as well as such standard measures as spot size, MTF, and the Strehl ratio.

The presentation of the design data and performance in this book is explained in Chap. 5, as is the evaluation of the aberration data in estimating the potential of a particular design for a specific task. Chapters 6 through 22 discuss the major lens-design types and the manner of their design. For each there is the prescription, a drawing of the lens, and ray aberration analysis plots (with ray intercept plots, longitudinal plots of spherical, spherochromatic, and field curvature, plus transverse plots of distortion and lateral color). Chapter 23 on tolerance budgeting and Chap. 24, the Formulary, conclude the book.

Two particular design forms, the telescope objective in Chap. 6 and the Cooke triplet in Chap. 8, are given special and extensive attention. The design techniques used for these relatively simple designs are explained at length. These same techniques can be applied to more complex designs as well as to these more basic lens types.

Chapter 2

Automatic Lens Design: Managing the Lens Design Program

2.1 Optimization

"A lens is about as nonlinear as anything in physics." RUDOLF KINGSLAKE

What is usually referred to as *automatic lens design* is, of course, nothing of the sort. The computer programs so described are actually optimization programs that drive an optical design to a local optimum, as defined by a *merit function* (which is not a true merit function, but actually a defect or error function). In spite of the preceding disclaimers, we will use these commonly accepted terms in the discussions that follow.

The lens design program typically operates this way: Each variable parameter is changed (one at a time) by a small increment whose size is chosen as a compromise between a large value (to get good numerical accuracy) and a small value (to get the local differential). The change produced in every item in the merit function is calculated. The result is a matrix of the partial derivatives of the defect items with respect to the parameters. Since there are usually many more defect items than variable parameters, the solution is a classical least-squares solution. It is based on the assumption that the relationships between the defect items and the variable parameters are linear. Since this is usually a false assumption, an ordinary least-squares solution will often produce an unrealizable lens or one that may in fact be worse than the starting design. The *damped least-squares* solution, in effect, adds the (suitably weighted) squares of the parameter changes to the merit function, heavily penalizing any large changes and thus limiting the size of the changes in the solution. The mathematics of this process is described in Spencer, "A Flexible Automatic Lens Correction Program," *Applied Optics*, vol. 2, 1963, pp. 1257–1264, and in many other publications; it will not be repeated here.

The damped least-squares optimization process will pursue *any* improvement, no matter how small, and no matter the cost. If one arrives at a design with an overly thick element or a very big airspace, it's wise to test to see if it's really doing much good. Reoptimizing with the offending dimension fixed at a reasonable value and comparing the results will usually yield the answer to this question.

If the changes of the solution are small, the nonlinearities will not ruin the process, and the solution, although an approximate one, will be an improvement on the starting design. Continued repetition of the process will eventually drive the design to the nearest local optimum.

One can visualize the situation by assuming that there are only two variable parameters. Then the merit function space can be compared to a landscape where latitude and longitude correspond to the two variables and the elevation represents the value of the merit function. Thus the starting lens design is represented by a particular location in the landscape and the optimization routine will move the lens design downhill until a minimum elevation is found. Since there may be many depressions in the terrain of the landscape, this optimum may not be the best there is; it is a local optimum and there can be no assurance (except in very simple systems) that we have found a "global" optimum in the merit function. This simple topological analogy helps us to understand the dominant limitations of the optimization process: the program finds the nearest minimum in the merit function, and that minimum is uniquely determined by the design coordinates at which the process is begun. The landscape analogy is easy for the human mind to comprehend; when it is extended to a 10- or 20-dimension space, one can realize only that it is apt to be an extremely complex neighborhood.

One version of adaptive optimization adjusts the targets in the merit function so that they are an ever-decreasing fraction of the distance to the desired solution. This has the virtue of keeping the steps in the optimization process relatively small (and less nonlinear) and the progress relatively constant. This may be incorporated in a program, or it can be accomplished manually by examining the calculated values of the individual operands in the merit function and adjusting the target values accordingly.

The *multiconfiguration* feature in most programs allows the optimization of a system in two or more configurations, wherein all the configurations share common elements of construction but differ in other aspects such as spacings and added components. The classic example is the zoom lens where each configuration uses the same optical elements, but several spaces differ (so as to change the focal length while maintaining focus and image quality). Other applications include: designing a system to operate at several different (or over a range of) object distances, as in a "macro" lens; increasing the depth of focus by designing configurations, each with slightly different image surface locations but the same optics; reducing sensitivity to tolerances by designing two configurations, each with a lens dimension (radius, thickness, alignment, and the like), which has a slightly different value in each configuration. Many programs perform a coarse search to determine the best value for the damping factor.

The result of a damped least-squares optimization run depends on the starting point given to the program. Obviously, changing the starting design form can produce a new result, for better or worse.

2.2 The Merit Function

The merit function is the most important aspect of an optical design program. As a result, the "care and feeding" of one's merit function may occupy more of the lens designer's time than anything else in the process.

Broadly speaking, the merit function can be described as a combination or function of calculated characteristics called "operands" or "targets," which is intended to completely describe, with a single number, the value and/or quality of a given lens design. This is obviously an exceedingly difficult thing to do. The typical merit function is the sum of the squares of many image defects; usually these image defects are evaluated for three locations in the field of view (unless the system covers a very large or a very small angular field). The squares of the defects are used so that a negative value of one defect does not offset a positive value of some other defect, and a large defect is doubly weighted by its size.

The defects may be of many different kinds; usually most are related to the quality of the image; however, any characteristic that can be calculated may be assigned a target value and its departure from that target regarded as a defect. Some rudimentary programs use only the third-order (Seidel) aberrations; these provide a rapid and efficient way of *adjusting* a design. These cannot be regarded as optimizing the image quality, but they do work well in *correcting* ordinary lenses. Another type of merit function traces a large number of rays from an object point. The radial distance of the image plane intersection of the ray from the centroid of all the ray intersections is then the image defect. Thus the merit function is effectively the sum of the root-mean-square (rms) spot sizes for several field angles. This type of merit function, while inefficient in that it requires many rays to be traced, has the advantage that it is both versatile and in some ways relatively foolproof. Some merit functions calculate the values of the classical aberrations, and convert (or weight) them into their equivalent wavefront deformations. (See Chap. 24 for the conversion factors for several common aberrations.) This approach is very efficient as regards computing time, but requires careful and expert design of the merit function. Still another type of merit function uses the variance of the wavefront to define the defect items. The merit function used in the various David Grey programs is of this type, and is certainly one of the best commercially available merit functions in producing a good balance of the aberrations.

Characteristics that do not relate to image quality can also be controlled by the lens design program. Specific construction parameters, such as radii, thicknesses, and spaces, as well as focal length, working distance, magnification, numerical aperture, required clear apertures, and the like, can be controlled. Some programs include such items in the merit function along with the image defects. There are two drawbacks that somewhat offset the neat simplicity of this approach. One is that if the first-order characteristics that are targeted are not initially close to the target values, the program may correct the image aberrations without controlling these first-order characteristics; the result may be, for example, a well-corrected lens with the wrong focal length or numerical aperture. The program often finds this to be a local optimum and is unable to move away from it. The other drawback is that the inclusion of these items in the merit function has the effect of slowing the process of improving the image quality. An alternative approach is to use a system of constraints outside the merit function. Note also that many of these items can be controlled by features that are included in almost all programs, namely *angle-solves* and *height-solves*. These algebraically solve for a radius or space to produce a desired paraxial ray slope or height.

In any case, the merit function is a summation of suitably weighted defect items, which is hoped to describe in a single number the worth of the system. The smaller the value of the merit function, the better the lens. The numerical value of the merit function depends on the construction of the optical system; it is a function of the construction parameters, which are designated as variables. Without getting into the details of the mathematics involved, we can realize that the merit function is an n-dimensional space, where n is the number of the variable constructional parameters in the optical system. The task of the design program is to find a location in this space (i.e., a lens prescription or a solution vector) that minimizes the size of the merit function. In general, for a lens of reasonable complexity there will a very large number of such locations in a typical merit function space. The automatic design program will simply drive the lens design to the nearest minimum in the merit function.

Optimally, the targets (or operands) in the merit function should be weighted so that all have a similar magnitude. This will avoid an implicit heavier weighting of items that are dimensionally larger. For example, the *optical path difference* (OPD) in millimeters would not be as significantly weighted as the OPD in waves, and the transverse measure of an aberration would be less significant than the longitudinal measure. One way of handling this, where appropriate, is to convert everything to OPD in wavelengths.

Note that a merit function based on the third- and fifth-order aberration coefficients will be very robust. This is because no trigonometric rays are traced; the paraxial rays used to calculate the third- and fifth-order contributions are not subject to failure due to a ray missing a surface or encountering total internal reflection (no matter how unrealistic the calculated paraxial ray path may be).

A minimum ray-set at each field point for an aberration merit function would include the following:

Principal ray (or axis)

Upper margin ray (vignetted)

Lower margin ray (vignetted)

Sagittal ray

The merit function can have specific targets for spherical, coma, and so on—these can be the complete merit function, or they can be added to the default merit function to shift the design emphasis to a problem area. When adding to the merit function, be sure not to duplicate any existing operands; this may lead to an ill-conditioned matrix.

A merit function *samples* the aperture, field, and spectrum. The sampling is often minimal in order to save time and simplify matters. But remember that high speed, or wide-angle systems, may need a finer sampling than an ordinary system does. The more rays used in the merit function, the better the correlation with performance, but the slower the calculation.

A system with a small aperture and/or a narrow field of view will need fewer rays and field points. For example, it is common to use three field points (including the axis); two are sufficient for a narrow field, but four or five may be necessary for a wide field.

A typical aberration merit function may include the following operands:

- First order: *Effective focal length* (EFL), magnification, total track length, image height, angle solves, height solves, and lens vertex length. Avoid conflicts from multiple specifications.
- Aberrations: Marginal spherical, coma, astigmatism and Petzval field curvature, distortion, axial chromatic, lateral color, chromatic variations, and secondary spectrum.

- Zonal aberrations: Rays at 0.7 of the aperture can be used for zonal aberrations. The third-order aberration sums can also be used to reduce zonal residuals if the marginal correction is held, as can the Conrady *D-d, offence against the sine condition* (OSC), and aperture coefficient of coma (S2T) operands.

Another possible approach is to begin the optimization with an all third-order merit function (or one with both third and fifth), and then progress to one that uses the actual aberrations.

A feature common to many programs is to calculate the rms spot size (radius) or the rms OPD using a form of quadrature. This allows these values to be calculated quite accurately using only a few rays, which are spaced and weighted according to the quadrature rules (as are the field points). The Gaussian quadrature calculated value of the function will be exact if twice the number of rays minus one exceeds the order of the function that defines the ray pattern. Thus five rays along a radius can accurately evaluate a ninth-order function, e.g., spherical up to the ninth order.

Usually the rays are specified by the intersection of rings and radial spokes in the aperture. In general, the more rays, the more accurate the calculation and the use of twice as many spokes as rings is customary. For example, six spokes and three rings is a common default.

When the operands are rms spot size, the program tends to produce the smallest possible spot; when rms OPD is used, the tendency is to produce a tight, hard core with flare or wings. Since *modulation transfer function* (MTF) is a function of the variance of the OPD, an rms OPD merit function often produces a better MTF than rms spot merit function. Note that a minimum rms spot is not the same as correcting the marginal spherical to zero. If you want TA_m equal to zero, rms spot won't do it; it tends toward undercorrected spherical accompanied by negative defocussing. To get $TA_m = 0.0$, use a target on the axial marginal ray.

Targets for first-order characteristics such as focal length, back focus, object to image distance, and magnification *may* be detrimental to the least-squares process. It may be beneficial to separate performance from such requirements. Consider using solves, Lagrange multiplier constraints, or even ad hoc intervention by the designer.

Systems that are not axially symmetric require more field points than the customary three. For example, a plane symmetric anamorphic system will need horizontal, vertical, and diagonal fields because the aberrations in each are different.

It is almost always necessary to tailor the design program's default merit function to suit the needs of the project at hand.

Solves available in almost all programs include:

1. The *angle solve*. This calculates (and adjusts) the curvature of a surface so as to produce a desired paraxial ray slope following the surface.

It can be applied to the axial ray or the chief ray. It is widely used to control focal length (or magnification) via the axial ray, and afocal magnification via the chief ray.

2. The *height solve*. This calculates (and adjusts) the distance to the next surface, which will produce the targeted ray height at that surface. It can be used on the chief ray to locate the pupil or on the axial ray to locate the paraxial focus position. A defocus from the paraxial focus is a much more stable variable than using the distance to the image surface as a variable.

3. The *pick up*. This will set the current curvature or thickness equal to a previous radius or thickness, with the same or a reversed sign. This is useful to maintain front-to-back symmetry or to use a surface twice, as in a mirror system.

Most programs also will allow a constant quantity to be added to the solved value.

Solves reduce the number of variables; this is good because it reduces the complexity of the solution. An angle solve to control focal length is exact only if the object distance is infinite; current programs use about 10^{20} units for the object distance. Most programs have a separate calculation for the exact value of the focal length, based on the data of two paraxial ray traces. If the tiny error in the focal length (approximately equal to f/d^2) because of the finite object distance is significant, this exact value can be used as a heavily weighted operand for the last step in the design process.

Some items that are commonly available for use in the merit function as operands or targets are as follows:

- Almost every constructional value, such as curvature, thickness, index, and aspheric coefficient
- Paraxial quantities, i.e., ray slopes and heights
- Aberration coefficients—chromatic, third- and fifth-order
- Ray trace values—intersections, angles (direction cosines and tangents), path length, field curvature (X_s and X_t), derivatives of height versus pupil or versus object shift, displacement of a ray relative to a reference ray, OPD, Conrady *D-d*, OSC, S2T, and MTF
- Mathematical combinations of previously defined operands: +, −, ×, ÷, **, ave., and rms

The designer should occasionally stop and consider just what each item in the merit function is actually doing, and how it relates to the other operands. Be sure that something in the merit function, in the constraints, in the variable bounds, or in the solves is not inadvertently limiting the progress toward a solution.

A system that is nominally intended for a zero field size (i.e., to be used on axis only) should be corrected for coma if at all possible. This is a good precaution against the possibility that in use, a misalignment will actually cause the system to be used off axis.

A good approach is to take a default merit function and then customize it. After adjusting, enlarging, customizing, reweighting and/or rebalancing the merit/function, you have created a merit function for your system, which is presumably capable of directing the program to a good solution. At this point you can

a. Start over again

b. Start over from a different starting design

c. Start over with a different design form (e.g., Tessar versus triplet)

d. Go "global"

Targets/operands should be weighted to be of about equal magnitude. For example, a given amount of spherical aberration expressed as a longitudinal aberration is a bigger number than when expressed as a transverse aberration, which is a bigger number than when expressed as an OPD. The relationship expressed as an equation is: $LA = TA/NA = 16 \cdot OPD/NA^2$. For an $f/3$ lens, $NA = 0.1667$ and $LA = 6 \cdot TA = 576 \cdot OPD$. Probably the simplest way to handle this in an aberration merit function is to convert the operands to an OPD expressed in units of the wavelength. See the equations in Chap. 24 for the conversions.

For a photographic application the corners should have a low weight; the objects in the corners are usually not of great interest and the fraction of the total area of the picture that is in the corner is a small one.

Don't try to control or fully correct an uncorrectable aberration, as for example, the astigmatism of a thin lens at the stop, which is totally independent of the lens shape or index.

Ideally, the merit function should correlate well with MTF (provided that a good MTF defines the desired state of correction). A merit function with this property is difficult to achieve. Some programs have the ability to use the MTF as an operand in the merit function, but this is slow at best, and unless the current design is close to the optimum, the process can be extremely volatile because the MTF is a very nonlinear function of the variables. An MTF merit function works well for a final balancing and adjustment of the design.

If non-image-quality items (e.g., EFL, back focal length (BFL), distances, thicknesses, and edge thicknesses) are put in the merit function as constraints (i.e., Lagrangian multipliers), this will eliminate any possible compromise between them and image quality.

An rms OPD merit function will usually produce a better MTF. The image is likely to have a good tight core with a flare or wings. A similar correction can be obtained by heavily weighting rays in the center of the aperture. An rms spot size merit function will usually not produce as good an MTF, but it will have less flare and a more compact total spot.

2.3 Local Minima

Figure 2.1 shows a contour map of a hypothetical two-variable merit function, with three significant local minima at points A, B, and C; there are also three other minima at D, E, and F. It is immediately apparent that if we begin an optimization at point Z, the minimum at point B is the only one the routine can find. A start at Y on the ridge at the lower left will go to the minimum at C; however, a start at X, which is only a short distance away from Y, will find the best minimum of the three, at point A. If we had even a vague knowledge of the topography of the merit function, we could easily choose a starting point anywhere in the lower right quadrant of the map, which would guarantee finding point A. Note

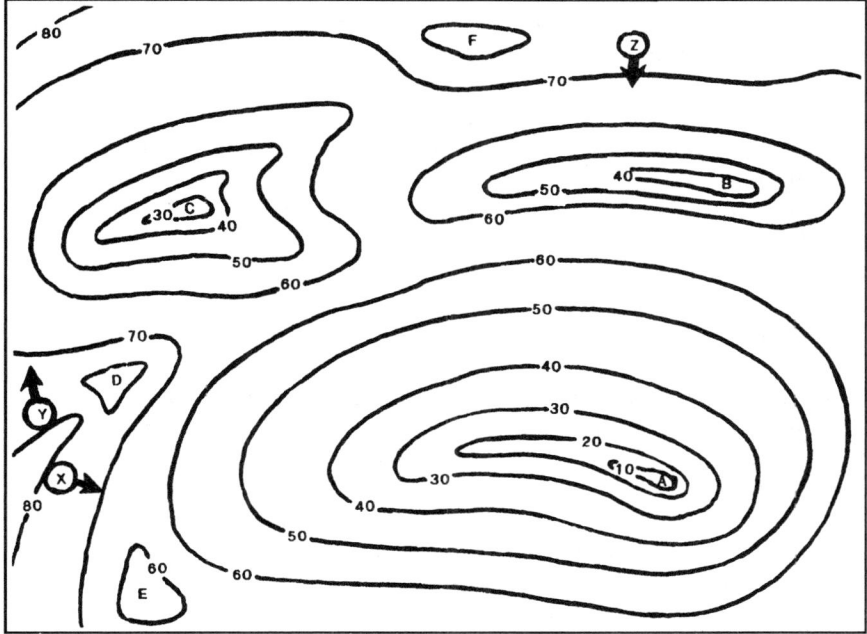

Figure 2.1 Topography of a hypothetical two-variable merit function, with three significant minima (A, B, C) and three trivial minima (D, E, F). The minimum to which a design program will go depends on the point at which the optimization process is started. Starting points X, Y, and Z each lead to a different design minimum; other starting points can lead to one of the trivial minima.

also that a modest change in any of the three starting points could cause the program to stagnate in one of the trivial minima at D, E, or F. It is this sort of minimum from which one can escape by jolting the design, as described below.

The fact that the automatic design program is severely limited and can find only the nearest optimum emphasizes the need for a knowledge of lens designs, in order that one can select a starting design form that is close to a good optimum. This is the only way that an automatic program can systematically find a good design. If the program is started out near a poor local optimum, the result is a poor design.

The mathematics of the damped least-squares solution involves the inversion of a matrix. In spite of the damping action, the process can be slowed or aborted by either of the following conditions: (*1*) A variable that does not change (or that produces only a very small change in) the merit function items. (*2*) Two variables that have the same, nearly the same, or scaled effects on the items of the merit function. Fortunately, these conditions are rarely met exactly, and they can be easily avoided.

If the program settles into an unsatisfactory optimum (such as those at D, E, and F in Fig. 2.1) it can often be jolted out of it by manually introducing a significant change in one or more parameters. The trick is to make a change that is in the direction of a better design form. (Again, a knowledge of lens designs is helpful.) Sometimes simply freezing a variable to a desirable form or changing the weights in the merit function can be sufficient to force a move into a better neighborhood. The difficulty is that too big a change may cause rays to miss surfaces or to encounter total internal reflection, and the optimization process may break down. Conversely, too small a change may not be sufficient to allow the design to escape from a poor local optimum. Also, one should remember that if the program is one that adjusts (optimizes) the damping factor, the factor is usually made quite small near an optimum because the program is taking small steps and the situation looks quite linear; after the system is jolted, it is probably in a highly nonlinear region and a big damping factor may be needed to prevent a breakdown. A manual increase of the damping factor can often avoid this problem.

Another often-encountered problem is a design that persists in moving to an obviously undesirable form (when you *know* that there is a much better, very different one—the one that you really want). Freezing the form of one part of the lens for a few cycles of optimization will often allow the rest of the lens to settle into the neighborhood of the desired optimum. For example, if one were to try to convert a Cooke triplet into a split-front-crown form, the process might produce either a form like the original triplet with a narrow airspaced crack in the front crown or a form with rather wild meniscus elements. A technique that will usually avoid these unfortunate local optima in this particular case is to freeze the front element to a plano-convex form by fixing the second surface to a plane for a few cycles of optimization. Again, one must know which lens forms are the good ones.

2.4 The Landscape Lens

As a real example to illustrate a simple merit function with two local optima, and to demonstrate the path of a typical damped least-squares optimization, we take a look at the *landscape lens*, which was probably the first real camera lens, i.e., a lens that covered a reasonably wide field. Its initial use was in the *camera obscura*, basically a dark room wherein the lens formed an image of the outside landscape on a wall or table for sketching, painting, or simply viewing.

The optical system of the landscape lens comprises a single positive element and a separate aperture stop. The stop is placed at the *natural* stop position where (assuming the lens has undercorrected third-order spherical aberration) the coma is zero and the tangential field is as backward curving (or the least inward curving) as possible. By bending the lens, one can determine the lens shape for which the field curvature is minimized when the coma is zero. There are two solutions: the lens is meniscus shaped in both, and the stop is located about 10 or 20 percent of the focal length away from the concave side of the lens. In the *front meniscus* the lens is convex toward the distant object and the stop is between the lens and the image; in the *rear meniscus* the lens is concave to the object and the stop is out in front of the lens.

To illustrate the effect of shifting the aperture stop in a lens with simple, undercorrected spherical aberration, Fig. 2.2 shows a ray intercept plot (i.e., an H-tan U curve) with three positions for the stop indicated by the circled numbers. (Note that shifting the portion of the plot under consideration from left to right is equivalent to moving the stop from right to left, because by convention the intersection points of the upper rays of a beam are plotted at the right.) In the plot, when a straight line is drawn connecting the intercepts for the upper and lower rays of a beam, the sag of the plot equals the tangential coma, and the slope of the line ($\delta H/\delta \tan U$) equals the tangential field curvature, X_t. Thus for stop position no. 1, we see overcorrected (positive) coma and an inward-curving field. Moving the stop to position no. 2 corrects the coma to zero, and the field is now backward curving. By inspection it is quite apparent that this is the stop location at which the field is most backward curving; this is called the natural stop position. At stop location no. 3 we once again have an inward-curving field and coma, but with the stop in this position the coma is undercorrected (negative).

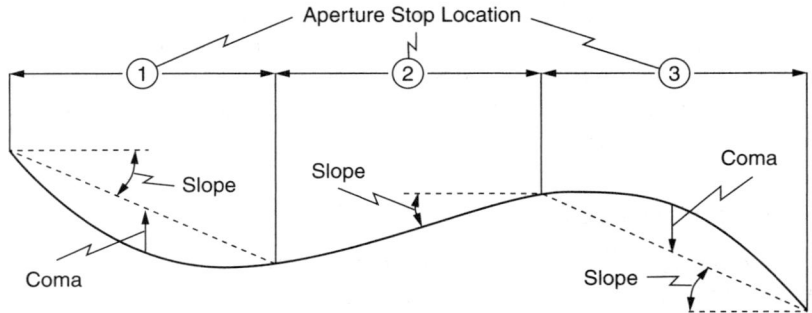

Figure 2.2 A ray intercept plot of off-axis, undercorrected, third-order spherical, illustrating how moving the stop will sample different portions of the plot. The different sections show the change of coma and field curvature as the stop is shifted. It also indicates that the stop position with zero coma is also the stop position that produces the most backward-curving field. This is the natural stop position.

The landscape lens is an excellent example with which to demonstrate the existence of multiple optima in lens design. Figure 2.3 is a contour map of an optical software default rms-spot-size merit function, plotted against the lens shape (cv 1) and the stop location. The data are for a BK7 lens of 100-mm focal length at a speed of $f/16.7$ (NA = 0.03) and a half-field of 20°. The evaluation is monochromatic.

The two local minima in the merit function are immediately apparent. The paths of two *damped least-squares* (DLS) optimizations are shown by the sequential circled numbers. For both paths, the starting lens shape was plano-convex with the aperture stop placed at the plano surface. When the starting lens was oriented with the convex surface facing the distant object, the DLS routine found the front meniscus form. When the plano side faced the object, the rear meniscus form was the result. Although the optimization paths appear almost random, note that each step in the process did reduce the merit function. Interestingly, from any starting point in the upper-right or lower-left quadrants of the plot, after the first few steps the optimization always followed a path very similar to these.

This is one of the very few lenses that can be designed analytically, using the third-order aberration contribution equations (either the surface- or the thin lens-contribution equations—see Chap. 24).

As can be seen from the MTF and ray intercept plots in Figs. 2.4 to 2.6, the rear meniscus is optically the better of the two. Its monochromatic image quality is 25 or 30 percent better than the front meniscus, its surface curvatures are weaker, and its distortion is barrel (and thus somewhat less objectionable to the eye than the pincushion of the front meniscus). In spite of all this, the front

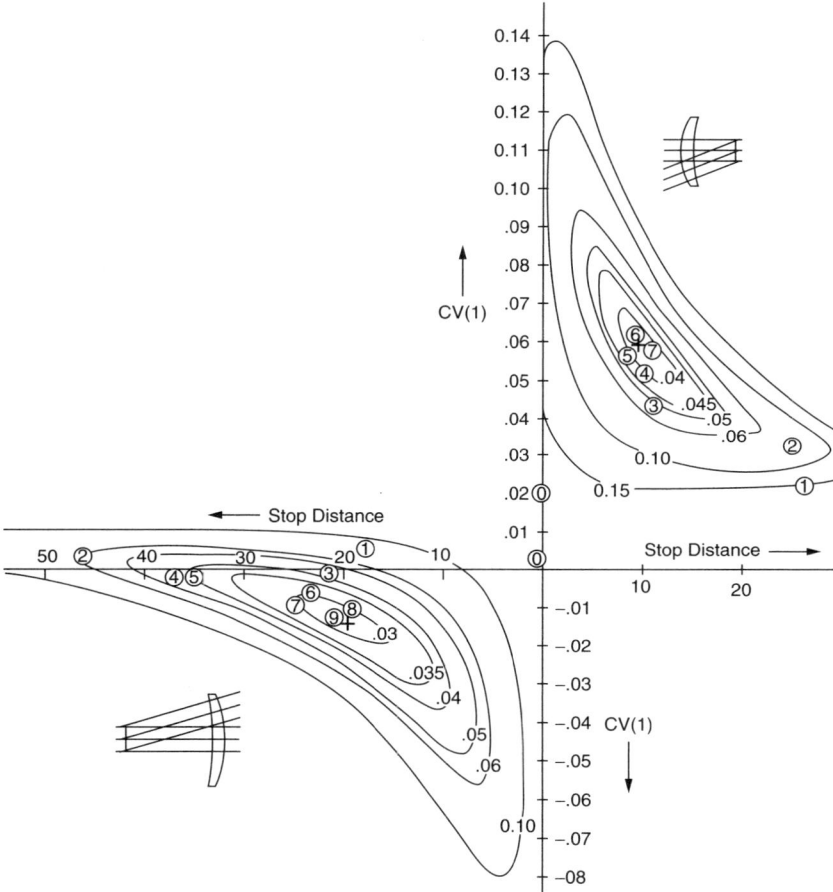

Figure 2.3 A contour plot of the merit function (rms spot size) as a function of the lens shape [cv (1)] and the stop position. The encircled numbers show the progress of the steps in the optimization process, starting from a plano-convex lens with the stop in contact with the plane surface.

meniscus has been the lens of choice for inexpensive cameras for 50 or 60 years.

The reasons for preferring the front meniscus form are threefold:

1. The overall length of the camera with the front meniscus is about 20 percent shorter than with the rear meniscus. (In the rear meniscus the length of the camera is determined by the lens focal length plus the distance to the stop.)

2. The shutter mechanism is protected and kept dirt-free because it's behind the front meniscus lens.

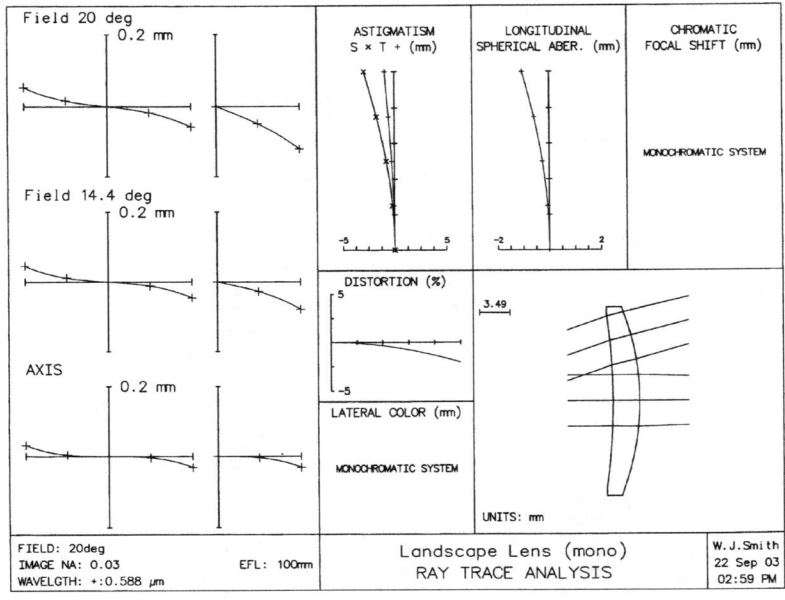

(a)

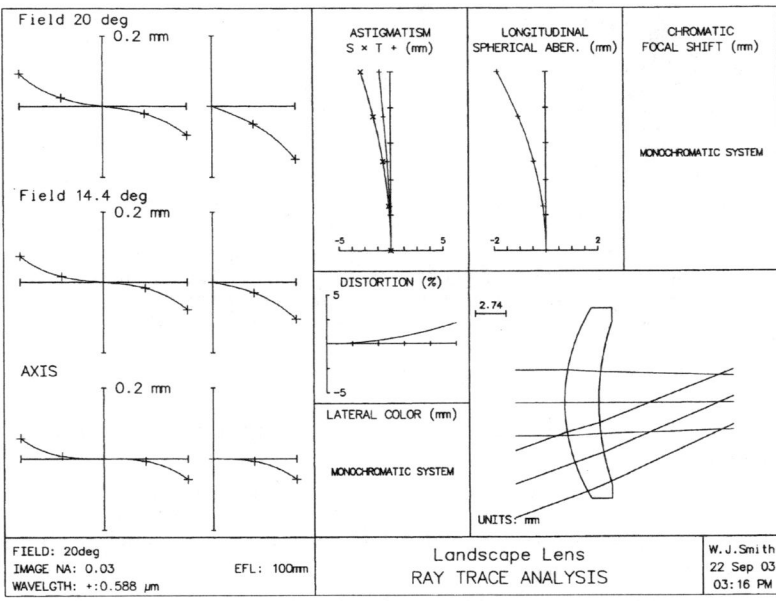

(b)

Figure 2.4 (a) The aberration plots for the rear meniscus solution, and (b) for the front meniscus solution.

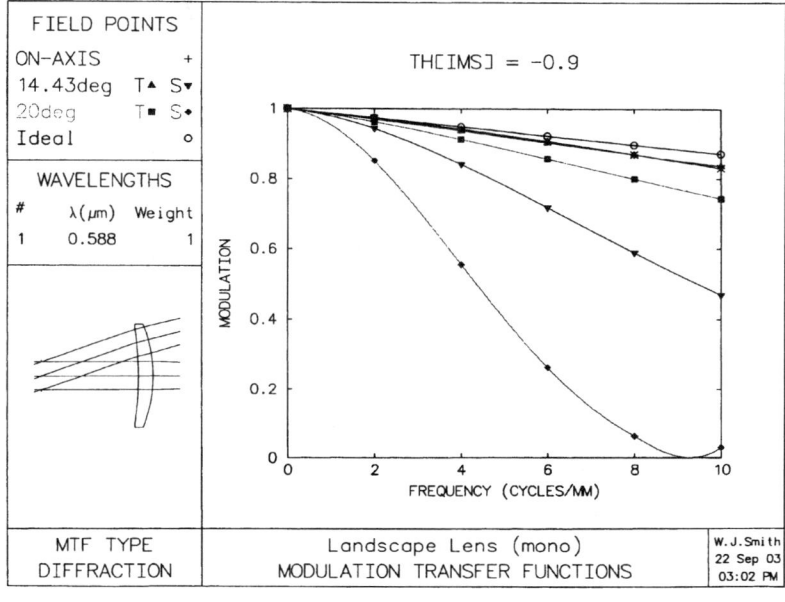

(a)

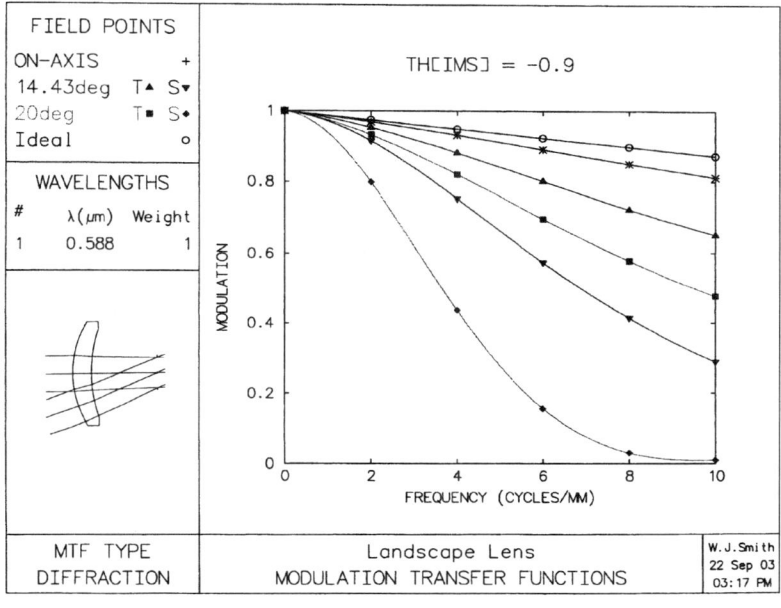

(b)

Figure 2.5 (a) The MTF for the rear meniscus solution, and (b) the MTF for the front meniscus solution. Both are defocused 0.9 mm from the paraxial focus.

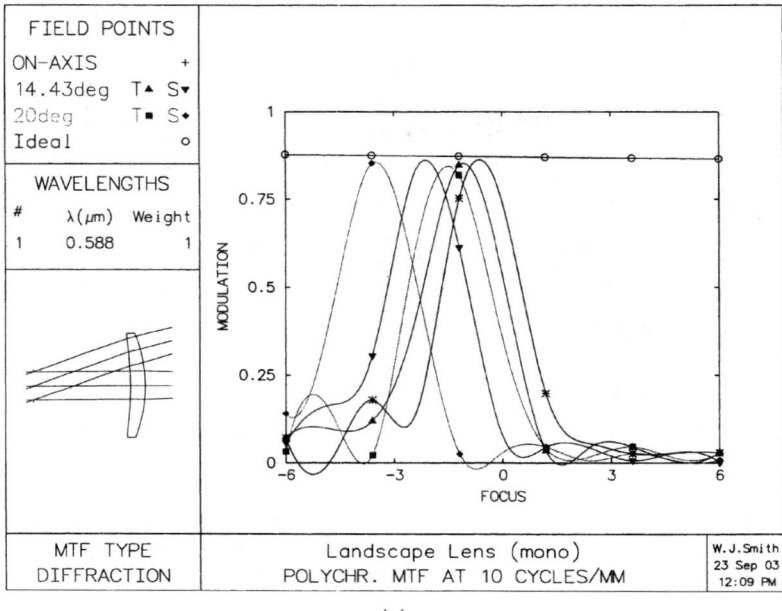

(a)

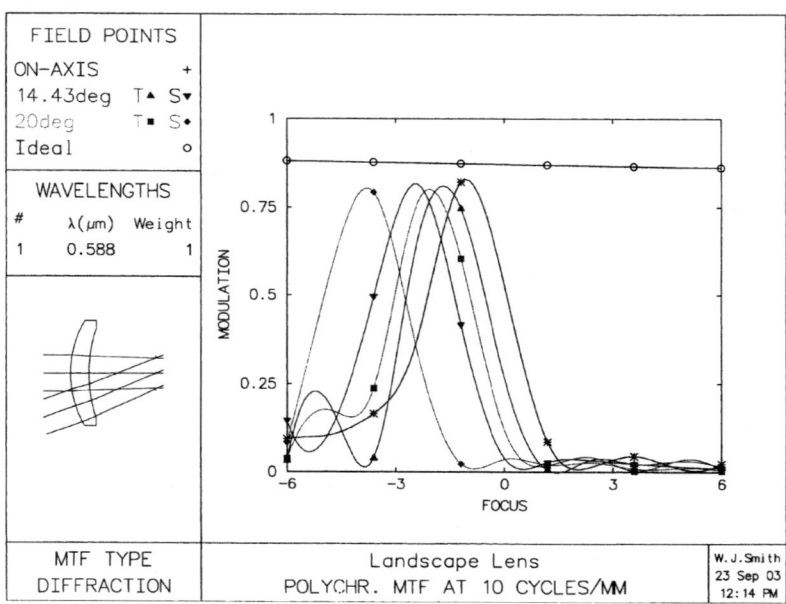

(b)

Figure 2.6 (a) The through-focus MTF plot at 10 lpm for the rear meniscus, and (b) for the front meniscus.

3. The customer sees the lens, which is larger and far more appealing than the shutter mechanism and the small aperture stop.

And in addition, since inexpensive camera lenses are injection molded out of plastic, the stronger surface curvatures of the front meniscus are no more expensive to make than the flatter surfaces of the rear meniscus.

This design form is quite forgiving in that several different merit functions that we tried all yielded quite similar (but definitely different) lenses with similar levels of the primary aberrations; however, even in simple designs, there is often more than meets the eye. It regularly turns out that, even in simple systems, many forms exist that have a similar level of performance. One must look carefully for real-world considerations that may differentiate one design from another. In addition to the basic default (rms spot size) merit function evaluated at the paraxial focus and for a single wavelength, we also tried:

a. Allowing defocus
b. Simply optimizing for zero coma and a flat tangential field
c. Optimizing with third-order aberration contributions
d. Optimizing with third-order plus fifth-order aberrations
e. Including C- and F-light spot sizes

While each different merit function produced a different design, all the designs had very similar performance as regards the monochromatic Seidel aberrations. There was, however, a real difference between the optimum monochromatic designs and those produced when the merit function included chromatic aberration. In this lens lateral color has a significant and detrimental effect on the image quality, and the longer the distance from lens to stop, the larger the lateral chromatic. (This is one of the very few cases where a spot diagram—showing the lateral color—is actually of value to the design process.) Thus, when a polychromatic optimization was done, the solutions had stop distances that were much smaller than those of the monochromatic designs. This is of course a realistic consideration, since the polychromatic design will have a much better off-axis white light performance. Another real world consideration is the effect on costs resulting from the smaller diameter required when the stop distance is reduced.

Incidentally, the "descendants" of the landscape lens include the symmetrical doublet *periscopic*, the *hypergon*, the *topogon*, and the thick double meniscus anastigmats of a century ago (which are discussed in Chap. 11).

2.5 Types of Merit Functions

Many programs allow the user to define or modify the merit function. This can be a valuable feature because it is almost impossible to design a truly *universal* merit function. As an example, consider the design of a simple Fraunhofer telescope objective: a simple four-target merit function that controls the spherical and chromatic aberrations of the axial marginal ray and the coma of the oblique ray bundle (plus the focal length) is all that is necessary. If the design complexity is increased by allowing the airspace to vary and/or adding another element, the merit function may then profitably include entries that will control zonal spherical, spherochromatism, and/or fifth-order linear coma. But as long as the lens is thin and in contact with the aperture stop, it would be foolish to include in the merit function entries to control field curvature and astigmatism. There is simply no way that a thin stop-in-contact lens can have any control over the inherent large negative astigmatism; the presence of a target for this aberration in the merit function will simply slow down the solution process. It would be ridiculous to use a merit function of the type required for a photographic objective to design an ordinary telescope objective. (Indeed, an attempt to correct the field curvature may lead to a compromise design with a severely undercorrected axial spherical aberration, which, in combination with coma, may fool the computer program into thinking that it has found a useful optimum.)

There are many design tasks in this category, where the requirements are effectively limited in number and a simple, equally limited merit function is clearly the best choice. In such cases, it is usually obvious that some specific state of correction will yield the best results; there is no need to *balance* the correction of one aberration against another.

More often, however, the situation is not so simple; compromises and balances are required and a more complex, suitably weighted merit function is necessary. This can be a delicate and somewhat tricky matter. For example, in the design of a lens with a significant aperture and field, there is almost always a (poor) local optimum in which (1) the spherical aberration is left quite undercorrected, (2) a compromise focus is chosen well inside the paraxial focus, (3) the Petzval field is made inward-curving, and (4) overcorrected oblique spherical aberration is introduced to balance the design. A program that relies on the rms spot radius for its merit function is sometimes likely to fall into this trap. A better design usually results if the spherical (both axial and oblique) aberrations are corrected, the Petzval curvature is reduced, and a small amount of overcorrected astigmatism is introduced. When one recognizes this sort of situation, it is a simple matter to adjust the weighting of the appropriate targets in the merit function to force the design into a form with the type of aberration balance that is desired. Another way to avoid this particular problem is to force the system to be evaluated/designed at the paraxial focus rather than at a compromise focus, i.e., to not allow (or to penalize) defocusing. As can be seen, the design of a general-purpose merit function that will optimally balance a wide variety of applications is not a simple matter.

Although it is not always necessary, there are occasions when it is helpful to begin the design process by controlling only the first-order properties (image size, image location, spatial limitations, and the like). Then one proceeds to control the chromatic and perhaps the Petzval aberrations. (Things may even go better if the first-order and the chromatic are fairly completely worked out by hand calculation before submitting the system to an automatic design process.) The next step in the sequence is to correct the primary aberrations (spherical, coma, astigmatism, and distortion), either directly or by using the Seidel coefficients, and finally proceed to balancing and correcting the higher-order residuals. This sort of ordered approach is sometimes useful (or even necessary) when one is exploring terra incognita, and, of course, it requires a user-defined merit function if it is to be implemented.

2.6 Stagnation

Sometimes the automatic design process will stagnate and the convergence toward a solution will become so slow as to be imperceptible. This can result from being in a very flat and broad optimum in the merit function. It can also result from an ill-designed merit function. Often first-order properties specified in the merit function are the cause of the problem. It is only too easy to require contradictory or redundant characteristics. This is especially true for zoom lenses or multiconfiguration systems, which can be confusingly complex. When stagnation occurs, or convergence is slower than you feel it should be, it is wise to stop and examine the merit function for problems. Look critically at every item in the merit function and consider what it is intended to be doing and what it *actually* does. Eliminate redundancies and try to make each entry in the merit function explicitly control its intended characteristic.

Another condition that may cause stagnation exists when the aberrations (or operands) of the initial design are extremely large. The small incremental changes in the variables used in the optimization process may produce changes in the merit function value, which are so tiny (relative to the large aberrations) that they are completely lost in the numerical noise of the computations. In this case a few manual changes to the initial design may reduce the merit function to a more reasonable level.

2.7 Generalized Simulated Annealing

The discussions above have centered on the standard DLS program or its equivalent. There have been several versions of *random search* programs proposed in the past. The most recent of these is quite sophisticated and is called *generalized simulated annealing*. In this, the computer *randomly* selects the lens dimensions (within a limited range and according to some probability distribution) and evaluates the resulting lens prescription. If the new version is better than the old, it is unconditionally accepted. If it is worse, it *may* be accepted, on the basis of random chance, weighted by a probability function, which reduces the chance of acceptance in proportion to the amount that the lens is worse than the original form. This sort of approach obviously allows the program an easy escape from the local minima described with Fig. 2.1, but it equally obviously requires a very large number of trials before a random chance can find a good combination of dimensions for the lens. Nonetheless, it does work, but not rapidly. As computers increase in speed, a program of this sort may displace or supplement the DLS as the routine of choice for automatic lens design.

In determining the acceptance of a new design that is worse than its predecessor, we define:

$$Z = \exp\left[\frac{-B \cdot \delta\phi}{(\phi_0 - \phi_{\text{best}})}\right]$$

where ϕ = value of the merit function
B = constant (e.g., ≈ 3.5)
$\delta\phi$ = increase in ϕ
$\phi_0 = \phi$ before the changes
ϕ_{best} = best possible value of ϕ

We randomly generate a number Γ between 0 and 1. Then the new design is accepted if $Z > \Gamma$. Thus Z approximates the chance of acceptance, and if:

$$\delta\phi/(\phi_0 - \phi_{\text{best}}) = 10.0 \quad \text{then} \quad Z \approx 10^{-15}$$
$$= 1.0 \quad\quad\quad Z \approx 0.03$$
$$= 0.1 \quad\quad\quad Z \approx 0.7$$
$$= 0.01 \quad\quad\quad Z \approx 0.97$$

These programs are often called *global optimization*. But consider that one can only say that such a solution is the optimum within a defined search area, after a certain number of hours of search.

Generalized simulated annealing in any of its forms is based on a specific merit function, a limited region of search, and a limited amount of search time. Given enough time, the probability is high that the global optimum within the search volume will be found.

2.8 Considerations about Variables for Optimization

The potential variables for use in optimization include: the surface curvatures, conic constants, and asphericities; the surface spacings; and the refractive characteristics of the materials involved. Occasionally tilts and decentrations are also included as variables.

2.8.1 Bounds on variables

There are several forms for bounds. The program may have a table (spreadsheet) of maximum and minimum values for the variables. There are different ways that these may be applied, i.e., soft versus hard versus elastic versus rebound. *Constraints* are handled by Lagrangian multipliers and are solved for (as exactly as the ever-present nonlinearities allow) with a priority over the minimization operands in the merit function. A variable bound may be included in the merit function as a least-squares target, as an inequality (< or >), or it may be channeled (< and >). A merit function bound on a variable allows the variable to penetrate the boundary a little, hopefully temporarily. The designer's choice of bounds, constraints, operands, and weights may affect the direction and progress toward a design solution, and may define a different local optimum.

2.8.2 Materials

Although the material characteristics are not continuous variables, for optical glasses at least the index and dispersion (or V value) can be varied within the boundaries of the glass map (Fig. 2.7) as if they were. The real glass nearest the optimized values can be substituted for the optimized glass to achieve nearly the same resultant design after another cycle or two of optimization with the real glass. Note that this is *not* true for partial dispersions, since there are relatively few glasses with partial dispersions unusual enough to be useful in the correction of secondary spectrum. Obviously, for applications outside the spectral regions where optical glass is usable, one cannot treat the refractive characteristics as variables, since the available materials tend to be few and far between.

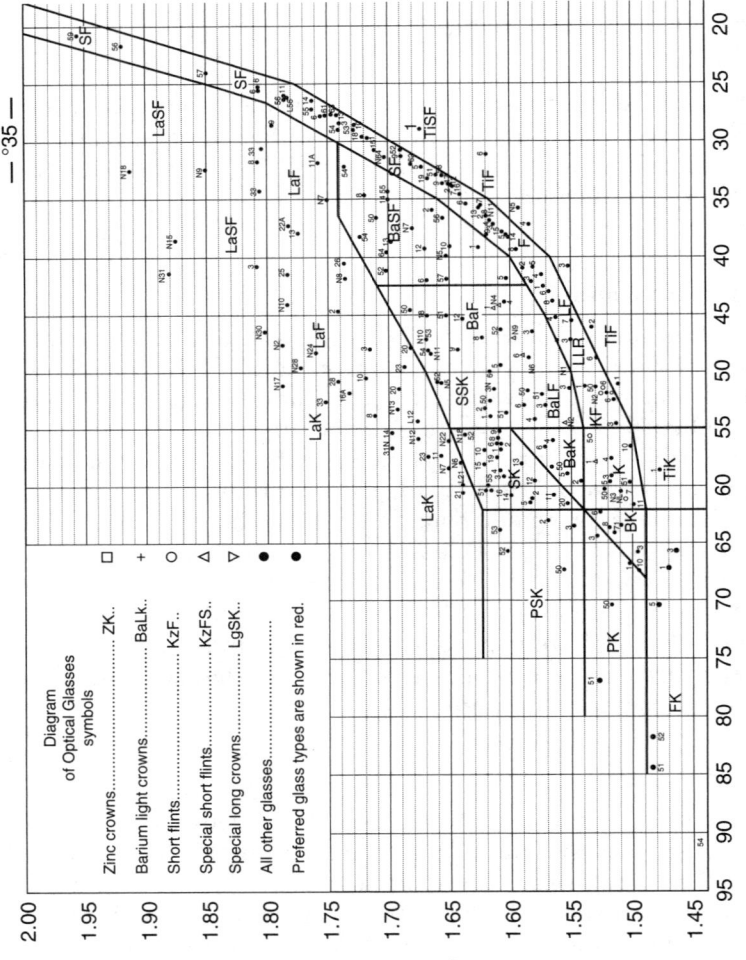

Figure 2.7 The glass map or "glass veil." Index (n_d) plotted against reciprocal relative dispersion (Abbe V value). The glass types are indicated by the letters in each area. The glass line is made up of the glasses of types K, KF, LLF, F, and SF, which are strung along the bottom of the veil. (Note that K stands for *kron*, German for crown, and S stands for *schuver*, or heavy or dense.) (*Courtesy of Schott Glass Technologies, Inc., Duryea, PA.*)

In many of the simpler types of designs it is absolutely essential to allow the glass characteristics to vary. In the Cooke triplet, for example, the relationship between the V-values of the crown and flint elements determines the overall length of the lens. As described in Sec. 8.3, the length of a triplet (and that of most airspaced anastigmats) determines the amount of higher-order spherical aberration and astigmatism; these in turn determine the aperture and field coverage capabilities of the lens. If the simpler types of lenses are to be fully optimized to suit the application at hand, the glass characteristics *must* be allowed to vary.

Some optimization programs have difficulty with the bounds of the glass map; if this is a problem, the optimization process is often facilitated by starting the variable glass well away from the boundary, so that it can find its best value before encountering the boundary problem.

It is often better to let the program vary the flint glasses rather than to vary the crowns. This is because the crowns usually tend to go toward the upper left corner (high index, high V-value) of the glass map. Flints head for the lower right corner, and are then, of course, constrained to lie on the glass line. The glasses along the glass line are numerous, inexpensive, and almost universally well behaved. On the other hand, the crown glasses in the upper left region of the map include in their number many that are expensive and/or easily attacked by the environment. Thus one might be willing to accept the computer's choice of a glass along the glass line, but would prefer to make a more discriminating selection from among the others.

Many programs allow the index to vary while holding the relative dispersion constant. For flint glasses, this amounts to varying the glass along the *glass line*. This is a very useful feature because this is often exactly the type of variation desired for flint glasses.

Be sure to bound the index and dispersion if they are allowed to vary; otherwise the index may well go to two or three, and the relative dispersion may go anywhere from more than one to less than zero.

2.8.3 Curvatures

In general, one would expect to want to make use of every available variable. This is almost always true regarding the curvatures, all of which, unless there is a reason to constrain the shape of an element, are usually allowed to vary. Unless the last surface is close to the image, the value of the last curvature can be "solved" to maintain μ'_k; this controls the focal length or the magnification of the system.

2.8.4 Airspaces

Ordinarily, airspaces may be regarded in the same light as curvatures, since they are continuously variable and are very effective variables.

2.8.5 Defocusing

Although the distance by which the design image plane departs from the paraxial focus is usually an airspace and can be regarded as a variable, its effects can be insidious. If the image surface is allowed to depart from the paraxial focus from the beginning of the optimization process, an unfortunate lens may result. In some lenses, and with some optimization merit functions, the tendency is to produce a lens with:

1. The image plane well inside the paraxial focus
2. A large undercorrected spherical aberration
3. A strongly inward-curving field
4. A heavy overcorrecting oblique spherical

Although this combination occupies a local optimum in the merit function, this is often not the best state of correction. It fools the optimization program because the undercorrected spherical causes the best axial focus to lie to the left of the paraxial focus and the overcorrected oblique spherical causes the best off-axis focus to lie to the right of the inward-curving field; the net result is that, to the program, the field seems flat. One can usually avoid this pitfall by not allowing any defocus in the early stages of the optimization and/or putting a heavy penalty on the defocusing. Note that for *some* lenses, such as non-diffraction-limited systems, the correction described above may in fact be a good one. See the comments on aberration balance in Sec. 3.8.

Note that the defocus variable, as a departure from the paraxial focus (which can be set by a height solve), is much more stable than simply using the distance to the image surface as a variable. Including the defocus, suitably weighted, in the merit function will put a brake on a runaway defocus variable.

2.8.6 Thickness

Element thicknesses must be regarded quite differently than airspaces. They must of course be bounded by the necessity for a practical edge thickness for the positive elements and a reasonable center thickness for the negative elements. In many designs, element thickness is an insignificant and ineffective variable (and one whose effects are easily duplicated by an adjacent airspace). In this circumstance one can arbitrarily select a thickness on the basis of economy or ease of fabrication. The elements in such designs are typically quite thin; for example, consider a telescope objective or an ordinary Cooke triplet.

There are, however, many systems in which the element thickness is not only an effective variable, but one that is essential to the success of the design type. The older meniscus lenses (Protar, Dagor, and the like) and the double-Gauss forms all depend on the separation of the concave and convex surfaces of their thick meniscus components to control the Petzval curvature and, in many instances, the higher-order aberrations as well. In lenses of this type it is absolutely essential that these glass thicknesses be allowed to vary.

One must be wary of, and skeptical toward, a thickness variable, which is very weakly effective. Occasionally an optimization program will produce a design with an overly thick element, where the large thickness produces only a very small improvement (which is not worth the added cost of producing the thick element). This occurs because the optimization routine will seek out any improvement that it can get, no matter how small, and without concern as to the cost. It is wise to test the value of a thick element if there is any doubt about its utility. This is readily accomplished with another optimization run, which fixes the thickness in question to a smaller value. Very often the performance of the thin version will not be noticeably different from that of the optimum thicker version. Although most significant with respect to lens thickness, this same rationale obviously applies to air-spaces as well. Another solution is to add the errant thickness to the merit function with a small or zero target.

2.8.7 Aspheric surfaces

Surface asphericity can be an extremely effective (if sometimes expensive) variable, but it is one that often requires a bit of finesse. On occasion, one may be ill-advised to begin an optimization with the conic constant and all the aspheric deformation coefficients used simultaneously as variables. The conic constant and the fourth-order deformation coefficient both affect the third-order aberrations in exactly the same way. Thus they are at least partially redundant. But more significantly, identical variables have an undesirable effect on the mathematics of the optimization process. It is often advisable to vary one or the other, but not both. A safe practice is to vary only the conic constant (or the fourth-order term) at first, and then add the higher-order terms (sixth, eighth, and tenth) one at a time, as necessary. The tenth-order term is, in many systems, totally unnecessary, adding little or nothing to the quality of the system; in fact, the eighth-order term is often something that can be done without.

A surface defined by a tenth-order polynomial can cause the spherical aberration to be corrected exactly to zero at four ray heights. If there are only four axial rays in the merit function, their ray intercept errors may all be brought to zero; the danger is that, between these rays, the residual aberration may be unacceptably large. A tenth-order surface can be a rather extreme shape. Thus the use of an aspheric surface sometimes calls for more rays in the merit function than one might otherwise expect to need. With a program that allows wavefront deformation or OPD targets in the merit function, the severity of this problem can be lessened somewhat.

Note that a conic constant is totally meaningless when applied to a plano surface. In order to aspherize a plane surface, a fourth-order deformation must be used. An aspheric located at the aperture stop can only (in the third order) affect the spherical aberration. If it is desired to affect coma, astigmatism, or distortion, the aspheric should be located well away from the stop. An aspheric does not affect Petzval or chromatic. An argument can be made that no more than three aspherics are necessary in a lens, one near the stop, plus one near each end of the lens. This is probably an oversimplification for complex lenses with significant higher-order aberrations. Note also that the third-order aberrations tend to be affected by the fourth-order surface deformations, the sixth-order deformations are effective against the fifth-order aberrations, and so on.

2.9 How to Increase the Speed or Field of a System and Avoid Ray Failure Problems

Very often, the lens designer is faced with the necessity of increasing the speed (i.e., relative aperture, numerical aperture, and the like) and/or the field of view of an existing optical system. There are two common reasons why this may be desirable. One may want to adapt an existing design (such as those in this book) to an application that requires a larger aperture or wider field than that for which the original lens has been configured. The other common situation is simply the creation of an entirely new system with a relatively large aperture and/or field. In either case the difficulty that can arise is that the rays that are needed to design the system may not be able to get through the initial lens prescription.

There are two reasons that a ray may not be able to get through. One reason is that the height of the ray at a surface may be greater than the radius of the surface; the ray simply misses the surface entirely and its path obviously cannot be calculated any further. The second reason is that the ray may encounter *total internal reflection* (TIR) in passing from a higher index to a lower; again, the ray path cannot be calculated. Each of these conditions represents a boundary that, if crossed, causes failure of the ray trace. Note also that as these boundaries are *approached*, the situation rapidly becomes very unstable. This is because close to the boundary, the angle of incidence (for the case of the ray height approaching the value of the radius) or the angle of refraction (for the case of the ray approaching TIR) is very near to 90°. Near this angle, Snell's law of refraction ($n \sin I = n' \sin I'$) becomes *very* nonlinear, producing a highly unstable situation, which often explodes as the lens construction parameters are incremented to calculate the partial derivative matrix in the course of the optimization.

A good way of dealing with this situation is simply to back off from the aperture and/or field requirement that is causing the problem. Most design programs have the capability to easily adjust or scale the aperture and field angle. If a change is made to smaller values of aperture or field, the rays will no longer be so near to the failure boundary. If the lens is now optimized, the program is very likely to adjust the lens parameters so as to reduce the angles of incidence, because this is usually a change that causes the aberrations to be reduced. The optimizing changes can thus be expected to pull the problem situation in the system further away from the ray failure boundary, *if this is possible*.

After the optimization has relaxed the problem, the field and/or the aperture can usually be adjusted to a moderately larger value without again encountering the failure boundary. Depending on just how sensitive the system is to the problem, an increase of about 10 to 50 percent may be appropriate. Now another cycle of optimization will strongly tend to again reduce the troublesome angles of incidence.

This process of adjusting the field or the aperture to larger values and then optimizing is continued until the desired aperture or field is attained without ray failures. This works well, *provided* that this desired result is possible for the lens configuration under study. It may be necessary to choose another configuration, usually one with more elements. When this is necessary, a drawing of the lens and rays (of the last design form that has successfully passed all the rays) will usually indicate which rays and which surfaces are causing the problem. One simply looks for angles of incidence or refraction that are large (and often near 90°). Then the offending element can be split into two (or more) elements shaped to reduce these angles. The scale-and-optimize process can now be repeated with a much improved chance of success. Note that the examination of the critical ray paths (typically those of the marginal rays) for large angles of incidence or refraction is a comple-tely general technique that will often indicate the source of a design problem.

2.10 Test Plate Fits, Melt Fits, Thickness Fits, and Reverse Aberration Fits

When the deleterious effects of fabrication tolerances become too large to bear, a technique commonly used to reduce these effects is to fit the lens design to the known values for the radius tooling and/or to the measured glass indices. The former is called a *test plate* (or *test glass* or *tooling*) fit; the latter is called a *melt fit*.

The *test plate fit* is begun by first obtaining a list of the available test plate radii from the shop that is scheduled to fabricate the lens. It is wise to ascertain that the radius values of the list are not just nominal values but are based on accurate and recent measurements of the test plates, since there is often a significant difference between the two.

The fit is carried out as follows: A surface is selected at which to begin the process. This selection is based on one of the following criteria: (1) the surface most sensitive to change,* (2) the surface with the shortest radius, (3) the strongest surface (i.e., with the largest surface power $[n' - n]/r$), or (4) the surface that shows the largest curvature difference from the nearest available test plate radius. Very often most, if not all four, of these criteria will indicate the same surface; if not, the choice of which one to use is almost a matter of taste. The nearest radius on the test plate list is substituted for the selected surface and the lens is reoptimized, allowing all the variable parameters to change (except of course, the radius that has been set to the test plate value).

Another surface is then chosen and fixed to the nearest test plate radius; the lens is reoptimized again. This process is repeated until all the surfaces have been fitted to test plates. It is usually wise to avoid fitting both surfaces of a singlet (or all surfaces of a component) until all components have at least one fitted surface each. This allows the unfitted surface to vary and adjust for power, chromatic, and Petzval aberrations and the like for as long as possible.

With a reasonably complete test plate list, all radii can usually be fitted without significantly degrading the image quality (i.e., the merit function). In fact, the merit function is often slightly improved by this process because of the additional cycles of optimization that have been performed plus the small shocks or jolts the radius changes produce. If the test plate list is limited or has a gap in it, it may not be possible to fit all the design radii to test plates without degrading the performance. In such a case, one must either fabricate new tooling and test plates or seek a new vendor for the parts. One should *always* try for a 100 percent fit.

A *melt fit* reoptimizes the lens design by using measured data for the material indices instead of the nominal values from the glass catalog. This measured data comes in the form of what is called a *melt sheet* provided by the glass manufacturer or supplier. For noncritical applications, this data is usually sufficient. A worthwhile elaboration of the process is to determine the difference between the measured index values and the catalog values. These differences are then plotted against wavelength. This plot should be a smooth, relatively level line. A data point that does

---

*Note that the relative sensitivity of any dimension of the system can be determined quite easily by making an incremental change in the dimension and noting the effect it produces on the merit function. This, of course, assumes a merit function that represents the quality of the image reasonably well.

not plot smoothly is suspect; the measured data may well be in error. A slope of the line indicates a *V*-value difference from the catalogue values. A curvature of the line indicates a change in the partial dispersion. Next, a smoothly drawn curve through the points is used to determine improved values for the index differences. These differences are then applied to the catalog values to arrive at better values for the measured melt data, and the improved values are used in the melt fit reoptimization. The smoothed curve of differences can also be used to determine the index for wavelengths that are not included in the melt sheet.

An ordinary melt sheet will list the indices for the wavelengths of d, e, C, F, and g light. These are usually not individually measured. Instead, the index is measured for d light and the index difference between C and F light is measured. These two measurements are fed into a computer program, which uses the known characteristics of the glass type to calculate the indices for e, C, F, and g light. This, while not ideal, is adequate for many, even most, applications; however, for some critical applications, and for all applications in which an attempt has been made to reduce or correct the secondary spectrum, it may be quite unsatisfactory. For an additional charge, the glass manufacturer can provide what is usually called a *precision* melt sheet, for which the indices have been measured individually and to a greater precision. Index values for wavelengths specific to the application can also be measured. It is, of course, wise to subject even this data to the difference test and smoothing process described in the preceding paragraph.

We called the third of these fits a *thickness fit* in the heading for this section. This is a process that is carried out during the final assembly of the lens. In sum, one reoptimizes the lens by varying the airspaces of the system, using accurately measured data for the radii, thicknesses, and indices of the fabricated elements. If the melt data and the test plate data are available, this just amounts to adding the measured element thicknesses and reoptimizing.

In a *reverse aberration fit*, the aberrations (or wavefronts) of an assembled system are measured. Then the nominal design is optimized to match the aberration (or wavefront) values that were measured. If the system is then adjusted by the opposite of the changes found by the optimization, the expectation is that the system performance will be improved. This technique has been used with success in multimirror systems, which are not axially symmetric and in which the alignment and positioning of the mirrors in space is both difficult and critical.

As can be seen, all of the above procedures are designed to almost completely eliminate the effects of the fabrication tolerances on the performance of the system. What is left, instead of the tolerances, is the uncertainty or inaccuracy of the various measurements on which the fits are based. The effect is usually quite modest and, therefore, acceptable;

however, these uncertainties are, in fact, the exact equivalent of tolerances in determining the performance of the fabricated system.

2.11 Spectral Weighting

In any ray-tracing process, the index of refraction of the material is, of necessity, that corresponding to a single specific wavelength. Most lens *design* programs use three wavelengths to represent the spectral bandpass for which the system is to be designed. Some allow more. *Analysis* programs may allow the use of five or ten suitably weighted wavelengths in calculating such things as MTF, point spread functions, radial energy distributions, and the like.

In either case, the question becomes, "What wavelengths should I use, and how should they be weighted?" For visual systems and just three wavelengths, the classical answer is to use d (or e), C, and F light, with weightings typically set at 1.0, 0.5, and 0.5, respectively, or perhaps with a uniform weighting. For other applications, this approximately corresponds to using the central wavelength and the wavelengths 25 percent from the extreme edges of the passband. If the results of an image analysis need not be especially precise, three wavelengths may be sufficient. For calculations done in the midst of the design process, this is often good enough to enable a judgment as to the relative merit of, or the rate of improvement between, two stages in the design process.

However, what should one do *in general?* To immediately dispose of the obvious, it is apparent that the more wavelengths used, the more accurate the results can be. That said, how do we select the wavelengths to be used? There are three obvious choices. In order to assess the full effects of the chromatic aberrations, one would like to include the extreme long and short wavelengths. One would also probably like to consider dividing the spectral weighting function into increments of equal power or response, so that each wavelength would represent an equally weighted sector. But if this is done, the extremes of the spectral passband are not included. Another possible option is to choose wavelengths that are evenly distributed across the spectral band, and to weight them according to the spectral response function. One might choose an even distribution on a wavelength scale, or what could be a bit better, an even distribution on a wave number (reciprocal wavelength) scale.

From this it should be apparent that most wavelength and weighting choices are based on some sort of compromise. Often the outer two wavelengths are chosen to be fairly close to the ends of the passband and the intermediate wavelengths and weights are a compromise between an even power distribution and an even wavelength spacing. If there are peaks or bumps in the spectral response function, wavelengths are often selected to be at or near the peaks. Note that, to a limited extent, one's choices will partially control the design process: heavier weighting at the ends of the spectrum will obviously emphasize both primary and secondary chromatic aberration, spherochromatism, and the like; heavy weights on the central wavelengths will emphasize the monochromatic aberrations at the expense of the chromatic aberrations.

2.12 How to Get Started

Experienced designers are often asked questions such as "How do you know where to start?" or "How did you decide that a (name of a design type) could meet the specs?" or "Why did you shift the components around?"

The answer to these questions is almost always "experience," which probably means that the full answer is different for every problem. It would be foolhardy to pretend to be able to give a complete, definitive answer or even a set of answers to such questions, but there are a few guidelines that should be reasonably dependable.

Figure 2.8 is, in effect, a compilation of some of that experience. If the system to be designed roughly corresponds to a photographic objective or to one of the other types indicated, the figure can be used as an easy guide to the selection of an appropriate form. In this plot the areas corresponding to various combinations of field and aperture are labeled to indicate the type of system commonly used there. Obviously the boundary of each area is fuzzy and ill-defined. In a presentation of this type there is also an implied level of performance within each area, which is presumably typical of each particular lens applications. In general, the performance (image quality, resolution, or whatever) is better when a given lens type is optimized for a smaller field and/or smaller aperture, that is, for a combination closer to the origin of the plot. Thus one can select the design type from Fig. 2.8 corresponding to the field-aperture combination required with reasonable assurance that it is an appropriate choice. But should the performance prove inadequate for the application at hand, one can move up to the design form above and to the right on the plot, which should have a greater performance capability. Of course, if the required level of performance is known in advance to be relatively high, one would select the more capable type to begin with. The same selection approach can be applied to the components of a more complex system. There are many design types and applications, which are not represented on this necessarily abbreviated chart; however, many of these are presented in the lens designs included in this book. The reader may wish to mark up Fig. 2.8 to add the types that are of particular interest to him or her.

For a more complex system, the approach must start on a more fundamental level. The first step is to collect and tabulate the requirements to be met; these may include such things as:

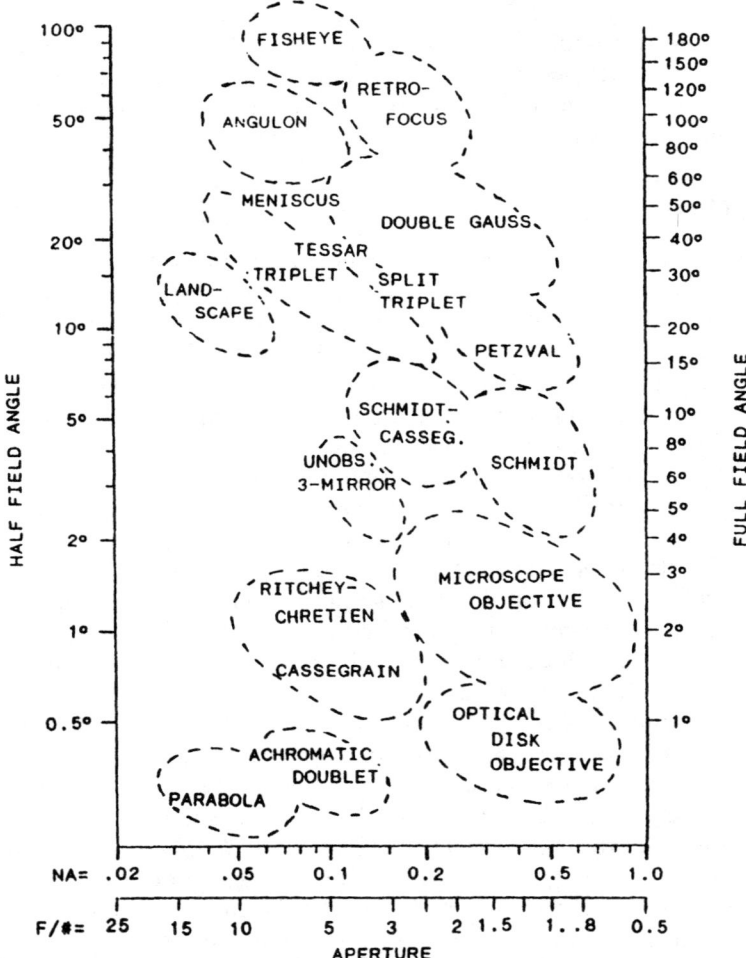

Figure 2.8 Map showing the design types commonly used for various combinations of aperture and field of view.

Resolution or performance (versus diffraction limit)

Wavelength

Fields of view

Image size

Aperture, numerical aperture (NA), and f number

Vignetting or illumination uniformity

Focal lengths or magnifications

Space limitations, clearance, windows, and so on

(See Chap. 1 for a more extensive listing of possible specifications.)

The next step is to make a first-order layout of the system that will satisfy the requirements. The first-order layout is simply an arrangement of component powers (or focal lengths) and spacings that will produce an image in the required location, in the required orientation, and of the required size. At this stage, no consideration is given to the design type of each component; one is concerned only with its power, aperture, and field as first-order, i.e., paraxial, characteristics. For systems (or portions of systems) consisting of two components, the equations of Chap. 24 can be extremely useful. For more complex systems, the component by component ray-tracing equations may be used. A general approach is to trace paraxial rays that will define the required characteristics. The ray-trace results (ray heights, ray slopes, intersection lengths, and so on) can be expressed as equations with the component powers and spacings as unknowns to be solved for. An important facet of this stage is that one should try to find a layout that minimizes the component powers, or minimizes (or equalizes) the "work" (ray height times component power). Doing this will almost always produce a system with less aberration residuals, one that is less expensive to fabricate and less sensitive to fabrication and alignment errors. This process can be carried out with an optical design program as well.

Sometimes one can leap directly from the first-order layout to choosing the component design types, and thence to the optimization stage; however, if this is not the case, the next step is usually to analyze and/or correct the chromatic aberrations. Equations (24.85) and (24.97) can be used for the whole system, or the components can be individually achromatized. To this end, the element powers for a thin achromatic doublet are given by

$$\phi_A = \frac{V_A}{(V_A - V_B)F} \tag{2.1}$$

$$\phi_B = \frac{V_B}{(V_B - V_A)F} = \frac{1}{F} - \phi_A \tag{2.2}$$

and the element surface curvatures are determined from

$$C_1 - C_2 = \frac{\phi}{n-1} = \frac{1}{r_1} - \frac{1}{r_2} \tag{2.3}$$

At this point a sketch of the system is often helpful. Simply make a scale drawing, showing each element as either a plano-convex or an equi-convex form (plano-concave or equi-concave for negative elements). If the elements look too fat, they should be split into two or more elements. Be sure that the element diameters are sized properly for the rays that they must pass. At this stage the system should begin to look like a good lens. Again, the computer can be used to make the drawings.

The next step is to give some consideration to the Petzval curvature. Choosing suitable anastigmat types for the components is one way to handle this. Another is to use a field flattener in an appropriate location (i.e., near an image) in the system. And, of course, the usual device of configuring the system or component with separated positive and negative elements or surfaces to reduce the Petzval sum can always be utilized if a new design must be created from scratch.

Often many of these steps can be handled conveniently and expeditiously by the automatic design program. The first-order layout can be done with zero-thickness plano-convex or plano-concave elements, allowing the spaces and the curvatures of the curved surfaces (but not the plano surfaces) to vary. The merit function is a simple one, configured to define the required first-order characteristics. Note well the comments in Sec. 2.6 regarding stagnation and contradictory or redundant first-order entries in the merit function.

Since the chromatic aberrations and the Petzval curvature depend on element power and not on element shape, the lens design program can also be used to find a layout that is a preliminary solution with the chromatic and Petzval adjusted to desired, reasonable values, which typically should be small and negative.

Sometimes it is useful as a next step to allow the elements to bend and to correct the third-order aberrations. This can be done by putting an angle-solve on the second surface of each element so that the axial ray slope is maintained. More often than not, however, the next step will skip over the third-order and go directly to a full-dress thick lens optimization run.

A design start may be based on: (*1*) previous work from one's files; (*2*) an idea for the general configuration necessary; (*3*) published prescriptions or scalable drawings; (*4*) lens libraries; and (*5*) imagination and intuition.

Some starting points that have been suggested:

a. A series of plano–plano plates.

b. A series of nondispersive aspheric singlets, to be later converted to achromatic doublets after the first-order and the monochromatic aberrations have been satisfied.

The starting point (b) has, under the ministrations of a good designer, been successful. But the unguided start from plane parallel plates is almost a parlor trick. Once the merit function is defined and the thickness, spacing, and glass of the plates are given, we must realize that, for a given program, the end result is then completely predetermined; there is no magic. A good design result is of course possible, and occasionally a pleasant surprise can surface, but this is unlikely unless there is a bit of chicanery or some steering of the process (the latter is, of course, why there are lens designers).

A general starting process:

1. Do a thin lens layout and drawing. Look for ridiculous powers, spacings, and the like.
2. Achromatize, do a drawing, and look again.
3. Revise the layout to reduce, add, or split elements.
4. Now start the program.

2.12.1 Hypothetical, generic design process

After selecting a design form or type:

1. Start by varying all the curvatures and the large air spaces. Manually adjust the center thicknesses of the positive elements for a suitable edge thickness. Check all bounds and first-order targets for compliance. Consider the general practicality of the result. Consider splitting elements that seem too strong.

2. Add all thicknesses and spaces to the variable list and optimize. Correct and temporarily freeze any problem areas (e.g., CT < 0.0, ETH < 0.0, or extreme thickness or spaces). If a ray fails (by missing a surface or by TIR), restart with only first-, third-, and fifth-order targets in the merit function. These cannot fail because they are not based on trigonometric (exact) rays.

3. Add glass to the variable list, either along the glass line, or vary both index and dispersion. Use tight bounds, because glass variables can easily get out of control. Sometimes varying only one or two glasses at a time is a useful limitation.

4. Adjust the target weights up or down or add targets as appropriate to control or attack trouble spots.

5. Fix the glasses, one at a time, to catalog types and vary all curvatures and thicknesses.

As the design progresses, occasionally check the MTF (both through focus and through frequency). Check the distortion and lateral color; it's easy to overlook them in the heat of battle. Draw the lens and the marginal rays at several field angles. Look for large incidence angles; enlarge the scale; and examine any trouble spots. Print out the third- and fifth-order surface contributions and look for any surface with larger than average contributions; these often signal trouble.

In the best of all worlds, one simply sets up what seems to be a likely layout and proceeds directly to the automatic lens design program, which promptly turns out an excellent design. "Experience?" Possibly. But certainly, "Lots of luck!"

Chapter 3

Improving a Design

3.1 Lens Design Tip Sheet: Standard Improvement Techniques

This is a listing of changes in the design form or the design approach that may significantly improve or change a lens design. Many changes can be incorporated without a major change in the system's power or magnification. They can be introduced blindly because they are expected to be beneficial.[*] Note that most of these changes require additional intervention, i.e., complementary changes to the balance of the system to restore (and hopefully improve) the design. In no particular order, the suggestions are:

1. Bend an element. Change its shape but maintain its power.
2. Add a field flattener or a concave surface close to the image.
3. Remove a weak element (add cv_1, cv_2, $cv_1 - cv_2$, and th to the merit function with zero targets, and gradually increase the weighting).
4. Split a strong troublesome element into two weaker elements. Beware of *total internal reflection* (TIR) at the split; hold the incidence angles to less than the critical angle, $I_c = \arcsin(n'/n)$.
5. Replace a thick element with two thin elements and an airspace.
6. Add a plane parallel plate in front or in back of the lens, or in a large airspace, and vary its curvatures and thickness.
7. Add an element and later remove it (gradually, by targeting it in the merit function).

---

[*]This list was inspired in part by Ellis Betensky in "Postmodern Lens Design," *Optical Engineering*, Vol. 32, No. 8, pp. 1750–1756, Aug. 1993.

8. Replace a singlet made of expensive high-index glass with two elements of cheap glass.
9. Raise the index of the materials, maintaining element power.
10. Introduce unusual partial dispersion materials if secondary spectrum is a problem.
11. Convert a singlet into a cemented doublet with glasses of: (*a*) different indices, (*b*) different *V*-values, or (*c*) both.
12. Cement two adjacent surfaces that have nearly equal curvatures.
13. Reverse, i.e., turn around, a cemented doublet.
14. Change a thick, strong element into a cemented doublet.
15. Split (airspace) a cemented doublet (beware of TIR).
16. Insert a thick, low-power meniscus element before, after, or within the lens. A concave surface near the image can act as a field flattener. A *concentric* shell field flattener, which is also concentric with the focus, will not change the efl, focus position, spherical, coma, axial chromatic, lateral chromatic, or sagittal field curvature. It will affect the Petzval curvature, the distortion, and the tangential field curvature.
17. Reverse or flip a meniscus element, or a low-powered *pair* of elements. A concentric shell flipped about its center of curvature will not change *anything*, but it may lead the design to a quite different form.
18. Add thickness to a positive power meniscus to improve the Petzval field curvature by holding the power constant, which will require a stronger concave surface.
19. Add an aspheric to a surface and later remove it gradually by targeting it to zero in the merit function.
20. Insert or add a *disposable* zero power aspheric element, which will be removed later on.
21. Aspherize a surface (near the stop for spherical, or away from the stop to also affect coma, astigmatism, and distortion).
22. Move the stop to a new location to use a better portion of the oblique beam (or avoid a worse portion).
23. Look for surfaces or elements with large third- or fifth-order contributions or a large angle of incidence. They are usually troublemakers.
24. Make a large change in a constructional parameter and freeze it for several optimization cycles, then release it. Choose a parameter that looks troublesome. Change it in the direction of reducing the angles of incidence.

25. Temporarily loosen (or remove) a bound or constraint, or reduce the weight on a target to allow the design to temporarily cross over the boundary to find a new solution, and then reimpose the limits.
26. Systematically vary the starting point for the optimization (in hopes of finding a new and better local optimum).
27. Reduce the weights on the off-axis targets in the merit function (or eliminate them entirely) for a few cycles, and then raise (or reapply) the weights.
28. Temporarily reduce the damping factor in *damped least squares* (DLS) to escape from a local optimum (by allowing changes that increase the value of the merit function—by a factor of 100 to 10,000 per Dave Shafer), then reoptimize.
29. After the DLS process has bottomed out, some hacking around can often squeeze a little more out of a design by making more or less random arbitrary changes and then reoptimizing; changing to a different merit function, e.g., OPD versus spot size versus an aberration merit function; and jolting the design and reoptimizing.
30. Try a reduced or limited set of variables—the important ones.
31. Remove the following: ineffective variables, variables that null each other, and targets that are redundant.
32. Modify the merit function to severely target the most bothersome aberrations by increased weighting or by added operands.
33. In general, add new rays and targets to the merit function to pressure the aberrations identified as problems in the ray intercept plots.
34. Add more rays, fields, and colors to the merit function toward the end of the design process to get more accurate evaluations and also to be sure that something has not slipped through the cracks.

A good part of the "art" in lens design consists of knowing when to add or delete an element. The simple, straightforward application of these techniques is no guarantee of improvement in a lens in that they do not automatically correct the defects that they are intended to address. In general, these changes tend to reduce the aberration contributions of the modified components; to take full advantage of this, the aberrations of the balance of the system must be reduced as well. The operative principle is this: if large amounts of aberrations are corrected or balanced by equally large amounts of aberrations of opposite sign, then the residual aberrations also tend to be large. Conversely, if the balancing aberrations are both small, then the residuals tend to be correspondingly small.

Although things may go smashingly well if you vary everything at once, it's not always the best approach to vary everything right from the start. If there are many variables, it is sometimes useful to first vary only the curvatures and then add the thicknesses and spacings. Finally, allowing only one or two glasses at a time to vary is often wise.

When you have located the source of a troublesome aberration (perhaps by checking the third-order contributions) it should be corrected as soon (i.e., as close) as possible. Allowing it to propagate through the lens is a source of higher-order aberrations.

Always save any promising result you find in the course of a design as soon as you find it. Save it as a file or as a printout. After a long session of design changes and investigations, you may never be able to find it again, even if you try to retrace your steps from the start of the day.

In a finite conjugate system, do not constrain more than two of: (*a*) focal length, (*b*) object to image distance, or TTL, and (*c*) the magnification. Probably the best one to leave out is the focal length. If all three are constrained, a simultaneous solution may still be possible, but the design form will be severely limited and a better solution, which meets the magnification and TTL specifications, may be missed.

A good way of deciding where to attack a design is to draw the lens with marginal rays (and the upper and lower rays of the oblique beams). Look for large angles of incidence or refraction. This, combined with an analysis of the surface contributions of the third- (and fifth-) order aberrations will often pinpoint the source of troubles. A large ray height, either for the marginal or principal ray, may also indicate the surfaces at risk. If the angle of incidence or refraction of the ray is detrimentally large, or if the slope change (i.e., $u' - u$) is extreme, it can be added to the merit function to drive it down. The Aldis total aberration (see Ref. 3) is available in some programs; it is an all-inclusive representation of the surface contribution.

In most lenses there are many small variations of the design form with nearly equivalent performance or *modulation transfer function* (MTF). Note that the merit function and the MTF are very rarely well correlated (if at all). Poke around and look for something of interest that is off the beaten (DLS) path.

Other than at unit magnification (1:1), a fully symmetrical construction is usually not the best solution, although it may be a useful starting point. In fact, even at 1:1, the computer may find an unsymmetrical solution it likes better. Bear in mind that the improvement from the $f/4$ Plasmat lens to the $f/2$ double-Gauss lens was made possible by abandoning symmetry (and thickening the inner doublets). Although it's not a sufficient condition for a good design, a small Seidel sum is a necessary condition for a good design.

To take care of a troublesome aberration, target it in the merit function with a higher weighting, add the appropriate third- and fifth-order targets to the merit function, and add a ray (or field) to get it under control.

3.2 Glass Changes: Index and V-value

The refractive characteristics of the materials used in a lens are obviously significant and important to the design. In general, for a positive element, the higher the index the better. The higher index reduces the inward Petzval curvature, which plagues most lenses. It also tends to reduce most of the other aberrations as well. As an example, see Fig. 3.1, which clearly

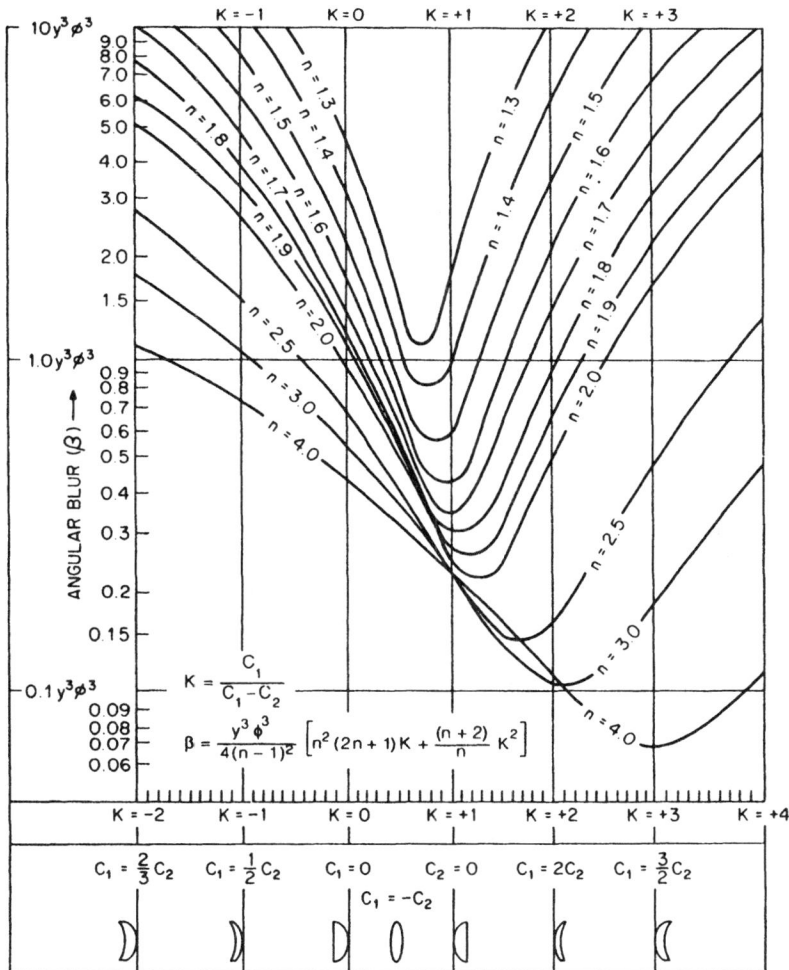

Figure 3.1 The angular spherical aberration blur of a single lens element as a function of lens shape, for various values of the index of refraction; ϕ is the element power and y is the semiaperture. The angular blur can be converted to longitudinal spherical aberration by $LA = 2\beta/y\phi^2$, or to transverse aberration by $TA = -2\beta/\phi$. (The object is at infinity.)

indicates the effect of higher index in reducing the spherical aberration of a single element. This sort of reduction is primarily a result of the fact that the surface curvature required to produce a given element power is inversely proportional to $(n - 1)$. The improvement also results from the reduction of the angles of incidence at the surfaces of the element.

In a negative element, the situation is less clear. From the standpoint of the Petzval correction, a low index would increase the overcorrecting contribution of a negative element. This can help to offset the (inward) undercorrection, which is a major problem in most lenses. On the other hand, a higher index would reduce the surface curvatures and have a generally desirable effect on the overall state of correction. The situation is usually resolved with the negative elements made from a glass along the "glass line" lower boundary of the glass map (Fig. 2.7).

A high V-value for the positive element and a low V-value for the negative element of an achromatic doublet reduce the element powers; this is ordinarily desirable. In lenses (such as the Cooke triplet) where the relative V-values of separated elements control the element spacing or the system length (see Chap. 8), this desideratum may be overridden by other concerns.

In general, for negative elements the flint glasses from along the glass line (see Fig. 2.7) are the best choice (except in a negative-power component). Again, in general, a higher index is better. Check the price and availability. Many glasses have recently been discontinued or reformulated to eliminate lead, arsenic, or cadmium. Some of the new "light" glasses are more attractive than their predecessors in many respects. Stain and weathering characteristics may be very important for some applications. Note that ordinarily bubbles are simply a cosmetic defect (unless the element is near an image). Striae are especially important in prisms or thick pieces. Incidentally, a system with many elements usually implies that a wide choice of glass types may be utilized.

Note that, as usual, when you are dealing with components of negative focal length, many of the considerations outlined above are reversed. In a negative achromatic doublet, the negative element is made of crown glass and the positive is made of flint. Here a high-index (flint) positive element will reduce the inward Petzval curvature, as will a low-index (crown) negative element.

3.3 Splitting Elements

Splitting an element into two (or more) approximately equal parts whose total power is equal to the power of the original element can reduce the aberration contribution by a significant factor. The reason why this reduces aberrations is that it allows the angles of incidence to be

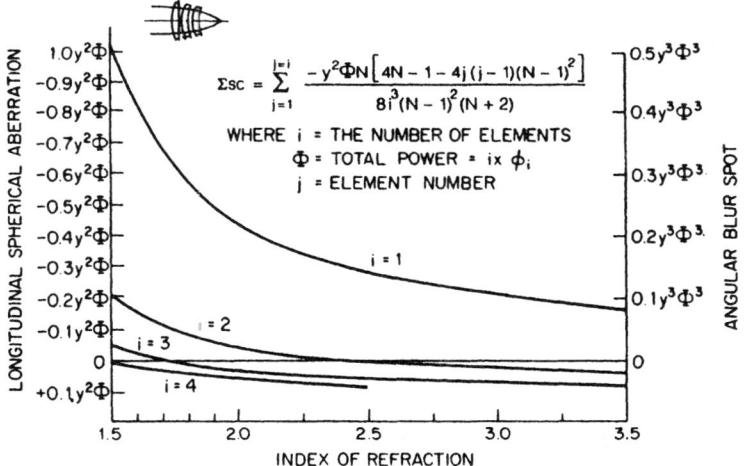

Figure 3.2 The spherical aberration of one, two, three, and four thin positive elements, each bent for minimum spherical aberration, plotted as a function of the index of refraction, and showing the reduction in the amount of aberration produced by splitting a single element into two or more elements (of the same total power). Each plot is labeled with i, the number of elements in the set. (The object is at infinity.)

reduced; the nonlinearity of Snell's law means that smaller angles introduce less aberration than do large ones. This technique is often used in high-speed lenses to reduce the zonal spherical residual and in wide-angle lenses to control astigmatism, distortion, and coma.

Figure 3.2 shows the thin lens third-order spherical aberration for spherical-surfaced positive elements shaped (or bent) to minimize the undercorrected spherical. The upper plot shows the spherical as a function of the index of refraction for a single element with a distant object. The curve labeled $i = 2$ shows the spherical for two elements whose total power is equal to that of the single element. The best split is 50-50, i.e., the split elements have equal power; this minimizes the spherical. (The same is true for a split into more than two elements, i.e., three, four, and so on, as shown in the curves labeled $i = 3$ and $i = 4$.) The improvement produced by splitting an element in two can be seen to be a factor of about 5 for lenses of index equal to 1.5. The higher the index, the greater the reduction; for an index of 1.8, the factor is about 7. At an index of 2.5 or higher, the spherical can be brought to zero or even overcorrected with just two positive elements. Most other aberrations are similarly affected by splitting, although it should be obvious that neither Petzval nor chromatic is changed by splitting.

In high-speed lenses this technique is frequently used to reduce the residual zonal spherical; the positive elements are split. This illustrates the basic idea. If the residual zonal spherical is negative (undercorrected), one splits a positive element; in the rare event that the zonal is positive (overcorrected), one would split a negative element. A similar philosophy can be applied for troublesome residuals of the other aberrations as well.

The choice of which element to split is often less apparent. The logical candidate would obviously seem to be the element that contributes most heavily to the problem aberration. (An examination of the third- and fifth-order surface contributions can often locate the source of the aberration. Alternately, look for large angles of incidence or refraction.) However, other considerations often become significant. For example, in the Cooke triplet, the rear element is the prime candidate for the split, and such a split is quite effective in reducing the zonal spherical, as Fig. 9.1 will attest. But the better choice for the split is the front element, not because it does a better job of reducing the zonal spherical, but because the resulting lens is better corrected for the other aberrations. This is because the meniscus-shaped second element reduces the Petzval sum and has very little spherical aberration. Figure 9.2 shows the simple split-front triplet. This is the ancestor of the Ernostar family of lenses; Figs. 9.6 to 9.10 illustrate designs that can be considered descendants from the split-front triplet. Although they have been largely superseded by the more powerful double-Gauss form, they are nonetheless excellent design types.

Many retrofocus and wide-angle lenses that use strong outer meniscus negative elements illustrate the use of this technique for the control of coma, astigmatism, and distortion by splitting these negative elements.

The implementation of this technique with an automatic design program is sometimes far from easy. For example, if one decides to split one of the crowns of a Cooke triplet and simply replaces one crown with two, after the computer optimization has run its course, the resultant lens may look like an ordinary triplet with a narrow cracklike airspace in the split element (a *cracked crown triplet*). The performance of the lens is the same as the original triplet; the split has not improved a thing. This is because the original lens was in a local optimum of the merit function. Aberrations other than the zonal spherical dominated the design; this caused the program to return the lens to its original design configuration. What is necessary in this situation is to force the split elements into a configuration that will accomplish the desired result.

Consider the split-front triplet. There are two ways to get to a design like Fig. 9.2. One approach is to make the lens so fast that the zonal spherical is by far the single, dominant aberration in the merit function. Then the program will probably choose a form that reduces the zonal spherical; the lens shapes in Fig. 9.2 are a likely result. A difficulty with this approach may be that you aren't interested in a very fast lens, or if you are, the rays may miss the surfaces of the initial design completely, or encounter TIR. The alternative approach is to constrain the front elements to a configuration in which the spherical is minimized. Simply fixing the first element to a plano-convex form (by not allowing the plano surface curvature to vary) or holding the second element to an aplanatic meniscus shape is usually sufficient to obtain a stable design that is different enough from the cracked crown triplet. When this has been accomplished the constraint can be released and the automatic design routine allowed to find what is (one hopes) a new and better local optimum. The problem here is that this approach presupposes a knowledge of the configuration that will produce a good result. Obviously a knowledge of both aberration theory and of successful design forms is a useful tool to the designer.

3.4 Separating a Cemented Doublet

Airspacing a cemented doublet can provide two additional degrees of freedom: two bendings instead of one, plus an airspace. While this technique does not have the inherent aberration reduction capability that many other modifications possess, the extra variables may indirectly make a design improvement possible. A difficulty in implementing this is that the refraction at the cemented surface is apt to become much more abrupt when it is split into two glass-air interfaces than when it was a cemented surface; in fact, rays may encounter TIR if a simple split is attempted without a concomitant reduction of the angle of incidence. Manual intervention in the form of adjusting the radii to reduce the angle of incidence is often necessary.

3.5 Compounding an Element

Compounding a singlet to a doublet can be viewed in two different ways:

1. As a way of simulating a desirable but nonexistent glass type
2. As a way of introducing a cemented interface into the element in order to control the ray paths

The longitudinal axial chromatic of a singlet is given by $LA_{ch} = -f/V$. Thus a fully achromatized lens (with $LA_{ch} = 0.0$) has the chromatic characteristic of a lens made from a material with a V-value of infinity; a partially achromatized doublet acts like a singlet with a very high V-value.

The Petzval radius of a singlet is given by $\rho = -nf$, where n is its index. An *old achromat* with a low-index crown and a higher-index flint has a shorter Petzval radius than a singlet of the crown glass. For example, an achromat of BK7 (517:649) and SF1 (717:295) has a Petzval radius $\rho = -1.37f$; in other words, in regard to Petzval field curvature, it behaves like a singlet with an index of 1.37. A *new achromat* has a high-index crown and a lower-index flint. A new achromat of SSKN5 (658:509) and LF5 (581:409) has a Petzval radius $\rho = -2.19f$ and a Petzval curvature that is characteristic of a singlet with an index of 2.19.

Thus an achromatized (or partially achromatized) doublet with a high-index positive element and a low-index negative element has many of the characteristics of a lens made of a high-index, high-V-value crown glass. (Note that for a negative focal length doublet, the reverse is true.) Both conditions are usually to be desired to flatten the Petzval field and to achieve achromatism.

Note that in almost all examples of Tessar-type lenses (and other types that utilize compounded elements), the doublets have positive elements with high index and high V-values, while the negative element of the doublet has both a lower index and V-value. See Chap. 10 for examples.

Figure 3.3 shows a singlet, an old achromat, and a new achromat, each with the same focal length. The equivalent V-value of each achromat is equal to infinity. The Petzval radius for each is given in the figure caption.

The cemented interface of the doublet can be used for specific control of specific rays. In a lens such as the Tessar, where the doublet is located well away from the aperture stop, the upper and lower rim rays of the

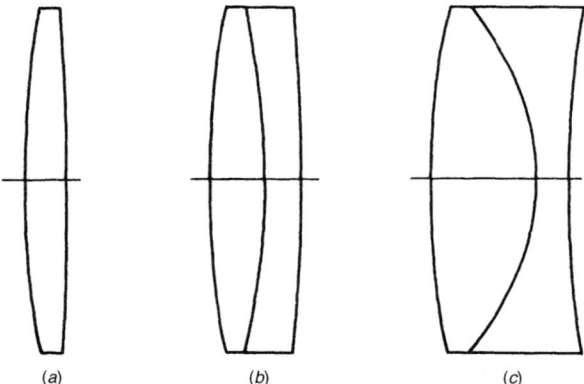

Figure 3.3 Three lenses, each with the same focal length f. (*a*) A singlet of BK7 (517:642) glass; Petzval radius equals $-1.52f$. (*b*) An old achromat of BK7 (517:642) and SF1 (717:295) glasses; Petzval radius equals $-1.37f$. (*c*) A new achromat of SSKN5 (658:509) and LF5 (581:409); Petzval radius is $-2.19f$.

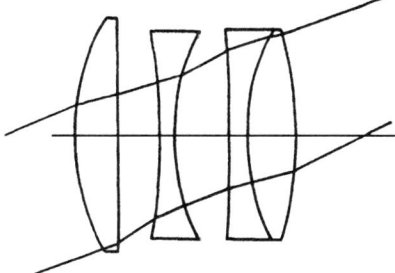

Figure 3.4 The upper and lower rim rays have significantly different angles of incidence at the cemented interface in the rear doublet of this Tessar design. Properly handled, this difference can be used to modify the correction of the coma-type aberrations.

oblique fan have very different angles of incidence at the cemented surface. In Fig. 3.4 it can be seen that the angle of incidence at this surface is much larger for the upper ray than for the lower. In this type of lens the cemented surface is typically a convergent one, and the (trigonometric) nonlinear characteristic of Snell's law means that the upper ray is, in this case, refracted downward more than it would be were the refraction linear with angle. Thus the upper ray is deviated in such a way as to reduce any positive coma of this ray. This illustrates the manner in which a cemented surface can be used for an asymmetrical effect on an oblique beam.

The cemented surface in the Tessar doublet:

1. Reduces the zonal spherical aberration
2. Reduces the overcorrected oblique spherical
3. Reduces the astigmatism of the zonal field (i.e., the "belly" in the X_s and X_t curves)

3.5.1 The Merté surface

A strongly curved, collective cemented surface with a small index break (to the order of 0.05 to 0.15) has an effect that can be used to reduce the undercorrected zonal spherical aberration. The central doublet of the Hektor lens shown in Fig. 3.5 illustrates this principle. The cemented surface is a collective one (in that $[n' - n]/r$ is positive) and contributes undercorrected spherical aberration. For rays near the axis, the spherical aberration contribution of the surface is modest; however, when the ray intersection height increases and the angle of incidence becomes large, as shown in Fig. 3.5, the trigonometric nonlinearity of Snell's law causes the amount of ray deviation to be disproportionately increased. This causes the undercorrection from this surface to dominate the spherical aberration. The result is a spherical aberration characteristic like that shown in Fig. 3.5. The spherical in the central part of the aperture appears

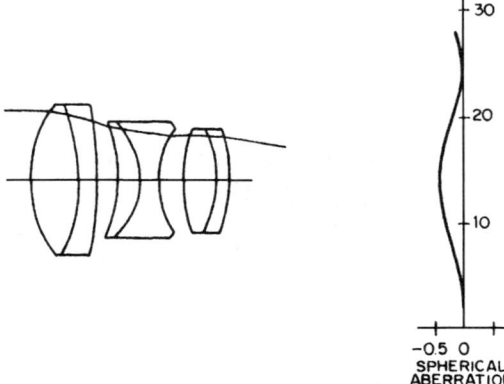

Figure 3.5 The cemented surface in the center doublet of this Hektor lens is what is called a Merté surface. The index break $(n' - n)$ across the surface is small, but at the margin of the aperture the angle of incidence for the axial ray becomes quite large. This combination produces an undercorrecting seventh-order spherical aberration which, as the plot shows, dominates the spherical aberration at the margin of the aperture, causing the marginal spherical to be negative rather than the usual positive value.

quite typical—the undercorrected third-order dominates close to the axis and the overcorrecting fifth-order causes the plot to curve back as the ray height increases. However, toward the edge of the aperture the undercorrection of the Merté surface becomes dominant and the aberration plot reverses direction again. The net result is the equivalent of a reduced zonal spherical aberration.

It is rare to see as extreme an example of the Merté surface as that illustrated in the Hektor of Fig. 3.5. Such a surface is very sensitive to fabrication errors and is thus expensive to make. It is also often best used close to the aperture stop because, if it is located away from the stop, the asymmetrical effects described two paragraphs above can become quite undesirable; however, it is well worth noting that even an ordinary collective cemented surface has a tendency to behave as a mild Merté surface and to reduce the spherical zonal, at least somewhat.

3.6 Vignetting and Its Uses

Vignetting, which is simply the mechanical limitation or obstruction of an oblique beam, is usually regarded primarily as something that reduces the off-axis illumination in the image; however, vignetting often plays an essential role in determining the off-axis image quality as well as the illumination. Of course there are many applications for which vignetting cannot be tolerated; the illumination must be as uniform as possible across the entire field of view. Thus for such applications, the complexity of the lens design must be sufficient to produce the required image quality at full aperture over the full field.

But for many applications, vignetting is quite tolerable. In commercial applications the clear apertures may well be established so as to be just sufficient to pass the full aperture rays for the axial image. It is not at all unusual for vignetting to exceed 50 percent at the edge of the field. For a camera lens, this vignetting will completely disappear when the iris of the lens is stopped down to an aperture below the vignetting level. Since camera lenses are most often used at less than full aperture, the vignetting is not as significant as it is in a lens that is always used at full aperture, such as a microscope or projection lens.

The *benefit* of vignetting is that it cuts off the upper and/or lower rim rays of the oblique tangential fan. Since these are ordinarily the most poorly behaved rays, the image quality may well be improved by their elimination. Most lenses that cover a significant field are afflicted with oblique spherical aberration, a fifth-order aberration that looks like third-order spherical aberration, but which varies as the square of the field angle. And since its magnitude is different for sagittal than for tangential rays, it can be seen to have characteristics of both astigmatism and spherical aberration. Oblique spherical aberration usually causes the rays at the edge of the oblique bundle to show strongly overcorrected spherical aberration; vignetting is a simple way to block these aberrant rays from the image.

Another factor favoring the use of vignetting is that it results from lens elements with small diameters. In general, one can count on a smaller-diameter lens being less costly to fabricate.

For a camera lens, one must be certain that the iris diaphragm is located centrally in the oblique beam so that, when the iris is closed down, the central rays of the beam are the ones which are passed. These are usually the best-corrected rays of the oblique beam. Also this location assures that the vignetting will be eliminated at the largest possible aperture.

A lens drawing with an oblique fan of 21 rays can yield a quick visual analysis of the vignetting; it can also show which parts of the system affect the aberrations and how they affect them. A fan of 21 rays is useful in checking the vignetting in increments of 10 percent of the semiaperture of the beam.

Note that vignetting is closely related to the overall length of the lens and if a vignetting problem seems intractable, a shorter lens may help. Unfortunately, lenses usually want to be long and large.

3.7 Eliminating a Weak Element—the Concentric Problem

Occasionally an automatic design program will produce a design with an element of very low power. Frequently this means that the element can be removed from the design without adversely affecting the quality of the design. Often a straightforward removal will not work; the design process may simply blow up. An approach that usually works (if anything will) is to add the thickness and the surface curvatures of the element to the merit function with target values of zero, allowing them to continue as variable parameters. Sometimes targeting the difference between the two curvatures is also useful. Usually, if the element isn't necessary to the design, a few cycles of optimization, possibly with gradually increasing weights on these targets, will change the element to a very thin, nearly plane parallel plate, which can then be removed without severe trauma to the design. If your design program will not accept curvatures and thicknesses as targets, an alternative technique is to remove the curvatures and the thickness as variables and to gradually weaken the curvatures and reduce the thickness (by hand) while continuing to optimize with the other variables.

An unfortunate form of the weak element is a fairly strongly bent meniscus, which the computer uses for a relatively important design function, such as the correction of spherical aberration or the reduction of the Petzval curvature. It is rarely possible to eliminate such an element because it is an integral part of the design. The unfortunate aspect of this situation is that, if the surfaces of the meniscus element are concentric or nearly so, the customary centering process used in optical manufacture is impossible or impractical, and the element is costly to fabricate. This situation can be ameliorated[*] by forcing the centers of curvature of the surfaces apart by a distance sufficient to allow the use of ordinary centering techniques. Again, including the required center-to-center spacing in the merit function and reoptimizing will usually modify the offending element to a more manufacturable form without significant damage to the system performance.

3.8 Balancing Aberrations

The optimum balance of the aberrations is not always the same in every case; the best balance varies with the application and depends on the size of the residual aberrations. In general, for well-corrected lenses, the aberrations should be balanced so as to minimize the optical path

---

[*]See: Smith, "The Problem of the Concentric Meniscus," *Optical Engineering*, Vol. 27, pp. 1039–1041, 1988.

difference (OPD), i.e., the wavefront variance, but there are significant exceptions.

3.8.1 Spherical aberration

If a lens is well-corrected and the high-order residual spherical aberration is small so that the OPD is to the order of a half- wave or less, then the best correction is almost always that with the marginal spherical corrected to zero, as illustrated in Fig. 3.6b; however, when the zonal spherical is large, there are two situations where one may want to depart from complete correction of the marginal spherical.

If the lens will always be used at full aperture (as a projection lens, for example), and if the spherical aberration residual is large (say to the order of a wave or so), the diffraction effects will be small when compared to the aberration blur; then the spherical aberration should be corrected to minimize the size of the blur spot rather than to minimize the OPD. This will produce the best contrast for an image with relatively coarse details, i.e., for a resolution well below the diffraction limit. As an example, at a speed of $f/1.6$, a 16-mm movie projection lens has a diffraction

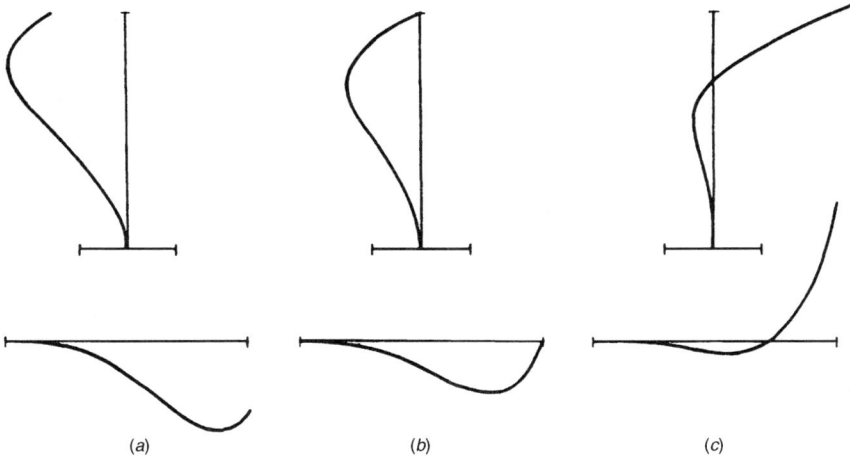

Figure 3.6 Three states of correction of spherical aberration are shown. Each has the same amount of fifth-order spherical, but different amounts of third-order. (a) Spherical aberration balanced to give the smallest possible size blur spot when refocussed. This correction may be optimum when the aberration is large and the required level of resolution is low compared to the diffraction limit. (b) Spherical aberration balanced for minimum OPD. This is optimum when the system is diffraction-limited. (c) Spherical aberration balanced to minimize the focus shift as the lens aperture is stopped down. This correction is used in camera lenses when the residual spherical is large. The upper row is longitudinal spherical versus ray height; the lower row is transverse ray intercept plots.

cutoff frequency of about 1100 *line pairs per millimeter* (lpm). But its performance is considered quite good if it resolves 100 lpm, an order of magnitude less than the diffraction limit. Such a lens can advantageously be corrected for the minimum diameter geometrical blur spot. This state of correction occurs (for third- and fifth-order spherical) when $LA_z = 1.5LA_m$, or $TA_z = 1.05TA_m$, and the defocus is $\delta = 1.25LA_m = 0.83LA_z$; the result is a high-contrast, but low-resolution, image. This correction is illustrated in Fig. 3.6a. See also the comments on defocusing in Secs. 2.4 and 2.7.

For a lens used at varying apertures, as in a typical camera lens, it is important that the best focus position not shift as the size of the aperture stop is changed. If the spherical aberration is corrected at the margin of the aperture, or corrected as described in the paragraph above, the position of the best focus will shift as the aperture is changed. The best focus will move toward the paraxial focus as the aperture is reduced. The state of correction that is often used in such a case is overcorrection of the marginal spherical, as shown in Fig. 3.6c (assuming an undercorrected zonal residual). For example, when $LA_m = -LA_z$ (or $LA_m + LA_z = 0$) the result is a design in which the focus is reasonably stable as the lens is stopped down. The *resolution* is better than it would be otherwise, but, at full aperture, the *contrast* in the image is quite low. This works out reasonably well, especially in a high-speed camera lens because camera lenses are only infrequently used at full aperture. Typically, photographs are taken with the lens stopped down well below the full aperture, and, when the camera is stopped down, this state of correction yields a much better photograph than that produced by a balance like Fig. 3.6a or Fig. 3.6b.

Note that designing a camera lens in two configurations (wide open and stopped down about 50 percent) is one way of getting this "stable focus" kind of balance. Other ways are to design at the paraxial focus (i.e., allow no defocus) or to simply optimize the lens stopped down a bit from wide open.

The three correction states shown in Fig. 3.6 also indicate the manner in which the spherical aberration is changed when the third-order aberration is changed. This is a typical situation often encountered in lens design—the fifth- (and higher-) order aberrations are relatively stable and difficult to change, but the third order is easily modified (by bending an element, for example). In the figure, all three illustrations have exactly the same amount of fifth-order spherical; the difference between the three is solely in the amount of third-order. Note that in the (upper) longitudinal plots, the change of the aberration from one illustration to the next varies as y^2, whereas in the (lower) transverse plots the differences vary as y^3.

Although patent designs are customarily shown with the marginal spherical corrected to zero, it is apparent that this is not always the best state of correction.

Note that sagittal fan ray intercept plots are often very similar to the axial-ray plot. This is because the sagittal cross section remains relatively constant compared to the tangential cross section. Thus, depending on which section has the worse oblique problems, one should consider the changes of the appropriate beam cross sections.

Oblique spherical is often the aberration that limits the useful field of view. One may consider oblique spherical to be the result of the change in the beam shape from circular on axis to elliptical as the field obliquity increases—this upsets the axial balance of the spherical. The tangential and sagittal fans are affected differently, thus their defocusing effects may be balanced by third-order astigmatism and Petzval.

3.8.2 Chromatic aberration and spherochromatism

Here the question is how to balance the spherochromatism, which typically causes the spherical aberration at short (blue) wavelengths to be overcorrected and that at the long (red) wavelengths to be undercorrected. If the aberration is small (diffraction-limited), the best correction is probably with the chromatic aberration corrected at about the 0.7 zone of the aperture. This means that the central half of the aperture area is undercorrected for color and the outer half of the aperture is overcorrected, as shown in Fig. 3.7a. But if the amount of the aberration is large, the spherical overcorrection of the blue marginal ray causes a blue flare and a low contrast in the image. In these circumstances the correction zone can advantageously be moved to (or toward) the marginal zone, as shown in Fig. 3.7b. This will probably reduce the resolution somewhat because it increases the size of the core of the image blur, but it improves the contrast significantly and yields a more pleasing image, free of the blue flare and haze. This state of correction is accomplished by increasing the undercorrection of the chromatic aberration of the paraxial rays. This is good for a system where the resolution is much less than the diffraction cutoff frequency.

For camera lenses, Kingslake recommends correcting the chromatic at 0.4 to 0.7 of the aperture because in use they are usually stopped down enough to cut off the marginal blue flare produced by this state of correction. In most situations, this will result in better definition.

Secondary spectrum is discussed in Chap. 6.

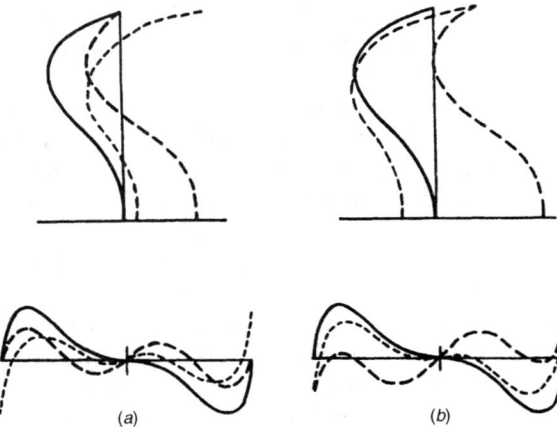

Figure 3.7 Spherochromatism. (*a*) Chromatic aberration balanced so that the outer half of the aperture is overcorrected and the inner half is undercorrected. This may be best if the amount of aberration is small. (*b*) Chromatic aberration balanced so that it is corrected at the margin. If the aberration is large, this correction eliminates the blue flare that can result from the type of correction in (*a*). Note that the state of correction is more easily perceived in the upper, longitudinal aberration plots, whereas the effect on the blur spot size and flare is much more apparent in the lower transverse ray intercept plots.

3.8.3 Astigmatism and Petzval field curvature

In a typical anastigmat lens the fifth-order astigmatism tends to become significantly undercorrected (i.e., negative) as the field angle is increased. In order to minimize the astigmatism over the full field, the third-order astigmatism is overcorrected enough to balance the undercorrected fifth-order astigmatism. The result is the typical field curvature correction with the sagittal focal surface located inside the tangential focal surface in the central part of the field because of the overcorrected third-order astigmatism, with the reverse arrangement in the extreme outer portions. The field angle at which the s and t fields cross (i.e., where the astigmatism is zero) is called the *node*. Usually the two fields separate very rapidly outside the node, and the image quality quickly deteriorates, often suddenly. The Petzval curvature is usually made somewhat negative, so that both fields are slightly inward-curving and the *effective* field is as flat as possible. A typical state of correction is shown in Fig. 3.8.

Note that a field correction with the s and t focal surfaces spaced equally on either side of the focal plane (so that the compromise "smallest circle

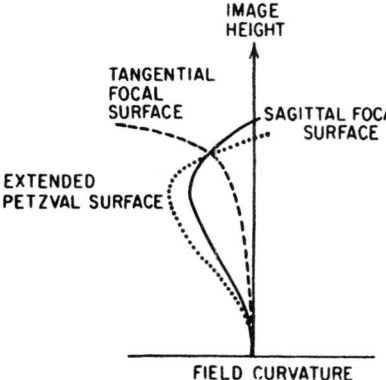

Figure 3.8 This is the typical balance of astigmatism and Petzval curvature in the presence of undercorrecting fifth-order astigmatism and overcorrecting high-order Petzval curvature. This is achieved by leaving the third-order Petzval slightly inward-curving and overcorrecting the third-order astigmatism by a small amount. This is the usual aberration balance for most anastigmats.

of confusion" focal surface is flat) is definitely *not* the best state of correction. In considering the correction of the field curvature, one should also bear in mind that the oblique spherical aberration (a fifth-order aberration that varies as the cube of the aperture and the square of the field angle) typically goes overcorrected with increasing field angle. In addition, the oblique spherical is usually more significant for the tangential fan of rays than for the sagittal. Thus the *effective* field curvature for the full fan of rays is usually more backward-curving than the X_s and X_t field curves indicate and the *effective* astigmatism is greater. These field curves indicate the imagery of a very small bundle of rays close to the principal ray, and do not take the oblique spherical of the full aperture into account. Thus, for most designs, the astigmatism and field curvature are usually arranged somewhere between the state at which the s and t curves are superimposed (i.e., zero astigmatism) and that at which the t field is approximately flat. Often a through-focus *modulation transfer function* (MTF) plot that includes both on-axis and off-axis plots will indicate quite clearly the effective field curvature; this is more informative than the X_s and X_t curves alone.

Note well that these discussions have assumed the type of higher-order residual aberrations one ordinarily finds—overcorrected fifth-order spherical aberration, undercorrected fifth-order astigmatism, and overcorrected spherochromatism for the shorter wavelengths. Although rare, the reverse is sometimes encountered. In such circumstances the obvious move is to apply the above advice in reverse.

As an additional consideration, note that the undercorrection of either the chromatic aberration or the Petzval curvature has the usually desirable side effect of reducing the power of the elements of the lens system. Thus a secondary benefit of this undercorrection is the reduction of residual aberrations in general because a lower-power element produces less aberration, which means that there is less higher-order residual aberration left when the aberrations are balanced out.

The usual field curvature plots of X_s and X_t like those in Fig. 3.8 are based on paraprincipal rays. These plots effectively indicate the imagery through a pinhole aperture at the stop. The through-focus plot of the MTF for several field angles (e.g., Fig. 13.37) can be used to determine the *effective* field curvature for the lens at full (or any other) aperture.

Sometimes increasing the weight on the zonal field targets in the merit function will reduce the "belly" at the zonal field.

3.8.4 Coma

Both of the two fifth order coma aberrations (linear and elliptical) can be balanced by third-order coma, but in different ways. Figure 3.9a shows a typical balance of linear coma (which varies as hy^4), and Fig. 3.9b shows the balance of elliptical coma (which varies as h^3y^2). Note that linear coma is balanced across the aperture and elliptical coma is balanced across the field.

Although it is ordinarily not a dominant problem, field curvature (X_s and X_t) can vary with wavelength. This can be detected by inspection of the off-axis ray intercept plots, or the field curvature can be plotted in different wavelengths. This is obviously a longitudinal chromatic which varies with field angle.

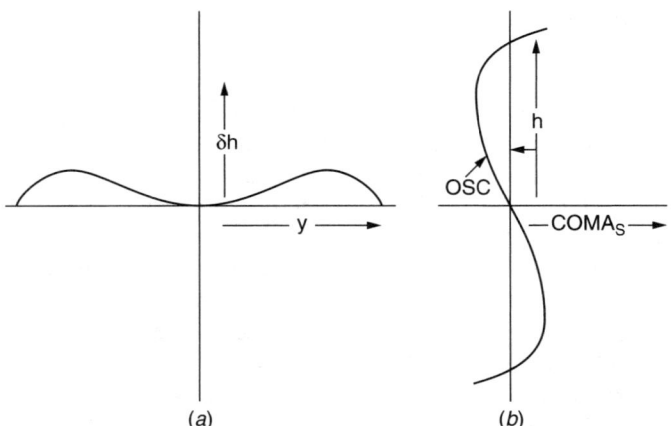

Figure 3.9 (a) A ray intercept plot showing undercorrected fifth-order linear coma (which varies as hy^4) balanced by overcorrected third-order coma (which varies as hy^2). (b) A plot of coma versus field height showing overcorrected fifth-order elliptical coma (which varies as h^3y^2) balanced by undercorrected third-order coma (which varies as hy^2). Note that OSC = (sagittal coma)/h, as h approaches zero.

3.9 The Symmetrical Principle

When an optical system has mirror (i.e., front-to-back) symmetry about the aperture stop (or a pupil), as shown in Fig. 3.10, the system is free of coma, distortion, and lateral color. This results from the fact that the components on one side of the stop have aberrations that exactly cancel the aberrations from the components on the other side of the stop. Obviously, to have mirror symmetry, the system must work at unit magnification, with equal object and image distances. For the symmetry to be *absolutely* complete, the object and image surfaces must be identical in shape; this then would imply separately curved sagittal and tangential surfaces at both object and image. (This is why the best 1:1 design is often not completely symmetrical.) However, the third-order coma, distortion, and lateral color are completely removed by symmetry, even with flat object and image surfaces.

Of course, most systems do not operate at unit magnification, and therefore a symmetrical construction of the lens will not completely eliminate these aberrations; however, even for a lens with an infinitely distant object, these aberrations are markedly reduced by symmetry, or even by an approximately symmetrical construction. This is why so many optical systems that cover a significant angular field display a rough symmetry of construction. Consider the Cooke triplet—it has outer crown elements that are similar but not identical in shape, and the center flint, while not equiconcave, is biconcave, and, except for slow-speed triplets, the airspaces are quite similar in size. The benefit of this is that the higher-order residuals of coma, distortion, and lateral color are markedly reduced by this symmetry. This is especially true for wide-angle lenses when good distortion correction is important.

A monocentric system, where all surfaces—refractive or reflective—have a common center and where the stop is located at the common center, has no coma, no astigmatism, and no lateral color. And if the object and image are both in a plane through the center, the spherical can also be zero (e.g., Offner and Dyson systems).

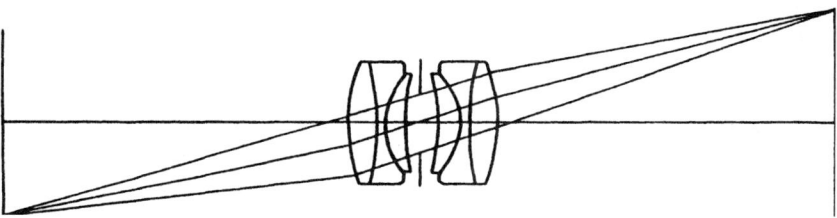

Figure 3.10 A fully (left to right) symmetrical system is completely free of coma, distortion, and lateral color because the aberration in one half of the system exactly cancels out the aberration in the other half.

Before you decide that symmetry or monocentricity is always a blessing, remember that the big steps in progressing from the $f/4$ Plasmat lens to the $f/2$ double-Gauss lens were (a) dropping symmetry and (b) increasing the doublet thicknesses.

3.10 Aspheric Surfaces

Many designs can be improved by the use of one or more aspheric surfaces. Except for the case of a molded or diamond-turned element, an aspheric surface is many times more expensive to fabricate than a spherical surface. A conic aspheric is easier to test than a general aspheric and is therefore apt to be somewhat less costly. For many systems, e.g., mirror objectives, aspheric surfaces are essential to the design and cannot be avoided.

One technique for introducing an aspheric into an optical system is to first vary only the conic constant or the fourth-order deformation coefficient. (Note that the conic constant and the fourth-order deformation term have exactly the same effect on the third-order aberrations. Thus, allowing both to vary in an automatic design program may cause a slowing of the convergence or, in extreme cases, a failure of the process. Occasionally the difference between the effect of the conic and the fourth-order term on the fifth- and higher-order aberrations may be useful in a design, but more often than not one of the two is redundant.) If the effect of varying the conic constant alone is inadequate, one can then allow the sixth-order term to vary, then the eighth-order, and so on. Some designs have aspherics specified to the tenth-order term when just the sixth or eighth would suffice. It is a good idea to calculate the surface deformation caused by the highest-order term used; if it is a fraction of a wave at the edge of the surface aperture, its utility may well be totally imaginary.

Occasionally one encounters a design specification or print in which the aspheric is specified by a tabulation of sagittal heights instead of an equation. The optimization program can be used to fit the constants of the standard aspheric surface equation (Eq. (5.1)) to the tabulated data. The specification table is entered in the merit function as the sag of the intersections of (collimated) rays at the appropriate heights. The surface coefficients are allowed to vary, and the result is a least-squares fit to the sag table.

The equations of Chap. 24, Sec. 11 indicate the effects of a conic or a fourth-order aspheric term on the third-order aberrations. Several points are worthy of note. The aspheric has no effect on the Petzval curvature, on power, or on axial or lateral chromatic, although it can change the chromatic variation of aberrations. Further, if it is located at the aperture stop or at a pupil, then the principal ray height y_p is zero and the aspheric has no effect on third-order coma, astigmatism, or distortion; it can only affect spherical. In the Schmidt camera (Fig. 18.15) the aspheric surface is located at the stop because the coma and astigmatism are already zero (the stop being at the center of curvature of the spherical mirror). The purpose of the aspheric is to change *only* the spherical aberration. Conversely, if the purpose of an aspheric is to affect the third-order coma, astigmatism, or distortion, then it must be located at a significant distance from the aperture stop. Even if located at the stop, an aspheric can affect the high-order off-axis aberrations.

It is also worth noting that the primary effect of the conic, or fourth-order, deformation term is on the third-order aberrations. The primary effect of the sixth-order deformation term is on the fifth-order aberrations, and so on.

Note that a conic constant is meaningless on a plane ($cv = 0.0$) surface; a fourth-order deformation term must be used instead. Conics and the fourth-order aspheric terms mostly affect the third-order aberrations. Do not vary the conic constant and the fourth-order coefficient at the same time; they have exactly the same effect on the Seidel aberrations.

Common uses of aspheric surfaces:

1. To correct spherical aberration, as in the Schmidt telescope
2. To control astigmatism and distortion in wide angle eyepieces
3. To correct distortion in mapping or measuring systems
4. To correct pupil spherical in complex visual systems

In using an aspheric, more rays are needed than for a system with only spherical surfaces. Shannon (Ref. 22) suggests rays at 0.7, 0.85, 0.9, and 0.95 for a 10th order aspheric. In any case, one must watch out for big residuals between the rays.

Chapter

4

Evaluation: How Good Is This Design?

4.1 The Uses of a Preliminary Evaluation

At some point in the process of designing an optical system, the designer must decide whether the design is good enough for the application at hand. With modern computing power, it is not a difficult matter to calculate the *modulation transfer function* (MTF) or the *point spread function* (PSF), and to accurately include the effects of diffraction in the calculations. The process does consume a finite amount of time (which, on a slow computer, may be a significant amount), and it is useful to be able to make an estimate of the system performance from a more limited amount of ongoing data. A reasonable estimate can avoid wasting time and computer paper in evaluating a clearly deficient design, or it can signal an appropriate point at which to conduct a full-dress evaluation.

An ongoing approximation of the MTF may be useful for comparisons or for judging the improvement between design stages; a three-wavelength evaluation, if consistent, is acceptable for this purpose. A five-wavelength MTF is apt to be more accurate.

Some common metrics of image quality are: *peak-to-valley optical path difference* (P-V OPD); rms OPD; point spread function (PSF); Strehl ratio; MTF; rms spot; radial energy density (RED); and encircled or ensquared energy.

4.2 OPD versus Measures of Performance

The distribution of illumination in the point spread function, particularly in the diffraction pattern of a reasonably well-corrected lens, is often used as a measure of image quality. The *Strehl ratio* (or *Strehl*

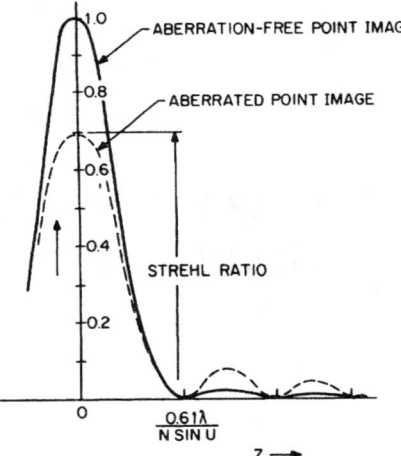

Figure 4.1 The Strehl ratio is the illumination at the center of the diffraction pattern of an aberrated image, relative to that of an aberration-free image.

definition) is the ratio of the illumination at the center of an aberrated point image to the illumination at the center of the point image formed by an aberration-free system. Figure 4.1 illustrates the concept.

There are several simple equations relating the rms OPD to the Strehl ratio. Although approximations, they are reasonably accurate when the aberrations are small (which is the only condition in which the Strehl ratio is meaningful). Here are three:

$$SR = \exp[-(2\pi w)^2]$$
$$SR = 1 - (2\pi w)^2$$
$$SR = \exp(-43w^2)$$

where w is the rms OPD in waves $\approx \dfrac{\text{P-V OPD}}{3.5}$.

The *Marechal criterion* (Strehl = 0.8; OPD ≈ 0.25 wave) may be satisfied for monochromatic image quality if:

$$[(\text{sph}/\text{sph-4})^2 + (\text{coma}/\text{coma-4})^2 + (\text{astig}/\text{astig-4})^2] \leq 1.0$$

where sph = amount of spherical
 sph-4 = amount of spherical corresponding to a quarter wave of OPD
 coma = amount of coma

coma-4 = amount of coma corresponding to a quarter wave of OPD
astig = amount of astigmatism
astig-4 = amount of astigmatism corresponding to a quarter wave of OPD

The relationship between rms OPD and P-V OPD depends on the type of aberration that produces the wavefront deformation. Here are some conversions:

1. For defocus: rms = 0.29 (P-V)
2. For refocused third-order spherical: rms = 0.075 (P-V)
3. For refocused, balanced third- and fifth-spherical: rms = 0.017 (P-V)
4. For third-order coma: rms = 0.12 (P-V)
5. For astigmatism at best focus: rms = 0.20 (P-V)
6. For smooth, random wavefront errors: rms ≈ 0.2 (P-V)

See Chap. 24 for the relationships between P-V OPD and the various aberrations.

Another measure of image quality uses the percentage of the total energy in the point image contained within the diameter of the Airy disk. This diameter remains relatively constant in size for small amounts of aberrations. Table 4.1 gives the relationships between the wavefront deformation (or OPD), the Strehl ratio, and the energy distribution.

TABLE 4.1 Tabulation of the Strehl Ratio and the Energy Distribution as a Function of the Wavefront Deformation

	Relation of image quality measures to OPD			
			% energy in	
P-V OPD	RMS OPD*	Strehl ratio	Airy disk	Rings
0.0	0.0	1.00	84	16
0.25RL† = λ/16	0.018λ	0.99	83	17
0.5RL = λ/8	0.036λ	0.95	80	20
1.0RL = λ/4	0.07λ	0.80	68	32
2.0RL = λ/2	0.14λ	0.4‡	40	60
3.0RL = 0.75λ	0.21λ	0.1‡	20	80
4.0RL = λ	0.29λ	0.0‡	10	90

*The table assumes that RMS OPD equals 0.29 times the P-V OPD.
†RL means the Rayleigh limit of one-quarter-wavelength peak-to-valley OPD.
‡The smaller values of the Strehl ratio do not correlate well with image quality.

The relationships between the basic aberrations and the OPD are given in Chap. 24, as are the relationships between rms OPD and P-V OPD and between rms OPD and the Strehl ratio.

4.2.1 The modulation transfer function or MTF

Another commonly utilized measure of performance is the modulation transfer function, which describes the image modulation or contrast as a function of the spatial frequency of the object or image. We define the contrast or modulation of a pattern of lines whose intensity varies as a sine function by

$$M = (\max - \min)/(\max + \min)$$

where max and min, are, respectively, the maximum and minimum brightness (or illumination) levels in the line pattern. If we compare the modulation of the object M_o to that of the image M_i, the ratio M_i/M_o is the modulation transfer factor for the spatial frequency of the line pattern. When expressed as a function of the spatial frequency v (usually in cycles or line pairs per millimeter), this is the modulation transfer function, or MTF (v). One can see why in the early days of its development MTF was referred to as frequency response, contrast transfer, or sine wave response. MTF is the *real* part of the complex optical transfer function, or OTF (v); the *imaginary* part is the phase transfer function, or PTF(v). The PTF is essentially a sideways shift of the image from where it should be. The negative MTF often seen in a defocussed, well-corrected lens is a 180° phase shift.

The calculation of MTF can be done by autocorrelating the aberrated wavefront against itself, shifted laterally by a fraction of the aperture, the shift corresponding to the spatial frequency. If the lens is perfect, with no wavefront deformation, then the process amounts to simply shifting the aperture contour (e.g., a circle) against itself. The normalized area common to both circles, for example, is then the MTF. This is shown in Fig. 4.2a for a circular aperture; Fig. 4.2b for a rectangular aperture; and Fig. 4.2c for an aperture with a central obstruction. For a rectangular aperture the common area will plot against displacement as a straight line. A central obscuration in the aperture will reduce the MTF at low frequencies and raise it slightly at high frequencies. If the aperture transmission varies, or if the intensity of the wavefront is not uniform (as in a laser beam), a similar technique will work with the aperture areas weighted appropriately. The MTF plots corresponding to these four cases are shown in Fig. 4.3. Obviously the displacement at which the common area is zero corresponds to the cutoff frequency.

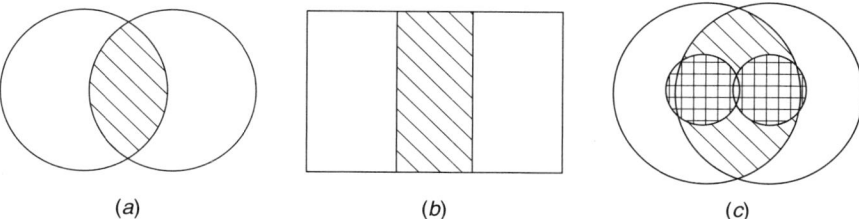

Figure 4.2 For a perfect lens, the MTF is equal to the (normalized) area common to the aperture and the aperture displaced laterally by a distance corresponding to the spatial frequency. The cutoff frequency v_0 corresponds to the shift where the common area is zero. (*a*) A circular aperture. (*b*) A rectangular aperture. (*c*) A circular aperture with a central obstruction. The MTF plots for these apertures are shown in Fig. 4.3.

In calculating MTF by the autocorrelation of the wavefront with itself, the spacing of the rays in the aperture must be close enough that the program does not loose track of the phase of the wavefront. The change in OPD from ray to ray should be small, to the order of a quarter or half wave or less. If the OPD is large (greater than one or two wavelengths), the results of a standard, default, diffraction MTF calculation may be meaningless, and a geometric MTF (Eq. (4.4)) may be much better.

The MTF of a perfect lens (with a uniformly transmitting circular aperture and imaging a wave front of uniform intensity) is given by Eqs. (4.1) and (4.2).

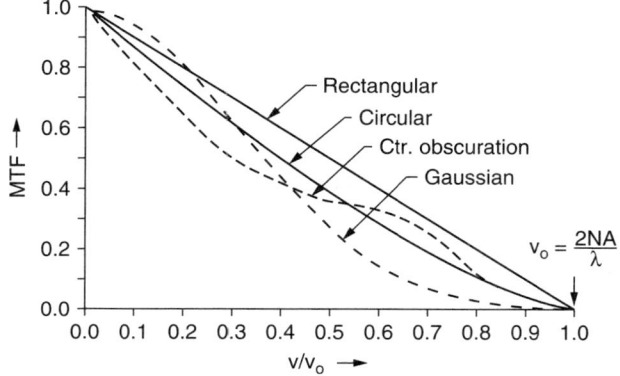

Figure 4.3 The MTF plots for a perfect lens with: (*a*) a circular aperture; (*b*) a rectangular aperture; (*c*) a circular aperture with a central obstruction; and (*d*) an apodized aperture, or one illuminated by a gaussian laser beam.

$$\text{MTF}(v) = \frac{2}{\pi}(\phi - \cos\phi \sin\phi) \tag{4.1}$$

where

$$\phi = \arccos\left(\frac{\lambda v}{2\text{NA}}\right) \tag{4.2}$$

This is plotted as curve A in Figs. 4.4 and 4.5. Figure 4.4 shows the effect on the MTF of defocusing an otherwise aberration-free lens. The spatial frequency in these plots is normalized to the cutoff frequency

$$v_0 = \frac{2\text{NA}}{\lambda} = \frac{1}{\lambda(f \text{ number})} \tag{4.3}$$

When the OPD is large (say more than one or two waves), the following geometrical approximation (derived from the geometric defocusing expression) can be used to calculate the MTF with reasonable accuracy:

$$\text{MTF}(v) = \frac{J_1[8\pi n \text{OPD}(v/v_0)]}{4\pi n \text{OPD}(v/v_0)} \tag{4.4}$$

where v_0 = cutoff frequency (Eq. (4.3))
n = index of the image medium
OPD = peak-to-valley wavefront deformation in waves

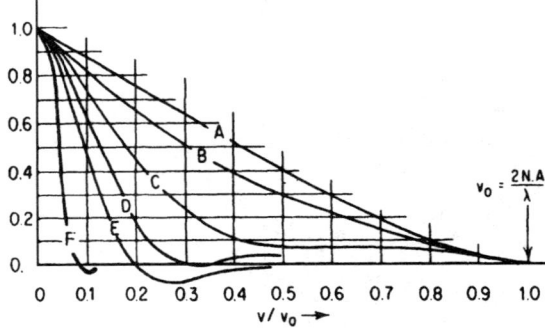

Figure 4.4 The effect of defocusing on the modulation transfer function of an aberration-free system. (A) In focus (OPD = zero); (B) Defocus = $\lambda/2n \sin^2 U$ (OPD = $\lambda/4$); (C) Defocus = $\lambda/n \sin^2 U$ (OPD = $\lambda/2$); (D) Defocus = $3\lambda/2n \sin^2 U$ (OPD = $3\lambda/4$); (E) Defocus = $2\lambda/n \sin^2 U$ (OPD = λ); (F) Defocus = $4\lambda/n \sin^2 U$ (OPD = 2λ). Curves are based on diffraction effects—not on a geometric calculation.

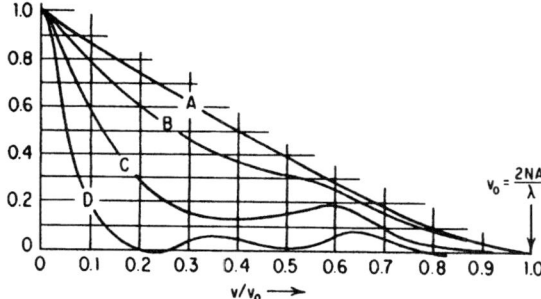

Figure 4.5 The effect of third-order spherical aberration on the modulation transfer function. (A) LA_M = zero OPD = 0; (B) $LA_M = 4\lambda/n \sin^2 U$ OPD = $\lambda/4$; (C) $LA_M = 8\lambda/n \sin^2 U$ OPD = $\lambda/2$; (D) $LA_M = 16\lambda/n \sin^2 U$ OPD = λ.

and

$$J_1[x] = \frac{x}{2} - \frac{(x/2)^3}{1^2 2} + \frac{(x/2)^5}{1^2 2^2 3} - \cdots$$

The diffraction cutoff frequency is given by

$$v_0 = \frac{2NA}{\lambda}$$

and the numerical aperture, NA, of a beam varies with the obliquity at which the beam strikes the image surface. The NA is likely to differ in the sagittal and tangential directions. It is affected by vignetting. Distortion of the exit pupil will change it. There is also an anamorphic factor for the tangential lines in an off-axis image; the tangential lines are spread out more, and the frequency is reduced to the sagittal lpm times the cosine of the obliquity.

Note that for visual work we can use 555 nm for the wavelength and get an easily remembered cutoff frequency of

$$v_0 = 1800/(f \text{ number}) \text{ lines per mm}$$

An optical system is a low-pass filter with an absolute cutoff limit given by v_0. An optical system cannot transmit information of spatial frequencies higher than the cutoff frequency.

Figure 4.5 shows the effect of simple third-order spherical aberration on the MTF. Note that, although the curves of Figs. 4.4 and 4.5 are not identical, they are quite similar. This similarity of effect is the basis for the common rule of thumb that a given amount of OPD will degrade the image by about the same amount, regardless of what type of aberration produced the OPD.

A convenient relationship to remember is that (per Chap. 24) a quarter wave of OPD corresponds to a transverse spherical aberration (either marginal or zonal) of about

$$\text{TA} = \frac{4\lambda}{\text{NA}} \qquad (4.5)$$

This is a useful way to make a quick and dirty evaluation from just the ray intercept plots.

Note that a bar target test for resolution or performance is significantly different than a test using a sine-wave-modulated target. The MTF plot "A" of Fig. 4.4 would be 10 to 15 percent higher in the midfrequency range if a bar target were used. A three-bar pattern (as used in the USAF 1951 target) has a lot of its power in the subharmonic frequencies and will have a 7 percent MTF at the cutoff frequency (where a sine wave target would show a zero contrast).

The MTF and the Strehl ratio are related in an interesting way. Remembering that the MTF plot may change as a function of the orientation of the line pattern, we can realize that, viewed in three dimensions the MTF plot is a surface, and the surface contains a volume. It turns out that the volume for an aberrated lens, normalized by the volume for a perfect lens, is equal to the Strehl ratio. Some workers look at the area under the ordinary two-dimensional MTF plot as a measure of relative quality.

Mouroulis[17] has reported that for visual purposes a quality factor equal to the area under the MTF plot between the frequencies of 5 and 25 cycles per degree in eye space correlates well with visual evaluations of image quality.

The Mouroulis visual quality rating is

$$\text{MTF}_* = \int_{v=5}^{v=20} \text{MTF}(v)$$

= the area under the MTF plot between 5 and 20 cycles

where the frequency v is in cycles per degree at the eye. This can be expanded to apply to both x and y; one simply uses the *volume* of the MTF solid between the frequency limits. Obviously this approach can be generalized by adjusting the frequency limits and the frequency metrics to suit the application.

The human visual response is plotted in Fig. 4.6 for both photopic (daylight) and scotopic (dark adapted) conditions. The figure also shows the integrated response, so that the fraction of the total response between two wavelengths can be estimated. This is one basis for establishing a spectral weighting scheme for use in calculating MTF, PSF, and so on.

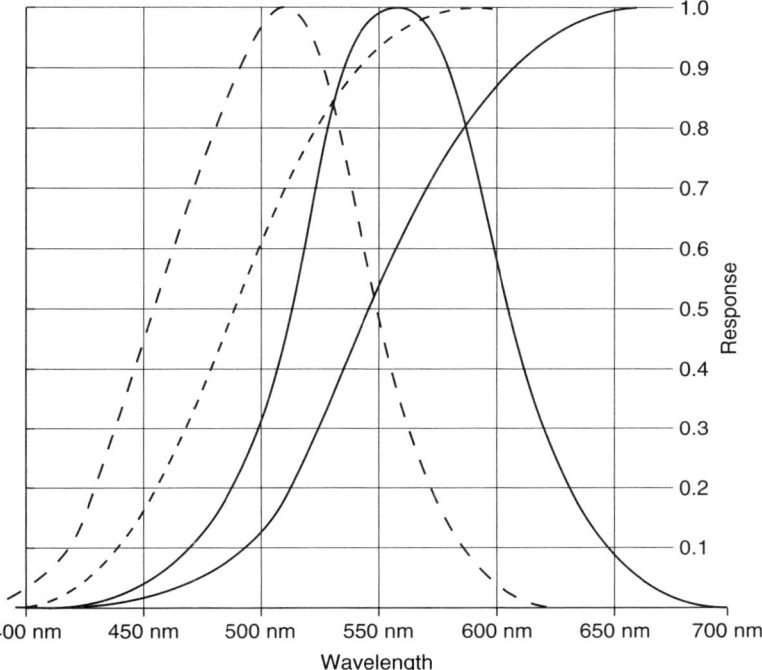

Figure 4.6 The spectral response of the human eye. The solid lines are the photopic (daylight) response, and the dashed lines are the scotopic (dark-adapted) response. There are two curves for each response: one is the relative response at the indicated wavelength; the other is the (integrated) fraction of the response for wavelengths shorter than that indicated.

A simple five wavelength spectral weighting derived from this plot for ordinary visual work might be:

Wavelength	Weight
486 nm (F)	0.25
510 nm	0.5
546 nm (e)	1.0
588 nm (d)	0.5
620 nm	0.25

For film or scotopic work the wavelengths can be shifted 40 or 50 nm toward the blue.

Most optical software does not show the negative values in the MTF plot. This is unfortunate because such an all-positive plot, when smoothed, may indicate a higher resolution than actually exists. Care is advisable when the plotted MTF is low, especially if, as shown in Fig. 4.7c, the plot

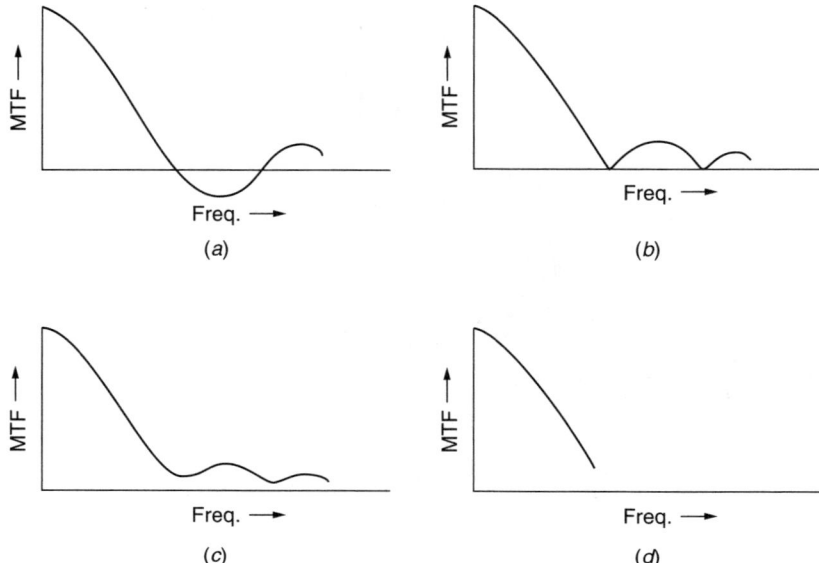

Figure 4.7 (a) The MTF plot for a lens with a negative MTF (i.e., a 180° phase shift) at certain frequencies. (b) Same as (a) except that the negative MTF is shown positive, as an all-positive plot. (c) What the computer program draws after the data of (b) is smoothed by a curves-fitting routine. The validity of the plotted MTF values at the intermediate frequencies is questionable. (d) The uncertainity is removed and the plot is correct if the MTF is plotted only up to the frequency at which the phase change occurs.

is undulating. Figure 4.7 shows: (a) the true MTF plot; (b) the all-positive version; and (c) the resulting plot when the curve fitting program has smoothed this all-positive data. Cutting off the plot just a little short of the frequency at which the MTF sign reversal occurs, as shown in Fig. 4.7d, can avoid the curve-fitting problem and indicate the real situation (because the curve-fitting program never gets the false positive data).

4.3 Geometric Blur Spot Size versus Certain Aberrations

Many times, the system characteristic of interest is the size of the blur (B) produced as the image of a point source. There are a few simple relationships that are useful in this regard. Note that these are based on geometrical optics (i.e., rays) and that diffraction effects are not included.

a. Third-order spherical at best focus (three-fourths of the way from paraxial to marginal focus):

$$B = 0.5 \text{ LA}_m \tan U_m = 0.5 \text{ TA}_m \quad (4.6)$$

If the geometric blur spot resulting from refocused third-order spherical (per Eq. (4.6)) is the same size as the Airy disk, then the OPD will be about 0.15 wavelengths. In light of this, the occasional use of the 100 percent blur spot size as a performance criterion does not seem too unreasonable.

b. Third- and fifth-order spherical (with the marginal spherical corrected, focused at 0.42 LA_z from the paraxial focus):

$$B = 0.84\ LA_z \tan U_m = 0.59\ TA_z \tag{4.7}$$

c. Third- and fifth-order spherical (corrected so that $LA_z = 1.5\ LA_m$, or $TA_m = 1.06\ TA_z$, and focused at 0.83 LA_z from the paraxial focus; this correction yields the smallest-diameter blur spot for a given amount of fifth-order spherical and is a good balance for a system with one or two waves of OPD from spherical aberration since it provides good contrast):

$$B = 0.5\ LA_m \tan U_m$$
$$= 0.5\ TA_m$$
$$= 0.33\ LA_z \tan U_m$$
$$= 0.47\ TA_z \tag{4.8}$$

Note that the above are based on the idea of the smallest spot containing 100 percent of the energy in the image of a point. For many applications this concept is valid and useful, but for best image quality there is usually another focus or correction at which the image has a smaller, brighter core and a larger flare; this is usually judged to be better for definition and pictorial purposes.

The effect of a large amount of defocusing on a well-corrected image is to produce a uniformly illuminated blur disk with a diameter of

$$B = 2(\text{defocus})\tan U_m \approx \frac{\text{defocus}}{f\ \text{number}} \tag{4.9}$$

Astigmatism and field curvature can be evaluated by applying Eq. (4.9) separately in the sagittal and tangential meridians.

Although ordinary axial chromatic is also defocusing, the blur it produces is not uniformly illuminated, but has the energy centrally concentrated. At the midway focus point, the diameter of the blur containing 100 percent of the energy is

$$B = LA_{ch} \tan U_m = TA_{ch} \tag{4.10}$$

However, the central concentration leads to a situation where 75 to 90 percent of the energy is in a spot only half this size, and 40 to 60 percent is in a spot one-quarter as large. (The smaller percentages apply for a uniform spectral response distribution and the larger percentages for a triangular spectral distribution.)

For third-order coma, the blur is the typical comet shape, and has a height equal to the tangential coma and a width (in the sagittal direction) two-thirds this size. Note, however, that about 50 percent of the energy in the coma patch is in the point of the figure, whose size equals about one-third of the tangential coma (i.e., it equals the sagittal coma).

4.4 Interpreting MTF—The Modulation Transfer Function

The interpretation of an MTF plot is often problematical; it shifts as the lens is refocused, and it is not the easiest thing in the world to decide how good an image is on the basis of an examination of its MTF plot.

The *limiting resolution* is easily determined if the system sensor can be characterized by an *aerial image modulation* (AIM) curve. The AIM curve is a plot of the threshold, or minimum, modulation required in the image for the sensor to produce a response. When plotted against spatial frequency, the intersection of the AIM curve and the MTF plot clearly indicates the limiting resolution, as shown in Fig. 4.8.

Although the AIM curve in Fig. 4.8 is shown as a straight line, it more typically curves upward as the frequency increases. There are big differences between the AIM curves for color and black-and-white film and even bigger differences between high-speed and low-speed film.

The Nyquist frequency is equal to $1/(2 \times \text{pixel spacing})$, and the corresponding period equals twice the pixel spacing. Thus a practical resolution limit with a pixelated sensor is a frequency a bit smaller than the Nyquist frequency.

The spherochromatic aberration and the secondary spectrum can be used to indicate the level of the monochromatic aberrations below which it is not worth pursuing, unless steps are taken to reduce the chromatic effects.

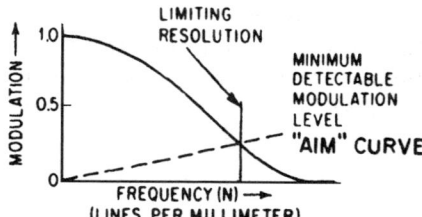

Figure 4.8 The intersection of the AIM curve and the MTF curve indicates the limiting resolution of the system.

Evaluation: How Good Is This Design?

A criterion for *excellent* performance (one that is often used as a design goal for top-of-the-line professional motion picture camera lenses) is to look for a 50 percent MTF at 50 lpm. Another criterion that has been presented for commercial 35-mm camera lenses is 20 percent MTF at 30 lpm over 90 percent of the field and yet another is 50 percent at 30 lpm. These criteria are applied at full aperture. These will give some idea of the range of the MTF values, which are more or less standard for this type of work.

Additional allowances suggested (by other authors) for a camera lens:

1. Distortion: < 2% or < 1%
2. Focus shift on stopping down: <0.02 mm
3. Vignetting: <20 or 50% at full field, zero at half field
4. Blur tolerance at 50 lpm: −0.02 mm
5. Defocus for 50 lpm: <0.02 · (f number) mm
6. MTF for a 35-mm camera lens on axis: 30% at 50 lpm; 50% at 30 lpm
7. MTF for an $f/1.8$ six element double Gauss: 65% at 40 lpm; 40% at 80 lpm

It is well to consider what the subject matter is likely to be in the corners of a rectangular field of view. For a typical snapshot, going clockwise from the upper-left corner we find: sky, sky, grass, grass. None of these would seem to require a high level of definition. Additionaly, when considered on a relative area basis, the significance of the corners is quite low.

A *final* check on the design should be made with more rays, more fields, and/or more wavelengths than used during the design process.

4.5 Fabrication Considerations

Here is a list of things to avoid in your lens designs. Of course, some may be unavoidable, but observing these caveats can lower costs, ease manufacturing, and possibly keep the optical shop from feeling that you should be lynched first thing tomorrow.

1. Avoid soft, easily abraded materials.
2. Avoid thermally weak materials that can be split by a mild thermal shock, such as encountered in blocking or washing. (They usually have a high coefficient of thermal expansion (CTE).)
3. Avoid easily stained materials and those easily attacted by moisture.
4. Avoid expensive materials.
5. Avoid thin elements that may warp or spring in fabrication (during blocking or polishing).

6. Avoid thin-edged elements. They are easily chipped, may be sharp edged at the blank diameter, and are difficult to handle and mount properly.
7. Avoid very thick elements. The material cost may be significant. For positive elements, thickness reduces the number of elements on a blocker, and there are big gaps between the blocked elements (Ref. 28,1980, p. 466).
8. Avoid strong curves, i.e., a large ratio of diameter to surface radius. Smith (Ref. 23, p. 552) indicates that the number of elements of diameter d that can be blocked on a tool of diameter T to be $N \approx 0.75\ (T/d)^2$.
9. Avoid meniscus elements with concentric (or nearly concentric) surfaces. They are not centerable.
10. Avoid *nearly* equi-convex or equi-concave elements. They are often assembled backward. Try for exactly equi-convex or equi-concave.
11. Avoid *nearly* plane surfaces. Go for truly plano.
12. Avoid bevels with odd angles, i.e., stick to 45°, 30°, and 60°.
13. Avoid precision bevels, in favor of simple, cheap protective chamfers. Mount or space from surfaces, not bevels.
14. Avoid cemented triplets or quadruplets.
15. Avoid tight cosmetic specs (scratch and dig).
16. Avoid thin, narrow airspaces between surfaces, especially if the ray slopes are large in the space. Note that the space may be correcting a high-order aberration created elsewhere in the system. Such spaces often need extremely tight fabrication tolerances (Ref. 28, 1980).
17. In general, just avoid tight tolerances if you can.
18. Another fabrication consideration arises in the course of the design process from the choice of the center thickness bounds. Obviously the thickness bound for a positive element must consider the edge thickness, and the bound for a negative element must not be so small that the element cannot be fabricated. But too large a value placed on the center (or edge) thickness bound may force the design toward elements of large diameter, volume, and cost (Ref. 28,1980). This occurs because the element thickness pushes the far surface further away from the stop; the ray beams spreading from the stop then require larger clear apertures for the elements, which then require increased thickness, and so on in a vicious circle. Violating an arbitrary minimum thickness limit by a bit may improve the situation markedly. It's a sort of snowball or domino effect; a thinner element allows a smaller diameter, which allows a thinner lens and smaller diameters for subsequent elements, which in turn can be thinner, and so on.

Chapter 5

Lens Design Data

5.1 About the Sample Lens Designs

One of the features of this volume is a fairly large set of lens designs, their prescriptions, and their aberration plots. This set is not intended to be a complete or extensive collection of all or even most of the published lens designs. We happily leave that to others. The set is intended to be a selection of lens designs that will serve two primary purposes. These are to serve as a set of suitable starting designs and also to serve as a set of designs that illustrate to the reader the principles and techniques of successful lens designs.

In this second edition of *Modern Lens Design* we have pruned out about half of the lens designs featured in the first edition. This was done (among other reasons) to make room for the addition of the descriptions of a good number of actual design projects, undertaken especially for this book as demonstrations of "how it's done." These descriptions report the initial specifications and "givens" plus the reasons for the design steps that were taken. They also report on the missteps; there was no papering over of goofs or redoing the job to make things (and the author) look better.

The designs in this book were drawn from many sources. Many are derived from the patent literature, or books that include patent references. Some of the designs are from the technical literature, such as journals, proceedings, or other books about lenses. Some are from private communications and some have never been previously published.

Note that the lens designs in patents (especially older patents) often do not show the best embodiment of the lens. This was probably done to protect the best (and secret) design from competitors. Today it is such a simple matter to optimize a not-so-good design that most patent embodiments now are quite good, and making minor design changes to conceal the best version is simply not worth the trouble (and there are legal reasons as well).

In most cases the published designs have been modified to some extent. For the majority of the designs, we have specified the optical glass as one of those from the Schott (Schott Glass Technologies, Inc.) catalog. We have chosen what we feel is the nearest Schott glass to that indicated in the source for the lens data. Occasionally this may constitute a significant change, but we have attempted to stay as close to the original data as possible. In a few designs, non-Schott glasses have been used. Note that many glass catalogs are currently in a state of flux; as the contents are changed, the number of glasses is reduced, and the glasses are reformulated to reduce density, improve durability, and eliminate lead, arsenic, and cadmium. For the newer designs we have used the standard six-digit code to identify glass types, rather than listing the index and V-value. For example, the six-digit code for LaK12 in Fig. 5.1 would be 678552, indicating an index of 1.678 and a V-value of 55.2.

In the first edition of this book the lens design figures were produced by a special, custom version of the lens design program GENII. This allowed us to produce all the significant aberration plots on one sheet. In this edition the new lens figures were done with the current version of OSLO, which includes all the features for which we needed a custom program just a decade ago. The new aberration plots are similar enough

F/4.5 25.2 deg Triplet US 1.987,878/1935 Schneider

radius	thickness	mat'l	index	V-no	sa
26.160	4.916	LAK12	1.678	55.2	11.7
1201.700	3.988	air			11.7
−83.460	1.038	SF2	1.648	33.8	10.2
25.670	4.000	air			10.2
	6.925	air	1.651	55.9	9.2
302.610	2.567	LAK22			10.3
−54.790	81.433	air			10.3

EFL	= 98.56	= Effective focal length
BFL	= 81.43	= Back focal length
NA	= −0.1127 (F/4.4)	= Numerical aperture (F-number)
GIH	= 46.33 (HFOV = 25.17)	= Image height (half field in degrees)
PTZ/F	= −2.831	= (Petzval radius) /F
VL	= 23.43	= Vertex length
OD	infinite conjugate	= Object distance

Figure 5.1 Sample lens prescription.

to those in the style of the first edition that the interpretation should be obvious.

The aperture and field that are indicated for any given lens design are more a matter of taste than anything else. What constitutes an acceptable level of aberration depends mostly on the application to which the lens is to be put. Thus the values for field and aperture that accompany each design in this book have often been selected somewhat arbitrarily to yield a level of correction we thought reasonable.

The choice of the clear apertures for the lens elements is equally arbitrary. Obviously, the clear aperture of an element cannot be so large that the edge thickness at that diameter becomes negative or impractically thin. We have selected what seemed to be reasonable values for the clear apertures, based on both edge thickness considerations and the choice of a vignetting factor that allows a reasonably sized oblique beam through the lens and also trims the oblique beam to eliminate the worst-behaved rays.

5.2 Lens Prescriptions, Drawings, and Aberration Plots

The lens design data and the associated graphics for this book have been produced by computer. While data input errors and other glitches are always possible, by producing the lens data table, the lens drawing, and the aberration plots all from the same lens data file, we hope to prevent most of the errors that have afflicted some other efforts of this type. For the new designs in this edition (which represent a variety of applications), we have departed from the practice of always scaling the focal length of each lens to 100 mm. We have instead used focal lengths appropriate to the application that the design process story is illustrating. We have also incorporated the design data into the text because, for the most part, the new design data is integral to the description of the design process in the text.

5.2.1 Lens prescription

A sample lens data table is shown in Fig. 5.1. The lens construction data are tabulated in a quite straightforward way. The columns are headed *radius, thickness, mat'l, index, V-no*, and *sa* (for semiaperture); the meanings should be apparent. The radius value follows the usual sign convention that a positive radius has its center of curvature to the right of the surface. Plano surfaces (i.e., with infinite radius) are indicated by a blank entry in the radius column. The thickness and material following a surface are presented on the same line as the surface radius, and have the same number.

With few exceptions, the material names are those of Schott Glass Technologies, Inc. The index and V number values correspond to the wavelengths given with the ray intercept plot (e.g., see Fig. 5.3); for most lenses we have used the d, F, and C lines. The location of the aperture stop is indicated by a blank in the radius column with air on both sides of the surface. Aspheric surfaces are specified by the conic constant kappa and/or the aspheric deformation coefficients. The equation for the surface is

$$z = \frac{cy^2}{1+[1-(1+\kappa)c^2y^2]^{1/2}} + ADy^4 + AEy^6 + AFy^8 + AGy^{10} \quad (5.1)$$

where c = surface curvature ($c = 1/r$)
y = radial distance from the axis
κ = conic constant kappa
AD, AE, ... = fourth, sixth, ... order deformation coefficients

The conic constant is described as follows:

$\kappa > 0$	An oblate spheroid; an ellipse with the focal points not on the axis
$\kappa = 0$	A sphere
$0 > \kappa > -1$	A prolate spheroid; an ellipse with the focal points on the axis
$\kappa = -1$	A paraboloid
$\kappa < -1$	A hyperboloid

Note that the first term of Eq. (5.1) is the equation of a conic section; if $\kappa = 0.0$, it is the equation of a sphere.

The data below the prescription tabulation has the following meanings:

EFL	Effective focal length
BFL	Back focal length (the distance from the last surface to the paraxial focal point)
NA	Numerical aperture (the corresponding f number is in parentheses)
GIH	Gaussian (paraxial) image height (half-field in degrees is in parentheses)
PTZ/F	Petzval radius as a fraction of EFL
VL	Vertex length from first to last surface
OD	Object distance

Equation (5.1) can simulate a cone surface (axicon) with an extreme hyperbola; use a very large curvature and set $\kappa = 1/\cos^2\theta$, where θ is the half angle of the cone.

5.2.2 Lens drawing

A sample lens drawing is shown in Fig. 5.2. The scale of the lens drawing is indicated by the dimensioned length of the line in the lens sketch. The rays in the sketch are the marginal and principal rays corresponding to the aperture and field angle, which are tabulated with the lens data. The aperture stop location is indicated by the point at which the principal ray crosses the optical axis. The lens elements are drawn to the clear apertures given in the prescription table as sa, the semiaperture.

Additional rays can easily be added to the lens drawing if desired by using a technique that is exact only for paraxial rays and is often accurate enough for use in estimating or drawing ray paths. A ray may be scaled by simply multiplying the heights at which it strikes the surfaces by a scaling constant. Also, rays may be added by adding their intersection heights together. In each case the result is a reasonable approximation to the path of another ray. Obviously, two rays can be scaled and then added. Thus any desired third ray can be drawn by determining its intersection heights from

$$Y_3 = AY_1 + BY_2 \tag{5.2}$$

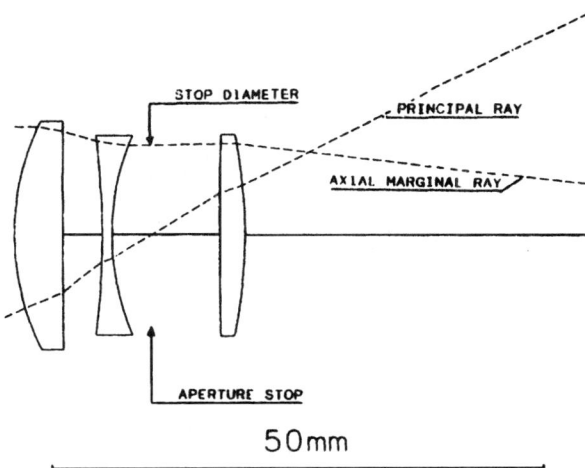

Figure 5.2 Sample lens drawing.

90   Chapter Five

where A and B are scaling factors (see Chap. 24, Sec. 4) and Y_1 and Y_2 are the ray heights of the rays in the lens drawing. If one defines the desired third ray by its intersection with any two surfaces (which may include the object or image surface), then a simultaneous solution for A and B may be found from the two equations that result when the appropriate values of Y_1, Y_2, and Y_3 are substituted into the equation above. See Eqs. 24.40 through 24.43.

5.2.3 Aberration plots

A sample aberration plot is shown in Fig. 5.3. The aberration plots include both tangential and sagittal ray intercept plots (sometimes called H-tan U curves) for the axis, 0.7 field, and full field. The ray displacements are plotted vertically, as a function of the position of the ray in the aperture. The vertical scale is given at the lower end of the vertical bar for the axial plot; the number given is the half length (i.e., from

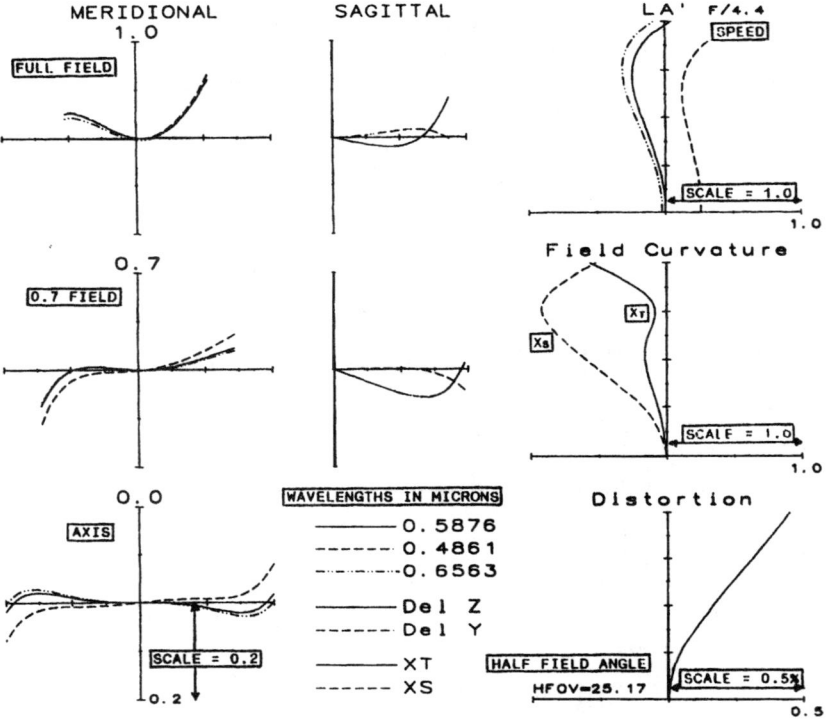

Figure 5.3 Sample aberration plot.

the origin to the end) of the vertical line in the plot. The horizontal scale is proportional to the tangent of the ray slope angle. Following the usual convention, the upper ray of the ray fan is plotted to the right. In the sagittal plots, the solid line is the transverse aberration in the x, or sagittal, direction and the dashed line is the ray displacement in the y direction (which is the sagittal coma).

The ends of the ray intercept plot are determined either by the aperture stop or by the vignetting. An arrowhead on the horizontal scale line of the plot indicates that the plot was terminated because the intersection height went offscale.

In addition to the ray intercept plots (which are, in general, probably the most broadly useful presentation of the aberration characteristics of a design), two aberrations are also presented as longitudinal plots. The longitudinal representations of spherical aberration and field curvature have been the classical, conventional presentation for decades, despite the fact that they give a very incomplete picture of the state of correction of the lens; however, a longitudinal plot of the spherical aberration in three wavelengths does allow a much clearer understanding of the spherochromatism, as well as the secondary spectrum. The scale factor for this plot is the number given at the right end of the horizontal axis; the number is the half length of the horizontal line. The vertical dimension of the plot is the height of the ray at the pupil; the f number is given at the top of the plot. This is the f number of the imaging cone and is equal to 1/2NA. The longitudinal field curvature plots yield an excellent picture of the correction of the Petzval curvature and the astigmatism. The scale for X_s and X_t is given at the right end of the horizontal axis; again the number is the half length of the horizontal line. The solid line is X_t and the dashed line is X_s. The vertical scale is the fraction of the *gaussian image height* (GIH); the half-field angle is given at the left side of the distortion plot. The scale for distortion is in percent, and the number indicates the half length of the horizontal line.

The field curvature values, X_s and X_t, indicate the imagery of an infinitesimal fan of rays about the principal ray, i.e., paraprincipal rays. These are analogous to the paraxial rays about the optical axis. The slope of the ray intercept plots indicates the longitudinal defocus of a pinhole-sized beam of rays about the ray that has been plotted in the ray intercept plot. Note that in the ray intercept plot, a left- or right-shift along the meridional fan plot corresponds to a shift of the stop. See Fig. 2.2.

Spherical aberration is sometimes plotted longitudinally as LA = δL = $(L' - l')$ versus the height of the ray in the pupil. With an object at infinity, the paraxial focal length is $f = -y_1/u_k'$. The "marginal focal length" is given by $F = -Y_1/\sin U_k'$. The variation of focal length with ray height, $\delta F = (F-f)$, can be plotted in the same diagram as spherical, $\delta L = (L'-l')$. The space between the δL and δF plots is a measure of OSC, which is equal to δ(sagittal coma)/δh, and indicates the amount of coma at small field angles. This plot was widely used to display coma when most aberrations were presented longitudinally.

5.3 Estimating the Potential of a Redesign

For an existing design it is relatively easy to estimate the effect on the aberration plots caused by a modest redesign if we apply a knowledge of third-order aberration theory. This is possible because the third-order aberrations of a lens are easily adjusted by changing the spaces or the shapes of the elements, whereas the amount of higher-order aberration tends to be relatively stable and resistant to change.

Equations (5.3) and (5.4) are a power series expansion of the relationships between the ray intersection with the image plane (x', y') as a function of the object height h and the ray position in the pupil (defined in polar coordinates s and θ), as shown in Fig. 5.4.

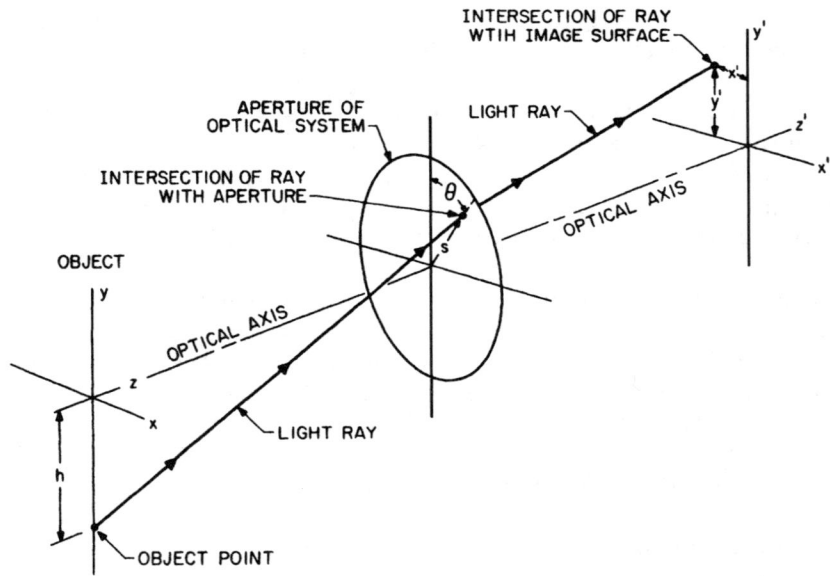

Figure 5.4 A ray from the point $y = h$, $(x = 0)$ in the object passes through the optical system aperture at a point defined by its polar coordinates (s, θ) and intersects the image surface at x', y'.

$$y' = A_1 s \cos \theta + A_2 h$$
$$+ B_1 s^3 \cos \theta + B_2 s^2 h(2 + \cos 2\theta) + (3B_3 + B_4)sh^2 \cos \theta + B_5 h^3$$
$$+ C_1 s^5 \cos \theta + (C_2 + C_3 \cos 2\theta) s^4 h + (C_4 + C_6 \cos^2 \theta) s^3 h^2 \cos \theta$$
$$+ (C_7 + C_8 \cos 2\theta) s^2 h^3 + C_{10} sh^4 \cos \theta + C_{12} h^5$$
$$+ D_1 s^7 \cos \theta + \cdots \quad (5.3)$$

$$x' = A_1 s \sin \theta$$
$$+ B_1 s^3 \sin \theta + B_2 s^2 h \sin 2\theta + (B_3 + B_4) sh^2 \sin \theta$$
$$+ C_1 s^5 \sin \theta + C_3 s^4 h \sin 2\theta + (C_5 + C_6 \cos^2 \theta) s^3 h^2 \sin \theta$$
$$+ C_9 s^2 h^3 \sin 2\theta + C_{11} sh^4 \sin \theta$$
$$+ D_1 s^7 \sin \theta + \cdots \quad (5.4)$$

Notice that in the A terms, the exponents of s and h are unity. In the B terms, the exponents total 3, as in s^3, $s^2 h$, sh^2, and h^3. In the C terms, the exponents total 5, and in the D terms, 7. These are referred to as the first-order, third-order, and fifth-order terms, and so on. There are 2 first-order terms, 5 third-order, 9 fifth-order, and $[(n+3)(n+5)/8 - 1]$ nth-order terms. In an axially symmetrical system there are no even-order terms; only odd-order terms may exist. If we depart from symmetry as, for example, by tilting a surface or introducing a toroidal or other nonsymmetrical surface, then there will be even-order terms.

It is apparent that the A terms relate to the paraxial (or first-order) imagery. A_2 is simply the magnification (h'/h), and A_1 is a transverse measure of the distance from the paraxial focus to our "image plane." All the other terms in Eqs. (5.3) and (5.4) are called *transverse aberrations*. If the image surface is at the paraxial focus, they represent the distance by which the ray misses the ideal image point as defined by the paraxial imaging equations.

The B terms are called the *third-order*, or *Seidel*, or *primary* aberrations. B_1 is spherical aberration; B_2 is coma; B_3 is astigmatism; B_4 is Petzval; and B_5 is distortion. Similarly, the C terms are called the *fifth-order* or *secondary* aberrations. C_1 is fifth-order spherical aberration; C_2 and C_3 are linear coma; C_4, C_5, and C_6 are oblique spherical aberration; C_7, C_8, and C_9 are elliptical coma; C_{10} and C_{11} are Petzval and astigmatism; and C_{12} is distortion.

The 14 terms in D are the seventh-order or tertiary aberrations; D_1 is the seventh-order spherical aberration. A similar expression for OPD, the wavefront deformation, is given in Sec. 5.4 (Eqs. (5.5) and (5.6)).

The terms with the B coefficients are the transverse third-order aberrations. The longitudinal aberrations are equal to the transverse aberrations divided by $-u_k'$; since u_k' is a direct function of the ray height s, one can convert the transverse into longitudinal aberrations by reducing the exponent of s by 1 (and adjusting the coefficients).

Note that the third-order spherical ($B_1 s^3 \cos\theta$), the fifth-order ($C_1 s^5 \cos\theta$), the seventh-order, and all the other spherical terms do not contain the field term h, and are thus constant over the whole field. Spherical is the only monochromatic aberration on the optical axis (where h = zero). Oblique spherical is like a cross between spherical and astigmatism aberrations. Consider that there are three constants (or, if you wish, two constants, $[C_4 + C_6]$ for y and $[C_5 + C_6]$ for x, indicating that oblique spherical can be different in x and y. It varies as h^2 like astigmatism and as a function of s^3 like spherical. The defocussing effects of oblique spherical can be balanced out by third-order astigmatism, but the third-order spherical cannot balance out the s^3 term except at one point in the field.

Since the high-order aberrations tend to be quite stable, one can assume that if the spherical aberration is adjusted by changing the third-order aberration so as to correct it at a given aperture, the change in the longitudinal aberration will be proportional to the square of the ray heights. For example, if the spherical at ray height Y_1 is to be changed by $d\text{LA}_1$, then at ray height Y_2 it will change by $d\text{LA}_2 = d\text{LA}_1(Y_2/Y_1)^2$. The change of the transverse spherical will vary as a cubic function, so that the transverse change $d\text{TA}_2 = d\text{TA}_1(Y_2/Y_1)^3$. Figure 3.6 shows the effect of changing the third-order spherical (assuming a constant fifth-order spherical).

If the axial chromatic is changed, the ray intercept plots for the different colors will simply rotate with respect to each other. The secondary spectrum and spherochromatism will change very little, so that one can readily estimate the ray intercept plots that will result from a simple change in the axial chromatic. Figure 3.7 shows two different balances of axial chromatic and spherochromatism. Similarly, a lateral chromatic change will change the relative heights of the different color ray plots, and the amount of the height change will be proportioned to the image height.

The change in the third-order longitudinal astigmatism and field curvature is proportional to the square of the image height or field angle. Thus if the field curvature is changed by dX_1 at image height H_1, the change at H_2 is $dX_2 = dX_1(H_2/H_1)^2$. The change in the tangential field curvature X_t is three times the change in the sagittal field curvature X_s if the change is produced by changing the amount of the astigmatism; however, if the change results from a change in the

Petzval curvature, both X_s and X_t are shifted by the same distance. Note that the slope of the ray intercept plot ($dH'/d \tan U$) at the principal ray is equal to the tangential field curvature (X_t) (or X_s for the sagittal plot).

Changes in the third-order coma produce a parabolic-shaped change in the ray intercept plot. If the plot is raised by an amount dH at the ends of the plot, it will be raised by $(0.7)^2 \, dH = 0.5 \, dH$ at the 0.7 zones of the aperture. The amount of the coma change for other field angles will vary directly with the field angle or image height.

The change in percent distortion will vary with the square of the field angle. The change in the lateral color varies directly with the field angle.

Thus one can look at the aberration plots for a given design and, by applying the techniques outlined above, easily visualize what they will look like after an adjustment has been made to fit the design to the application at hand.

Figure 5.5 illustrates the relationships between the four ways that aberrations are conventionally depicted. The longitudinal measure (LA) defines the longitudinal distance by which a ray misses the ideal axial intersection. The *transverse measure* (TA) indicates the transverse (normal to the axis) distance by which the ray misses the ideal intersection with the image surface. The *angular aberration* (AA) is simply the transverse TA divided by the distance l' from the second principal point to the image. The wavefront aberration is the distance (along the ray) that the actual wavefront departs from an ideal spherical reference surface.

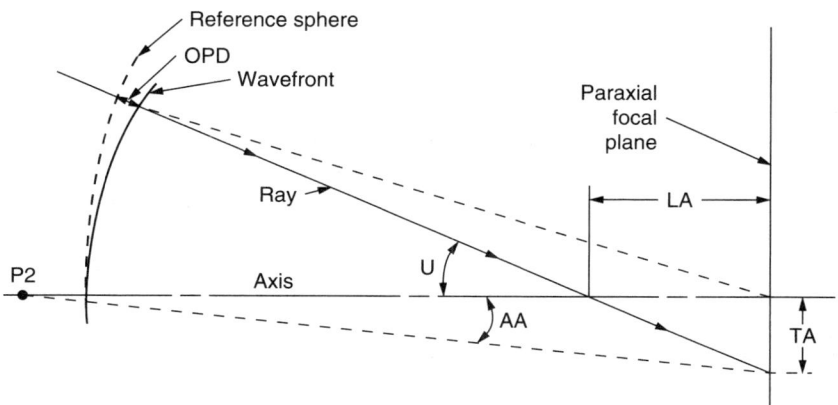

Figure 5.5 Showing the relationships between the several ways of describing the size of an aberration. LA = longitudinal aberration; TA = transverse aberration = TA · tan U; AA = angular aberration = TA/(P2-to-focus); OPD = ∫AA.

These aberrations are related by:

$$TA = LA \cdot \tan U$$
$$AA = TA/l'$$
$$OPD = \int AA$$

The *optical path* (OP = $\Sigma n \cdot d$) is related to the time of travel of light, which is equal to $\Sigma n \cdot d/c$. Ideally the OP from the object point to a reference sphere centered on the image point (and often located at the exit pupil—or at infinity) should be constant over the full aperture. The *optical path difference,* OPD = (OP$_{ray}$ − OP$_{ref}$), where OP$_{ray}$ is the path along a ray and OP$_{ref}$ is the path along the axis or along the principal ray. The *pupil function* is OPD (x, y); the *wave (front) function* is $w(x, y)$ = OPD $(x, y)/\lambda$ in waves; and the *phase function* $\Phi(x, y)$ is $2\pi w (x, y)$ in radians.

Aberrations may be *intrinsic* or *induced*. The intrinsic aberrations are those of a surface (or element) that are unaffected by the aberrations of the other surfaces. Induced aberrations are created by the aberrations (i.e., changes in the ray heights or angles) of the other elements. Usually the lower-order aberrations of the other surfaces cause induced higher-order aberrations. For example, the third-order aberrations of preceding surfaces will induce fifth-order spherical in following surfaces. See Chap. 6, Secs. 6.3 and 6.4 for an example of how the third-order spherical and first-order chromatic aberration in the first element affect the zonal (fifth-order) spherical and spherochromatic of the lens.

5.4 Scaling a Design, Its Aberrations, and its Modulation Transfer Function

A lens prescription can be scaled to any desired focal length simply by multiplying all of its dimensions by the same constant. All of the *linear* aberration measures will then be scaled by the same factor. Note, however, that percent distortion, *chromatic difference of magnification* (CDM), the numerical aperture or *f* number, the aberrations expressed as angular aberrations, and any other *angular* characteristics remain completely unchanged by scaling.

The exact *diffraction modulation transfer function* (MTF) cannot be scaled with the lens data. The diffraction MTF, since it includes diffraction effects that depend on wavelength, will not scale because the wavelength is not ordinarily scaled with the lens. A *geometric* MTF can be scaled by dividing the spatial frequency ordinate of the MTF plot by the scaling factor. Of course, because it neglects diffraction, the geometric

MTF is quite inaccurate unless the aberrations are large, with OPD to the order of one or two wavelengths (and the MTF is correspondingly poor).

A diffraction MTF can be scaled *very* approximately as follows: Determine the OPD that corresponds to the calculated MTF value of the lens for several spatial frequencies. This can be done by comparing the MTF plot for the lens to Figs. 4.5 and 4.6, which relate the MTF to OPD. Then multiply the OPD by the scaling factor and, again using Figs. 4.5 and 4.6, determine the MTF corresponding to these scaled OPD values. Obviously the accuracy of this procedure depends on how well the simple relationships of Figs. 4.5 and 4.6 represent the mix of aberrations in a real lens.

In the event that a proposed change of aperture or field is expected to produce a change in the amount of the aberrations, one can attempt to scale the MTF as affected by aberration. This is done by determining the type of aberration that most severely limits the MTF, then scaling the OPD according to the way that this aberration scales with aperture or field, in a manner analogous to that described in Sec. 5.3. In general, OPD as a function of aperture varies as one higher exponent of the aperture than does the corresponding transverse aberration. For example, the OPD for third-order transverse spherical (which varies as Y^3) varies as the fourth power of the ray height. In a form analogous to Eqs. (5.3) and (5.4) which indicate a power series expansion of the transverse aberrations as a function of aperture and field, Eq. (5.5) gives the relationship for OPD. As in Sec. 5.3, the terms of the equation refer to Fig. 5.4.

$$\begin{aligned}
\text{OPD} = {} & A'_1 s^2 + A'_2 sh \cos\theta \\
& + B'_1 s^4 + B'_2 s^3 h \cos\theta + B'_3 s^2 h^2 \cos^2\theta + B'_4 s^2 h^2 + B'_5 sh^3 \cos\theta \\
& + C'_1 s^6 + C'_2 s^5 h \cos\theta + C'_3 s^4 h^2 + C'_5 s^4 h^2 \cos^2\theta + C'_7 s^3 h^3 \cos^3\theta \\
& + C'_8 s^3 h^3 \cos^3\theta + C'_{10} s^2 h^4 + C'_{11} s^2 h^4 \cos^2\theta + C'_{12} sh^5 \cos\theta \\
& + D'_1 s^8 + \cdots
\end{aligned} \quad (5.5)$$

Note that although the constants here correspond to those in Eqs. (5.3) and (5.4), they are not numerically the same and are primed to so indicate; however, (because rays are normal to wavefronts) the expressions are related by

$$y' = \text{TA}_y = \frac{l}{N}\frac{\partial \text{OPD}}{\partial y} \quad \text{and} \quad x' = \text{TA}_x = \frac{l}{N}\frac{\partial \text{OPD}}{\partial x} \quad (5.6)$$

where l is the pupil-to-image distance and N is the image space index.

Note that the exponent of the semiaperture term s is larger by 1 in the wavefront expression than in the ray-intercept equations. Fourth-order OPD terms correspond to third-order transverse aberration terms.

5.5 Notes on the Interpretation of Ray Intercept Plots

5.5.1 Intercept plots

When the image plane intersection heights of a fan of meridional rays are plotted against the slope of the rays as they emerge from the lens, the resultant curve is called a ray intercept curve, an $H' - \tan U'$ curve, or sometimes (erroneously) a rim ray curve. The shape of the intercept curve not only indicates the amount of spreading or blurring of the image directly, but also can serve to indicate which aberrations are present.

In Fig. 5.6 an oblique fan of rays from a distant object point is brought to a perfect focus at point P. If the reference plane passes through P, it is apparent that the $H' - \tan U$ curve will be a straight horizontal line; however, if the reference plane is behind P (as shown) then the ray intercept curve becomes a tilted straight line since the height H' decreases exactly as $\tan U'$ decreases. Thus it is apparent that shifting the reference plane (or focusing the system) is equivalent to a rotation of the $H' - \tan U'$ coordinates. A valuable feature of this type of aberration representation is that one can immediately assess the effects of refocusing the optical system by a simple rotation of the abscissa of the

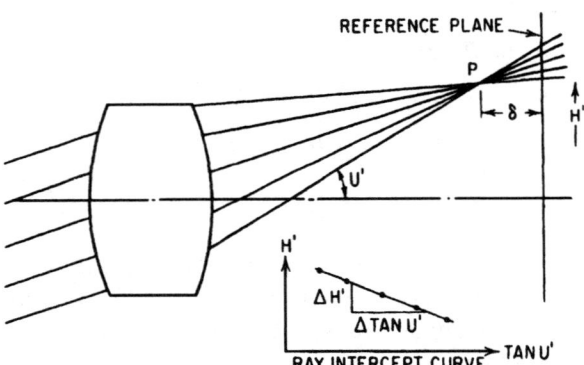

Figure 5.6 The ray intercept curve (H' versus $-\tan U'$) of an image point that does not lie in the reference plane is a tilted straight line. The slope of the line ($dH'/d \tan U'$) is mathematically identical to the distance from the reference plane to the point P. Note that this distance is equal to X_t, the tangential field curvature (if the reference plane is the paraxial focal plane).

figure. Notice that the slope of the line ($\Delta H'/\Delta \tan U'$) is exactly equal to the distance δ from the reference plane to the point of focus, so that, for an oblique ray fan, the tangential field curvature is equal to the slope of the ray intercept curve. Obviously, this relationship will also hold for sagittal ray fans.

Note that many, if not all, programs plot the ray intercept height H versus the relative position of the ray in the pupil, rather than versus tan U as is shown here. Besides simplicity, one reason for this is that the H-tan U plot can be nonmonotonic for some lenses (such as very wide angle or fish-eye lenses). The difference between the two is almost always invisible. One visible difference is that for a defocussed, high NA perfect lens (e.g., a paraboloid mirror) the plot is not a straight line when plotted against the ray height; it's a cubic parabola like third-order spherical. In some programs there is an option between the plotting versus tan U and plotting versus y.

Note that the convention for ray intercept plots is to plot the off-axis curves for an image located a positive distance above the axis, and that rays from the top of the exit pupil are plotted at the right of the plot. This may not be in accord with the sign convention for ray slope, but these conventions allow one to recognize immediately the sign of the underlying aberrations.

Figure 5.7 shows a number of intercept curves, each labeled with the aberration represented. The generation of these curves can be readily understood by sketching the ray paths for each aberration and then plotting the intersection height and slope angle for each ray as a point of the curve. Distortion is not shown in Fig. 5.7; it would be represented as a vertical displacement of the curve from the paraxial image height h'. Lateral color would be represented by curves for two colors that were vertically displaced from each other. The ray intercept curves of Fig. 5.7 are generated by tracing a fan of meridional or tangential rays from an object point and plotting their intersection heights versus their slopes. The imagery in the other meridian can be examined by tracing a fan of rays in the sagittal plane (normal to the meridional plane) and plotting their x-coordinate intersection points against their slopes in the sagittal plane.

It is apparent that the ray intercept curves that are "odd" functions, that is, the curves that have a rotational or point symmetry about the origin, can be represented mathematically by an equation of the form

$$y = a + bx + cx^3 + dx^5 + \cdots$$

or $$H' = a + b \tan U' + c \tan^3 U' + d \tan^5 U' + \cdots \qquad (5.7)$$

All the ray intercept curves for *axial* image points are of this type. Since the curve for an axial image must have $H' = 0$ when $\tan U' = 0$, it

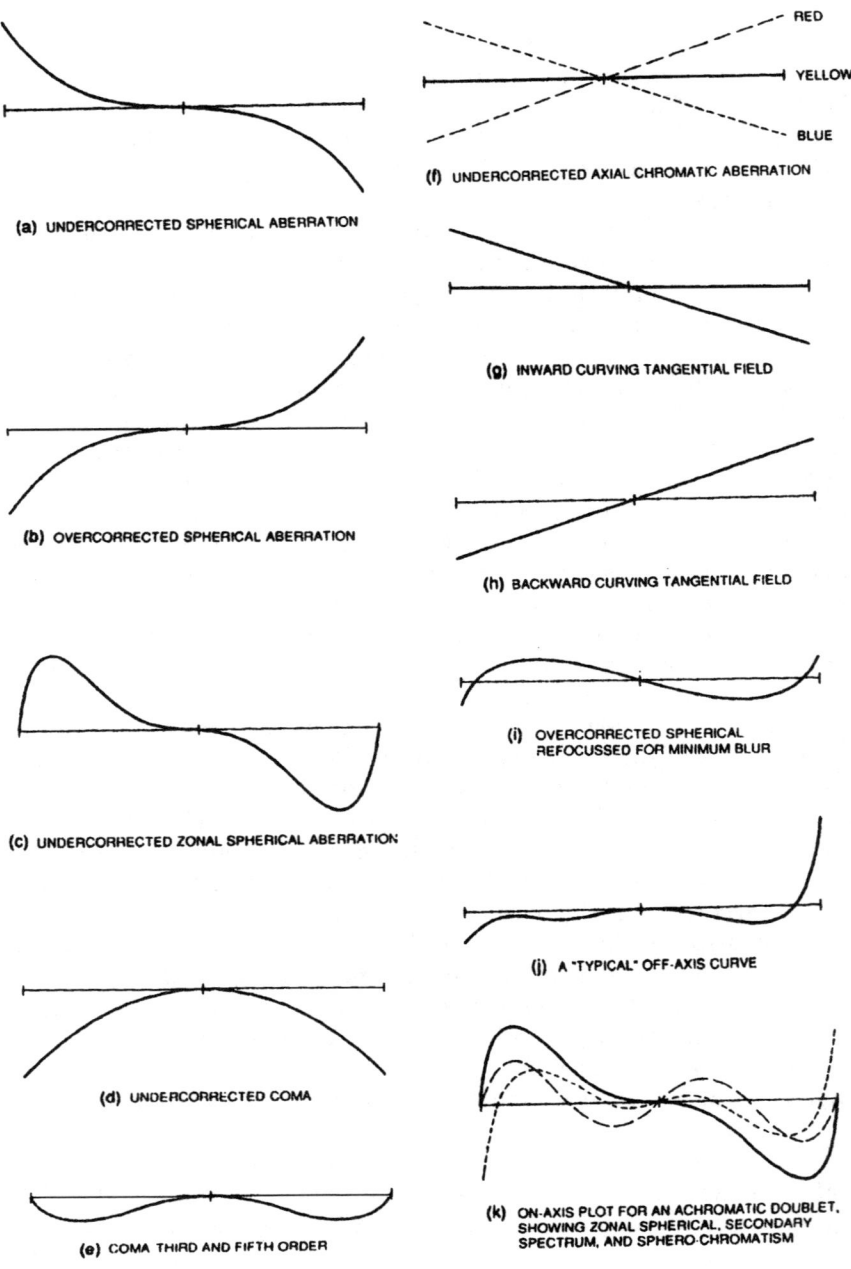

Figure 5.7 Sample ray intercept plots for various aberrations. The ordinate for each curve is the height at which the ray intersects the (paraxial) image plane; usually H is plotted relative to the principal ray height, which is set to zero. The abscissa is tan U, the final slope of the ray with respect to the optical axis. Note that, regardless of the sign convention for the ray slope, it is conventional to plot the ray through the top of the exit pupil at the right of the figure, and that the curves for image points above the axis are usually shown. Observance of these conventions makes it much easier to interpret the plots.

is apparent that the constant a must be a zero. It is also apparent that the constant b for this case represents the amount the reference plane is displaced from the paraxial image plane. Thus the curve for lateral spherical aberration plotted with respect to the paraxial focus can be expressed by the equation

$$TA' = c \tan^3 U' + d \tan^5 U' + e \tan^7 U' + \cdots \tag{5.8}$$

It is, of course, possible to represent the curve by a power series expansion in terms of the final angle U', or $\sin U'$, or the ray height at the lens (Y), or even the initial slope of the ray at the object (U_0) instead of $\tan U'$. The constants will, of course, be different for each.

For simple uncorrected lenses, the first term of Eq. (5.8) is usually adequate to describe the aberration. For the great majority of corrected lenses the first two terms are dominant; in a few cases three terms (and rarely four) are necessary to satisfactorily represent the aberration. As examples, Figs. 5.6a and b can be represented by $TA' = c \tan^3 U'$, and this type of aberration is called *third-order spherical*. Figure 5.6c, however, would require two terms of the expansion to represent it adequately; thus $TA' = c \tan^3 U' + d \tan^5 U'$. The amount of aberration represented by the second term is called the *fifth-order aberration*. Similarly, the aberration represented by the third term of Eq. (5.8) is called the *seventh-order aberration*. The fifth-, seventh-, ninth-, and so on, order aberrations are collectively referred to as *higher-order aberrations*.

The ray intercept plot is subject to a number of interesting interpretations. It is immediately apparent that the top-to-bottom extent of the plot gives the size of the image blur. Also, a rotation of the horizontal (abscissa) lines of the graph is equivalent to a refocusing of the image and can be used to determine the effect of refocusing on the size of the blur.

Figure 5.6 shows that the ray intercept plot for a defocused image is a sloping line. If we consider the slope of the curve at any point in an H–$\tan U$ ray intercept plot, the slope is equal to the defocus of a small-diameter bundle of rays centered about the ray represented by that point. In other words, this would represent the focus of the rays passing through a pinhole aperture so positioned as to pass the rays at that part of the H–$\tan U$ plot. Similarly, since shifting an aperture stop along the axis is, for an oblique bundle of rays, the equivalent of selecting one part or another of the ray intercept plot, one can understand why shifting the stop can change the field curvature.

The OPD or wavefront aberration can be derived from an H–$\tan U$ ray intercept plot. The area under the curve between two points is equal to the OPD between two rays that correspond to the two points. Ordinarily, the reference ray for OPD is either the optical axis or the principal ray (for an oblique bundle). Thus the OPD for a given ray is usually the area under the ray intercept plot between the center point and the ray.

Mathematically speaking, then, the OPD is the integral of the H–tan U plot and the defocus is the first derivative. The coma is related to the curvature or second derivative of the plot, as a glance at Fig. 5.7d will show.

It should be apparent that a ray intercept plot for a given object point can be considered as a power series expansion of the form

$$H' = h + a + bx + cx^2 + dx^3 + ex^4 + fx^5 + \cdots \quad (5.9)$$

where h is the paraxial image height, a is the distortion, and x is the aperture variable (e.g., $\Delta\tan U'$). Then the art of interpreting a ray intercept plot becomes analogous to decomposing the plot into its various terms. For example, cx^2 and ex^4 represent third- and fifth-order coma, while dx^3 and fx^5 are the third- and fifth-order spherical. The bx term is due to a defocusing from the paraxial focus and could be due to curvature of field. Note that the constants a, b, c, and so on will be different for points at differing distances from the axis.

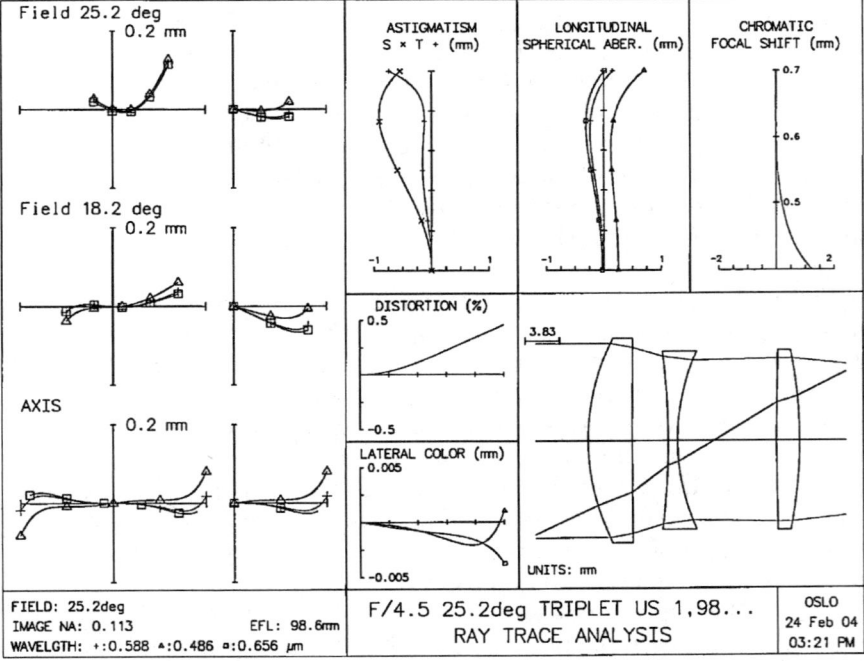

Figure 5.8 A complete aberration analysis on one page, with ray intercept plots, field curvature, distortion, lateral color, spherical and chromatic plus, a sketch of the lens system. Lens is Fig. 5.1.

5.6 Various Evaluation Plots

Here are some examples of the many evaluation plots that are available in a full scale lens design program. Believe it or not, this list does not exhaust the variety of plots available; the type and style of the plots will, of course, vary from program to program. The plots are based on the U.S. Patent no. 1,987,878 (1935) Cooke triplet shown in Figs. 5.1 and 5.3, which is apparently not as well corrected as one might expect.

Figure 5.8 is a basic "one-sheet" presentation of ray intercept plots; it also includes a lens drawing and individual plots of field curvature, spherical aberration, distortion, and lateral color. (This is similar to the presentation in Fig. 5.3.) Figure 5.9 is a through-focus MTF plot at a specific frequency for three field points, and Fig. 5.10 is the through-frequency plot at the nominal focus setting. Figure 5.11 plots the MTF at three frequencies against the field, and Fig. 5.12 plots the monochromatic rms spot size and rms OPD against the field. Figure 5.13 is a lens drawing with marginal and principal rays; more rays can be

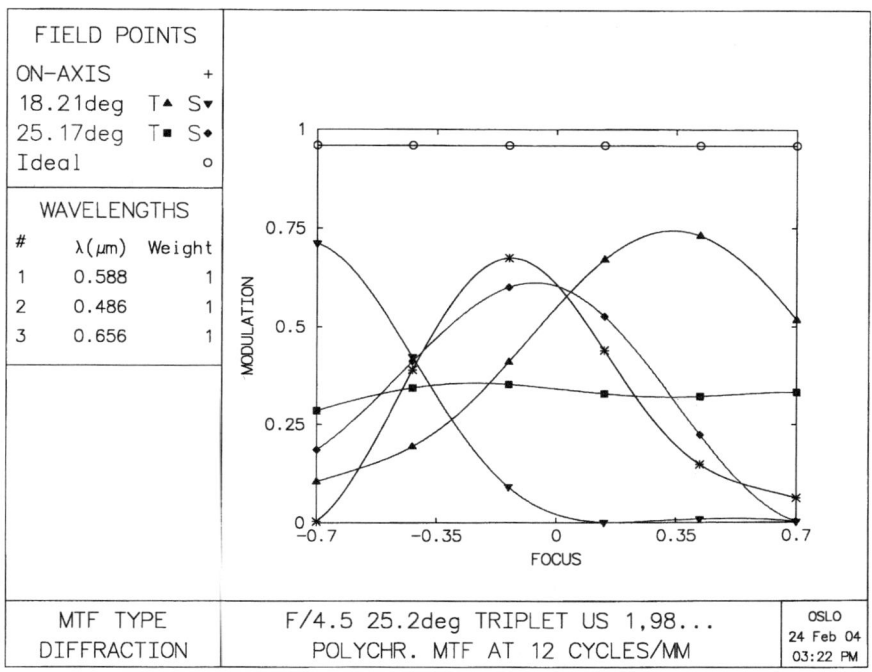

Figure 5.9 A through-focus of plot of the MTF for three field points. Lens is Fig. 5.1.

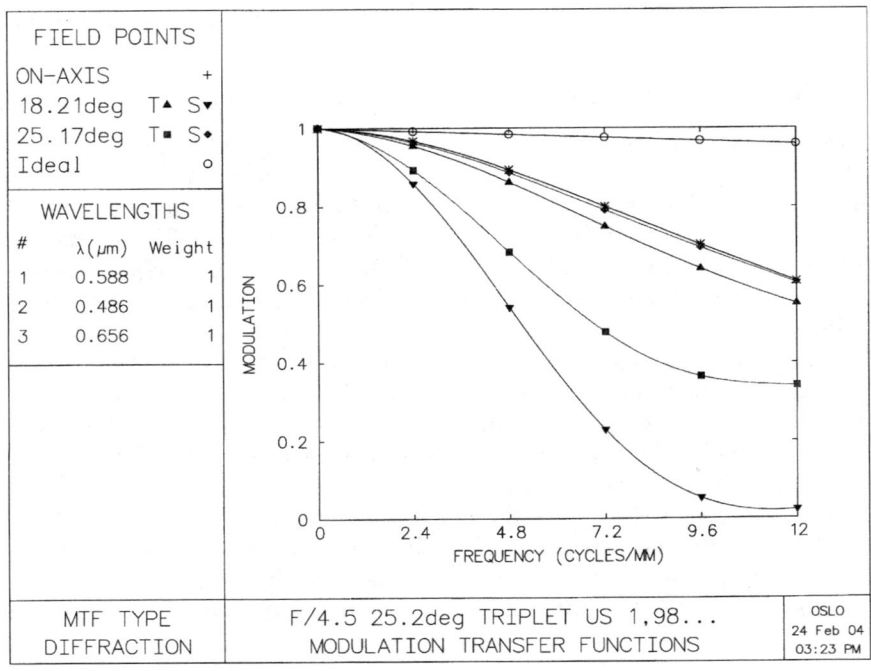

Figure 5.10 A through-frequency plot of MTF for three field points. Lens is Fig. 5.1.

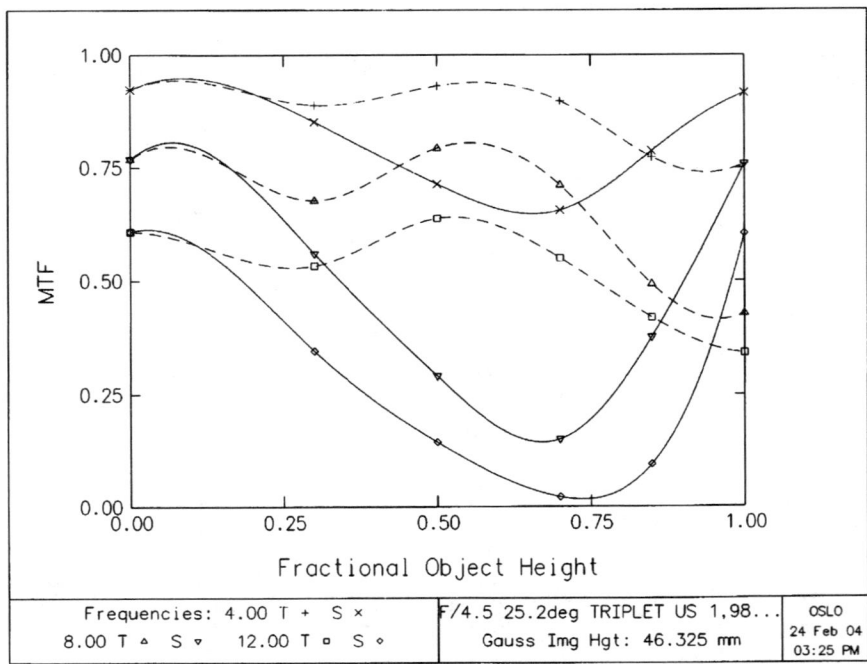

Figure 5.11 A plot of the MTF versus relative field for three spatial frequencies. Lens is Fig. 5.1.

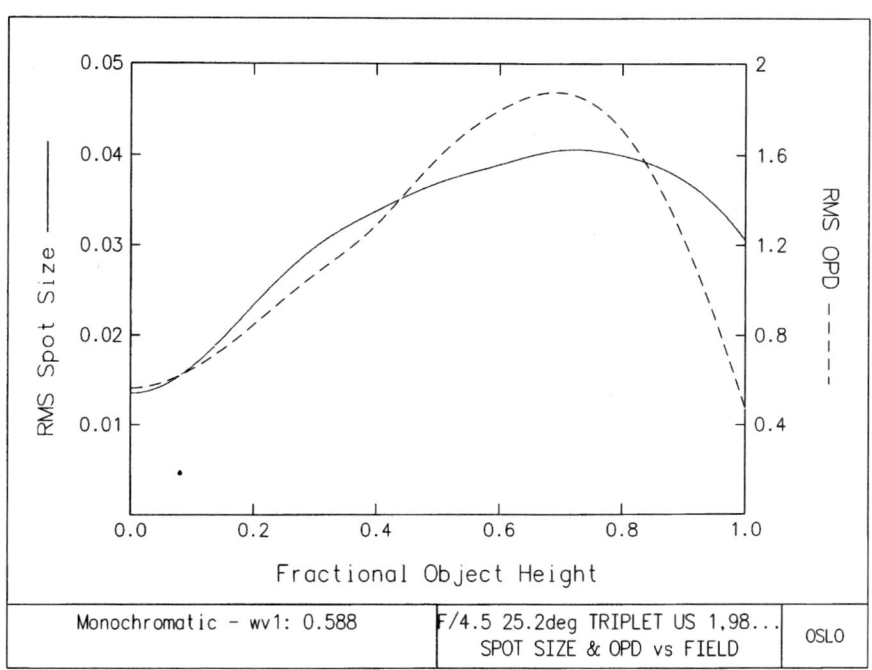

Figure 5.12 Plots of rms spot size and rms OPD versus relative field position. Lens is Fig. 5.1.

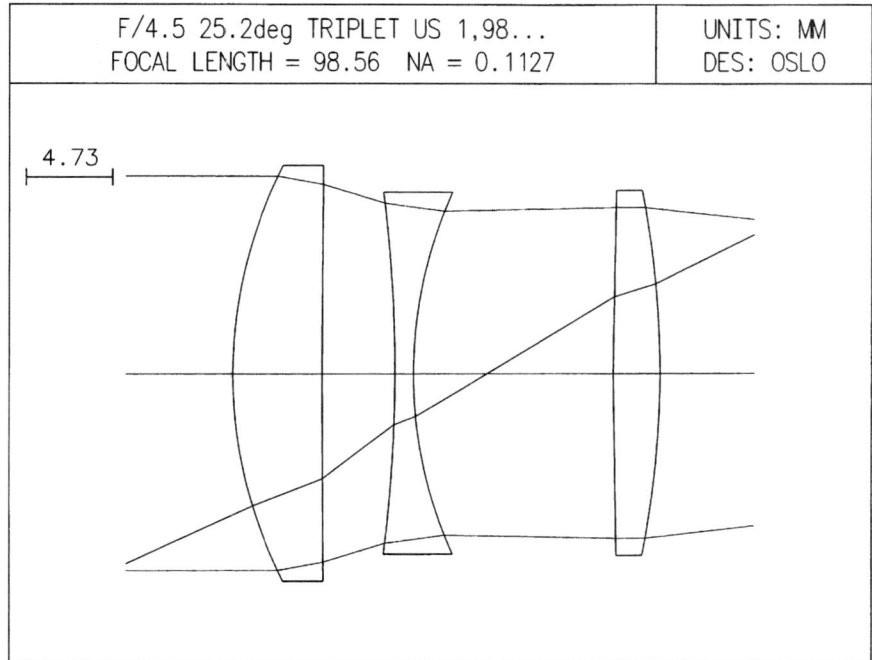

Figure 5.13 A drawing of the lens with marginal and principal rays. Lens is Fig. 5.1.

106   Chapter Five

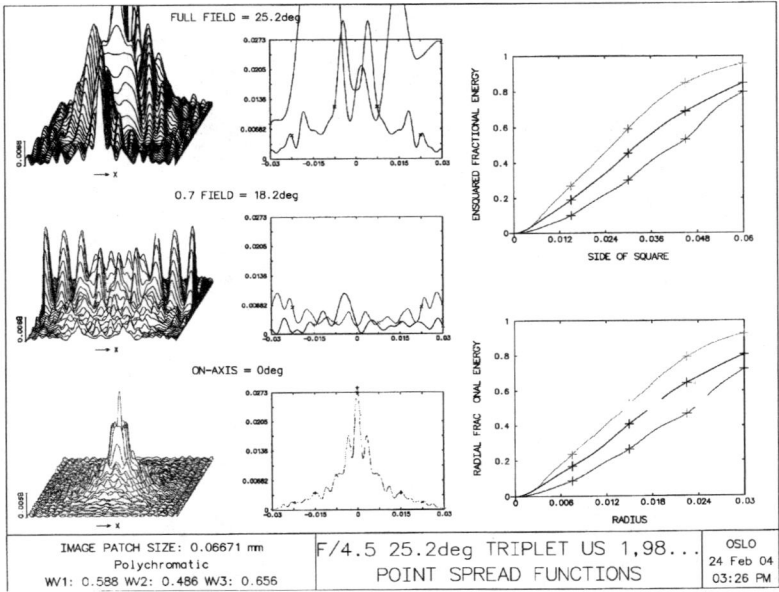

Figure 5.14 Point spread functions (PSF), both three dimensional and sectional, plus encircled and ensquared energy plots. Lens is Fig. 5.1.

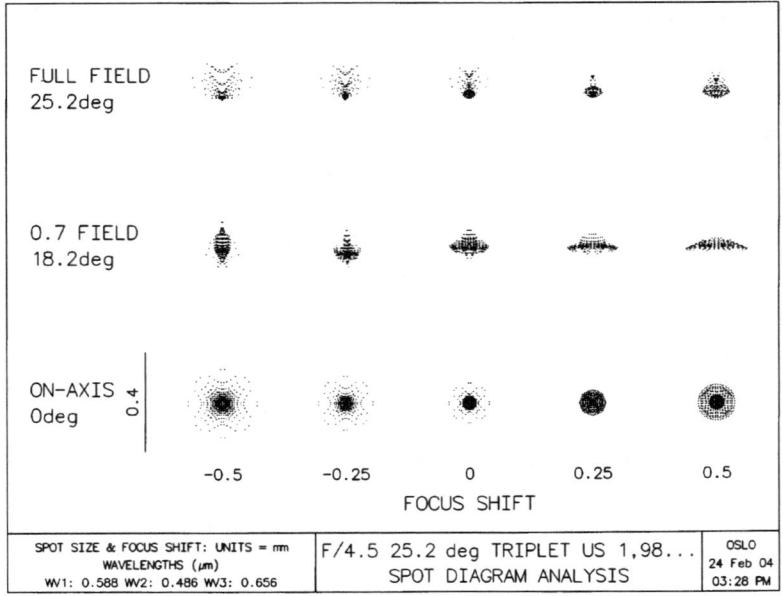

Figure 5.15 Through-focus spot diagrams for three field points. The original of this is in color (which will clearly indicate the significance of lateral color). Lens is Fig. 5.1.

Lens Design Data   107

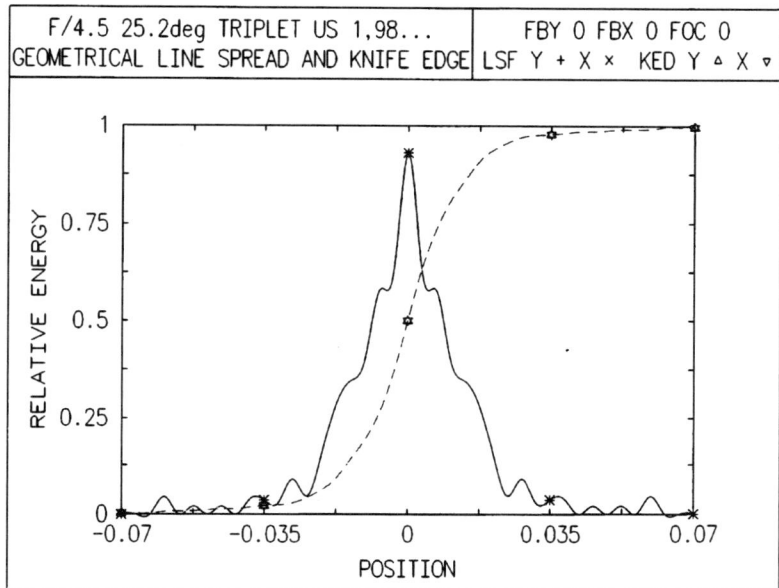

Figure 5.16 The on-axis geometric line spread function (LSF) and knife edge scan. Lens is Fig. 5.1.

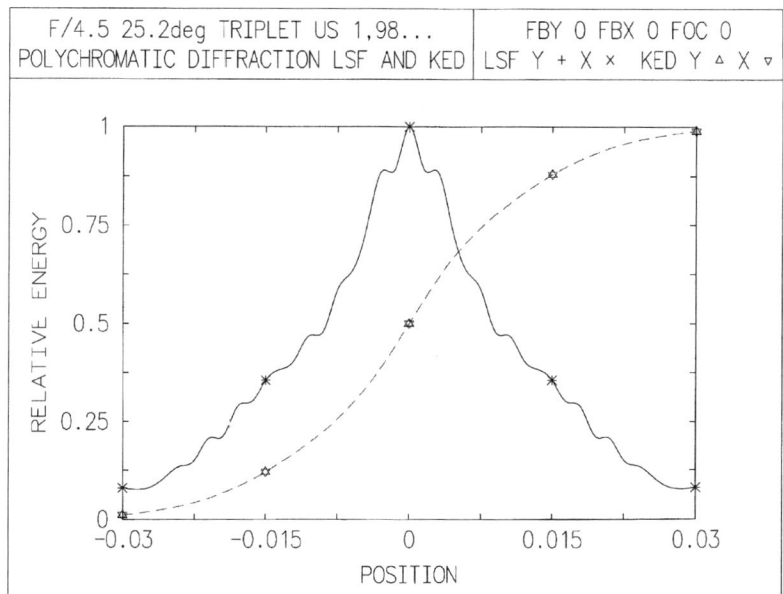

Figure 5.17 The on-axis diffraction line spread function (LSF) and the knife edge scan. Lens is Fig. 5.1.

added as needed. Figure 5.14 plots three-dimensional and sectional point spread functions as well as the fractional monochromatic encircled and ensquared energy. Polychromatic versions of the latter are also available, as are spot diagrams in Fig. 5.15. Geometric (Fig. 5.16) and diffraction (Fig. 5.17) polychromatic line spread functions and knife-edge scans are also available—as are many, many more.

Chapter 6

Telescope Objectives

6.1 The Thin Airspaced Doublet

The telescope objective can be considered in many ways to be a classic example of a thin lens. A thin aplanatic (i.e., free of both coma and spherical aberration) lens cannot be corrected for astigmatism. A thin lens has astigmatism as given by Eq. (24.83), *transverse astigmatism contribution* (TAC) = $h^2 \phi' u_{k'}/2$. For an aplanat this astigmatism cannot be changed by moving the stop. The stop shift equation (Eq. (24.95)) indicates that, if spherical and coma are zero, the astigmatism remains the same regardless of the location of the stop.

Therefore, since the astigmatism must be large and negative per Eq. (24.83), the image quality away from the optical axis is destined to be poor, and the useful field is limited to a few degrees; however, given sufficient effective degrees of freedom, one *can* control the focal length and can easily correct the coma, chromatic, and spherical aberrations. Because the lens is thin and has only a small field, distortion and lateral color are usually completely negligible.

The element powers for a thin achromat of power $\Phi = 1/f$ are given by

$$\phi_A = \frac{\Phi V_A}{V_A - V_B} \tag{6.1}$$

$$\phi_B = \frac{\Phi V_B}{V_B - V_A} = \Phi - \phi_A \tag{6.2}$$

where $V = (n_d - 1)/(n_F - n_C)$.

In a thin airspaced doublet, there are four degrees of freedom—the curvatures of the four surfaces. This is just sufficient for the task outlined above. Since the spherical aberration is a quadratic function of lens shape, for a thin lens there are two solutions to the problem. These are the Gauss form and the Fraunhofer form (or Steinheil, if the flint element faces the distant object).

Several solutions are illustrated here. Figure 6.1 shows a Gauss lens; Fig. 6.2, a Fraunhofer; and Fig. 6.3, a Steinheil, all at a speed of $f/7$. Note that the residual aberrations of the Fraunhofer and the Steinheil forms are quite similar, but the Steinheil form has stronger surface curvatures than the Fraunhofer.

The Gauss form has low spherochromatic because of the increasing airspace toward the edge of the aperture. This works the same way as the airspace in Fig. 6.7, which was introduced to reduce the spherochromatic aberration. It should also reduce the spherical zonal, but the extreme bendings create so much spherical aberration that any improvement is lost. The *Gauss space*, as it's sometimes called, is also noticeable in the double-Gauss lens (Chap. 12). The Gauss form has somewhat less spherochromatic but much more zonal spherical aberration, high-order coma, and slightly more secondary spectrum. The Gauss solution does not exist if the relative aperture of the lens is too large; the increased element thickness required for the large-diameter elements increases the effective spacing between the elements, and the lowered ray heights on the flint preclude a simultaneous solution for chromatic, coma, and spherical.

As can be seen, these designs have residuals of zonal spherical aberration and spherochromatism, plus secondary spectrum. Which of these constitutes the dominant limitation to performance depends on the aperture, focal length, and spectral bandpass. If the field of view is large enough, the fifth-order coma may also be a problem. There are specific techniques that can be used to attack each of these aberrations; they are outlined in the following sections.

6.2 Merit Function for a Telescope Objective

The merit function for a telescope objective can be, and should be, very simple. A "default" merit function, which has targets for astigmatism and field curvature, is definitely not optimal when, as in this case, spherical, coma, and axial chromatic are the only aberrations to be corrected (because there is no way that astigmatism and field curvature can be controlled in a thin, compact aplanat). A default merit function may work out if the design field is kept small enough so that the coma outweighs the field curvature. For an ordinary-cemented or edge-contacted telescope objective the merit function should be limited to targets (operands) for spherical, coma, and axial chromatic. The chromatic

Telescope Objectives 111

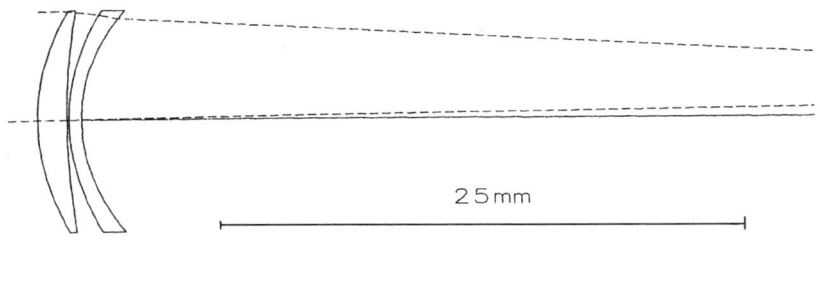

F/7 1degHFOV GAUSS TELESCOPE OBJECTIVE

radius	thickness	mat'l	index	V-no	sa
17.654	1.400	BK7	1.517	64.2	7.2
53.898	0.100	air			7.2
16.530	0.600	SF1	1.717	29.5	7.2
13.075	94.931	air			7.2

EFL = 100
BFL = 94.93
NA = -0.0713 (F/7.0)
GIH = 1.75
PTZ/F = -1.583
VL = 2.10
OD infinite conjugate

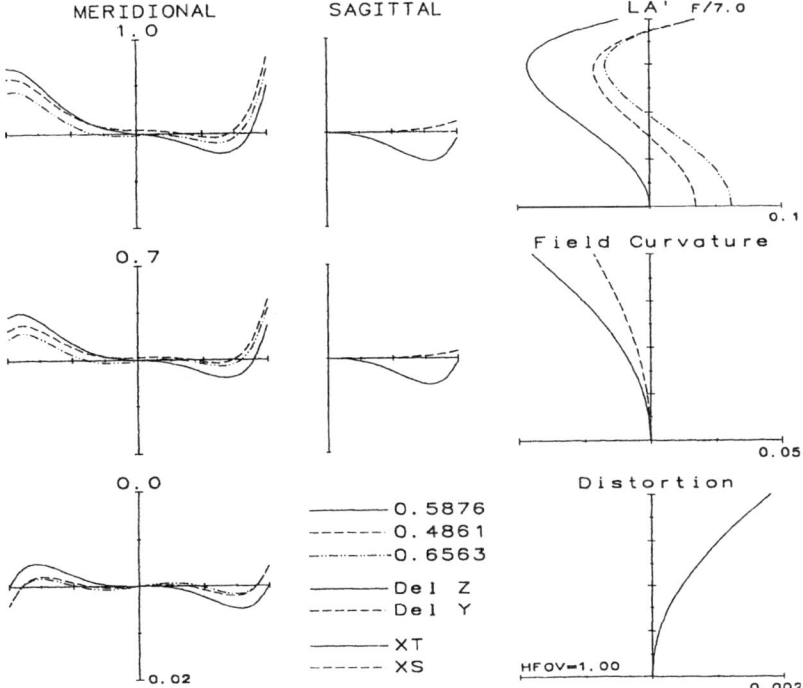

Figure 6.1 *f*/7 Gauss telescope objective (BK7 and SF1).

112 Chapter Six

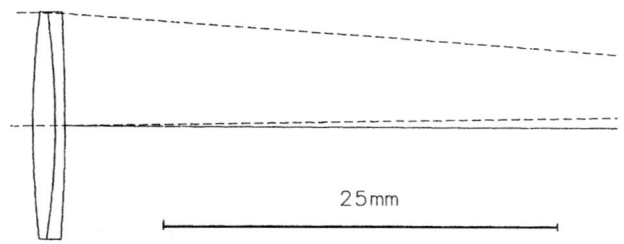

F/7 1degHFOV FRAUNHOFER OBJECTIVE

radius	thickness	mat'l	index	V-no	sa
60.415	1.400	BK7	1.517	64.2	7.2
-52.830	0.012	air			7.2
-51.552	0.600	SF1	1.717	29.5	7.2
-128.126	99.017	air			7.2

EFL = 99.98
BFL = 99.02
NA = -0.0714 (F/7.0)
GIH = 1.75
PTZ/F = -1.38
VL = 2.01
OD infinite conjugate

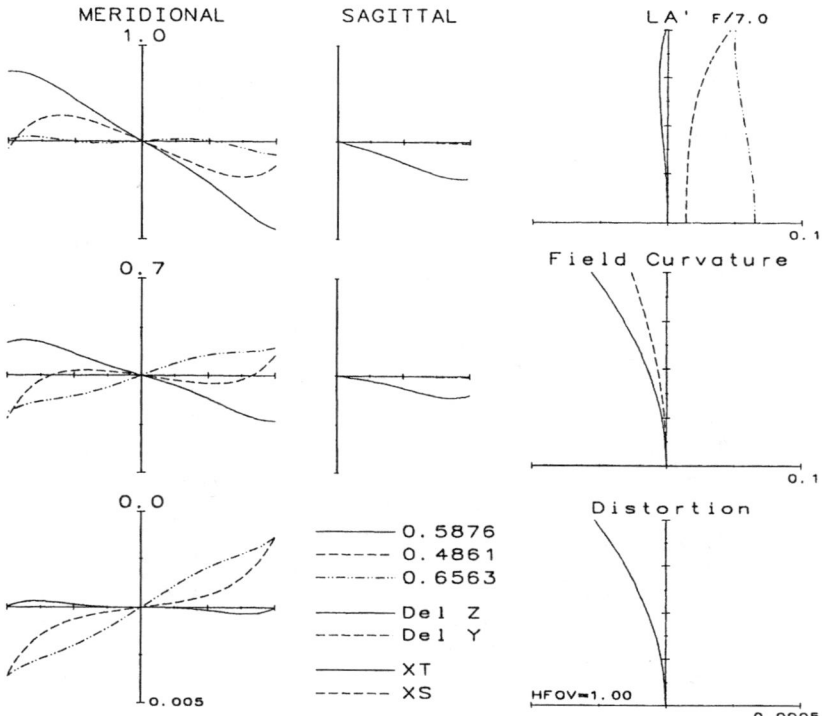

Figure 6.2 $f/7$ Fraunhofer telescope objective (BK7 and SF1).

Telescope Objectives 113

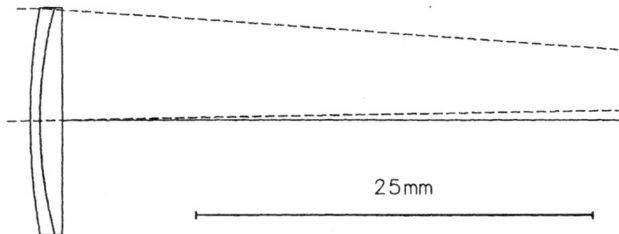

F/7 1degHFOV STEINHEIL OBJECTIVE

radius	thickness	mat'l	index	V-no	sa
41.582	0.600	SF1	1.717	29.5	7.2
27.865	0.019	air			7.2
28.425	1.400	BK7	1.517	64.2	7.2
-1936.302	98.514	air			7.2

EFL = 100
BFL = 98.51
NA = -0.0714 (F/7.0)
GIH = 1.75
PTZ/F = -1.385
VL = 2.02
OD   infinite conjugate

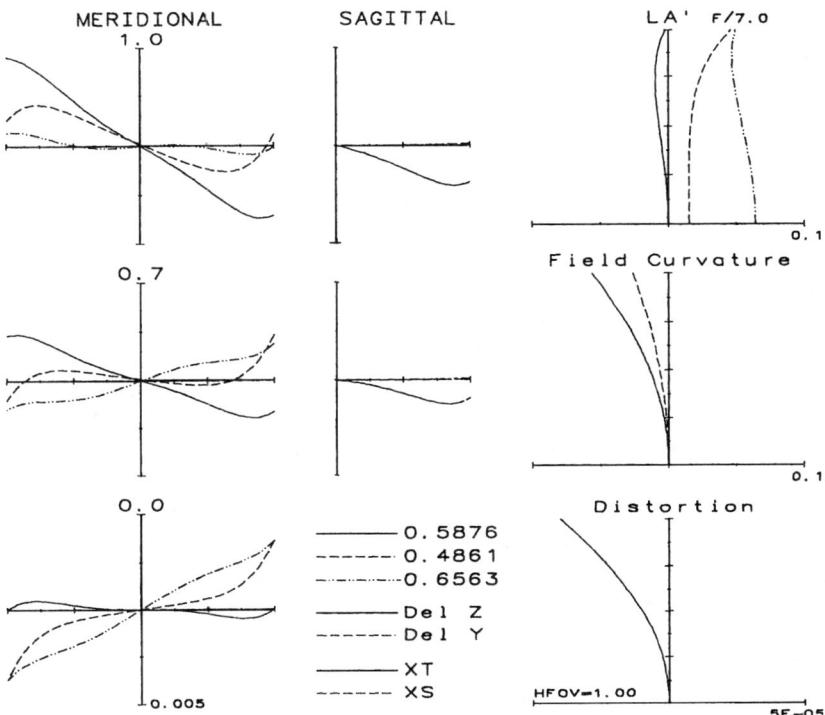

Figure 6.3 $f/7$ Steinheil telescope objective (BK7 and SF1).

may be controlled by a Conrady $\Sigma(D - d)\delta n$ target on the marginal ray; this will correct the chromatic of the (approximately) 0.7 zonal ray as shown in Fig. 3.7a.

6.2.1 The Conrady $D - d$ target; coma; focal length

Consider two wave fronts of different wavelengths passing through a lens. The physical path taken by each along a specific ray from the object point is, to a close approximation, the same. The optical path is $OP_i = \Sigma D_i n_i$, and the difference between the optical paths for the two wavelengths is simply $\Sigma D_i(n_1 - n_2) = \Sigma D_i \delta n$. Taken along the axis (or along a principal ray) the chromatic path difference is $\Sigma d_i \delta n$. Now if the chromatic path differences are the same along the ray and the axis, then $\Sigma(D - d)\delta n$ will be zero. If the ray is the marginal ray, then the wavefronts will be coincident at both the axis and the margin, and the two wavefronts must be parallel at about the 0.7 zone of the aperture (depending on the exact nature of the spherochromatism). Since rays are perpendicular to wavefronts, the chromatic aberration of the rays at this zone height will be zero.

If a different balance of correction is needed, then rays in C and F light (for example) at the desired height in the aperture can be traced and the difference of their heights in the focal plane can be targeted to zero. Alternately the ray height used in the $(D - d)\delta n$ operand can be changed.

The coma of a narrow field can be controlled with a target for *offence against the sine condition* (OSC), which is equal to the sagittal coma divided by the image height. Alternately, targeting the sum of the intersection height differences (δY from the principal ray) for the upper and lower rim rays to equal zero will control the tangential coma of the marginal rays for any size field.

The focal length can be controlled by an angle solve on the last surface, or possibly by a direct target in the merit function. A super simple merit function might include only targets for spherical, OSC, and $\Sigma(D - d)\delta n$; these require the trace of only one ray, the marginal ray. A three-color *rms* spot size merit function with only one field point (the axis) plus an OSC target could also do the job quite simply. Note that even for a nominal field of zero, coma should always be controlled because misalignment may cause the system to operate slightly off axis. Another possible coma target is called S2T in some programs. It is basicly the curvature (i.e., second derivative) of the ray intercept plot, projected (as h^2) to the full aperture. If calculated on the principal ray it gives the coma near the principal ray. Many of the designs in this chapter used this operand, as is indicated by the straightness of the ray intercept plot at the principal ray.

If the lens is to be airspaced, the zonal spherical and/or the spherochromatism can be corrected. In this case additional targets for these aberrations are necessary. If the chromatic of the C- and F-light marginal rays is corrected, the Conrady $(D-d)\delta n$ sum on the marginal ray is then a convenient measure of the spherochromatism, and makes a good operand for the purpose of controling or reducing it. Alternately, the difference between the intersection heights of the *zonal* C and F rays will indicate the spherochromatism (again assuming the marginal chromatic aberration is corrected). In the presence of spherochromatism, a minimum chromatic blur spot size should result if the sum of the marginal and zonal chromatic aberrations is targeted to zero (so they are equal and oppositely signed).

A 0.707 zonal ray with its intersection height targeted to zero can control the zonal spherical. But note that when both the marginal and zonal spherical are targeted, the least-squares program will strike a compromise, ordinarily with the zonal spherical undercorrected and the marginal spherical overcorrected. Obviously the weight on the marginal spherical should be heavier than on the zonal when a corrected marginal ray is desired. In some cases a target on zonal coma may also be useful.

6.3 The Design of an *f/7* Cemented Doublet Telescope Objective

The cemented form for a doublet is preferred over an airspaced lens because

a. The elements can be better trued and centered with respect to each other in the optical shop than they can be in a mounting cell.
b. There are only two air-glass surfaces to reflect the light instead of four.
c. In a high-speed system the edge rays may have very large incidence angles or even encounter total internal reflection (TIR) as they enter the airspace. Tolerances are very tight.

In the cemented doublet telescope objective there are not enough construction degrees of freedom to control the focal length and also correct the spherical, coma, and chromatic aberrations. In addition to the three surface curvatures, glass choice must be used as the fourth variable. It is well known that when bending a cemented achromatic doublet (Ref. 23, pp. 405, 406), the spherical aberration will plot as a vertical parabola and the coma will plot as a straight line, which is zero at the shape where the spherical plot peaks. The spherical may never reach zero, or it may have two bendings at which spherical is zero but where coma is not corrected.

TABLE 6.1 The Transverse Marginal Spherical Aberration for Various Combinations of Glass Types in a 100 mm f/7 Doublet Corrected for Coma and Axial Chromatic Aberrations

		K7 511604	K50 523602	BaK2 540597	SK11 564608	SK12 583595	SK14 603606	SK16 620603	LaK21 640601
SF57	847238	−0.020	−0.020	−0.020	−0.021	−0.020	−0.020	−0.020	−0.020
SF6	805254	−0.016	−0.016	−0.012	−0.018	−0.017	−0.019	−0.019	−0.019
SF4	755276	−0.008	−0.009	−0.010	−0.014	−0.014	−0.016	−0.017	−0.018
SF10	728284	−0.006	−0.007	−0.008	−0.013	−0.013	−0.016	−0.018	−0.019
SF8	689312	+0.007	+0.005	**+0.001**	**−0.007**	−0.008	−0.014	−0.016	−0.019
SF2	648339	+0.021	+0.016	**+0.010**	**−0.004**	−0.007	−0.016	−0.021	−0.026
F2	620364	+0.038	+0.030	**+0.019**	**[−0.001]**	−0.009	−0.022	−0.030	−0.038
F8	596392	+0.062	+0.049	**+0.030**	**−0.002**	−0.018	−0.037	−0.051	−0.067
LF8	564438	+0.120	+0.088	**+0.040**	**−0.032**	−0.086	−0.121	−0.160	−0.205
LLF6	532488	+0.157	**+0.054**	**−0.126**	−0.319	−0.708	−0.753	−1.14	−4.65

The ideal situation is one where the spherical plot just reaches a zero value at the same bending for which the coma is zero. To reach this condition the spherical plot can be raised by selecting a new flint glass with a lower index and a higher V-value, or a new crown glass with a higher index and a lower V-value (or lowered by the reverse changes).

This is a bit easier said than done, glass variables being notoriously tricky in practice. In an effort to clarify this situation we chose eight crown glasses with V-values close to 60 (which seems to be a reasonable value for a telescope objective crown) and eight flints spaced along the glass line. Allowing cv(1) and cv(2) of the doublet to vary, using an angle solve on the last surface to hold the focal length, with a flint center thickness (CT) of 1.5, and controlling the crown edge thickness to 1.5, we optimized to get the marginal coma and axial chromatic both equal to zero for a 100 mm focal length at a speed of $f/7.0$ and a field of $\pm 1°$. The results are presented in Table 6.1, which gives the resulting marginal spherical, TA_m, for each of the 64 possible glass combinations.

If we plot these results (as TA_m versus the crown index) or if we simply examine the boldface data in the table, it is evident that, with our chosen crown V-value of 60, if we are to get an aplanatic achromatic doublet we must select a crown with an index between 1.53 and 1.565 and pair it with a flint from along the glass line between LLF6 (523488) and SF8 (689312). A different crown V-value would probably produce somewhat different but similar results.

Choosing SK11 (564608) and F2 (620364), indicated in the table by brackets, we get the following design:

$R1 = +63.4648$	$T1 = 2.732$	SK11 564608
$R2 = −36.1200$	$T2 = 1.5$	F2 620364
$R3 = −228.969$	BFL = 97.767	

Telescope Objectives 117

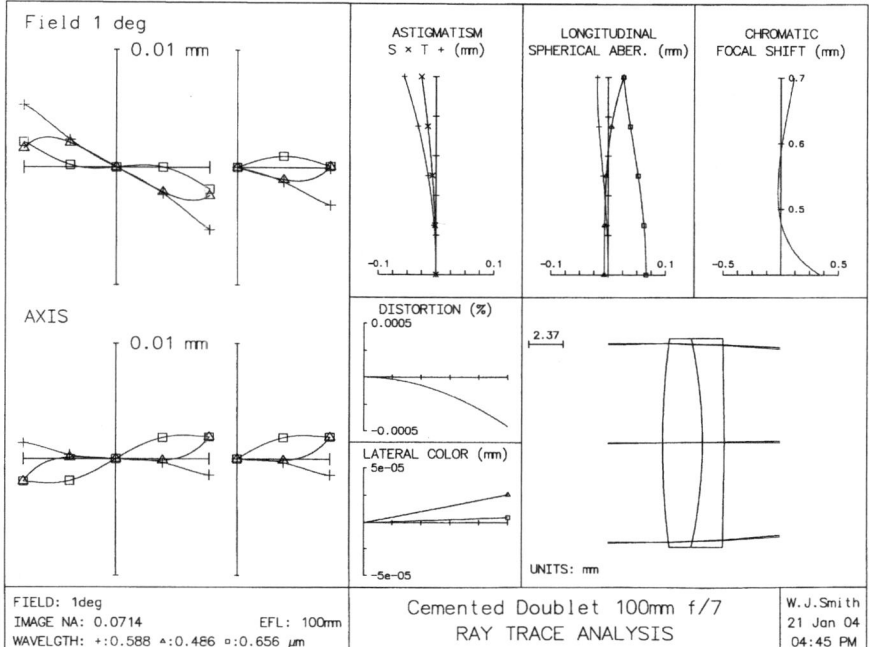

Figure 6.4 $f/7$ cemented Fraunhofer telescope objective (SK11 and F2).

The ray intercept plots and the through-focus *modulation transfer function* (MTF) are shown in Figs. 6.4 and 6.5. The inward-curving field can easily be seen in the through-focus MTF plot, where the off-axis best focus is shifted inward with little degradation (at its best focus).

At a focal length of 100 mm the aberrations are very small and the chromatic might be slightly better balanced by correcting it at the 0.7 zone of the aperture with a Conrady $D - d$ operand, rather than correcting it at the margin as done here. At a longer focal length the present correction might be preferred, as indicated in Chap. 3, Sec. 3.8.

Figures 6.6 and 6.7 illustrate similar lenses at a speed of $f/3.0$, the first cemented and the other airspaced (so as to reduce spherochromatism and zonal spherical as discussed below). Note that the residual aberrations of the design shown in Fig. 6.7 represent a compromise between the correction of the spherochromatism and the correction of the zonal spherical. The airspace is larger than the optimum for the zonal spherical and smaller than optimum for spherochromatism. Possibly another set of glasses could be found where both aberrations would be corrected at the same spacing.

Also worth noting in Fig. 6.7 is the high-order coma. This is primarily fifth-order linear coma, which varies as y^4h. (Fifth-order elliptical coma

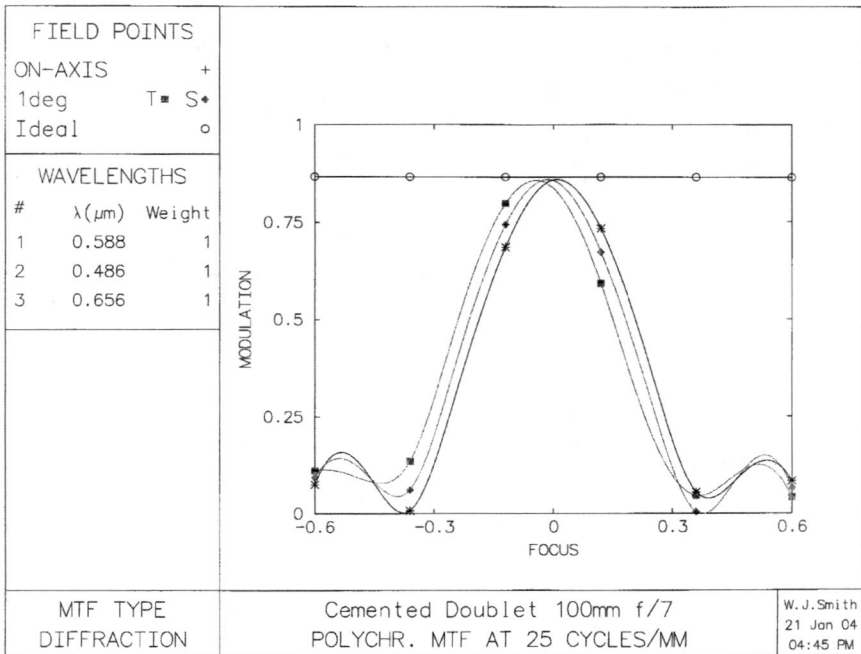

Figure 6.5 The MTF for the lens of Fig. 6.4.

varies as y^2h^3, and third-order coma varies as y^2h.) Fifth-order linear coma is also apparent in both Fig. 6.7 and the Gauss lens of Fig. 6.1.

Although seldom used in ordinary telescope objectives, higher-index glasses can reduce the spherical zonal and spherochromatism by a factor of 2 or 3. In general, even for an airspaced doublet there seems to be an optimum mate for a given glass. Figure 6.8 is an airspaced doublet of high-index glasses; compare its aberrations with Fig. 6.9, which uses low-index glass types.

6.4 Spherochromatism

Changing the spacing between two components (of an achromatic system) that are individually and oppositely uncorrected for axial chromatic aberration will change the spherochromatism of the system. In Fig. 6.10 the paths of a marginal ray in red light (solid line) and in blue light (dashed line) are shown (with their differences exaggerated). When the airspace is increased, the ray heights at element B are reduced; they are reduced more for blue light than for red light, because element A has undercorrected chromatic aberration. This reduces the spherical overcorrection produced by element B more for blue light than for red.

Telescope Objectives 119

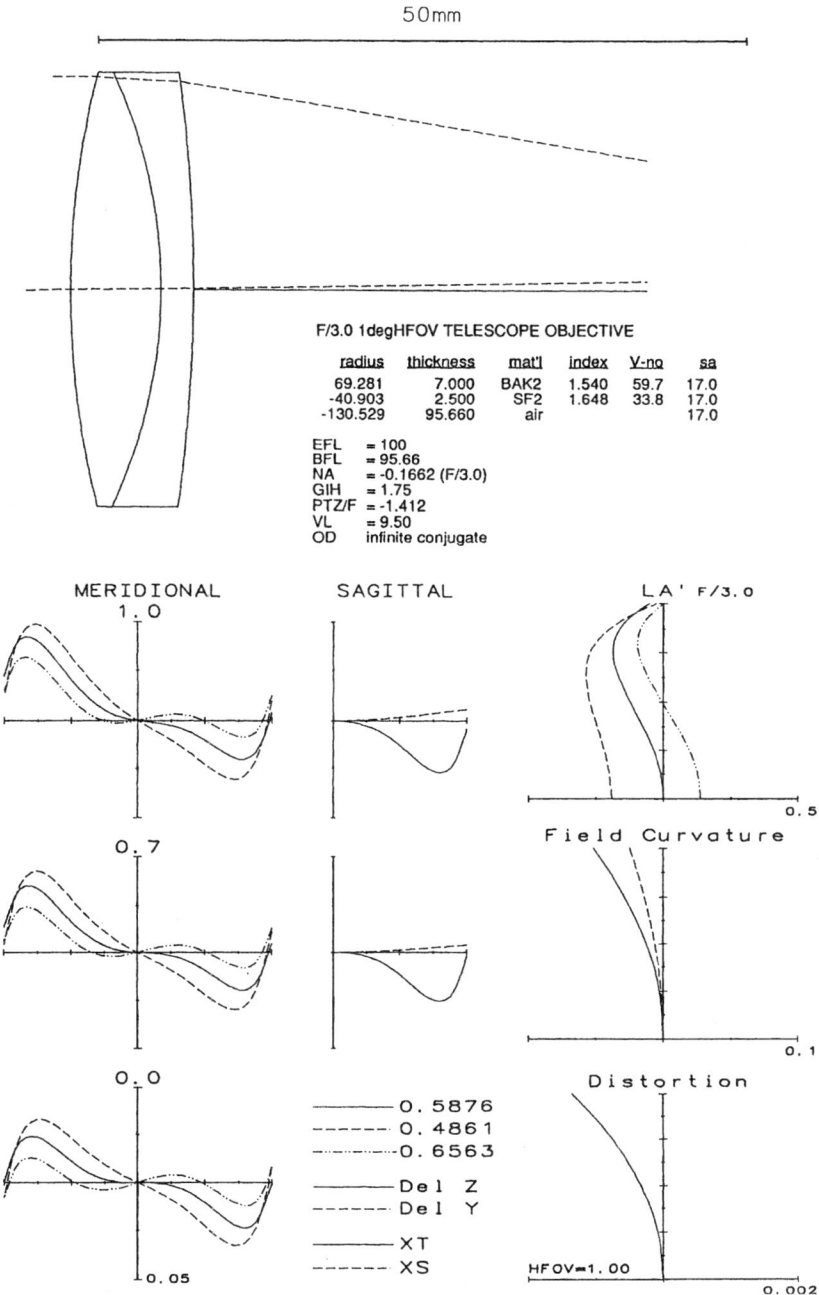

Figure 6.6 *f*/3 cemented doublet.

120 Chapter Six

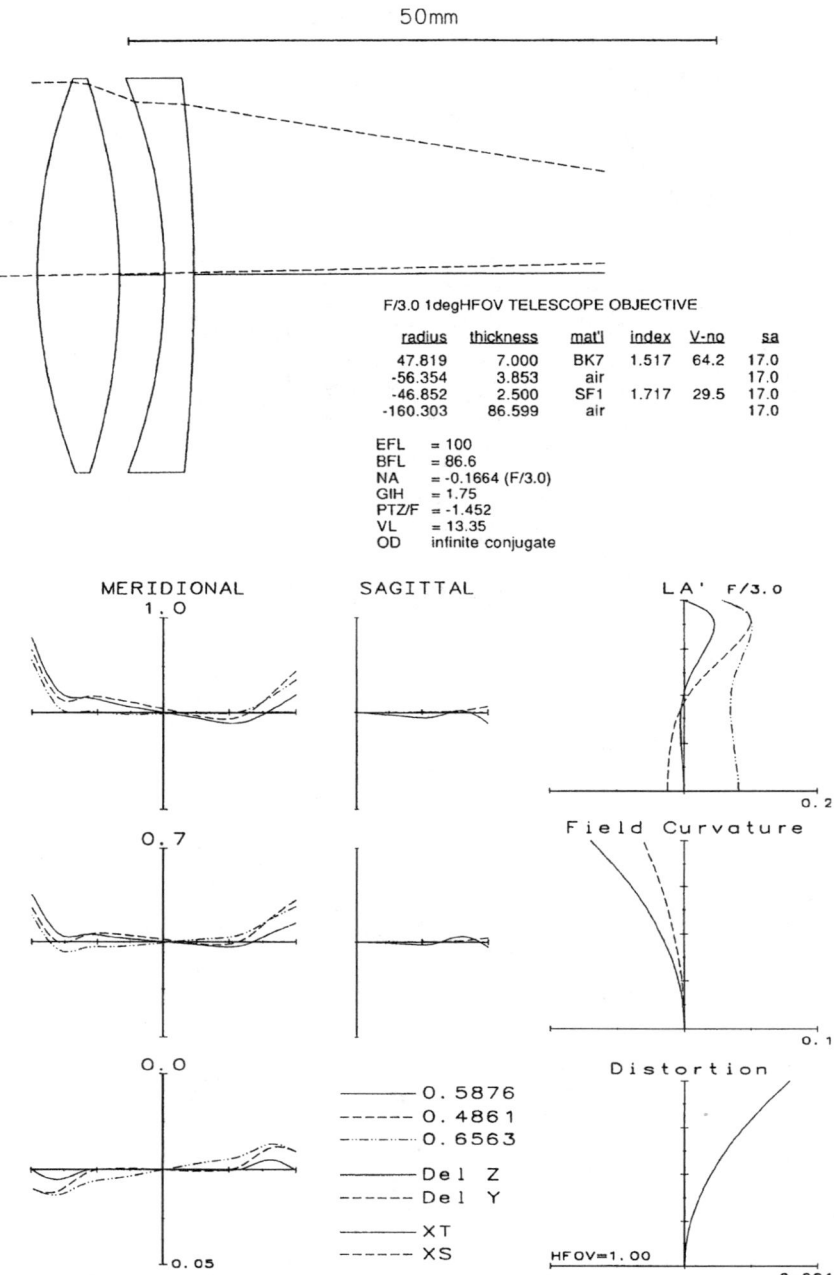

Figure 6.7 $f/3$ doublet, airspaced to control spherochromatism and zonal spherical aberration.

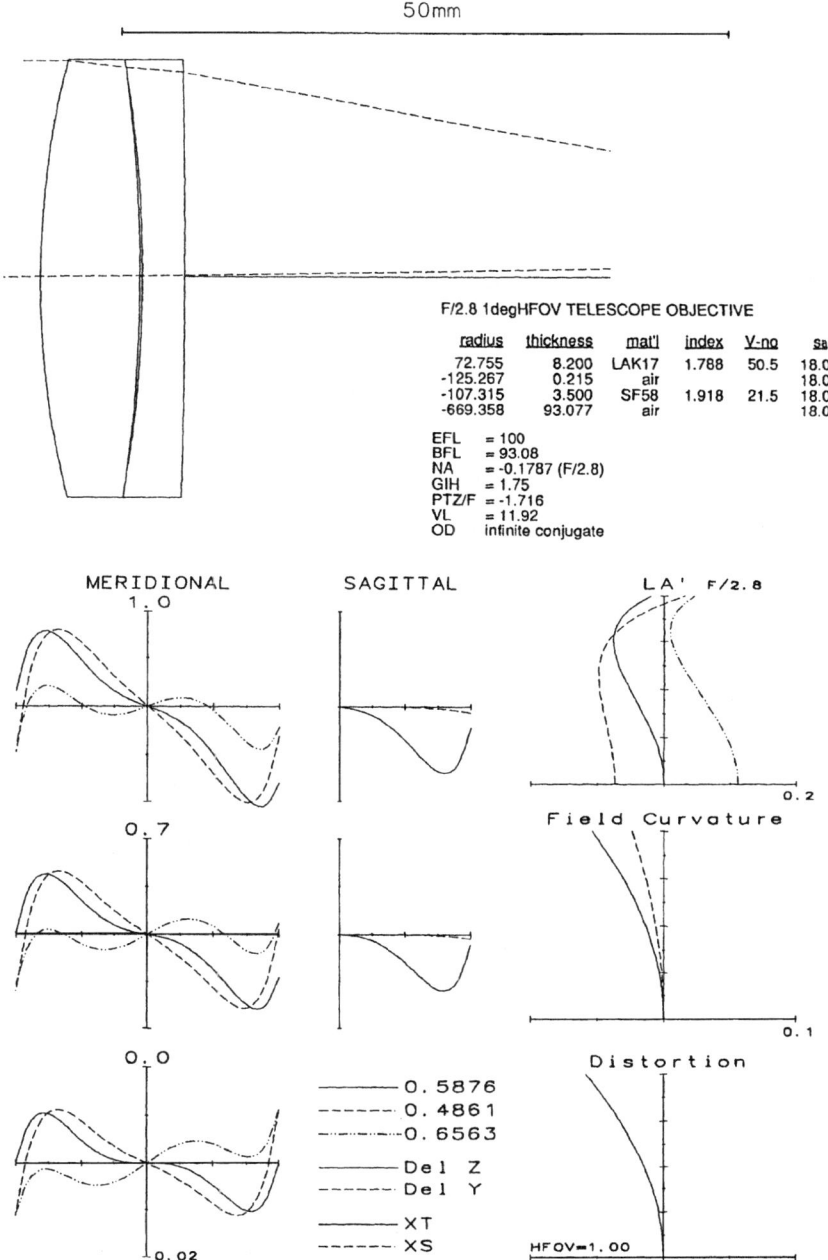

Figure 6.8 High index doublet (LAK17 and SF58).

122 Chapter Six

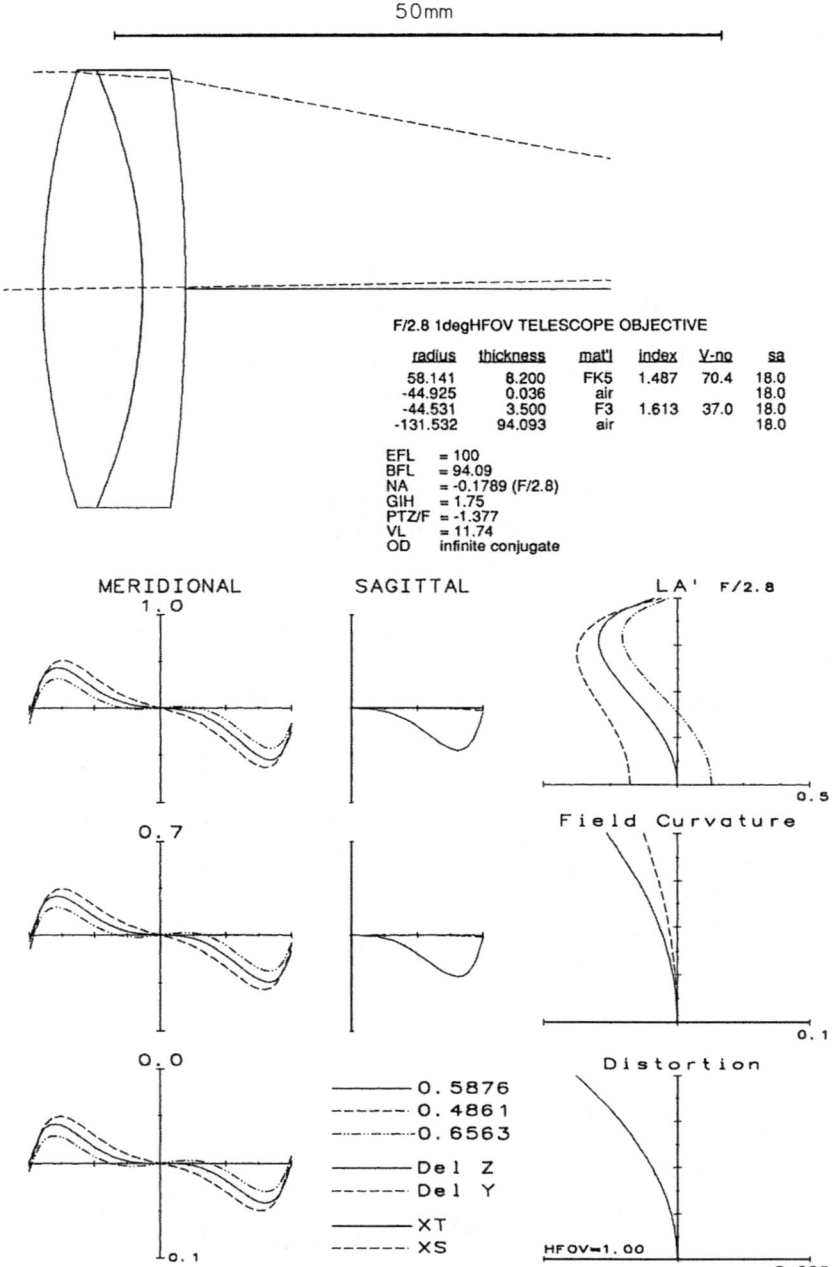

Figure 6.9 Low index doublet (FK5 and F3).

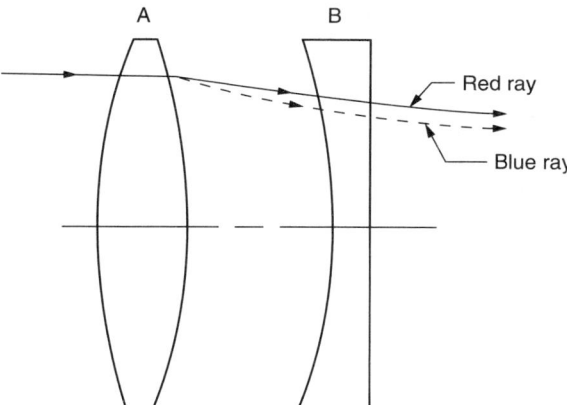

Figure 6.10 Ray diagram (exaggerated) showing the correction effect of spacing on spherochromatic.

Since in ordinary spherochromatism blue light shows overcorrected spherical and red light shows undercorrected, the change introduced by spacing will reduce the spherochromatism. This effect is illustrated by comparing Fig. 6.6 with Fig. 6.7. *Note well that this principle is not limited to doublets or telescope objectives; it can be applied quite generally.*

Note that spherochromatism is often *balanced* by adjusting the primary chromatic aberration, as described in Chap. 3, Sec. 3.8. This does not correct the spherochromatic, as the spacing change does; it simply minimizes its effects.

6.5 Zonal Spherical Aberration

Changing the spacing between two components that are individually and oppositely uncorrected for spherical aberration will change the zonal residual aberration. The principle works a lot like a spacing change does with spherochromatism.

The spherical aberration contribution of element B in Fig. 6.11 is overcorrection. The exaggerated ray paths shown between A and B indicate that both the marginal and zonal ray heights are reduced at element B, and that the reduction is proportionately larger for the marginal ray (solid line) than for the zonal ray (dashed line). This is the result of the undercorrected spherical aberration in element A. Thus the overcorrected spherical aberration contribution for each ray at B is reduced, but the overcorrection of the zonal ray is reduced proportionately less than that of the marginal ray. When the elements are reshaped to restore the spherical correction, the relative increase in the

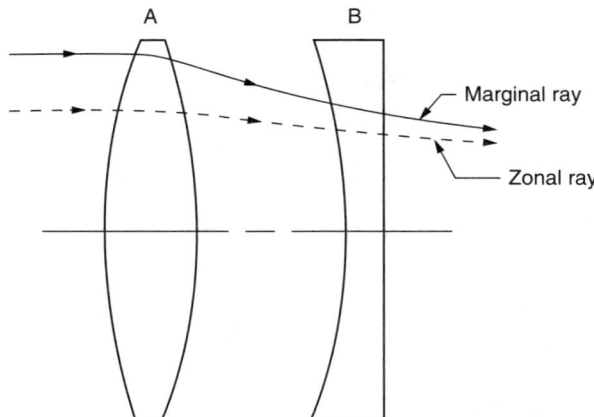

Figure 6.11 Ray diagram (exaggerated) showing the correction effect of spacing on zonal spherical aberration.

overcorrection of the zonal ray will reduce the (negative) zonal spherical aberration.

These correction-by-spacing-change improvements of spherochromatism and zonal spherical do not come without a price, however. A cemented doublet can be aligned during the cementing operation to a very tight alignment tolerance. In an airspaced doublet the alignment is determined by how closely the lens diameter fits the inside diameter of the mount. An edge-contact doublet is preferred to a more widely spaced doublet because the edge bevel that affects the element tilt is quite precise, whereas a larger air space is usually effected by a thin metal spacer, which is difficult to fabricate with sufficient precision. Mounting (cementing) the elements trued in precision individual cells is one way of minimizing this problem.

6.6 Induced Aberrations

The two preceding sections illustrate quite clearly the idea of induced versus intrinsic aberrations. The third-order aberration contributions are based solely on the paraxial angles and ray heights that occur at the surface or element contributing the aberration. The higher-order aberrations, however, are also affected by the aberrations of the other surfaces or elements. In the two cases described above, the change of the doublet airspace affects the spherochromatism and the zonal spherical because the individual elements of the doublet are uncorrected for chromatic and spherical aberrations, respectively. If the elements were fully corrected, the effect of the spacing change would be negligible.

6.7 Three-Element Objectives

The idea of splitting elements to reduce aberrations can be applied to the telescope objective by dividing the crown element into two or more elements. Note that, in general, splitting a positive element will reduce an undercorrected zonal spherical residual and splitting a negative element will reduce a positive or overcorrected zonal. A simple split can reduce the zonal; however, it does nothing for the spherochromatism, which must be controlled as described above by respacing components that are chromatically uncorrected. The triplet telescope objective can be executed as three airspaced elements, or as a singlet crown plus a cemented doublet. The latter form is somewhat easier to fabricate because it is less sensitive to misalignments between the components. Either form has enough effective degrees of freedom to control both spherochromatism and zonal spherical, as well as the primary aberrations: chromatic, coma, and spherical. Figures 6.12 to 6.14 show a series of telescope objectives, all using BK7 (517:642) and SF1 (717:295) glass and all at a speed of $f/2.8$. This is too fast for most applications but is used here to clearly display the aberrations. Figure 6.12 is the basic edge-contacted doublet, shown here for comparison. In Fig. 6.13 the crown element is split to reduce the zonal and the flint is spaced away to correct the spherochromatism. The result is a lens whose axial aberrations are practically eliminated except for secondary spectrum (which could be reduced by using glasses with unusual partial dispersions).

Figure 6.14 shows one of the many arrangements for a singlet plus a cemented doublet. Here the crown split has been quite effective in reducing the spherical (by about a factor of about 17x compared to that in Fig. 6.12). Although the spherochromatism is not as well controlled as in Fig. 6.13, because the airspace is less effective, it is only one-third as big as in Fig. 6.12. Splitting the crown has reduced the aberrations of the singlet and thus the effectiveness in changing the ray heights on element B.

6.8 Secondary Spectrum (Apochromatic Systems)

First things first. The words *achromat* and *apochromat* should be pronounced in a similar fashion. That is, achromat is pronounced with the accent on the first syllable, i.e., "A-chro-mat." The word apochromat should be pronounced "AP-o-chro-mat," not "a-POCH-ro-mat."

The secondary spectrum contribution of a thin lens is given by Eq. (24.89). The secondary spectrum of a doublet (or of any thin achromat using just

126 Chapter Six

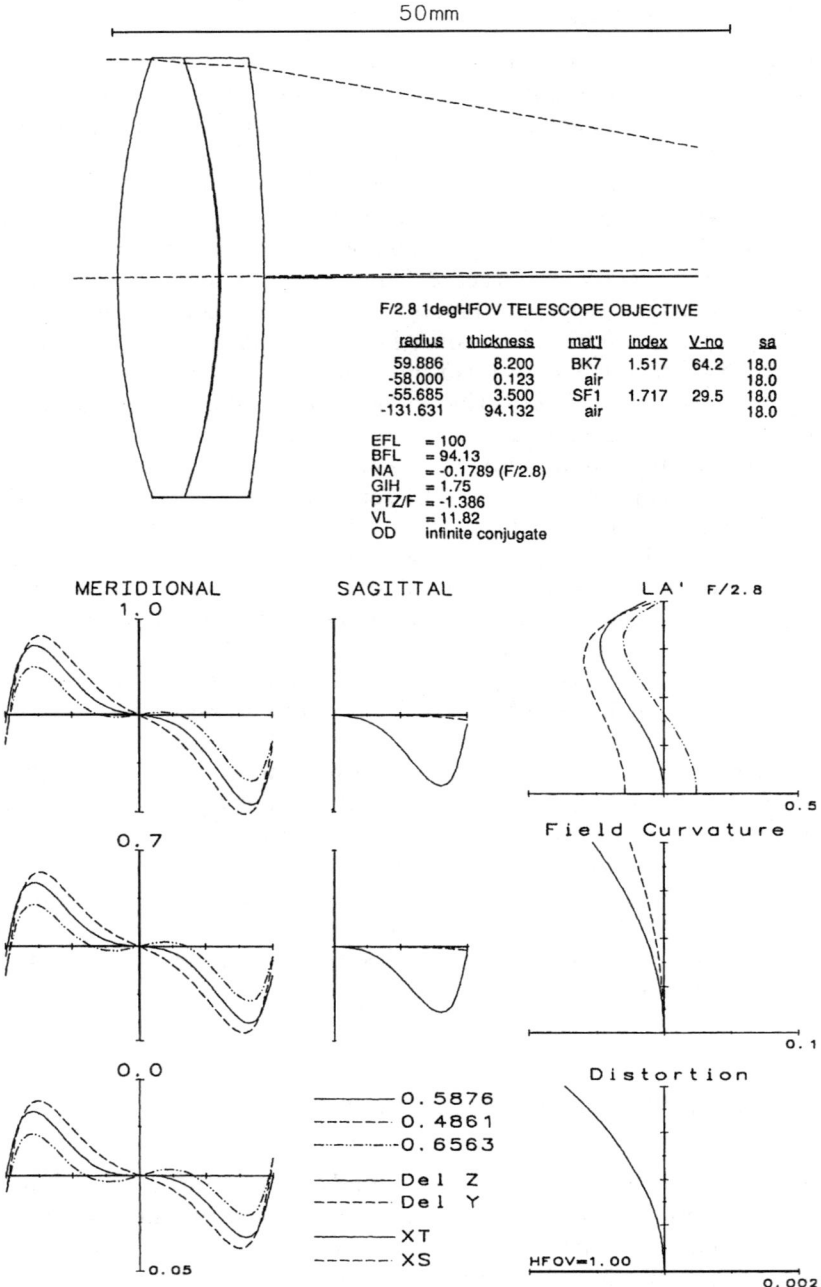

Figure 6.12 An $f/2.8$ airspaced doublet telescope objective.

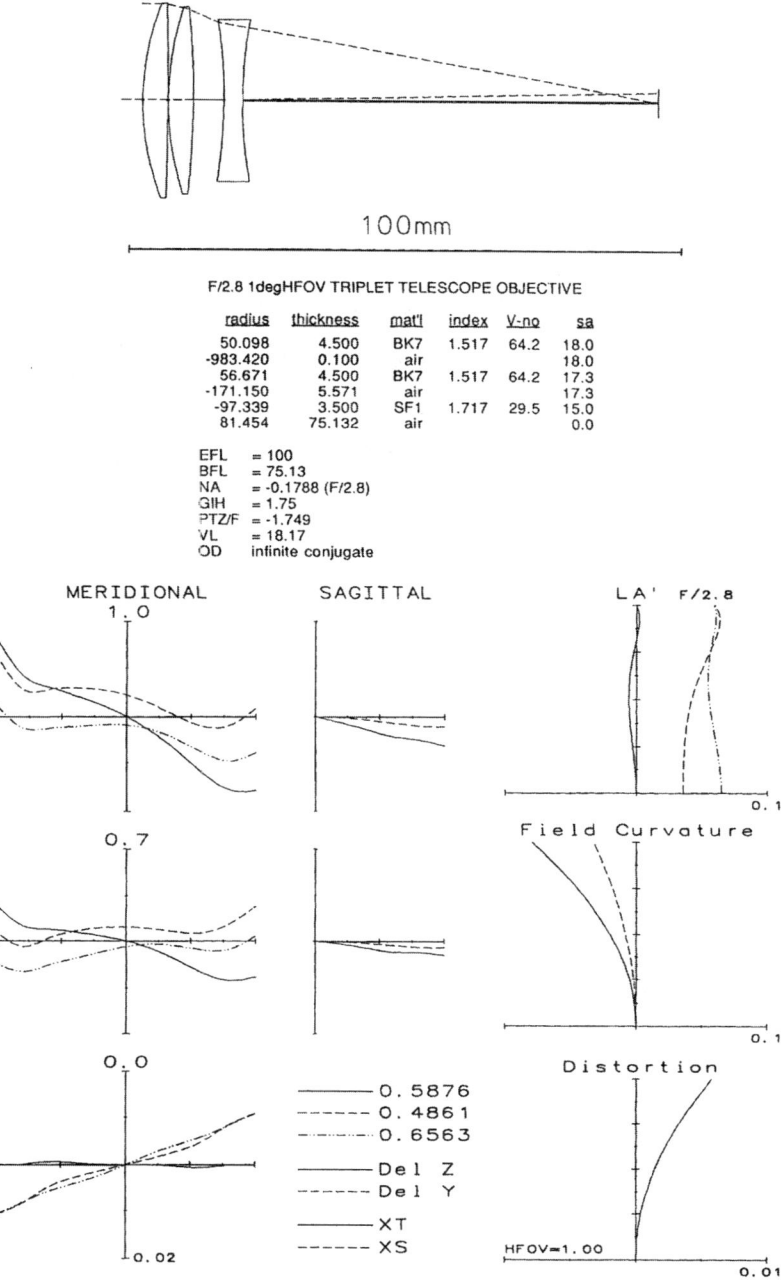

Figure 6.13 A three element $f/2.8$ airspaced telescope objective. Compare with Fig. 6.26.

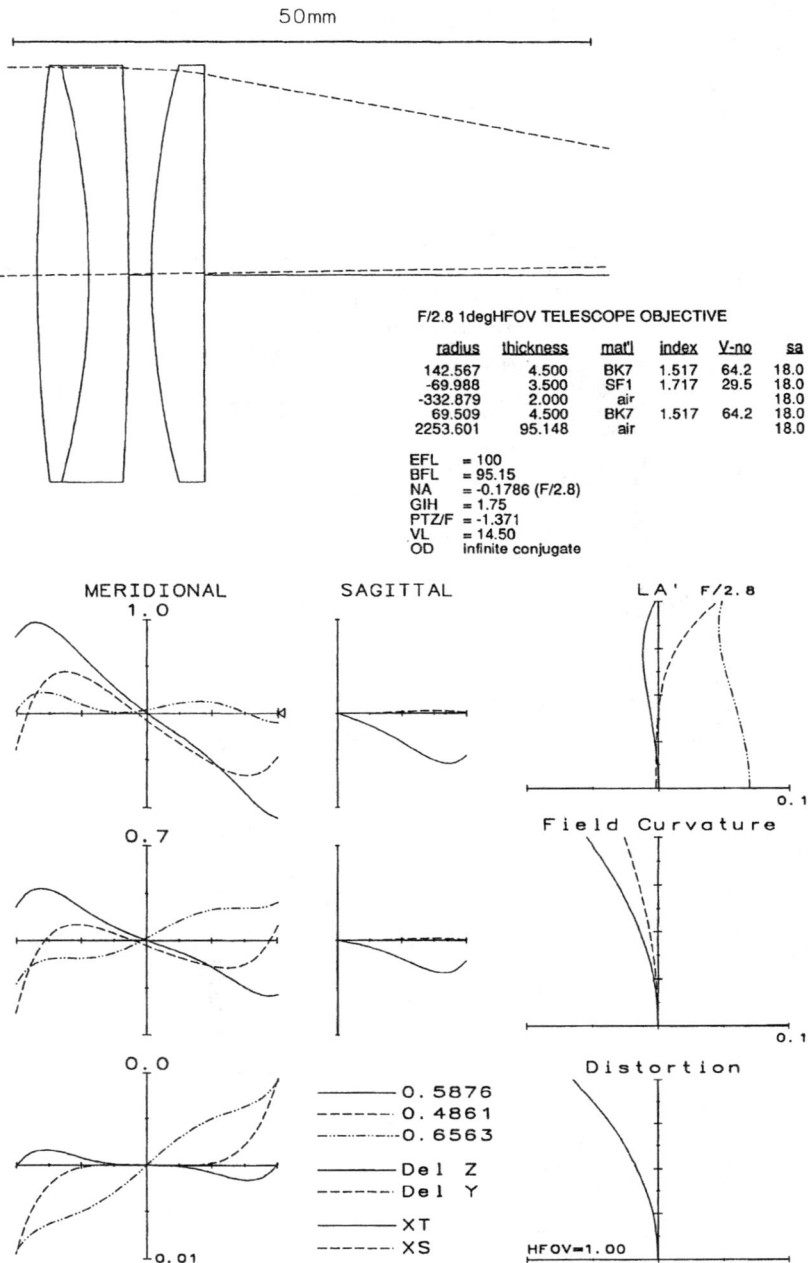

Figure 6.14 A three element $f/2.8$ objective with a cemented doublet, which reduces alignment sensitivity.

two glass types) is given by

$$SS = \frac{f(P_A - P_B)}{V_A - V_B}$$
$$= f\left(\frac{\Delta P}{\Delta V}\right) \qquad (6.3)$$

where f = focal length
P = partial dispersion = $(n_F - n_d)/(n_F - n_C)$
V = Abbe V number = $(n_d - 1)/(n_F - n_C)$

The secondary spectrum can only be reduced by using a combination of glasses that have a lower $\Delta P/\Delta V$ than ordinary glass types. Glasses with unusual partial dispersions can be selected from glass catalogs. Some manufacturers include a separate list of their glasses tabulating the amount by which the partial dispersion departs from the normal run of glass. For positive elements, glasses such as FK51, FK52, FK54, PSK53, PK51, LgSK2, the higher-index SF, and the TiF glasses (the last two for use in negative achromatic components), and many crystals such as CaF_2 will tend to reduce the secondary spectrum in any sort of lens. For negative elements, the short flints (KzF and KzSF types) and some lanthanum glasses are useful. Many, if not most, of these materials are characterized by poor resistance to atmospheric attack, poor working characteristics in the shop, and frequently by a high price. As glasses, they tend to be more difficult to manufacture ("melt") than the ordinary glasses. As a result, their optical quality is sometimes lower than that of normal glasses. In addition, most glass pairs that have well-matched partial dispersions, so that $(P_A - P_B)$ is very small, also tend to have small V-value differences; per Eqs. (6.1) and (6.2), this means that the element powers required to produce achromatism will be correspondingly large. The result is that although such glasses can reduce or eliminate secondary spectrum, the large element powers cause greatly increased amounts of spherochromatism and spherical zonal. Such a lens is shown in Fig. 6.15; compare its residuals to the ordinary doublets of the same speed in Figs. 6.2 and 6.3. Obviously, then, such lenses must be used at slower speeds, or designed with more elements to reduce these residuals. Note that the lens of Fig. 6.15 has been spaced to reduce spherochromatic and zonal spherical aberrations, and that the lenses of Figs. 6.2 and 6.3 are not; they are edge contacted. Even so, the difference in correction is quite apparent.

Figure 6.16 is a plot of partial dispersion versus V-value; each dot represents an optical glass. The slope of a line connecting any two

Chapter Six

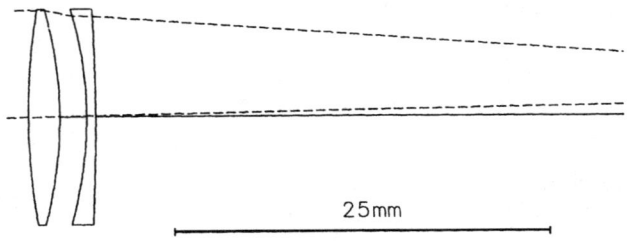

F/7 1degHFOV FRAUNHOFER OBJECTIVE

radius	thickness	mat'l	index	V-no	sa
40.679	2.100	FK54	1.437	90.7	7.2
-27.252	1.789	air			7.2
-25.887	0.600	KZFS2	1.558	54.2	7.2
-141.761	93.440	air			7.2

```
EFL   = 100
BFL   = 93.44
NA    = -0.0714 (F/7.0)
GIH   = 1.75
PTZ/F = -1.364
VL    = 4.49
OD    infinite conjugate
```

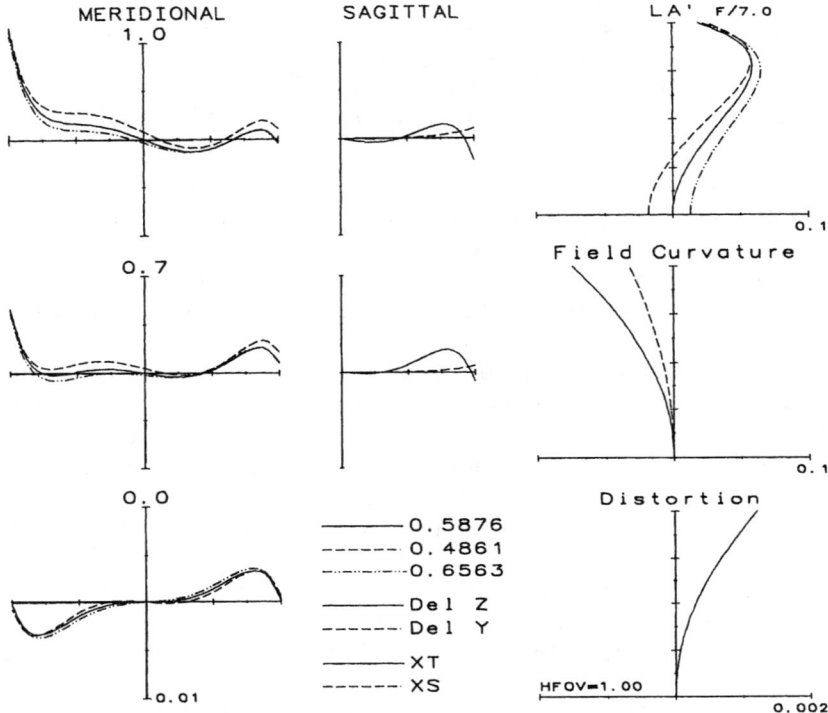

Figure 6.15 $f/7$ apochromatic doublet, airspaced for spherochromatism (FK54 and KZFS2).

Telescope Objectives   131

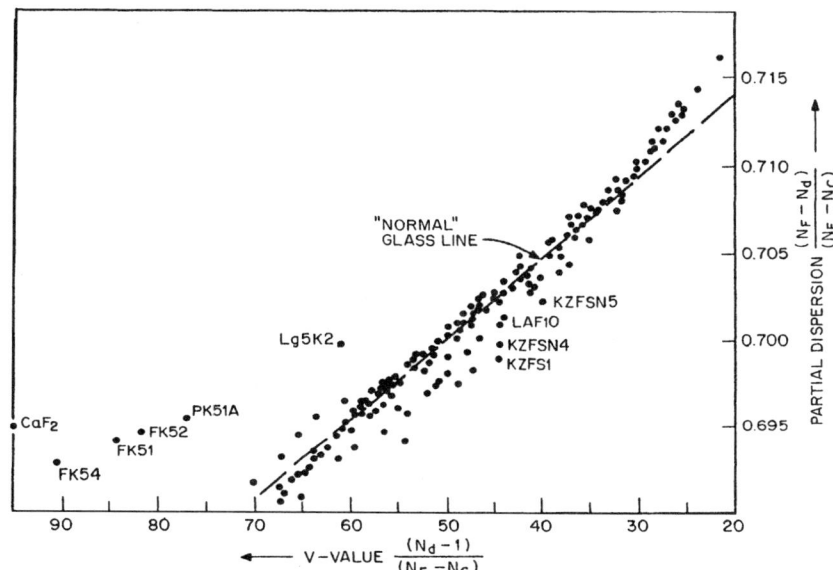

Figure 6.16 Plot of the partial dispersion $P = (n_F - n_d)/(n_F - n_C)$ versus Abbe V number $V = (n_d - 1)/(n_F - n_C)$ for the Schott glass catalog, showing several glasses with unusual partial dispersions that are useful in reducing secondary spectrum.

glasses is equal to $(P_A - P_B)/(V_A - V_B)$, which, as indicated in Eq. (6.3), determines the amount of secondary spectrum in a thin lens. It is apparent that any pair of glasses selected from along the "normal line" will have the same secondary spectrum as any other pair. It is also apparent that the glasses with unusual partial dispersions will form doublets with reduced secondary, and that the further they lie away from the normal line, the greater the difference between their V-values, the lower the element powers, and the lower the residual aberrations the lenses will have.

On a P versus V plot such as Fig. 6.16, one can simulate a nonexistent glass anywhere along a straight line connecting the points representing two real glasses. Thus, if a telescope objective is made as a triplet rather than a doublet, it is possible to reduce the $(P_A - P_B)/(V_A - V_B)$ to zero by using two of the glasses to simulate one whose partial dispersion exactly matches that of the third glass. Figure 6.17 illustrates such a glass combination in an airspaced (coma-corrected) version. Note that this technique does not eliminate the need for glasses with unusual partials; it merely facilitates a good match for the partials. By suitable glass choice, it is possible to bring four or even five wavelengths to a common focus; for most applications this is an unnecessary refinement.

132 Chapter Six

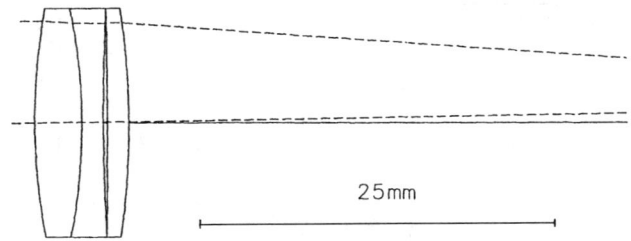

F/7.0 1degHFOV TELESCOPE OBJECTIVE

radius	thickness	mat'l	index	V-no	sa
44.144	3.300	PK51	1.529	77.0	8.0
-39.524	1.500	LSF18	1.913	32.4	8.0
158.460	0.278	air			8.0
-418.801	1.500	SF57	1.847	23.8	8.0
-54.845	97.720	air			0.0

EFL = 100
BFL = 97.72
NA = -0.0714 (F/7.0)
GIH = 1.75
PTZ/F = -1.142
VL = 6.58
OD infinite conjugate

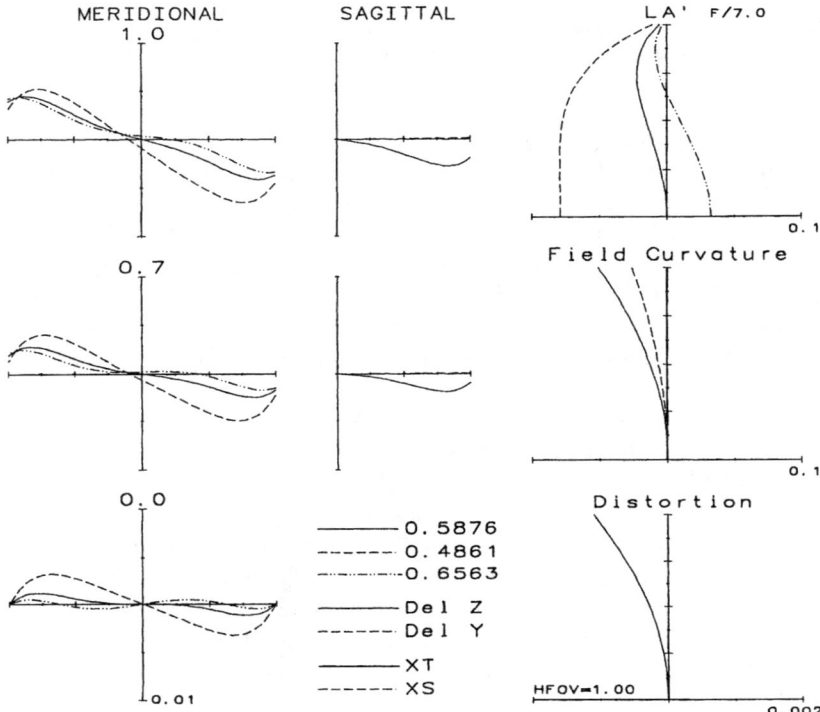

Figure 6.17 An airspaced $f/7$ apochromatic triplet.

6.9 The Design of an f/7 Apochromatic Triplet

6.9.1 Partial dispersion, secondary spectrum, and wavelengths

The *normal line* in the plot of partial dispersion versus V-value (Fig. 6.16) is anchored by K7 (n_d = 1.51112, V = 60.41, $P_{F,d}$ = 0.695) and F2 (n_d = 1.62004, V = 36.37, $P_{F,d}$ = 0.706). Thus $\delta P/\delta V$ = 0.000461, and a thin achromat made from K7 and F2 glass (or from any pair of glasses from along the normal line) would have a secondary spectrum equal to 0.000461f, or f/2168. If we accept the catalog values, an achromat of PK51A (529770) and LAK8 (713538) will have a secondary spectrum of 0.0000038f, or f/261,587. The total element powers add up to 5.65Φ, some 40 percent larger than the 4.02Φ for K7 and F2. But before getting too excited over this, we should note that the catalog index values are given to only five places, which seriously limits the accuracy of the calculations above. Kingslake suggests that seven-figure accuracy may be necessary (Ref. 11, p. 133).

The *short flints*, whose name usually begins with the letters KZ (from *kurz*, German for "short"), have a short, blue end of the dispersion spectrum, and the "long crowns" have a stretched blue end. In KzFS, the S stands for heavy, indicating a high index.

Note well that the shapes or contours of the glass map and the partial dispersion map change as the wavelength region changes from the visual region. The change is especially pronounced in the near infrared, where crown and flint glasses switch their relative dispersions.

Some spectral lines used in lens design, their sources, and wavelengths are as follows:

t	Hg	1013.98 nm	d	He	587.5618
s	Cs	852.11	e	Hg	546.0740
r	He	706.5188	F	H	486.1327
C	H	656.2725	F'	Cd	479.9914
C'	Cd	643.8469	g	Hg	435.8343
	He-Ne	632.8	h	Hg	404.6561
D	Na	589.2938	i	Hg	365.0146

6.9.2 Apochromatic triplets

The element powers for a thin lens apochromatic triplet of unit power ($\Phi_{abc} = 1.0$) are given by the following equations:

$$X = V_a\,(P_b - P_c) + V_b\,(P_c - P_a) + V_c(P_a - P_b)$$

$$\phi_a = V_a\,(P_b - P_c)/X$$

$$\phi_b = V_b\,(P_c - P_a)/X$$

$$\phi_c = V_c(P_a - P_b)/X = 1.0 - \phi_a - \phi_b$$

$$V = (n_d - 1)/(n_F - n_C) \qquad P_{dx} = (n_d - n_x)/(n_F - n_C)$$

The denominator term in these equations, X, is equal to twice the area of the triangle defined by the three glasses in the P versus V map. The larger this area, the lower the element power, and the better the lens is likely to be. This makes a reasonable, practical, and simple basis for glass selection.

For a three-glass apochromat we have six possible arrangements of the elements: *abc, acb, bca, bac, cab,* and *cba.* Usually the arrangement is chosen so that the strong positive element (low index, high V-value) is first, the negative element (high index, medium V-value) is second, and the weak (positive or negative) element (higher index, low V-value) is last. This arrangement tends to yield a rational looking set of curves. If we now include the cemented and airspaced combinations, e.g., *abc, a · bc, ab · c,* and *a · b · c,* where *a·b* indicates an airspace, there are 6 × 4 = 24 possible arrangements for a compact apochromatic triplet (once the glasses have been selected). In addition, just as in the doublet, the airspace size may be used as a variable as described in Secs. 6.4 and 6.5. Obviously a complete exploration of the apochromatic triplet can cover a good bit of territory even after the glass selection is made.

Most apochromats tend to be slow speed (i.e., large f number) because the high powers of the elements necessary for color correction cause heavy residual aberrations, i.e., zonal spherical, spherochromatism, and other high-order aberrations. Often the improvement in image quality provided by the apochromatic color correction is completely swamped by these residuals, and a more complex design form is needed to take full advantage of this improvement in the color correction. A compromise design with a less-than-complete correction of the secondary spectrum is often useful.

Once the glass types have been selected, the design process for the apochromatic triplet is similar to that for the ordinary doublet telescope objective, except that a target for apochromatism must be added to the usual spherical, coma, spherochromatic, and primary chromatic targets in the merit function. One possible set of targets for this purpose would seek to make the marginal (or zonal) ray focal plane intersection heights equal for all three wavelengths (for example, $Y_d - Y_C = 0$ and $Y_F - Y_C = 0$).

In designing an apochromat for fabrication (as opposed to a paper design study) one should check the index of the glasses very carefully, by the most precise data determination available (such as extra-precise measurements) rather than relying on catalog values. A very small error in the index values can ruin the apochromatic correction. As noted, Kingslake has suggested that seven-place accuracy for the index is necessary; the index used in computer programs is usually calculated from a dispersion equation, and to more than seven places. Recently the glass manufacturers have switched to a new Sellmeier equation (Ref. 23, p. 176), which is said to be an order of magnitude more accurate than the equation previously used. The measurement accuracy is still a bit problematic, however.

Some glass combinations that have been suggested in the literature for apochromatic triplets are:

FK6–KzFS1–SF15

BK7–KzFS1–BaFN10

PK51–LaF21–SF15

PK51–LSF18–SF57

PSK53A–KzFS1–TiF6

An infrequently encountered apochromatic device is the use of a zero-power cemented doublet (or triplet) composed of glasses with nearly the same index and V-value but with different partial dispersions. Such a component is inserted into an airspace of a lens in order to reduce the secondary spectrum without having much effect on the other aberrations (which are presumably already well corrected). There are several dense flints that can be used for this purpose.

6.9.3 The design process

The glass types cited in several published designs were collected and plotted on a P versus V map. The three glasses that made the triangle with the largest area were selected, as indicated in Fig. 6.18. These glasses are tabulated below with their V-values, partial dispersions and the element powers needed to produce an apochromat of unit power ($\Phi = 1.0$).

	V-value	P_{Fd}	Power ϕ	Curvature = $\phi/(n-1)$
PK51A	76.981	0.6955	+3.09507	+5.85573
KZFSN2	54.157	0.6941	−2.03137	−3.63809
SFL57	23.622	0.7146	−0.06370	−0.07524

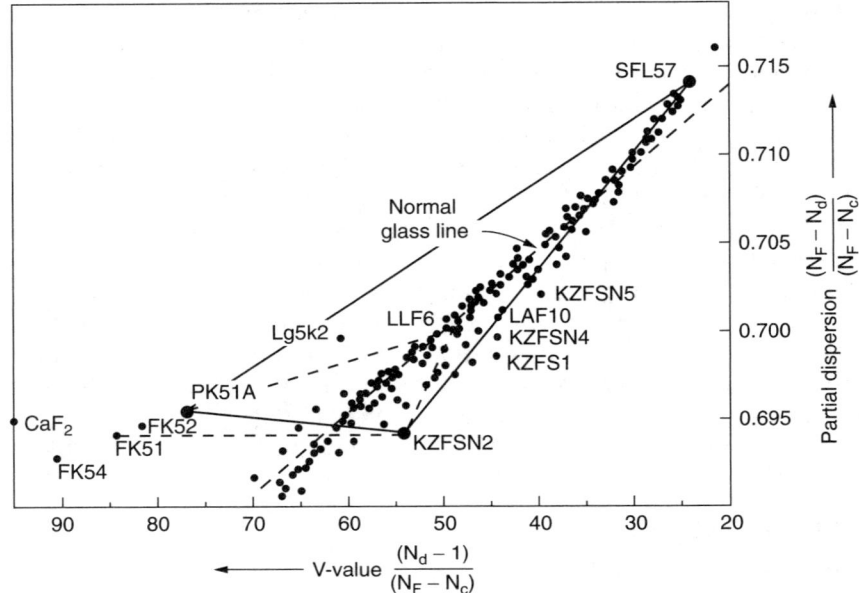

Figure 6.18 Plot of the partial dispersion $P = (n_F - n_d)/(n_F - n_C)$ vs. Abbe V number $V = (n_d - 1)/(n_F - n_C)$ for the Schott glass catalog, showing several glasses with unusual partial dispersions useful in reducing secondary spectrum. The glasses used in the sample design project are indicated by the triangles.

We elect to try a cemented triplet and (selecting the most common arrangement) begin with the leading element as a strong positive equiconvex. Scaling for a focal length of 100, adding suitable thicknesses, and using an angle solve on the last surface, the starting prescription for an $f/7$ lens becomes:

$R1 = +34.15$	$T1 = 2.5$	PK51A	529770
$R2 = -34.15$	$T2 = 1.5$	KZFSN2	558542
$R3 = +140.8$	$T3 = 1.5$	SFL57	847236
$R4 = +121.62$	BFL = 94.71		

The basic merit function that we used included targets for:

Marginal transverse spherical

Marginal transverse F-C chromatic

Marginal transverse F-d chromatic

Tangential coma = $3 \cdot H \cdot OSC$

TABLE 6.2 Prescription of the Apochromatic Triplet Objective Shown in Figs. 6.19 and 6.20

$R1 = +61.2926$	$T1 = 2.5$	PK51A	529770
$R2 = -27.8097$	$T2 = 1.5$	KZFSN2	558542
$R3 = -153.503$	$T3 = 1.5$	SFL 57	847236
$R4 = -193.063$	BFL = 97.396		

All curvatures were allowed to vary, with an angle solve on the fourth curvature. The data of the optimized lens after a few iterations is shown in Table 6.2.

The aberrations are plotted in Fig. 6.19 and the MTF in Fig. 6.20. The performance on axis is near perfect. The off-axis performance is poor but it's about as good as can be, considering the astigmatism inherent in a thin aplanat. Although we have corrected the coma, the field curvature and astigmatism are large, and as noted before, unavoidable. The secondary spectrum is negligible, but the spherochromatism is normal (i.e., the F-light spherical is overcorrected and the C-light is undercorrected) and this is the limiting factor on the axial performance.

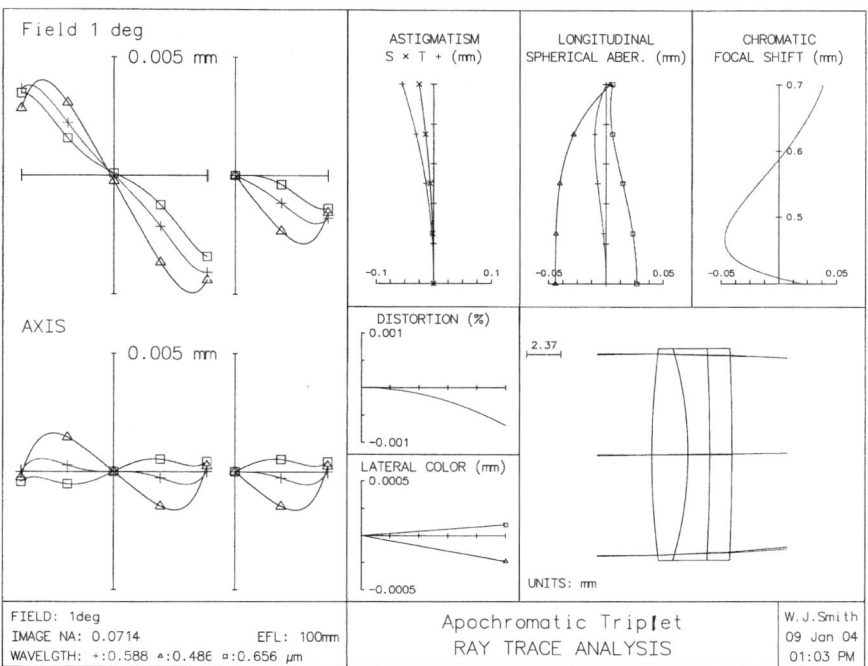

Figure 6.19 Ray intercept plot for cemented apochromat of Table 6.2.

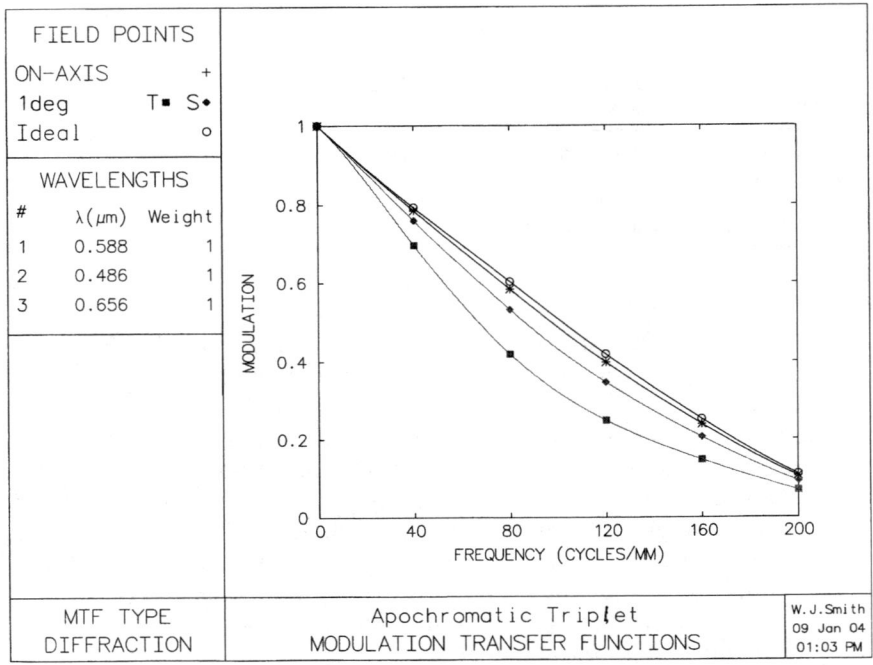

Figure 6.20 MTF plot for cemented apochromat of Table 6.2.

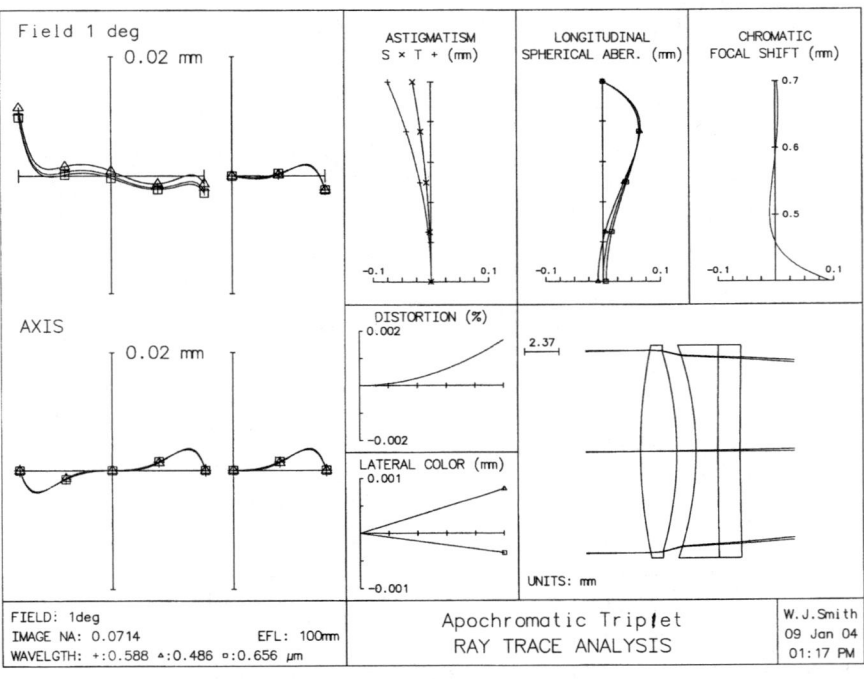

Figure 6.21 Ray intercept plot for the airspaced apochromat of Table 6.3.

In the doublet telescope objective of Fig. 6.7 we corrected the spherochromatism by spacing apart elements that were oppositely uncorrected for chromatic aberration. In this design we should be able to space the first element away from the other two and correct the spherochromatism in the same way. We add a marginal-ray Conrady $(D-d)\delta n$ target to the merit function and allow the airspace and all curvatures to vary. With the marginal chromatic aberration corrected, the Conrady $(D-d)$ sum will be a measure of the area between the C- and F-ray intercept plots, and is thus a direct measure of the spherochromatism. The aberrations and MTF are shown in Figs. 6.21 and 6.22 and the resulting lens shown in Table 6.3.

Now the spherochromatism has been eliminated and *at the best focus* the axial MTF is effectively perfect. Although the coma is only a bit overcorrected, the off-axis MTF has become worse. This is because the overcorrected zonal spherical has shifted the best focus away from the lens and in effect increases the inward-field curvature (which is the result of the negative astigmatism.)

Adding an off-axis marginal rim-ray coma target to the merit function (in addition to the OSC target) yields a better balance for the coma,

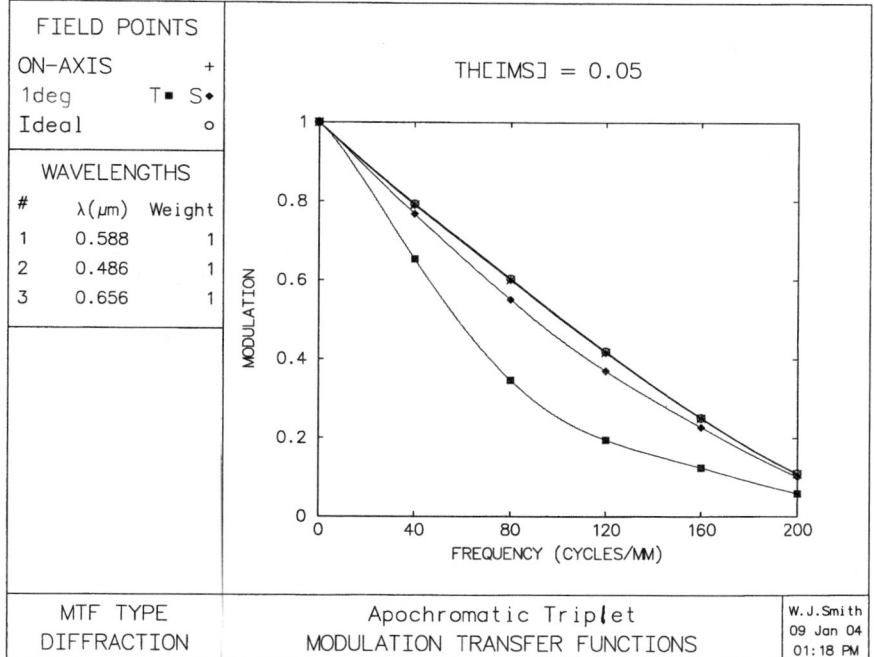

Figure 6.22 MTF plot for the airspaced apochromat of Table 6.3.

TABLE 6.3 Prescription for an Airspaced Apochromatic Triplet with Corrected Spherochromatic, as Shown in Figs. 6.21 and 6.22

$R1 = +38.0181$	$T1 = 2.5$	PK51A	529770
$R2 = -26.0148$	$T2 = 1.3086$	air	
$R3 = -23.8169$	$T3 = 1.5$	KZFSN2	558542
$R4 = +499.250$	$T4 = 1.5$	SFL57	847236
$R5 = +319.550$	BFL = 91.156		

and surprisingly also reduces the zonal spherical, but does allow some secondary spectrum to return. The axial performance is very slightly degraded and the off-axis MTF is noticeably better. The lens data are shown in Table 6.4 and the aberrations and MTF are shown in Figs. 6.23 and 6.24.

When we break the contact between the second and third elements and allow both airspaces to vary, this produces a quite meniscus third element and an unreasonably large airspace, all without any significant improvement. It would appear that at this point we have exhausted the possibilities in our triplet. But we have studied only one set of glasses and only one sequence of the elements, so we have no guarantee that there is not a better triplet somewhere, even if we restrict ourselves to a limited set of glasses.

Just for the fun of it we return to the P versus V glass map and select an ordinary glass near the normal glass line from about the middle of the plot, LLF6 (532488), as a possible substitute for the more extreme SFL57 glass. When we try the cemented triplet version of this lens, the zonal spherical and the spherochromatism are about 30 percent worse with LLF6 than with the SFL57, and the MTF at 80 lpm is down about 10 percent.

The airspaced triplet version is shown in Table 6.5, and Figs. 6.25 and 6.26 show the aberrations and MTF of the LLF6 lens. Compared to the SFL57 triplet the secondary spectrum is about the same, but the other axial aberrations are three to five times larger. The tangential field curvature, X_t, is about 10 percent less and this (plus a reduced high-order coma) yields about a 5 percent improvement in the off-axis MTF at 80 lpm.

TABLE 6.4 Prescription for an Airspaced Apochromatic Triplet with Improved Coma Correction (See Figs. 6.23 and 6.24)

$R1 = +40.2701$	$T1 = 2.5$	PK51A	529770
$R2 = -36.3204$	$T2 = 2.4631$	air	
$R3 = -31.1800$	$T3 = 1.5$	KZFSN2	558542
$R4 = -326.924$	$T4 = 1.5$	SFL57	847236
$R5 = +2961.4$	BFL = 89.216		

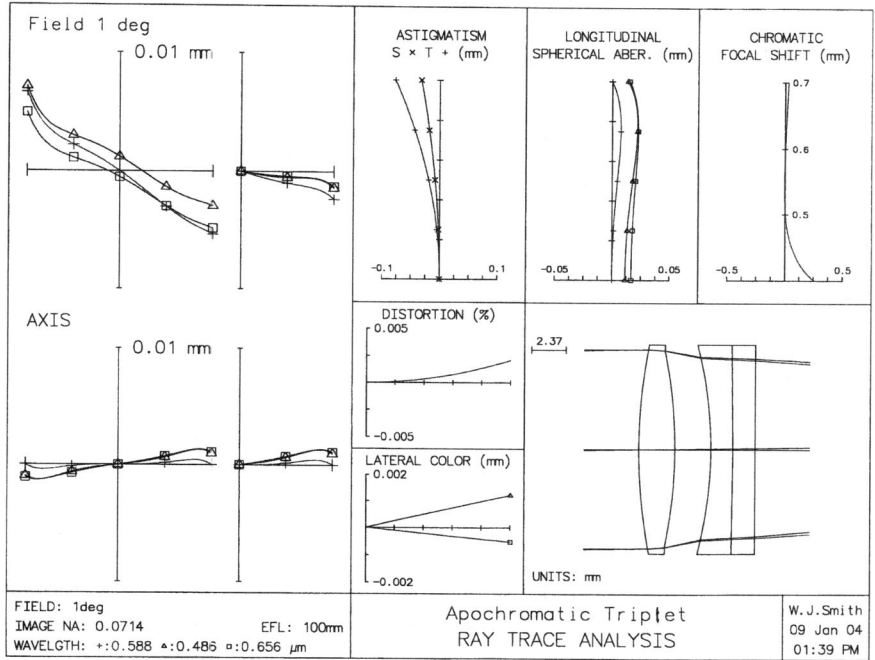

Figure 6.23 Ray intercept plot for the airspaced apochromat of Table 6.4.

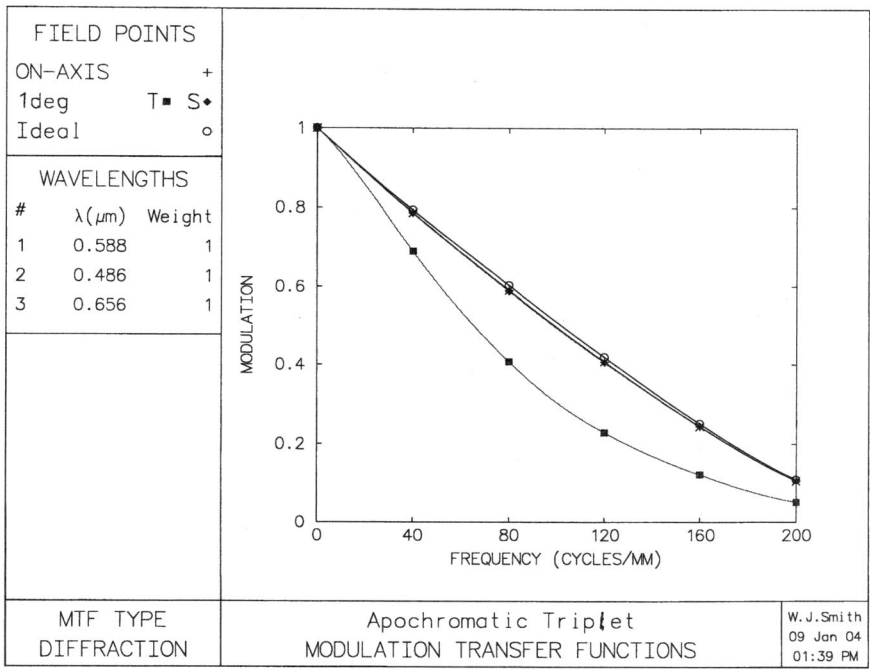

Figure 6.24 MTF plot for the airspaced apochromat of Table 6.4.

TABLE 6.5 Prescription for the Airspaced Apochromatic Triplet Which Used an Ordinary Light Flint to Replace the SFL57 Glass of the Previous Designs (See Figs. 6.25 and 6.26)

$R1 = +39.4654$	$T1 = 2.5$	PK51A	529770
$R2 = -28.5866$	$T2 = 1.4133$	air	
$R3 = -25.9508$	$T3 = 1.5$	KZFSN2	558542
$R4 = -36.3578$	$T4 = 1.5$	LLF6	532488
$R5 = +234.436$	$BFL = 91.083$		

Again looking at the P versus V glass map that we used at the start, we notice that two of the glasses we used, PK51A and KZFS2, have nearly equal values of the partial dispersion P. While not a perfect match, they should produce a doublet with very low secondary spectrum. Removing the third (SFL57) element and also removing the secondary spectrum target from the merit function, plus later on triple weighting the zonal spherical target and doubling the weight on the marginal spherical target, we manage to get some very weird doublet configurations of quite undistinguished quality; however, remembering a very old technique taught by Conrady and Kingslake[1,2,11] we set $R3$ equal to $R2$ (via a curvature pickup) to maintain a reasonable shape and arrive at the following airspaced doublet. See Table 6.6.

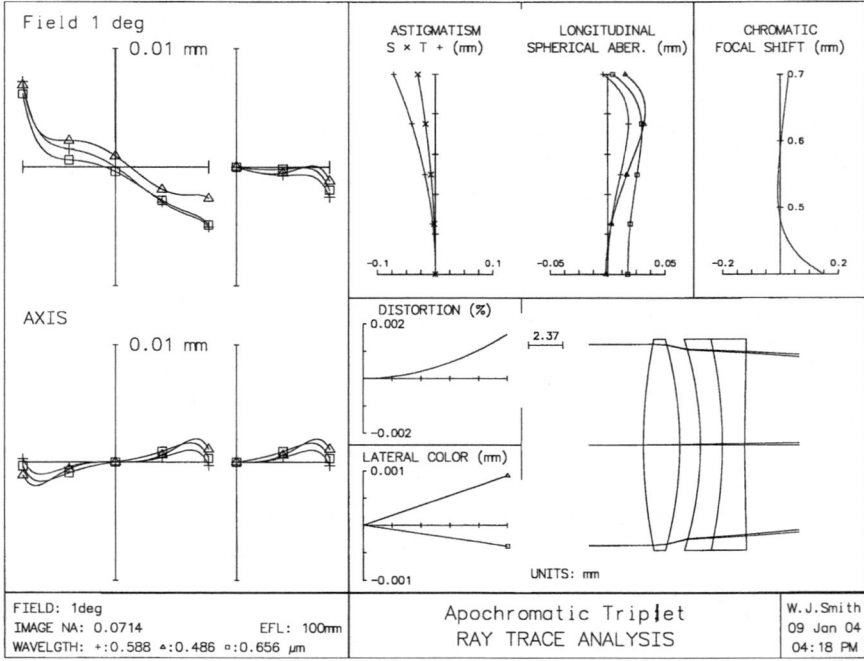

Figure 6.25 Ray intercept plot for the airspaced apochromat of Table 6.5.

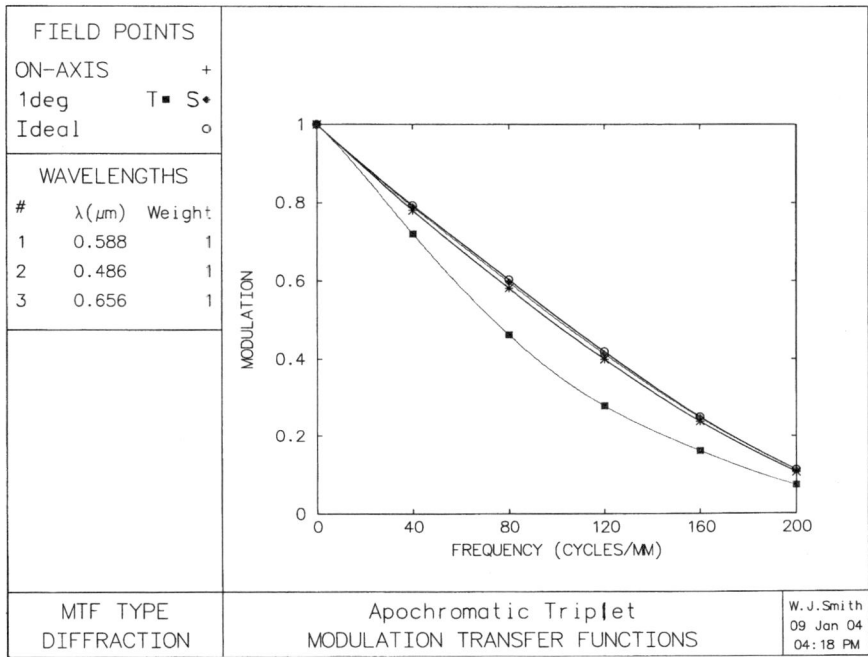

Figure 6.26 MTF plot for the airspaced apochromat of Table 6.5.

Somewhat surprisingly, *on the whole* this doublet is a trifle better than our best triplet. The aberrations and the MTF are shown in Figs. 6.27 and 6.28. There is some secondary spectrum but it is modest; however, the zonal spherical is less and the astigmatism is also slightly less. As a result we have an axial MTF, which is not quite as good as the triplet, but an off-axis MTF that is notably better.

The P versus V plot of Fig. 6.18 also indicates that FK51 and KZFSN2 should form a nearly perfect pair. We try them, and although they are slightly better than the PK51A/KZFSN2 doublet, the very small improvement in this design hardly seems worth the increased glass cost and fabrication difficulties that accompany the FK51 glass.

TABLE 6.6 The Prescription for a Nearly Apochromatic Doublet, Shown in Figs. 6.27 and 6.28

$R1 = +56.9160$	$T1 = 2.5$	PK51A
$R2 = -21.8717$	$T2 = 0.1057$	air
$R3 = -21.8717$	$T3 = 1.5$	KZFSN2
$R4 = -276.466$	BFL = 97.358	

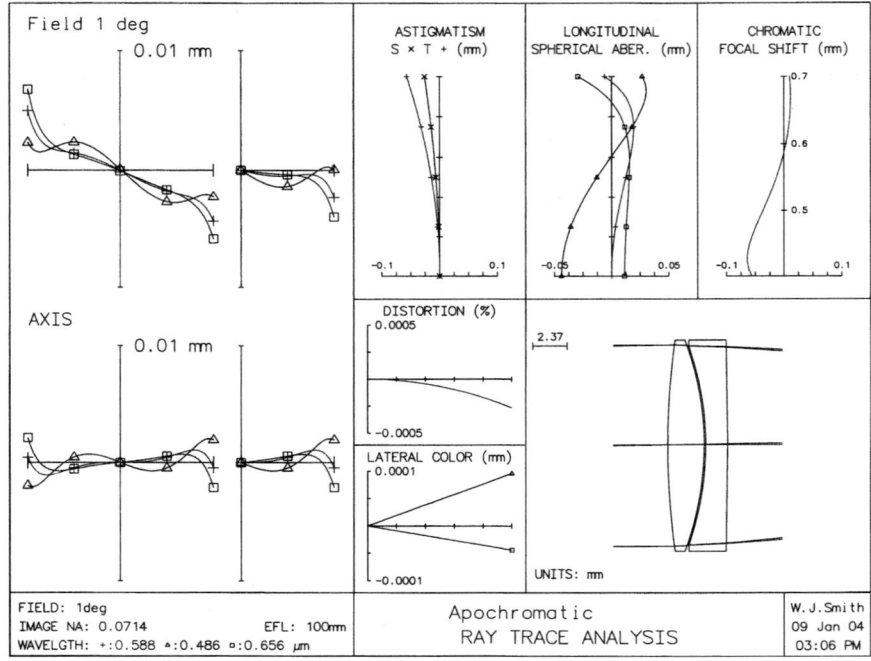

Figure 6.27 Ray intercept plot for the doublet apochromat of Table 6.6.

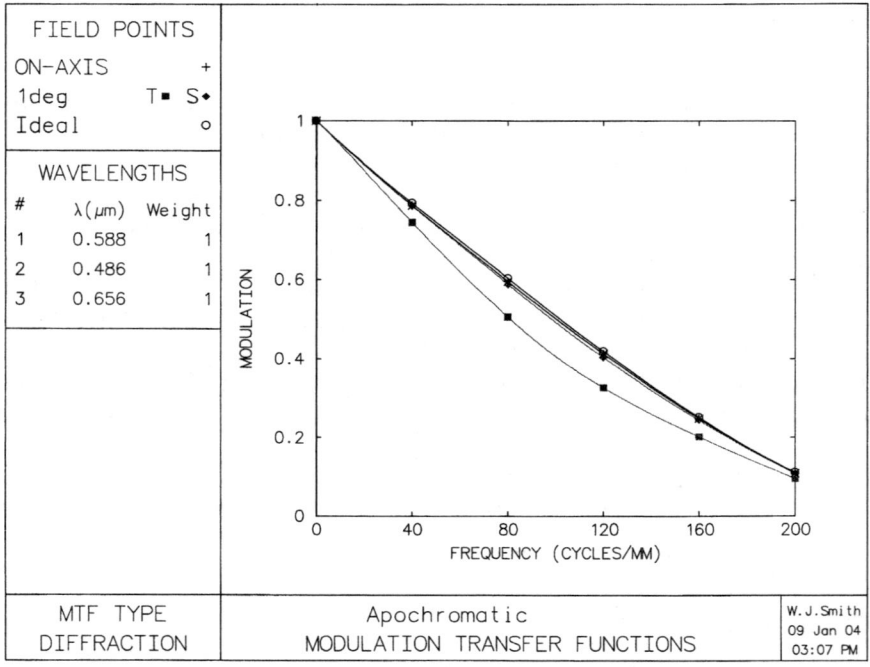

Figure 6.28 MTF plot for the doublet apochromat of Table 6.6.

The reality is that in many lenses secondary spectrum is less important than spherochromatism or the chromatic variations of astigmatism, coma (especially in telephoto designs), distortion, and so on.

When the chromatic aberration is corrected for three wavelengths, i.e., as in an apochromat, the tertiary spectrum may be out of balance, for example, with the positive tertiary chromatic at one end of the spectrum larger than the negative tertiary chromatic at the other end. This can be rebalanced by shifting the center wavelength used in solving the achromatic equations toward the end of the spectrum which has the larger tertiary spectrum.

Note. Pronounce apochromat the same way you pronounce achromat; think A-chromat and APO-chromat. It is not pronounced a-POCH-romat!!

6.10 The Diffractive Surface in Lens Design

A *diffractive surface* as used in lens design is a *fresnel* surface "modulo 2π." In other words, it is a fresnel surface where the height of each step is such that the wave front is retarded or stepped by exactly one wavelength. Thus the step height is $\lambda/(n - 1)$, assuming that the surface is bounded by air. For a glass or plastic surface ($n \approx 1.5$), this is a step height of about two wavelengths, as opposed to a step height on the order of tenths of a millimeter or more for an ordinary plastic fresnel. The slope and shape of the fresnel facets can be defined by a sphere or an aspheric. Note that similar results can be obtained with a local variation of the index of refraction.

6.10.1 Diffraction efficiency

The term *kinoform* indicates a surface with smooth facets. A curved-surface kinoform theoretically can have 100 percent efficiency. A linear (cone-shaped) kinoform can be 99 percent efficient. A "binary" surface approximates the smooth fresnel facets with a stair-step contour produced by a high-resolution photolithographic process. The surface relief is created by exposure through a series of masks. The number of levels produced equals 2^n, where n is the number of masks used, hence the name *binary*. The efficiency (i.e., the percentage of light that goes in the desired direction) of a binary surface is limited by the number of levels used to approximate the ideal smooth contour of the fresnel facet. A one-mask, two-level surface is 40.5 percent efficient; a two-mask, four-level surface is 81.1 percent efficient; a three-mask, eight-level surface is 95.0 percent efficient; a four-mask, 16-level surface is 98.7 percent efficient; and an M-level surface is $[\sin(\pi/M)/(\pi/M)]^2$ efficient. The theoretical efficiency of any diffraction surface, whether kinoform or binary, will be reduced by any fabrication departures from the ideal shape, such as rounding of sharp corners, and so on.

Since the wave front is stepped or retarded at each diffractive fresnel step by exactly one wavelength for the nominal wavelength, it is apparent that the coherent behavior of the system is preserved only for the nominal wavelength. At this wavelength, the phase from the top of one zone exactly matches that from the bottom of the preceding zone. The surface is less efficient for other wavelengths, and thus the spectral bandwidth over which a diffractive surface is useful is limited. This limitation may show up as inefficiency or as unwanted diffractive orders, ghosts, stray light, low contrast, and so on. The efficiency at other than the nominal wavelength (λ_0) is

$$E = [\sin \pi(1-\lambda_0/\lambda)/\pi(1-\lambda_0/\lambda)]^2$$

Over a bandwidth of ($\Delta\lambda$), the average efficiency is

$$\text{ave } E \approx 1 - [\pi (\Delta\lambda)/6\lambda_0]^2$$

6.10.2 Manufacturability

The following expressions allow an estimate of the practicality or manufacturability of a diffractive lens. As indicated above, the step height is $\lambda/(n-1)$. The radial spacing distance from one fresnel step to the next is approximately

$$\text{Spacing} \approx R\lambda/Y(n-1) = F\lambda/Y$$

where R is the diffractive surface radius of curvature, F is its focal length, and Y is the radial distance from the axis. The minimum spacing (at the edge of the diffractive lens) is

$$\text{Min spacing} \approx 2\lambda (f/\#) = \lambda/\text{NA}$$

where $f/\# = F/2Y_{max}$ = the relative aperture, and NA = $n \sin u$ = the numerical aperture. The total number of fresnel steps or zones is

$$\text{Number of steps} \approx D^2/8\lambda F$$

where D is the lens diameter. It is apparent that the longer the wavelength and the weaker the power of the diffractive surface, the wider and deeper are the steps, and the easier is the fabrication task. Techniques used for fabrication include single-point diamond turning (especially good for long-wave IR), ion-beam machining, electron-beam writing, laser-beam writing, and photolithography (which is extremely difficult on curved surfaces but effective on plano surfaces). For large commercial quantities, injection-molded plastic elements are an economical choice. Another useful process is epoxy replication. Applications of diffractive optics include hybrid (combined refractive and diffractive) lenses, microlens (size about 50 µm) arrays, anamorphic arrays, prisms, beamsplitters, beam multiplexers, filters, and the like.

6.10.3 The Sweatt model

From a lens design standpoint, an easy way to handle and understand the use of a diffractive surface is through the *Sweatt model*. W. C. Sweatt[*] showed that a raytrace model consisting of a very high index, zero-thickness lens could be used to predict the effect of a diffractive surface; the higher the index, the closer the results of the raytrace come to matching the exact diffraction results. An index of about 10,000 is a reasonable value to use. Since the diffractive effect is a direct function of wavelength, the index of the model should vary as the wavelength, and

$$n(\lambda) = 1 + (n_0 - 1)(\lambda/\lambda_0)$$

where λ_0 and n_0 are the nominal wavelength and index, respectively.

Thus, for the visual region, using d, F, and C light, we have for d-light at 0.5875618 µm,

$$n_d = 10{,}001.00$$

F-light at 0.4861327 µm,

$$n_F = 8{,}274.73$$

C-light at 0.6562725 µm,

$$n_C = 11{,}170.42$$

and the Abbe V-value,

$$V = (n_d - 1)/(n_F - n_C) = -3.45$$

The negative V-value results from the fact that the index rises with wavelength instead of dropping as in ordinary refractive materials. The partial dispersion is $P = (n_F - n_d)/(n_F - n_C) = 0.5962$. These extremely unusual values make the diffractive surface a most singular optical

---

[*]*J. Opt. Soc. Am.*, vol. 67, 1977, p. 803, vol. 69, 1979, p. 486; *Appl. Opt.*, vol. 17, 1978, p. 1220.

material. This low-V-value (i.e., high dispersion) characteristic of a diffractive device indicates that there will be very large amounts of chromatic aberration when a diffractive surface is used over a significant spectral bandwidth.

6.10.4 The achromatic diffractive singlet

If we assume a single element of BK7 (n_d = 1.5168, V = 64.2, P = 0.6923), we can apply Eqs. 6.1 and 6.2 to determine the powers of the singlet and the diffractive element that will produce an achromat. The result is a power of $\phi_a = V_a \Phi/(V_a - V_b) = +0.949\Phi$ for the BK7 element and $\phi_b = +0.051\Phi$ for the diffractive element (where Φ is the desired power of the achromat). The negative V-value of the diffractive surface produces an achromat where both elements have positive power. If we allow the diffractive surface to be aspheric (in the actual surface this is done by making the slope and shape of the fresnel facets correspond to those of an aspheric surface), we can produce a singlet of the desired power, which is corrected for spherical aberration, chromatic aberration, and coma. The necessary four degrees of freedom are the power and bending of the singlet and the power and fourth-order asphericity (or conic constant) of the diffractive surface.

The resulting design is shown in Fig. 6.29. The residual aberrations (zonal spherical, spherochromatism, and secondary spectrum) can be compared with those of the ordinary achromatic doublet shown in Fig. 6.6. Note that the sign of the secondary spectrum is reversed from that of an ordinary doublet (because of the unusual P and V of the diffractive surface) and that the spherochromatism is large, more than twice that of the doublet of Fig. 6.6 (and is also of reversed sign). The spherochromatism can be corrected by aspherizing the first surface with a fourth-order deformation term in a manner analogous to adjusting the airspace of the doublet in Fig. 6.7 (i.e., we change the relative heights at which the red and blue rays strike the diffractive surface). The zonal spherical can be removed with a sixth-order deformation term on the first surface. The use of an aspheric surface is an economically practical move, assuming that the lens is to be injection-molded from plastic. The result is a lens whose only axial aberration is about 0.17 mm of secondary spectrum.

Alternately, because photolithographic fabrication is most conveniently done on a flat surface, one might want to limit the lens shape to a plano-convex form and use as degrees of freedom the lens index, its radius, the power of the diffractive surface, and its asphericity. The optimal index is about 1.55 for this lens. If the lens material is acrylic (n = 1.492), and if we elect to control focal length, spherical, and chromatic

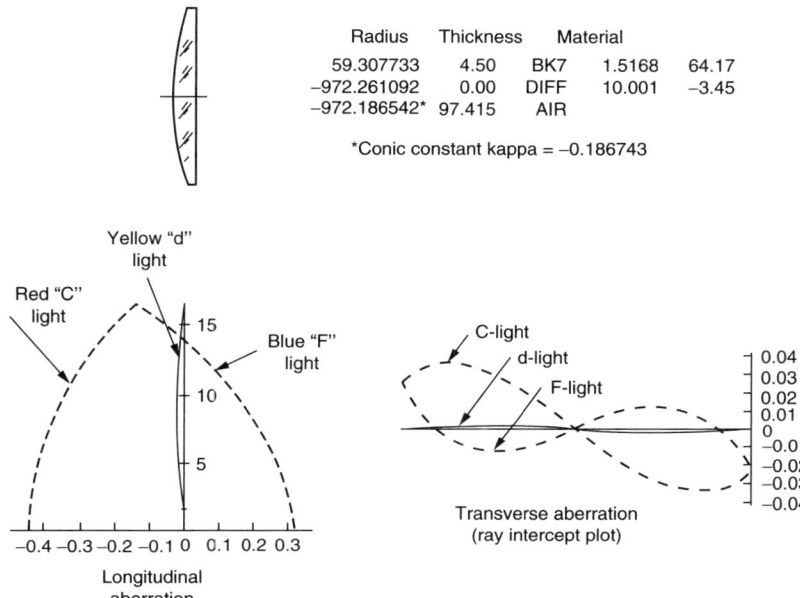

Radius	Thickness	Material		
59.307733	4.50	BK7	1.5168	64.17
−972.261092	0.00	DIFF	10.001	−3.45
−972.186542*	97.415	AIR		

*Conic constant kappa = −0.186743

Figure 6.29 The spherical aberration and spherochromatism of a hybrid refractive-diffractive singlet, efl = 100, $f/3.0$. Compare with the doublet of Fig. 6.6 (but note that the scales for LA are different). Both the spherochromatism and secondary spectrum are larger and of opposite sign from the lens of Fig. 6.6. As indicated in the text, the spherochromatism and zonal spherical can be eliminated easily by aspherizing the first surface (which would be quite a feasible option if the lens were injected-molded from acrylic).

(neglecting coma), the tangential coma at one degree off axis is −0.0156; if the material is polystyrene ($n = 1.590$), it is +0.0101.

Achromatic diffractive singlets have been very satisfactorily used in eyepieces, magnifiers, zoom camera lenses, and many other applications where the object field is of relatively uniform brightness. Their compactness and light weight as compared with a glass achromat make them very desirable for many applications such as head-mounted displays. Diffractive surfaces sometimes have proven less satisfactory for systems where there is a high brightness source in (or near) the field of view or a wide spectral bandwidth.

6.10.5 The apochromatic diffractive doublet

Since the unusual V-value and partial dispersion of the diffractive surface are so far from the line of normal glasses in a P versus V plot, we can easily produce an apochromatic lens using two ordinary materials plus a diffractive surface to eliminate the secondary spectrum.

The element powers for a three-element apochromat can be found using the following equations from Sec. 6.8:

$$X = V_a (P_b - P_c) + V_b (P_c - P_a) + V_c (P_a - P_b)$$

$$\phi_a = \Phi V_a (P_b - P_c)/X$$

$$\phi_b = \Phi V_b (P_c - P_a)/X$$

$$\phi_c = \Phi V_c (P_a - P_b)/X$$

where Φ is the power of the apochromatic triplet, V_i is the V-value, and P_i is the partial dispersion of the ith element.

If we use acrylic ($n = 1.4918$, $V = 57.45$, $P = 0.7014$) and polystyrene ($n = 1.5905$, $V = 30.87$, $P = 0.7108$) for elements a and b, and the diffractive surface ($n = 10,001$, $V = -3.45$, $P = 0.5962$) for element c, we get the following starting powers for the elements:

$$\phi_a = +1.9544\Phi \quad \text{(acrylic)}$$

$$\phi_b = -0.9640\Phi \quad \text{(polystyrene)}$$

$$\phi_c = +0.0096\Phi \quad \text{(diffractive)}$$

The lens can be corrected for marginal and zonal spherical aberration, coma, chromatic, spherochromatic, and secondary spectrum using the techniques outlined above. A drawback for this particular lens is that the secondary spectrum varies with aperture and is corrected only at one zone.

6.11 A Final Note

Note well that many of the sample telescope objective designs in this chapter are shown at speeds that are quite high for a telescope objective, e.g., $f/2.8$ or $f/3.0$. This has been done so that the aberrations and the techniques used to reduce them are clearly demonstrated. Most telescope objectives are used at speeds that are considerably slower and at which the aberrations are much smaller.

Figures 21.1, 21.2, and 21.3 in Chap. 21 show telescope objectives designed for use in the 8- to 12-μm region of the infrared. Figure 22.9 in Chap. 22 is effectively a telescope objective designed for use in monochromatic (laser) light; both elements are of flint glass.

Chapter 7

Eyepieces and Magnifiers

7.1 Eyepieces

The eyepiece of a telescope or microscope is an unusual optical system in that its pupil must be located completely outside the system. The aperture stop is usually located at the objective lens. The exit pupil is the image of the stop, and it must lie a suitable distance away from the eyepiece so that the eye can be placed at the pupil (to see the full field of view). This distance, or *eye relief*, must be at least 9 or 10 mm, just long enough to clear the eyelashes; an additional 5 mm eases the situation considerably. An eye relief of 20 mm or so is about the minimum necessary for comfortable use by spectacle wearers. Many systems depart from these guidelines. Any system subject to sudden motion (e.g., a rifle scope, which recoils) needs a much longer eye relief to prevent injury to the eye. For a 22-caliber rifle, the eye relief should be 2 in or more; for a high-powered rifle, the eye relief is typically about 4 or 5 in. It should be obvious that, for a given apparent field, the longer the eye relief, the larger the diameter of the eye lens must be to pass the edge-of-the-field rays. At the other extreme, the eyepieces of surveying telescopes, laboratory equipment, microscopes, and many high-power eyepieces often have an uncomfortably short eye relief.

Because there is absolutely no symmetry about the stop, distortion, coma, and lateral color tend to be problematic in an eyepiece. Most eyepieces have an amount of distortion that would be considered intolerable in a camera lens. Several percent is typical, and, in wide-angle eyepieces, a distortion of 8 to 12 percent is not uncommon. The distortion is usually of the f-theta, or f-sin θ type, however, and this is more easily tolerated than is that of the opposite sign. Coma and lateral color are ordinarily well corrected, but spherical and axial color are not; they are typically small and are often balanced by compensating aberrations in the objective and/or the erector lenses.

Spherical aberration of the pupil and distortion are directly related. For example, if an eyepiece has undercorrected spherical aberration of the chief or principal ray, that ray is usually bent toward the axis at a greater angle than it should be; the greater slope means that the apparent image of this part of the field will be too large, i.e., too far from the axis, and pincushion distortion is the result.

Pupil spherical means that the exit pupil position changes for different field angles. Spherical aberration of the pupil is usually undercorrected, so that the pupil image from the outer field is closer to the eyepiece than is the on-axis pupil. Since the rotation of the eyeball to view the edge of the field shifts the eye's pupil off the axis, this undercorrected pupil aberration is undesirable and simply exacerbates the situation. (Note that overcorrected pupil spherical would tend to match up with the shift of the pupil of the eye.) Vignetting between the exit pupil and the eye pupil may be severe with under-corrected pupil spherical. If the eye is placed closer to the lens (i.e., at the edge-of-the field pupil) to view the edge of the field, the result can be what is called the *kidney bean effect*, where both the axis and the edge of the field are unvignetted, but a dark, bean-shaped area caused by zonal vignetting appears in the zonal area of the field.

The eye rotation itself makes a wide angle eyepiece difficult to use, unless the exit pupil is very large and able to accomodate the motion of the eye pupil when the eye rotates.

Aspheric surfaces can be used to correct the pupil spherical, especially when located near the focus or on a curved surface concave to the pupil/stop. In an eyepiece the aspheric surface need not be very precisely finished because only a small area of the surface (the size of the pupil of the eye) is used to view a given field point. For an eyepiece, a surface accuracy equivalent to that of window glass is usually adequate.

The following tabulation lists a range of eyepiece characteristics:

	High performance	More typical
1. Axial rms blur spot	<0.3 mr	<1.0 mr
2. Off-axis rms blur	<0.7 mr	<1.5 mr
3. Field curvature and astigmatism	<1 Diopter	<3 Diopter
4. Axial color* (C-F)	<3 mr	<6 mr
5. Lateral color (C-F)	<0.5 mr	<1.5 mr
6. OSC	<0.0015	<0.0025
7. Distortion	<2%	<5%
8. Transverse pupil spherical	<0.4 mm	<1.0 mm
9. Eye relief	>25 mm	>15 mm

*The eye has about one diopter of undercorrected axial chromatic, so that overcorrected color in an eyepiece would offset this.

In reference to the table:

a. A 4 mm diameter pupil is assumed
b. "mr" stands for milliradian ≈ 3 min arc
c. OSC = (sagittal coma)/h (near the axis)
d. VA (visual acuity): 20/10 = 3 cycles/mr; 20/20 = 1.5 cycles/mr

Note that these are intended to indicate a common range of values, and that these values are often exceeded in either direction.

Remember that eyepieces are often used in systems that include prisms, and these prisms must be included in the design, usually with the eyepiece. When an eyepiece is a part of a telescope design, consider that the longer the telescope, the weaker the components, which means (for a given objective diameter) a lower *numerical aperture* (NA) for the components, less Petzval and astigmatism, and a longer eye relief, all to the good. But the cost and the weight of the scope will increase, and the internal image size and eyepiece dimensions will scale up. Also note that distortion calculated (in reverse, with the object at infinity) as barrel is actually seen as pincushion.

Field curvature and astigmatism are important in eyepieces. For many eyepieces, there is simply no correction of the Petzval curvature; the eyepiece is effectively composed of only positive components, and the result is a strongly inward-curving Petzval surface. Eyepieces that have flatter Petzval fields usually achieve them with thick meniscus components (which function just as described in Chap. 11). In some, a concave surface located near the focal plane has a strong field-flattening effect. This arrangement is the equivalent of a negative field lens; in addition to flattening the Petzval field it also lengthens the eye relief (and necessitates a correspondingly larger-diameter eye lens). A separate negative field lens is occasionally encountered when the concomitant increase in the size and weight of the eyepiece is an acceptable trade-off for a long eye relief, as in a tank sight, for example.

For an eyepiece afflicted with an inward-curving Petzval field, the astigmatism should definitely *not* be negative; usually some positive, overcorrecting astigmatism is desirable to artificially flatten the field. Often the factor that limits the extent of the angular field of the eyepiece is an overcorrected higher-order astigmatism, which produces both an abruptly backward-curving tangential field and too much astigmatism. In many eyepieces the overcorrected astigmatism arises at a diverging cemented surface. An increase in either the surface curvature or the index difference across the surface will increase the overcorrection of the astigmatism; a suitable balance between these two factors is necessary to both flatten the tangential field and restrain the high-order astigmatism.

A high-power microscope objective with an aplanatic front has lateral color, which, given the limitations of this classical design form, cannot be controlled. A *compensating eyepiece* is one that is designed to have a matching amount of lateral color so that the final image presented to the eye is free of lateral color.

The field curvature of an eyepiece (or telescope) is best evaluated in diopters of defocusing at the eye. In an eyepiece designed or analyzed separately from the telescope, the ray trace is conventionally done in reverse, with the object located at infinity on the eye side. Therefore, the field curvature (X_s and X_t) that is calculated this way must be converted from the short conjugate to diopters at the long conjugate. Using the newtonian imaging equation, we get $x' = -f^2/x$, and the reciprocal of x' (in meters) is the field curvature in diopters. Thus the field curvatures X_s and X_t can be converted to diopters by

$$D = \frac{X}{f^2} \qquad X \text{ and } f \text{ in meters}$$

$$= \frac{1000 X}{f^2} \qquad X \text{ and } f \text{ in millimeters}$$

$$= \frac{39.37 X}{f^2} \qquad X \text{ and } f \text{ in inches}$$

If D is positive, the eye can accommodate to focus on the image; if it is negative the normal eye cannot focus on the image. In most cases a field curvature of about 1 diopter (of either sign) is acceptable. A negative D of 3 diopters is about the largest tolerable curvature.

The off-axis image quality of most eyepieces is relatively poor. As a result, the outer portions of the field are used primarily to identify and locate objects of interest, which are then brought to the center of the field for a closer examination. Interestingly enough, the human eye shares this characteristic and mode of operation, so this seems quite natural to most observers.

Note that the focusing of a telescope eyepiece to accomodate nearsightedness or farsightedness is given by:

$$\delta = D \cdot f^2/1000 \text{ mm}$$

where D is the diopter correction required, and both δ and f are in millimeters. Typically the focus adjustment range is ±4 diopters or less.

Note that the common practice of raytracing from the exit pupil toward the objective lens may introduce errors in the analysis or design of an eyepiece. In the presence of spherical aberration of the pupil, the paraxial pupil position may be quite incorrect for the oblique beams, and this can lead to the calculation of incorrect paths.

7.2 A Pair of Magnifier Designs

The first sample lens of this chapter, Fig. 7.1, is a magnifier, suitable for use as a slide viewer, an optical comparitor, or a general-purpose magnifier. Unlike a telescope eyepiece, a good magnifier must be relatively insensitive to the position of the eye, because in a magnifier there is no exit-pupil-image of the objective lens to define the location of the user's eye. This requires a better level of correction, especially for spherical aberration, in the magnifier, so that the image doesn't swim or distort as the eye is moved about. The symmetrical eyepiece (Figs. 7.23 to 7.27) and the orthoscopic eyepiece (Fig. 7.21) also work well as magnifiers.

The second sample lens, Fig. 7.2, is also a magnifier, but of a totally different type. It can be considered either as a magnifying glass placed in front of a galilean telescope or as a reverse telephoto construction. The result is a working distance that is (in this sample) 50 percent longer than the effective focal length. This type of magnifier can also be executed in a much simpler construction than that of Fig. 7.2.

The ordinary 10x magnifier has a focal length of 1 in; it is quite convenient to hold and use. A 20x magnifier has a focal length of only 0.5 in. A simple 20x magnifier is about the smallest and most powerful magnifier that can conveniently used in the hand. A 20x magnifier must be held uncomfortably close to both the eye and the subject for a reasonable field to be seen. On occasion one's nose prevents getting this close. The Brueke style magnifier or a compound microscope are almost necessary for powers above 20x.

A viewing eyepiece for a head/helmet mounted display is essentially a magnifier with a somewhat limited allowable motion of the pupil. A plastic achromatic hybrid singlet with a diffractive surface (Chap. 6, Sec. 6.10) can be a good eyepiece for this application. It is attractive because of its very light weight and compactness.

7.3 The Simple, Classical Eyepieces

The *Huygens, Ramsden,* and *Kellner* eyepieces are shown in Fig. 7.3. They are simple and inexpensive, but, as you might expect, they are not outstanding performers. They use common, low-index glass; both elements of the Huygens and the Ramsden are often made from window glass, as is the Kellner field lens. Often only the convex surface is ground and polished. There is little advantage in departing from the plano-convex form.

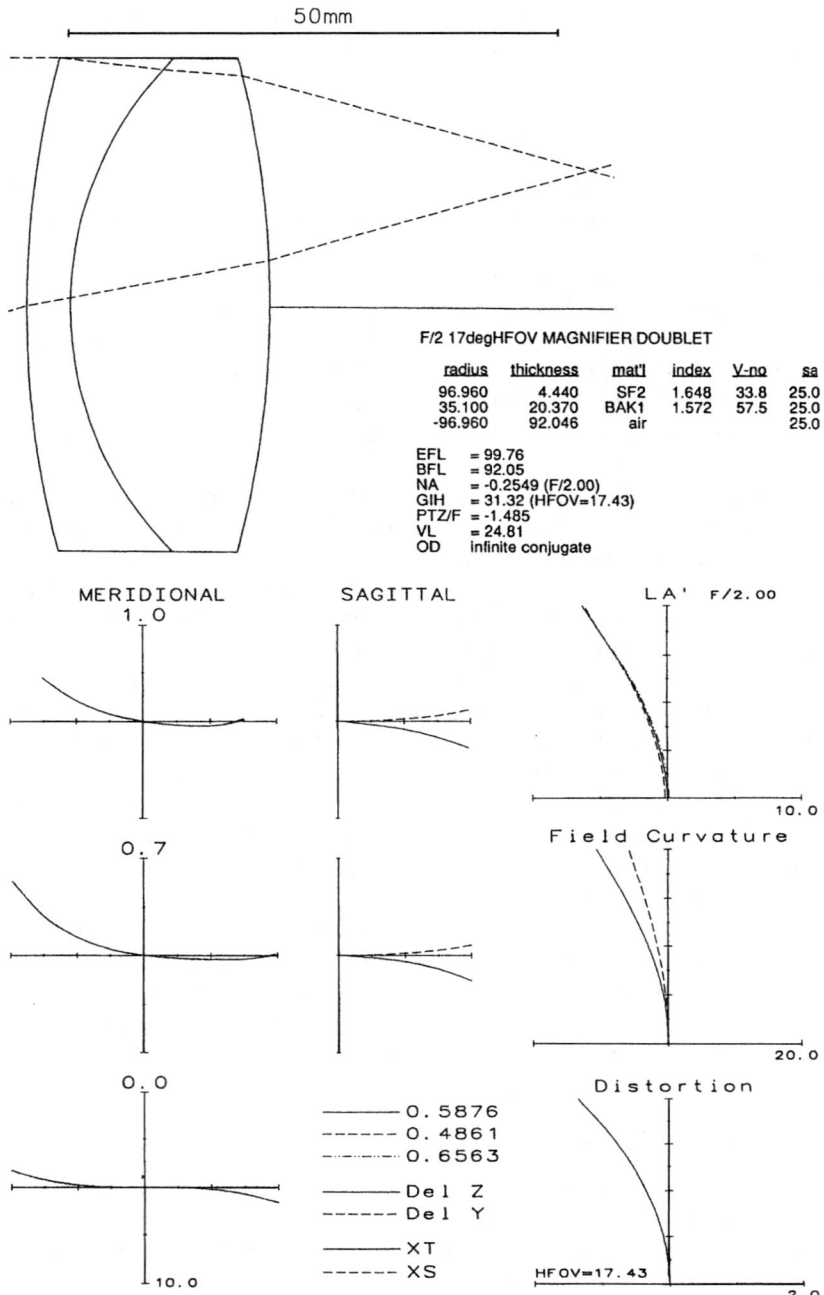

Figure 7.1 An excellent general purpose magnifier.

Eyepieces and Magnifiers 157

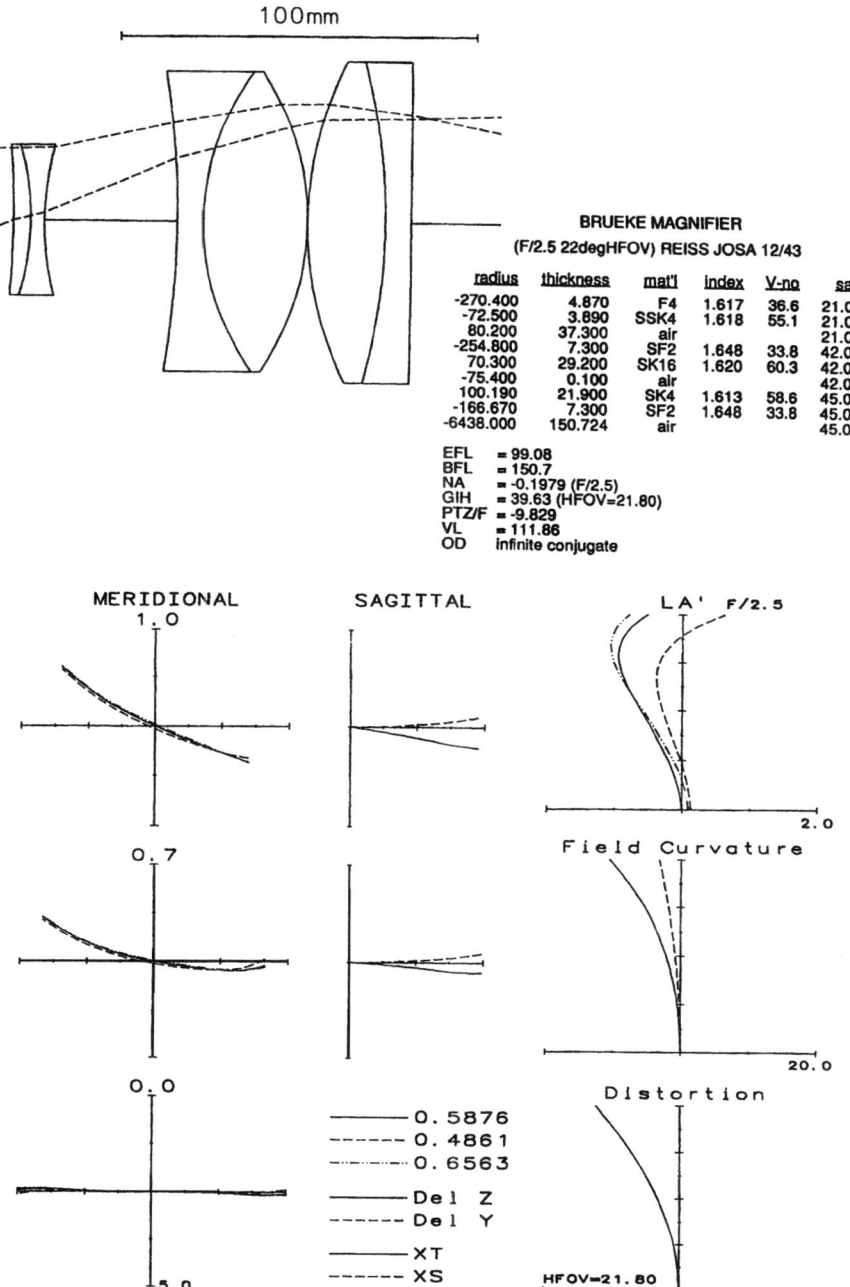

Figure 7.2 A long working-distance magnifier. This can be considered either as a reversed telephoto lens or as a galilean microscope. This type is good for higher powers.

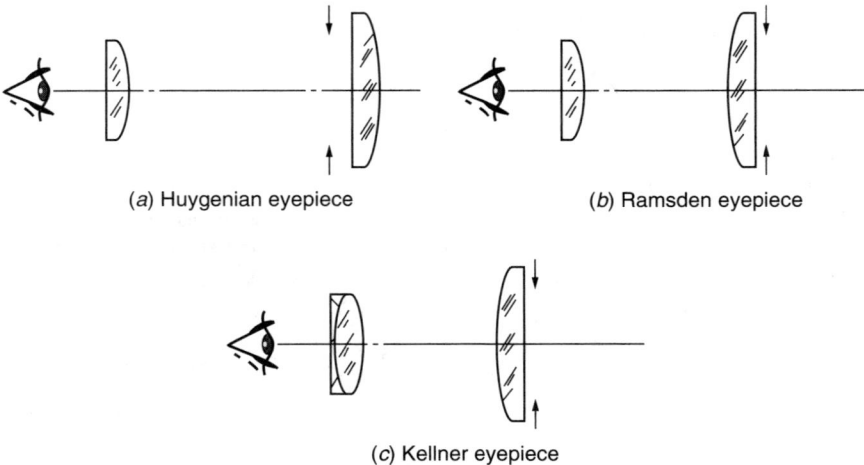

Figure 7.3 Three simple eyepieces.

7.3.1 The Huygens eyepiece

In this eyepiece (Fig. 7.3a) the lateral color is corrected by choosing a spacing such that the lateral chromatic of one element exactly balances that of the other. This spacing is approximately equal to the average of the focal lengths of the two elements. The coma is corrected by adjusting the ratio of the element powers. With the coma corrected for a given objective distance, the stop/pupil is in what is called the *natural* stop position. In a system that is undercorrected for spherical aberration, putting the stop in the *natural* coma-correcting location means that the astigmatism is as overcorrected as possible. In other words, the field curvature is as backward curving (or as little inward curving) as possible. The Huygenian lens, as shown in Fig. 7.3a, is one where the focus distance is negative and there is an internal image. The image at this location is not corrected, so this eyepiece is not suitable for use with a reticle. The eye relief is usually uncomfortably short.

7.3.2 The Ramsden eyepiece

In the Ramsden (Fig. 7.3b) the airspace is reduced sufficiently from that of the Huygens to provide an accessible, external focal plane. The lateral color is no longer fully corrected, but the eye relief is greatly improved, and there is an accessible, corrected, image surface that permits the use of a reticle or cross hair. As in the Huygens, the coma can be corrected by adjusting the ratio of the element powers. The field lens is turned around with the plano facing outward. If desired, this orientation would allow a design with a reticle on this surface.

7.3.3 The Kellner eyepiece

This is basically a Ramsden with an achromatized eye lens. It is sometimes used in inexpensive binoculars. In the Kellner (Fig. 7.3c) it is not uncommon for the components to depart from the plano-convex form. The doublet eye lens not only reduces the lateral color, but the index break across the cemented surface can be used as a degree of freedom. This can be effective in controlling the astigmatism since a divergent surface is usually the source for overcorrected astigmatism. A barium crown positive element can provide an index break to the order of $\delta n = 0.075$.

Table 7.1 lists the *relative* characteristics of these three eyepieces. Note well that these are typical values and that glass thickness and the like can modify this data. The table also lists data taken from the lens designs given in MIL-141,[7] which were designed for a 10x scope (and which apparently incorporated some compromises in balancing the aberrations).

TABLE 7.1 The Relative Characteristics of the Three Simple Eyepieces Shown in Fig. 7.3

	Huygens	Ramsden	Kellner
Relative			
Spherical aberration	1.0	0.2	0.2
Axial chromatic	1.0	0.5	0.2
CDM* = Lateral chromatic/h	0.00	0.01	0.003
Distortion	1.0	0.5	0.2
Coma	0.0	0.0	0.0
Field curvature (Petz)	1.0	0.7	0.7
Eye relief	1.0	1.5–3.0	1.5–3.0
ϕ_e/ϕ_f (low power)	2.3	1.4	0.8
ϕ_e/ϕ_f (high power)	1.3	1.0	0.7
Field	±15°	±15°	±18–20°
Eyepieces from MIL-141 (Ref. 7)			
Spherical aberration	1.0	0.23	0.20
Axial chromatic	1.0	0.64	0.15
CDM*	0.0	0.010	0.007
Distortion	1.0	0.5	0.4
Coma (OSC)	0.0	0.0023	0.0003
Field curvature (Petz)	1.0	0.64	0.66
Eye relief	1.0	4.1	2.5
$\phi_e/\phi_f^\dagger$ (10x)	1.41	1.13	1.19

*Chromatic difference of magnification.
†ϕ_e and ϕ_f are the powers of the eye lens and field lens, respectively.

Another simple eyepiece, which may be viewed as a modification of the Kellner, consists of a closely spaced singlet and doublet. It is quite common in binoculars. It can be found either with the doublet as the eye lens or as the field lens. As a field lens, the doublet is sometimes made quite thick and meniscus, close and concave to the focal plane. This arrangement is beneficial in its effect on both the Petzval curvature and the eye relief. The design exercise in Sec. 7.4 of this chapter is based on this configuration. Higher-index glasses are sometimes used.

7.4 Design Story of an Eyepiece for a 6 × 30 Binocular

This project can be a surprisingly complex one. The reason is that there are not sufficient variables in a simple eyepiece to do a good job of controlling all of the aberrations. Efforts to flatten the Petzval field (by a negative field lens or an Erfle-type construction, for example) result in more costly, larger diameter elements. In addition, the objective lens, because it is thin and at the aperture stop, will have both a negative astigmatism and a negative Petzval curvature. If a reticle must be imaged sharply there is an obvious conflict between producing a flat eyepiece field and balancing the inward-curving objective field with a large amount of overcorrected astigmatism (which is, in and of itself, a conflict).

Complicating the problem is the fact that the image of the aperture stop (the objective) is the exit pupil of the binocular, which must be a sufficient distance from the final eyepiece element to allow a comfortable clearance for the eye (the eye relief). This total lack of front-to-back symmetry about the stop means that we cannot count on symmetry to reduce the distortion, coma, or lateral color. So the design of a simple eyepiece usually devolves into establishing the least undesirable balance of the aberrations.

We take our guidelines from the external dimensions of an inexpensive pair of binoculars labeled "6 × 30 8° field." We measure the eye relief with a scale at a very approximate 15 mm. The eye lens clear diameter is about 15 mm. From a crude tracing of the measured external binocular dimensions we conclude that the objective focal length must be about 150 mm; the eyepiece focal length is thus 150/6 = 25 mm, and the Porro no. 1 type erecting prisms[23] have a total glass path of about 120 mm.

We elect to use an eyepiece type that has largely replaced the classic Kellner in inexpensive binoculars—a compact *inverted Kellner*. This form consists of a singlet eyelens plus a cemented doublet with the flint element facing the objective. The Edmund RKE eyepiece and the Konig USP 1,159,233 are of this general type. In any eyepiece with cemented components, and especially in the simple ones, the index difference or break between the cemented crown and flint elements can be quite significant in controlling the astigmatism and distortion. We look up a few designs of this type and find that the average index break is about 0.16. On the basis that we expect a higher index to produce a better design, we elect to use a medium barium crown SK4 (613586) for the crowns and dense flint SF11 (785258) for the flint. Both are preferred types, low in price, and fairly stable. This gives us an index break of 0.172.

This type of eyepiece can be expected to have a longer eye relief than the 15 mm of the binoculars we used as a model. A field of 8° (±4°) at the objective means an angular field at the eyepiece of $\pm\arctan(6 \times \tan 4°)$; this works out to about ±22.8°. If the eye relief is about 25 mm, we will need a clear semiaperture of about 13 mm, so we assume a 14 mm semiaperture for the eyepiece elements to start.

For a starting point we choose a plano-convex eye lens and a plano-convex doublet with an equi-convex crown. We start with all radii equal to 25 mm and guess at some thicknesses. We adjust the radius of the singlet eye lens to 29.5 mm to get a focal length of 25 mm. The entrance beam radius is 2.5 mm (= 0.5 × 30 mm/6) and the field angle is ±22.8°. We must include a block of glass to represent the prisms; they significantly affect the aberrations of the binocular (by overcorrecting the spherical, axial color, and astigmatism, and undercorrecting the coma, distortion, and lateral color.) Our rough starting layout is shown in Table 7.2 and Figs. 7.4 and 7.5.

Note that the through-focus *modulation transfer function* (MTF) plot of Fig. 7.5 neatly and separately indicates the field curvature, astigmatism, and image quality.

Surface no. 1 is the exit pupil of the binocular. We will follow the usual practice of designing and raytracing from the long conjugate to the short, so surface no.1 will be our entrance pupil and aperture stop. Surfaces 8 and 9 represent the block of BK7 glass of the prism system. The equivalent air path of $T8$ (equal to $t/n = 120/1.5168$) plus airspaces $T7$ and $T9$ add up to 150, the objective focal length.

TABLE 7.2 The Rough Starting Layout for the 25 mm Binocular Eyepiece, as Shown in Figs. 7.4 and 7.5

$R1$ = plano	$T1$ = 21.4	(aperture stop)
$R2$ = plano	$T2$ = 8.0	SK4 (613586)
$R3$ = −29.5	$T3$ = 1.0	
$R4$ = +25.0	$T4$ = 10.0	SK4 (613586)
$R5$ = −25.0	$T5$ = 3.0	SF11 (785258)
$R6$ = plano	BFL = 15.885	(solved for paraxial focus)
$R7$ = plano	$T7$ = 10.0	
$R8$ = plano	$T8$ = 120.0	BK7 (517642) (the prism)
$R9$ = plano	$T9$ = 60.89	
$R10$ = plano	$T10$ = −150.008	(solved for paraxial focus)
$R11$ = image		

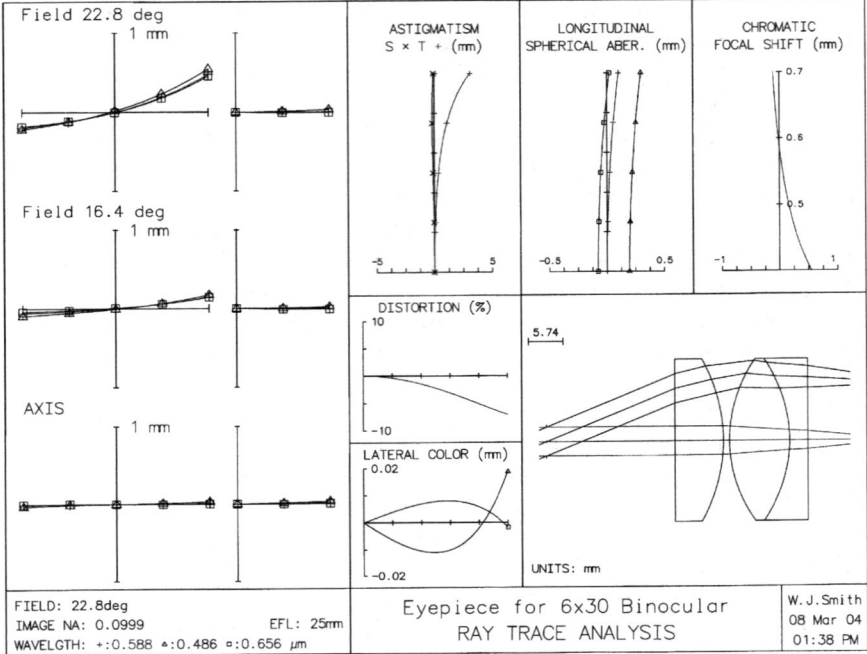

Figure 7.4 The ray aberrations of the starting eyepiece design, Table 7.2.

Surprisingly (or perhaps not), our rough guesses have given us a fair lens. The distortion is a bit large at −7 percent, the lateral color is well balanced (with +0.02 mm at the edge of the field and −0.02 at the zone), but the coma is quite large at about +0.14 mm, and the tangential field is about +3.1 mm over corrected, with a flat sagittal field. At 21.4 mm the eye relief is a little short of our expectations, and the crown edge thicknesses need adjustment. But all in all, this does demonstrate that some rough guesses, or perhaps even a pickup collection of elements thrown together, can make a reasonable eyepiece.

We now allow $R2$, $R3$, $R4$, $R5$, and $R6$ to vary, as well as $T1$ (the eye relief), $T2$, and $T4$ (the crown thicknesses). We set up a simple merit function to allow us to control and balance the system characteristics. Our custom merit function and targets are:

1. The eye relief ($T1$) = 24 mm (later 25 mm)
2. The principal ray height at the objective ($R10$) = 0
3. The focal length = 25 mm
4. The edge thickness of crown 2-3 = 2.0 mm

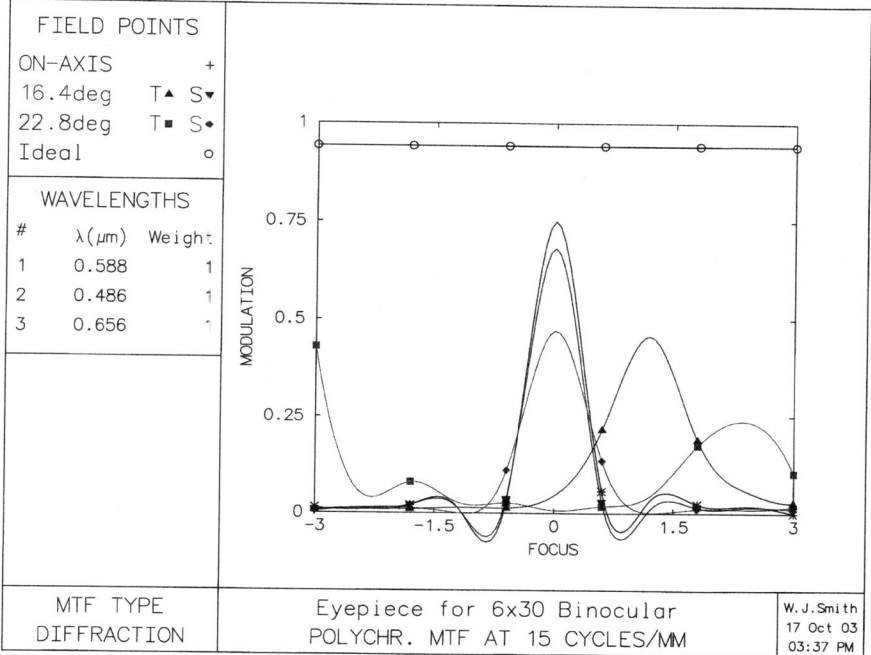

Figure 7.5 The through-focus MTF of Fig. 7.4, Table 7.2.

5. The edge thickness of crown 4-5 = 2.0 mm
6. The lateral chromatic = 0
7. The distortion = −7%
8. The coma = 0
9. The sagittal field curvature = −0.7 mm
10. The tangential field curvature = +2.0 mm
11. The astigmatism = +2.5 mm

The optimization, by *damped least squares* (DLS), was allowed to proceed only one iteration at a time. At the start and then after each iteration, the targets for eye relief, distortion, and field curvature were adjusted to values that seemed suitable and achievable, based on a consideration of the immediately preceding raytrace analysis. After four iterations, the weights were also adjusted to emphasize the troublesome aberrations. The benefit of this gradual, *adaptive* approach is that it tends to prevent the DLS process from chasing after wildly unsuitable design forms.

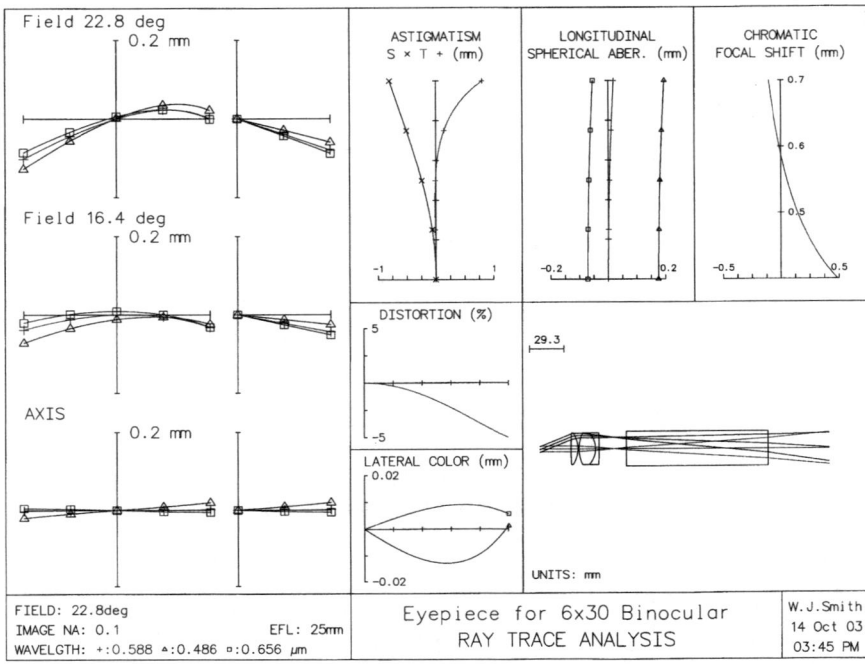

Figure 7.6 The eyepiece aberrations after the first run, Table 7.3.

After the fourth iteration the weights on Y_{pr} at $R10$, lateral color, and coma were doubled. After the sixth, the eye relief ($T1$) was dropped as a variable and fixed at 25 mm. The distortion target was gradually reduced to −5 percent, the X_s to −0.6 mm, the X_t to +0.7 mm, and the astigmatism to −1.6 mm. Finally, when things seemed to have stablized, the distortion target was set to zero with a 75x weight, the weights on coma and lateral color were increased to 5x, the edge thicknesses were freed, and the DLS iterations were allowed to continue until there was no change. The resulting lens is shown in Figs. 7.6 and 7.7 and seems to be quite a reasonable design. The design data is shown in Table 7.3.

7.4.1 Other possibilities—Equi-convex crowns

The well-regarded Edmund RKE eyepiece is distinguished by the fact that both of its crown elements are equi-convex. This feature can yield some economies in fabrication, so we decide to investigate this form. We stick with the SK4 and SF11 glasses, and use minus curvature pickups to make the crowns equi-convex. Using our very first starting layout and

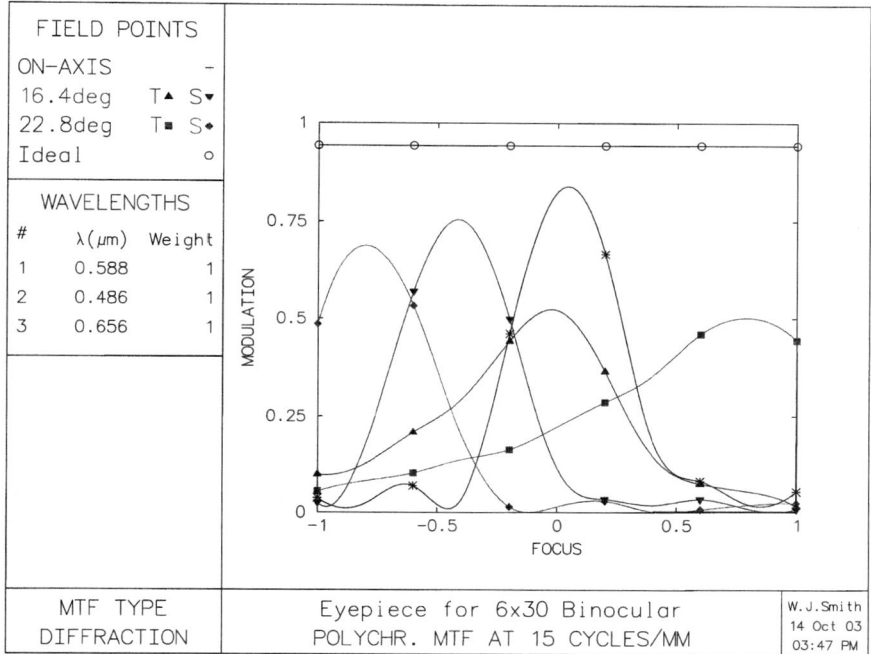

Figure 7.7 The MTF of Fig. 7.6, Table 7.3.

making its eye lens equi-convex, we do a bit of juggling and get the version as shown in Table 7.4.

As is apparent from Figs. 7.8 and 7.9, this start is not quite as fortuitous a beginning as the starting form we originally tried. The coma, astigmatism, and lateral color are acceptable, but the eye relief is short, and the distortion is a deal breaking −10.7 percent. We follow the procedure used in the preceding design, adjusting the merit function

TABLE 7.3 The Prescription for the Initial Optimized 23 mm Eyepiece Design, as Shown in Figs. 7.6 and 7.7

$R1$ = plano	$T1$ = 25.0	(stop)	
$R2$ = +232.155	$T2$ = 6.0	SK4	613586
$R3$ = −26.0616	$T3$ = 1.0		
$R4$ = +33.7713	$T4$ = 14.0	SK4	613586
$R5$ = −20.9068	$T5$ = 3.0	SF11	7785258
$R6$ = −290.362	BF = 13.0655	(solved)	
$R7$ = plano	$T7$ = 10.0		
$R8$ = plano	$T8$ = 120.0	BK7	517642
$R9$ = plano	$T9$ = 60.89		
$R10$ = plano	$T10$ = −150.0039	(solved)	
$R11$ = image			

TABLE 7.4 Rough Starting Point for an Eyepiece with Equi-Convex Crown Elements (See Figs. 7.8 and 7.9)

$R1$ = plano	$T1$ = 21.1	(stop)
$R2$ = +54.4	$T2$ = 8.0	SK4 613586
$R3$ = −54.4	$T3$ = 1.0	
$R4$ = +25.0	$T4$ = 10.0	SK4 613586
$R5$ = −25.0	$T5$ = 3.0	SF11 785258
$R6$ = plano	$T6$ = 14.479	(solved)
$R7$ = plano	$T7$ = 10	
$R8$ = plano	$T8$ = 120.0	BK7 517642
$R9$ = plano	$T9$ = 60.89	
$R10$ = plano	$T10$ = −150.004	(solved)
$R11$ = image		

targets after each iteration. After the fourth iteration we fix the eye relief to 25 mm, and target the distortion at zero (it has gotten to −11 percent), increasing its weight to 10x. For the next two iterations we increase the weight to 100x, then 300x, but the distortion is incorrigible. Stuck at −10.5 percent, it will not budge. We are in a bad local optimum. The other

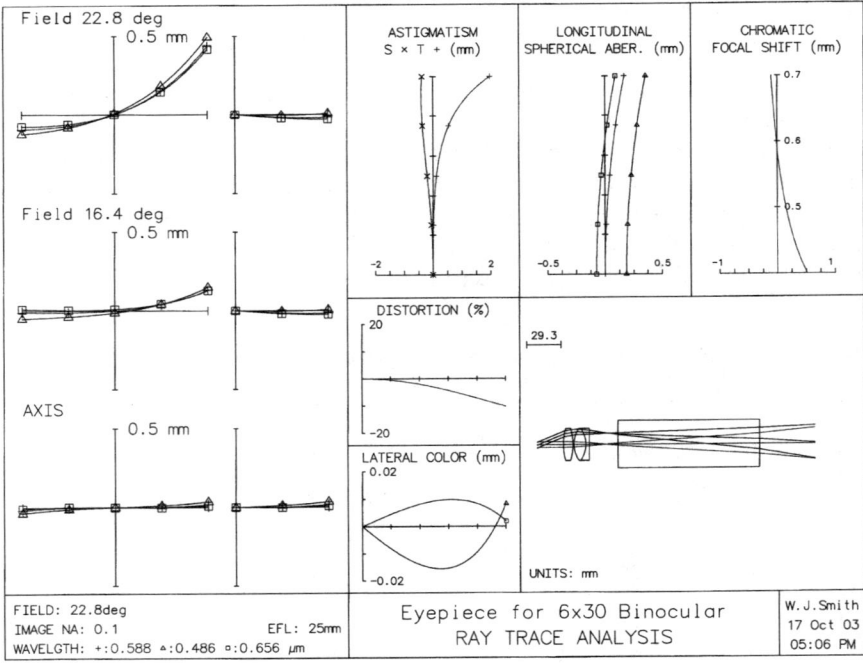

Figure 7.8 The equi-convex crown eyepiece starting form of Table 7.4.

Eyepieces and Magnifiers 167

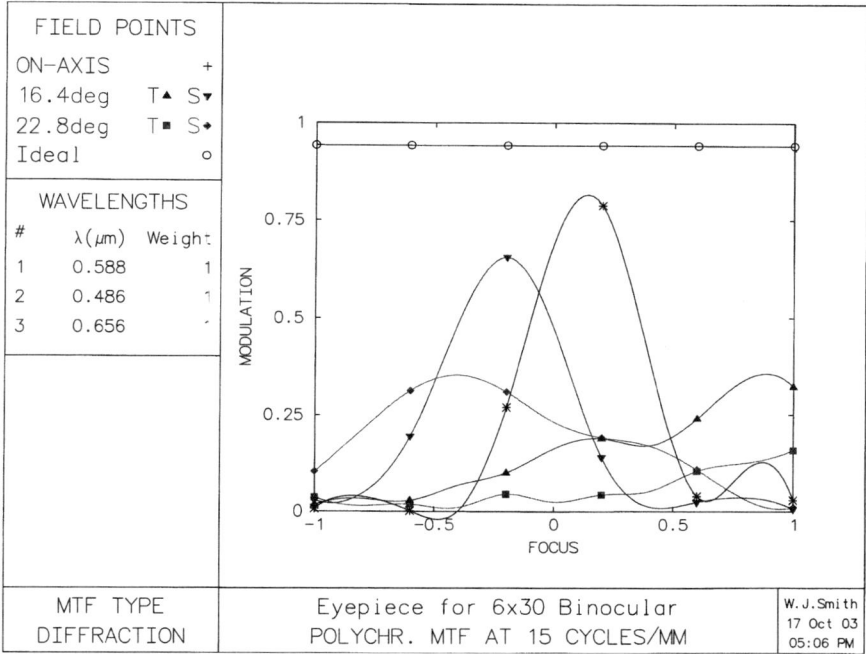

Figure 7.9 The MTF of Fig. 7.8, Table 7.4.

aberrations are OK, so we must accept the fact that (with the glasses we have chosen) the price for equi-convex crowns is heavy distortion, and a smaller field coverage might be advisable. Perhaps a different selection of glasses might help. The performance is shown in Figs. 7.10 and 7.11, and the final design data are shown in Table 7.5.

TABLE 7.5 Prescription for the Optimized Eyepiece with Equi-Convex Crowns (See Figs. 7.10 and 7.11)

$R1$ = plano	$T1$ = 25.0	(stop)	
$R2$ = +41.7316	$T2$ = 6.872	SK4	613586
$R3$ = −41.7316	$T3$ = 1.0		
$R4$ = +25.3740	$T4$ = 10.409	SK4	613586
$R5$ = −25.3740	$T5$ = 3.0	SF11	785258
$R6$ = +61.0636	$T6$ = 12.6598	(solved)	
$R7$ = plano	$T7$ = 10.0		
$R8$ = plano	$T8$ = 120.0	BK7	517642
$R9$ = plano	$T9$ = 60.89		
$R10$ = plano	$T10$ = −150.004	(solved)	
$R11$ = image			

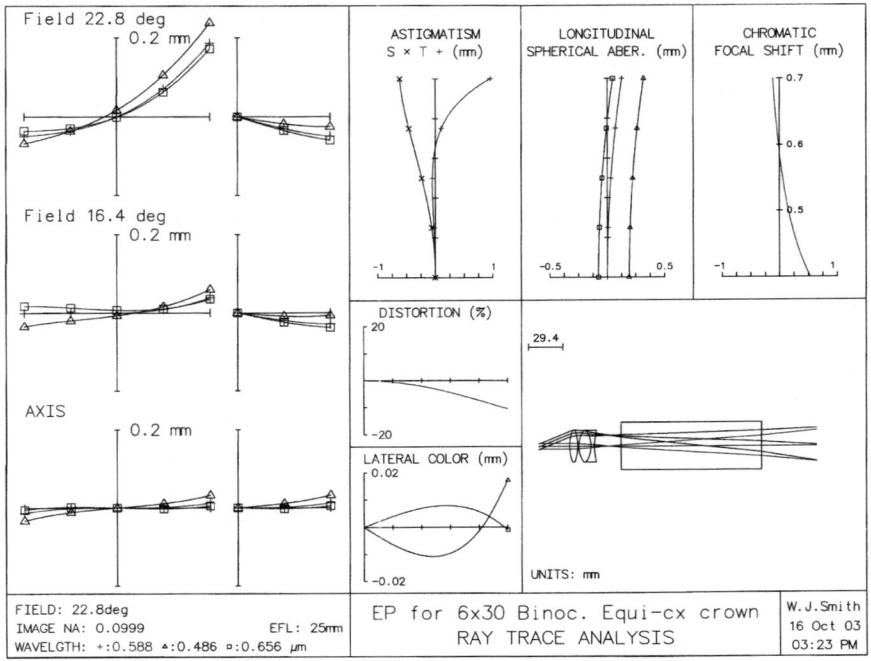

Figure 7.10 The optimized version of Figs. 7.8 and 7.9, Table 7.5.

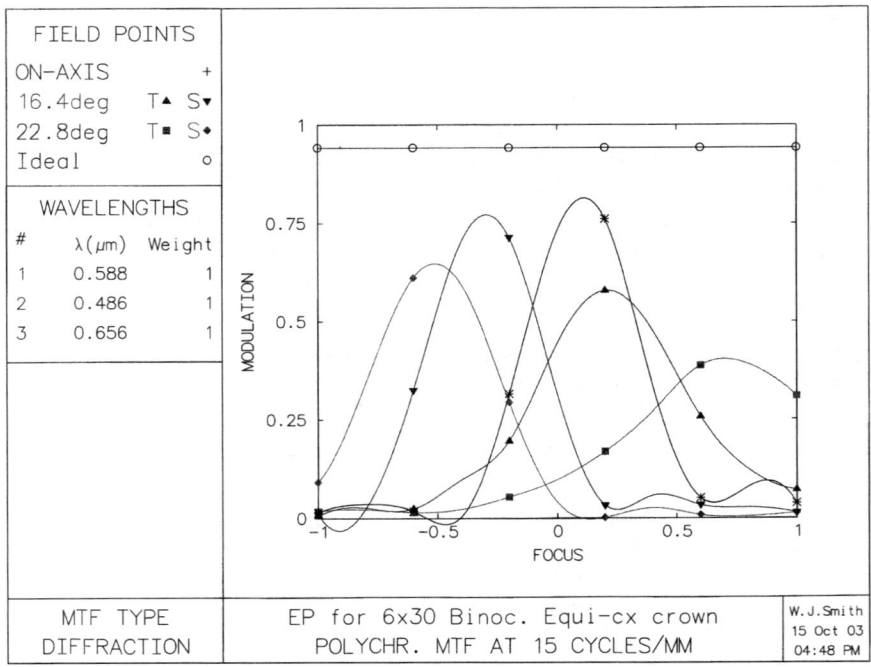

Figure 7.11 The MTF for Fig. 7.10, Table 7.5.

7.4.2 Other possibilities—plano surfaces

We next decide that while we are on an economy hunt, we should try some plano surfaces. We use the preceding design as a start, fix $R2$ as a plano and adjust $R3$ to get a 25 mm focal length. After a few iterations, we have the design below with better distortion (–6.4 percent) and the other aberrations at reasonable values. Figures 7.12 and 7.13 show the aberrations and MTF for the lens as shown in Table 7.6.

It would appear that $R3$, $R4$, and $R5$ are close enough that they could all have the same radius (to reduce tooling costs) without much difficulty.

Looking for something better, we decide to abandon the equi-convex crown idea and let both $R4$ and $R5$ vary independently. We start with the lens above. After an iteration the distortion is only –3.2 percent but the coma has grown. We reduce the distortion weight from 300x to 50x. Now the distortion is –5.0 percent and the other aberrations are acceptable, as indicated by the data shown in Table 7.7 and Figs. 7.14 and 7.15.

Again, as an interesting commercial idea: make $R3 = R5$.

The long radius on the last surface ($R6 = +366.110$) also looks to be a likely candidate for flattening, so we set $R6$ = plano. The first iteration

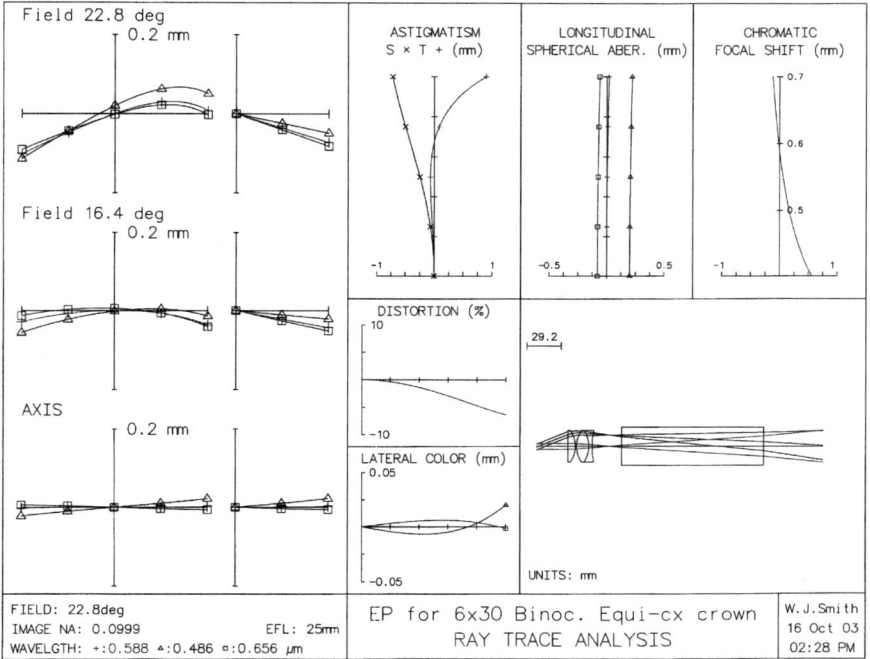

Figure 7.12 Plano first surface ($R2$) starting form, Table 7.6.

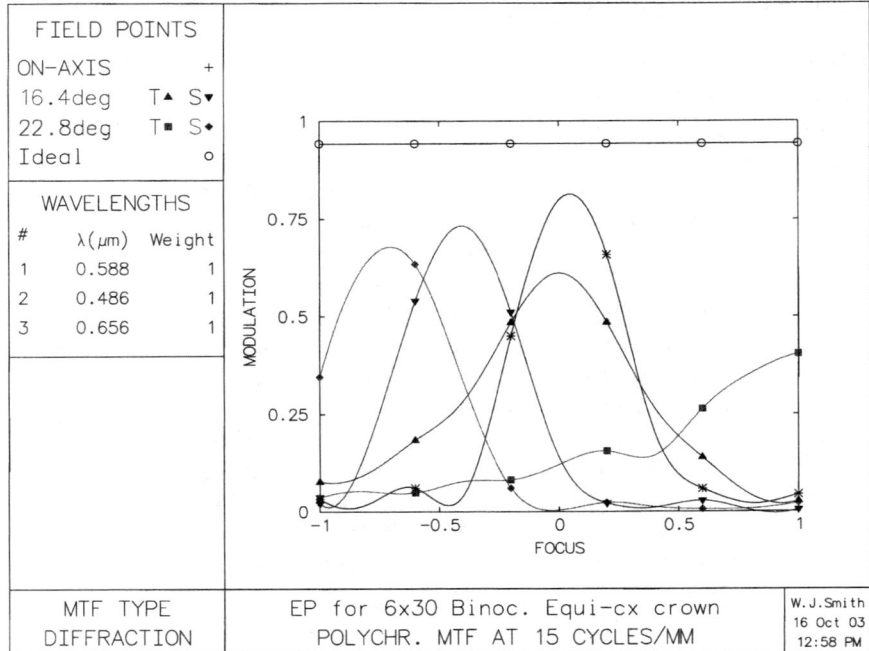

Figure 7.13 The MTF of Fig. 7.12, Table 7.6.

is promising with the distortion down to −4.5 percent, so we reduce the distortion weight from 50x to 1.0x, iterate again and get Figs. 7.16 and 7.17 for the data shown in Table 7.8.

7.4.3 Comparison with the symmetrical (Plössl) eyepiece

Finally, by way of comparison, we design a symmetrical eyepiece. This classic eyepiece consists of two identical achromatic doublets facing crown to crown. It is a simple, versatile, and widely used eyepiece.

We stay with the same glasses (SK4 and SF11) and use the same merit function as with our original effort. We soon discover that the eye relief must be shortened to about 20. The final design is shown in Table 7.9, and the aberrations and MTF are plotted in Figs. 7.18 and 7.19.

Given the robustness of the symmetrical form, it seems likely that $R2$ and $R7$ could be made plano, or perhaps the crowns could be made equiconvex, or, if we adjust the glass types to maintain achromatism, we might make a plano- and equi-convex form work. Another possibility

Eyepieces and Magnifiers

TABLE 7.6 Eyepiece Prescription for the Design wih a Plano-Convex Eyelens and an Equi-convex Crown Element in the Doublet (See Figs. 7.12 and 7.13)

$R1$ = plano	$T1$ = 25.0	(stop)	
$R2$ = plano	$T2$ = 6.478	SK4	613586
$R3$ = −24.1374	$T3$ = 1.0		
$R4$ = +25.5267	$T4$ = 10.343	SK4	613586
$R5$ = −25.5267	$T5$ = 3.0	SF11	785258
$R6$ = +106.187	$T6$ = 14.690	(solved)	
$R7$ = plano	$T7$ = 10.0		
$R8$ = plano	$T8$ = 120.0	BK7	517642
$R9$ = plano	$T9$ = 60.89		
$R10$ = plano	$T10$ = −150.004	(solved)	
$R11$ = image			

would be to abandon symmetry and to allow the two doublets to differ. As with our *compact inverted Kellner*, there are many avenues through which we may exercise our imagination.

7.4.4 Summary

The targeted characteristics of the designs above are summarized in Table 7.10 below. In the table E.R. stands for eye relief, Eth is edge thickness, Lat. Chr. is lateral chromatic, Dist. is distortion, Coma is tangential coma, X_s and X_t are the sagittal and tangential field curvatures, Astig. is the difference between the two, and $\Sigma|cv|$ is the sum of the (absolute) surface curvatures.

With the exception of the figures in boldface, these designs are pretty much the same. Thus one might choose to go with the "2plcx" column design with two plano surfaces and the lowest total curvature.

TABLE 7.7 Eyepiece Prescription for an Improved Lens Derived from Table 7.6, Allowing R4 and R5 to Vary Independently (See Figs. 7.14 and 7.15)

$R1$ = plano	$T1$ = 25.0	(stop)	
$R2$ = plano	$T2$ = 6.543	SK4	613586
$R3$ = −23.7440	$T3$ = 1.0		
$R4$ = +29.0817	$T4$ = 10.438	SK4	613586
$R5$ = −22.6787	$T5$ = 3.0	SF11	785258
$R6$ = +366.110	$T6$ = 15.157	(solved)	
$R7$ = plano	$T7$ = 10.0		
$R8$ = plano	$T8$ = 120.0	BK7	517642
$R9$ = plano	$T9$ = 60.89		
$R10$ = plano	$T10$ = −150.004	(solved)	
$R11$ = image			

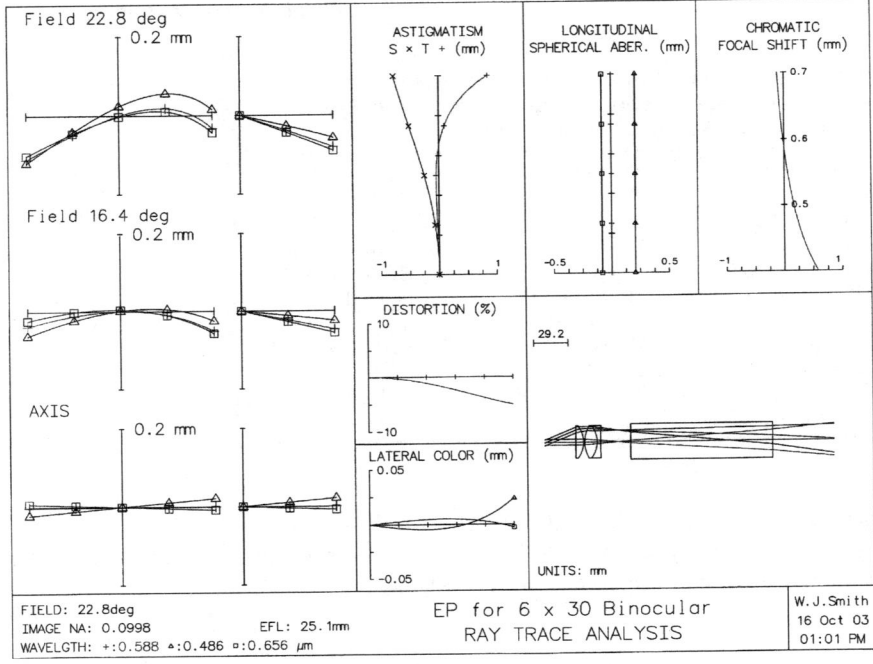

Figure 7.14 The Plano first surface eyepiece after optimization, Table 7.7.

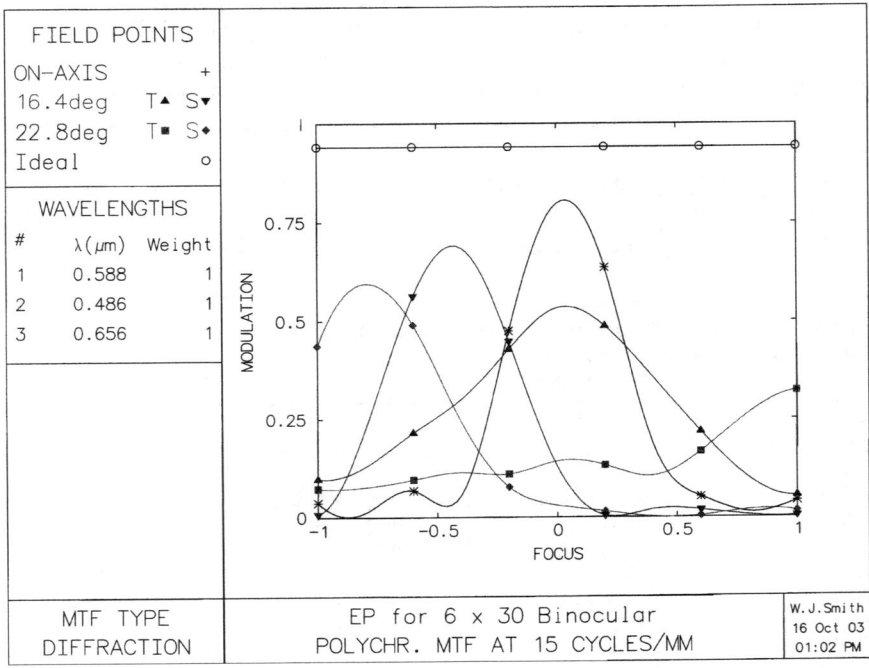

Figure 7.15 The MTF of Fig. 7.14, Table 7.7.

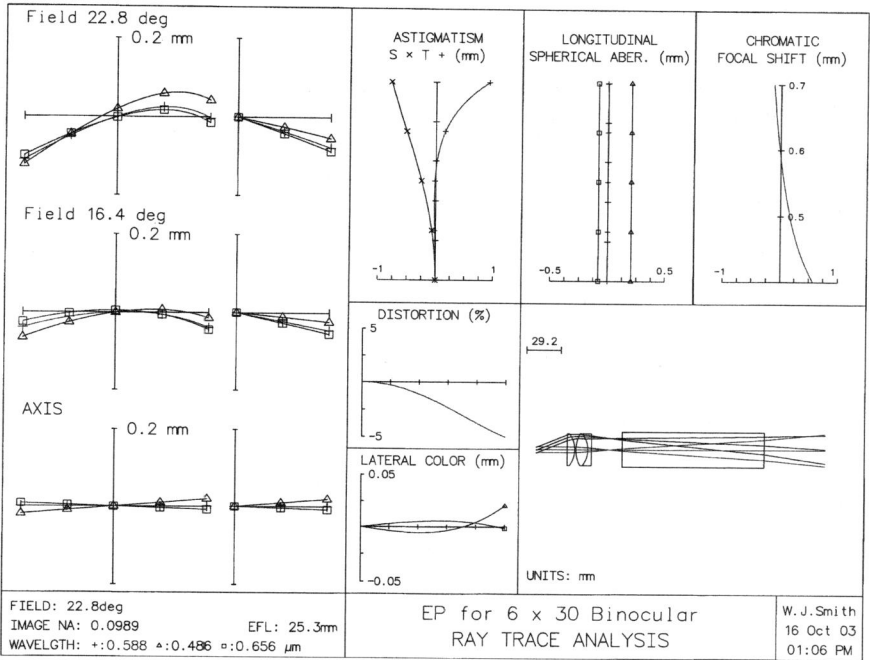

Figure 7.16 Plano surfaces on $R2$ and $R6$, optimized, Table 7.8.

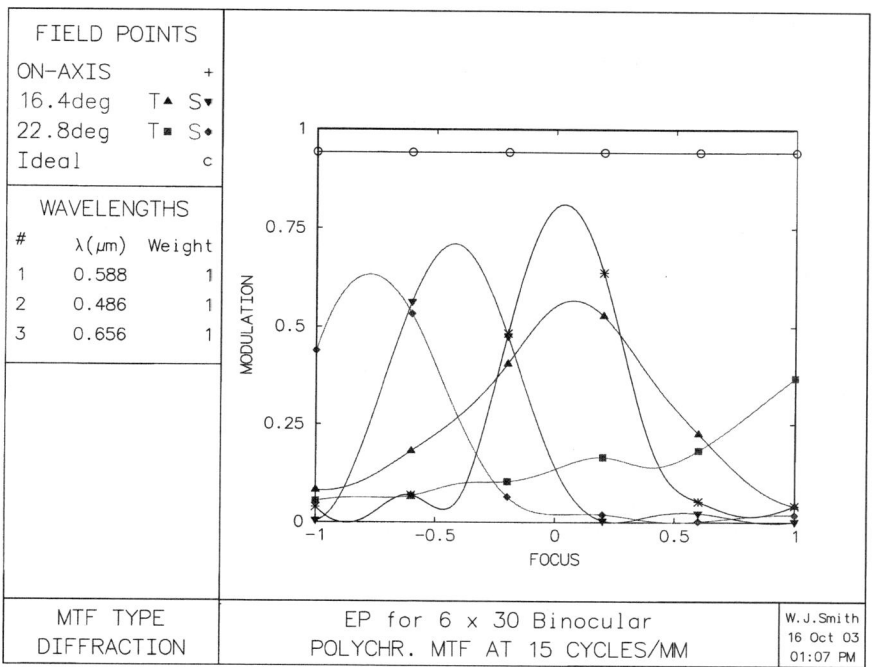

Figure 7.17 The MTF for Fig. 7.16, Table 7.8.

TABLE 7.8 Prescription for the Eyepiece with Plano Forms of Both the Eyelens and the Doublet (See Figs. 7.16 and 7.17)

$R1$ = plano	$T1$ = 25.0	(stop)	
$R2$ = plano	$T2$ = 6.334	SK4	613586
$R3$ = −24.1819	$T3$ = 1.0		
$R4$ = +30.8199	$T4$ = 10.137	SK4	613586
$R5$ = −23.0555	$T5$ = 3.0	SF11	785258
$R6$ = plano	$T6$ = 15.861	(solved)	
$R7$ = plano	$T7$ = 10.0		
$R8$ = plano	$T8$ = 120.0	BK7	517642
$R9$ = plano	$T9$ = 60.89		
$R10$ = plano	$T10$ = −150.004	(solved)	
$R11$ = image			

7.4.5 Final comments

This will conclude our excursion into this eyepiece form. But note well that, *inter alia*, we have left totally unexplored the possible effects of:

a. The eye relief
b. The element thicknesses
c. The airspace
d. Changing glasses to higher-index types
e. An aspheric surface or two
f. Reversing the doublet
g. Reversing the entire eyepiece

Among other material factors that affect the design, we have assumed that a 5 percent distortion was acceptable and that minimum X_s and X_t values were desirable. This latter is a compromise

TABLE 7.9 Symmetrical (Plössl) Binocular Eyepiece, as Shown in Figs. 7.18 and 7.19

$R1$ = plano	$T1$ = 19.598	(stop)	
$R2$ = +92.6992	$T2$ = 2.0	SF11	785258
$R3$ = +33.1098	$T3$ = 8.101	SK4	613586
$R4$ = −34.2162	$T4$ = 1.0		
$R5$ = +34.2162	$T5$ = 8.101	SK4	613586
$R6$ = −33.1098	$T6$ = 2.0	SF11	785258
$R7$ = −92.6992	$T7$ = 10.0		
$R8$ = plano	$T8$ = 120.0	BK7	517642
$R9$ = plano	$T9$ = 60.89		
$R10$ = plano	$T10$ = −150.004	(solved)	
$R11$ = image			

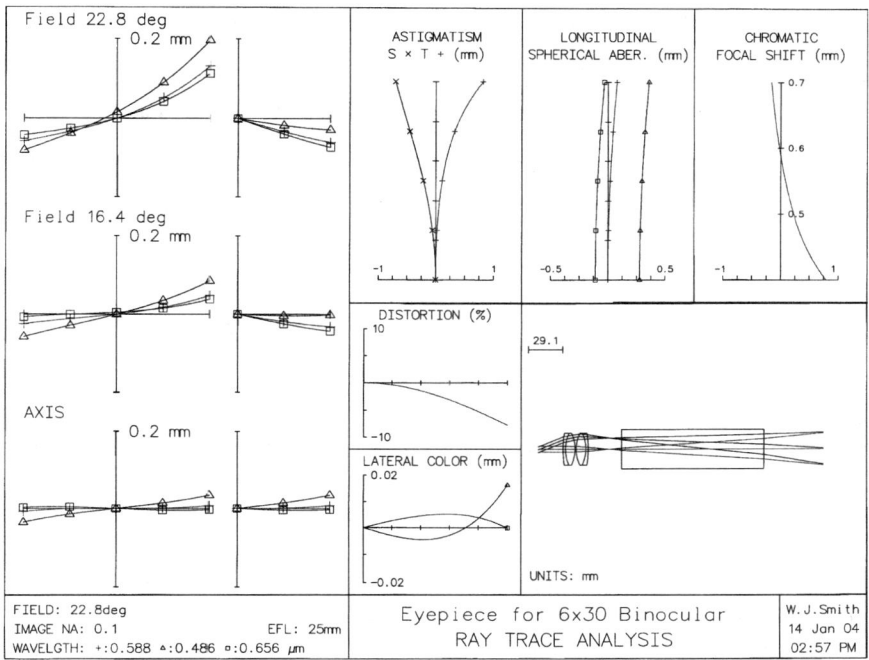

Figure 7.18 Symmetrical or Plössl eyepiece aberrations, Table 7.9.

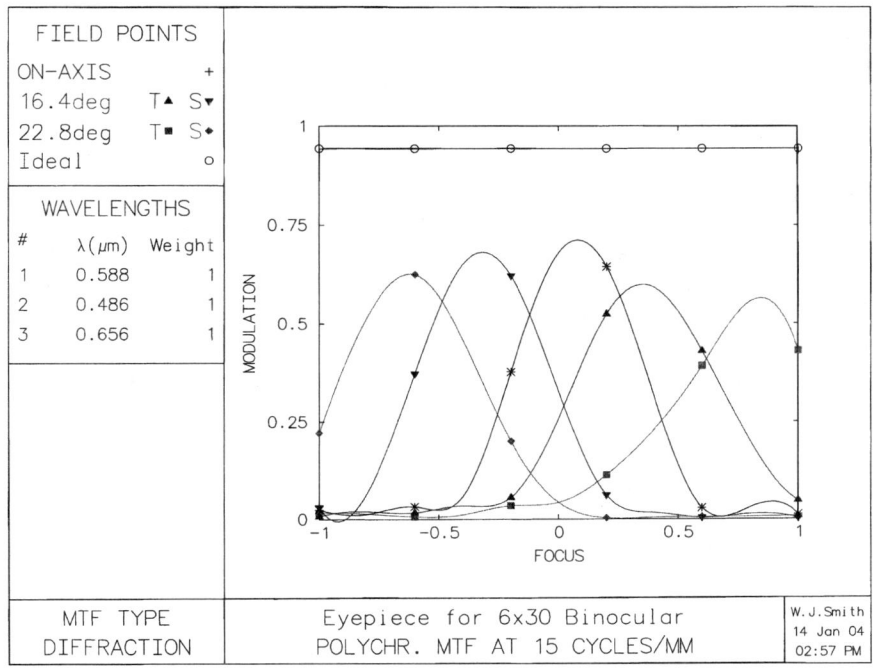

Figure 7.19 The MTF for Fig. 7.18, Table 7.9.

TABLE 7.10 Table Comparing the Eyepiece Characteristics of the Designs Undertaken in Sec. 7.5 (The Boldface Entries Indicate the Values which are not Satisfactory)

Fig.	7.4	7.6	7.10	7.12	7.14	7.16	7.18
Type	start	base	2eqcx.	1eqcx	1plcx	2plcx	**Symm**
E.R.	21.4	25	25	25	25	25	**19.6**
Eth 2, 3	**4.47**	1.5	2.04	2.00	1.98	1.87	2.0
Eth 4, 5	**1.43**	5.6	1.99	1.98	2.01	2.04	2.0
Lat. Chr.	+.010	−.005	+.018	+.022	+.026	+.022	+.017
Dist.	**−7.0%**	−5.0%	**−10.5%**	**−6.4%**	−5.0%	−5.0%	**−7.7%**
Coma	**+.142**	−.050	+.069	−.047	−.072	−.057	+.037
X_s	−.09	−.80	−.62	−.72	−.78	−.77	−.70
X_t	**+3.1**	+.80	+.64	+.90	+.85	+.92	+.82
Astig.	**+3.18**	+1.61	+1.57	+1.62	+1.63	+1.70	+1.52
$\Sigma\|cv\|$	0.114	0.124	0.143	0.129	0.123	0.117	0.140

between minimizing the astigmatism and setting X_s equal to zero. With regard to the field curvatures, we are faced with the fact that the 150 mm objective will have a Petzval field curvature of $X_{\text{ptz}} \approx -0.27$ mm and a thin lens AC = −0.37 mm. The objective field curvatures are thus $X_s \approx -0.63$ mm and $X_t \approx -1.37$ mm. Our eyepiece has a Petzval curvature X_{ptz} of about −1.6 mm. To match the objective field curvature, the eyepiece tangential field curvature would have to be $X_t = +1.37$, and its astigmatism would be two-thirds of the difference, or 0.67(1.6 + 1.37), or about + 2.0 mm. Most of the designs above have an astigmatism of about + 1.6 mm, which seems a reasonable compromise value.

From all this one might conclude that eyepiece design is an endless swamp. To some extent this is true; however, we should realize that designs are often concluded when an acceptable form is located, and that, while an academic exploration of the entire design volume is a lot of fun, it is seldom actually pursued to completion. But what we have seen is a simple example of what might be called adjustable or adapting optimization, where the targets and weightings in the merit function are not fixed but are adjusted as the design develops, based on the previous progress and what seems reasonably possible. This is a technique often used by experienced designers and tends to produce reasonable and steady progress toward the final result.

7.5 Four-Element Eyepieces

The first of these, Fig. 7.20, is often used as a long-eye-relief microscope eyepiece. It can also be executed with identical meniscus singlets and/or with an equi-convex crown in the cemented doublet. The glasses here are typical, although SF1 (717-293) is often used for the flint.

Eyepieces and Magnifiers 177

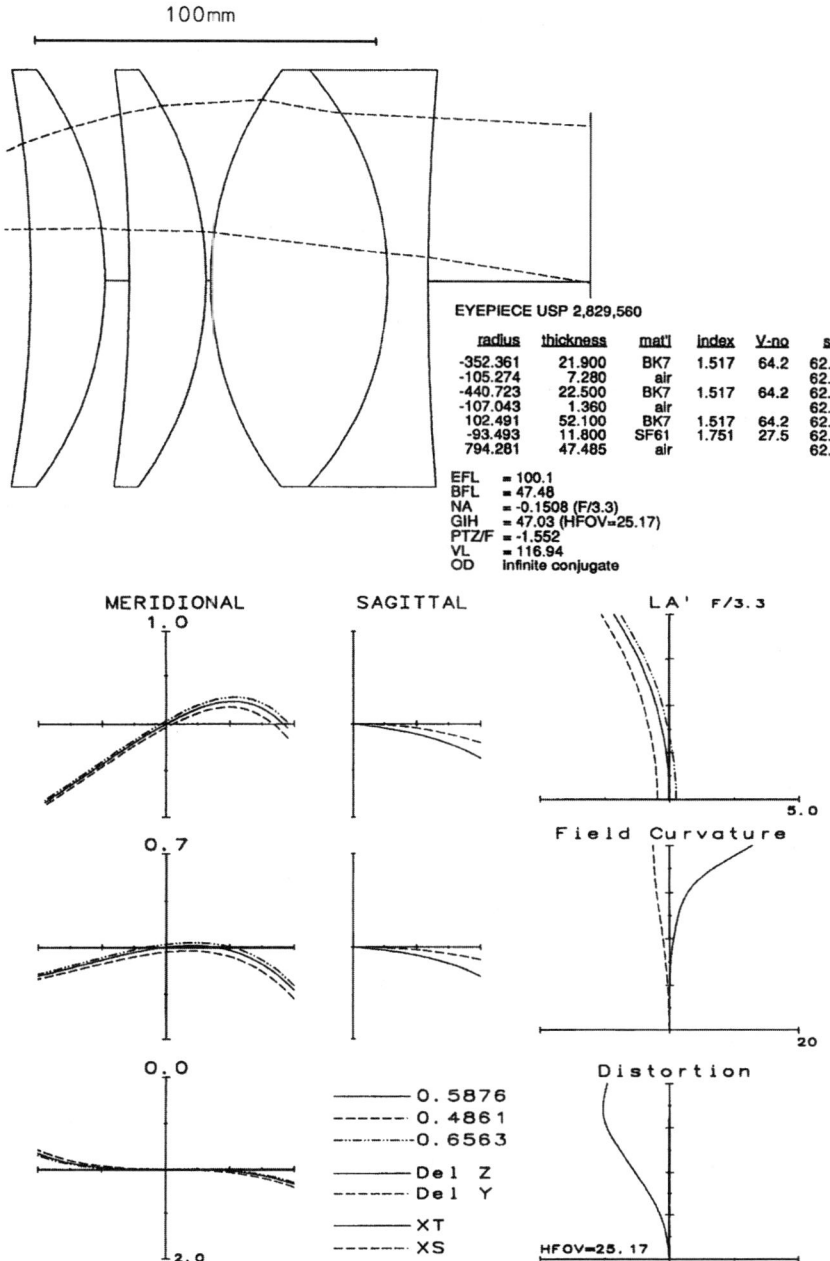

Figure 7.20 A four element 50° eyepiece.

The classical orthoscopic eyepiece is shown in Fig. 7.21. This construction is quite typical; a plano-convex eye lens of a light barium crown (or a light flint) and a symmetrical cemented triplet. The orthoscopic is noted for its freedom from distortion, although this example taken from MIL-141 is obviously not outstanding in this regard (Ref. 7). Like Fig. 7.20, it has a long eye relief. A heavier flint and BK7 in the triplet might help.

The eyepiece of Fig. 7.22 can be regarded as a simplified Erfle eyepiece. The thick meniscus doublet helps the Petzval curvature, but is not as effective in correction of the lateral chromatic as the doublet in Fig. 7.20.

The next five lenses are examples of the *symmetrical* or *Ploessl* eyepiece. This is an excellent, versatile, general-purpose, medium-field eyepiece. It has a long eye relief and is relatively insensitive to pupil shift. The angular field coverage is usually limited by higher-order overcorrected astigmatism, which causes the tangential field to become very strongly backward curving toward the edge of the field. Of course, some overcorrected astigmatism is necessary to offset the inward Petzval curvature; the amount is determined by both the shape of the components and the index break at the cemented surfaces. This form (two facing achromats) is extremely tolerant of modification; almost anything goes!

Figure 7.23 is a classic example, executed in BK7 (517-642) and SF12 (648-338) glasses. Figure 7.24 utilizes higher-index glasses (SK1, 610-567, and SF61, 751-275) to achieve a modestly improved performance. Figure 7.25 uses the classical glasses, but features economical equiconvex crown elements. Figures 7.26 and 7.27 are departures from strict symmetry: the first keeps a symmetrical arrangement of the glass types, but has plano-concave flints and one equi-convex crown; the second uses four different glasses, but manages identical doublets both equiconvex crowns and plano-concave flints. This design would call for very careful production control in the shop to avoid mixing the elements, unless the doublets are made to different diameters.

The symmetrical eyepiece is simple, versatile, and robust, with a long eye relief. One can take almost any pair of achromats, combine them face to face and get a reasonable eyepiece. In fact, the achromats don't even have to be a symmetrical pair, or even be oriented face to face to find something useful.

The last eyepiece of this section (Fig. 7.28) is composed of two identical, nearly plano-convex doublets. The design data for the pupil position is obviously incorrect, as can be seen by the path of the principal ray in the drawing (obviously the eye relief should be longer, i.e., ≈80), but this configuration, as well as a version with both elements reversed, is often a useful design form for rifle scopes.

Eyepieces and Magnifiers 179

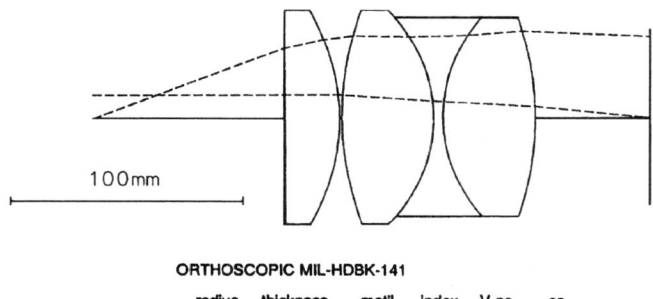

ORTHOSCOPIC MIL-HDBK-141

radius	thickness	mat'l	index	V-no	sa
	82.560	air			10.0
	23.280	BAK1	1.572	57.5	40.0
-90.950	1.100	air			46.0
129.490	39.680	KF3	1.515	54.7	46.0
-63.690	4.050	F3	1.613	37.0	43.0
63.690	39.680	KF3	1.515	54.7	43.0
-129.490	49.185	air			43.0

EFL = 100.1
BFL = 49.19
NA = -0.1001 (F/5.0)
GIH = 36.44 (HFOV=20.00)
PTZ/F = -1.257
VL = 190.35
OD infinite conjugate

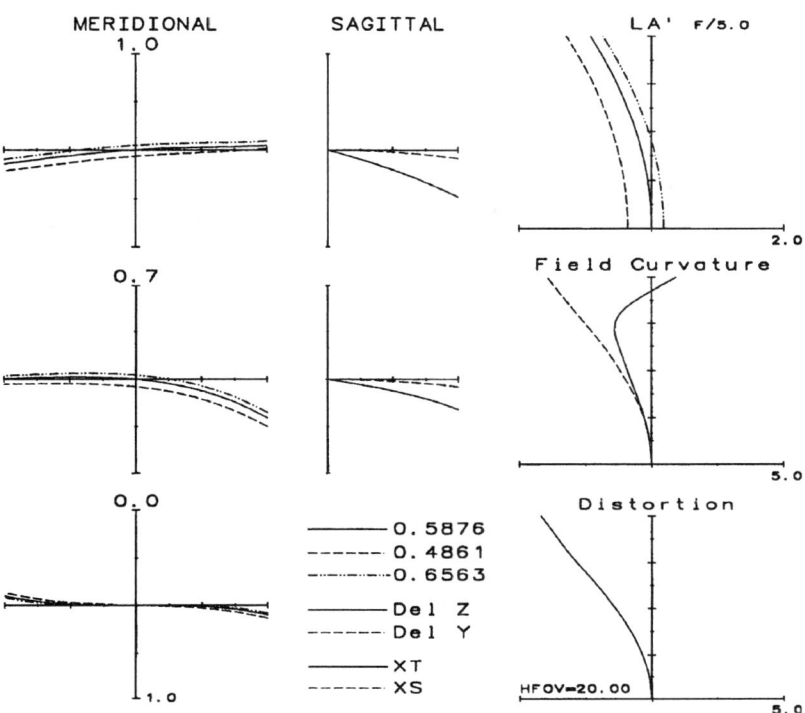

Figure 7.21 Orthoscopic eyepiece.

180   Chapter Seven

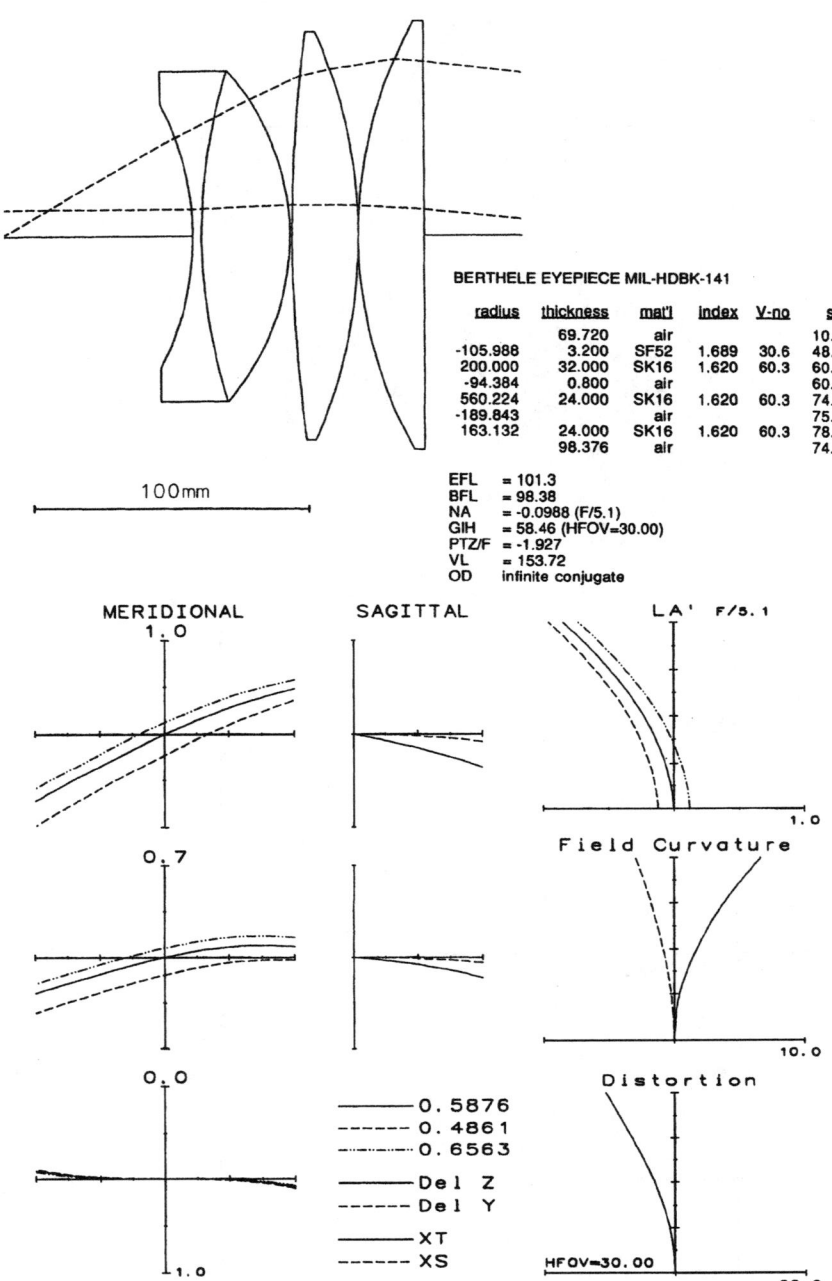

Figure 7.22   Berthele 60° eyepiece.

Eyepieces and Magnifiers 181

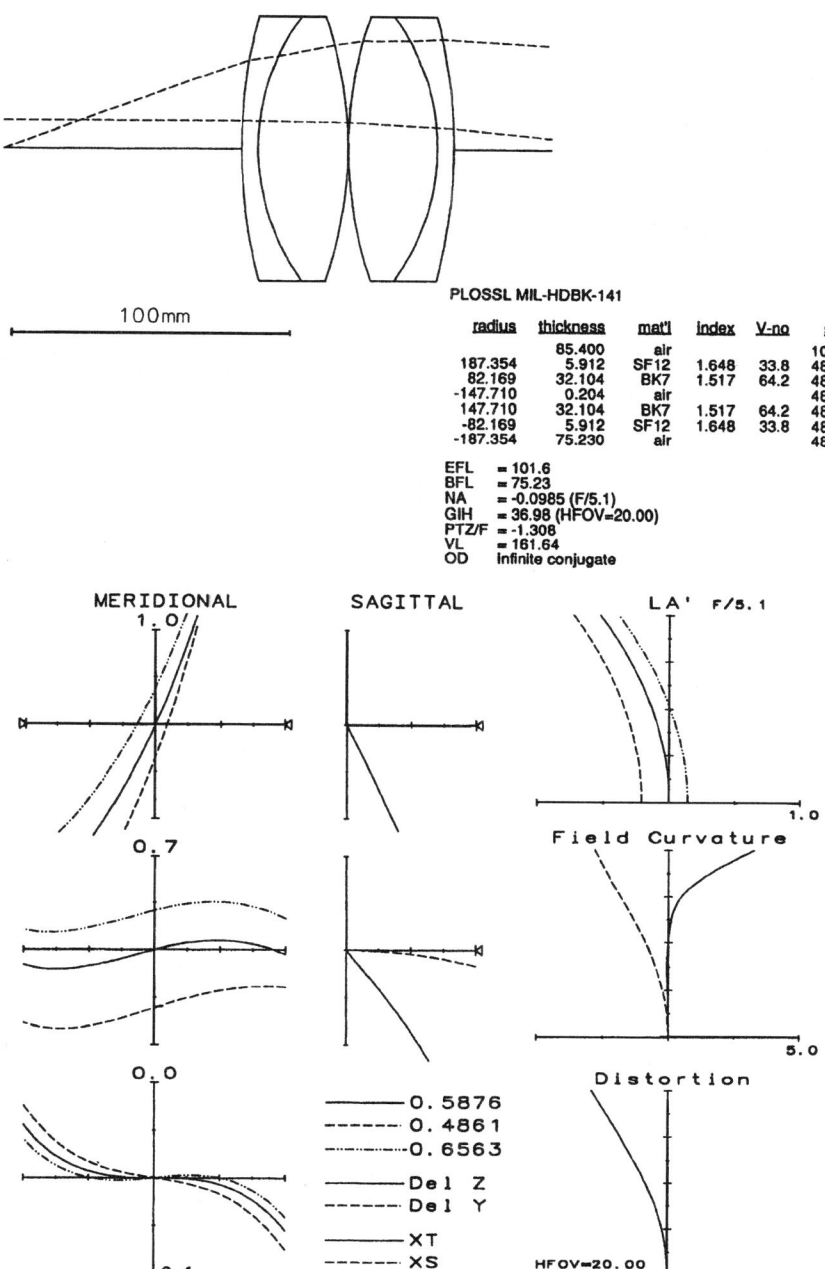

Figure 7.23 Symmetrical/Plössl 40° eyepiece.

182 Chapter Seven

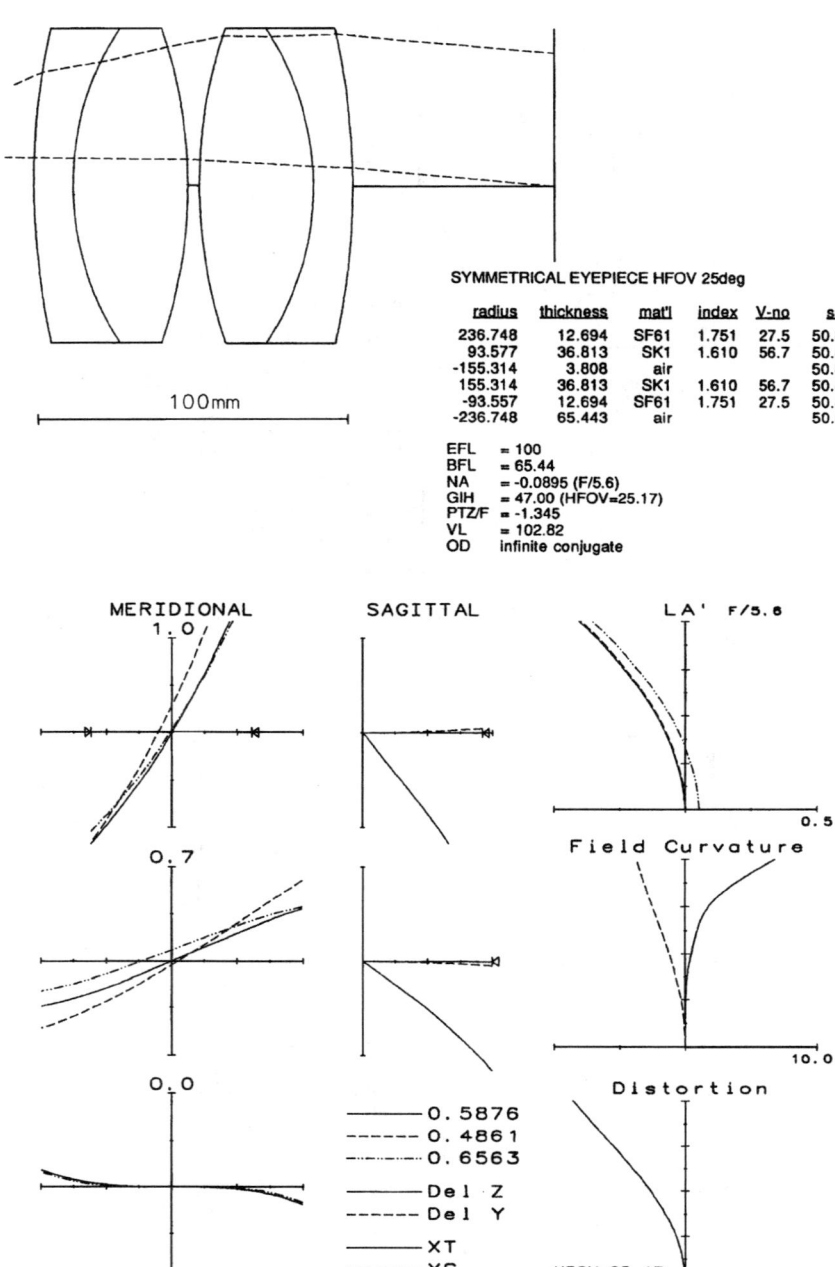

Figure 7.24 Symmetrical/Plössl 50° eyepiece.

Eyepieces and Magnifiers

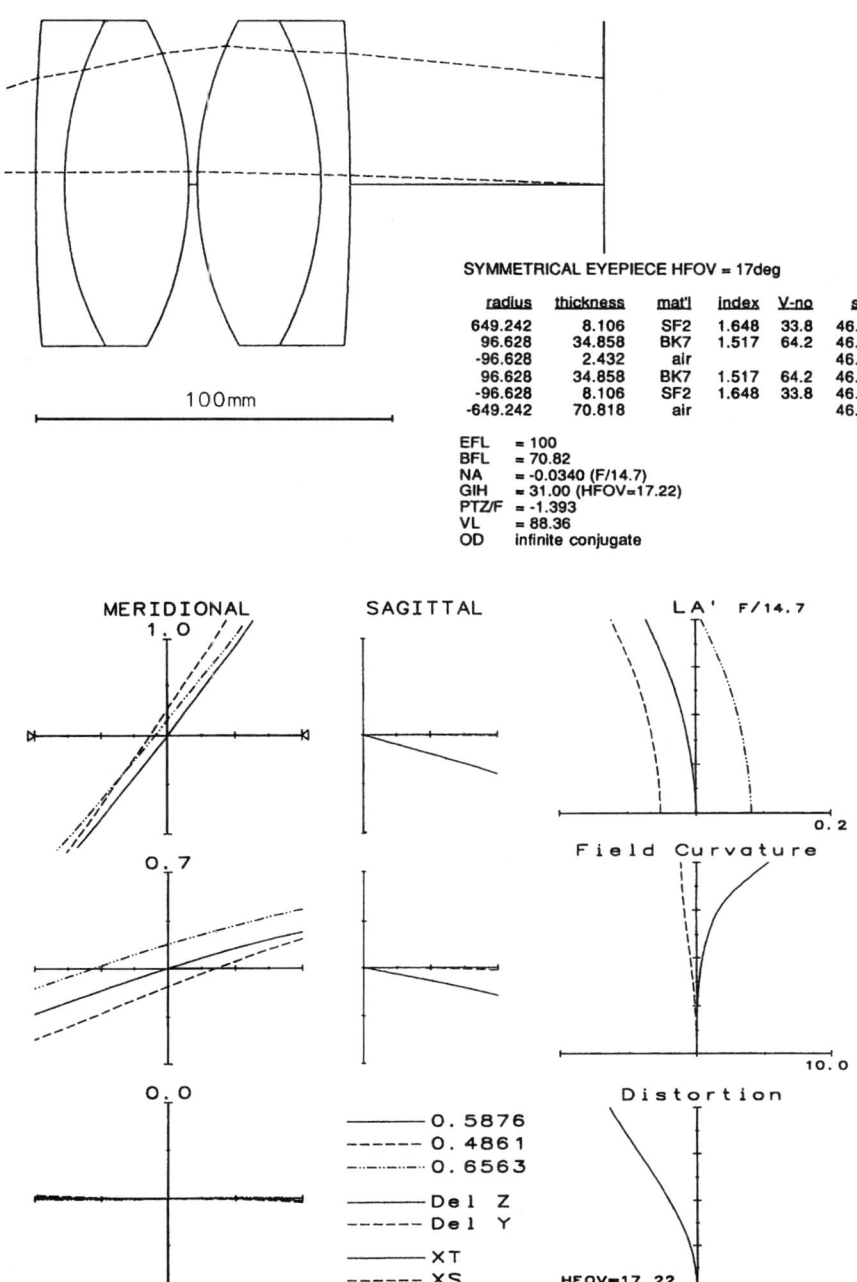

Figure 7.25 Symmetrical/Plössl 35° eyepiece.

184 Chapter Seven

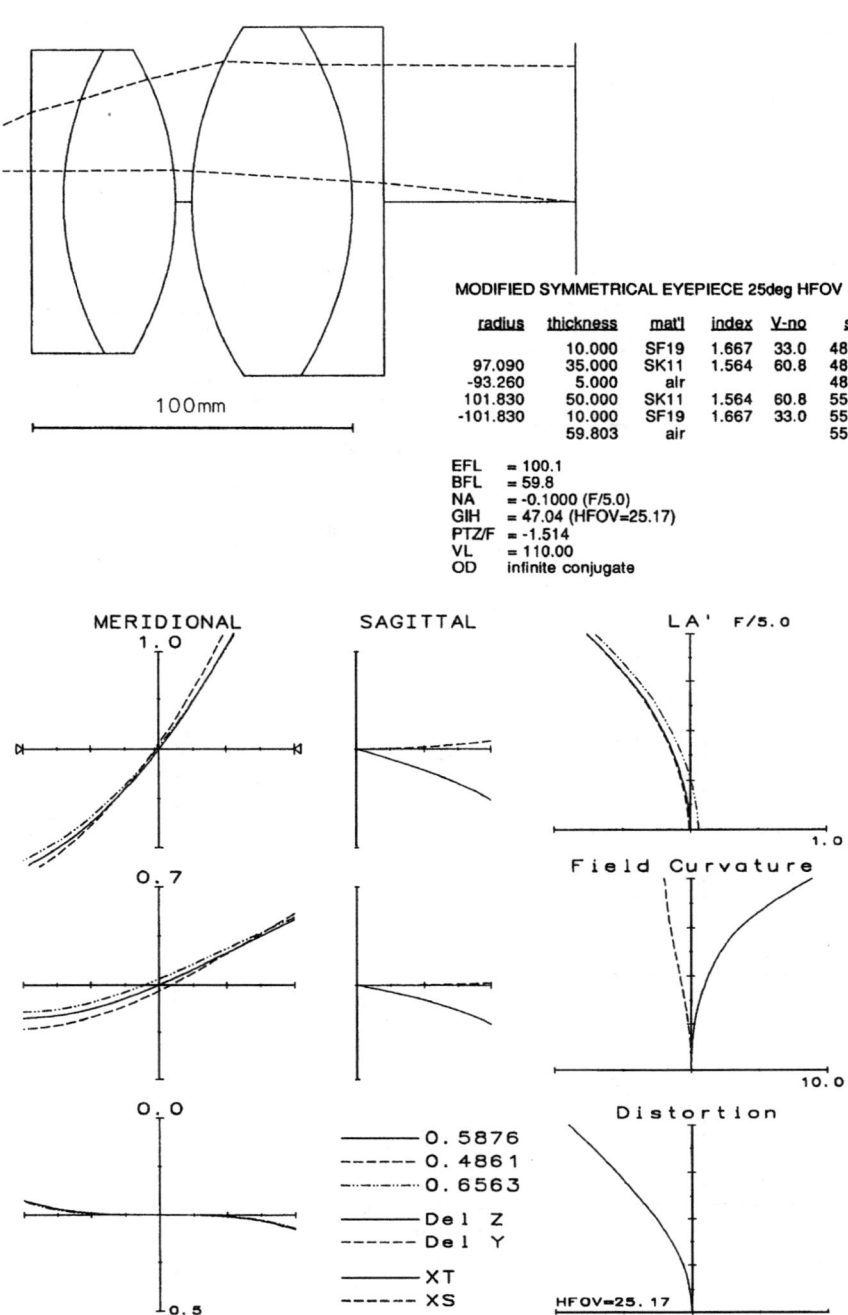

Figure 7.26 Unsymmetrical, two doublet, 50°, higher-index eyepiece.

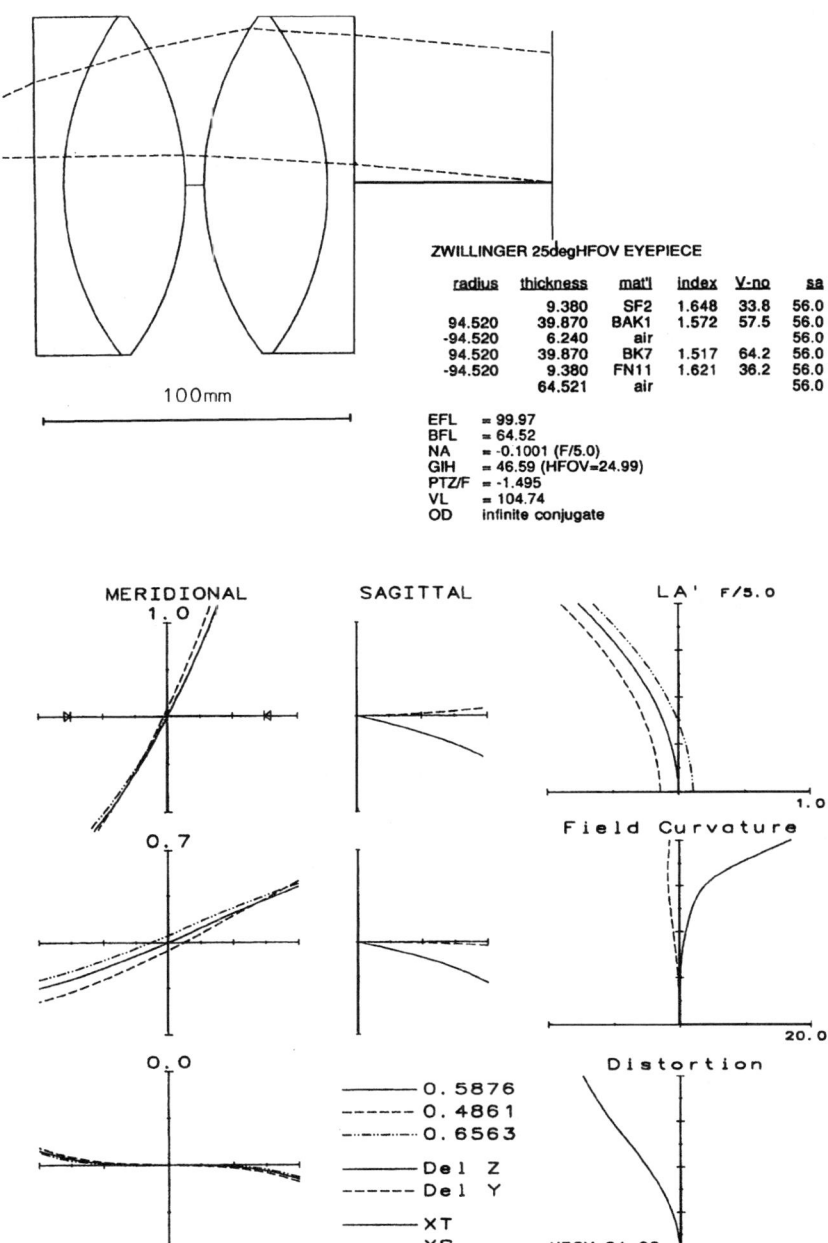

Figure 7.27 Symmetrical radii, but unsymmetrical glass, 50° eyepiece.

186 Chapter Seven

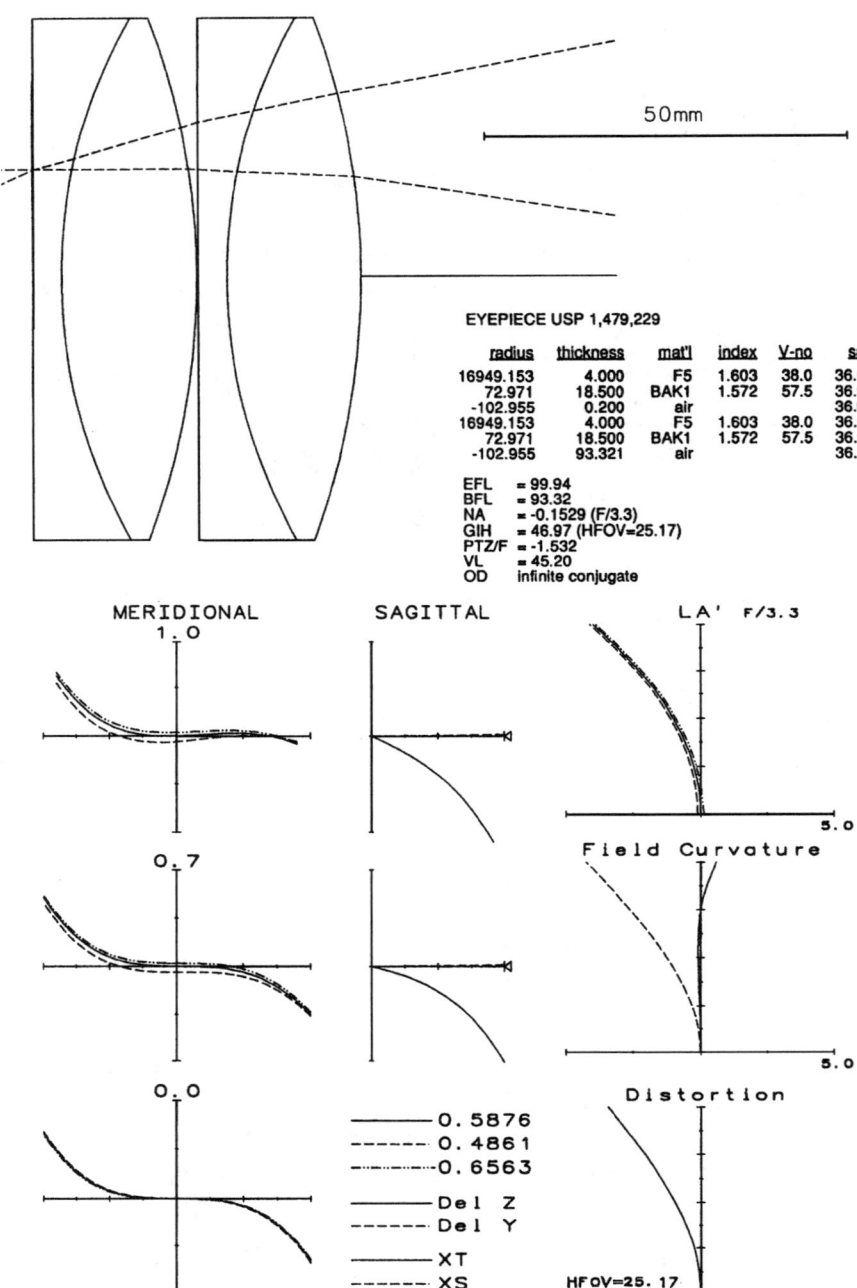

Figure 7.28 Two doublet eyepiece. Note the incorrect pupil location.

7.6 Five-Element Eyepieces

The basic Erfle eyepiece, shown in Figs. 7.29 and 7.30, is a workhorse wide angle eyepiece capable of covering fields exceeding 60–65°. Distortion and field curvature both tend to be large at these angles. The Erfle utilizes what is, in effect, a negative field lens near the focal surface to flatten the Petzval field and increase the eye relief. It does this by locating the strong concave surface of a meniscus doublet very close to the focal plane. Note that Fig. 7.29 uses higher-index glass, as contrasted with the ordinary glasses of Fig. 7.28. Section 7.7 discusses two Erfle-type magnifiers with extremely high-index glasses.

7.7 Very High Index Eyepiece/Magnifier

The two following "eyepieces" utilize very high-index glasses. The crown (positive) elements are N-LaSF44 (804465) and the flints are SFL57 (847236). These are reasonable glasses, although the LaSF glass does cost 20 or 30 times as much as BK7 and the SF glass cost is about six times BK7. The lenses were designed to view a diffuse image directly, and thus have no fixed exit pupil as there is for a telescope eyepiece. Thus, as used, they are really magnifiers (as discussed in Sec. 7.2). The specifications called for performance with a pupil 7 mm in diameter located 35 mm from the lens. A clearance of 5 mm from the object (the image here) was required for focusing and mechanical clearance. The focal length was specified indirectly as the viewing angle for a specified object size; as a result the required focal length was a function of the distortion. In these two designs the necessary focal length came out to be 22.7 mm for the first and 23.48 mm for the second.

The first system was designed to specifications that called for performance over a field of ±18.3°. The design is a fairly standard six-element Erfle form with a rather thick $T8$ of 10.1 mm. The performance was specified as the MTF at 33 line pairs per millimeter (lpm) (the Nyquist frequency for 15 μm pixels); for completeness we show the MTF out to 70 lpm. The aberrations and MTF are shown in Figs. 7.31 and 7.32, and the lens prescription is shown in Table 7.11.

An interesting feature of all cemented doublets that use high-index glasses is that the cement layer reflects about 10 percent when the large index break at the two surfaces of the low-index cement layer are taken into account. This problem can be ameliorated by low reflection coating the cementing surfaces; a quarter-wave layer of sapphire (Al_2O_3) does pretty well.

Looking over the drawing of this lens, there are two design modifications that could possibly prove to be advantageous. The third element ($R5$-$R6$) is relatively weak (efl = –281) and would seem to be a likely

188 Chapter Seven

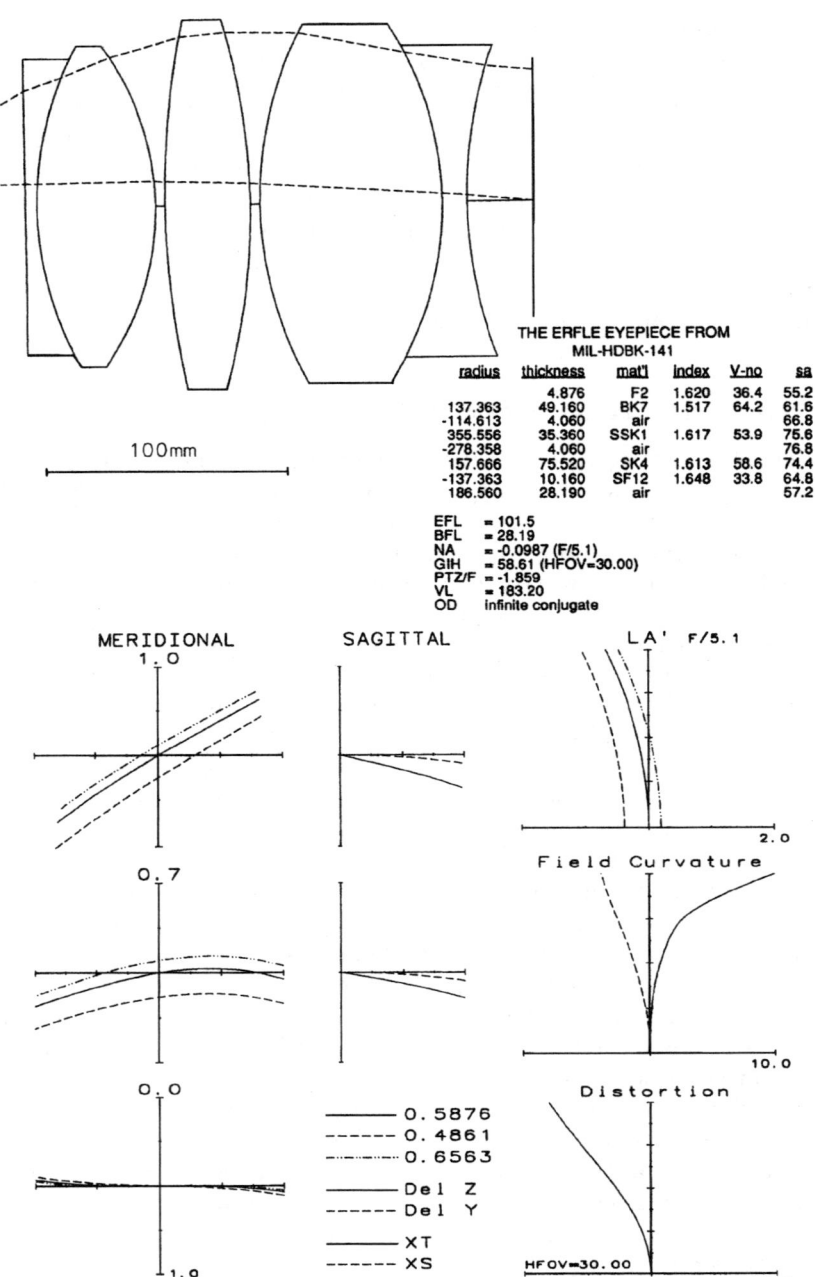

Figure 7.29 Five-element Erfle 60° eyepiece, from MIL-141 (Ref. 7).

Eyepieces and Magnifiers 189

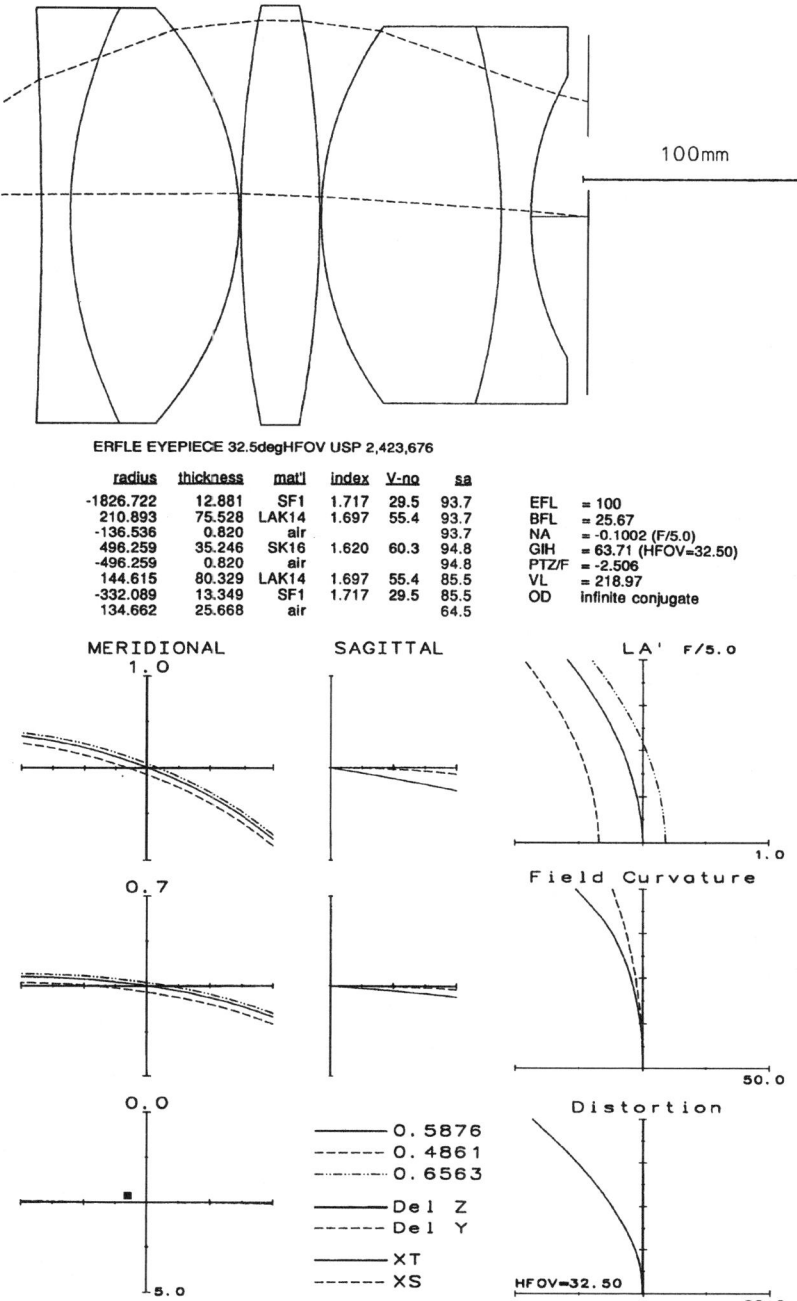

ERFLE EYEPIECE 32.5degHFOV USP 2,423,676

radius	thickness	mat'l	index	V-no	sa
-1826.722	12.881	SF1	1.717	29.5	93.7
210.893	75.528	LAK14	1.697	55.4	93.7
-136.536	0.820	air			93.7
496.259	35.246	SK16	1.620	60.3	94.8
-496.259	0.820	air			94.8
144.615	80.329	LAK14	1.697	55.4	85.5
-332.089	13.349	SF1	1.717	29.5	85.5
134.662	25.668	air			64.5

EFL = 100
BFL = 25.67
NA = -0.1002 (F/5.0)
GIH = 63.71 (HFOV=32.50)
PTZ/F = -2.506
VL = 218.97
OD infinite conjugate

Figure 7.30 Five-element Erfle 65° eyepiece, with high-index glass.

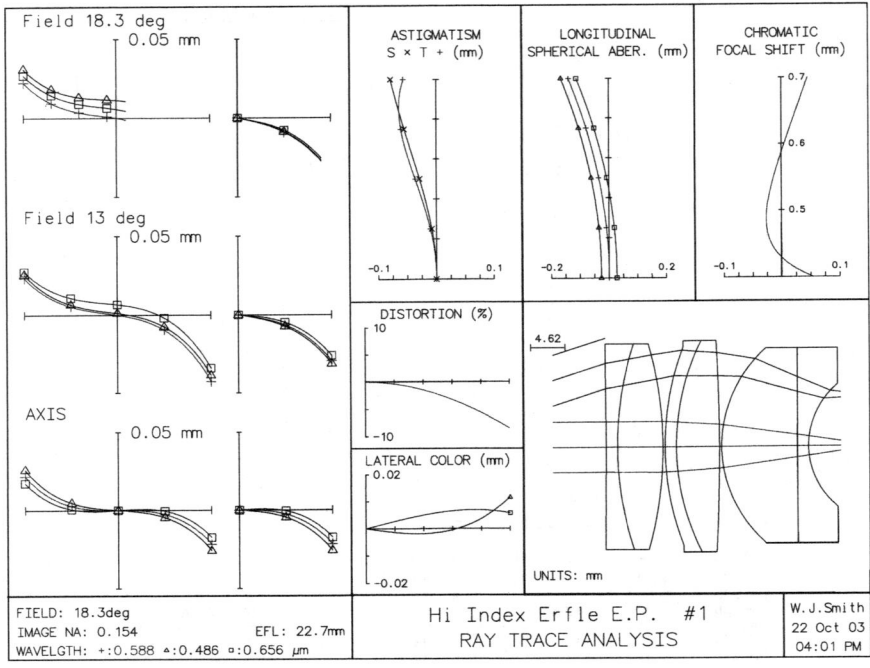

Figure 7.31 Aberrations of a six element, 37°, very high-index Erfle magnifier, Table 7.11.

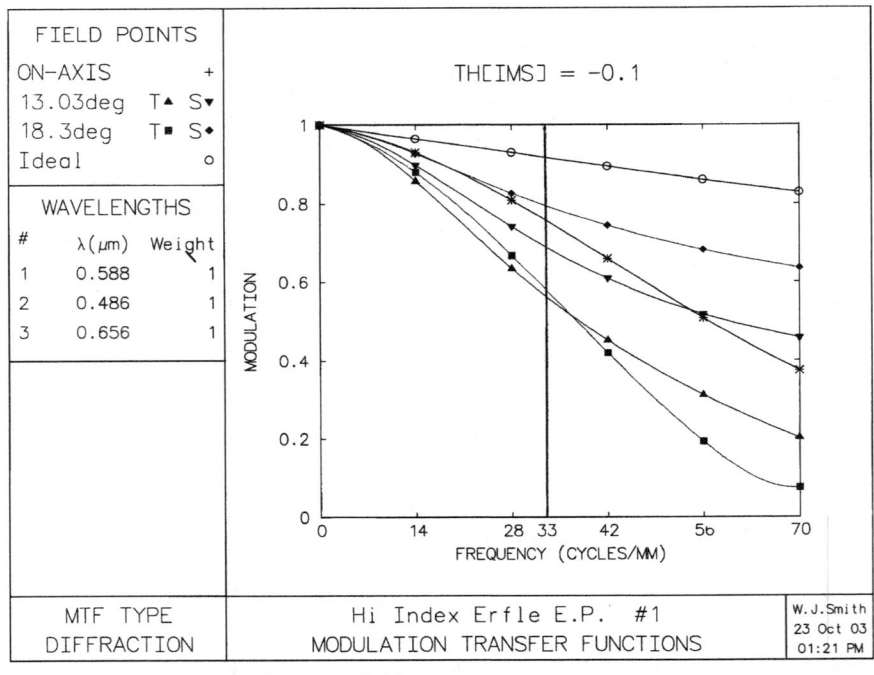

Figure 7.32 The MTF for Fig. 7.30, Table 7.11.

TABLE 7.11 Prescription for a High-Index, Six Element Erfle Style Magnifier, as Shown in Figs. 7.31 and 7.32

$R1$ = plano	$T1$ = 35.0	(pupil)	
$R2$ = −932.63	$T2$ = 1.5	SFL57	847236
$R3$ = +46.829	$T3$ = 6.1040	N-LaSF44	804465
$R4$ = −54.210	$T4$ = 0.3		
$R5$ = +39.986	$T5$ = 1.5	SFL57	847236
$R6$ = +33.643	$T6$ = 5.7284	N-LaSF44	804465
$R7$ = −264.11	$T7$ = 0.3		
$R8$ = +18.410	$T8$ = 10.067	N-LaSF44	804465
$R9$ = +1181.7	$T9$ = 1.5	SFL57	847236
$R10$ = +11.298	BFL = 8.866		
$R11$ = image			

EFL = 22.700
NA = 0.1542 (F/3.24)
GIH = 7.51 (±18.3°)
PTZ/F = −11.37
VL = 27.
OD = Inf.

candidate for removal. The fifth element ($R8$-$R9$) is thick; here one might consider replacing the thick glass element by a thinner element plus an airspace (to reduce the weight while still retaining the spacing between $R8$ and $R10$, and also to gain two more design variables). In the course of the project the specifications were modified; the required field of view was reduced from ±18.3° to ±15°. The design incorporating these changes and the smaller field is given in Table 7.12 and the performance plots are in Figs. 7.33 and 7.34.

TABLE 7.12 Prescription for a High-Index, Five Element Erfle Style Magnifier, with an Air Spaced Final Doublet (See Figs. 7.33 and 7.34)

$R1$ = plano	$T1$ = 35.0	(pupil)	
$R2$ = −56.205	$T2$ = 1.5	SFL57	847236
$R3$ = +71.419	$T3$ = 5.3	N-LaSF44	804465
$R4$ = −35.009	$T4$ = 0.25		
$R5$ = +22.449	$T5$ = 5.35	N-LaSf44	804465
$R6$ = plano	$T6$ = 0.25		
$R7$ = +17.013	$T7$ = 4.85	N-LaSf44	804465
$R8$ = +46.323	$T8$ = 1.651		
$R9$ = plano	$T9$ = 1.5	SFL57	847236
$R10$ = +11.436	BFL = 11.15		
$R11$ = image			

EFL = 23.475
NA = 0.149 (F/3.35)
GIH = 6.29 (±15°)
PTZ/F = −57.6
VL = 20.65
OD = Inf.

192   Chapter Seven

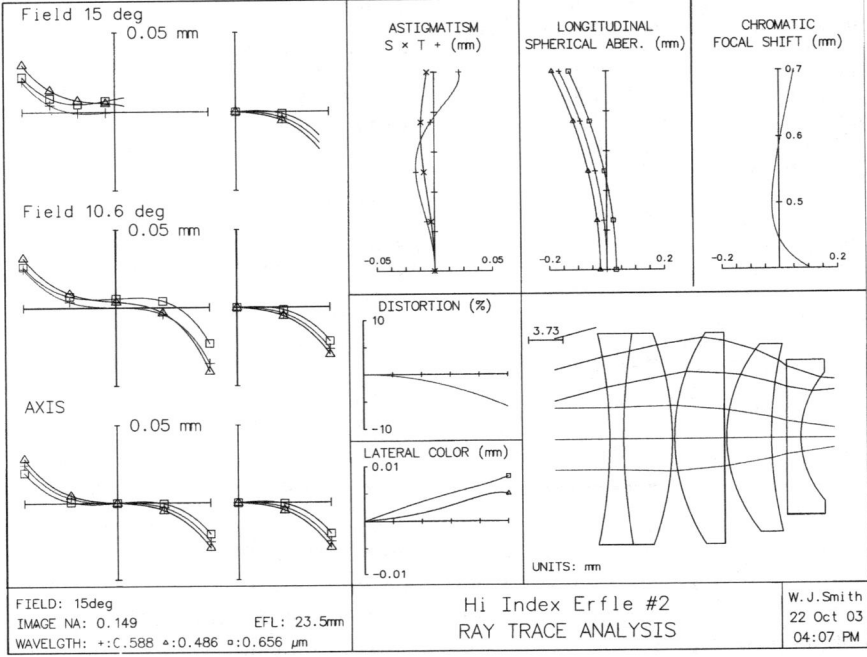

Figure 7.33 Aberrations of a split doublet five element Erfle 30° very high-index magnifier, Table 7.12.

This design has only five elements and also incorporates two inexpensive plano surfaces. The smaller field allows element diameters that are smaller than in the first design. It easily met the performance specifications, having a slightly better performance at 33 lpm than the six-element design. Of course this is made possible by the reduction of the angular field from 36.6 to 30°.

The custom merit function used in these designs is interesting (and typical) in that it includes: (a) the various physical limits imposed both by the specification and reasonable fabrication practice; (b) the design program's default rms spot size merit function with Conrady $(D - d)\delta n$ chromatic control, plus a distortion target; and (c) various targets that were added to control or correct the more troublesome aberrations that popped up in the course of the designs. The final merit function included targets for the following:

1. EFL
2. Working distance >5 mm

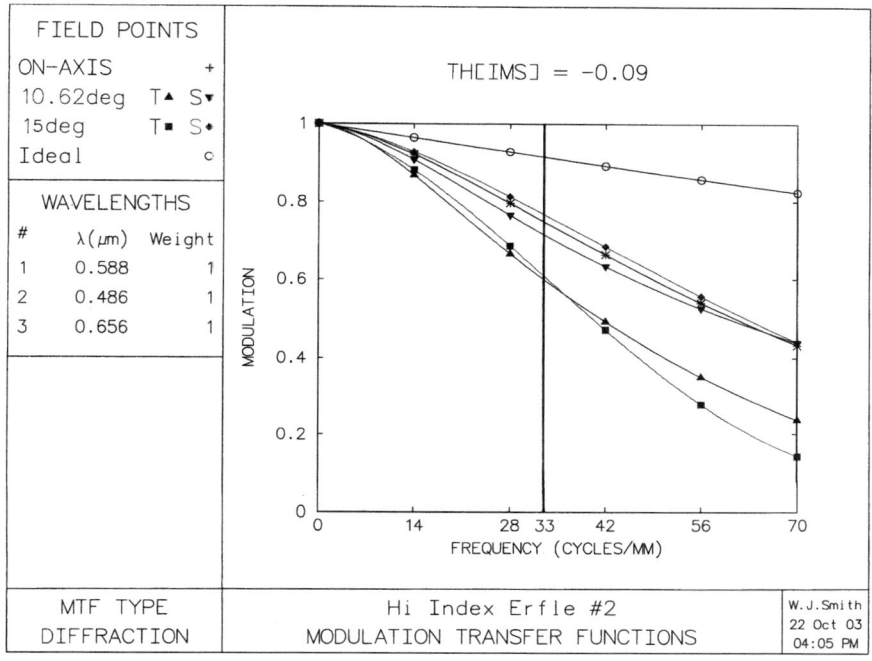

Figure 7.34 The MTF for Fig. 7.32, Table 7.12.

3. Overall length <44 mm
4. Edge thickness of positive elements >2 mm
5. X_s, X_t and astigmatism at full field
6. X_s and X_t at 0.7 field
7. Tangential coma at full field
8. Tangential coma at 0.7 field
9. Paraxial lateral chromatic
10. Lateral chromatic at full field
11. Lateral chromatic at 0.7 field
12. Rms spot size at all three fields
13. Conrady $(D - d)\delta n$ at all three fields
14. Distortion at full field

In the final merit function: items 1, 2 and 3 were constrained, all the others were minimized; items 5, 6, 7, and 14 were weighted 100x; item 8 was weighted 300x; and items 10 and 11 were weighted 1000x.

194 Chapter Seven

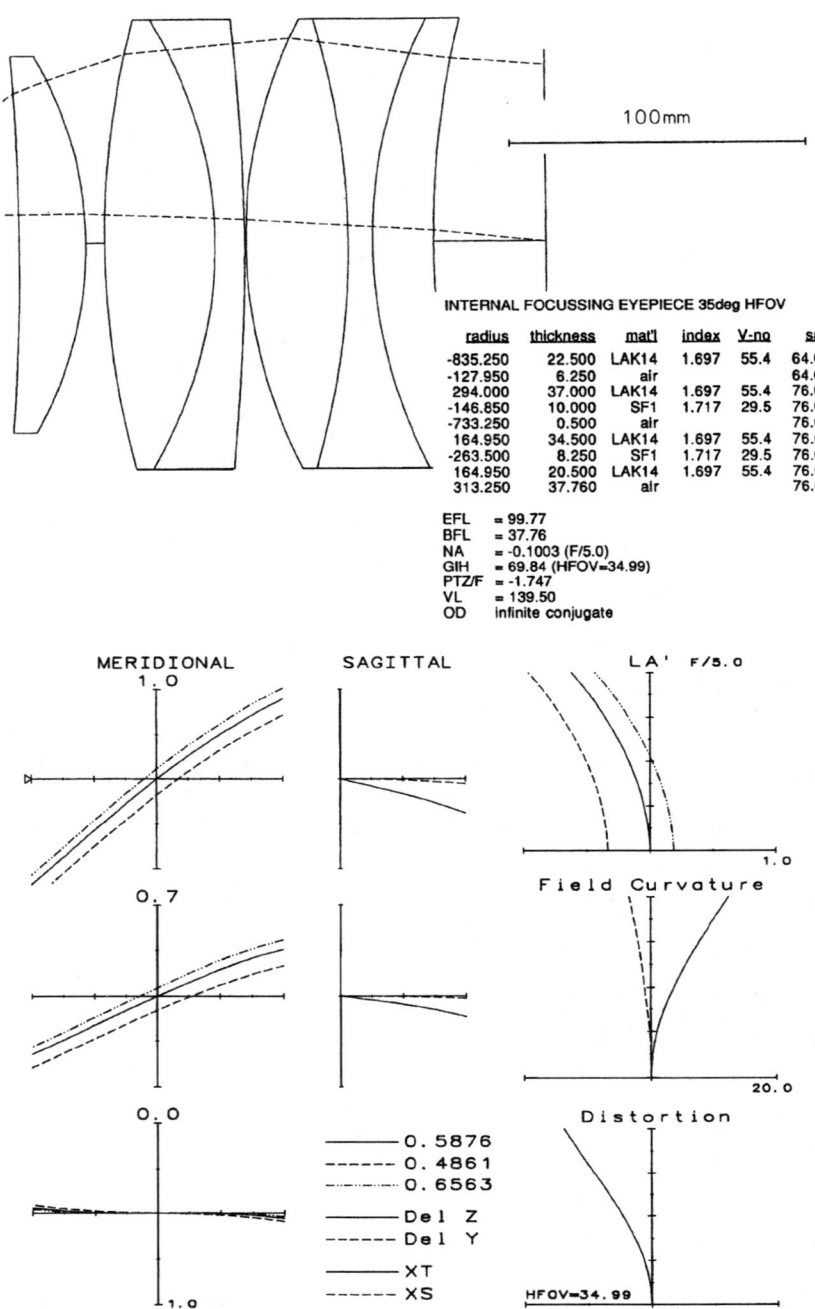

Figure 7.35 Internal focusing 70° eyepiece.

Eyepieces and Magnifiers

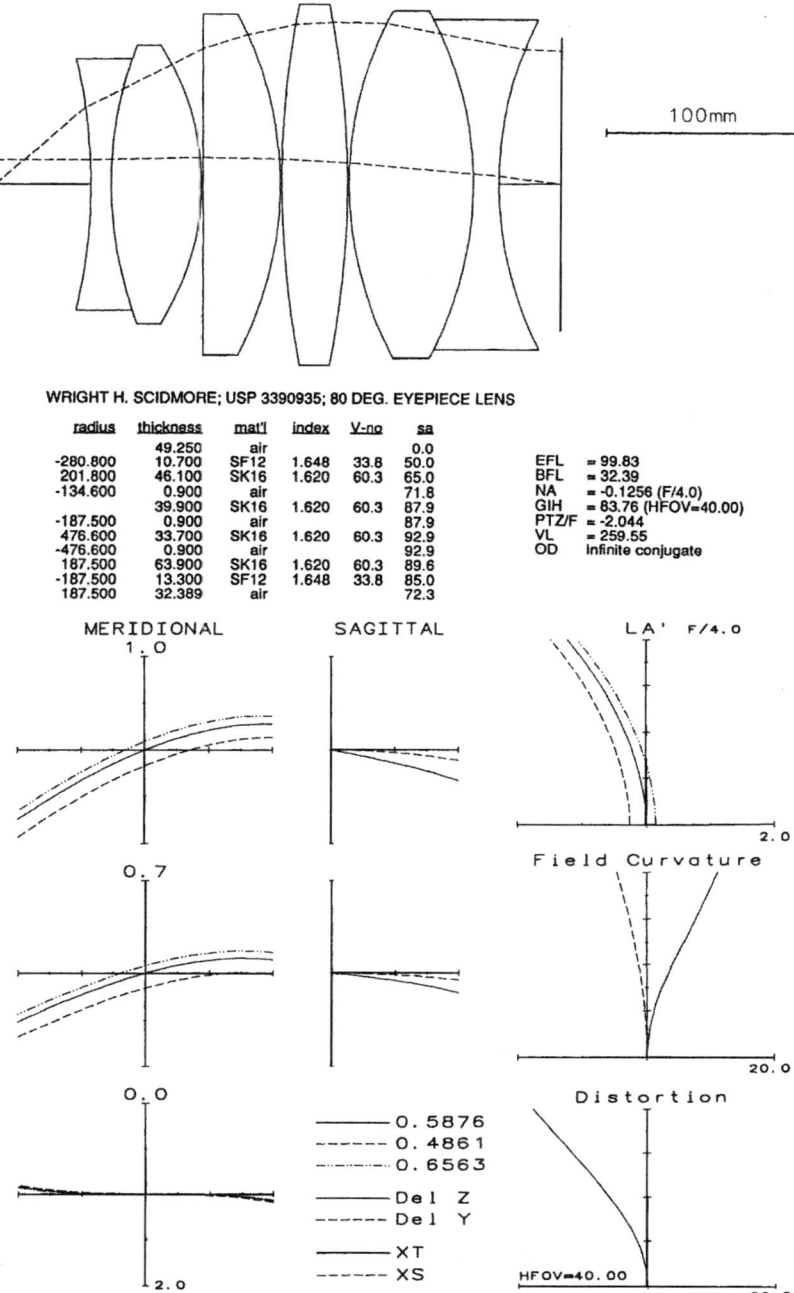

WRIGHT H. SCIDMORE; USP 3390935; 80 DEG. EYEPIECE LENS

radius	thickness	mat'l	index	V-no	sa
	49.250	air			0.0
-280.800	10.700	SF12	1.648	33.8	50.0
201.800	46.100	SK16	1.620	60.3	65.0
-134.600	0.900	air			71.8
	39.900	SK16	1.620	60.3	87.9
-187.500	0.900	air			87.9
476.600	33.700	SK16	1.620	60.3	92.9
-476.600	0.900	air			92.9
187.500	63.900	SK16	1.620	60.3	89.6
-187.500	13.300	SF12	1.648	33.8	85.0
187.500	32.389	air			72.3

EFL = 99.83
BFL = 32.39
NA = -0.1256 (F/4.0)
GIH = 63.76 (HFOV=40.00)
PTZ/F = -2.044
VL = 259.55
OD Infinite conjugate

Figure 7.36 Six-element split singlet 80° Erfle eyepiece.

196 Chapter Seven

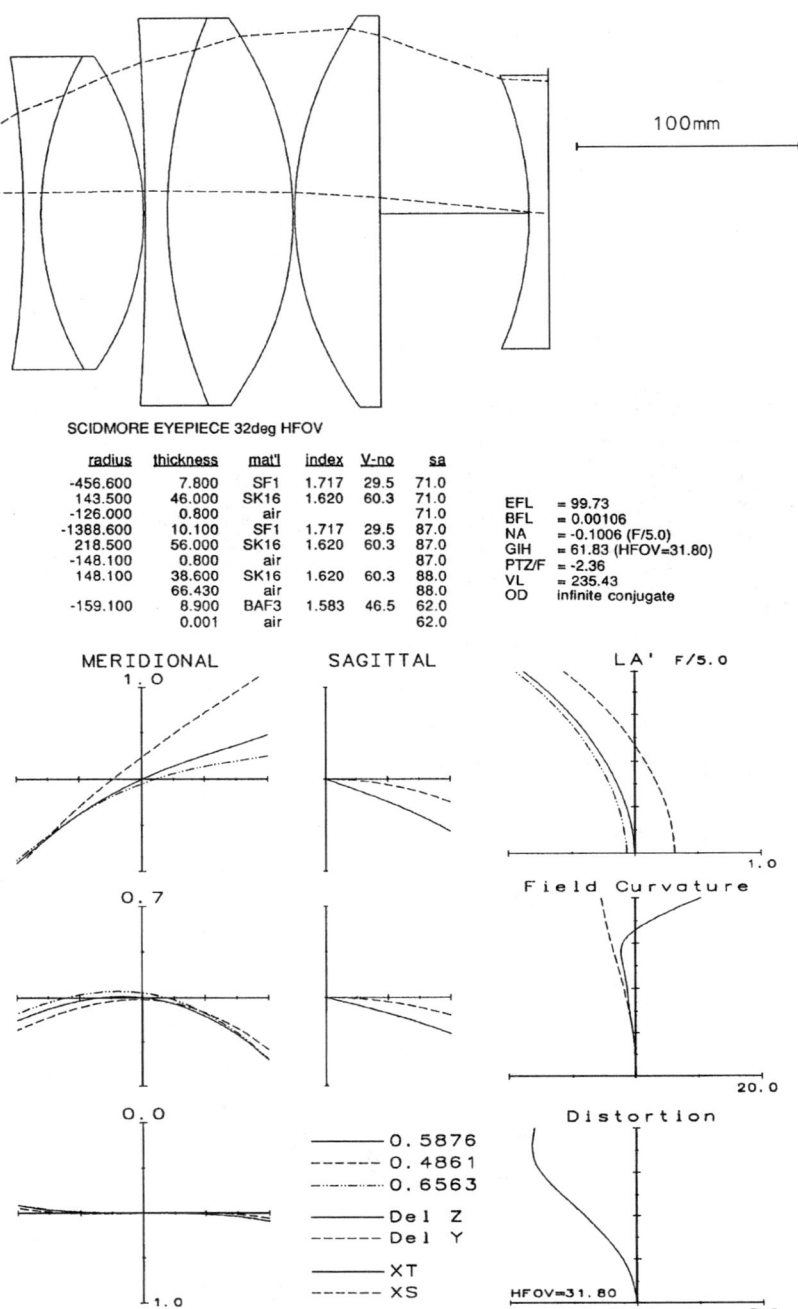

Figure 7.37 Modified 64° Erfle eyepiece with a field flattener.

Eyepieces and Magnifiers 197

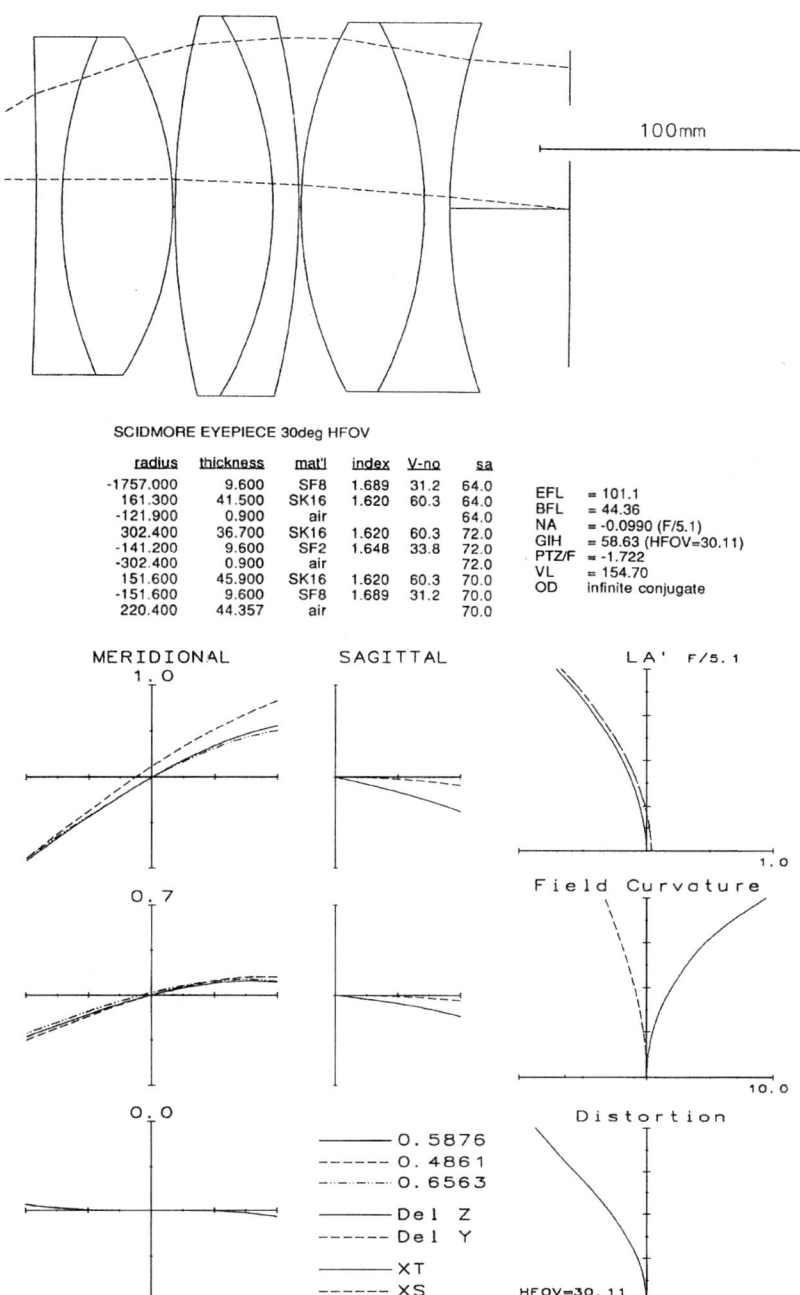

SCIDMORE EYEPIECE 30deg HFOV

radius	thickness	mat'l	index	V-no	sa
-1757.000	9.600	SF8	1.689	31.2	64.0
161.300	41.500	SK16	1.620	60.3	64.0
-121.900	0.900	air			64.0
302.400	36.700	SK16	1.620	60.3	72.0
-141.200	9.600	SF2	1.648	33.8	72.0
-302.400	0.900	air			72.0
151.600	45.900	SK16	1.620	60.3	70.0
-151.600	9.600	SF8	1.689	31.2	70.0
220.400	44.357	air			70.0

EFL = 101.1
BFL = 44.36
NA = -0.0990 (F/5.1)
GIH = 58.63 (HFOV=30.11)
PTZ/F = -1.722
VL = 154.70
OD infinite conjugate

Figure 7.38 Six-element Erfle 60° eyepiece.

198 Chapter Seven

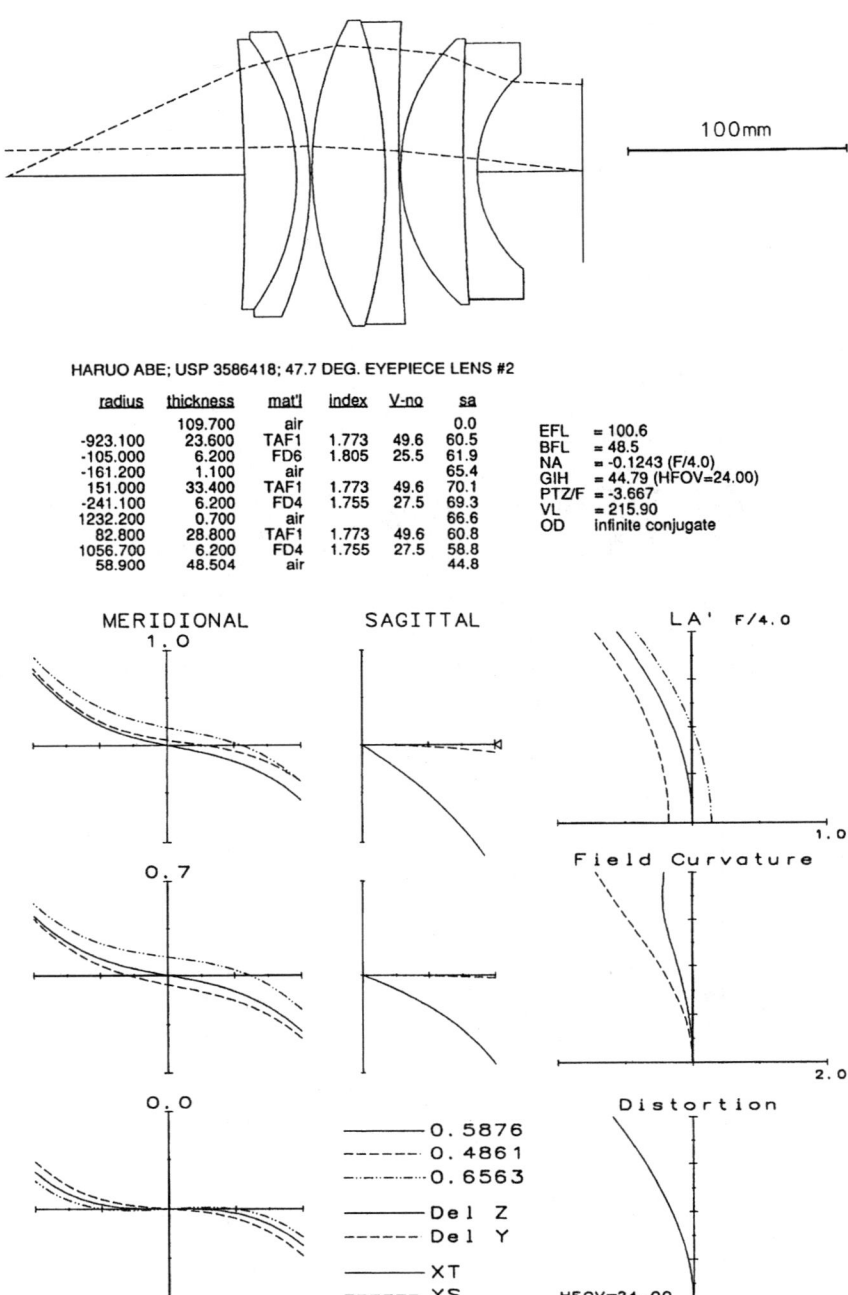

HARUO ABE; USP 3586418; 47.7 DEG. EYEPIECE LENS #2

radius	thickness	mat'l	index	V-no	sa
	109.700	air			0.0
-923.100	23.600	TAF1	1.773	49.6	60.5
-105.000	6.200	FD6	1.805	25.5	61.9
-161.200	1.100	air			65.4
151.000	33.400	TAF1	1.773	49.6	70.1
-241.100	6.200	FD4	1.755	27.5	69.3
1232.200	0.700	air			66.6
82.800	28.800	TAF1	1.773	49.6	60.8
1056.700	6.200	FD4	1.755	27.5	58.8
58.900	48.504	air			44.8

EFL = 100.6
BFL = 48.5
NA = -0.1243 (F/4.0)
GIH = 44.79 (HFOV=24.00)
PTZ/F = -3.667
VL = 215.90
OD infinite conjugate

Figure 7.39 Six-element high-index 48° Erfle eyepiece.

Eyepieces and Magnifiers 199

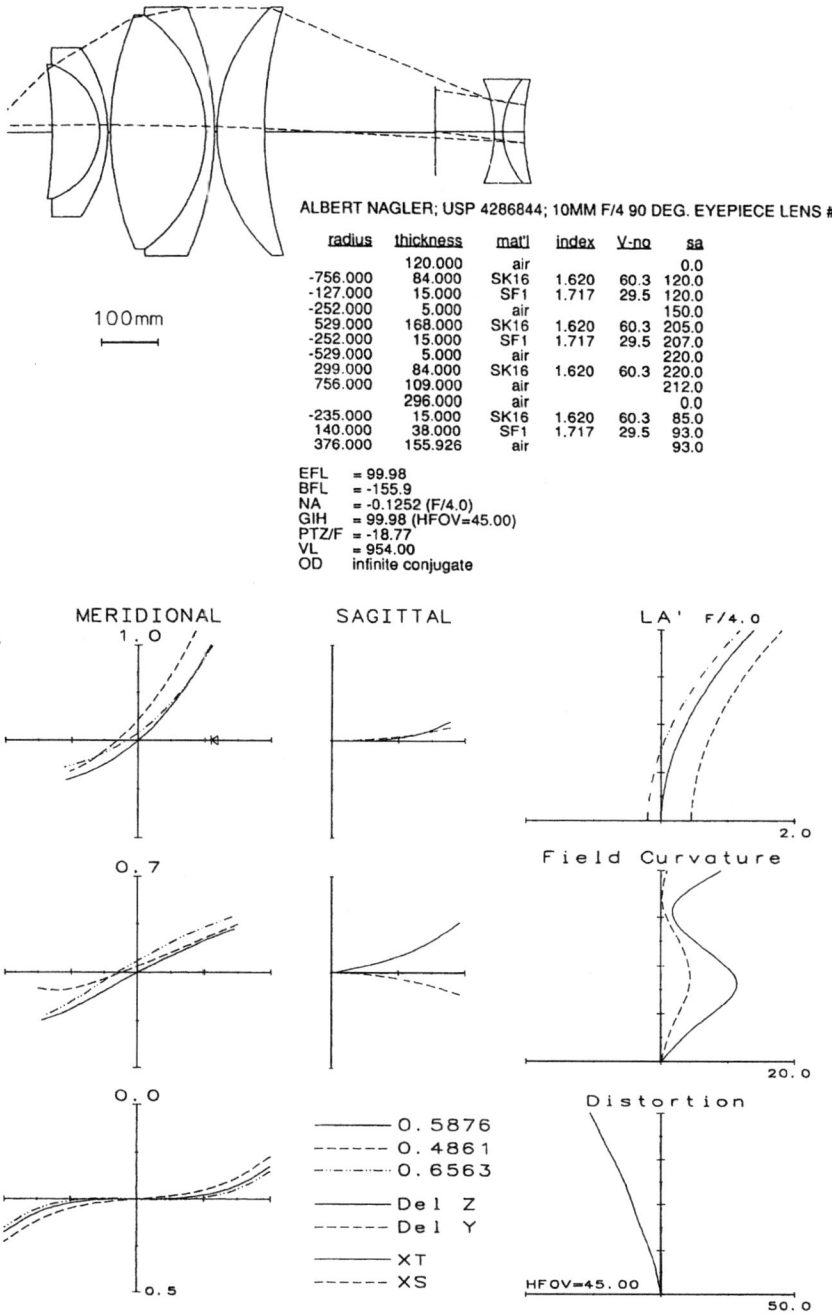

Figure 7.40 Nagler 90° eyepiece, with a negative achromat field lens.

A final note. Because the five-element design was more sensitive to tolerance problems, it lost out to a ±15° version of the first lens when it came time to fabricate the product. We thank Kaiser Electro-Optics, Inc. of Carlsbad, CA for permission to include these interim designs.

7.8 Six- and Seven-Element Eyepieces

The internal focusing eyepiece of Fig. 7.35 is focused, not by moving the whole eyepiece, but by moving just the doublet and the triplet, while maintaining the eye-lens-to-focal plane spacing constant. This allows a more easily sealed construction for the focusing mechanism of the assembly. At the scale of the data, the two components can be shifted about 1 unit toward the eye and about 23 units away; the amount of focusing that this produces is obviously a function of the focal length to which the design is scaled. At a focal length of about 1.5 in, this shift will focus through about 4 diopters at the eye. The internal focusing idea can of course be applied to almost any of the eyepiece designs. A multiconfiguration design approach could be used to minimize the aberration/performance changes as the system is respaced to refocus.

The Fig. 7.36 design covers a rather astonishing 80° total field of view and is an elaboration of the five-element Erfle, with a split or doubled inner singlet. Note that in this design both the field lens and the eye lens are meniscus. Figure 7.37 has an explicit negative field lens added to flatten the field and lengthen the eye relief. Its plano side also serves as a reticle surface, eliminating the need for a separate element to carry the reticle pattern. The negative field lens also means large lens diameters.

Figures 7.38 and 7.39 are six-element variations on the basic Erfle eyepiece with relatively moderate fields of about 60° and about 50°. In practice, a field as large as 60° or more tends to be rather difficult to use since, as the eye rotates in its socket, it is not easy to keep the pupil of the eye and the exit pupil of the instrument in proper alignment. Note also that spherical aberration of the pupil becomes both more important and more of a design problem as wider fields are covered.

The last eyepiece of this chapter, the Nagler eyepiece of Fig. 7.40 achieves a 90° field and a similar level of correction, but uses a unique construction, incorporating a negative achromatic field lens beyond the focal plane. Pupil aberration can be a problem in this eyepiece also, although more recent designs are improved. Note that locating the negative field lens beyond the focus actually shortens the focal length of the eyepiece.

Chapter 8

Cooke Triplet Anastigmats

8.1 Airspaced Triplet Anastigmats

The Cooke triplet (Fig. 8.1) is an especially interesting design form for several reasons. It possesses just enough effective degrees of freedom to control or correct all the primary aberrations. Because of the nonlinearity of the relationships between the aberrations and the design variables, there are (theoretically) at least eight potential solutions for the primary aberrations, two or three of which may be useful.

The basic method used to flatten the Petzval field curvature in *all* anastigmats is the longitudinal separation of positive power from negative power. This separation may be between surfaces, between elements, or between components. Since the contribution to the Petzval curvature is TPC = $(n' - n)ch^2 n_k' u_k'/2nn'$ for a surface (per Eq. (24.67)), or TPC = $h^2 \phi u_k'/2n$ for a thin element (per Eq. (24.84)), it is obviously independent of the height at which the rays strike. The contribution to the system power, however, is $y(n' - n)c$ for a surface, or $y\phi$ for a component. Thus, increasing the spacing of positive power away from negative power, which lowers the relative ray height y on the negative, will reduce the (negative) power contribution of the surfaces without changing their Petzval contribution. The result is an effective net positive power without the undesirable excess of inward Petzval curvature, which would otherwise accompany positive power.

All anastigmats make use of this principle. Some incorporate thick meniscus components to separate positive convex surfaces from negative concave surfaces. Examples are the older anastigmats such as the Protar and Dagor (Chap. 11), the Ernostar and Sonnar types (Chap. 9), and the widely used and powerful Biotar or double-Gauss form (Chap. 12). Others, such as the Cooke triplet and its modifications, or the retrofocus forms

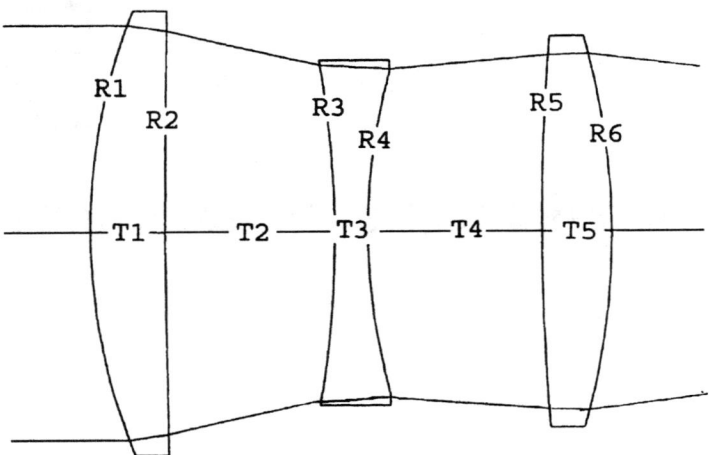

Figure 8.1 The Cooke triplet.

(Chap. 14), utilize spaced-apart components to achieve the same ends. The Cooke triplet can be viewed as the trunk of a family tree of airspaced anastigmats.

For the Cooke triplet the powers and spaces must satisfy the following relationships (which are the shape-independent aberrations):

For lens power (efl):

$$\phi = \frac{1}{y_A}\Sigma y_i\phi_i = \frac{1}{\text{EFL}}$$

For axial chromatic:

$$\text{TA}_{ch}\text{A} = \frac{1}{u'_k}\Sigma\frac{y_i^2\phi_i}{V_i}$$

For lateral chromatic:

$$\text{T}_{ch}\text{A} = \frac{1}{u'_k}\Sigma\frac{y_i y_{Pi}\phi_i}{V_i}$$

For Petzval curvature:

$$\text{TPC} = \frac{h^2}{2u'_k}\Sigma\frac{\phi_i}{n_i}$$

Since the triplet has three element powers and two airspaces that are effective constructional variables (in regard to the *primary* aberrations, the element thicknesses are only weakly effective and duplicate the airspaces), these requirements are easily satisfied. Two element powers and the two spaces are dedicated to this end. The remaining element power and the three shapes (bendings) of the elements are then available to control the primary spherical, coma, astigmatism, and distortion.

Actually, as will be discussed below, the choice of the glass for the crowns (and the subsequent selection of the flint glass to suit the field and aperture of the application) is a very important factor, although in practice the crown glasses are not often treated as free variables. Usually both positive elements are made of the same glass. From a general design standpoint, it's difficult to see any reason for them to differ; what's good in one location is also good in the other.

Regarding the nonlinear equations for the primary aberration contributions versus shape, the spherical, astigmatism, distortion (and coma, if the element is not at the stop) equations are all quadratic. *Quadratic* tells us that two element shapes may each yield some given amount of aberration. (Alternatively, there may be no shape that will have the desired amount.) Thus with three elements there are $2^3 = 8$ possible solutions (assuming good choices have been made for the glass types and the desired aberration sums). With a simple merit function made up of only primary aberration targets, this means that there may be eight local minima. Add glass variables and higher-order aberration merit function targets to the mix, and the number of local minima becomes enormous indeed.

A condensed description of an analytic, aberration theory approach to the Cooke triplet design is as follows: The four shape-independent aberrations (EFL, axial color, lateral color, and Petzval sum) are controlled by the power and space layout for the elements. The four shape-dependent aberrations (spherical, coma, astigmatism, and distortion) are controlled by the element shapes or bendings.

The element powers and spaces are adjusted to satisfy the four equations above. R.E. Stephens in "The Design of a Triplet Anastigmat of the Taylor Type," *JOSA*, Vol. 38, No. 12, Dec. 1948, pp. 1032–1039, R.E. Hopkins in "Third and Fifth Order Analysis of the Triplet," *JOSA*, Vol. 52, No. 4, Apr. 1962, pp. 389–394, and also in MIL-141,[7] and Smith[23] give either equations or iterative procedures to accomplish this. At this stage there is an extra degree of freedom because there are five variables (two spaces and three powers) and only four equations (above) to satisfy. The value of one of these variables is arbitrarily chosen and reserved for later use as a variable.

As regards the starting values for these aberration goals, note that if the Petzval surface is made too flat, the element powers will tend to be large, and the high-order residuals will also be large. An inward-curving Petzval radius equal to two to four times the focal length is reasonable (two times for high NA types and four (or more) times for slow, wide angle lenses). Note also that if $s_1 = s_2$ and if $y_1\phi_1 = y_3\phi_3$, then the lateral color will be corrected. Also a typical value for the length of a triplet $L = (s_1 + s_2)$ is about 1.3 EFL/ (f number).

The element shapes can be found by an iterative process similar to the following:

1. A shape is chosen for the first element.

2. The stop is put at the second (middle) element. Note that with the stop at this element, its astigmatism is fixed.

3. The shape of the rear element is solved to get the desired astigmatism contribution sum. Since the astigmatism is a quadratic function of this element shape, there are usually two shapes for the third element that satisfy the requirement; usually the better looking is chosen.

4. Then the shape of the second element is adjusted to correct the coma. Note that with the stop at the second element, its coma is a linear function of bending, and there is only one solution for its shape.

5. The shape chosen for the first element can be varied (repeating the above steps each time) to correct the spherical aberration.

6. The leftover variable from the power and space solution can then be used to control distortion.

It is possible that there will be two real, useful solutions for the primary aberrations. One of the two may be better as regards the higher-order aberrations, i.e., oblique spherical (y^3h^2), elliptical coma (y^2h^3), or linear coma (y^4h). The so-called right-hand solution is the more common. Usually in this form the rear airspace is larger than the front, and the front radius of the second element, $R3$, is longer than the rear, $R4$. In the left-hand solution, the front airspace is larger than the rear, and the rear radius of the flint, $R4$, is larger than the front, $R3$. The first surface radius, $R1$, is usually longer for the left-hand version than the right. And in general the first radius, $R1$, tends to be longer than the sixth ($R6$) for both left and right forms.

With respect to the astigmatism balance:

1. The fifth-order astigmatism is usually negative.

2. The third-order astigmatism is made positive to balance the fifth order; the result is undercorrected astigmatism at the edge of the field, and overcorrected astigmatism at the "belly," or zone of the field.

3. The Petzval curvature is made inward curving to offset the overcorrected oblique spherical as well as the inward defocus caused by the negative spherical aberration zonal residual.

With an optical design program available to balance the Petzval, astigmatism, spherical, and the length effect on the higher-order aberrations, there is not much point in manipulating the third-order aberration equations. The balance of the aberrations can readily be adjusted by changing the weightings in the merit function, or by adding a few new rays and targets to the merit function.

Despite the neat congruence of eight effective variables and eight aberrations (including efl), the satisfactory solution to the Cooke triplet problem, i.e., the simultaneous correction or control of the eight characteristics, requires a delicate balance between the third-order and higher-order aberrations. For example, the Petzval curvature must be slightly inward curving with the Petzval radius $\rho = -K(\text{efl})$, where K is ordinarily in the range of about 2 to 6. In general, K should be near the small end of this range for high-speed, narrow-angle triplets and near the large end for low-speed, wide-angle lenses. In addition, the third-order astigmatism must be slightly overcorrected to offset the undercorrected fifth-order astigmatism. Too small a Petzval sum leads to a large belly $(X_s - X_t)$ at the zonal field. To make a rough preliminary estimate of the size and length of an anastigmat, for the vertex length you can use

$$L \approx (0.6 + 12/\theta) \cdot f/(f\text{ number}) \quad \text{with HFOV } \theta \text{ in degrees}$$

or

$$L \approx K \cdot f/(f\text{ number})$$

where K is about 1.3 (i.e., $K =$ from 1.0 to 1.7)

8.2 Glass Choice

The choice of glass types is an extremely important degree of freedom and has a significant effect on the characteristics of the triplet design. The glass for the positive elements should be a dense barium or lanthanum crown type; an index of 1.6 or more is almost essential. A triplet using an ordinary low-index crown glass (or even acrylic plastic) for the positive elements is, of course, possible, but the result is poor unless the aperture or field (or both) is small or aspheric surfaces are utilized. The other important factor in glass choice is the use of the relative difference in V values between the crown and flint elements as a means to adjust the vertex length of the triplet to its optimum value for the aperture and field.

8.3 Vertex Length and Residual Aberrations

In general, the longer an anastigmat is (i.e., the greater its vertex length), the smaller the field angle it will cover and the less zonal spherical aberration it will have. High-speed lenses of small angular coverage tend to have a large vertex length, whereas slow-speed, wide-angle lenses tend to be relatively short. Figure 8.2 is a plot of the zonal spherical and the angular coverage of typical anastigmats, as a function of vertex length for a large number of published lens designs. Each dot in the figure represents an ordinary anastigmat lens. Note that the figure is not limited to Cooke triplets. Lenses with three, four, five, six, and seven elements, with no restrictions on glass type, designed by about 50 different designers, are included. The correlation, while not exact, is obvious. The lines in the figure show the results of a study* in which the design characteristics were carefully controlled, so that the effects of vertex length on the lens characteristics were clearly demonstrated.

In Cooke triplets (and lenses derived from them) the vertex length can be controlled by the choice of glass types used in the design. The axial chromatic aberration contribution of a thin element varies with $y^2\phi/V$, as in Eq. (24.86). If we require a specific amount of chromatic aberration from an element (to correct the chromatic for the entire lens), it is apparent that y, the marginal-axial ray height, must be smaller if we use a glass with a low V-value than if we use a higher V-value glass. In the triplet, the ray height at the center flint will be reduced if the spacing between the elements is increased. Thus a greater vertex length produces a lower ray height at the flint and requires a lower V-value for the flint element (or a higher V-value for the positive crown elements) to maintain chromatic correction.

Therefore, to produce a long, high-speed, narrow-field (triplet) anastigmat, glasses with a large V-value difference between (positive) crown and (negative) flint elements are appropriate. For short, slow, wide-angle systems, a small V-value difference is suitable. By allowing the glass types to vary during the optimization process, an automatic design program can select the appropriate V values to produce a vertex length that is optimally suited to the aperture and field desired. Obviously, the starting glasses for the design should be chosen with this relationship in mind. It is usually preferable to allow the flint glass to vary and to manually select the crown glass from those in the upper left region of the glass map (Fig. 2.2) on the basis of cost, workability, and stability. The optimum flint glass is always on the glass line, because a low index in a negative element is beneficial to the Petzval correction. These

---

*W. Smith, "Control of Residual Aberrations in the Design of Anastigmat Objectives" *J. Opt. Soc. Am.*, vol. 48, 1958, pp. 98–105.

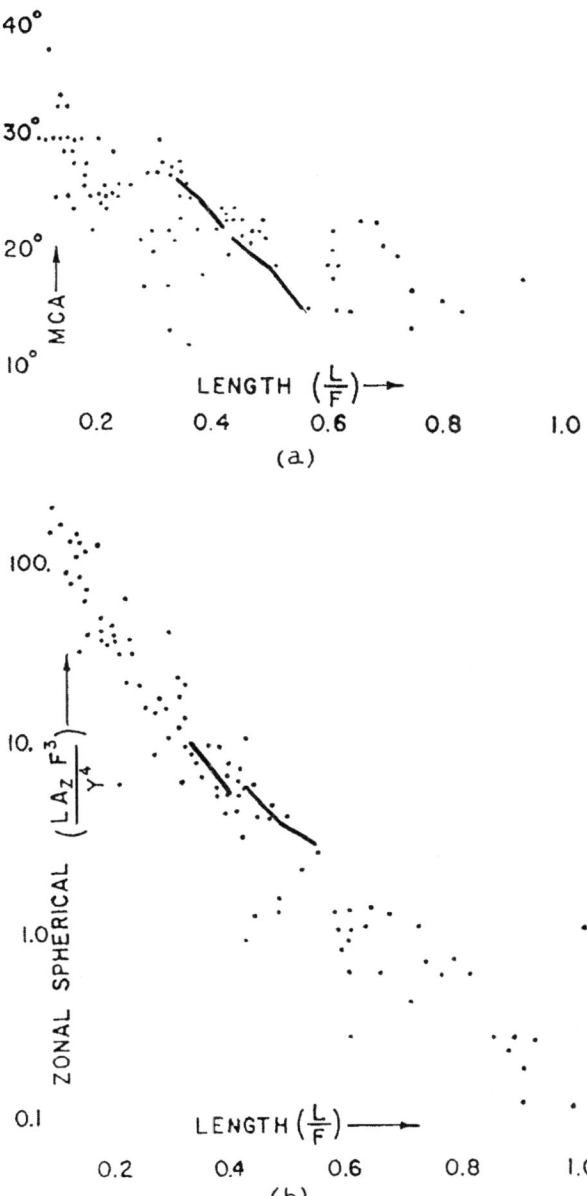

Figure 8.2 (a) Plot of angular field coverage or maximum coverage angle (MCA) as a function of vertex length (L/f) for about 85 different anastigmat designs. Lenses with three to seven elements, without restriction as to glass types, age, application or design configuration, are included. (b) Plot of fifth-order spherical (zonal) aberration as a function of the vertex length for the same lenses as in a. The solid lines are for a four-element Dogmar/Celor-type design in which the conditions (Petzval, chromatic, etc. correction) were carefully controlled so that the vertex length was the primary variable.

glass-line glasses are mostly inexpensive, durable, available, and workable types. It is usually quite safe to allow the design program to select from among these glasses.

Figure 2.8 gives a rough idea of the limit of the design capabilities of the Cooke triplet, as well as other design forms. In general, as with most lenses, when the speed of the triplet is increased, the angular coverage must be reduced (and vice versa) to maintain a given level of image quality.

In a long triplet (usually resulting from a flint glass with a low V-value and a high index) it can be seen that the principal ray heights at the outer elements will be large. The result is large high-order, off-axis aberrations (astigmatism, coma, and oblique spherical); the acceptable field is limited to a small angle. In addition to this, if the system is very long, the power of, and the axial ray height at, the flint element may be so small that it cannot correct the spherical aberration when all the other primary aberrations are corrected.

As regards the vertex length of a triplet:

1. Inward Petzval leads to a long system.
2. A ray bending ratio ($y_1\phi_1/y_3\phi_3$) equal to one produces a shorter system than a ratio of two or one-half.
3. Index has little effect on length, but high-index crowns lead to lower element powers.

8.3.1 A small field triplet anastigmat

It can be a bit tricky to design a lens of moderate speed and a small field (e.g., $f/3.5$ and $\pm 4°$) when the project at hand requires a much better off-axis performance than can be obtained with a telescope objective. A Cooke triplet is the obvious choice for this situation. But unfortunately the lens design program will likely go to a configuration with very good on-axis performance and very bad off-axis aberrations. In other words it may design a three-element telescope objective. One approach to avoid this could be to target a suitable length for the system, and/or reduce the weight on the axial targets. Another solution is to start out with a Cooke triplet covering a wider field angle than you need (to establish a stable anastigmat form), then gradually reduce the field, optimizing at each step in the reduction. This will hold the triplet anastigmat configuration and yield a flat field with much better off-axis performance (due to the control of astigmatism that a compact telescope objective lacks). Of course the axial performance will not compare with the excellent correction represented by the telescope objective.

An example of such a design is given in Table 8.1 and shown in Figs. 8.3 and 8.4.

TABLE 8.1 The Prescription for the f/3.5, +/−4 deg Field Triplet Shown in Figs. 8.3 and 8.4

$R1 = +33.134$	$T1 = 5.831$	SSKN5	658509
$R2 = +622.92$	$T2 = 14.692$		
$R3 = -52.103$	$T3 = 2.471$	SF63	748277
$R4 = +26.897$	$T4 = 1.977$		
$R5 = $ stop	$T5 = 12.588$		
$R6 = +90.454$	$T6 = 5.831$	SSKN5	658509
$R7 = -39.756$	BFL $= 76.952$		

Should a higher level of performance be necessary, a split front crown triplet (Chap. 9) design could be used.

8.4 Other Design Considerations

A number of other generalizations also apply to the Cooke triplet:

1. The higher the index of the positive elements, the better. Actually, it is primarily the *difference* between the index of the crown and flint that

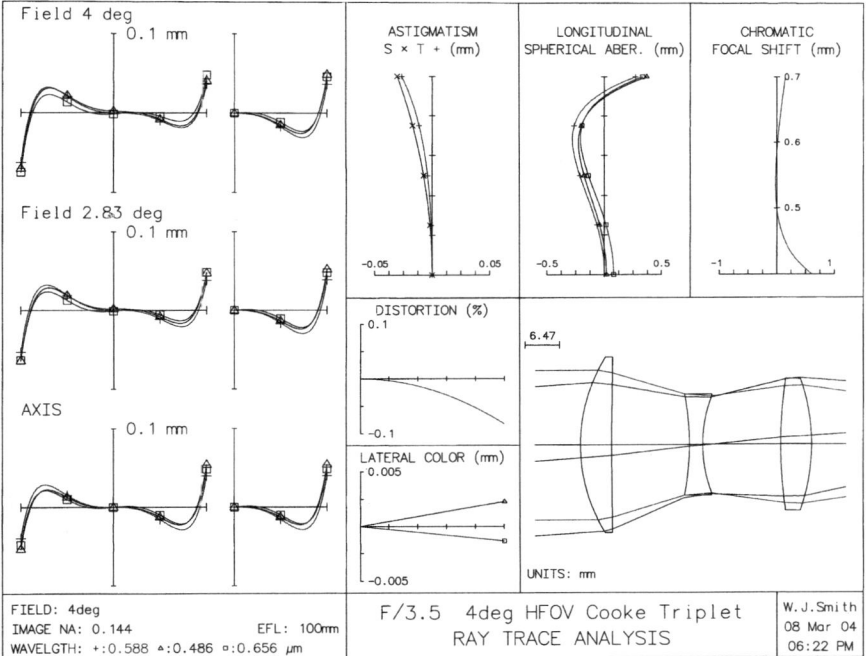

Figure 8.3 $f/3.5$ ±4° HFOV, Table 8.1, triplet form achieves more uniform performance over a small field than the compact telescope objective type.

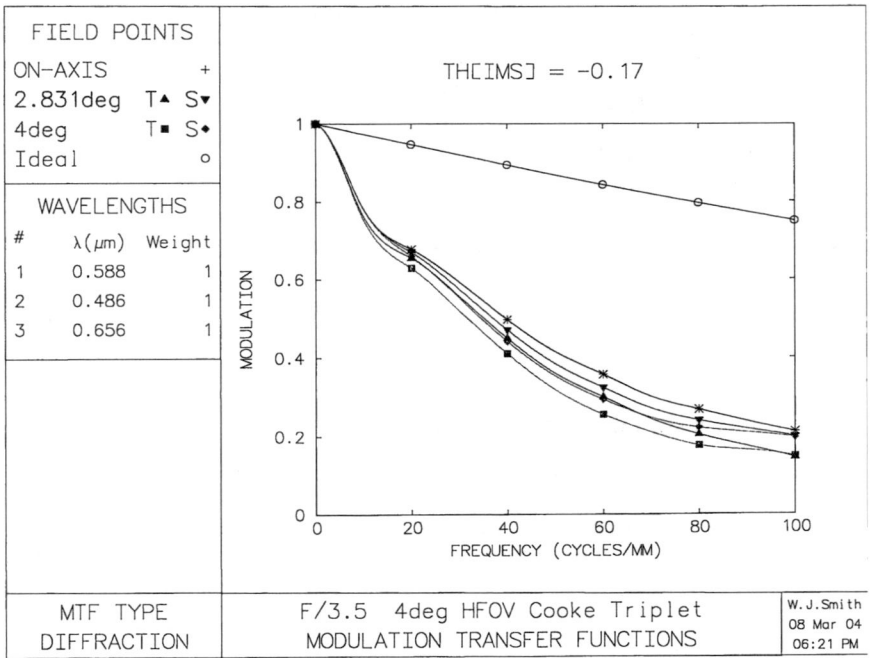

Figure 8.4 MTF for Fig. 8.3, Table 8.1.

affects the lens performance, but since the index of the flint cannot be reduced significantly below the glass line, and since the flint V-value should be chosen to produce the optimum vertex length for a given aperture and field as indicated in Sec. 8.3, the flint index is almost predestined once the crown glass type is chosen; thus the index difference is effectively determined by the index (and V-value) of the crown elements.

2. Allowing a more inward-curving Petzval field will result in lower powers for all the elements when the other primary aberrations are corrected. The effect of the lower element powers is that the residual aberrations are smaller—there is less zonal spherical aberration, less high-order coma, astigmatism, and the like.

3. Allowing the axial chromatic to be somewhat undercorrected has the same beneficial effect as described above for the Petzval undercorrection.

Figures 8.5 to 8.8 show a series of four Cooke triplets with various combinations of field and aperture ($f/6.3$, 27°; $f/4.0$, 23°; $f/3.0$, 19°; $f/2.5$, 16°), which were chosen to lie along a line well above and approximately parallel to the triplet area of Fig. 2.8. All four have SK4 crown elements

Cooke Triplet Anastigmats 211

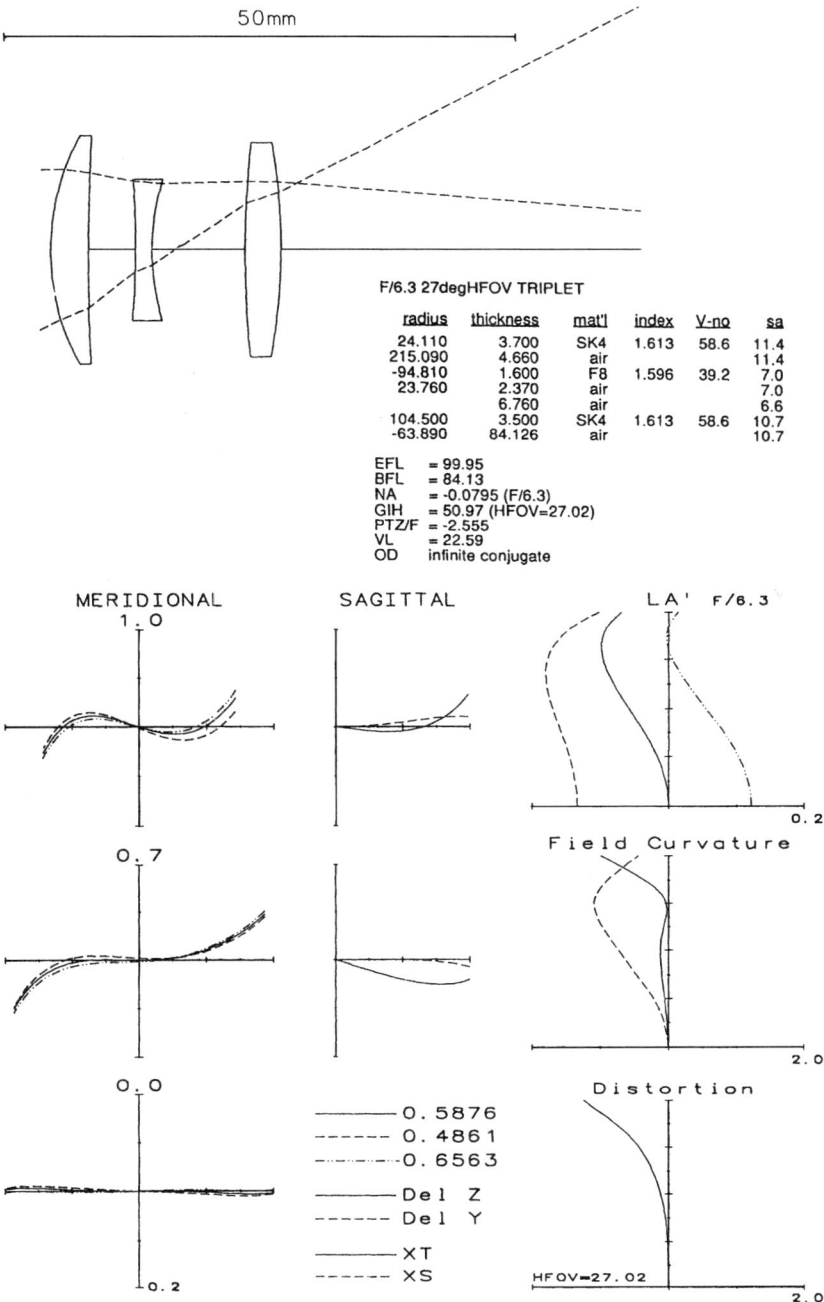

F/6.3 27degHFOV TRIPLET

radius	thickness	mat'l	index	V-no	sa
24.110	3.700	SK4	1.613	58.6	11.4
215.090	4.660	air			11.4
-94.810	1.600	F8	1.596	39.2	7.0
23.760	2.370	air			7.0
	6.760	air			6.6
104.500	3.500	SK4	1.613	58.6	10.7
-63.890	84.126	air			10.7

EFL = 99.95
BFL = 84.13
NA = -0.0795 (F/6.3)
GIH = 50.97 (HFOV=27.02)
PTZ/F = -2.555
VL = 22.59
OD infinite conjugate

Figure 8.5 $f/6.3$ ±27° triplet. Compare vertex length and V-value of flint with Figs. 8.6 to 8.8.

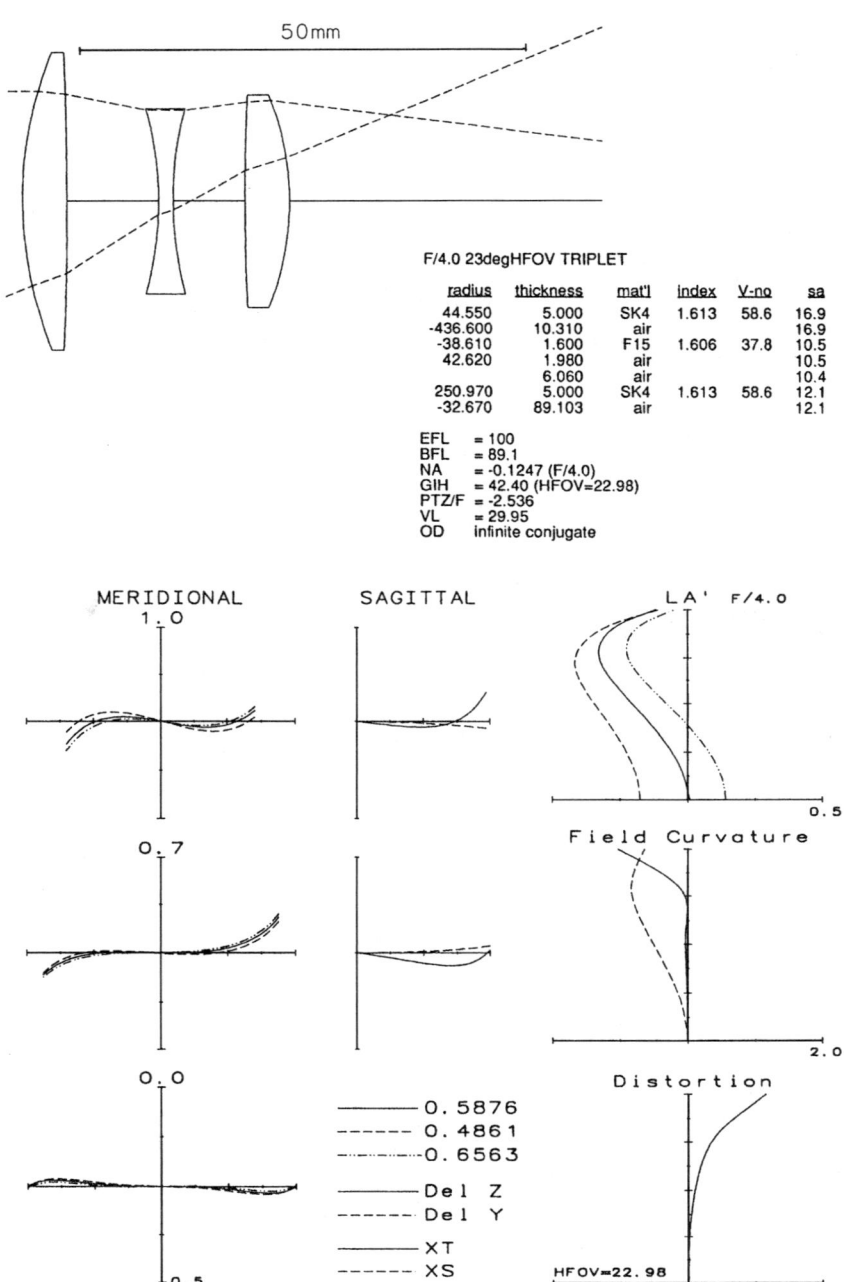

Figure 8.6 f/4.0 ±23° triplet. Compare vertex length and V-value of flint with Figs. 8.5 to 8.8.

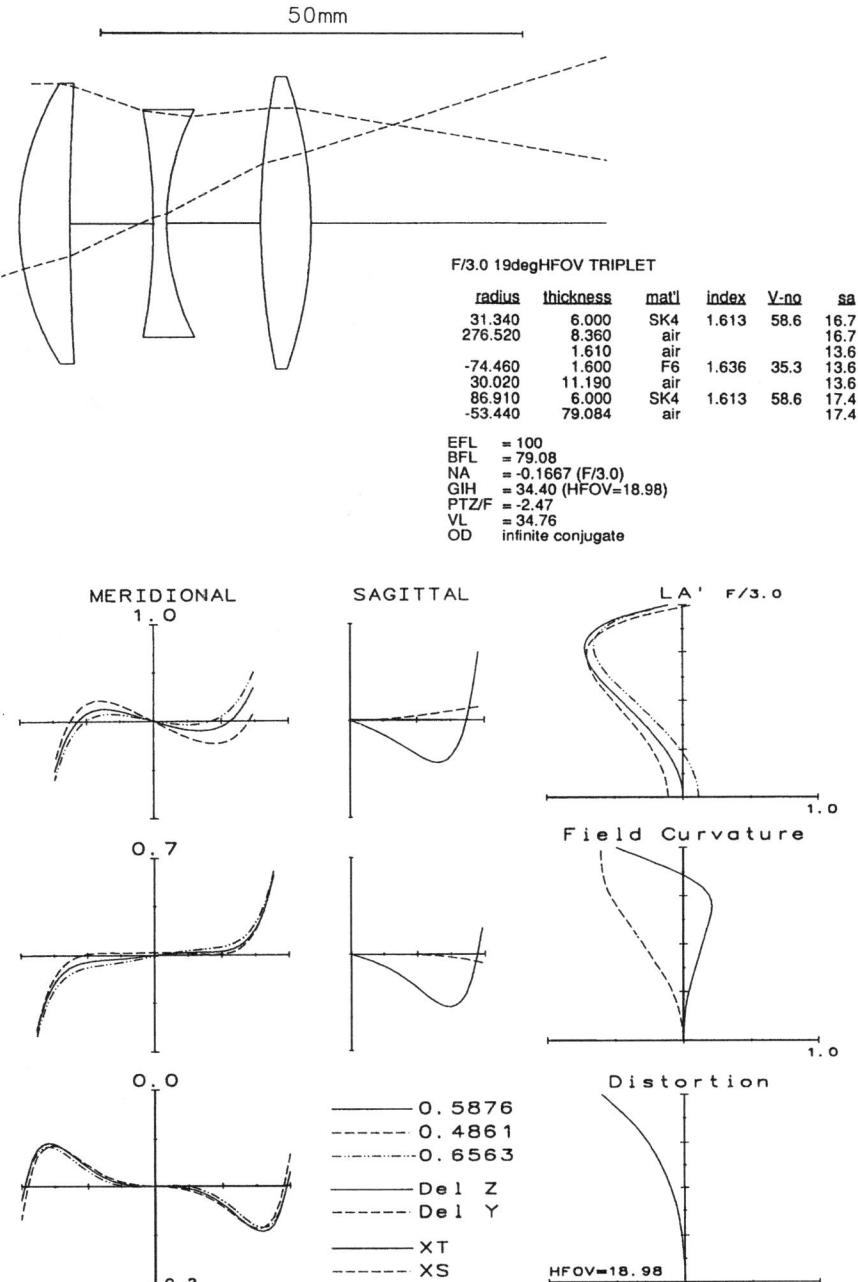

Figure 8.7 $f/3.0$ ±19° triplet. Compare vertex length and V-value of flint with Figs. 8.5 to 8.8.

214   Chapter Eight

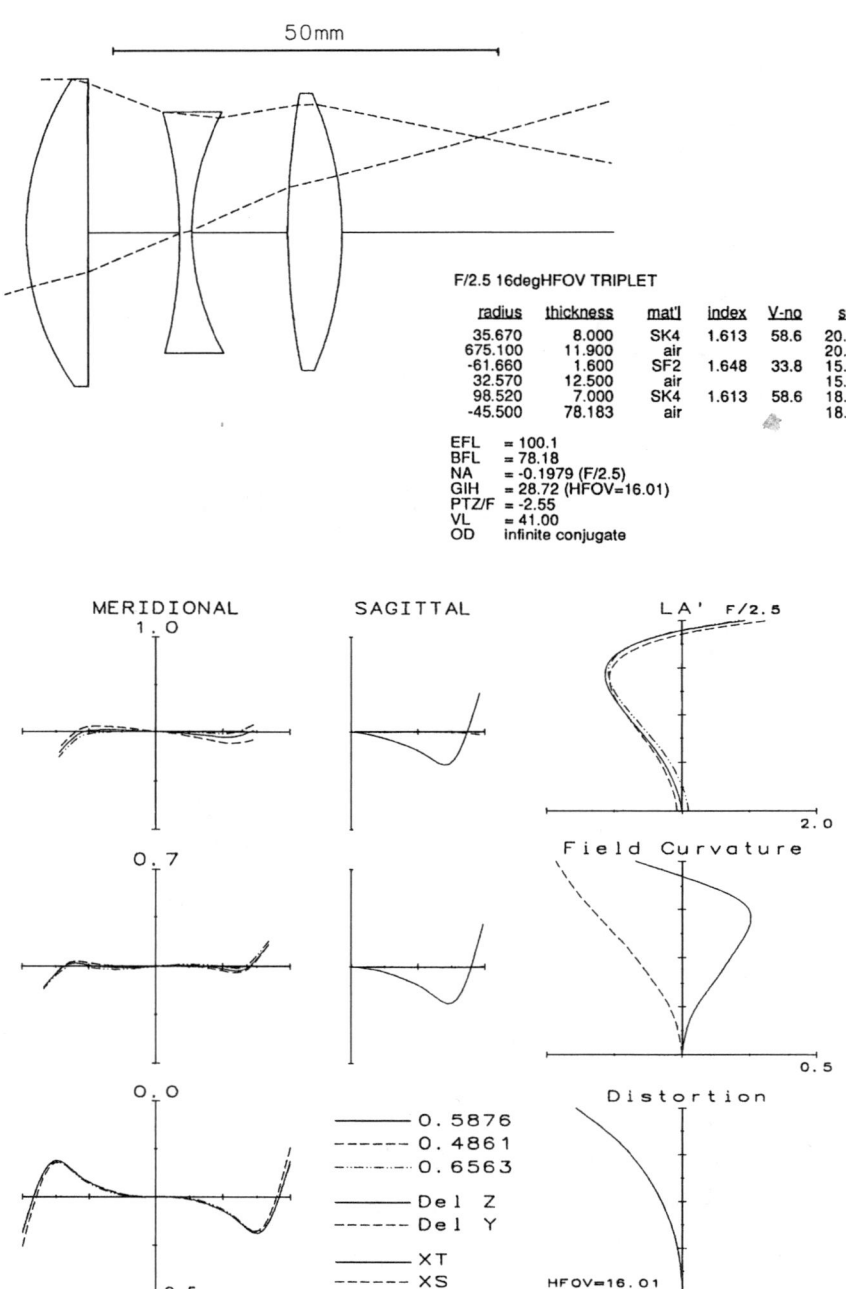

Figure 8.8 f/2.5 ±16° triplet. Compare vertex length and V-value of flint with Figs. 8.5 to 8.7.

and all were designed with the same optimization program and merit function. (Because the combinations of field and speed of these lenses are well beyond the normal capabilities of the triplet form, these lenses have large aberration residuals.) Notice the relationships between:

a. The vertex length
b. The flint glass V value
c. The field and aperture

which clearly illustrate the correlations shown in Fig. 8.2 and the discussion in Sec. 8.3.

Figures 8.9 to 8.14 present a selection of Cooke triplets of various aperture and field combinations. Figures 8.9 and 8.10 show two lenses with similar, but not identical, combinations of aperture and field. Note that the speed of $f/2.5$ in Fig. 8.9 has produced a very long lens (55 percent of the focal length) and the lens has achieved a good balance of correction. Note the unusually thick elements; it would be an interesting exercise to attempt to reduce the thickness of the two front elements and simultaneously maintain the same level of correction. Figure 8.10 indicates the improvement in field curvature (while maintaining the axial correction), which results from the use of a higher-index crown glass as compared to the lower-index designs.

Figure 8.11 shows a triplet of a speed and angular coverage that is typical of 35-mm slide-projection lenses.

Figure 8.12, at a speed of $f/4.4$ and a normal coverage angle is typical of a triplet camera lens. Figure 8.13 is an extremely well-corrected lens designed for microfilm work; note the high-index glasses and the fact that the field angle is about 35 percent that of Fig. 8.12; these factors combine to produce a level of correction that is well above that of any of the other lenses in this chapter. Note also that the front airspace is longer than the rear and that $R3$ is shorter than $R4$; this is the reverse of most Cooke triplets, i.e., the left-hand solution. Note the focal length of the example is 50.7 mm.

Figure 8.14 covers 68° at a speed of $f/6.3$. This general type of lens was for years the staple in inexpensive folding cameras. At this end of the speed and field range the front airspace tends to be quite small—this seems to allow the wide field of view. There is an old design form where the first two elements are cemented together, which might be considered a version of the triplet.

8.5 A Plastic, Aspheric Triplet Camera Lens

Through the courtesy of Ellis Betensky, arguably one of the world's best lens designers, we show an example of what can be done with aspheric surfaces. Normally a triplet made of plastic elements is a very poor example

216   Chapter Eight

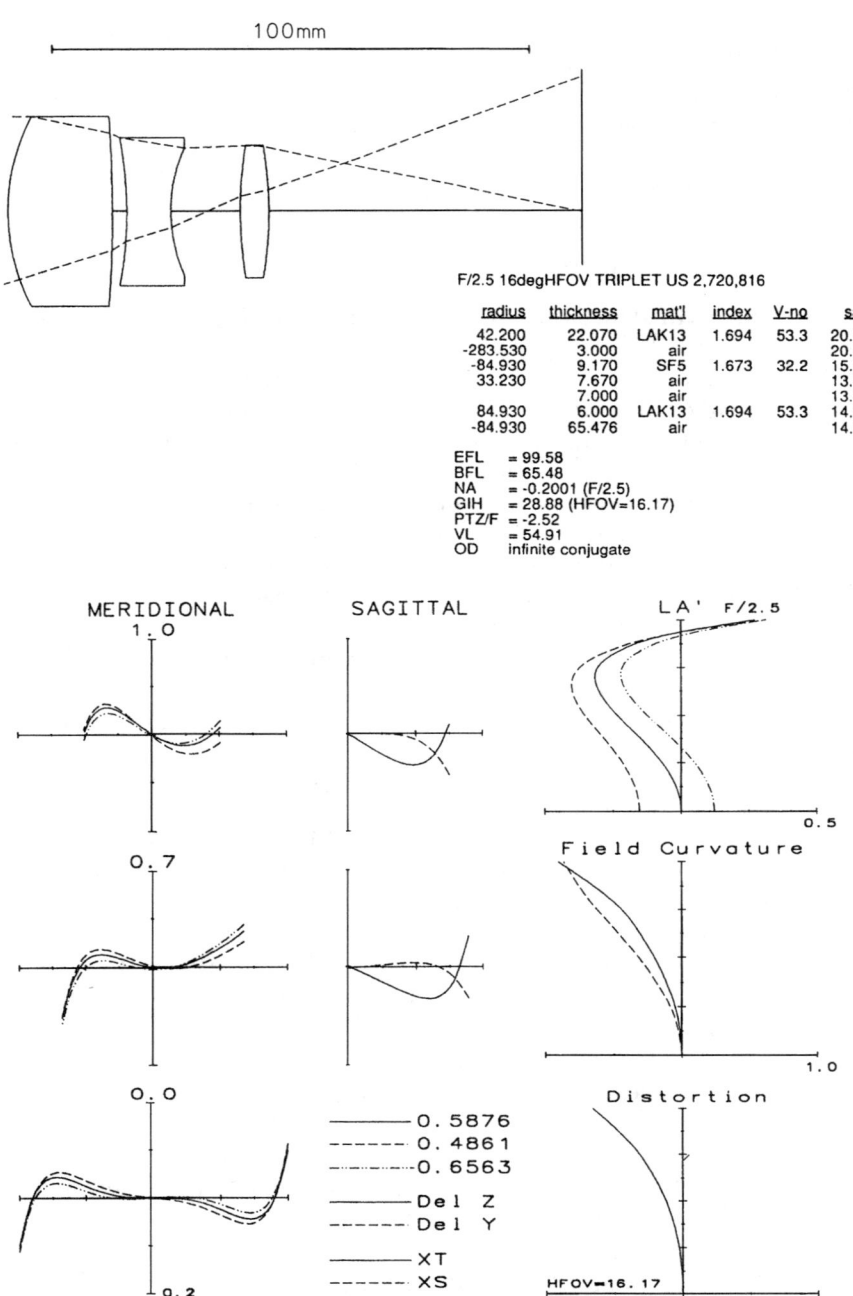

Figure 8.9 *f*/2.5 ±16° triplet with high-index crowns and thick elements yields better performance than Fig. 8.8.

Cooke Triplet Anastigmats 217

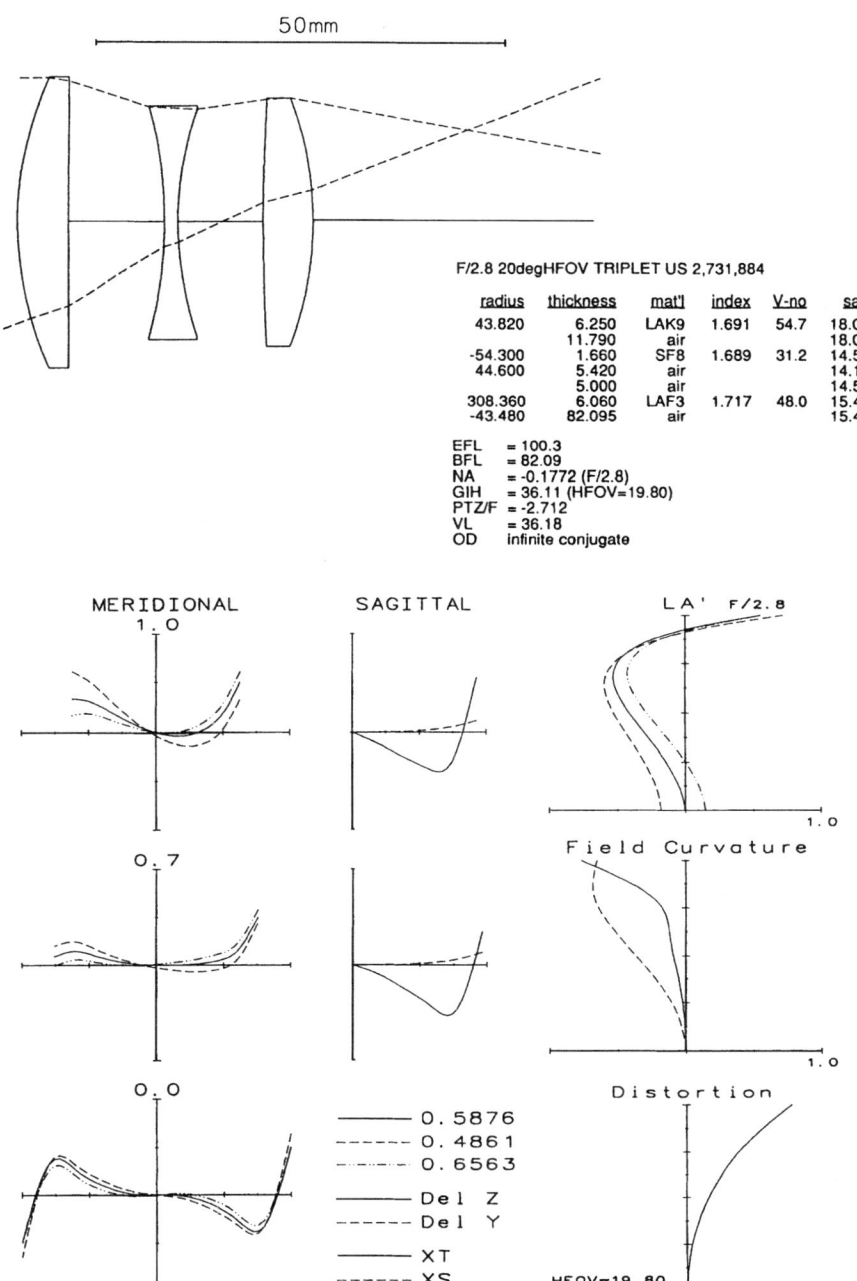

F/2.8 20degHFOV TRIPLET US 2,731,884

radius	thickness	mat'l	index	V-no	sa
43.820	6.250	LAK9	1.691	54.7	18.0
	11.790	air			18.0
-54.300	1.660	SF8	1.689	31.2	14.5
44.600	5.420	air			14.1
	5.000	air			14.5
308.360	6.060	LAF3	1.717	48.0	15.4
-43.480	82.095	air			15.4

EFL = 100.3
BFL = 82.09
NA = -0.1772 (F/2.8)
GIH = 36.11 (HFOV=19.80)
PTZ/F = -2.712
VL = 36.18
OD infinite conjugate

Figure 8.10 f/2.8 ±20° triplet with high-index crowns.

218 Chapter Eight

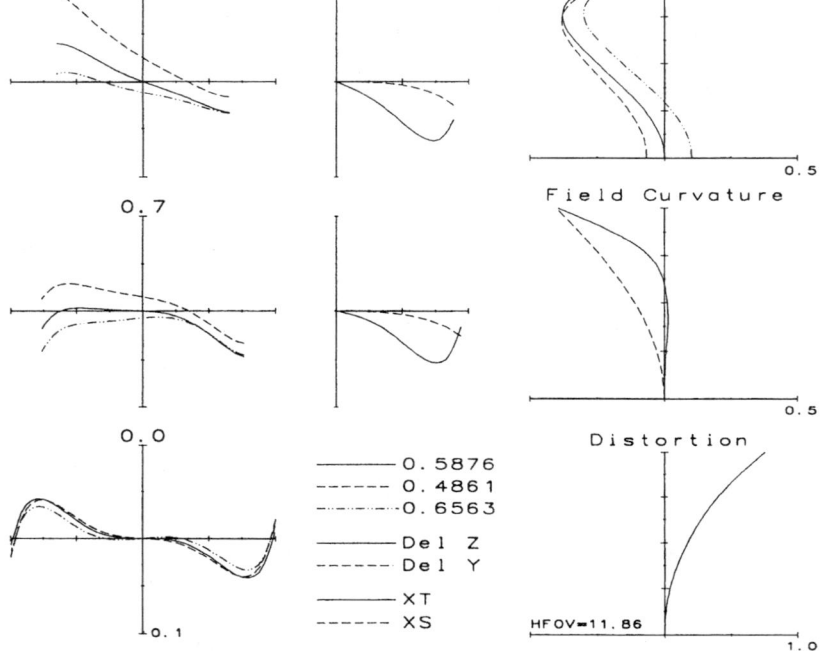

Figure 8.11 $f/3.5$ ±12° triplet of the type used in 35-mm slide projectors.

Cooke Triplet Anastigmats 219

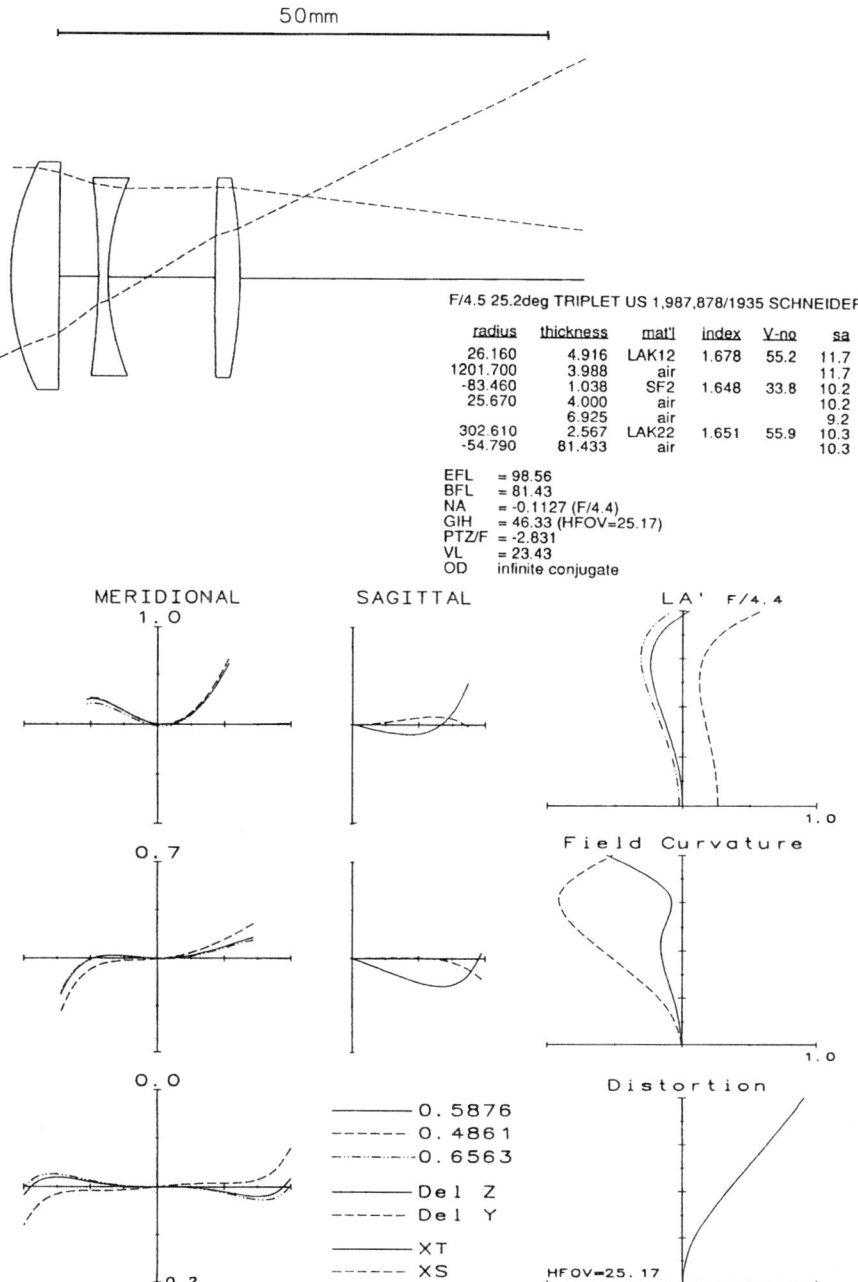

F/4.5 25.2deg TRIPLET US 1,987,878/1935 SCHNEIDER

radius	thickness	mat'l	index	V-no	sa
26.160	4.916	LAK12	1.678	55.2	11.7
1201.700	3.988	air			11.7
-83.460	1.038	SF2	1.648	33.8	10.2
25.670	4.000	air			10.2
	6.925	air			9.2
302.610	2.567	LAK22	1.651	55.9	10.3
-54.790	81.433	air			10.3

EFL = 98.56
BFL = 81.43
NA = -0.1127 (F/4.4)
GIH = 46.33 (HFOV=25.17)
PTZ/F = -2.831
VL = 23.43
OD infinite conjugate

Figure 8.12 $f/4.4$ ±25° triplet with high-index glass in a form typical of an inexpensive camera lens.

220 Chapter Eight

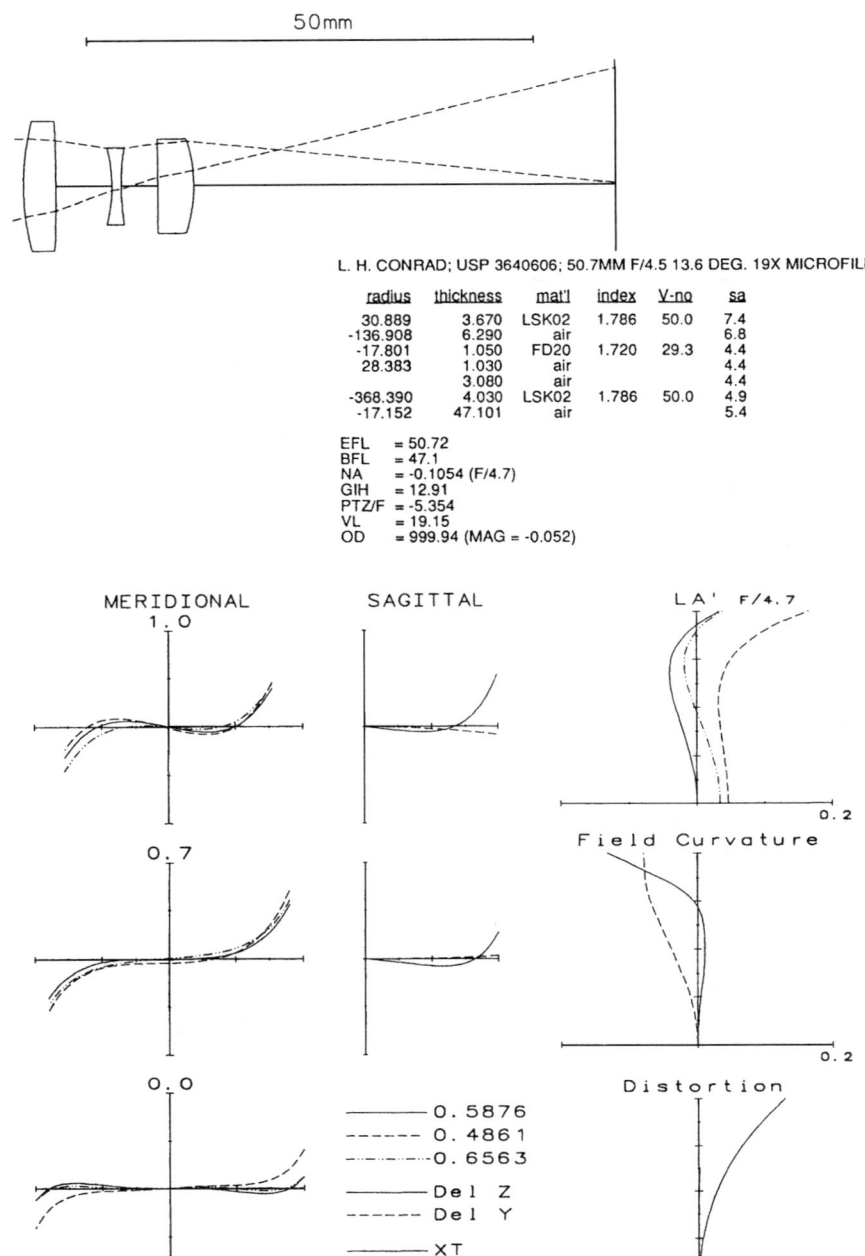

Figure 8.13 $f/4.7$ ±14° high-performance triplet for microfilm work.

Cooke Triplet Anastigmats   221

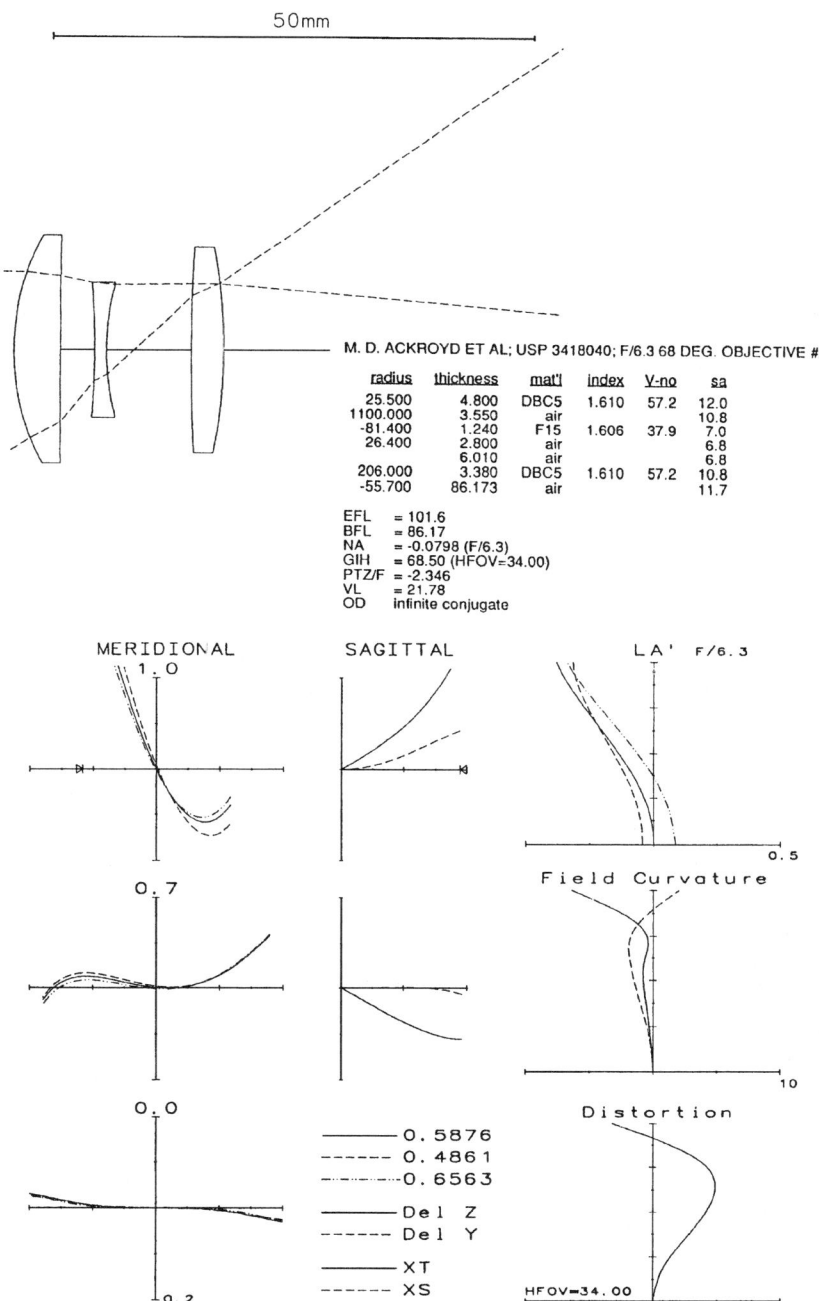

M. D. ACKROYD ET AL; USP 3418040; F/6.3 68 DEG. OBJECTIVE #2

radius	thickness	mat'l	index	V-no	sa
25.500	4.800	DBC5	1.610	57.2	12.0
1100.000	3.550	air			10.8
-81.400	1.240	F15	1.606	37.9	7.0
26.400	2.800	air			6.8
	6.010	air			6.8
206.000	3.380	DBC5	1.610	57.2	10.8
-55.700	86.173	air			11.7

EFL = 101.6
BFL = 86.17
NA = -0.0798 (F/6.3)
GIH = 68.50 (HFOV=34.00)
PTZ/F = -2.346
VL = 21.78
OD infinite conjugate

Figure 8.14 $f/6.3$ ±34° triplet of the type widely used in inexpensive folding cameras.

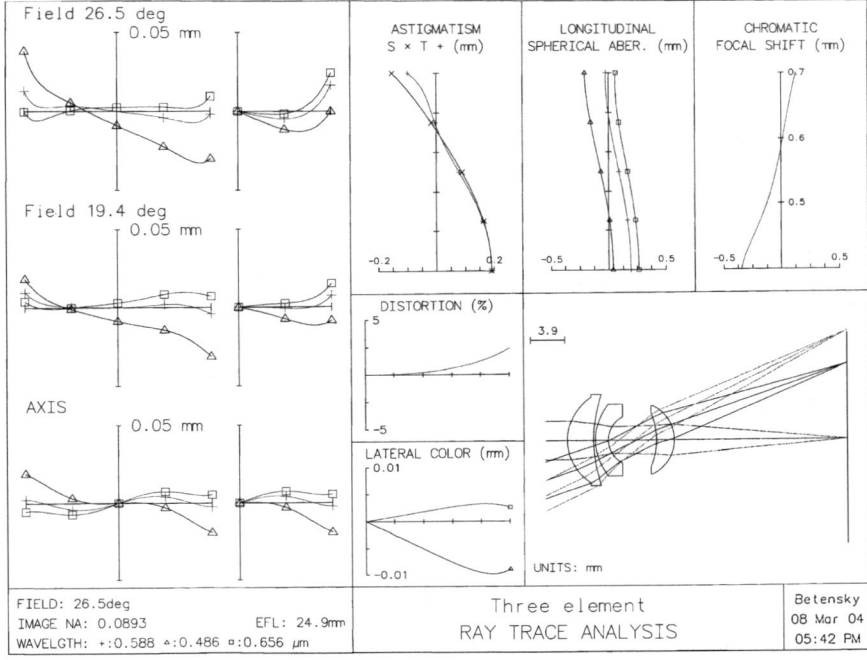

Figure 8.15 Aberration plots for the Betensky three element aspheric plastic camera lens, f/5.6 ±26°, Table 8.2.

of the breed. Figures 8.15 and 8.16 show the aberrations and *modulation transfer function* (MTF) of a triplet that Betensky designed a number of years ago. Table 8.2 gives the design data for this lens. (Another Betensky plastic aspheric lens is in Chap. 11, *Double Meniscus Anastigmats*.)

TABLE 8.2 The Prescription for the 25 mm, f/5.6, +/−26 deg Field Aspheric Plastic Triplet Shown in Figs. 8.15 and 8.16

$R1 = +6.069$	$T1 = 2.932$	Acrylic	490579
$R2 = +13.875$	$T2 = 0.200$		
AD = −.000285;	AE = 1.7067e − 06;		
AF = 1.1374e − 09;	AG = −8.9508e − 11		
$R3 = +6.039$	$T3 = 1.472$	Polycarb	585303
AD = −.000402;	AE = −1.3476e − 05		
AF = −4.3318e − 07;	AG = −6.7247e − 09		
$R4 = +3.061$	$T4 = 3.265$		
$R5 = $ Stop	$T5 = 2.240$		
$R6 = -8.670$	$T6 = 1.988$	Acrylic	490579
AD = −.000231;	AE = −3.2680e − 05		
AF = 2.9558e − 05;	AG = −1.3003e − 07		
$R7 = -4.540$	$T7 = 19.779$		
$R8 = $ plano	$T8 = -.191$ (defocus)		

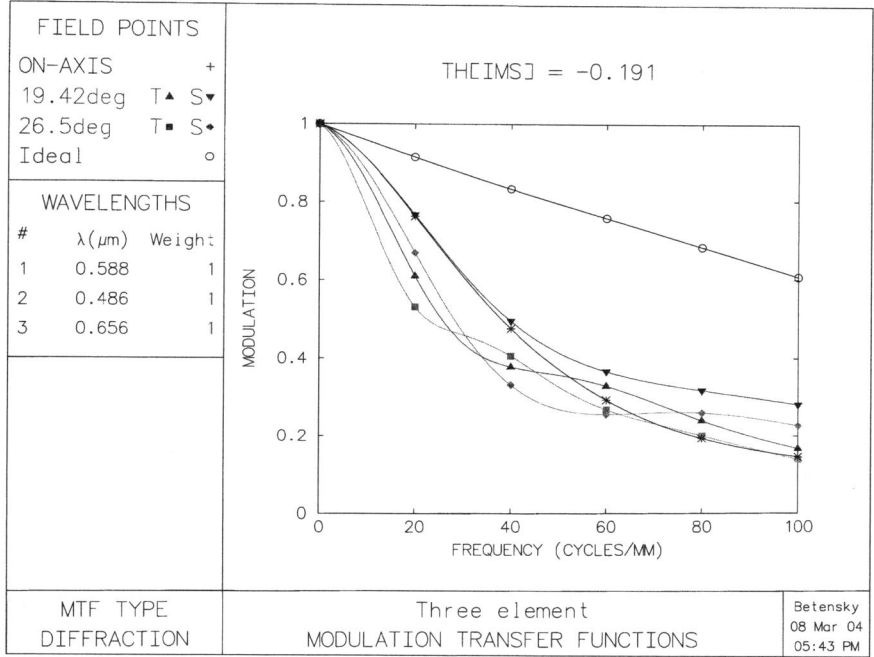

Figure 8.16 MTF plot for Fig. 8.15 lens, Table 8.2.

8.6 Camera Lens Anastigmat Design "from Scratch"—The Cooke Triplet

8.6.1 Project: Design a 2 in. *f*/4 lens for a 35-mm camera

As a vehicle to demonstrate anastigmat design, we elect to work with an objective lens for a 35-mm camera at *f*/4, covering ±23.1° (these are admittedly modest requirements, selected for ubiquity and generality). In the section of the story in this chapter, we will investigate the Cooke triplet, beginning with ordinary glasses (SK16 and F2), then investigating the range of ordinary (i.e., not lanthanum) glasses, selecting SSKN5 and SF15 as the best of this limited batch, then proceed to the rare earth glasses and on to aspheric surfaces.

The project will continue in Chap. 11 with the design of the Dogmar (Celor), and in Chap. 10 with the Tessar and Heliar types.

The Cooke triplet. We begin the project with the classic Cooke triplet, probably the simplest anastigmat suitable for the task. As shown in Fig. 8.1, it consists of two outer positive crown elements, significantly spaced from a central negative flint element. If we regard the Cooke triplet as three thin elements, we can consider that we have eight variable construction parameters—six surface curvatures plus two airspaces. Another, and more insightful, ordering is to count three powers, three shapes (or bendings), and two airspaces.

For a satisfactory design we must control the first- and third-order characteristics to appropriate values. These are the focal length, the two chromatic aberrations, and the five Seidel aberrations. It is interesting that we have just eight variable parameters to control eight characteristics. One's first (naive) thoughts run to "simply solve eight equations in eight unknowns."

Looking at the thin lens aberration contribution equations in Chap. 24, it becomes apparent that, once the glasses are selected, the *effective focal length* (EFL), the two chromatic aberrations, and the Petzval curvature are independent of the (thin) element shapes, and depend only the element powers and spacings (which control ray heights), whereas the spherical, coma, astigmatism, and distortion aberrations are all affected by the bending or shape of the element as well.

The thin lens aberration contribution equations can be set up and solved to control the shape-independent aberrations. The equations that result from this solution are complex enough that most prefer to choose the glass types and then use an iterative, converging, approximation technique to get the powers and spaces. A second iterative technique is then used to determine the element shapes. The quadratic nature of the thin lens third-order contribution equations indicates that if the glasses are well chosen, there may be as many as eight possible thin lens solutions; experience and observation tell us that only one or two of these solutions will be practical.

In using the above-mentioned iterative techniques it often turned out (given the initial glass choices) that because the final relationships in the iteration were quadratic, there were two significantly different final solutions (which were often dubbed the *right-hand* and *left-hand* solutions.) These solutions usually differed (a) in the relative size of the front and rear airspaces, (b) in the relative strength of the front and back curvatures of the central flint, and (c) in the relative strength of the curvatures of the outer crown surfaces (almost as if the lens were turned end for end in changing from the left solution to the right). As one might expect, there are differences in the higher-order aberrations, especially coma and astigmatism. The best designs seem to occur when the glasses are chosen in such a way that, given the angular field of view and the relative aperture required, the two solutions are almost identical, i.e., they approach the situation where there is only one solution to the final quadratic.

This tiresome algebraic approach does have the advantage that it nicely illustrates some of the interrelationships and limitations of the airspaced triplet configuration. But instead of struggling with the complexity of the aberration contribution equations to solve this problem for a fixed set of glasses, we will include glass as a variable and make use of the *damped least-squares* (DLS) capability of our optical software to find a solution. If we proceed in a logical and methodical way, the process should be simple, quick, and easy.

Goal. EFL = 2 in (50.8 mm), $f/4.0$ for 35 mm film, format 24 × 36 mm, diag. = 43.3 mm (HFOV = ±23.1°).

Assumptions.

1. We will use ordinary glasses, which are inexpensive, stable, and available (i.e., not LaK, LaF, or LaSF).

2. We start with SK16 620603 for both outer crowns, this being a promising glass and there being no advantage in using different glasses for the two crowns.

3. For the flint, we will optimize from along "the glass line," e.g., F2 (620364), SF2 (648338), and the like. These are good, available, and economical glasses. (We start with F2.)

4. These initial glass choices are based on knowledge of existing designs, which saves us a lot of stumbling around.

5. We put the aperture stop at the center flint (on $R4$) so that we can benefit from a rough front-to-back symmetry.

(For the sake of brevity and compactness, we will ignore the fact that in a camera lens the stop must be located within an airspace, allowing room for the iris diaphragm. For the same reasons we will also ignore the fact that camera lenses are usually designed to work best when *stopped down* one or two stops. This usually means that the marginal spherical aberration is overcorrected, so that $LA_m \approx -LA_z$.)

6. Many camera lenses are roughly *square*, that is, the lens diameter is approximately equal to its overall length. For our starting length, this *very* approximate rule of thumb gives us:

$$\Sigma T_i = \text{EFL}/(f/\text{no}) = 50.8/4 = 12.7 \text{ mm}$$

7. We set airspaces $T2 = T4 = 6.3$. (Note that if the product $y \cdot \phi$ is the same at both crowns, setting the airspaces equal will make y_p the same on both and will yield zero lateral color because $y_i \, y_{pi} \, \phi_i$ will be equal and oppositely signed for elements one and three).

8. Based on published triplet design data we can make an *extremely* rough guess at the radii:

$R1 = +18$; $R2 = $ plano; $R3 = -18$; $R4 = +18$; $R5 = $ plano; $R6 = -18$

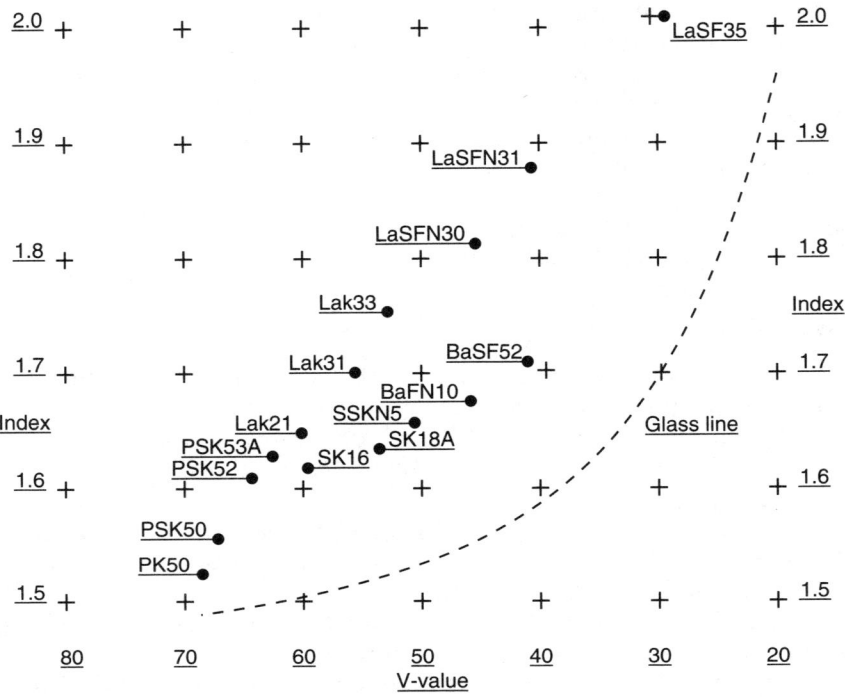

Figure 8.17 Index versus V-value glass map, showing the crown glasses used in Sec. 8.6.

The crown glass types discussed in this project are shown in the glass map of Fig. 8.17.

Solely for the purpose of illustration, we will duplicate the algebraic solution process outlined above, first determining the powers and spaces to satisfy the thin lens focal length, chromatic, and Petzval requirements. Then we will add element thicknesses and adjust the element shapes to control all the primary aberrations (first- and third-order) plus the fifth-order aberrations as well. Only then will we move on to the correction and balancing of the full pantheon of all the aberrations (to the extent that the simple triplet construction allows). And actually, this simple path, using first, third, and fifth order, is often a quite good general design approach, especially when starting a design from scratch.

9. We will vary CV1, CV4, CV6 and $T2$ (=$T4$). Alternatively we could fix the airspaces and vary the flint [no. 3] glass.

10. We create a custom merit function (MF) with targets for EFL = 50.8, TAch = 0.0, TchA = 0.0, and the Petzval radius = $-3.5 \times$ EFL.

A few DLS (Damped Least Squares) cycles yield the data as shown in Table 8.3.

TABLE 8.3 The Prescription of the First "Thin Lens" Triplet with the Focal Length = 50.8, Axial and Lateral Color = zero and the Petzval Radius = −177.8

$R1 = +16.94v$	$T1 = 0.0$	SK16	EFL = 50.8
$R2$ = plano	$T2 = 6.16v$		TAch = 0.0
$R3 = -18.00$	$T3 = 0.0$	F2	TchA = 0.0
$R4 = +17.85v$	$T4 = 6.16p$		Petz $R = -177.8$
$R5$ = plano	$T5 = 0.0$	SK16	
$R6 = -14.88v$	BF = 44.64		

NOTE: v indicates a variable and p means the value is picked up from a preceding thickness (no. 2 in this case).

Recommendation: After each stage of a design procedure such as this, it is very wise to save the design to disk. Then if something goes awry during the next step one can always back up and repeat the step without returning to the beginning and repeating everything all over again. Needless to say, an orderly scheme of naming and numbering the lens files is imperative.

11. Now we manually add element thicknesses to get the *edge thickness* (ET) for the crowns to about 1mm, and set $T3 = 1.0$ mm for the flint.

12. We add targets to the MF for ET = 1.0 at $y = 6.8$ mm for surfaces 1–2, and at $y = 6.0$ mm for 5–6, choosing these semiapertures to pass the axial marginal ray (after making a few trial drawings on the computer). This is a common and economical commercial choice for lens diameters.

A few DLS cycles with $T1$ and $T5$ added as variables yield the data shown in Table 8.4.

This is a thick lens triplet with the desired focal length, chromatic aberrations, and Petzval radius.

13. Next we add targets for (third- plus fifth-order aberrations) spherical, coma, astigmatism, and distortion to the MF to drive the (third-plus fifth-order) aberrations to zero, and add CV2, CV3, and CV5 as variables.

TABLE 8.4 Prescription for the First Thick Element Triplet with Aberrations Controlled as in Table 8.3

$R1 = +16.77v$	$T1 = 2.44v$	SK16	620603
$R2$ = plano	$T2 = 4.92v$		
$R3 = -18.00$	$T3 = 1.0$	F2	620364
$R4 = +17.96v$	$T4 = 4.92p$		
$R5$ = plano	$T5 = 2.24v$	SK16	620603
$R6 = -15.10v$	BF = 43.30		

After a few cycles of DLS iteration we meet all our targets and get the following data:

Prescription for the Triplet Lens with the First, Third, and Fifth Order Aberrations Controlled

$R1 = +20.19v$	$T1 = 2.26v$	SK16	620603
$R2 = -301.5v$	$T2 = 6.81v$		
$R3 = -21.97v$	$T3 = 1.00$	F2	620364
$R4 = +17.26v$	$T4 = 6.81p$		
$R5 = +85.47v$	$T5 = 2.21v$	SK16	620603
$R6 = -18.58v$	$BF = 42.40$		

Now a trigonometric raytrace analysis gives us:
Longitudinal spherical $\approx +0.15$
Longitudinal chromatic ≈ -0.15
Tangential Coma $\approx +0.1$
$X_s \approx 0$ and $X_t \approx -5$
Distortion and lateral color are close to zero.

Considering that we have only corrected the third- and fifth-order aberrations, at this stage these results are not too bad. But the vignetting is a rather horrific 80 to 90 percent at the edge of the field.

Nonetheless, we elect to forge on (saving this intermediate result to disk just in case). We drop the third- and fifth-order targets and use the software default rms spot size merit function (with added targets for EFL and ET). We choose to achromatize with a Conrady $(D-d)\delta n$ target, and add a penalty for any distortion in excess of 1 percent. To be careful, we iterate only once, and when everything seems OK, we add as variables the flint glass and the thickness $T4$ (dropping the $T2 = T4$ constraint). After two iterations the lens has become shorter and the flint is now 625358, i.e., $n = 1.625$ and $v = 35.8$ (it started at 620364). The vignetting has been reduced to about 50 percent (from 90 percent). After two more iterations the flint glass type is still 626358.

14. We fix the flint glass as F1 (626357) and optimize until there is no further improvement to the merit function. This gives the design shown in Figs. 8.18 and 8.19 with the data as shown in Table 8.5.

TABLE 8.5 Prescription for the Optimized Triplet Derived from the Previous Table, using the Software Default rms Spot Size Merit Function (Performance is Shown in Figs. 8.18 and 8.19)

$R1 = +18.441v$	$T1 = 2.255v$	SK16	620603
$R2 = +519.57v$	$T2 = 5.196v$		
$R3 = -29.616v$	$T3 = 1.000$	F1	626357
$R4 = +18.218v$	$T4 = 5.453v$		
$R5 = +68.949v$	$T5 = 2.069v$	SK16	620603
$R6 = -22.700v$	$BF = 42.509$		

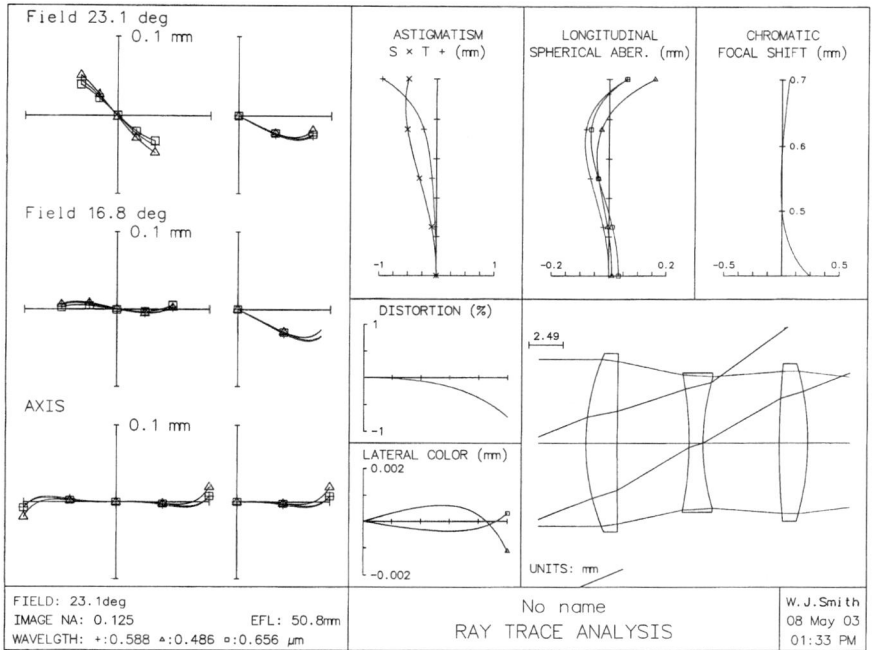

Figure 8.18 Aberrations of the $f/4.0 \pm 23°$ triplet resulting from the initial design run, Table 8.5.

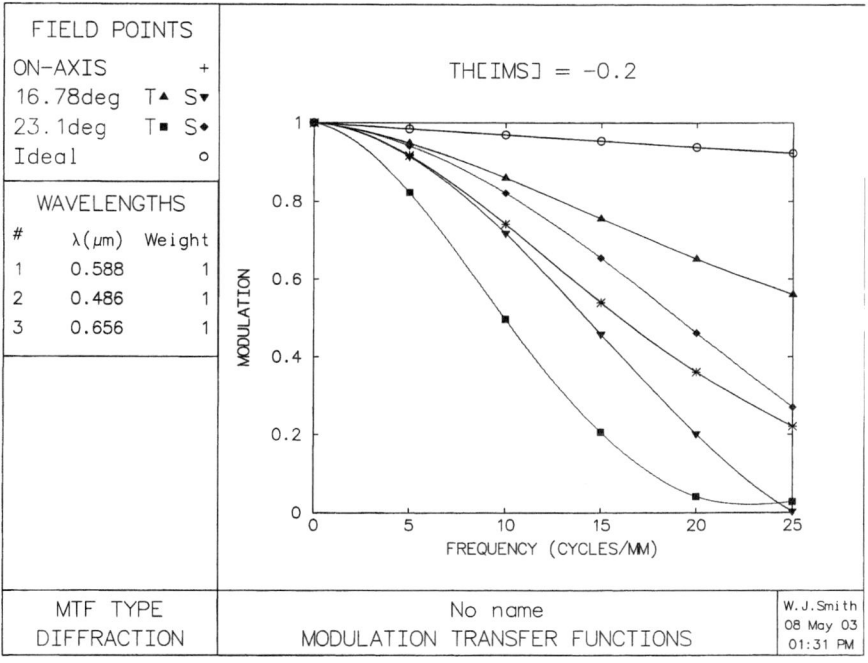

Figure 8.19 MTF for Fig. 8.18 lens, Table 8.5.

230 Chapter Eight

TABLE 8.6 Prescription of the Triplet Lens Derived from Table 8.5 After the Correctly Vignetted Rays were used in the Merit Function (The Performance is Shown in Figs. 8.20 and 8.21)

$R1 = +20.92$	$T1 = 2.14$	SK16	620603
$R2 = -19937$	$T2 = 5.86$		
$R3 = -22.60$	$T3 = 1.00$	F4	617366
$R4 = +20.00$	$T4 = 4.54$		
$R5 = +78.89$	$T5 = 2.22$	SK16	620603
$R6 = -18.63$	$BF = 44.01$		

Now, as a perfect example of the kind of thing that actually happens fairly often in the course of an "unrehearsed" design effort such as this, we suddenly remember that the default rays in our merit function did not take vignetting into account. We adjust the rays to more nearly represent the actual vignetting and repeat the optimization. The flint goes to 615368, which we fix to F4 (617366), and the MTF is much better. Reoptimized, we now have the data as shown in Table 8.6 (with dimensions rounded to two decimal places).

The ray aberrations and MTF are shown in Figs. 8.20 and 8.21.

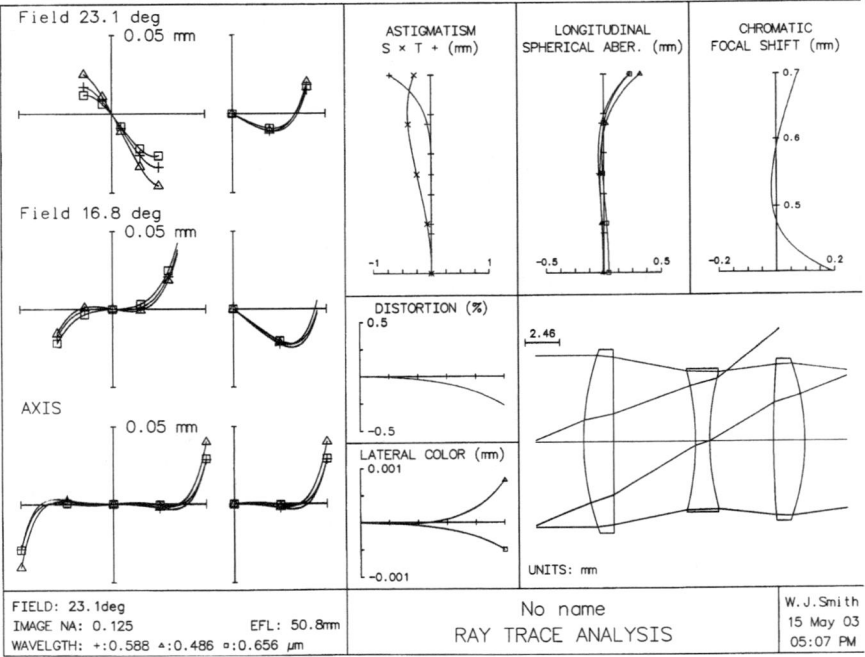

Figure 8.20 Left-hand triplet of Table 8.6; using a corrected design approach from the lens of Fig. 8.18.

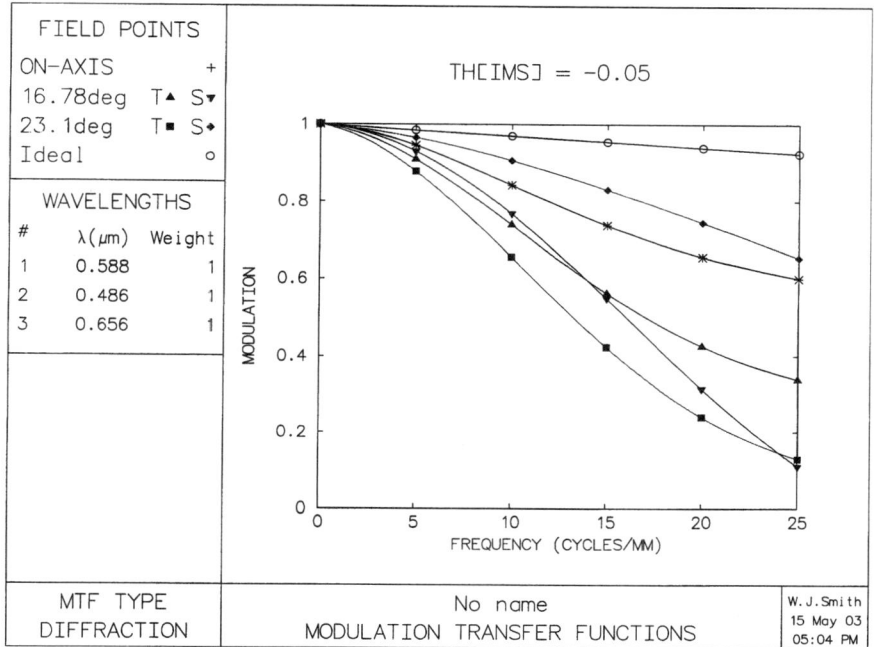

Figure 8.21 MTF for Fig. 8.20 lens, Table 8.6.

Note that somehow we have found a modestly different form, with the relative curvatures of $R1$ and $R6$ reversed, and the front airspace is noticeably larger than the rear one. These two design forms apparently represent what we earlier referred to as the left- and right-hand solutions.

At this point, we could, among many other options, continue this effort with even more accurately vignetted rays in our merit function (and undoubtedly get a slightly improved MTF). We could also better sample the performance with more rays, more fields, and more wavelengths in the merit function. But we elect to explore the initial format (with the larger rear airspace) by forcing the design toward this type of configuration. We could add a $T2 < T4$ target to the MF, or simply switch $T2$ and $T4$ and not allow them to vary for a few cycles. We try the latter and quickly find a design of the original type ($T2 < T4$ and $R1 < [-R6]$) whose performance in Figs. 8.22 and 8.23 looks just as good as the design immediately above.

The fact that these two designs are equivalent in performance indicates that we have found a solution region near the point where the left- and right-hand types merge into a single solution to the quadratic, as indicated above. This is because we allowed the flint to vary during the optimization. The separate left- and right-hand solutions only occur

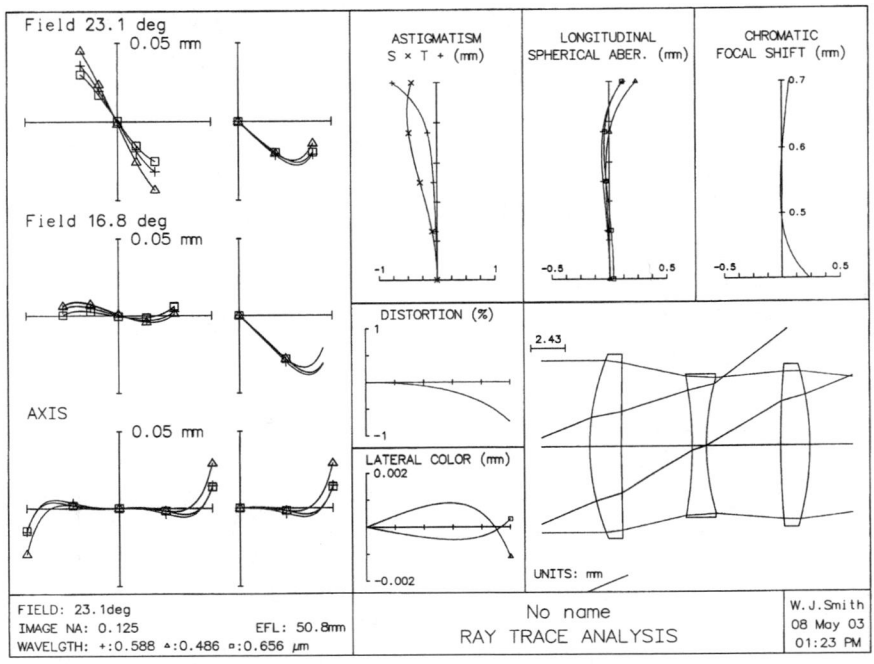

Figure 8.22 Right-hand triplet design forced from the left-hand lens of Fig. 8.20.

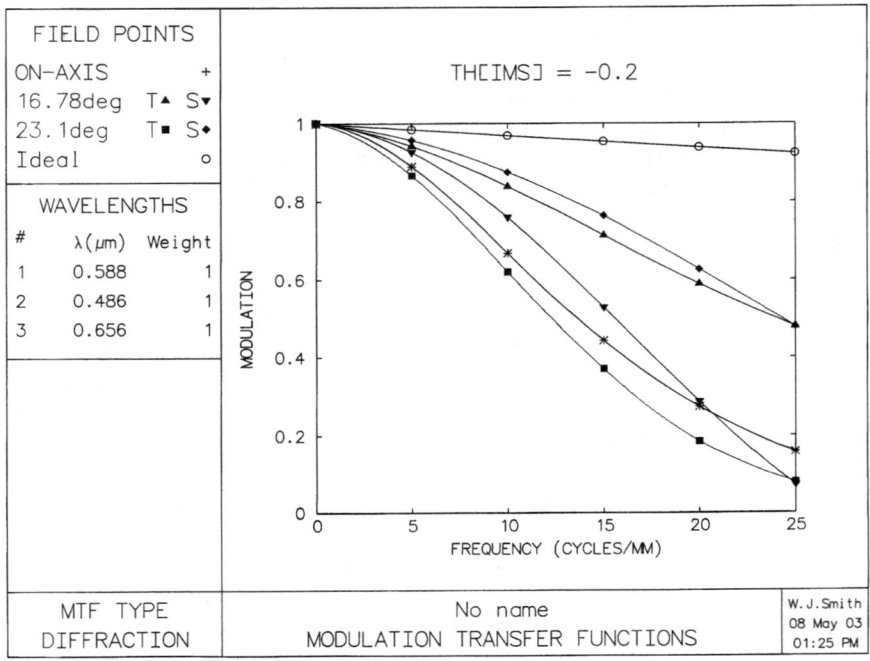

Figure 8.23 MTF for Fig. 8.22 lens.

when the glass types are fixed. When the glass types are fixed the vertex length of the triplet cannot vary to find the length that is optimum for the given aperture and field as required for the application.

By now it should be apparent that the design of the simple Cooke triplet is more than just "solving eight equations in eight unknowns." This is because the triplet performance is ultimately limited by the higher-order aberrations and not the third-order. Control of the third-order aberrations is a necessary but not sufficient condition for a good design.

Q. Is there a better choice than SK16 (620603) (from among the ordinary glasses) for the outer elements of our triplet? Note that our answer to this question may apply only to our particular combination of aperture and field, i.e., *f*/4 and 23.1° HFOV. The answer may be different for another combination.

A. We can do a quick and dirty survey of some other crown glasses, limiting ourselves to the PK, PSK, SK, SSK, BaF, and BaSF types. On the glass map (Fig. 8.17) these glasses lie along what might be considered the top of the "ordinary" region. We can proceed by starting with the last design, simply changing the crown glass type (remembering to again allow the flint to vary). To start, we change the crowns in our base design from SK16 (620603) to SK18 (639554) and optimize with the image surface at the paraxial focus; we then optimize again, allowing the image plane to be defocused. (Optimizing first at the paraxial focus before allowing defocusing is often a good procedure.)

15. Using the resulting SK18 design as a new starting point, we now change from SK18 to SSKN5 (658509) and optimize. We continue on with BaFN10 (670471) and then BaSF52 (702410). Note that we are taking modest steps along the sequence of glasses, rather than jumping from the base SK16 design to each new glass. This has the effect of keeping us in the same general solution region, avoiding radical departures from our general form. Next we return to our base SK16 design and change the crowns to PSK52 (603654), then to PSK53A (620635), PSK50 (557673), and finally to PK50 (521697). To save time we don't bother to fix the flint glass to a real catalog type for each design (although in practice, fixing and reoptimizing is a wise move since lens design programs will on occasion come up with some very unreal glasses). Also note that the BaSF, PSK, and PK glasses we have used are not exactly inexpensive and readily available. We justify these irregularities on the basis that we are only very quickly exploring the possibilities.

The resulting merit function values are shown in Table 8.7.

There are two irregularities in this table that show up if the data is plotted. The SK16 design was rounded to two places and reoptimized—this probably degraded the design slightly. The PSK53A (620635) glass is a bit above a smooth line through the other glasses on the glass chart—this should (and does) produce a better design.

TABLE 8.7 Showing the Results of Changing the Crown Glass Type and Reoptimizing. Columns 1 and 2 are the Crown Glass Type, Columns 3 and 4 are the Optimum Flint Glass Type and the rms Spot Size Merit Function when Optimized at the Paraxial Focus. Columns 5, 6, and 7 are the Flint Glass, the rms Merit Function, and the Focus Shift when Defocusing was Allowed

Crown		Flint	MF@parax	Flint and MF@defocus = δ		
BaSF52	702410	792261	0.000338	775267	0.000285	$\delta = -0.460$
BaFN10	670471	711295	0.000323	699302	0.000272	$\delta = -0.461$
SSKN5	658509	679314	0.000314	667322	0.000264	$\delta = -0.449$
SK18A	639554	644340	0.000319	637346	0.000270	$\delta = -0.439$
SK16	620603	617366	0.000328	610374	0.000277	$\delta = -0.448$
PSK53A	620635	603382	0.000318	596391	0.000268	$\delta = -0.438$
PSK52	603654	592396	0.000339	587404	0.000287	$\delta = -0.462$
PSK50	557673	580414	0.000417	575422	0.000354	$\delta = -0.550$
PK50	521697	566438	0.000493	563443	0.000419	$\delta = -0.635$

Examination of the ray intercept plots, longitudinal aberration plots, and graphs of spot size versus field angle indicates that the state of correction of all the lenses is quite reasonable and that the rms spot sizes are all comparable, except for the two low-index crowns, PK50 and PSK50, which have significantly larger sized spots. This is a good indication that the invention of the dense barium glasses was essential to the creation of the triplet and other anastigmats. Other than this the differences among the designs are relatively minor.

Thus, based on the rms spot sizes (MF in the table above) given by this particular merit function, SSKN5 seems (by a rather narrow margin) to be the best glass choice for the crown elements. The corresponding flint, 693305, is between SF8 (689312) and SF15 (699301). These are all acceptable glasses. The performance is indicated in Figs. 8.24 and 8.25. Our final design, rounded to three decimals, is shown in Table 8.8.

8.7 Possible Improvements to Our "Basic" Triplet

As possible improvements (while maintaining the triplet configuration) we can consider:

TABLE 8.8 Prescription for the Triplet Design with the Optimum Crown Glass (SSKN5, 658509), Shown in Figs. 8.24 and 8.25

$R1 = +18.343v$	$T1 = 2.228v$	SSKN5	658509
$R2 = +293.96v$	$T2 = 5.065v$		
$R3 = -31.660v$	$T3 = 1.000$	SF15	699301
$R4 = +18.879v$	$T4 = 5.239v$		
$R5 = +74.126v$	$T5 = 2.048v$	SSKN5	658509
$R6 = -22.758v$	$BF = 42.656$		

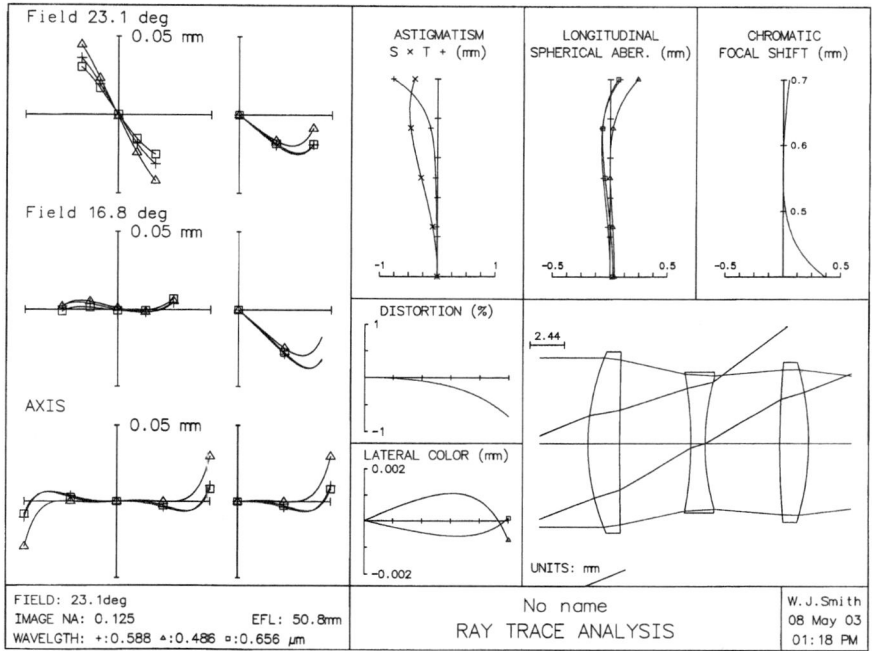

Figure 8.24 f/4.0 ±23° triplet design with the best of the ordinary crown glasses, Table 8.8.

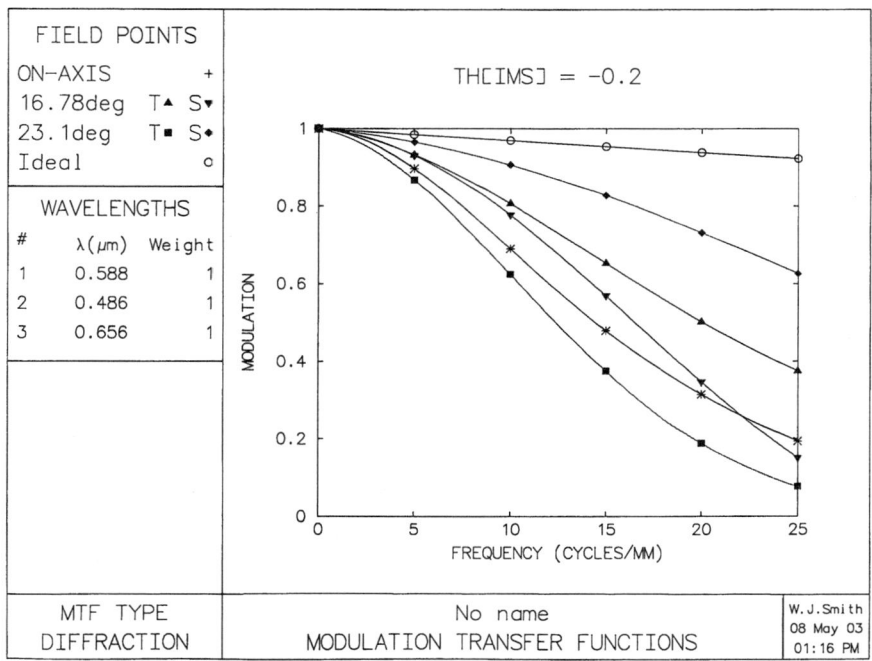

Figure 8.25 MTF for Fig. 8.24 lens, Table 8.8.

1. Changes, additions, or adjustments to the merit function to achieve a better balance of the aberrations. Different field weightings, additional rays, specific targets for the most troublesome aberrations, and the like are often useful. A totally different type of merit function may locate an improved design, especially if the current design is trapped in an inferior local optimum. Realistically, however, there is a limited amount of quality in a design as simple as the triplet, and about all we would be able to do in this regard right now is to spread it around differently. This might not be true for a more complex design with more degrees of freedom.

2. Change the crown glasses to one of the high-index lanthanum types. Raising the index of the glass in a design flattens the surface curvatures and strongly tends to reduce the aberrations. Which of the glasses from along the top of the glass map is the best choice is the subject of a discussion that follows. Perusal of a glass catalog will indicate that, in practice, one's choice should certainly include considerations of price, resistance to environmental attack, and availability.

3. Aspherizing one or more surfaces. Here, cost is a major consideration, as is the feasibility and availability of the necessary aspheric fabrication technology.

4. Increasing the thickness of the elements. Here, the cost factor depends not only on the cost of the increased volume of glass used in the elements, but also on the increased fabrication costs resulting from fewer elements on a blocker in the grinding and polishing operations. In the particular application at hand ($f/4 \pm 23.1°$) we find that increasing the element thickness is not effective in improving the design.

5. Failing these steps, consider changing to a better form, perhaps (for this example) a four-element design such as a Tessar or Dogmar, or if necessary, even a five element Heliar as discussed in Chaps. 10 and 11.

8.8 The Rare Earth (Lanthanum) Glasses

To evaluate the effect of using the rare earth glasses we can simply repeat the Table 8.7 glass survey in the LaK, LaF, and LaSF space on the glass map. We again start from our base SK16 design and substitute glasses from along the top of the glass map in Fig. 8.17. As before, we optimize at the paraxial focus, then reoptimize with defocus allowed. The results are shown.

Based on the merit function values, LaSFN31 at 881410 with an 870240 flint comes out on top. The flint is between SF57 (847238) and SF58 (918215). The LaSFN31 price is about 60 times the price of BK7, and the flints cost about 6 and 20 times BK7. These flints are environmentally sensitive to staining and hazing, and all three glasses tend to be yellowish in greater or less degree, having poor transmission in the blue. The cost and blue absorption decrease toward the top of the table, but we note that our previous SSKN5 design had merit function values

Table Showing the Merit Function Values when High Index, Rare Earth Crown Glasses are used (with, and without Defocus)

Crown		Flint	MF@parax	MF@defocus	$= \delta$
LaK21	640601	630353	0.000465	0.000355	$\delta = -0.488$
LaK31	697564	661327	0.000376	0.000291	$\delta = -0.397$
LaK33	754524	701301	0.000315	0.000247	$\delta = -0.329$
LaSFN30	803464	771269	0.000288	0.000235	$\delta = -0.287$
LaSFN31	881410	870240	0.000262	0.000226	$\delta = -0.231$
LaSF35	1022291	1340185	0.000428	0.000420	$\delta = -0.145$

of 0.000467 and 0.000366, close to those for LaK21 (0.000465 and 0.000355) (at the top of the table here), and SSKN5 costs only about half as much as LaK21.

It's quite apparent that moving toward the top of the glass map for the crown glass is beneficial for image quality, but one might profitably "cherry pick" for glasses in the area somewhat below the top of the map, using factors such as cost, color, and stability, in addition to index and dispersion, as criteria for selection.

With SF57 (847238) as the nearest flint glass, our reoptimized high-index design is shown in Table 8.9 and Figs. 8.26 to 8.28.

8.9 Aspherizing the Surfaces

Stop shift theory tells us that, for the third-order aberrations, an aspheric located at the aperture stop can affect *only* the spherical aberration. To affect coma, astigmatism, or distortion, the aspheric surface must be away from the stop so that the principal ray strikes the surface at a significant height. For our demonstration design we elect to go "all-out" and aspherize all three elements of our triplet. This gives us leverage on all the aberrations, including the higher orders. The lens must be completely redesigned for this to be effective, allowing all the curvatures and spacings to vary. We cannot simply aspherize the surfaces of an existing design and expect much in the way of results. Simple conic sections may not do too much either, but we can easily try them.

TABLE 8.9 Prescription for the Triplet with the Optimum High Index, Rare Earth Crown Glass (The Performance is Shown in Figs. 8.26 to 8.28)

$R1 = +22.727v$	$T1 = 1.810v$	LASFN31	881410
$R2 = +100.27v$	$T2 = 6.137v$		
$R3 = -42.707v$	$T3 = 1.000$	SF57	847238
$R4 = +24.474v$	$T4 = 5.603v$		
$R5 = +126.86v$	$T5 = 1.772v$	LASFN31	881410
$R6 = -28.901v$	$BF = 42.661$		

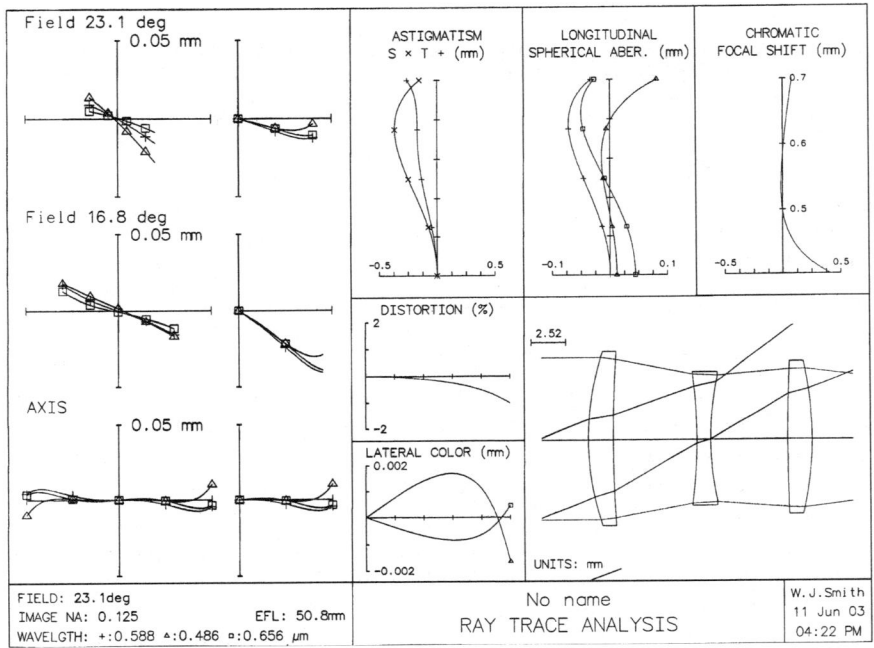

Figure 8.26 $f/4.0$ ±23° triplet design with the best of the Lanthanum glasses, Table 8.9.

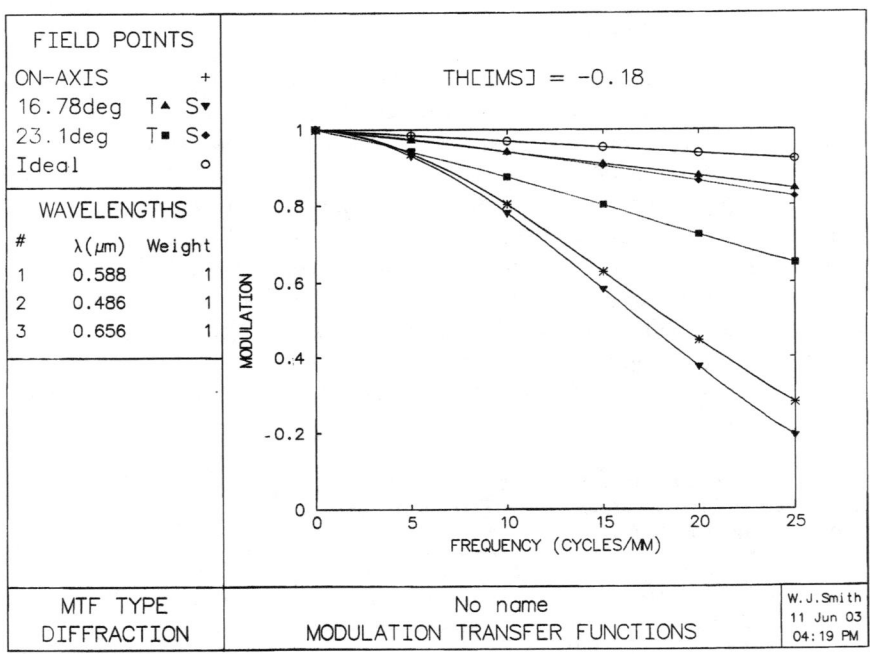

Figure 8.27 MTF versus frequency for Fig. 8.26 lens, Table 8.9.

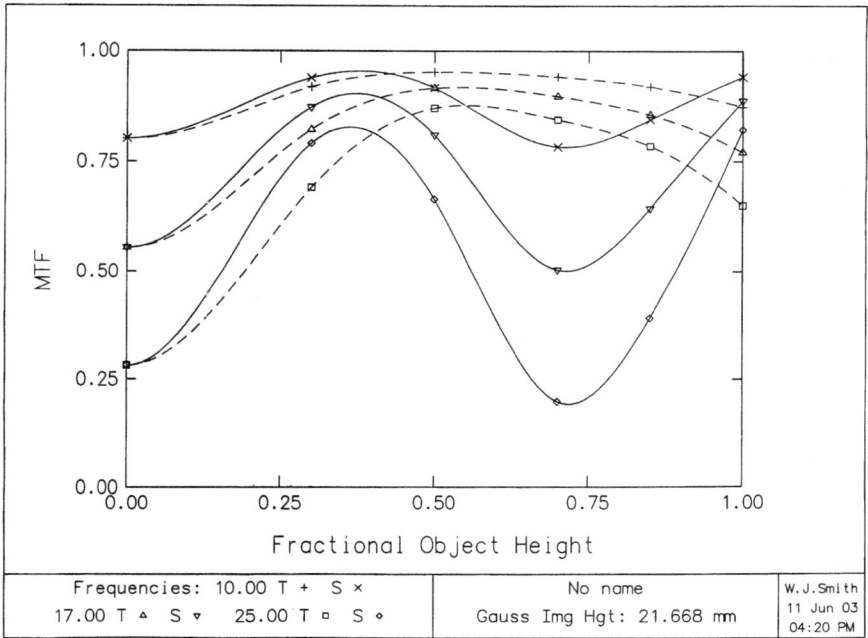

Figure 8.28 MTF versus field at 10, 17, 25 lpm for Fig. 8.26 lens, Table 8.9.

We can use the best of our ordinary-glass triplets with SSKN5 crowns as the starting point, aspherizing all three elements. Surface no. 4 (at the stop) is an intuitive choice, but whether to aspherize the inner or outer surfaces of the crowns is an open question. So we consider four cases: aspherize with either conics or general aspherics, and aspherize either the inner or outer crown surfaces. For the general aspherics we vary the fourth-, sixth-, and eighth-order coefficients. The results are shown in Table 8.10.

The asphericities (for surface no. 1 [or no. 2 for inner], surface no. 4, and surface no. 5 [or no. 6 for outer]) are shown in Table 8.11.

The conic constants for the inner lens surfaces are very large because they are on surfaces with small curvatures (i.e., long radii). (This brings up the point that a conic is meaningless on a plane surface.) For some reason the inner conics don't appear to be very effective; the merit function is not much smaller than the all spherical lens, and the optimized focus shift is large. The other three aspherizations are all clearly improvements, and of the same magnitude. Not too suprisingly, the general aspherics are somewhat more effective than the conics.

Aspheric surfaces usually prove more effective on a highly stressed lens. To demonstrate, we can start with the 100 mm $f/2.5$, 16° HFOV

TABLE 8.10 RMS Spot Size Merit Function Values Obtained When the Triplet is Redesigned Using Three Aspheric Surfaces

	MF@parax	MF@defocus = δ	
All spherical	0.000467	0.000366	$\delta = -0.474$
Conic outer	0.000263	0.000220	$\delta = -0.383$
Conic inner	0.000425	0.000307	$\delta = -0.606$
Asph. outer	0.000204	0.000203	$\delta = -0.269$
Asph. inner	0.000199	0.000197	$\delta = -0.300$

triplet of Fig. 8.8. This speed and field clearly qualify this triplet as a stressed lens. To make this a fair comparison, before starting we will reoptimize the design of Fig. 8.8 using the same merit function that we have used for all the preceding triplets (but holding EFL = 100.0 and the crown edge thicknesses at 2.0). The reoptimized lens data is in Table 8.12 and the aberrations are shown in Figs. 8.29 to 8.31.

In this lens the aspherics should prove to be extremely effective. After reoptimizing with spherical surfaces, we make surfaces 1, 4, and 6 general aspherics and allow the fourth-, sixth- and eighth-order aspheric coefficients (AD, AE, and AF) to vary during the optimization process. The performance is shown in Figs. 8.32 to 8.34 and the prescription is given in Table 8.13. The merit function values tell the tale:

	MF@parax	MF@defocus = δ	
Spherical surfaces	0.002529	0.001133	$\delta = -1.917$
Aspheric surfaces	0.000381	0.000379	$\delta = -0.283$

The raytraced aberrations agree with the merit function values, showing that aspherizing has produced a 7x improvement for zonal spherical, 4x improvement of field curvature, 2x improvement of distortion, and a 4x to 7x (depending on the focus chosen) improvement for the directly calculated rms spot size.

TABLE 8.11 The Conic Constants and Aspheric Coefficients of the Lenses Listed in Table 8.10

Conic outer	cc (1) = −1.128	cc (4) = −2.169	cc (6) = −0.232
Conic inner	cc (2) = −840.9	cc (4) = +0.257	cc (5) = +276.1
Asph. outer	AD (1) = −9.84 − 6	AE (1) = −5.34 − 8	AF (1) = +9.46 − 10
	AD (4) = −2.37 − 5	AE (4) = −9.38 − 7	AF (4) = +1.49 − 8
	AD (6) = −1.36 − 5	AE (6) = +1.19 − 7	AF (6) = −4.49 − 9
Asph. inner	AD (2) = +2.44 − 5	AE (2) = −2.67 − 7	AF (2) = +6.85 − 10
	AD (4) = −8.48 − 5	AF (4) = +1.43 − 7	AF (4) = +2.30 − 9
	AD (5) = −1.09 − 5	AF (5) = +1.66 − 7	AF (5) = −4.01 − 10

Cooke Triplet Anastigmats

TABLE 8.12 Prescription for the Optimized All-Spherical 100 mm, f/2.5, +/−16 deg Triplet used as the Base Lens for the Aspheric Study (The Lens is Shown in Figs. 8.29 to Fig. 8.31)

$R1 = +37.382$	$T1 = 7.819$	SK4	613586
$R2 = -10728$	$T2 = 12.115$		
$R3 = -59.259$	$T3 = 1.60$	Var.	642343
$R4 = +33.340$	$T4 = 13.687$	(stop)	
$R5 = +107.616$	$T5 = 7.253$	SK4	613586
$R6 = -45.220$	BFL = 78.565		

The ray intercept plots of the aspherized design in Fig. 8.32 show that there is heavy spherochromatism in the outer part of the aperture, with the blue ray significantly overcorrected at the margin. To improve this we drop the Conrady $(D-d)\delta n$ chromatic correction from the merit function and instead optimize on the rms spot size in all three colors. This chromatic rebalancing eliminates the blue flare in the image by undercorrecting the paraxial chromatic, and (somewhat surprisingly) also produces a worthwhile improvement in the MTF as shown in Figs. 8.35 to 8.37. Note that switching to a merit function based on rms OPD (rather than the rms spot size merit function we have used here)

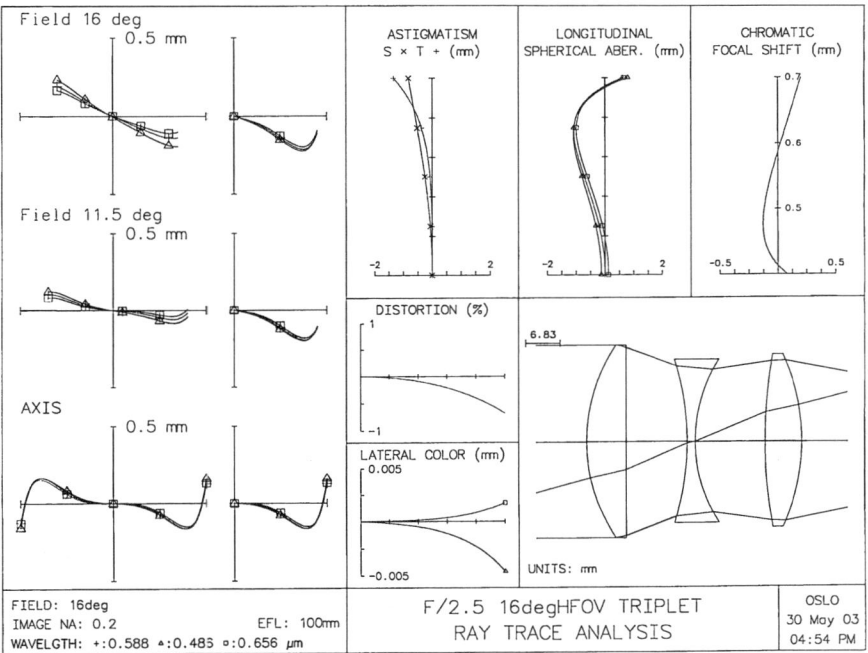

Figure 8.29 $f/2.5$ ±16° triplet with spherical surfaces, Table 8.12.

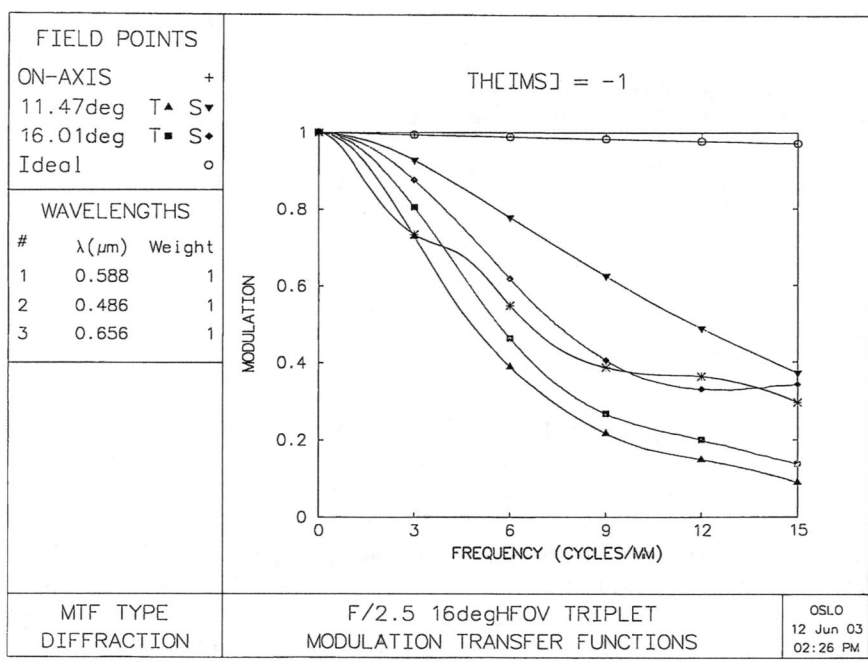

Figure 8.30 MTF versus frequency for Fig. 8.29 lens, Table 8.12.

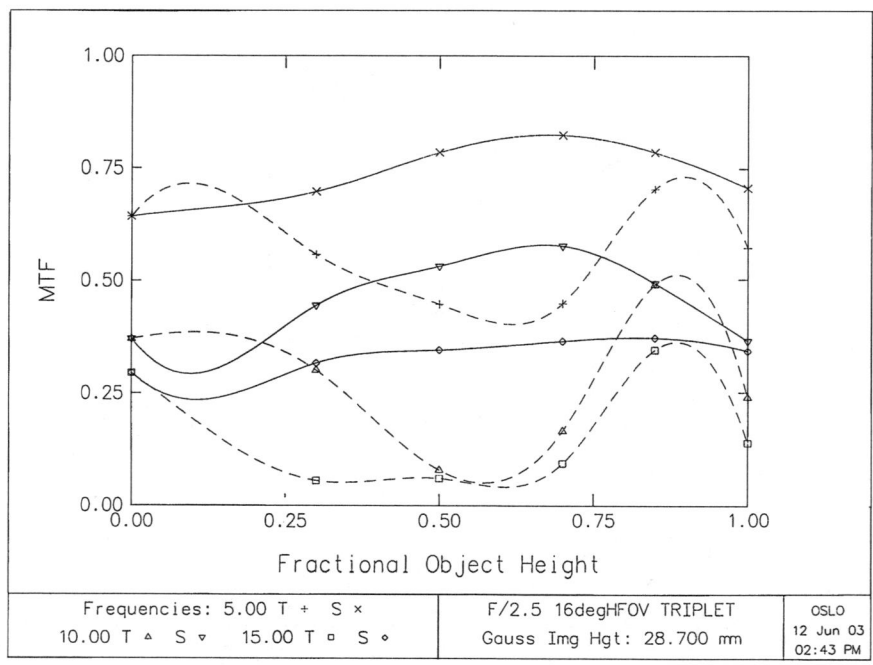

Figure 8.31 MTF vs. field at 5, 10, 15 lpm for Fig. 8.29 lens, Table 8.12.

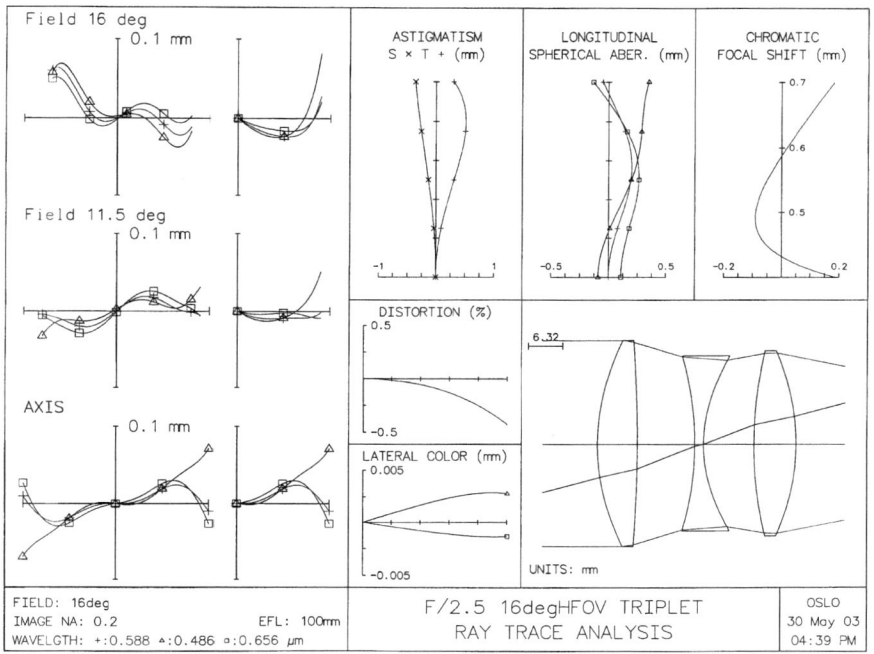

Figure 8.32 $f/2.5$ ±16° triplet with three aspheric surfaces, Table 8.13.

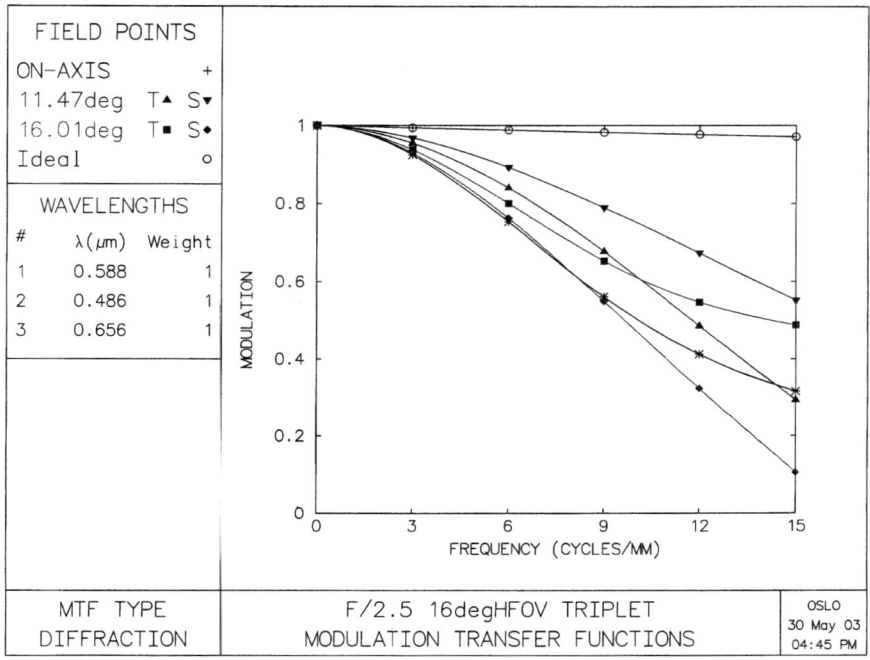

Figure 8.33 MTF versus frequency for Fig. 8.32 lens, Table 8.13.

243

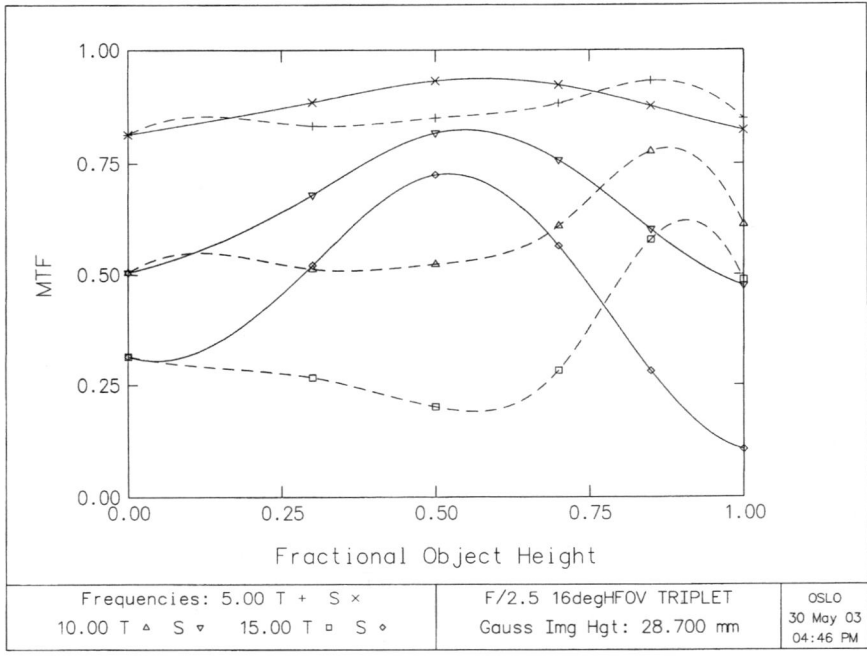

Figure 8.34 MTF versus field at 5, 10, 15 lpm for Fig. 8.32 lens, Table 8.13.

usually will produce a better MTF. The final aspherized design is shown in Table 8.14.

8.10 Increasing the Element Thickness

With this lens (the SSKN5 design at $f/4$ and $\pm 23.1°$) and with this merit function, changing the element thickness does not produce an improvement. This may or may not be true for other applications, other lens

TABLE 8.13 Prescription for the Initial Aspherized 100 mm, f/2.5 +/−16 deg Triplet Shown in Figs. 8.32 to 8.34

$R1 = +42.385$	$T1 = 7.302$	SK4 613586
AD = −2.4340e − 06	AE = −4.0604e − 10	AF = −1.1177e − 12
$R2 = -274.30$	$T2 = 10.555$	
$R3 = -63.721$	$T3 = 1.60$	Var. 622362
$R4 = +28.508$	$T4 = 9.151$	
AD = −9.2929e − 06	AE = −3.4034e − 09	AF = −8.3036e − 12
$R5 = 82.182$	$T5 = 7.636$	SK4 613586
$R6 = -45.159$	BFL = 82.288	
AD = +8.9979e − 07	AE = −8.1861e − 10	AF = +3.1951e − 12

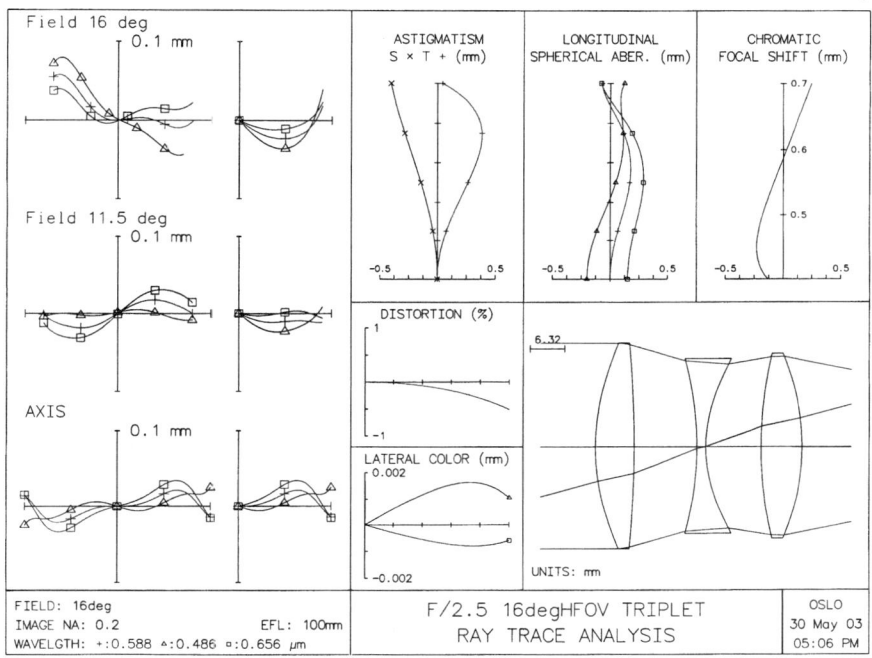

Figure 8.35 $f/2.5$ $\pm 16°$ aspheric triplet, improved design, Table 8.14.

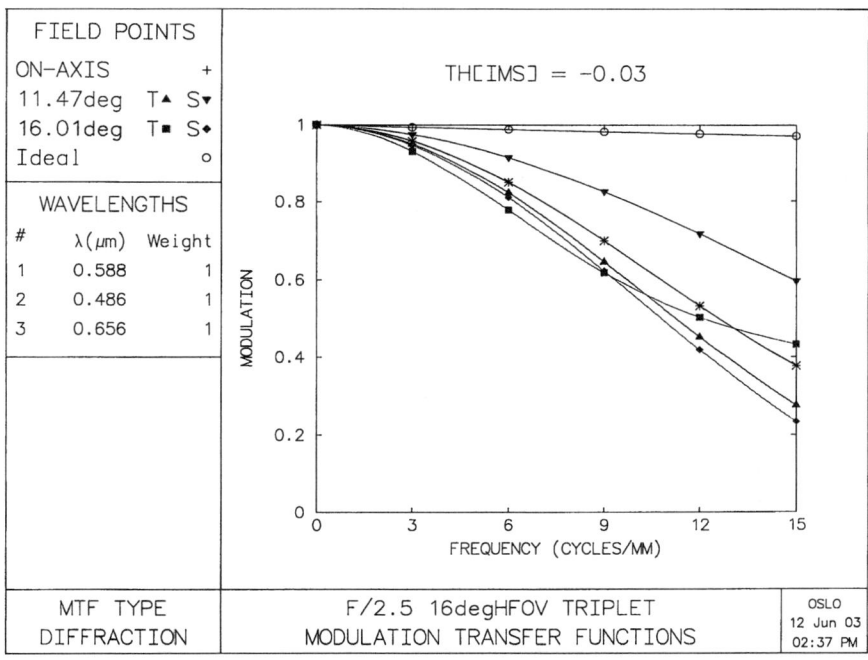

Figure 8.36 MTF versus frequency for Fig. 8.35 lens, Table 8.14.

245

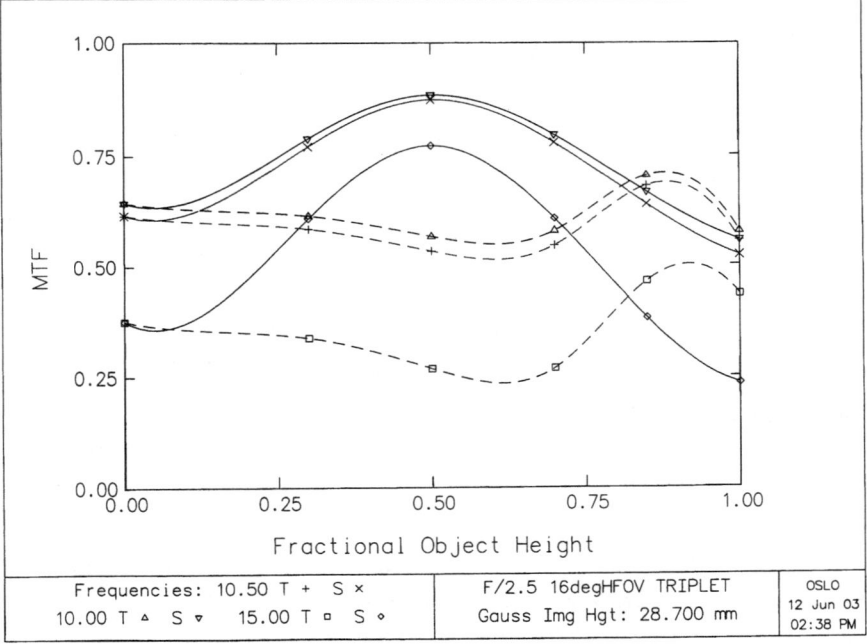

Figure 8.37 MTF versus field at 5, 10, 15 lpm for Fig. 8.35 lens, Table 8.14.

configurations, other glass types, or other merit functions. See, for example, Fig. 8.9 where, at $f/2.5$ and $\pm 16°$, lanthanum glass combined with thick first and second elements produced a marked reduction in the zonal spherical aberration of a triplet.

Q: Was this improvement due to the high-index glass, the thick elements, or both?

This camera lens study continues in Chap. 10 with the Tessar and Heliar forms, and in Chap. 11 with the Dogmar lens.

TABLE 8.14 Prescription for the Final Improved Aspherized Triplet Design, Shown in Figs. 8.36 to 8.37

$R1 = +45.125v$ $T1 = 7.078v$ SK4 613586
 AD $= -2.4843e - 06$ AE $= -6.1508e - 10$ AF $= -6.7307e - 13$
$R2 = -233.64v$ $T2 = 11.539v$
$R3 = -66.545v$ $T3 = 1.600$ 616369v
$R4 = +28.726v$ $T4 = 10.028v$
 AD $= -9.5160e - 06$ AE $= -5.0816e - 09$ AF $= -5.1165e - 12$
$R5 = +82.069v$ $T5 = 7.517v$ SK4 613586
$R6 = -46.284v$ BF $= 81.994$
 AD $= +9.5799e - 07$ AE $= -1.3183e - 10$ AF $= +2.6314e - 12$

Chapter 9

Split Triplets

As discussed in Sec. 3.3, splitting an element into two elements allows the aberrations of the two to be substantially reduced from that of the original single element. If the balance of the system can be adjusted to take advantage of this reduction, a significant improvement in performance can be achieved.

In the Cooke triplet, the outer crowns are often split to reduce the zonal spherical aberration and thus allow the speed of the lens to be increased. An example of the split-rear crown type is shown in Fig. 9.1. Note that, as the speed is increased, the angular field coverage is reduced and the vertex length is increased. The split-rear crown form is often used in camera lenses for 8-mm, 16-mm, small-format TV, and charge-coupled device (CCD) cameras, typically at a speed of about $f/2$ and a telephoto ratio near 1.0.

The split-front crown type has been widely used for (pseudo) telephoto camera lenses and for 35-mm slide projection lenses (Fig. 9.2). Figures 9.2 to 9.5 present an assortment of split-front triplets with glasses of low, medium, and high index, with thick and thin elements, with speeds from $f/2.8$ to $f/1.5$, and angular fields ranging from 20° to 35°. Figure 9.2 has identical front crowns for economy in manufacture. Another commercial version has a plano-convex front element and a more strongly bent meniscus second element. The two forms are quite equivalent as to performance and cost.

The rear element, on the basis that it is the larger contributor of undercorrected spherical aberration in the triplet, is the obvious candidate for a split; however, it turns out that splitting the front element is a better choice. The reason is that the shape of the second element in the split-front type is meniscus. In a truly thin lens element this would be immaterial, but with a thick meniscus the ray height drops significantly in

Chapter Nine

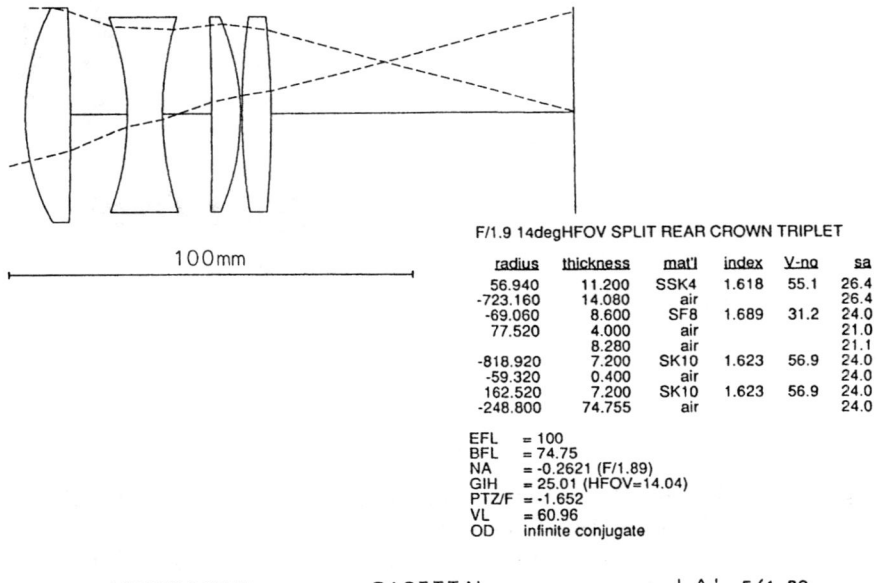

F/1.9 14degHFOV SPLIT REAR CROWN TRIPLET

radius	thickness	mat'l	index	V-no	sa
56.940	11.200	SSK4	1.618	55.1	26.4
-723.160	14.080	air			26.4
-69.060	8.600	SF8	1.689	31.2	24.0
77.520	4.000	air			21.0
	8.280	air			21.1
-818.920	7.200	SK10	1.623	56.9	24.0
-59.320	0.400	air			24.0
162.520	7.200	SK10	1.623	56.9	24.0
-248.800	74.755	air			24.0

EFL = 100
BFL = 74.75
NA = -0.2621 (F/1.89)
GIH = 25.01 (HFOV=14.04)
PTZ/F = -1.652
VL = 60.96
OD infinite conjugate

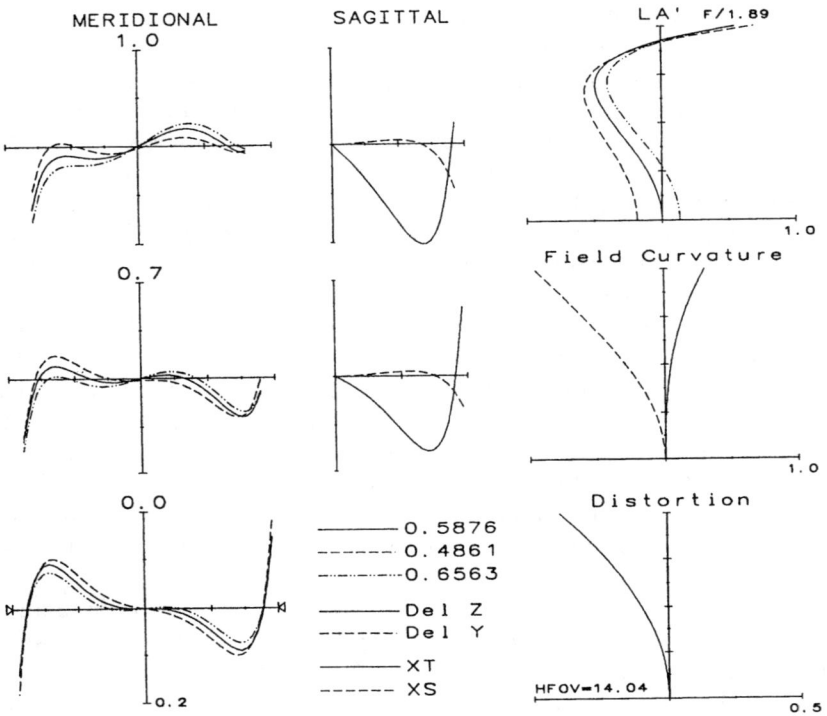

Figure 9.1 $f/1.9$ ±14° split-rear crown triplet.

Split Triplets 249

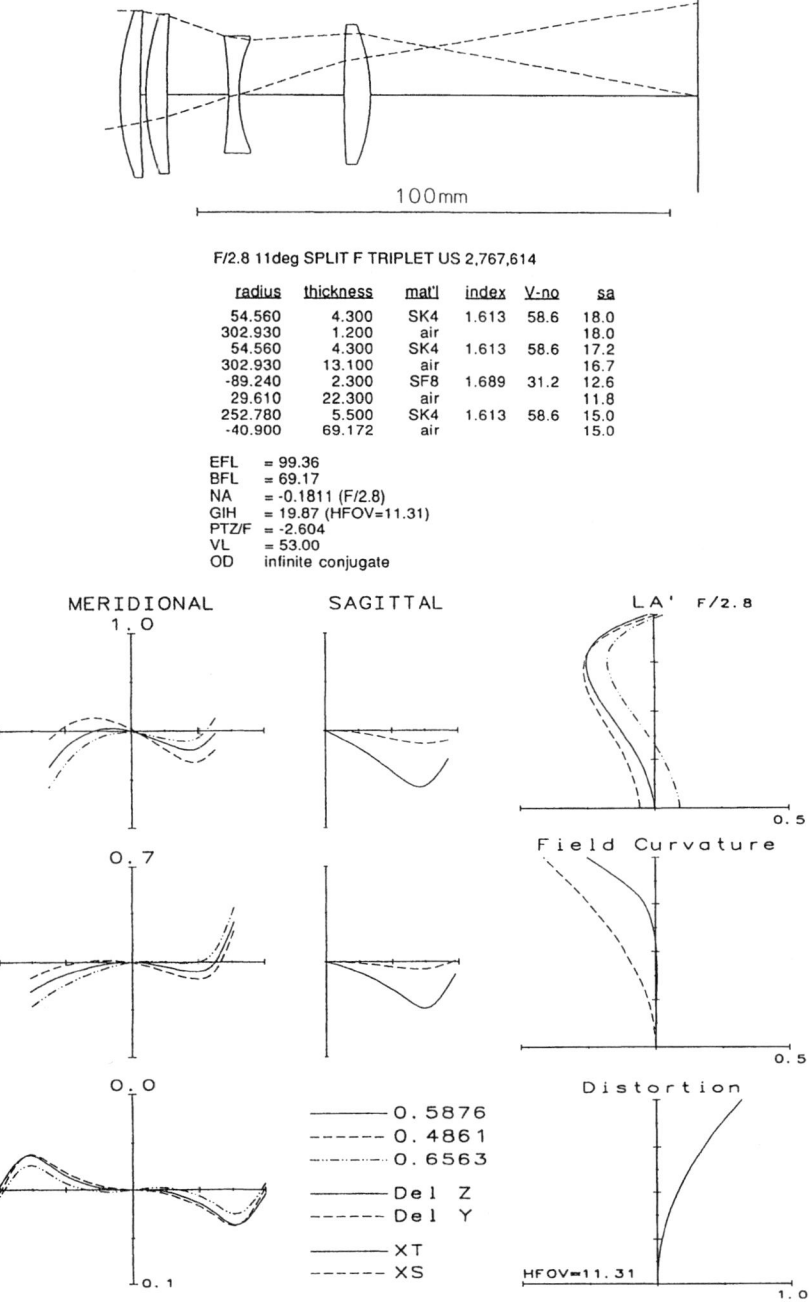

F/2.8 11deg SPLIT F TRIPLET US 2,767,614

radius	thickness	mat'l	index	V-no	sa
54.560	4.300	SK4	1.613	58.6	18.0
302.930	1.200	air			18.0
54.560	4.300	SK4	1.613	58.6	17.2
302.930	13.100	air			16.7
-89.240	2.300	SF8	1.689	31.2	12.6
29.610	22.300	air			11.8
252.780	5.500	SK4	1.613	58.6	15.0
-40.900	69.172	air			15.0

EFL = 99.36
BFL = 69.17
NA = -0.1811 (F/2.8)
GIH = 19.87 (HFOV=11.31)
PTZ/F = -2.604
VL = 53.00
OD infinite conjugate

Figure 9.2 $f/2.8$ ±14° split-front crown for a 35-mm slide projector.

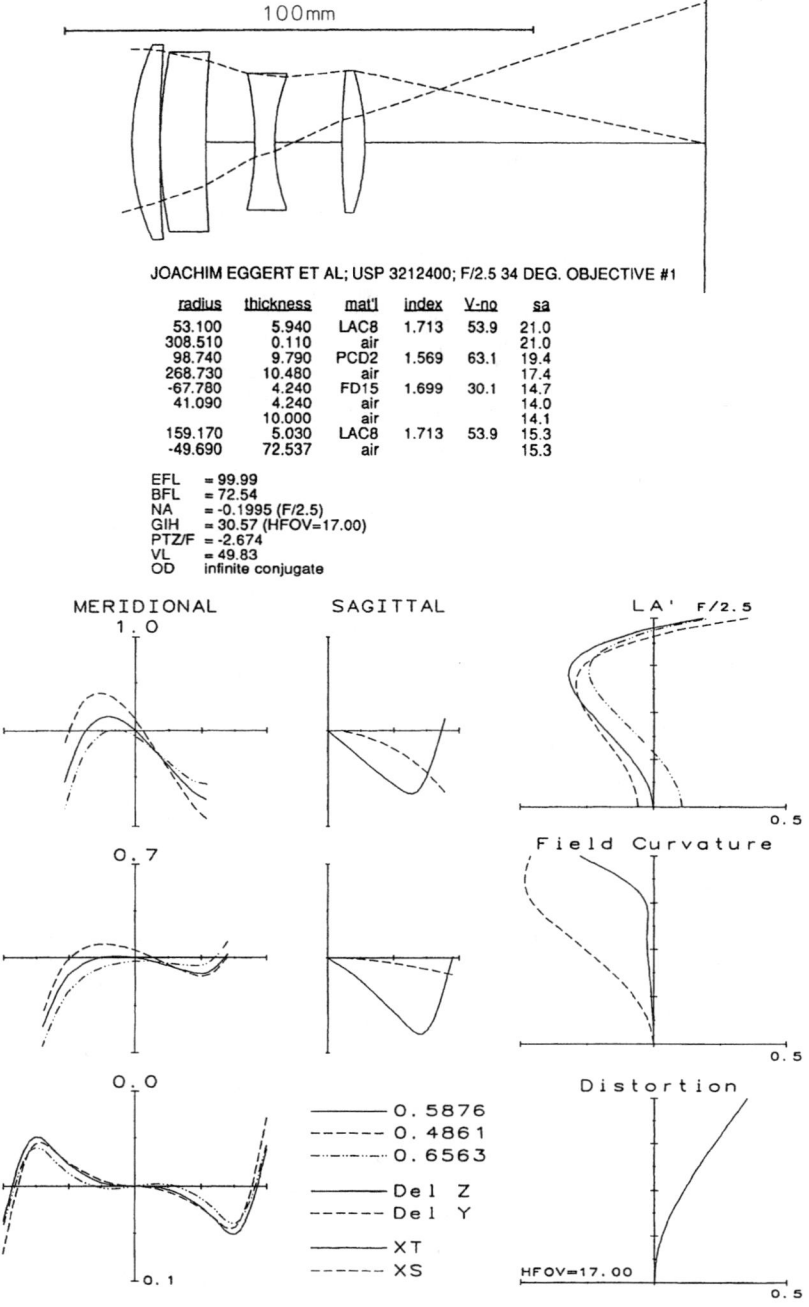

Figure 9.3 f/2.5 ±17° split-front crown triplet with lanthanum glass.

Split Triplets   251

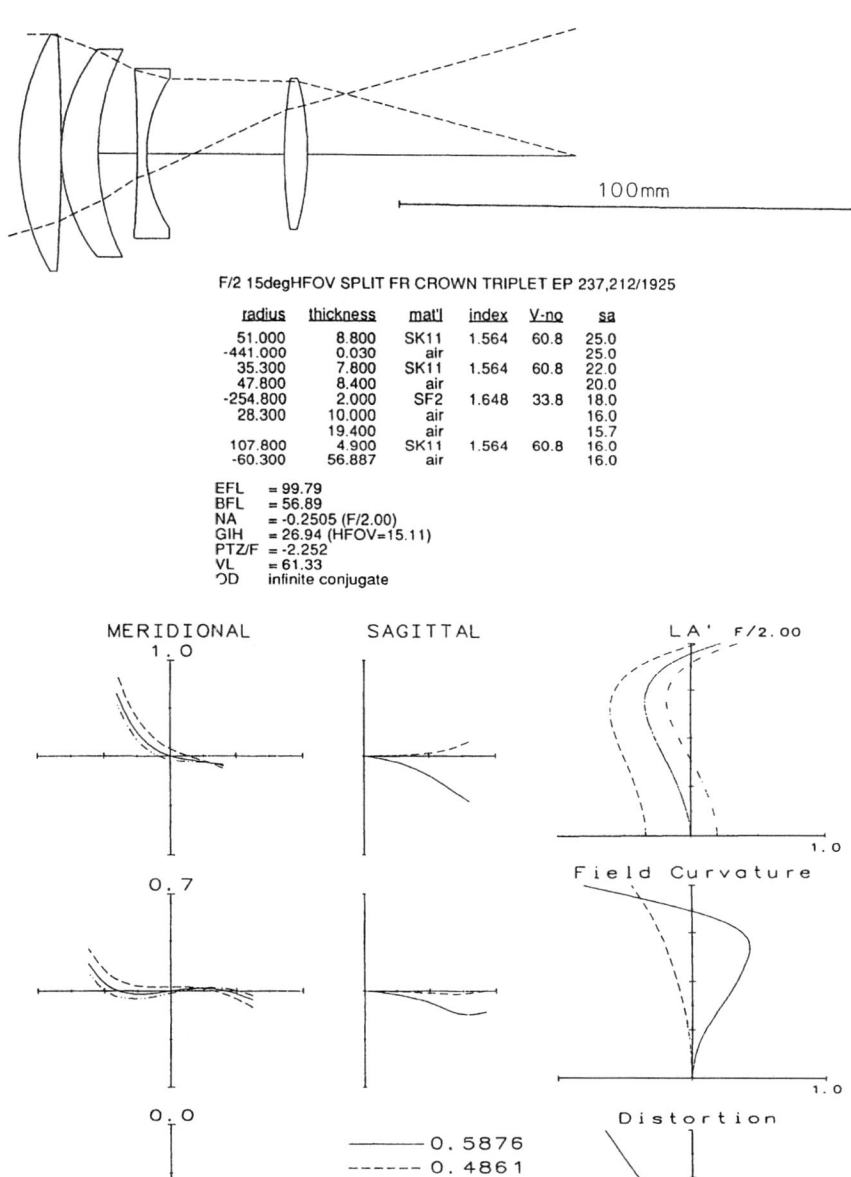

F/2 15degHFOV SPLIT FR CROWN TRIPLET EP 237,212/1925

radius	thickness	mat'l	index	V-no	sa
51.000	8.800	SK11	1.564	60.8	25.0
-441.000	0.030	air			25.0
35.300	7.800	SK11	1.564	60.8	22.0
47.800	8.400	air			20.0
-254.800	2.000	SF2	1.648	33.8	18.0
28.300	10.000	air			16.0
	19.400	air			15.7
107.800	4.900	SK11	1.564	60.8	16.0
-60.300	56.887	air			16.0

EFL = 99.79
BFL = 56.89
NA = -0.2505 (F/2.00)
GIH = 26.94 (HFOV=15.11)
PTZ/F = -2.252
VL = 61.33
OD infinite conjugate

Figure 9.4 $f/2.0$ ±15° split-front crown triplet.

252　Chapter Nine

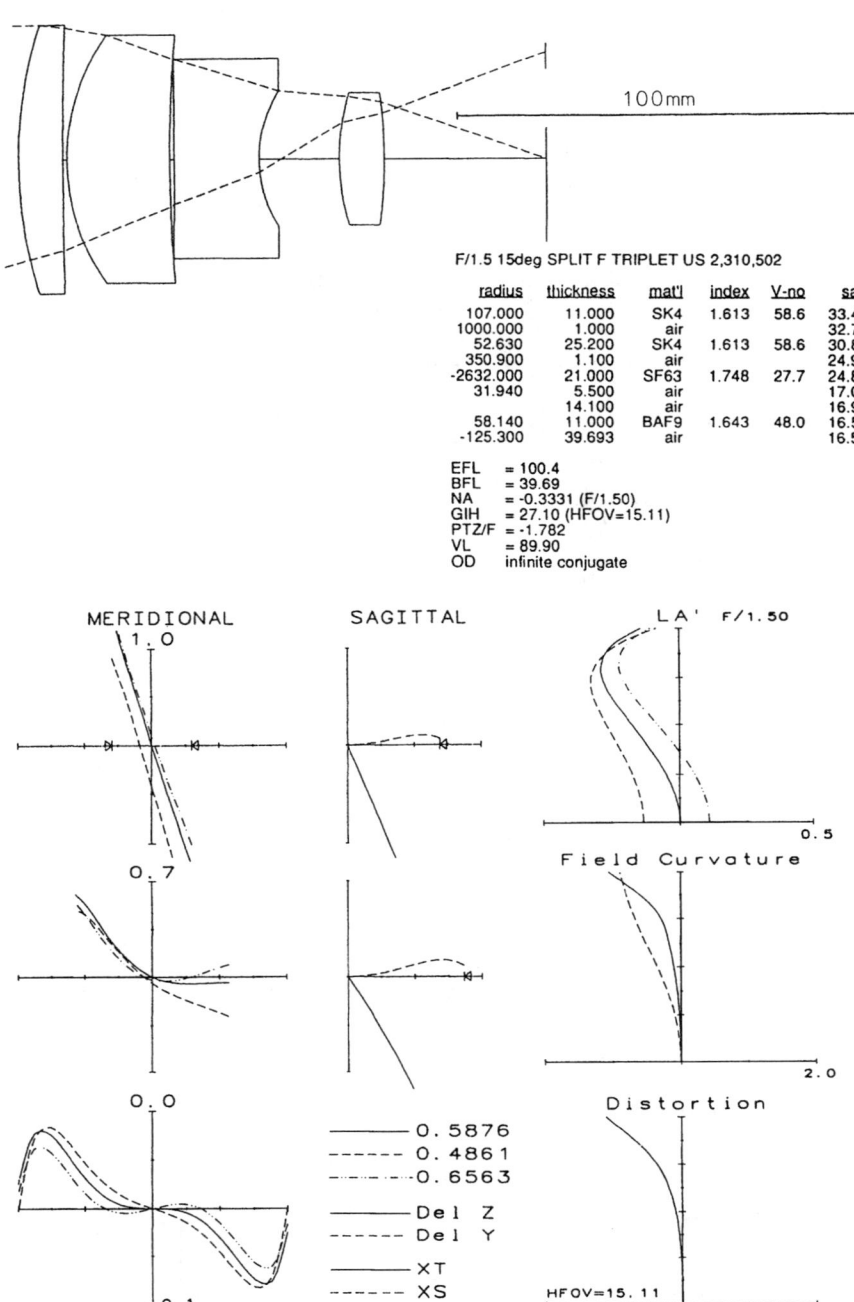

Figure 9.5 $f/1.5$ ±15° split-front crown triplet with thick elements.

passing from the first surface to the second. This means that to maintain the power of the meniscus element, we must increase the curvature of the concave second surface. This reduces the Petzval field curvature from this element. Once we realize this, the reason for the popularity of the split-front version and a thick meniscus second element (Fig. 9.5) is immediately apparent. In addition, a meniscus element in a converging beam is like the aplanatic lens discussed in Chap. 17, Sec. 17.2. It has a very small spherical contribution, and some shapes and powers have zero (or even positive) spherical aberration contributions.

The split-front triplet has also developed into two interesting and powerful, if currently rarely used, forms. The second element of the split front is meniscus in shape. If made thick, as discussed above, it has a field-flattening effect of its own, and the Sonnar form, Figs. 9.6 to 9.8, makes use of this to eliminate the need for the center flint negative element. These lenses begin to look like the front half of a Biotar/double-Gauss combined with the rear of a triplet or Tessar. The tremendous variety of Sonnar configurations and characteristics is only sampled here. They all cover reasonable angular fields, with speeds ranging from $f/2.9$ to $f/1.2$.

Traditionally the weak point of the Sonnar is considered to be its Petzval field curvature, although the examples here don't really illustrate that point. The lenses in this chapter are all relatively fast, and it is common for fast, narrow-field lenses to have a large Petval sum (which reduces element powers and the associated high-order aberrations).

The Sonnar was primarily used as a medium focal length objective for a 35-mm camera lens. It can be corrected to cover a $\pm 20°$ to $23°$ half field of view. Note the front part of the Sonnar is like a double-Gauss lens, and the rear may be a singlet, doublet, or triplet.

Another variation is the Ernostar type, Figs. 9.9 and 9.10, which uses the split triplet form as the basis for improvement by compounding the elements, beginning with the conversion of the new second element into a meniscus doublet. Both the Sonnar and the Ernostar types had a brief vogue as 35-mm camera lenses of normal and long focal length, but they have been largely superseded by other constructions, especially the double-Gauss lens.

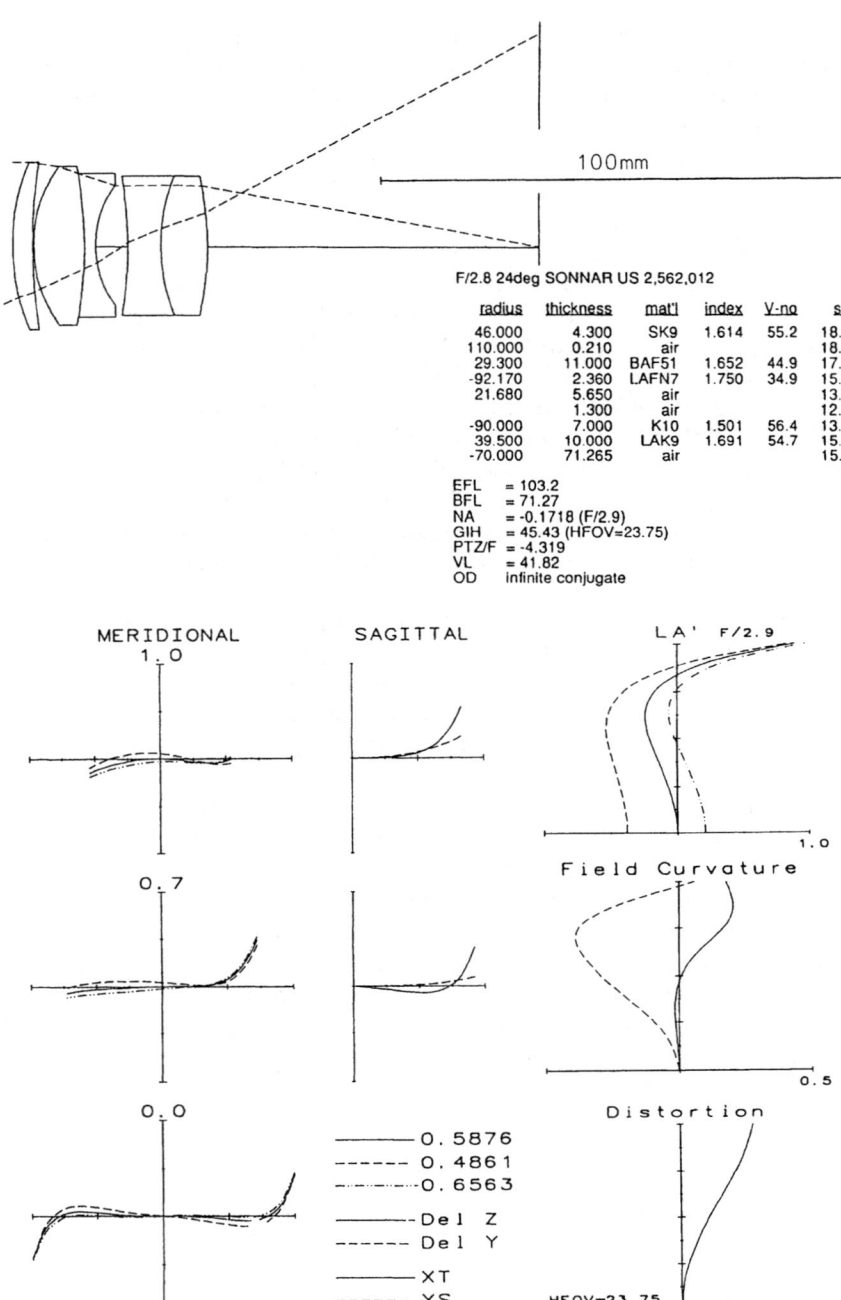

Figure 9.6 $f/2.9$ ±24° Sonnar.

Split Triplets 255

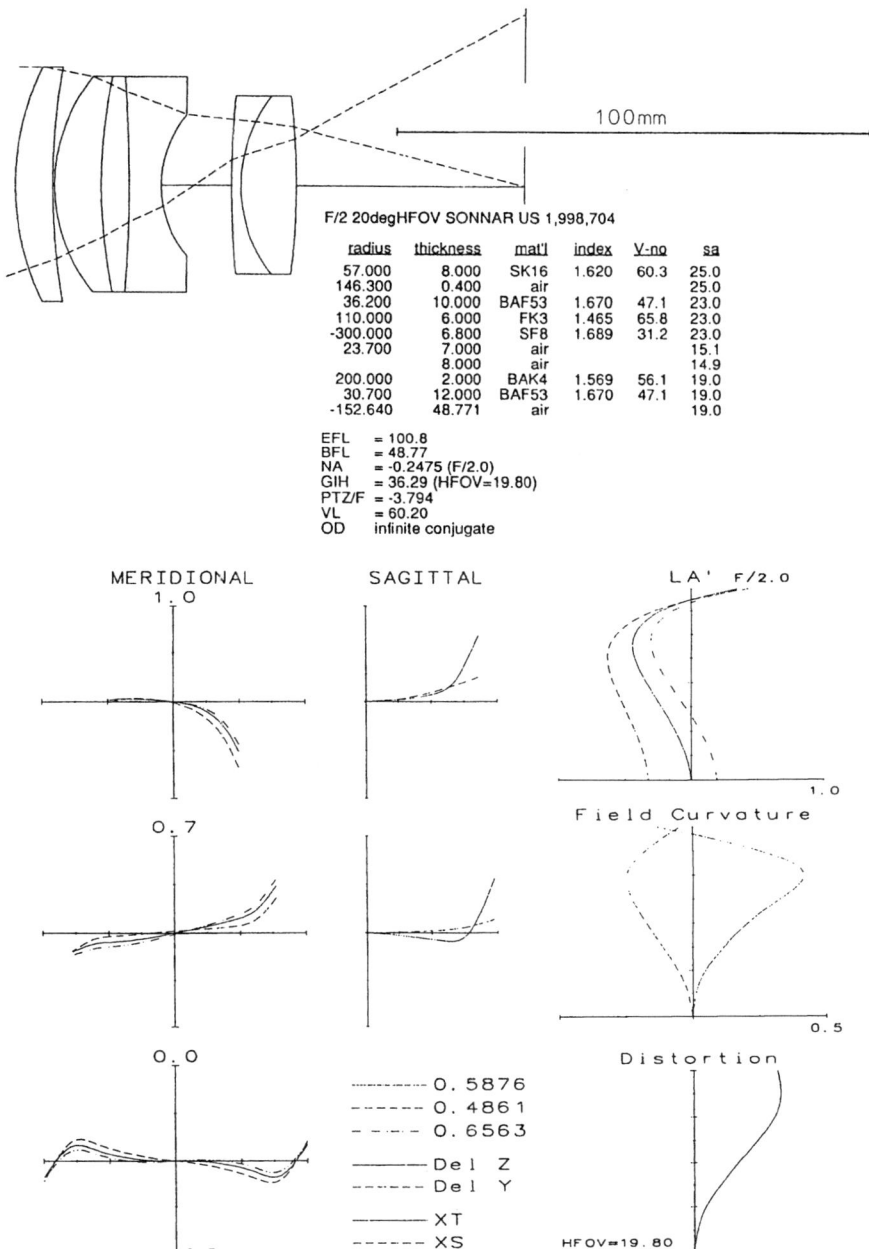

F/2 20degHFOV SONNAR US 1,998,704

radius	thickness	mat'l	index	V-no	sa
57.000	8.000	SK16	1.620	60.3	25.0
146.300	0.400	air			25.0
36.200	10.000	BAF53	1.670	47.1	23.0
110.000	6.000	FK3	1.465	65.8	23.0
-300.000	6.800	SF8	1.689	31.2	23.0
23.700	7.000	air			15.1
	8.000	air			14.9
200.000	2.000	BAK4	1.569	56.1	19.0
30.700	12.000	BAF53	1.670	47.1	19.0
-152.640	48.771	air			19.0

EFL = 100.8
BFL = 48.77
NA = -0.2475 (F/2.0)
GIH = 36.29 (HFOV=19.80)
PTZ/F = -3.794
VL = 60.20
OD infinite conjugate

Figure 9.7 $f/2.0$ ±20° Sonnar.

256 Chapter Nine

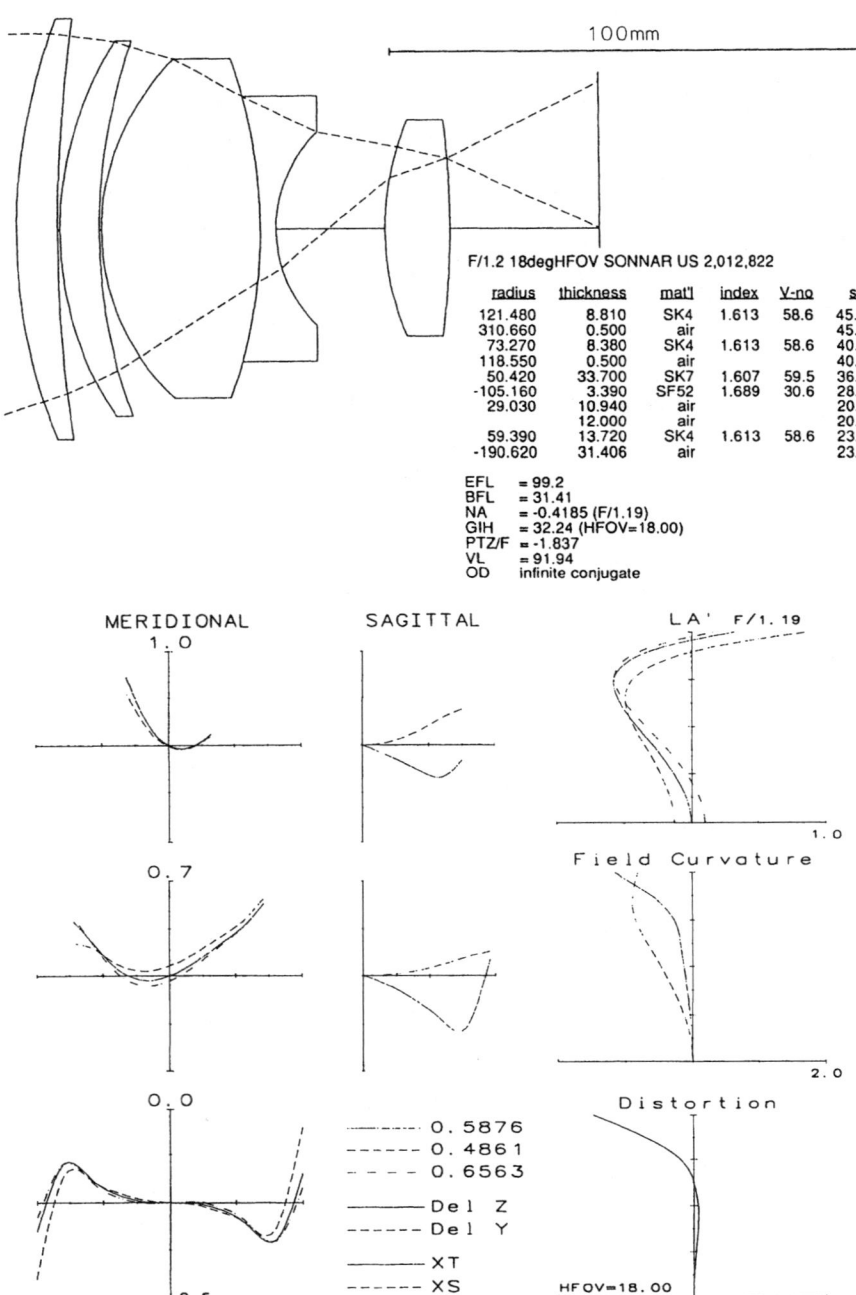

Figure 9.8 $f/1.2$ ±18° Sonnar.

Split Triplets

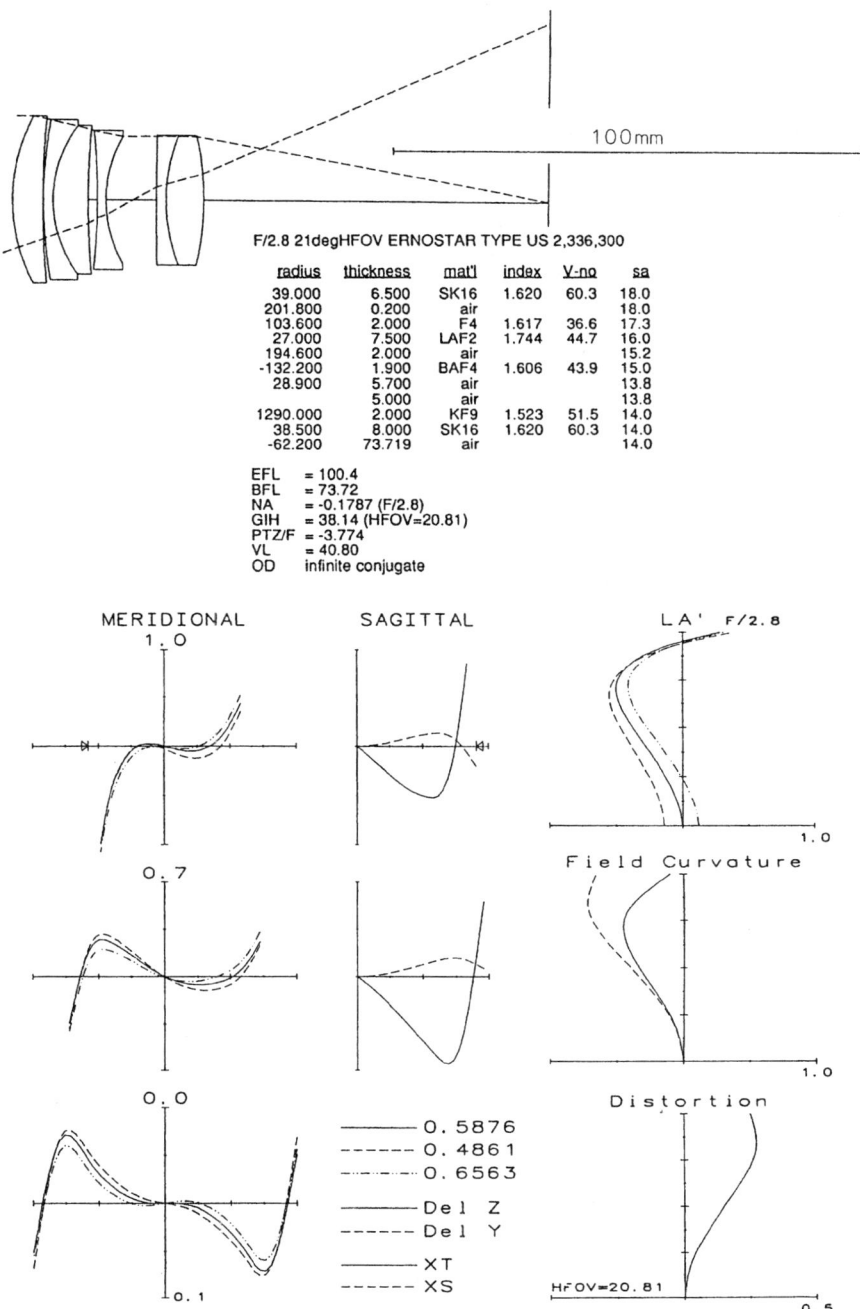

Figure 9.9 f/2.8 ±21° Ernostar.

258 Chapter Nine

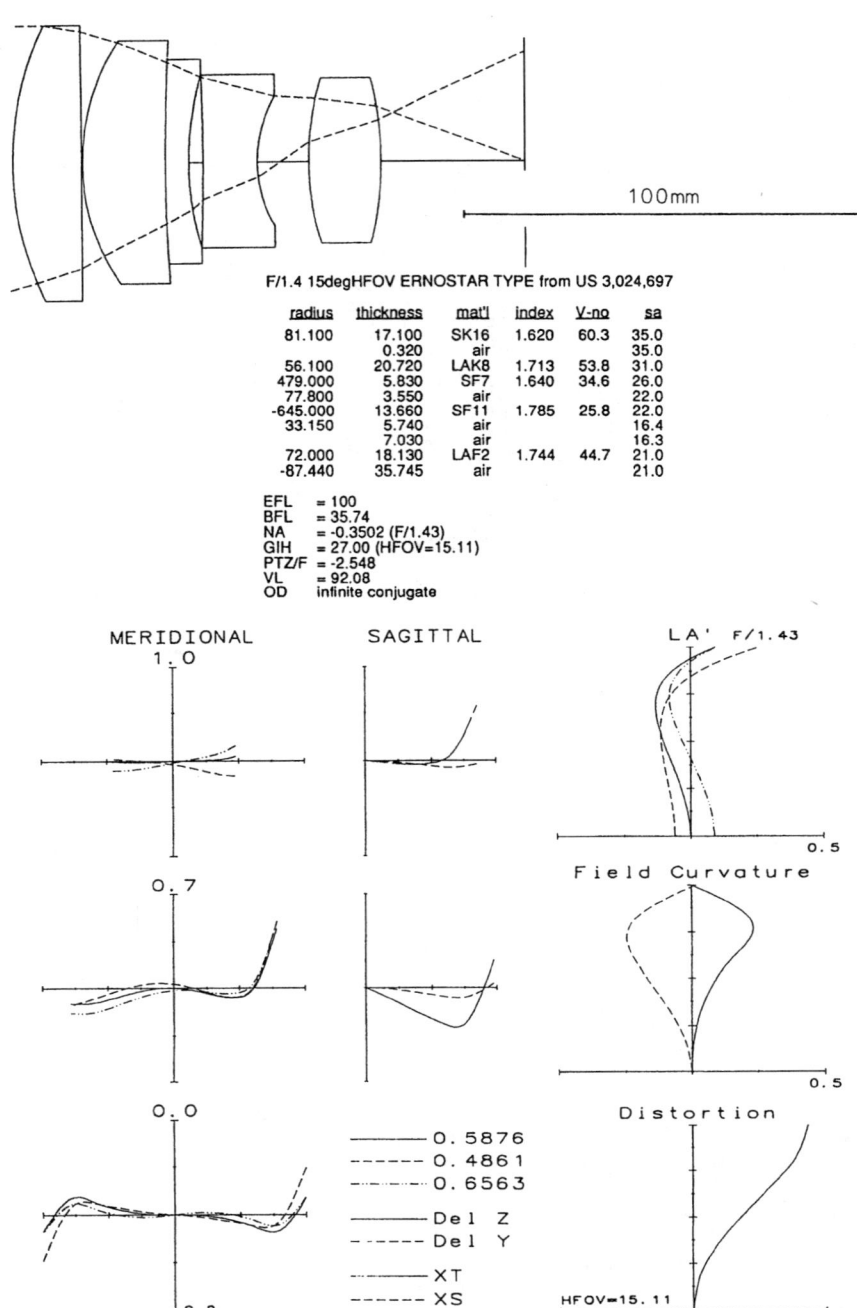

F/1.4 15degHFOV ERNOSTAR TYPE from US 3,024,697

radius	thickness	mat'l	index	V-no	sa
81.100	17.100	SK16	1.620	60.3	35.0
	0.320	air			35.0
56.100	20.720	LAK8	1.713	53.8	31.0
479.000	5.830	SF7	1.640	34.6	26.0
77.800	3.550	air			22.0
-645.000	13.660	SF11	1.785	25.8	22.0
33.150	5.740	air			16.4
	7.030	air			16.3
72.000	18.130	LAF2	1.744	44.7	21.0
-87.440	35.745	air			21.0

EFL = 100
BFL = 35.74
NA = -0.3502 (F/1.43)
GIH = 27.00 (HFOV=15.11)
PTZ/F = -2.548
VL = 92.08
OD infinite conjugate

Figure 9.10 $f/1.4$ ±15° Ernostar.

Chapter 10

The Tessar, Heliar, and Other Compounded Triplets

10.1 The Classic Tessar

Although the Tessar is actually a descendant of the double-meniscus anastigmat form, as described in Chap. 11, it can also be regarded as a modification of the Cooke triplet. As explained in more detail in Sec. 3.5, the substitution of a new achromat doublet for a crown in the triplet is the equivalent of using a high-index, high-V-value glass, and allows the possibility of utilizing the cemented surface of the doublet to control coma and oblique spherical aberration.

In the Tessar rear doublet, the convergent cemented surface:

a. Reduces the zonal spherical

b. Reduces the overcorrected oblique spherical

c. Reduces the "belly" in the astigmatism plots at the zonal field

d. Corrects the aberration of the upper oblique ray (see Fig. 3.4)

As a design approach, Kingslake has suggested using an index break of about $\delta n = 0.1$, fixing the radius of the cemented surface, and varying everything else.

The new achromat doublet of the Tessar (see Sec. 3.5) reduces the Petzval sum by using a crown that has a higher index than the flint.

In the Tessar there is a tendency toward negative high-order coma (hy^4). See Fig. 10.3 where the high-order (hy^4) is balanced by the third-order coma (hy^2)—this flattens the central part of the ray intercept plot.

The classical Tessar form is illustrated in Figs. 10.1 to 10.6. Figure 10.1 is an early Tessar by Rudolph, using relatively standard glasses and covering a relatively standard (for its day) 60° field at a speed of $f/6.3$.

260 Chapter Ten

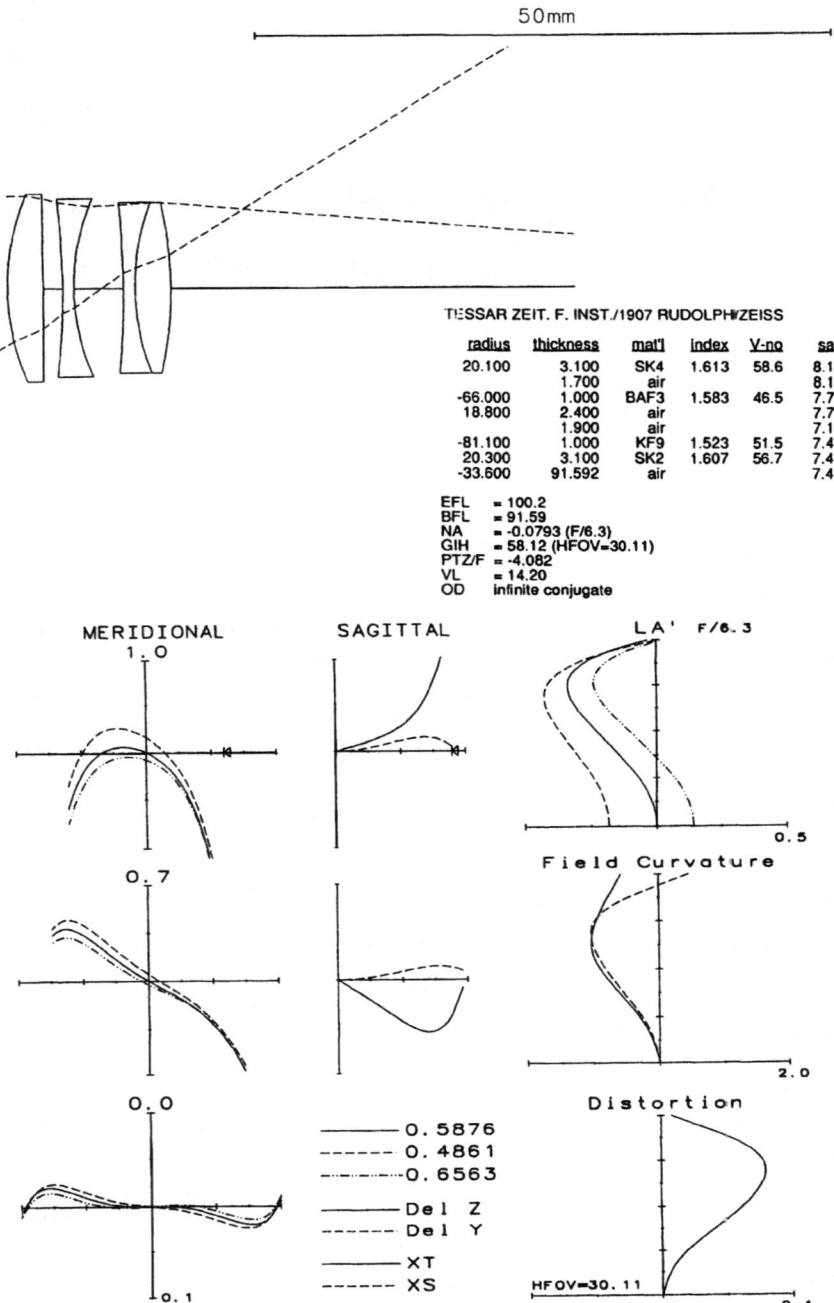

Figure 10.1 $f/6.3$ ±30° Rudolph Tessar.

The Tessar, Heliar, and Other Compounded Triplets 261

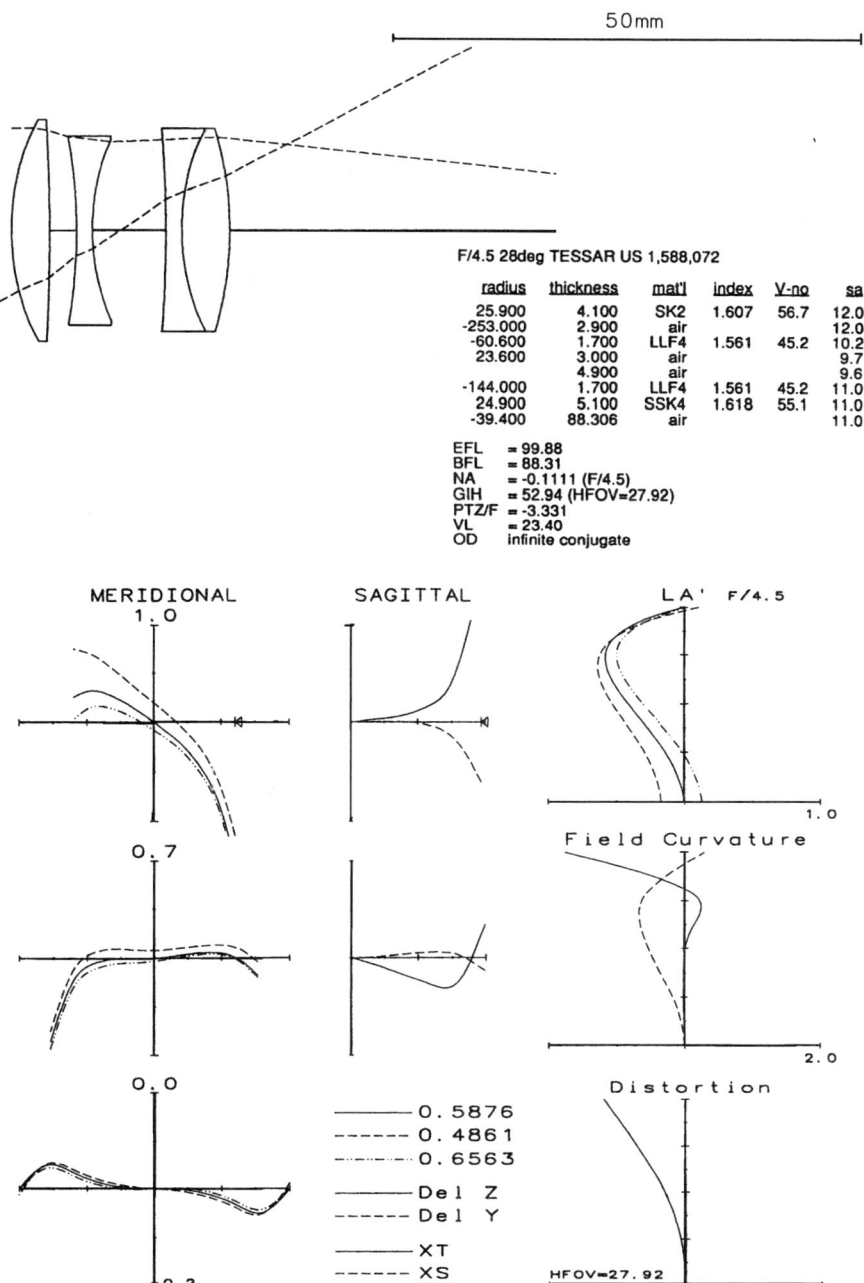

Figure 10.2 $f/4.5$ ±28° Tessar.

262 Chapter Ten

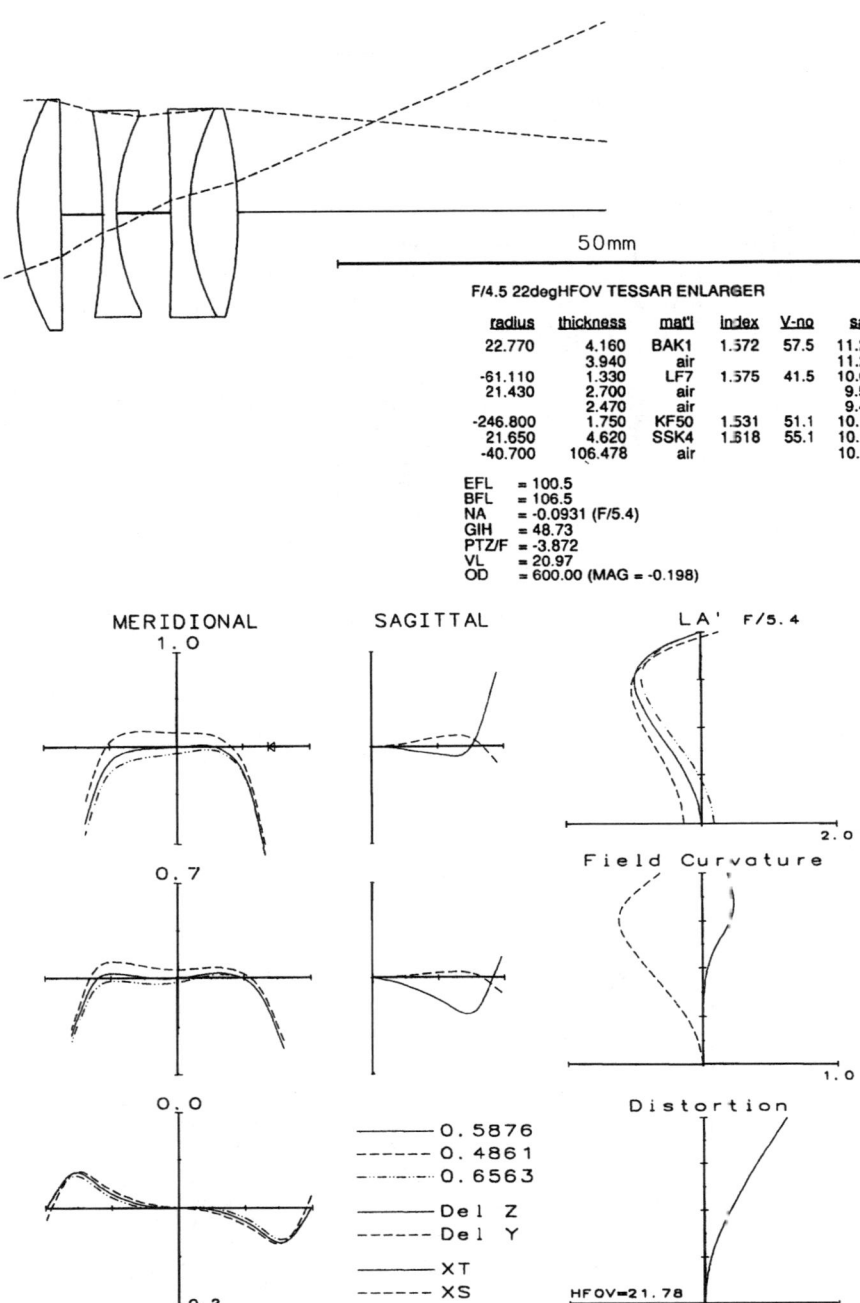

Figure 10.3 $f/4.5$ ±22° Tessar enlarger lens.

The Tessar, Heliar, and Other Compounded Triplets   263

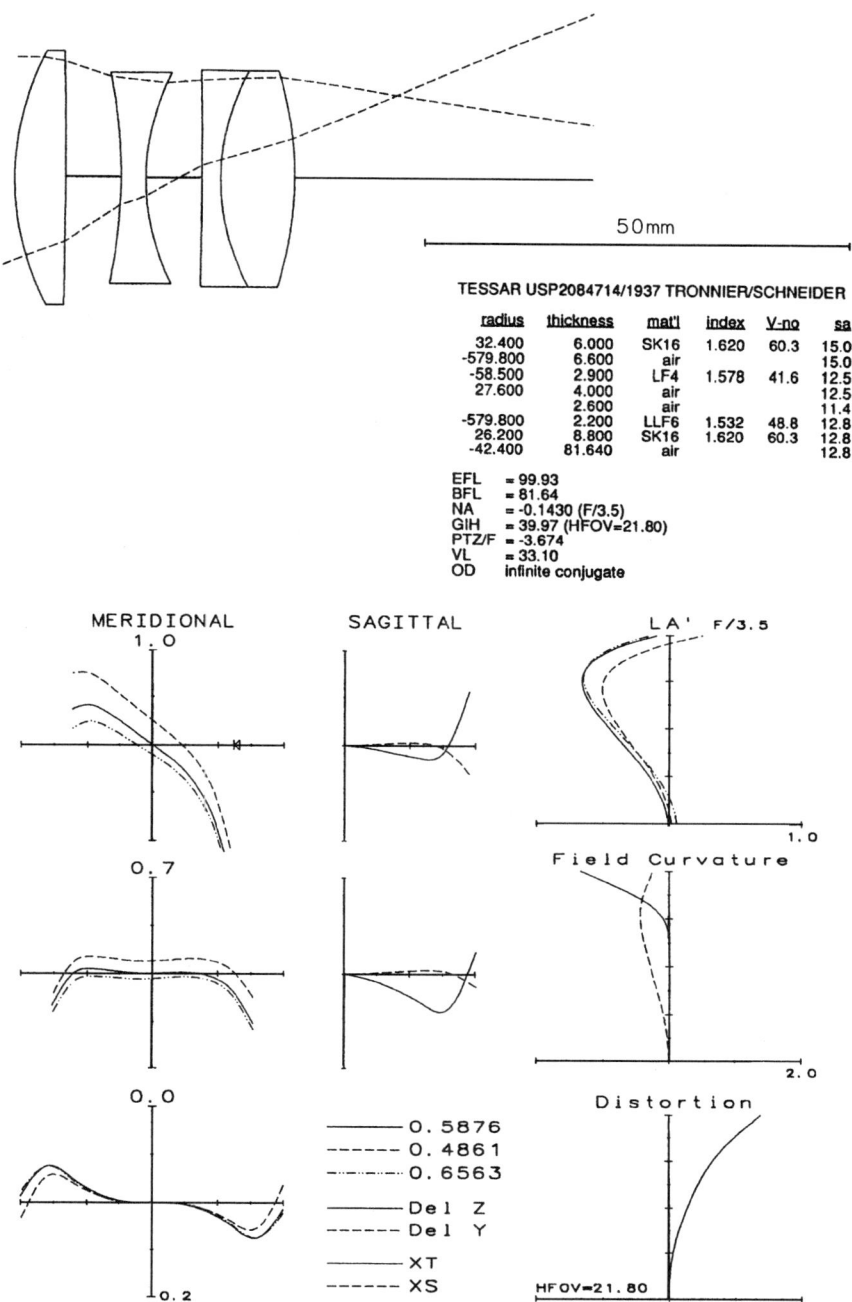

Figure 10.4 $f/3.5$ ±22° Tessar.

264 Chapter Ten

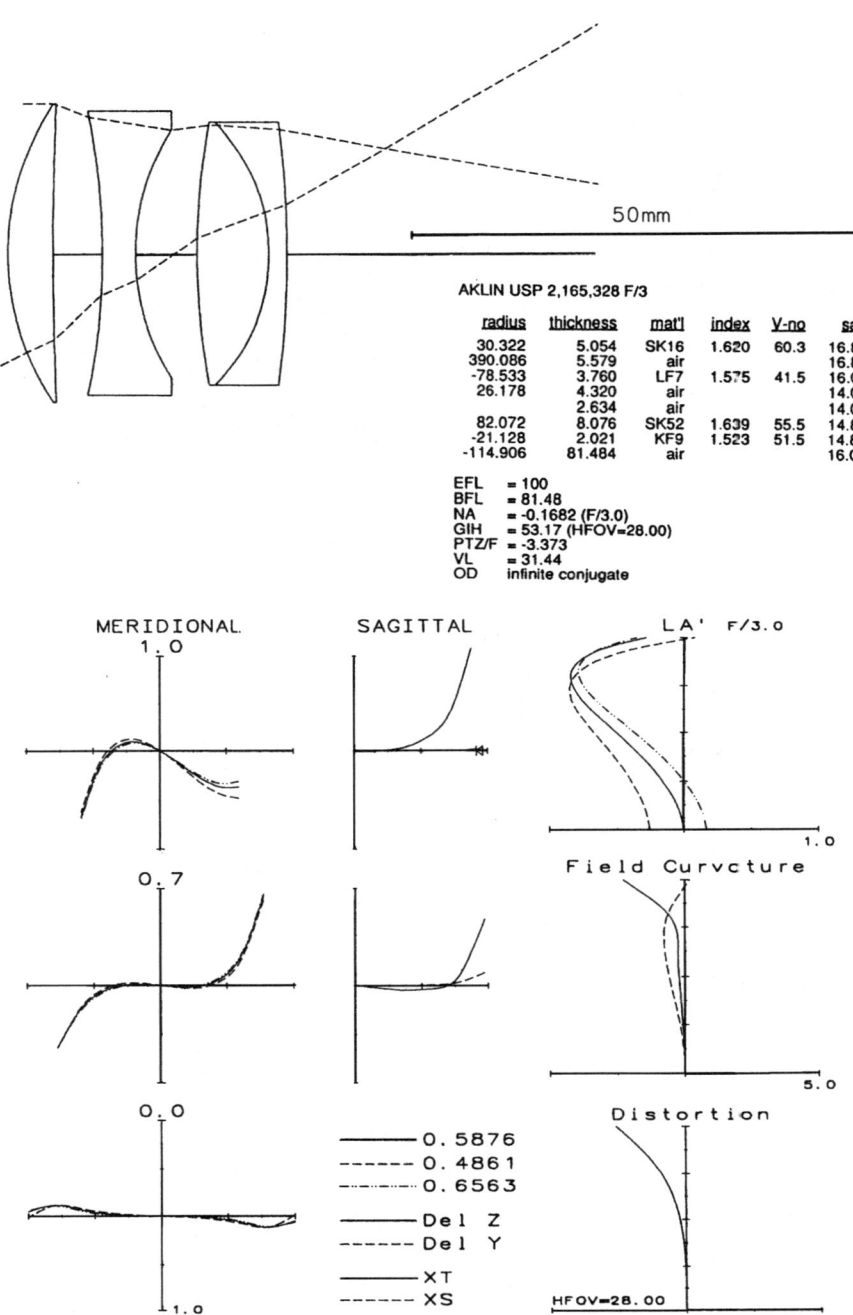

Figure 10.5 $f/3.0$ ±28° Tessar with reversed doublet.

The Tessar, Heliar, and Other Compounded Triplets

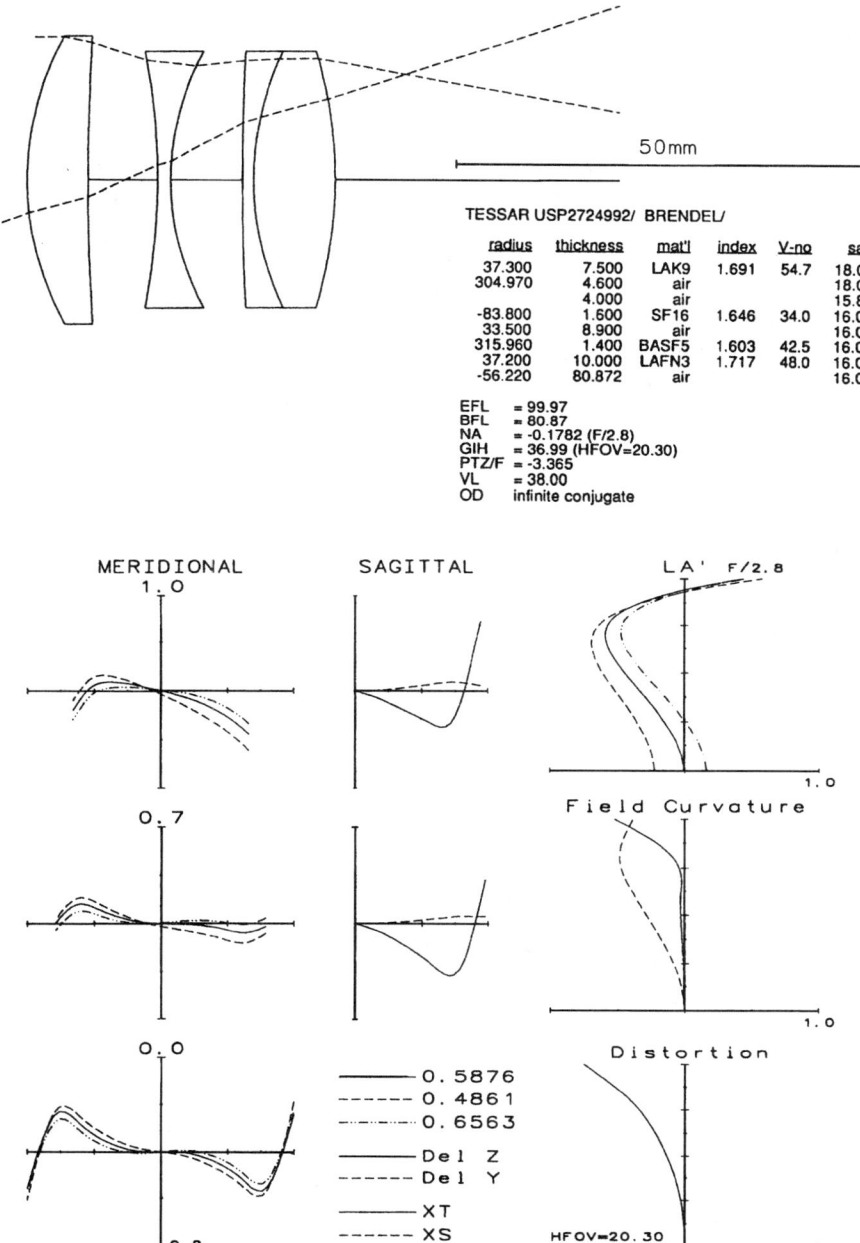

Figure 10.6 $f/2.8$ ±20° high-index Tessar.

Figure 10.2 covers a 56° field at $f/4.5$. Figure 10.3 is a lens similar to Fig. 10.2, but has been corrected for use as an enlarger lens—it is shown at a magnification of 1:5.

Figure 10.4 is a lens of ordinary glass, at a speed of $f/3.5$ with an angular field of 44°. Figure 10.5 covers a 56° field, is balanced for an aperture of $f/3.0$, and has its rear doublet arranged in reversed order. According to Kingslake, this orientation is to be preferred when high-index glass is used.

A still higher speed ($f/2.8$) is illustrated in Fig. 10.6 at an angular field of 41°; Fig. 10.6 uses higher-index glasses all around.

A Tessar-type lens with the *front* member compounded (instead of the rear) is shown in Fig. 10.7 at a speed of $f/2.8$; it uses glass of high index. It can be compared to Fig. 10.6, although the fields of view are different. Although rarely encountered, it is also possible to compound the center element of the Cooke triplet. See Figs. 10.12 and 10.13 for designs incorporating a compounded center component.

10.2 The Heliar/Pentac

The Heliar design uses the same concept of the *new achromat* doublet, but it has the advantage of having two of them. In addition, it has the benefit of an approximately symmetrical construction to help with coma, distortion, and lateral color. The result is an improvement over the Tessar. This design form has been executed with the doublets in the flint-out arrangement, as in Fig. 10.8 (in an early, fully symmetrical design), and in the (superior) crown-out arrangement, as in Figs. 10.9 and 10.10. Two other orientations of the general Heliar format are possible. In Fig. 10.11 both crowns face the short conjugate. The opposite orientation, with both crowns facing the object, is also possible.

10.3 The Portrait Lens and the Enlarger Lens

In a *portrait lens* a flattering softness of the image is achieved by correcting all the aberrations except spherical (and/or sometimes chromatic). If the front airspace of the Tessar is made adjustable, the degree of softness can be varied.

An *enlarger lens* must be insensitive to conjugate change since it is used at magnifications ranging from 1x to 10x or 15x. In designing an enlarger lens one should remember that an enlarger is usually focused with the iris wide open but is used stopped way down. The multiconfiguration feature of the design software can be used to optimize over a range of magnifications and apertures.

The Tessar is often used as an inexpensive enlarger lens, in a form, which is corrected for a magnification of about 5x. The Heliar and the

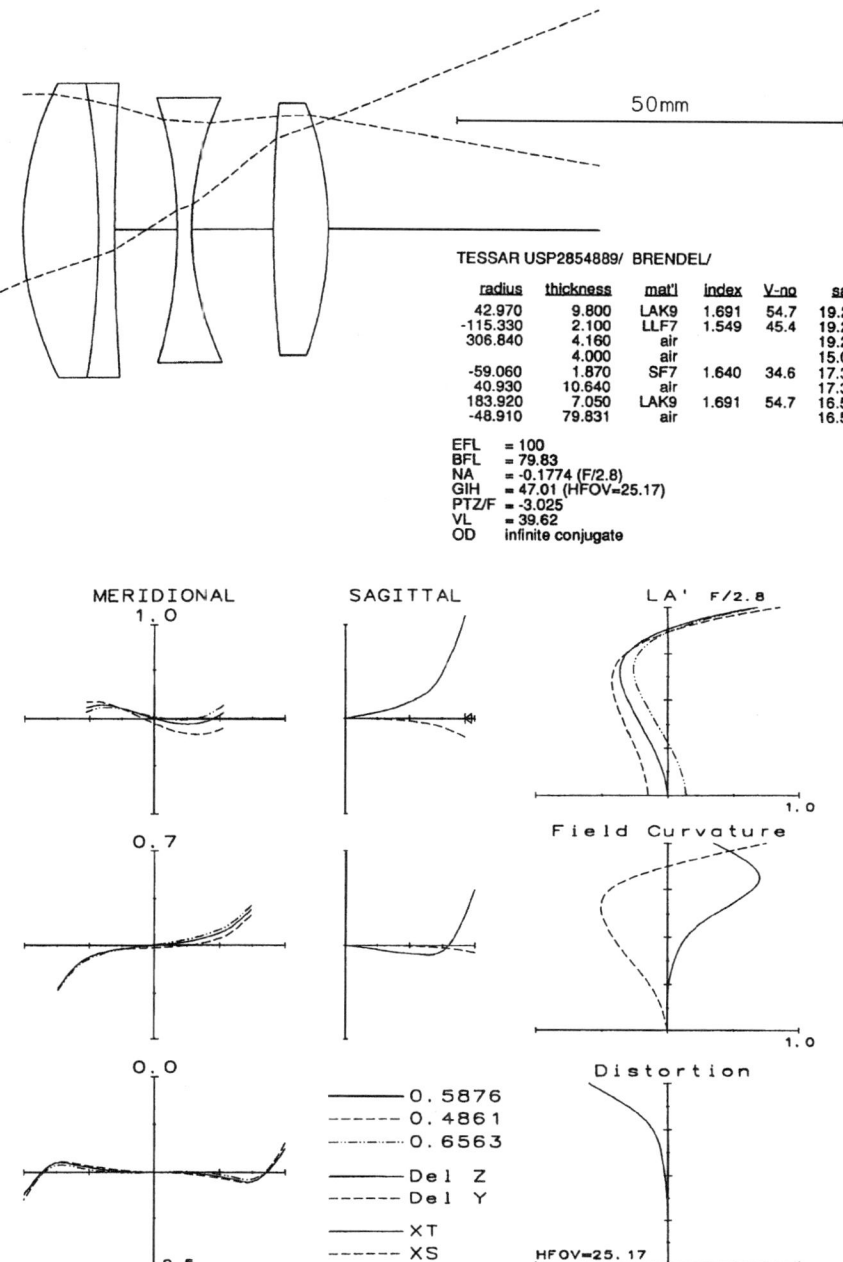

Figure 10.7 f/2.8 ±25° Tessar with doublet in front.

268 Chapter Ten

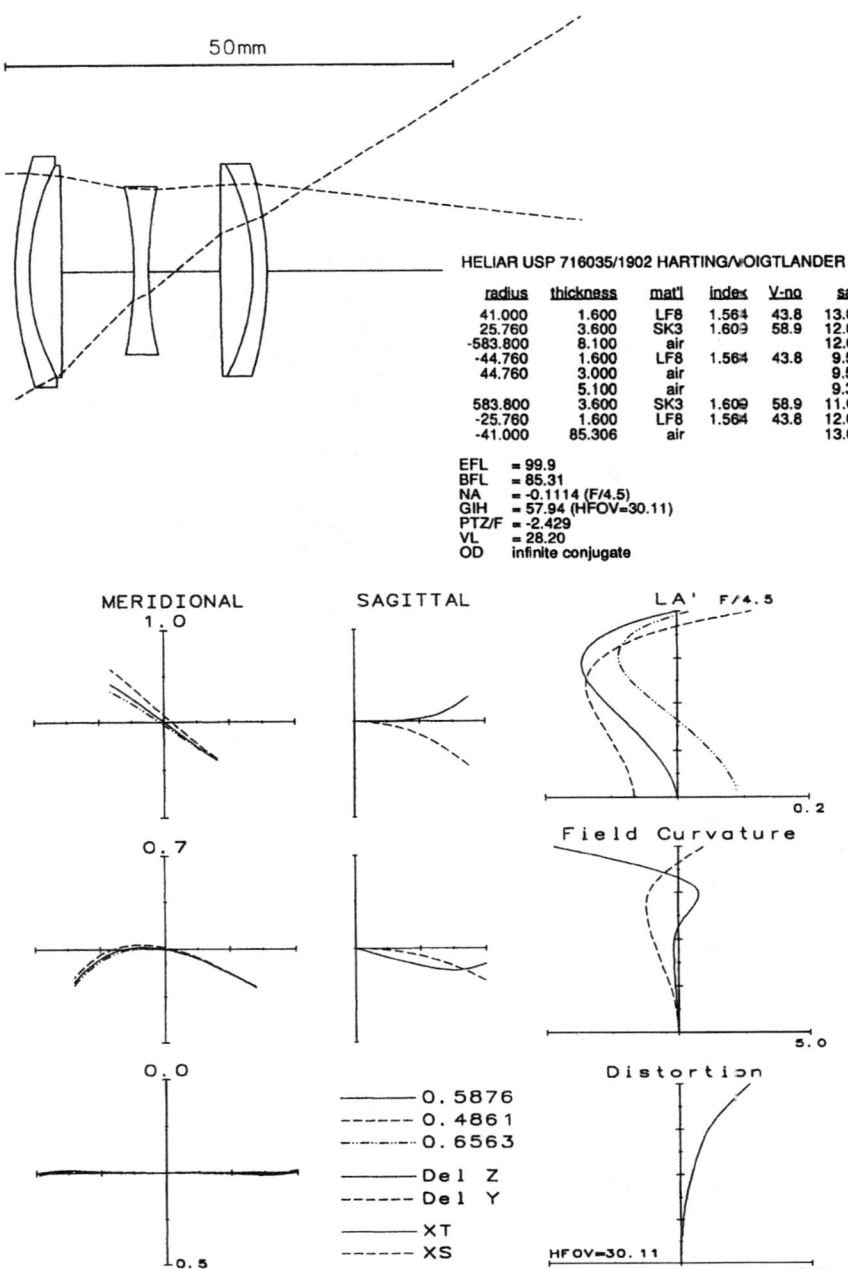

Figure 10.8 f/4.5 ±30° early Heliar with doublet flints facing outward.

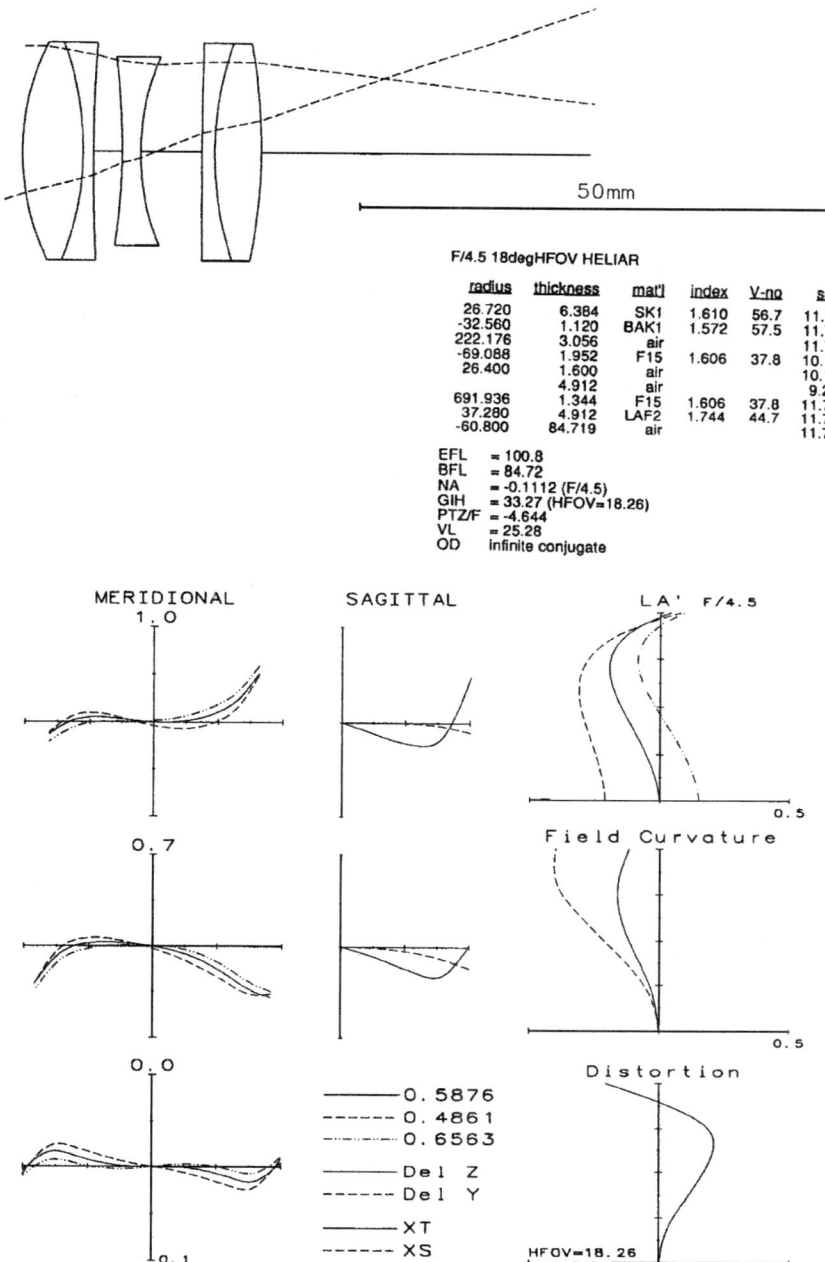

Figure 10.9 f/4.5 ±18° Heliar with crowns facing outward.

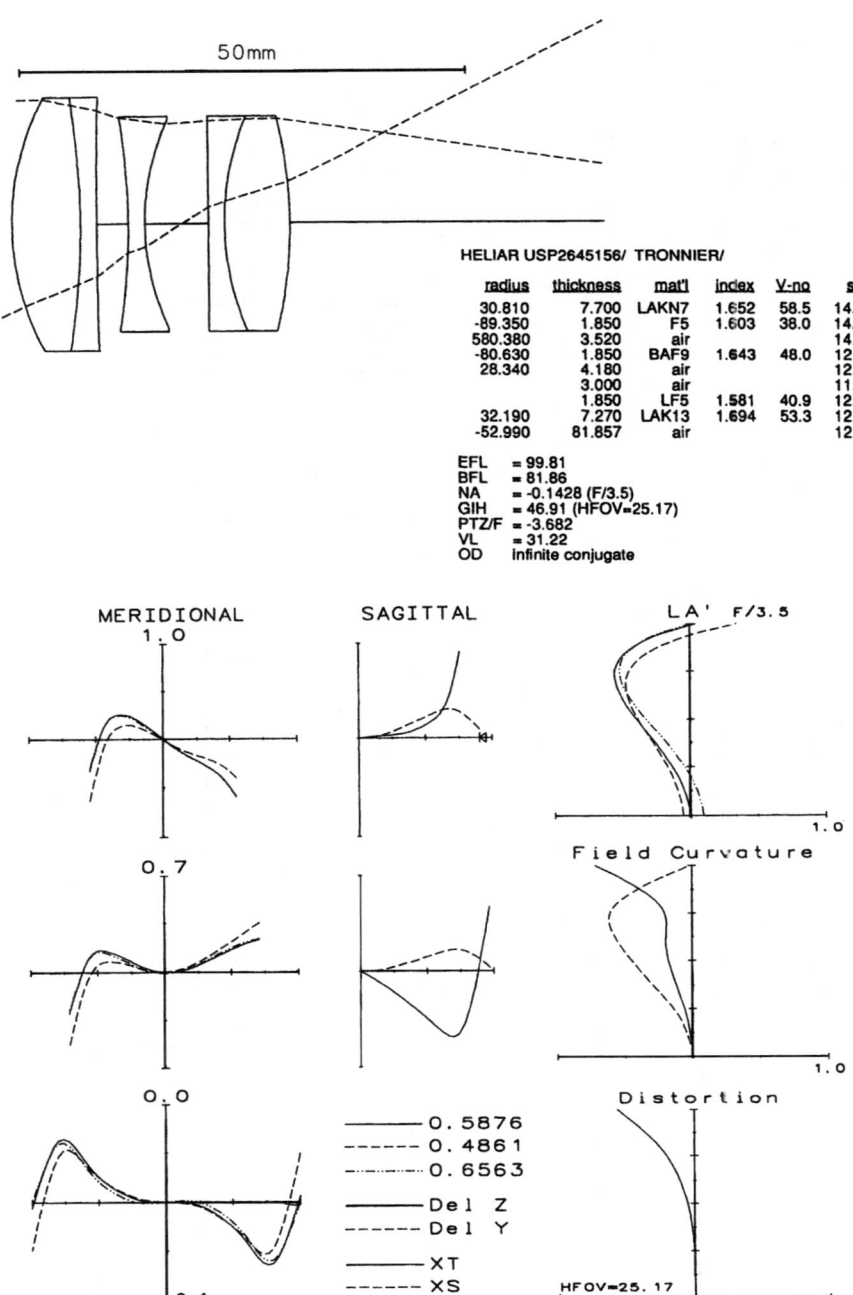

Figure 10.10 f/3.5 ±25° Heliar.

The Tessar, Heliar, and Other Compounded Triplets

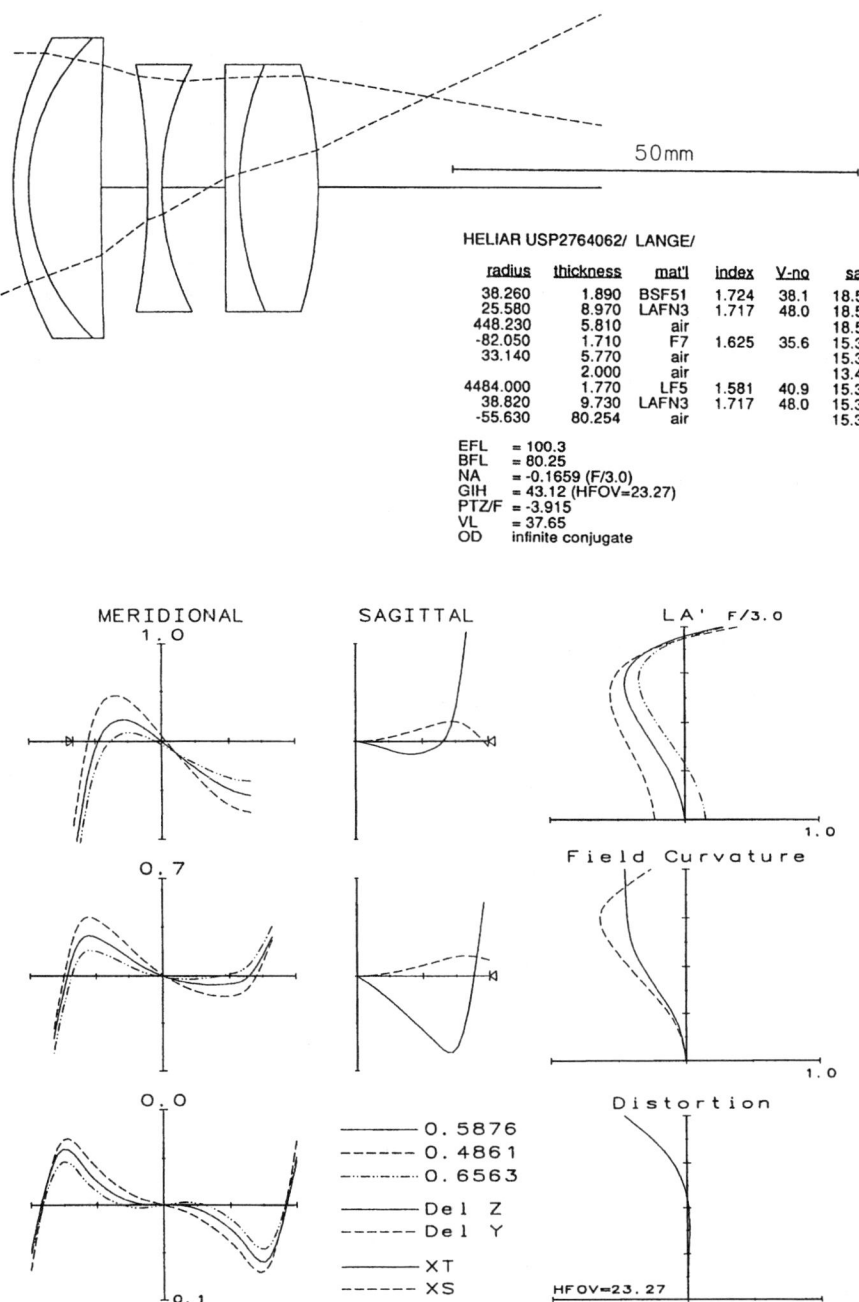

HELIAR USP2764062/ LANGE/

radius	thickness	mat'l	index	V-no	sa
38.260	1.890	BSF51	1.724	38.1	18.5
25.580	8.970	LAFN3	1.717	48.0	18.5
448.230	5.810	air			18.5
-82.050	1.710	F7	1.625	35.6	15.3
33.140	5.770	air			15.3
	2.000	air			13.4
4484.000	1.770	LF5	1.581	40.9	15.3
38.820	9.730	LAFN3	1.717	48.0	15.3
-55.630	80.254	air			15.3

EFL = 100.3
BFL = 80.25
NA = -0.1659 (F/3.0)
GIH = 43.12 (HFOV=23.27)
PTZ/F = -3.915
VL = 37.65
OD infinite conjugate

Figure 10.11 f/3.0 ±23° Heliar with reversed front doublet.

Dogmar (Chap. 11) are relatively stable as the enlarger magnification is changed, because of their roughly symmetrical construction. Other types occasionally found as enlarger lenses are triplets, the double Gauss, and the split Dagor (the latter in copy machines).

10.4 Other Compounded Triplets

Both the central and rear elements are compounded in Fig. 10.12, which covers about 32° at a speed of $f/2.7$. This lens could work well as a long-focal-length 35-mm camera lens; when it is stopped down, the off-axis ray intercept plots are quite flat.

The Hektor of Fig. 10.13 is an example of what can be accomplished by compounding all three elements of the triplet. This high-speed ($f/1.8$) example also makes use of a strong cemented Merté surface (see Sec. 3.5) in the center doublet (as does Fig. 10.12). This surface introduces undercorrected seventh-order spherical aberration, which has the effect of reducing the zonal spherical by causing the normally overcorrected spherical at the margin to reverse direction and go undercorrected.

10.5 Camera Lens Anastigmat Design "from Scratch"—The Tessar and Heliar

10.5.1 Project: Design a 2-in f/4 lens for a 35-mm camera

Continuing our design story from Chap. 8, we consider modifications of the Cooke triplet that depart from the three-element form. For the combination of speed and field we are investigating, four element forms that are suitable include the *Tessar* where one element of the triplet is replaced by a *new achromat*. Another four-element design is the *Dogmar*, also known as the *Celor* or *Aviar*, which we will take up in Chap. 11. In this chapter we will investigate the *Tessar*, and also go on to the five-element *Heliar*.

Compounded elements: New and old achromats. The Tessar and Heliar both incorporate cemented doublets in a *new achromat* form. In fact, almost all anastigmat lenses that utilize achromatic (or partially achromatic) cemented doublets use the new achromat form. In a new achromat the crown element has a higher index than the flint element. This was made possible by the development of the increased index barium crown glasses late in the nineteenth century. In the new achromat the cemented surface is converging, i.e., $c(n' - n)$ is positive, and it contributes undercorrected spherical aberration. Thus there is no mechanism to balance the undercorrected spherical produced by the outer surfaces of the doublet. In the *old achromat*, the cemented surface is divergent, i.e., $c(n' - n)$ is

The Tessar, Heliar, and Other Compounded Triplets 273

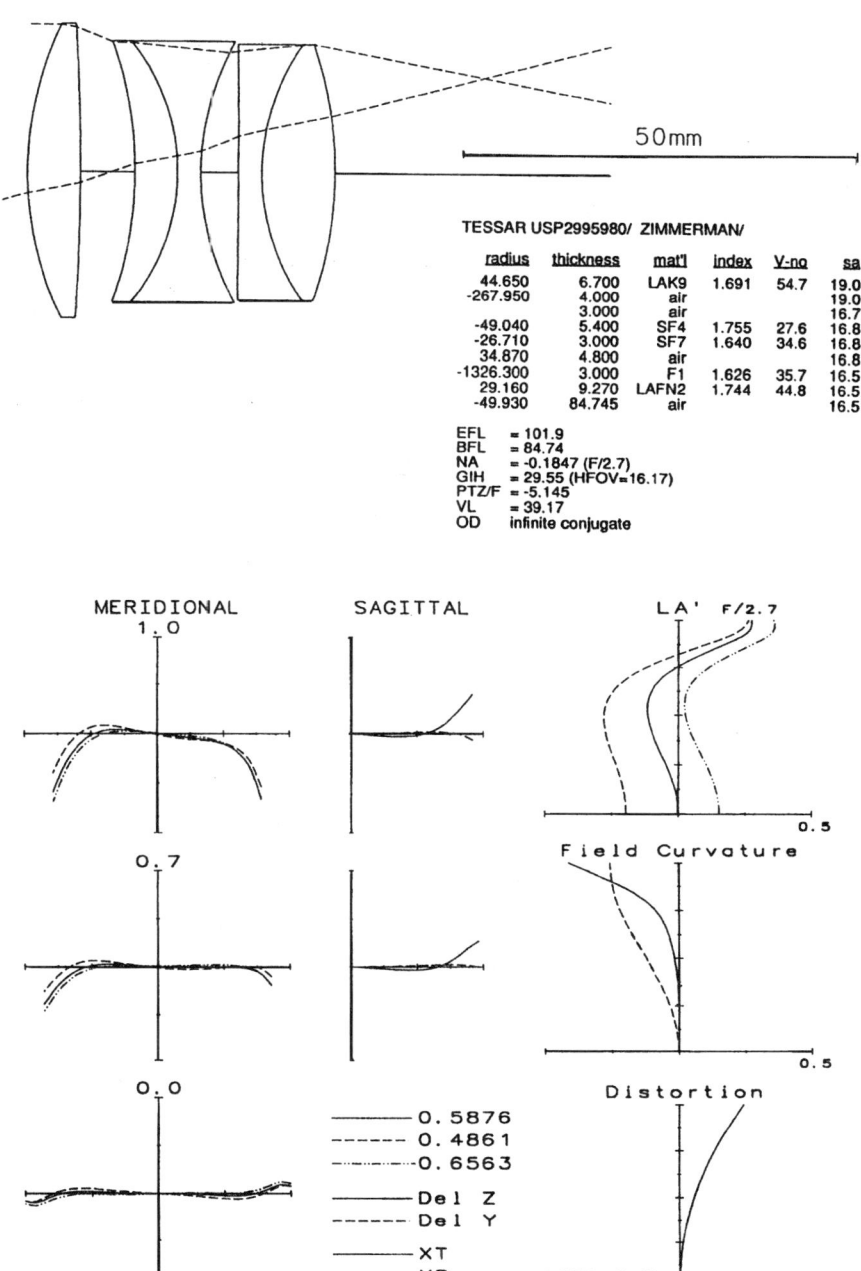

Figure 10.12 $f/2.7$ ±16° Tessar with compound middle lens.

274 Chapter Ten

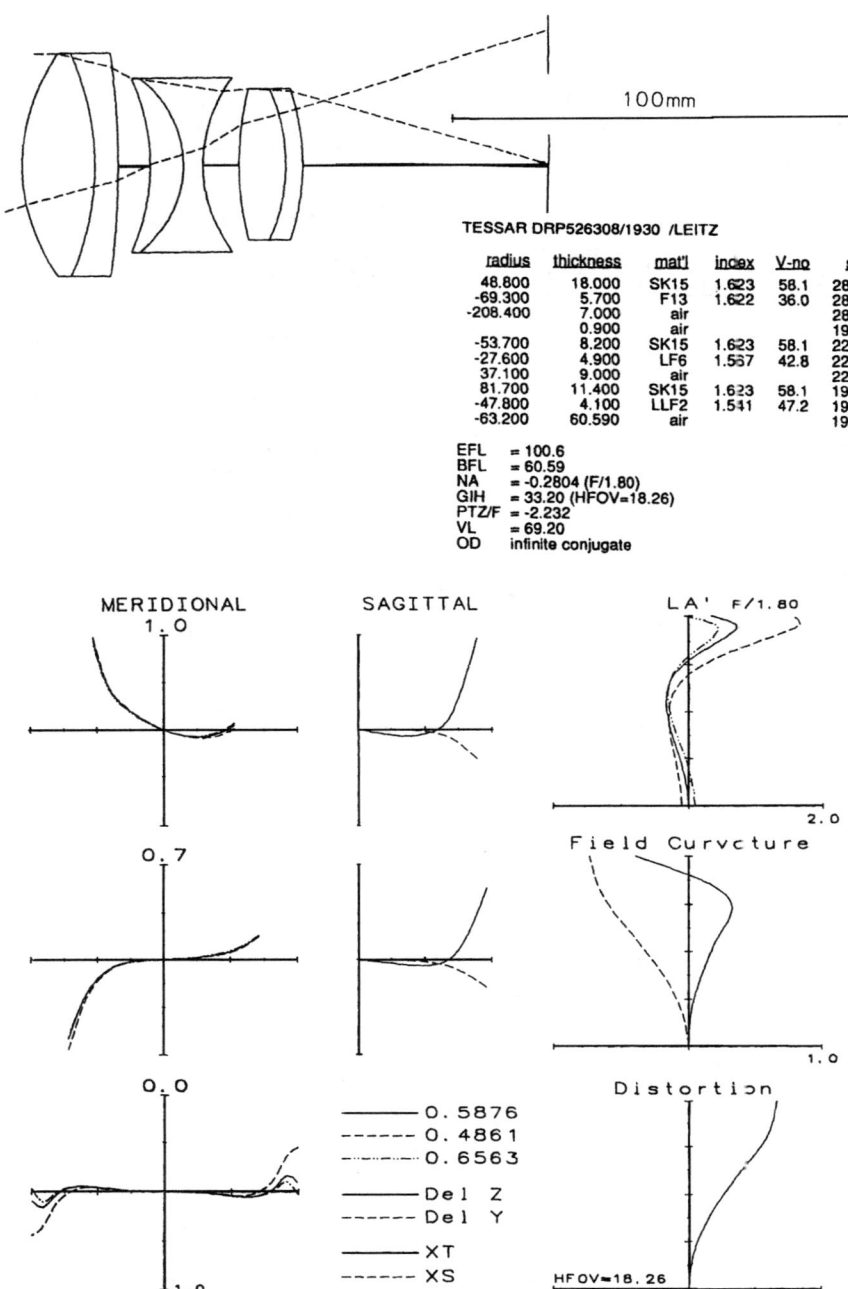

Figure 10.13 f/1.8 ±18° Hektor with all three components compounded.

negative, and it contributes overcorrected spherical. This is what makes a spherically corrected cemented achromatic doublet possible.

The appeal of the new achromat is that it has a reduced Petzval sum. Being at least partially achromatic, it is thus much like a high V-value, high-index glass singlet (as far as its chromatic and Petzval characteristics are concerned). In replacing a singlet the new achromat also possesses a cemented interface whose index break $(n' - n)$ and curvature can be used to balance other aberrations such as coma and astigmatism (provided that the doublet is located away from the aperture stop).

Typically, when used in a camera lens, a new achromat doublet has an index break at the cemented surface to the order of 0.05 to 0.12, with a larger break in some Tessars and a bit smaller break in the two doublets of the Heliar form. In the classic designs the doublet glasses are typically barium crowns and light flints. In newer designs the crowns are often a high-index lanthanum type and the flint glass has a correspondingly higher index. The usual orientation for the doublet in the Tessar is with the crown facing outward and the flint facing the stop. (With some higher-index glasses the reverse orientation is said to be better, but the curves are stronger.)

The Tessar lens. Let's see what we can do if we start with the SSKN5 Cooke triplet design and convert the rear crown to a new achromat doublet. (Note that there are equally good Tessar designs with the doublet as the front component.) We will simply add a cemented surface inside the rear crown to make a biconvex rear element, and convert the negative element to an appropriate flint glass. We increase the crown thickness to avoid a sharp edge, and also reduce the rear airspace and flint thicknesses to compensate for the added glass. The index for SSKN5 is 1.658, so a flint with an index in the range of 1.54 to 1.61 for the doublet would seem appropriate. Scanning the glass map we see that LF7 575415 is in the center of this range. Finally, we manually adjust $R5$ to get the right focal length.

Thus our starting layout is shown in Table 10.1.

We use the software default (rms spot size) *merit function* (MF) with added targets on focal length (50.8) and on *edge thickness* (ET, 0.8) for the

TABLE 10.1 The Starting Prescription for the Tessar Design Project

$R1 = +18.41$	$T1 = 2.22$	SSKN5	658509
$R2 = +287.9$	$T2 = 5.02$		
$R3 = -31.72$	$T3 = 0.8$	SF15	699301
$R4 = +18.91$	$T4 = 4.0$		
$R5 = +95.0$	$T5 = 0.8$	LF7	575415
$R6 = +22.90$	$T6 = 2.5$	SSKN5	658509
$R7 = -22.91$			

crowns. Perhaps surprisingly, the large number of variables often becomes something of a headache when we try to systematically investigate the Tessar. We have a choice of four glasses, and as continuous variables there are seven radii and six spacings, totaling a possible 17 parameters (or 21, if we count both index and dispersion). After we select SSKN5 and set the flint thicknesses at 0.8, we still have 13 variables, five more than we need to control the power and the primary aberrations. Controling the crown thicknesses to get an ET of 0.8 still leaves us with three extras. We seem to have an "embarasment of riches."

In considering the Cooke triplet from an aberration theory point of view, it was apparent that, because of the quadratic relationships between the third-order aberrations and the variable parameters, once the glass types were chosen there were eight potential thin lens third-order solutions to the primary aberrration problem. Now if we consider a merit function, which includes the effects of the higher-order aberrations, obviously many more local optima will be possible, each optimum representing a different balance or level of aberrations. The situation is even more complex in the Tessar than in the triplet. Fortunately we have about a century of Tessar design history to guide us, and the published standard design types are pretty good route markers.

The lens design solution space is n-dimensional, where n is the number of variables. When we think in terms of a simple two-dimensional "landscape" space, we may visualize the merit function terrain as a smooth bowl with a single optimum; however, in the much more complex n-dimensional space, while the solution area may still be broad and relatively level, there are many local depressions representing differing *optimum* designs of roughly equivalent quality. The one that our *damped least-squares* (DLS) software will go to is the one nearest to the starting design. Bearing in mind that the n dimensions are all quite different in numerical magnitude (e.g., surface curvature, surface spacing, glass index of refraction, and dispersion), it should be apparent that which local optimum is the nearest depends on how the variables are expressed or weighted. As a simple example, consider the relative magnitudes of the numbers representing radius, curvature, spacing, index, and dispersion. A seemingly simple difference such as whether the design calculations are executed in inches or millimeters will change the relationships greatly.

It should be apparent that our "embarassment of riches" in regard to variable parameters will extend to the number of *optimum* designs. We can expect that there will be many nearly equivalent designs clustered together, and there probably will be several basic clusters that differ in both form and performance. This is why a *global optimum* is a such a difficult thing to find.

After quite a few DLS optimization cycles we have this design as shown in Table 10.2. The aberrations and *modulation transfer function* (MTF)

TABLE 10.2 Prescription for the First Optimized Tessar Design (Using an rms Spot Size Merit Function). See Figs. 10.14 to 10.16

$R1 = +15.606v$	$T1 = 2.360$	SSKN5	658509
$R2 = -7.6e + 5v$	$T2 = 2.943v$		
$R3 = -33.701v$	$T3 = 0.800$	$658327v$	
$R4 = +15.076v$	$T4 = 4.115v$		
$R5 = -147.28v$	$T5 = 0.800$	$535501v$	
$R6 = +19.585v$	$T6 = 2.596$	SSKN5	658509
$R7 = -21.504v$	$BF = 43.473$		

are shown in Figs. 10.14 to 10.16. Figure 10.16 is a plot of the MTF versus field angle for 10, 17, and 25 lpm, and gives a good idea of how uniform the contrast is across the field.

The variable glasses look reasonable; no. 3 is between SF9 and SF19, while no. 5 is close to LLF6. The unusual thing about this design is that the program produced an index difference of +0.103 in the doublet but chose a flint with a V-value difference of only +0.8. Apparently it didn't need an achromatic rear doublet, but made good use of the converging cemented interface.

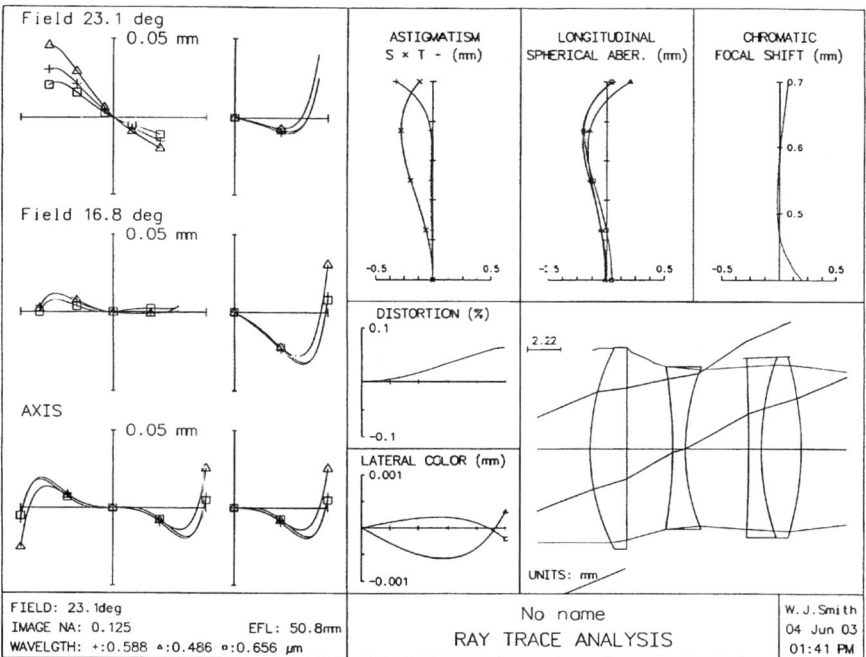

Figure 10.14 Ray aberrations for the first Tessar design, Table 10.2.

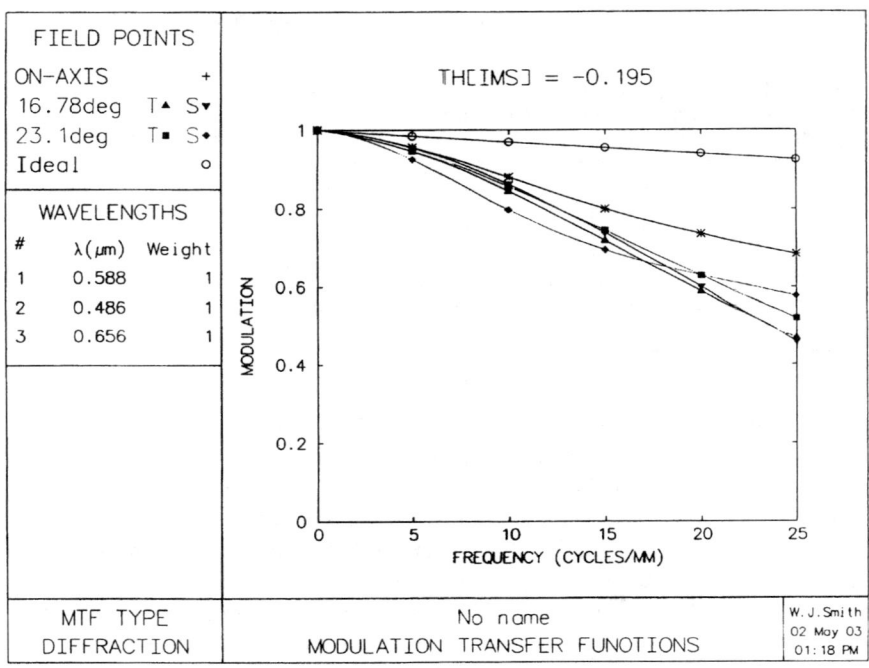

Figure 10.15 MTF versus frequency for the lens of Fig. 10.14, Table 10.2.

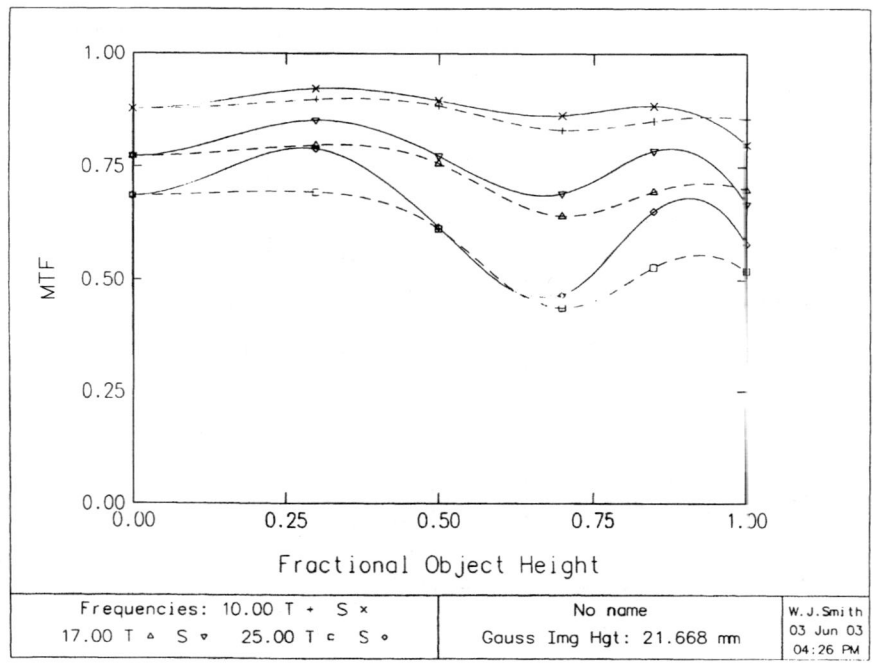

Figure 10.16 MTF versus field for 10, 17, 25 lpm for Fig. 10.14, Table 10.2.

TABLE 10.3 Prescription for the Second Optimized Tessar Design (using an Aberration Based Merit Function). See Figs. 10.17, 10.18, and 10.19

$R1 = +15.799v$	$T1 = 2.457v$	SSKN5	658509
$R2 = -198.17v$	$T2 = 2.078v$		
$R3 = -31.251v$	$T3 = 0.800$	$626356v$	
$R4 = +15.007v$	$T4 = 3.964v$		
$R5 = -63.774v$	$T5 = 0.800$	$554454v$	
$R6 = +18.830v$	$T6 = 2.708v$	SSKN5	658509
$R7 = -19.909v$	$BF = 44.766$		

We then changed the merit function to one based on aberrations rather than rms spot size. We also pushed the rear doublet flint to a lower V-value type. As a result we got the data as shown in Table 10.3. The aberrations and MTF are shown in Figs. 10.17 to 10.19.

Again the variable glasses are quite reasonable, with no. 3 near F1 and no. 5 between LLF1 and LLF7. This aberration-based merit function produced a more typical glass pair in the rear doublet with an index difference of +0.104 just as in the prior design, but with a V-value difference of +5.5 compared to +0.8. This lens is close, but not quite as good as the preceding design.

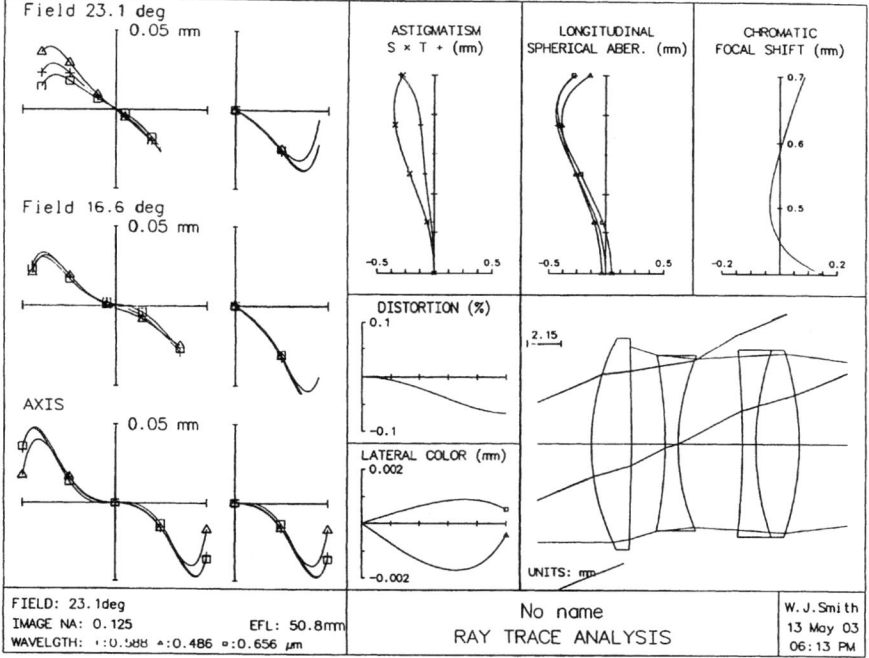

Figure 10.17 Ray aberrations for the second Tessar design, Table 10.3.

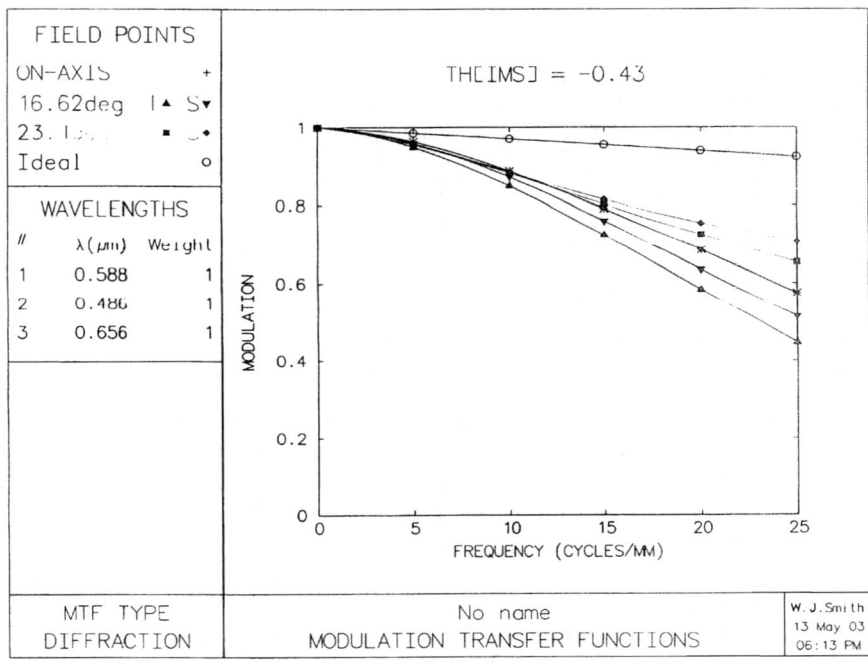

Figure 10.18 MTF versus frequency for Fig. 10.17, Table 10.3.

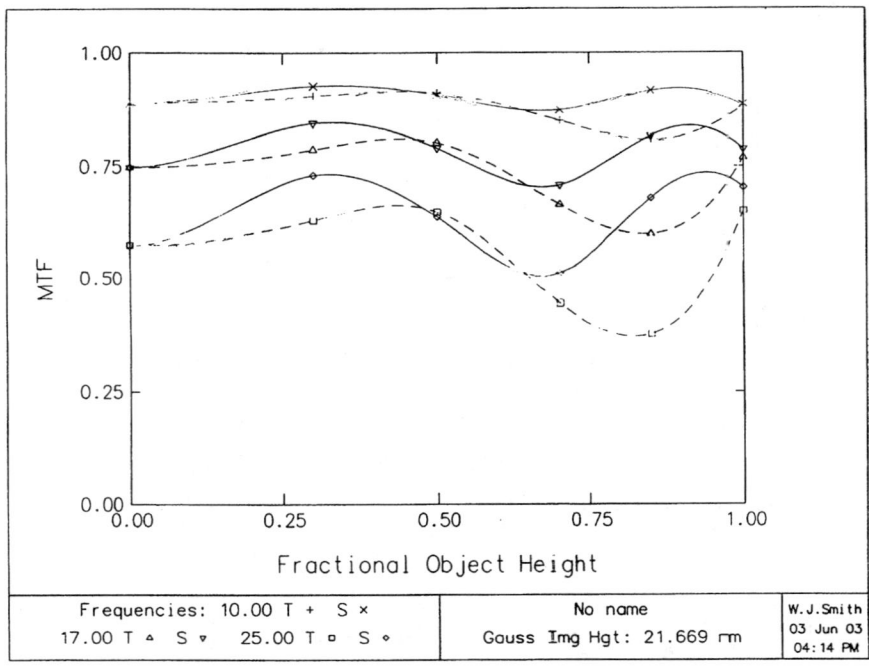

Figure 10.19 MTF versus field for 10, 17, 25 lpm for Fig. 10.17, Table 10.3.

The Tessar, Heliar, and Other Compounded Triplets   281

TABLE 10.4 Prescription for the Third Optimized Tessar, with Fixed Flint Glasses (See Figs. 10.20 to 10.22)

$R1 = +15.68$	$T1 = 2.38$	SSKN5	658509
$R2 = -891.19$	$T2 = 2.57$		
$R3 = -34.10$	$T3 = 0.80$	F6	636353
$R4 = +14.84$	$T4 = 4.34$		
$R5 = -121.17$	$T5 = 0.80$	LLF1	548457
$R6 = +18.96$	$T6 = 2.62$	SSKN5	658509
$R7 = -21.65$	$BF = 44.054$		

We return to the rms-spot-size merit function and freeze the rear doublet flint to LLF7 (548457) to see what this will give us. We later fix the center flint to F6 (636353) and round off the dimensions to two places. This rounding probably costs us a bit in image quality. The result is shown in Table 10.4. The characteristics of this design are plotted in Figs. 10.20 to 10.22.

The performance of this design is about the same as that of our first effort, and like the first it is slightly better than the immediately preceding

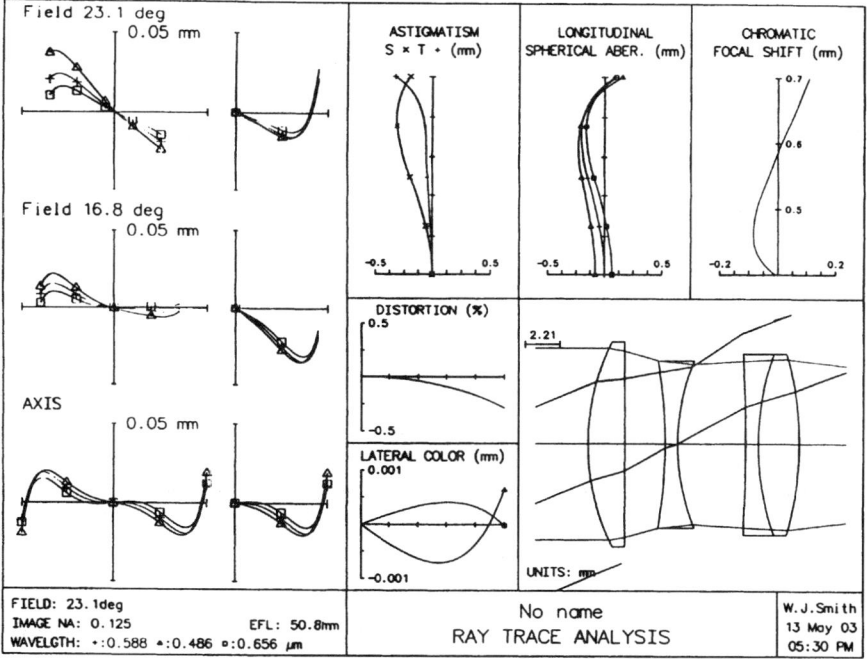

Figure 10.20 Ray aberrations of the third Tessar design, Table 10.4.

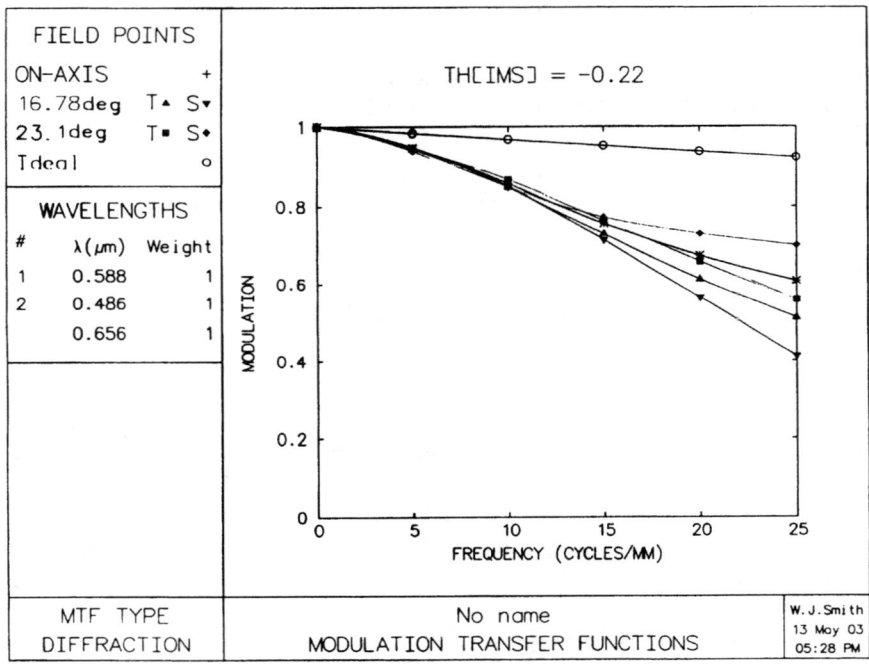

Figure 10.21 MTF versus frequency for Fig. 10.20, Table 10.4.

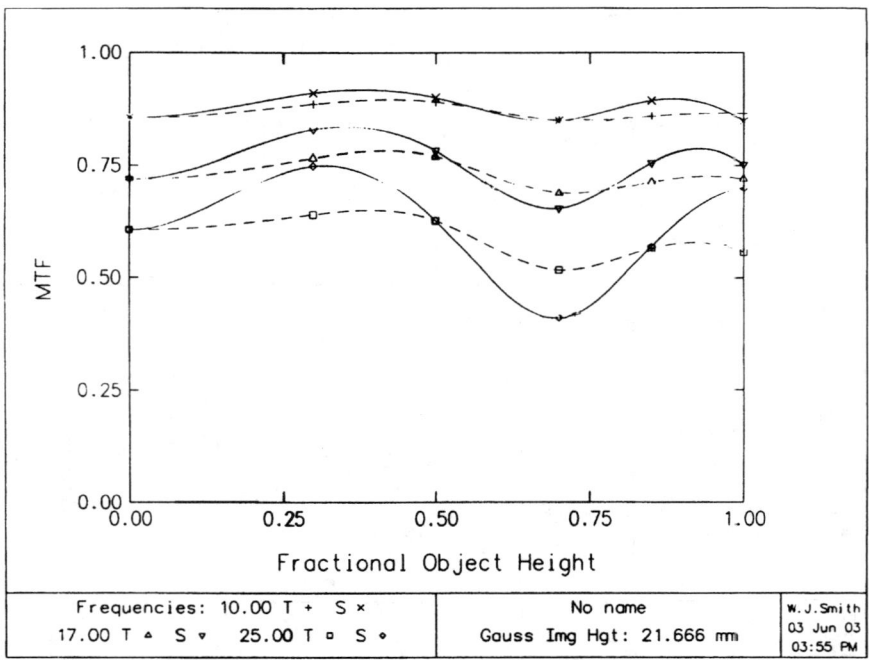

Figure 10.22 MTF versus field for 10, 17, 25 lpm for Fig. 10.20, Table 10.4.

TABLE 10.5 Prescription for the Fourth Optimized Tessar, using Lanthanum Crowns (See Figs. 10.23 to 10.25)

$R1 = +16.76$	$T1 = 2.16$	LAF3	717480
$R2 = +276.89$	$T2 = 2.93$		
$R3 = -45.27$	$T3 = 0.80$	SF5	673322
$R4 = +15.91$	$T4 = 5.46$		
$R5 = -129.68$	$T5 = 0.80$	KF3	515547
$R6 = +27.13$	$T6 = 2.20$	LAK31	697564
$R7 = -25.25$	$BF = 43.692$		

design. It would seem that, in this particular set of circumstances, the rms-spot-size merit function slightly outdid the aberration-based one.

We have used SSKN5 (658509) in all out Tessars so far, so now we allow the crowns to wander about (but we keep them on a short leash) and get the following rare earth design as shown in Table 10.5, again rounded to two places past the decimal point. Figures 10.23 to 10.25 show the aberrations and MTF for this design.

We can fairly compare this lens with the immediately preceding one since they are both rounded to two places past the decimal point. This

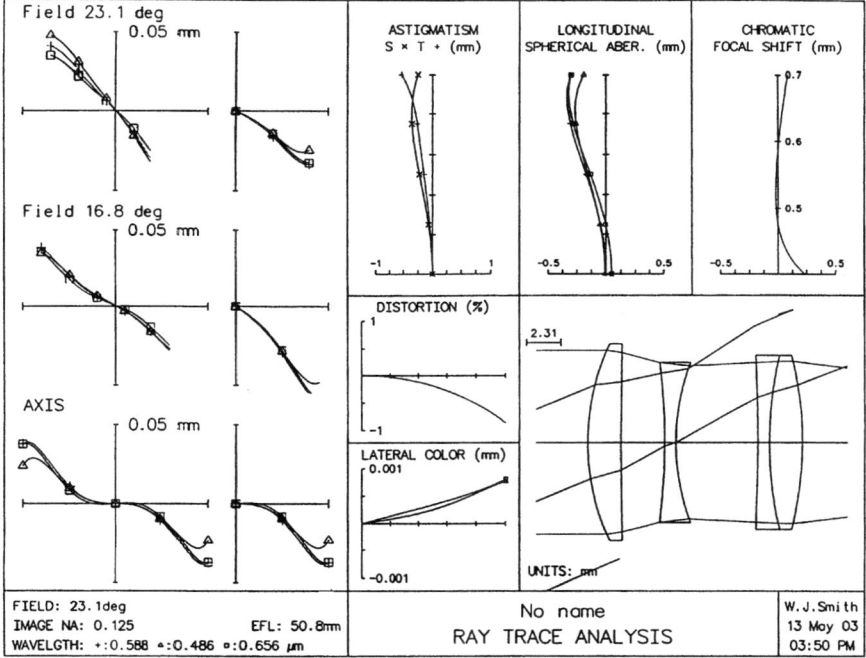

Figure 10.23 Ray aberrations for the fourth Tessar design, with higher-index glasses, Table 10.5.

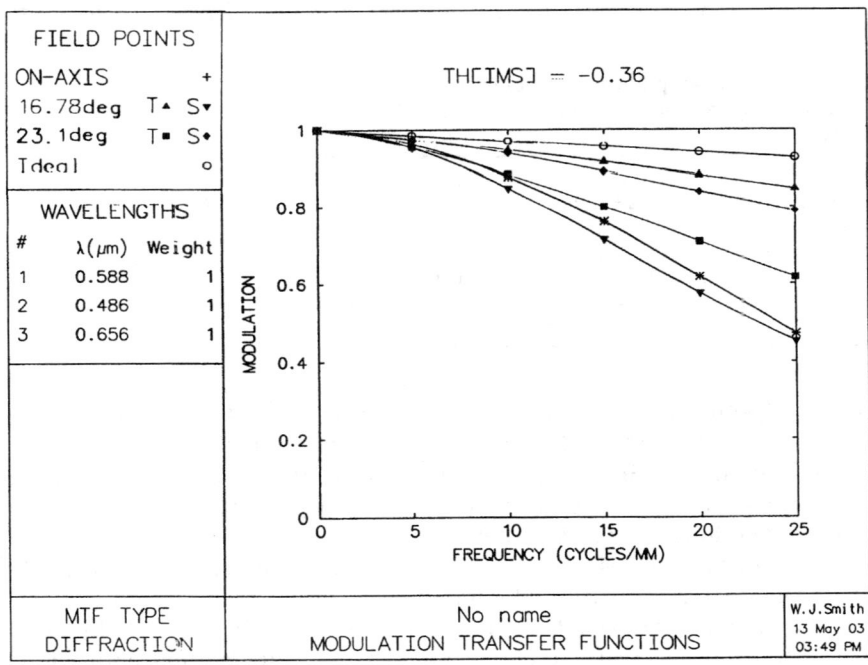

Figure 10.24 MTF versus frequency for Fig. 10.23, Table 10.5.

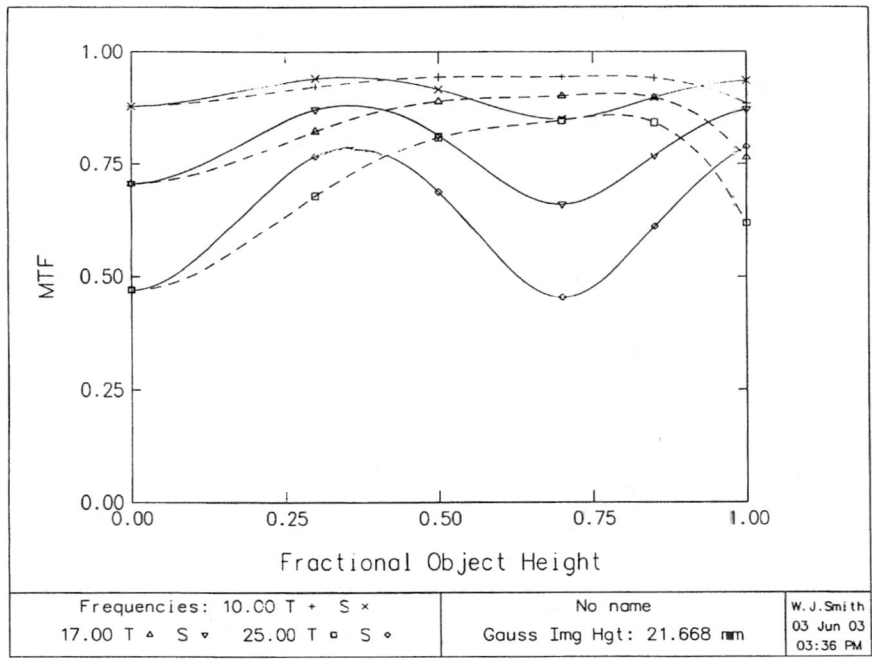

Figure 10.25 MTF versus field for 10, 17, 25 lpm for Fig. 10.23, Table 10.5.

one is clearly better. The surface curvatures are much flatter. It has less astigmatism, and the oblique spherical is well controlled. The sagittal MTF is only slightly better, but it is better, and in all other regards this one is clearly the winner. Again we note that this rms-spot-size merit function likes a large index difference (+0.182) and accepts a relatively small V-value difference (+1.7) in the rear doublet. An obvious drawback to this design is that the price of LAF3 is twice that of SSKN5, and LAK31 is nine times as much; however, the glasses are comparable for chemical stability. Possible replacements for LAK31 (697564) are LAK9 (691547) and LAKN14 (697554), which, respectively, cost only 29 and 42 percent of the cost of LAK31.

For brevity we have only used the rms-spot-size merit function for most of this process. We should note that often the rms *optical path difference* (OPD) merit function will produce a lens with a better MTF, but with more flare in the image.

Of course we have not completely explored our Tessar. Of necessity, this has been a relatively quick survey. There are still many variations left untried. Different glass combinations and modifications to the merit functions immediately come to mind. For the Tessar the stop should be well into the second airspace rather than at the fourth surface where we have placed it in both our triplet and Tessar designs. And of course, any general conclusions that we may draw are limited to the application that we have chosen (EFL = 50.8, $f/4.0$, ±23.1°).

The Heliar lens. For the Heliar we can begin with the same layout that we worked up for the start of the Tessar, except that we will use the doublet from the Tessar for both the front and rear doublets in our Heliar. We bend the flint to an equi-concave shape and set the two airspaces equal. We use pickups to maintain complete front-to-back symmetry. The resulting lens has a focal length of 48.9, so we adjust $R3$ (and $R6$) from −95 to −158 to get the focal length to 50.8. Note that we choose to fix the doublet glasses, but follow our usual custom of allowing the center flint to vary (so that the vertex length of the lens can be optimized to suit our field and f number). Our starting form is shown in Table 10.6.

We use the default rms-spot-size merit function with added targets for effective focal length (EFL) and (ETH), and iterate the DLS about 40 times. The MF drops from an initial value of 0.013341 to 0.000563 and is still dropping, but the performance leaves a lot to be desired and progress seems slow. So we release the symmetry constraints (still holding the doublet glasses as SSKN5 and LF7) and quickly get a big improvement, with the MF reduced to 0.000161. The design is now shown in Table 10.7. The performance curves are given in Figs. 10.26 to 10.28.

We now allow the flint glass of the doublets to vary and find that the design becomes much better but has significantly departed from symmetry.

TABLE 10.6 Rough, Fully Symmetrical Starting Prescription for the Heliar Design

$R1 = +22.9v$	$T1 = 2.50v$	SSKN5	658509
$R2 = -22.9v$	$T2 = 0.80$	LF7	575415
$R3 = -158.v$	$T3 = 4.51v$		
$R4 = -23.7v$	$T4 = 0.80$	SF15v	699301
$R5 = +23.7p$	$T5 = 4.51p$		
$R6 = +158.p$	$T6 = 0.80$	LF7	575415
$R7 = +22.9p$	$T7 = 2.50p$	SSKN5	658509
$R8 = -22.9p$	BFL = 42.13		

The front airspace is reduced to about 1.3 mm and the rear space is increased to about 4.7 mm. With the aperture stop on surface no. 5 the vignetting is now quite unbalanced. So we move the stop 2 mm (and later only 1 mm) into the rear airspace. The contrast (MTF) of the sagittal fan at the 0.7 field appears to be the biggest problem spot, so we increase the merit function weight on these rays to try to get a better balance. There is improvement, so we fix the glasses, reoptimize, and get the following data as shown in Table 10.8.

The ray aberrations and MTF are presented in Figs. 10.29 to 10.32. Notice that Fig. 10.32 plots the ray aberrations at a defocus of 0.25 mm; compare with Fig. 10.29. As an aside, we could have explored the possibility of pushing the two airspaces to more nearly equal values. This would probably take us down a completely different road than the one we are following.

There remain several degrees of freedom we have not tested. They are the doublet crown glasses and the symmetry of the doublet glass types. As usual, one would expect that crowns from the upper left area of the glass map should produce improvements; the patent literature seems to confirm this. The literature also seems to indicate that an unsymmetrical arrangement of glass types has found favor among lens designers. (Although one should always bear in mind that the improvements claimed

TABLE 10.7 Prescription for the Optimized Heliar Design, After Discarding the Front-to-Back Symmetry and Fixing the Doublet Glasses (See Figs. 10.26 to 10.28)

$R1 = +17.593v$	$T1 = 3.099v$	SSKN5	658509
$R2 = -25.287v$	$T2 = 0.800$	LF7	575415
$R3 = +83.861v$	$T3 = 2.822v$		
$R4 = -26.937v$	$T4 = 0.800$	586403v	
$R5 = +16.988v$	$T5 = 3.361v$		
$R6 = +383.75v$	$T6 = 0.800$	LF7	575415
$R7 = +16.838v$	$T7 = 2.731$	SSKN5	658509
$R8 = -22.200v$	BFL = 43.51		

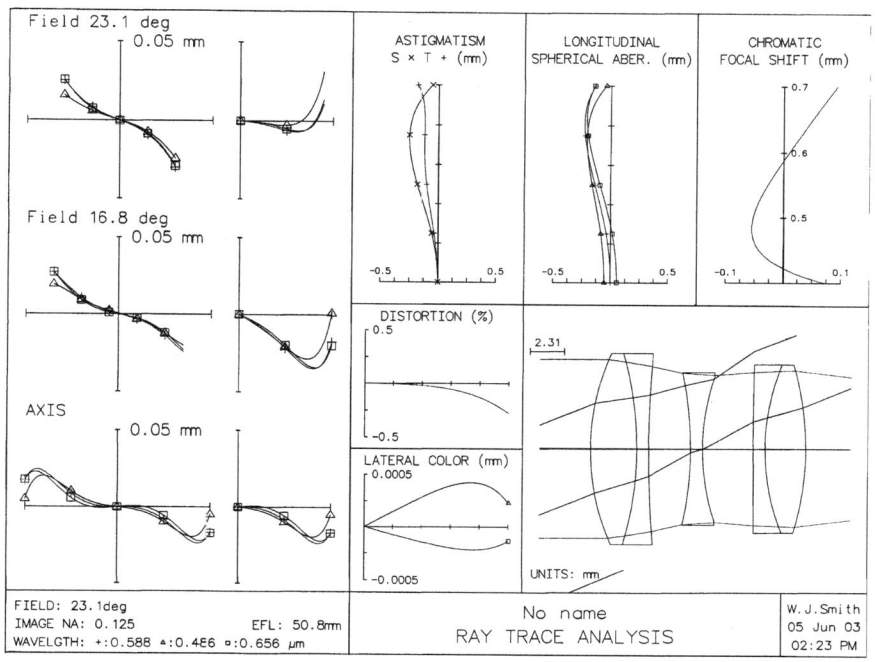

Figure 10.26 Ray aberrations for the first Heliar design, Table 10.7.

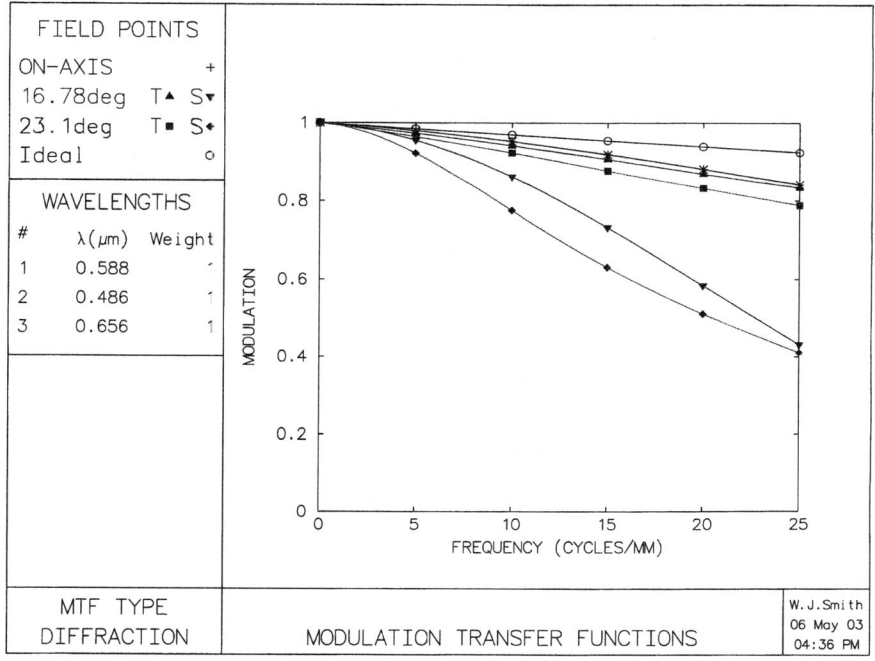

Figure 10.27 MTF versus frequency for Fig. 10.26, Table 10.7.

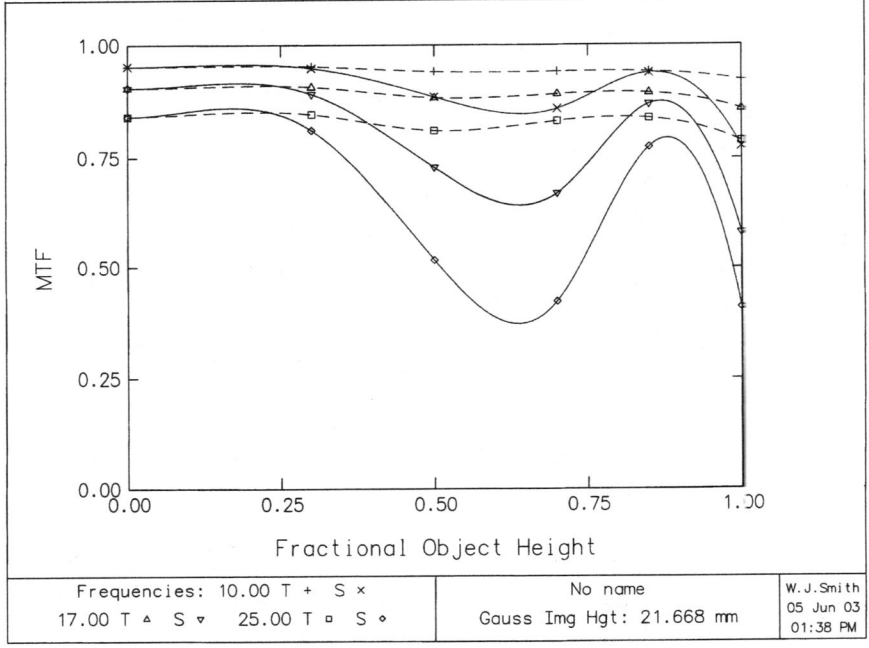

Figure 10.28 MTF versus field for 10, 17, 25 lpm for Fig. 10.27, Table 10.7.

in patents may consist primarily of the avoidance of the claims of prior art.)

Before leaving the Heliar we make a few quick trials of several crown glasses. We try BAFN11 (667484), SK16 (620603), and LAK3 (713538) without any great effect, even after switching to an aberration-based merit function. But when we try LASFN30 (803464) we get a surprise. The flint element of the front doublet changes from a strong negative

TABLE 10.8 Prescription of the Optimized Heliar Design with All Flint Glasses Allowed to Vary (See Figs. 10.29 to 10.32)

$R1 = +16.988v$	$T1 = 3.458v$	SSKN5	658509
$R2 = -19.298v$	$T2 = 0.800$	F8	596392
$R3 = +424.48v$	$T3 = 1.653v$		
$R4 = -29.925v$	$T4 = 0.800$	LLF2	541472
$R5 = +14.947v$	$T5 = 1.000$		
$R6 = $ Stop	$T6 = 3.874v$		
$R7 = -233.27v$	$T7 = 0.800$	F8	596392
$R8 = +16.237v$	$T8 = 2.693v$	SSKN5	658509
$R9 = -24.562v$	BFL = 42.98		

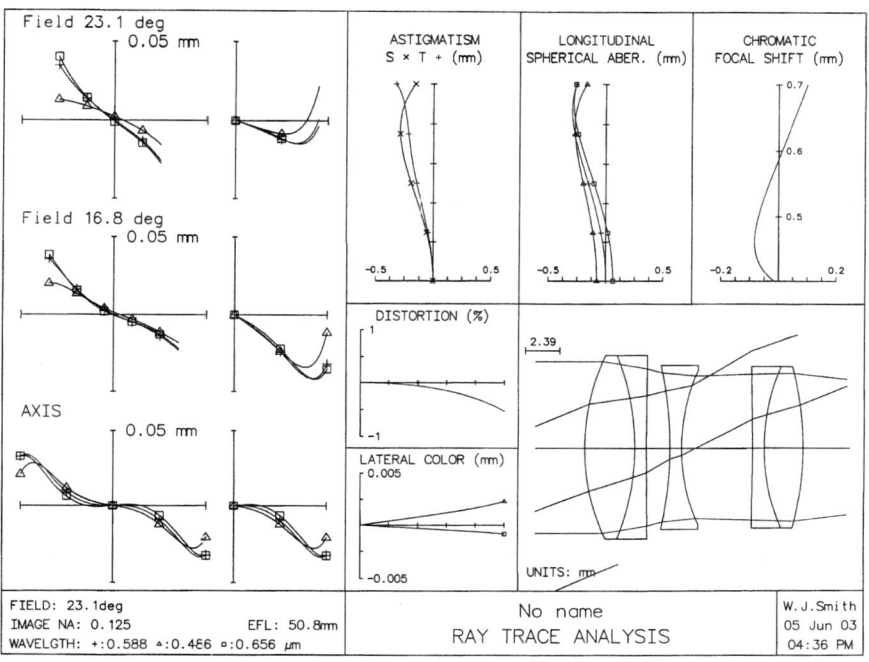

Figure 10.29 Ray aberrations for the second Heliar design, Table 10.8.

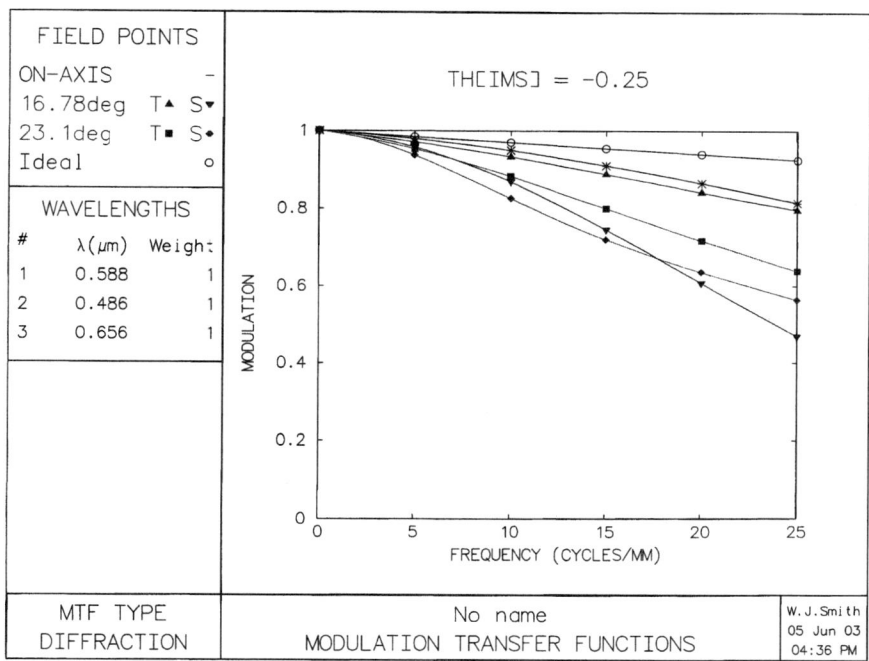

Figure 10.30 MTF versus frequency for Fig. 10.29, Table 10.8.

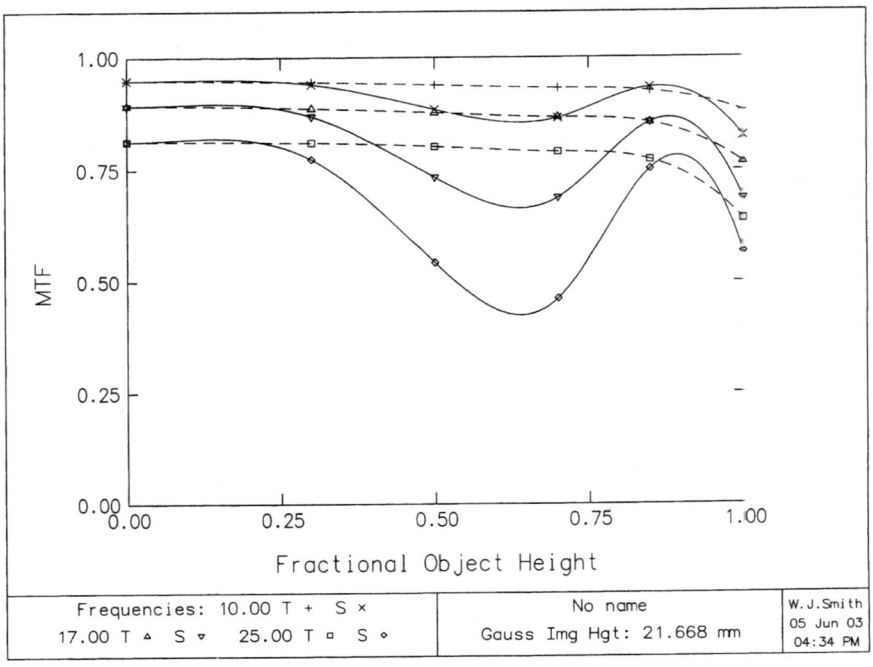

Figure 10.31 MTF versus field at 10, 17, 25 lpm for Fig. 10.30, Table 10.8.

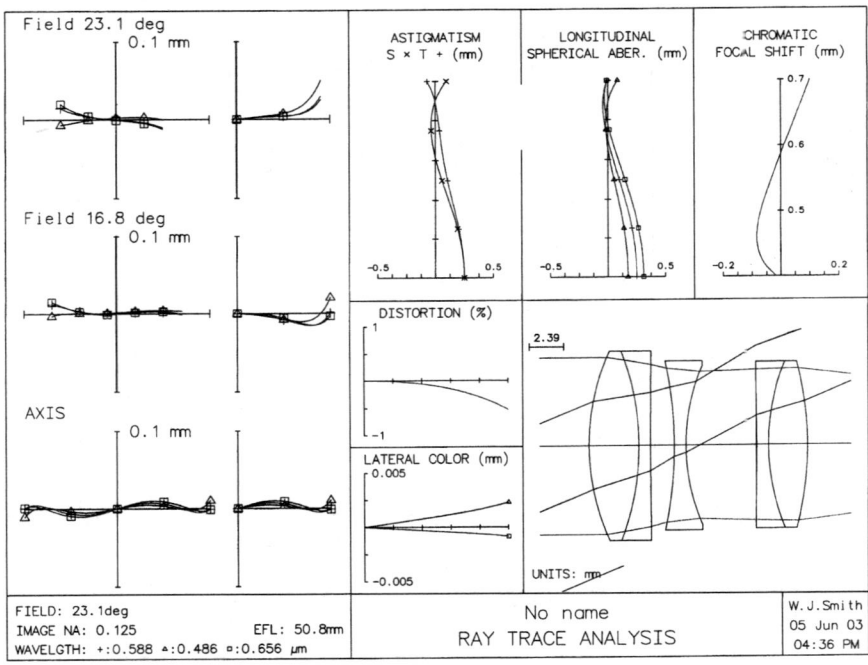

Figure 10.32 Ray aberrations for Fig. 10.29, Table 10.8 with a defocus of 0.25 mm.

TABLE 10.9 Prescription of the Optimized Tessar Design using Very High Index Crown Glass (See Figs. 10.33 to 10.35)

$R1 = +18.226v$	$T1 = 2.056v$	LASFN30	803464
$R2 = +391.54v$	$T2 = 0.800$	SF2	648338
$R3 = -213.45v$	$T3 = 1.425v$		
$R4 = -44.638v$	$T4 = 0.800$	SF2	648338
$R5 = +15.705v$	$T5 = 1.000$		
$R6 = $ Stop	$T6 = 6.519v$		
$R7 = -56.039v$	$T7 = 0.800$	SF2	648338
$R8 = +37.946v$	$T8 = 2.002v$	LASFN30	803464
$R9 = -25.841v$	BFL = 43.97		

element to a weakly positive, biconvex one! After double-checking for an error, we find that it's a good (even if quite odd) design, showing significant improvement at the zonal field over the previous design. (See Table 10.9.) See Figs. 10.33 to 10.35 for the aberration and MTF plots.

This arrangement is, to say the least, quite unusual. There seems to be little or no reason to have a front doublet of this peculiar form. Suspecting that a singlet front might do just as well, we take steps to

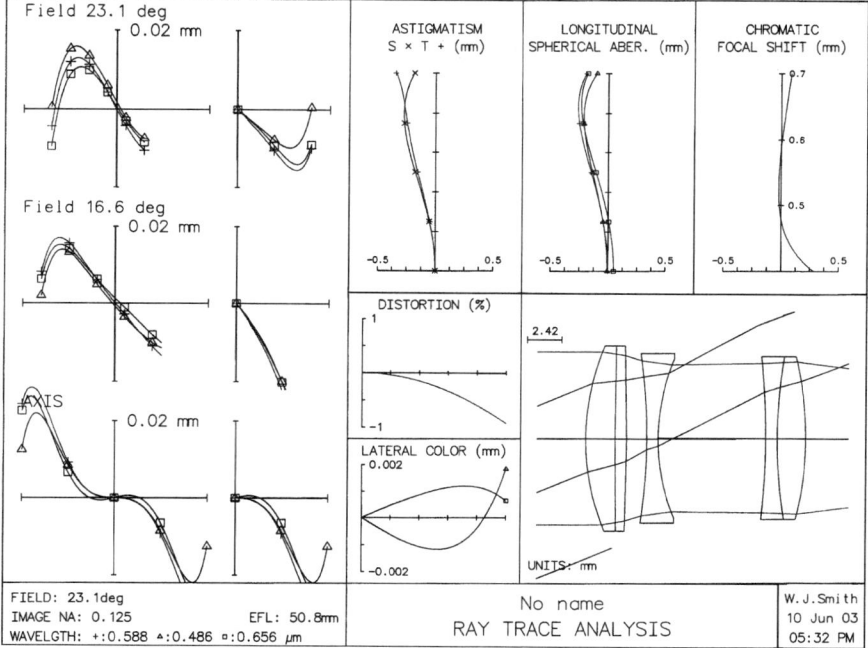

Figure 10.33 Ray aberrations for the third Heliar design with high index crowns, Table 10.9.

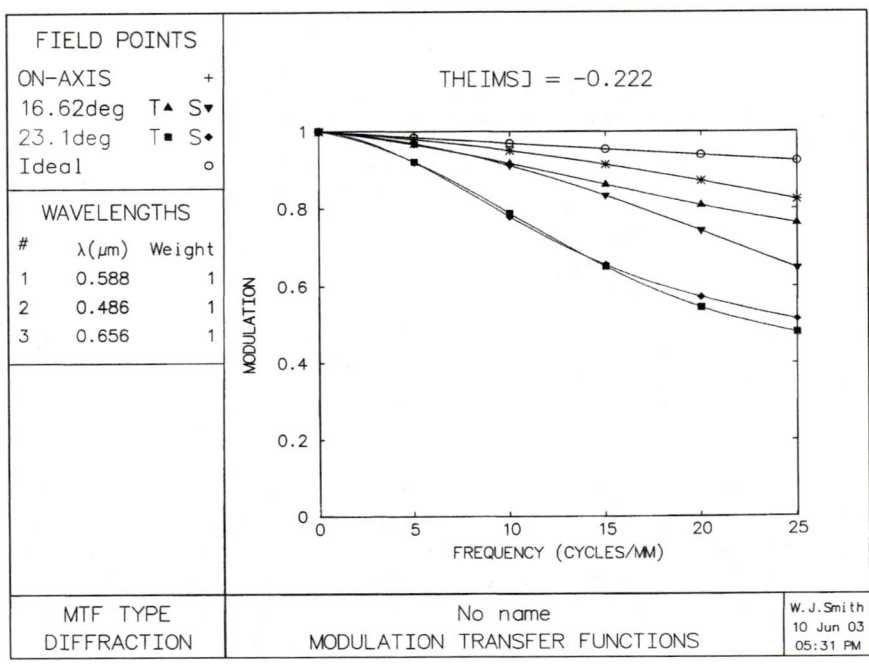

Figure 10.34 MTF versus frequency for Fig. 10.33, Table 10.9.

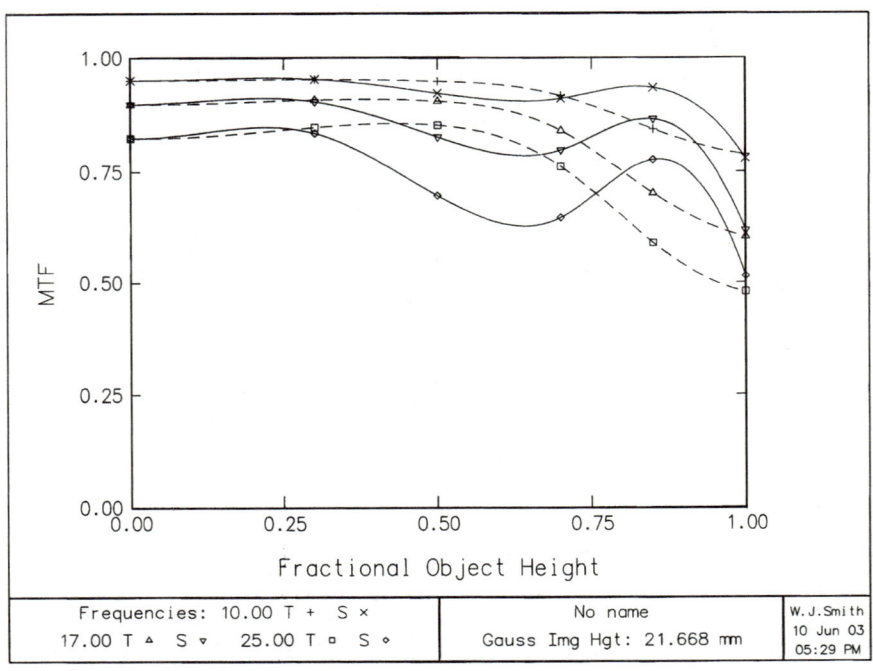

Figure 10.35 MTF versus field for 10, 17, 25 lpm for Fig. 10.33, Table 10.9.

TABLE 10.10 Prescription of the Tessar Design Which Resulted from Removing the Front Flint from the Lens of Table 10.9 (See Figs. 10.36 to 10.38)

$R1 = +16.327v$	$T1 = 2.218v$	LASFN30	803464
$R2 = +482.28v$	$T2 = 1.633v$		
$R3 = -66.448v$	$T3 = 0.800$	SF2	648338
$R4 = +14.099$	$T4 = 3.000$		
$R5 = $ Stop	$T5 = 5.482v$		
$R6 = -94.225v$	$T6 = 0.800$	SF2	648338
$R7 = +36.180v$	$T7 = 1.884v$	LASFN30	803464
$R8 = -31.741$	BFL = 42.803		

eliminate the flint in the front doublet. We add two targets to the merit function—(CV2-CV3) and TH2. We weight them very heavily and optimize. An examination of the lens drawing indicates that the stop is too far forward. Allowing its location to vary is unproductive so we manually move it further back. A few trials indicate that a TH4 of 3.0 is about right, and further optimization gives us the data as shown in Table 10.10.

The aberrations and the MTF are plotted in Figs. 10.36 to 10.38.

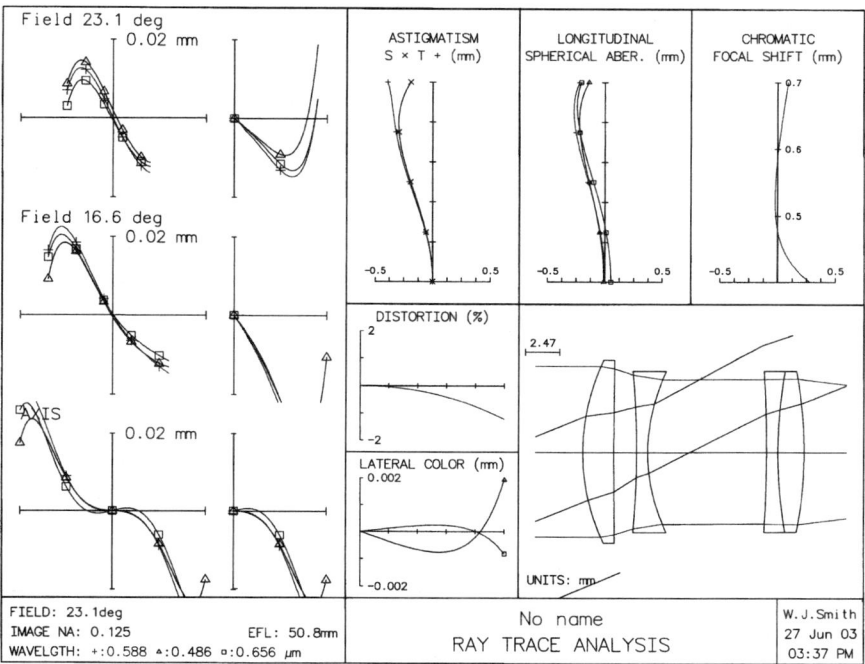

Figure 10.36 Ray aberrations for the Tessar derived from a Heliar, Table 10.10.

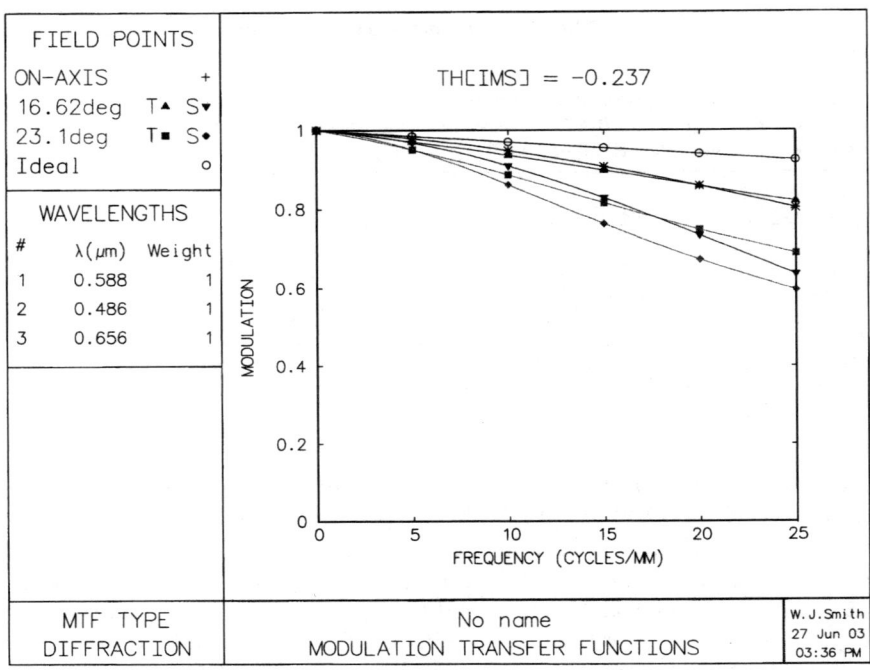

Figure 10.37 MTF versus frequency for Fig. 10.36, Table 10.10.

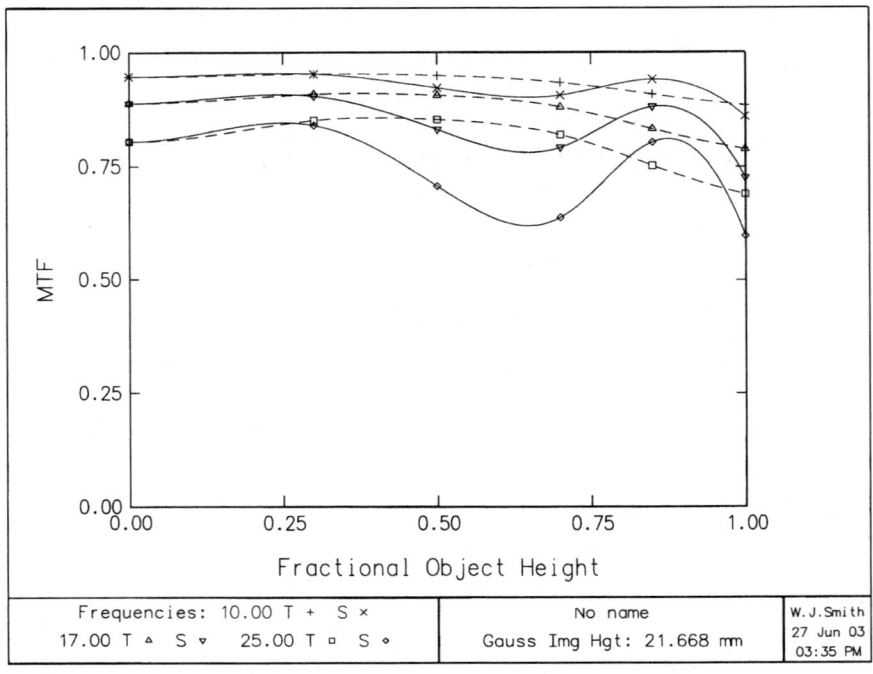

Figure 10.38 MTF versus field for 10, 17, 25 lpm for Fig. 10.36, Table 10.10.

Of course we are now back to the Tessar form. This lens is significantly better than any of the Tessar lenses we have previously designed. This result is not too surprising when we compare the crown glasses in this design with those used in our Tessar studies. Here we have a much higher index in LASFN30 (803464) than the LAF3 (717480) and LAK31 (647564) glasses, which were the highest-index glasses that we considered for the Tessar designs.

Some final thoughts. In conclusion, we should note that all of these designs have been optimized to get the best performance at full aperture. This is due to the nature of the default merit functions we used. This actually violates our stated goal of designing a camera lens. When a design has a lot of fifth-order spherical aberration (so that the residual aberration blur overwhelms the diffraction blur), optimization at full aperture usually produces a result that features undercorrected spherical aberration and a corresponding defocus as most of our lenses do. A balance of $LA_z = 1.5\ LA_m$ and a defocus of $\delta = 1.25\ LA_m = 0.83\ LA_z$ produces the smallest (geometrical) blur spot for a given amount of fifth-order spherical. An example of this type of correction is shown with Figs. 10.30, 10.31, 10.32, and 10.34 for the lens of Table 10.8, which show the ray intercept plots at both the paraxial focus and the "best focus."

The negative axial spherical aberration also helps offset the overcorrected fifth-order oblique spherical, which becomes troublesome toward the edge of the field in many designs. While this state of correction may be ideal for a lens that is always used at full aperture (such as a projection lens), it is not always a good choice for a camera lens. This is because as the aperture is stopped down, the best focus will shift toward the paraxial focus. [Another thing to bear in mind in evaluating lens designs is that most published design examples (including those in this text) have at least the marginal spherical aberration (and often some of the other aberrations as well) corrected to zero. The advantage of this type of presentation (with $LA_m = 0.0$) is that it allows an easy comparison of the residual aberrations between different designs.]

A typical camera lens with an iris is often designed so that the spherical aberration is overcorrected at the margin by roughly the same amount that the 0.7 zone is undercorrected, i.e., $LA_m \approx (-LA_z)$. A few suitable entries in the merit function can easily produce this balance, even in a default MF, if the entries are sufficiently weighted. When this is done the best focus does not shift when the lens is stopped down (as it would with the correction balance we have used in the designs above). The justification for this state of correction is that a camera lens is infrequently used at full aperture. The marginal overcorrection, and the accompanying reduction in the zonal spherical, yield a better performance at the small aperture opening where a camera lens is most often used. Stopping down the iris cuts off the overcorrected rays in the outer portion of the aperture, and the stopped down performance is many times better than it would be if the marginal spherical were undercorrected (or even fully corrected).

The penalty that we pay for this overcorrected spherical is a loss of contrast at full aperture. This is somewhat offset by the fact that, even with the lens wide open, a photograph is often underexposed because of a dark scene or a high shutter speed. The spherical aberration flare is then quite tolerable because the overall haze it produces acts as a preexposure to push the film past the toe of its H and D curve. This has the effect of speeding up the film. And often the photographer is happy just to get any picture at all under such difficult circumstances.

If nothing else this section should have indicated the many, many possible combinations possible in this type of lens, far more than we have space for in this volume. Further design explorations of the compounded three-component anastigmat (and we have left more than a few avenues open) are left to the reader.

Chapter 11

Double-Meniscus Anastigmats

11.1 Meniscus Components

All anastigmats achieve a flat field by the longitudinal separation of positive and negative power (surface, element, or component power). The earliest anastigmats flattened the field by using a thick meniscus component, which separated the positive and negative outer surfaces. At the time there were no antireflection coatings, and designers tried to minimize the number of air-glass surfaces so that the surface reflections would not cause unacceptable ghost images or reduced contrast. For this reason, most of the early anastigmats were limited to two cemented meniscus components (with only four air-glass surfaces).

11.2 The Hypergon, Topogon, and Metrogon

The *Hypergon* lens of Fig. 11.1 can be considered the progenitor of this class.* It consists of two identical meniscus elements, symmetrical about a central stop. The concave and convex radii differ by less than 0.7 percent (0.5 percent in some versions) so that the Petzval contributions of the convex surfaces are almost completely offset by the Petzval of the concave surfaces. The astigmatism is controlled by the distance of the elements from the stop, and the symmetrical construction almost completely eliminates the coma,

---

*The symmetric two-meniscus *periscopic* lens consists of a pair of landscape lenses (Sec. 2.4) oriented with their concave surfaces toward a central stop. Despite its superficial resemblance to the Hypergon, the periscopic lens cannot be considered to belong in this class of lenses. The elements are relatively thin, so that there is only a tiny effect on the Petzval sum (which is completely offset by the fact that there are two positive elements spaced apart from each other). In the *Hypergon* the two surfaces have nearly equal radii so their Petzval contributions almost completely offset each other, and the power of the element derives from the relatively large thickness separating the surfaces.

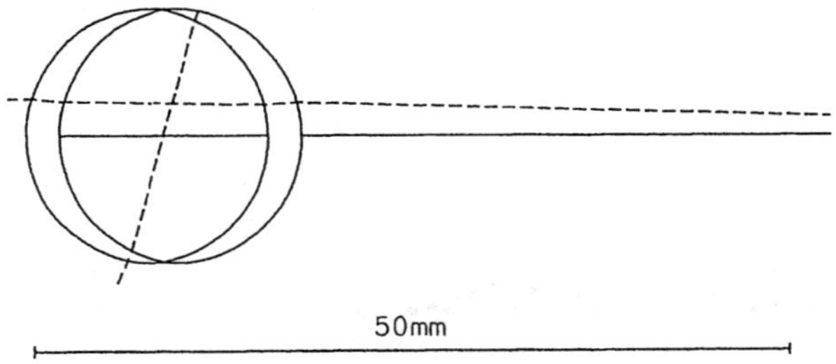

298 Chapter Eleven

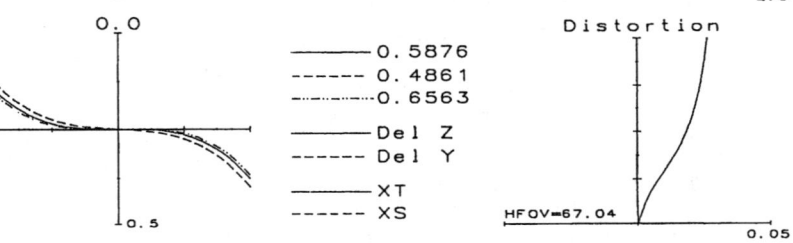

67degHFOV F/20 HYPERGON

radius	thickness	mat'l	index	V-no	sa
8.570	2.200	BK1	1.510	63.5	8.5
8.630	6.900	air			8.5
	6.900	air			2.2
-8.630	2.200	BK1	1.510	63.5	8.5
-8.570	92.925	air			8.5

EFL = 103.2
BFL = 92.93
NA = -0.0267 (F/20.6)
GIH = 243.44 (HFOV=67.04)
PTZ/F = -21.46
VL = 18.20
OD infinite conjugate

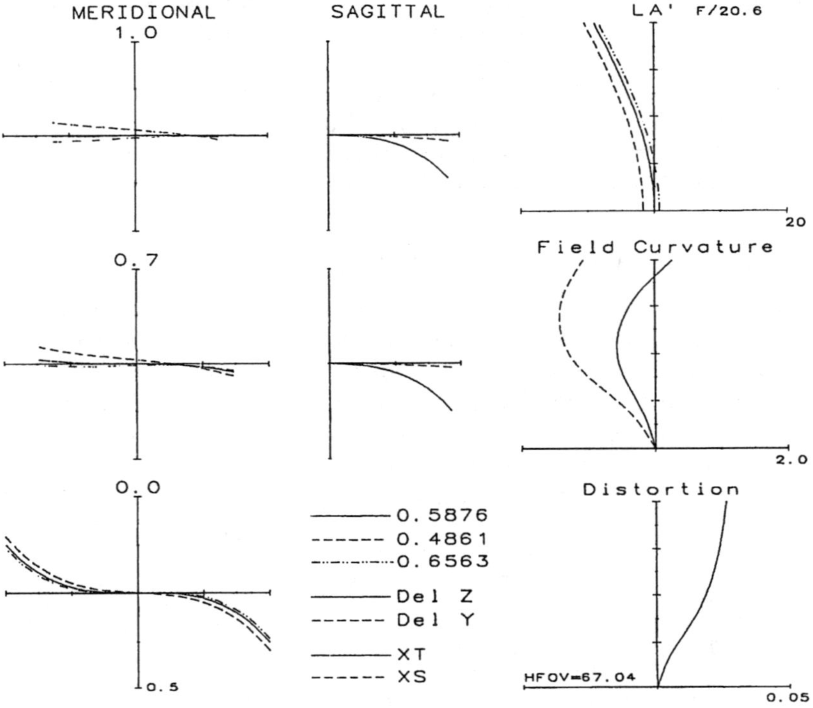

Figure 11.1 The Hypergon lens covers a 135° field (but at a speed of f/20 or f/30). Its Petzval sum is very low because the concave and convex radii differ by less than 0.7%.

distortion, and lateral color. The lens covers an astonishing field of 135°. Of course, the spherical aberration of these strongly bent meniscus elements is tremendous, and for this reason the lens must be used at a very low speed. The other major defect of this lens is the falloff in illumination with field angle. The cosine-fourth effect at 67° reduces the edge-of-field illumination to less than 2.5 percent of that at the center. (See Chaps. 14 and 15 for the improvement in illumination uniformity that results from the use of negative—rather than positive—outer elements in a wide-angle lens.) The illumination can be evened out either by a radial gradient filter or by a rotating star-pinwheel device introduced in front of the lens part way through the long (in the early days of photography) exposure.

The obvious way to improve the Hypergon is to add negative flint elements to correct the spherical aberration and axial chromatic aberration. The symmetrical *Topogon* of Fig. 11.2 covers a field of about 100° at a speed of $f/6.3$, using dense barium crown and extra-dense flint glasses, retaining the symmetrical construction and the strong meniscus configuration for all of the elements. In the *Metrogon*, the front crown is split into two elements, and the design achieves an excellent correction for distortion over a 100° field. These lenses were superceded by lenses of the type shown in Chap. 15 (which had more uniform illumination).

11.3 A Two Element Aspheric Thick Meniscus Camera Lens

We are again indebted to Ellis Betensky for this design, which might be appropriate for a disposable APS camera. Working at a speed of $f/8$, this system uses the thickness of the elements to flatten the Petzval field by separating the concave and convex surfaces. The element thickness is also used to space the outer surfaces away from the stop, where their asphericities can be effective in the control of astigmatism, coma, and distortion. The design data follows (see Table 11.1), and the aberrations and *modulation transfer function* (MTF) are shown in Figs. 11.3 and 11.4.

TABLE 11.1 Design Data for the Betensky Aspheric Plastic Doublet Shown in Figs. 11.3 and 11.4

$R1 = +3.9888$	$T1 = 1.1422$		Polycarb 585303
AD = −0.002530	AE = −0.000756 AF = +7.6893e−05	AG = −1.4527e−05	
$R2 = +2.8125$	$T2 = 0.3593$		
AD = −0.000726	AE = +0.000754 AF = −0.001741	AG = +0.000599	
$R3 = $ Stop	$T3 = 0.1$		
$R4 = +412.02$	$T4 = 2.0848$		Acrylic 490579
AD = +0.001002	AE = +0.004949 AF = −0.003314	AG = +0.000914	
$R5 = -6.0464$	BFL = 24.148		$\delta = 0.30$
AD = −0.002047	AE = +0.000115 AF = −0.000121	AG = +7.2364e−06	

300 Chapter Three

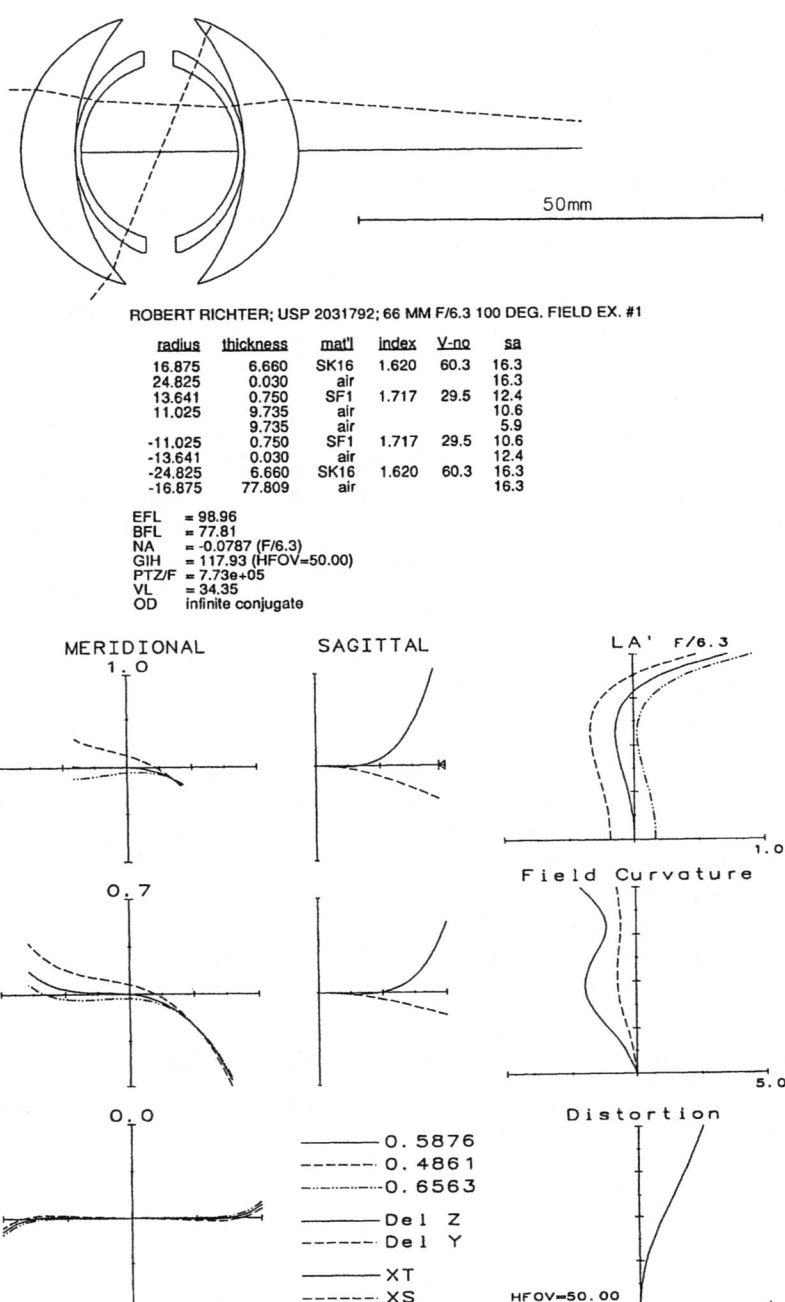

ROBERT RICHTER; USP 2031792; 66 MM F/6.3 100 DEG. FIELD EX. #1

radius	thickness	mat'l	index	V-no	sa
16.875	6.660	SK16	1.620	60.3	16.3
24.825	0.030	air			16.3
13.641	0.750	SF1	1.717	29.5	12.4
11.025	9.735	air			10.6
	9.735	air			5.9
-11.025	0.750	SF1	1.717	29.5	10.6
-13.641	0.030	air			12.4
-24.825	6.660	SK16	1.620	60.3	16.3
-16.875	77.809	air			16.3

EFL = 98.96
BFL = 77.81
NA = -0.0787 (F/6.3)
GIH = 117.93 (HFOV=50.00)
PTZ/F = 7.73e+05
VL = 34.35
OD infinite conjugate

Figure 11.2 The Topogon is a logical step from the Hypergon. The meniscus elements are thicker, and the negative elements are added to correct the chromatic and spherical aberrations. The thin shell elements are difficult to fabricate.

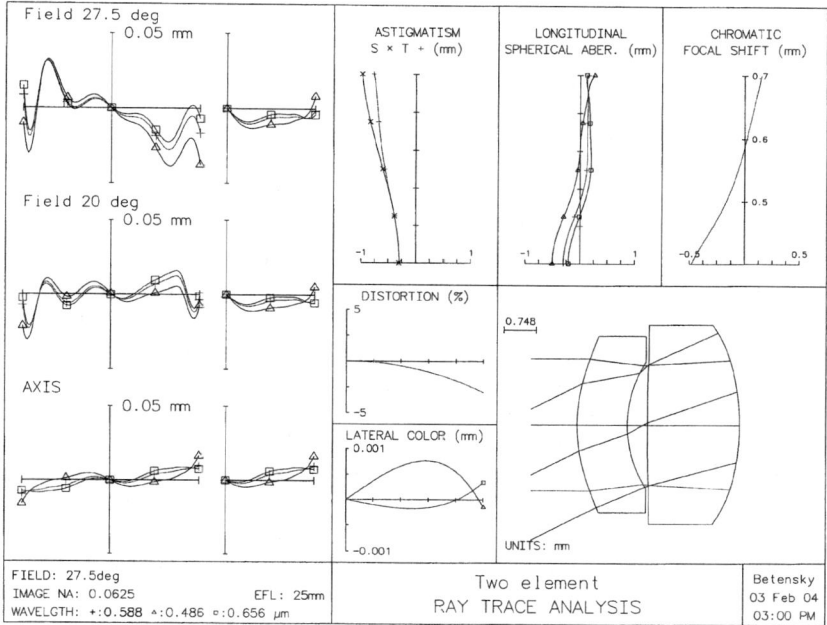

Figure 11.3 The ray aberrations of the Betensky two element aspheric thick meniscus lens of Table 11.1.

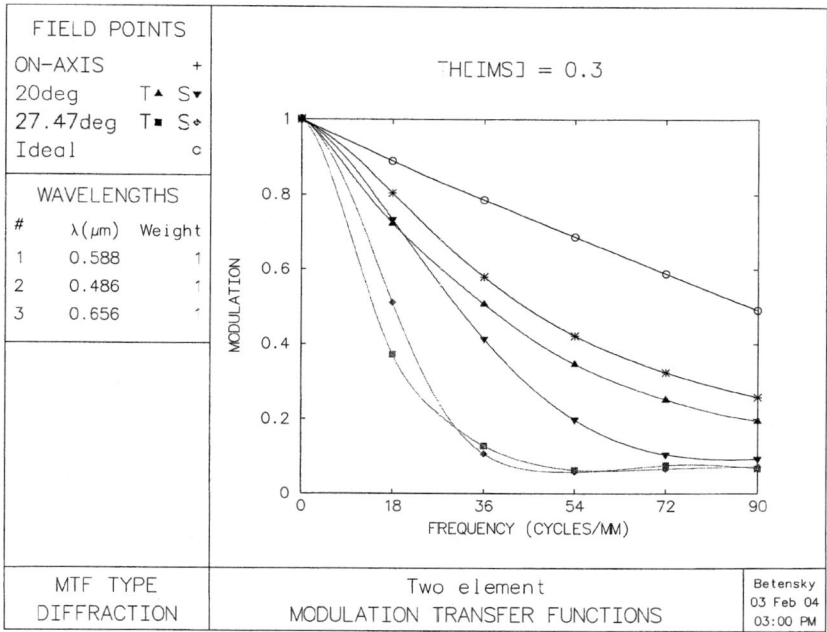

Figure 11.4 The MTF for the lens of Fig. 11.3, Table 11.1.

11.4 Protar, Dagor, and Convertible Lenses

The *Protar* (Fig. 11.5) is based on the combination of a weak, meniscus, *old achromat* front member and a strong, meniscus, *new achromat* rear member. See Sec. 3.5 for a discussion of *old* versus *new* achromats. The diverging cemented surface of the old achromat front is used to correct the spherical aberration; in the rear doublet, the low-index flint, high-index crown combination of the new achromat, along with the thick meniscus construction, is used to flatten the Petzval surface. The cemented interfaces and the stop position are used to control the astigmatism. Note that the divergent cemented surface in the front doublet is concave to the stop, and that the convergent cemented surface in the rear is convex to the stop. This is also true in the Dagor lens of Fig. 11.6.

The *Dagor* (Fig. 11.6) combines both the old achromat and the new achromat into a cemented triplet construction. If we visualize the central negative element of the triplet split into two parts, then the outer high-index element and the outer part of the (medium-index) middle element can be seen to make up a new achromat. The inner low-index crown element and the other part of the middle element make up the old achromat. The symmetrical construction about the stop reduces the coma and distortion, while the spacing from the stop and the cemented surfaces control the astigmatism.

Since either the front or the rear half of this sort of lens can be designed to perform reasonably well by itself, a *convertible lens* can be constructed to yield two different focal lengths when one uses either both components or one alone. If a hemisymmetrical construction (with the halves similar, but of different focal lengths) is used, then three different focal lengths are available. This was probably the original multifocal (not quite a zoom) lens.

Many elaborations on this theme were designed, with as many as five elements cemented together on each side of the stop, for a total of 10 elements. Little was gained by adding elements, and one cannot help but suspect that many elaborations were more for the purpose of evading patent coverage than for the improvement of image quality.

There are many designs that can be regarded as descendants of these early thick meniscus anastigmats. The *Tessar* (Chap. 10), although it looks like a modification of the Cooke triplet, was actually designed as a rear new achromat combined with an airspaced front pair replacing the old achromat of the Protar. The front airspaced pair can be regarded as a thick meniscus triplet component with a center lens made of air.

The *Biotar* or double-Gauss lens (Chap. 12) is also a (very distant) descendant of this general form, as indicated by the thick-meniscus inner doublets. The Biotar has evolved into what is arguably the most versatile and powerful of the standard design types.

Double-Meniscus Anastigmats 303

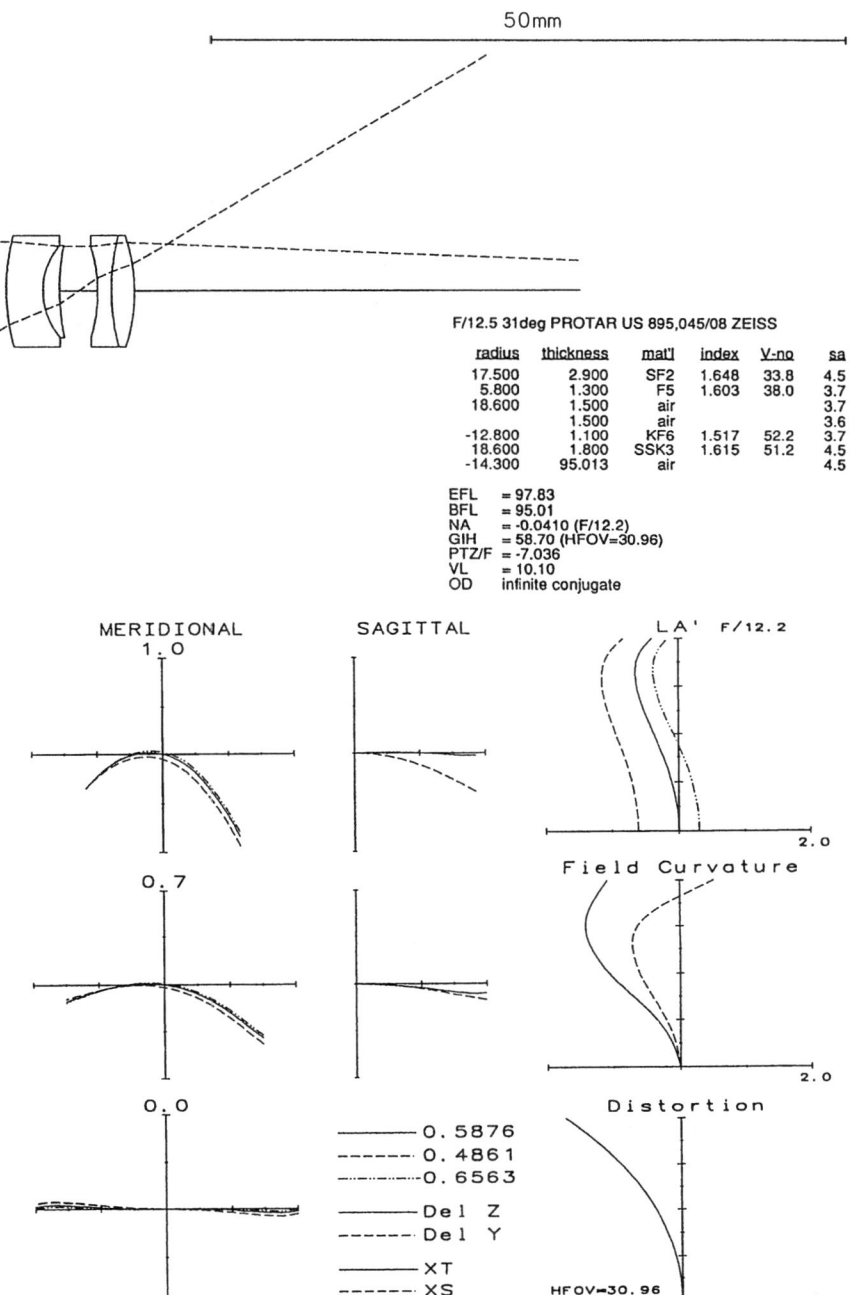

Figure 11.5 The classic Protar combined a thick meniscus old achromat front with a thick meniscus new achromat rear. The front cemented surface is diverging and the rear converging.

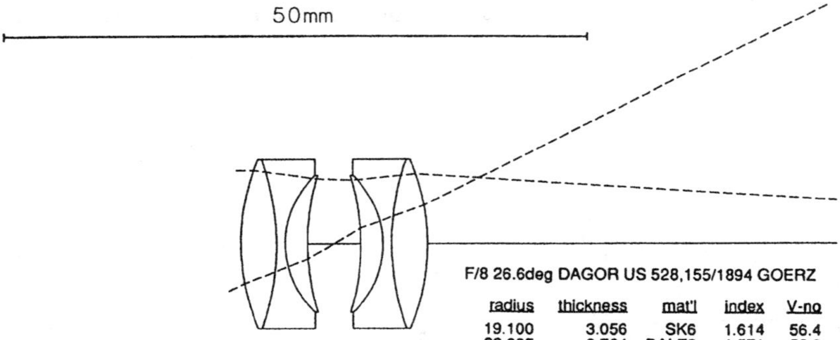

F/8 26.6deg DAGOR US 528,155/1894 GOERZ

radius	thickness	mat'l	index	V-no	sa
19.100	3.056	SK6	1.614	56.4	7.4
-22.635	0.764	BALF3	1.571	52.9	7.4
8.272	1.910	K4	1.519	57.4	6.0
20.453	2.292	air			6.0
	2.292	air			5.6
-20.453	1.910	K4	1.519	57.4	6.0
-8.272	0.764	BALF3	1.571	52.9	6.0
22.635	3.056	SK6	1.614	56.4	7.4
-19.100	96.267	air			7.4

EFL = 103.3
BFL = 96.27
NA = -0.0622 (F/8.0)
GIH = 51.67 (HFOV=26.57)
PTZ/F = -3.706
VL = 16.04
OD infinite conjugate

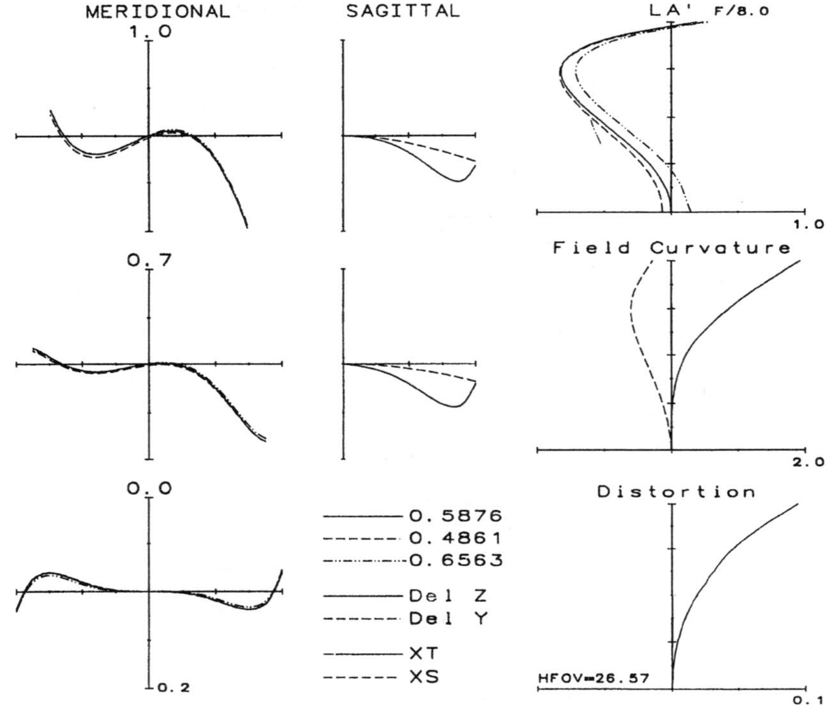

Figure 11.6 The symmetrical Dagor combined the two components of the Protar into a cemented triplet. If one mentally splits the flint down the middle, the first element (614564) and half of the second element (571529) form a new achromat; the balance of the cemented triplet forms an old achromat (571529 and 519574).

11.5 The Split Dagor

In the *Orthometar, Plasmat, W. A. Express,* and *Euryplan* (Fig. 11.7) types, the inner meniscus elements of the Dagor were split off, a higher-index glass was used, and complete symmetry was abandoned. The freedom to independently bend these inner elements allowed the designer to correct the spherical aberration without being limited to the low-index glass of the original old achromat part of the Dagor in order to do so. Of course, this split doubled the number of air-glass surfaces from four to eight, but this design was a significant improvement and reduced the zonal spherical of the Dagor. A fully symmetrical version of this form has been widely utilized in xerographic copiers, working at near-unity magnifications. Eventually low reflection coatings reduced the ill effects of the two added surfaces.

The data for a fully split Dagor are given in Table 11.2 for a focal length of 100, a field of ±30°, and a speed of $f/3.3$, and the aberrations are shown in Fig. 11.8

11.6 The Dogmar

The *Dogmar* (or *Celor*) lens form (Fig. 11.9) is also a member of the double-meniscus family, as can be realized if one considers each half to be a triplet with a center air lens. The Dogmar form is used as an excellent general-purpose camera lens, and its symmetry and stability of correction make it eminently suitable for an enlarger lens. The performance of the Dogmar can be improved by the use of higher-index crowns. For use as a process lens, glasses with unusual partial dispersions can be chosen (see Sec. 6.8) to reduce the secondary spectrum.

The original double Gauss consisted of two Gauss-type telescope objectives (Fig. 6.1) with the flints facing a central stop in a symmetric arrangement. Because the meniscus elements are wrapped around the stop, this form can cover a wider field than the Dagor. The oblique beam angles of incidence are small; this form might lead one to the Topogon design of Fig. 11.2.

11.7 Camera Lens Anastigmat Design "from Scratch"—The Dogmar Lens

We have previously followed the design of the Cooke triplet (Chap. 8) and the Tessar and Heliar lenses (Chap. 10). As a final design in this series we will work out the Dogmar/Celor/Aviar airspaced four element (+ − − +) design as our 2-in $f/4$ camera lens.

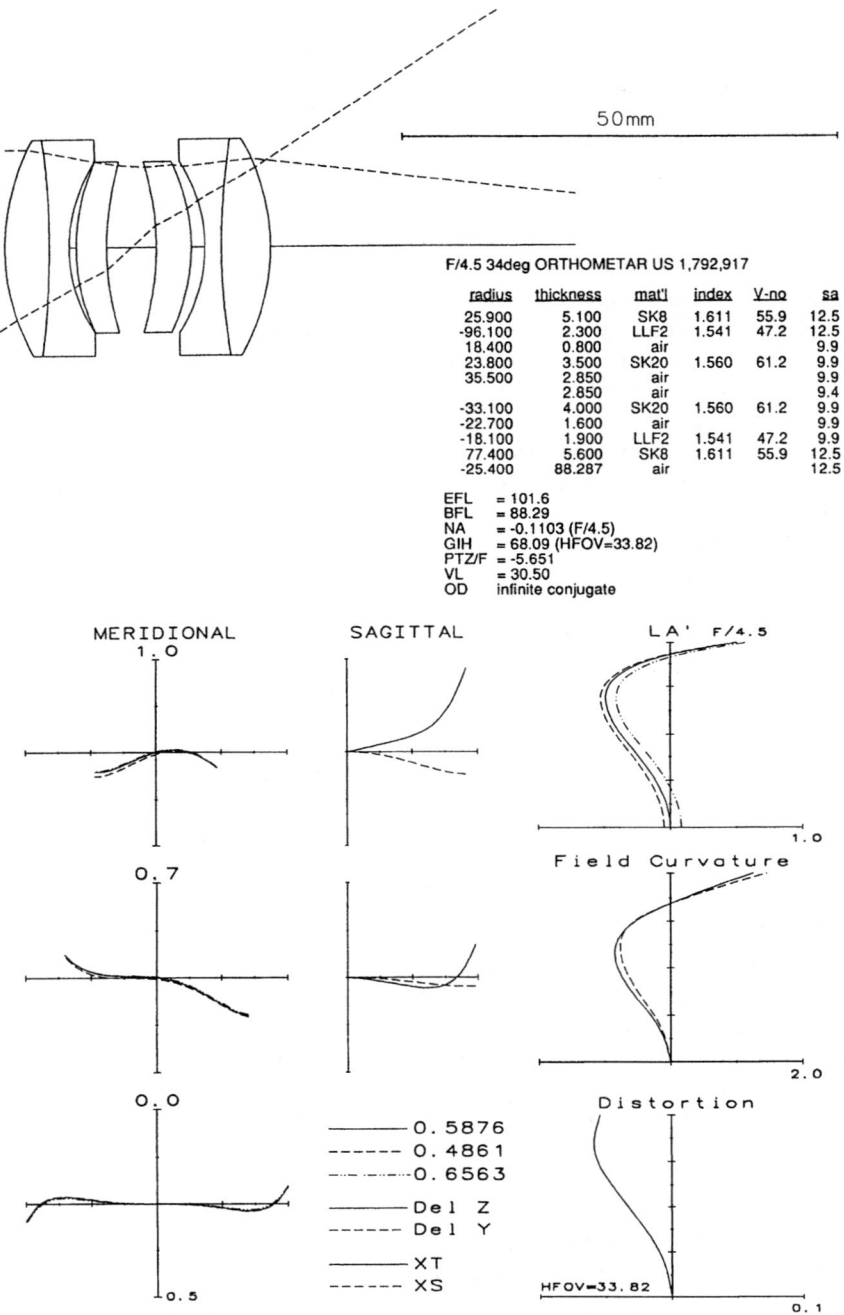

Figure 11.7 In the split Dagor, the airspace corrects the spherical, and a low-index crown third element is no longer necessary for this task. Abandoning symmetry produced an excellent general purpose lens. A more (or completely) symmetrical version is popular as a photocopy lens.

TABLE 11.2 Prescription for a Fully Split Dagor, with Six Airspaced Singlets (See Fig. 11.8)

$R1 = +24.291$	$T1 = 6.020$	SSKN5	658509
$R2 = +55.253$	$T2 = 3.512$		
$R3 = +50.167$	$T3 = 3.010$	F1	656357
$R4 = +17.077$	$T4 = 2.408$		
$R5 = +29.769$	$T5 = 3.010$	SK4	613586
$R6 = +50.416$	$T6 = 3.311$		
$R7 =$ Stop	$T7 = 4.013$		
$R8 = -90.391$	$T8 = 2.308$	SK8	611559
$R9 = -40.134$	$T9 = 2.609$		
$R10 = -19.766$	$T10 = 3.010$	F4	617366
$R11 = -85.755$	$T11 = 3.010$		
$R12 = -91.213$	$T12 = 5.017$	SSKN5	658509
$R13 = -24.281$	BFL = 78.061		

11.7.1 The Dogmar design

At first glance the Dogmar looks like a Cooke triplet with the flint element split in two. This is a fair enough assumption, even though the conceptual origin of the Celor design probably was as a symmetrical pair of thick meniscus triplets, with the center element of each triplet being

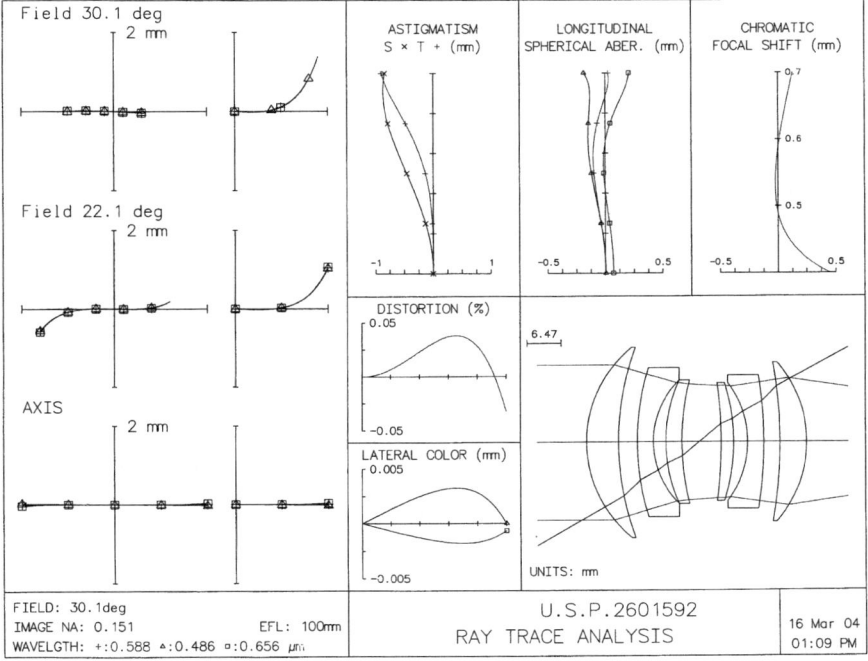

Figure 11.8 A fully split Dagor.

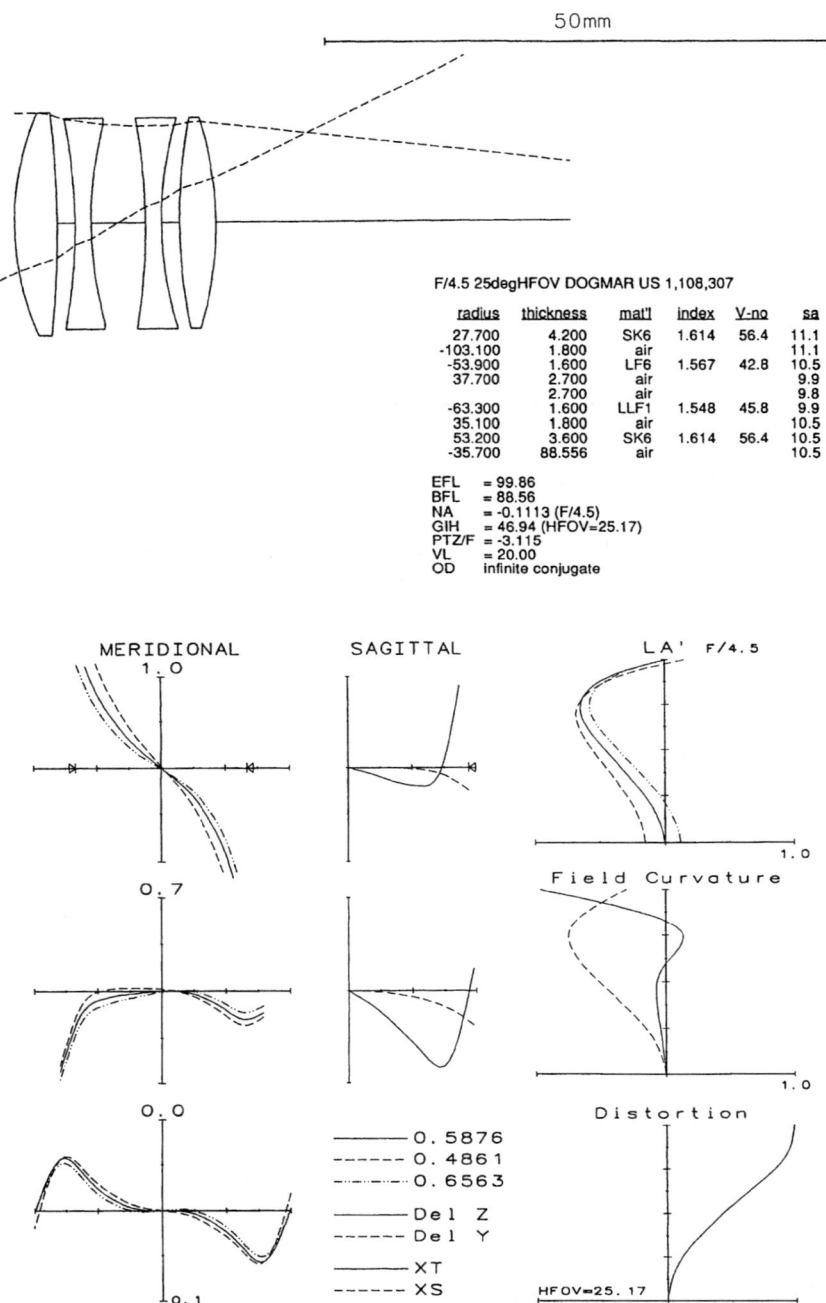

Figure 11.9 The Dogmar may be considered a pair of meniscus triplets, with the middle lens of the triplet made of air. Originally a fully symmetrical form, it was improved by a non-symmetrical configuration.

an air lens. If we start the design in the same way as we did the Cooke triplet, we might use the initial thin lens form, but split the flint in two elements, separated from the crowns by 2.0 and with the central airspace equal to 2.8. These add up to a little thinner system than the previous one, but we allow for the addition of element thickness later on. We put the stop in the middle of the central airspace and use curvature, thickness, and glass pickups to maintain complete front-to-back symmetry in the initial stages. Most lenses of this type use a light flint so that the doublets are close coupled, so we'll try LF5 (581409) as our starting flint, and SSKN5 (658509) for the crowns. Our start looks like this:

Rough Symmetrical Starting "Thin Lens" Prescription for the Dogmar Design

$R1 = +18v$	$T1 = 0.0$	SSKN5	658509
$R2$ = plano	$T2 = 2.0v$		
$R3 = -36v$	$T3 = 0.0$	LF5	581409
$R4 = +36p$	$T4 = 1.4$		
$R5$ = Stop	$T5 = 1.4p$		
$R6 = -36p$	$T6 = 0.0$	LF5p	581409
$R7 = +36p$	$T7 = 2.0p$		
$R8$ = plano	$T8 = 0.0$	SSKN5p	658509
$R9 = -18p$			

NOTE: v indicates a variable and p a pickup.

The independent variables are $R1$, $R3$, and $T2$. We use a simple merit function with targets for effective focal length EFL = 50.8, TAch = 0.0, TchA = 0.0, and the Petzval radius = $-3.5 \times$ EFL (just as we did in the triplet). The pickups indicated above maintain complete front-to-back symmetry. Note also that there are only three variables to control four targets; we are implicitly relying on symmetry to keep the lateral color small. The focal length of this starting prescription is 82, far from the desired 50.8, so we can expect some substantial changes as we proceed.

A run of several *damped least-squares* (DLS) cycles gives us:

Prescription of the Symmetrical Thin Lens Dogmar, Adjusted to Control EFL, TAch, TchA, and Petzval Radius

$R1 = +10.52v$	$T1 = 0.0$	SSKN5	EFL = 50.8
$R2$ = plano	$T2 = 0.98v$		TAch = 0.0
$R3 = -21.12v$	$T3 = 0.0$	LF5	TchA = 0.002
$R4 = +21.12p$	$T4 = 1.4$		Petz $R = -177.7$
$R5$ = Stop	$T5 = 1.4p$		$= -3.5$ EFL
$R6 = -21.12p$	$T6 = 0.0$	LF5	
$R7 = +21.12p$	$T7 = 0.98p$		
$R8$ = plano	$T8 = 0.0$	SSKN5	
$R9 = -10.52p$			

Adding element thicknesses, an edge thickness target for the first element, and allowing the flint glass to vary, we optimize again and obtain:

The Starting Thick Element Prescription for the Rough Symmetrical Dagor

$R1 = +12.42v$	$T1 = 2.83v$	SSKN5	658509
$R2 = $ plano	$T2 = 0.1v$		
$R3 = -25.64v$	$T3 = 0.8$	$597387v$	(near F8)
$R4 = +25.64p$	$T4 = 1.4$		
$R5 = $ Stop	$T5 = 1.4p$		
$R6 = -25.64p$	$T6 = 0.8$	$597387p$	
$R7 = +25.64p$	$T7 = 0.1p$		
$R8 = $ plano p	$T8 = 2.83p$	SSKN5p	
$R9 = -12.42p$			

We note that $T2$ and $T7$ are at the default minimum (0.1) of the design program for airspace variables, and a drawing of the lens shows that surfaces 2–3 and 7–8 interfere at the edges. All of our merit function targets are met, except for lateral chromatic at –0.005. This target is stymied by the need for a fourth variable and by our reliance on front-to-back symmetry.

We now allow the first four curvatures to vary, still maintaining symmetry, and use our third- and fifth-order aberration merit function. The flint changes slightly to 583405 (near LF5), and all the targets, although not at zero, are very close to being met.

Next we drop the symmetry constraint and optimize using the default rms spot size merit function; the results are not too promising. We try SK16 glass for the crowns, accepting the lower index in exchange for a higher V-value—it's better, but not great. We try allowing $T4 = T5$ to vary and find that it wants to go to zero, again without much progress. So we fix $T4 = T5 = 1.3$ and switch to a different default merit function, one which targets the raytraced aberrations. We still hold $T2 = T7$ and wind up with the design in Table 11.3, which is shown in Figs. 11.10 to 11.12.

TABLE 11.3 Prescription for the Dogmar Lens of Figs. 11.10 to 11.12, where the Glass Types were Changed and Symmetry was Abandoned

$R1 = +16.303v$	$T1 = 2.376v$	SK16	620603
$R2 = -85.456v$	$T2 = 1.474v$		
$R3 = -31.521v$	$T3 = 0.80$	LF8	564438
$R4 = +25.873v$	$T4 = 1.30$		
$R5 = $ Stop	$T5 = 1.30p$		
$R6 = -46.330v$	$T6 = 0.80$	LF8p	564438
$R7 = +19.980v$	$T7 = 1.474p$		
$R8 = +34.819v$	$T8 = 2.145v$	SK16p	620603
$R9 = -19.795v$	BF $= 44.763$		

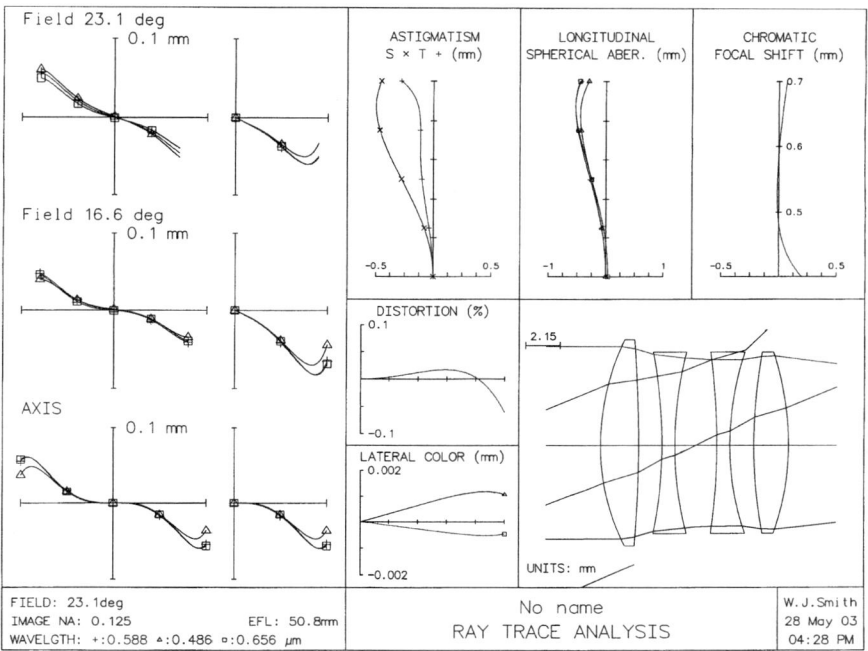

Figure 11.10 The ray aberrations of the first Dogmar type design, Table 11.3.

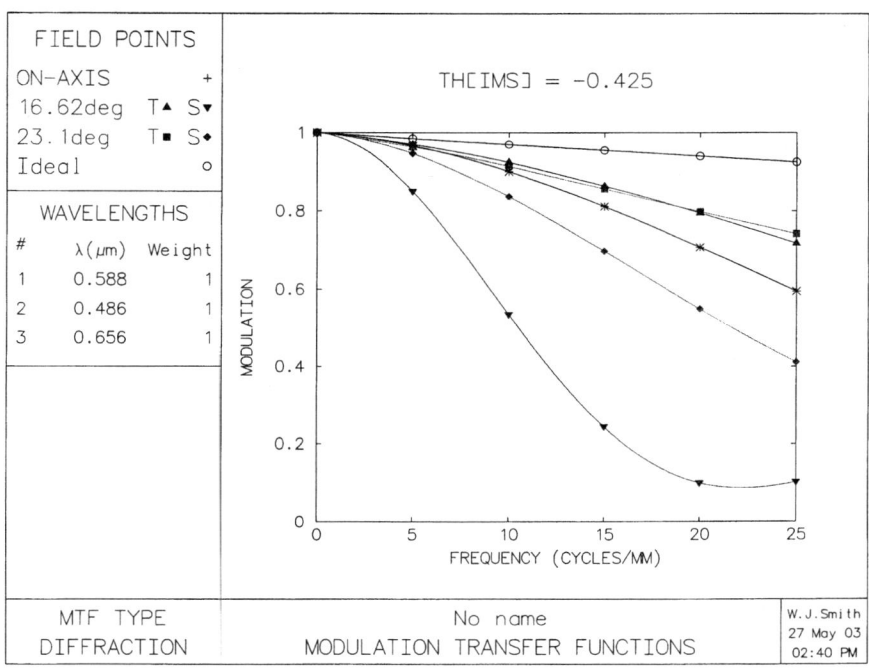

Figure 11.11 The MTF versus frequency for the lens of Fig. 11.10 and Table 11.3.

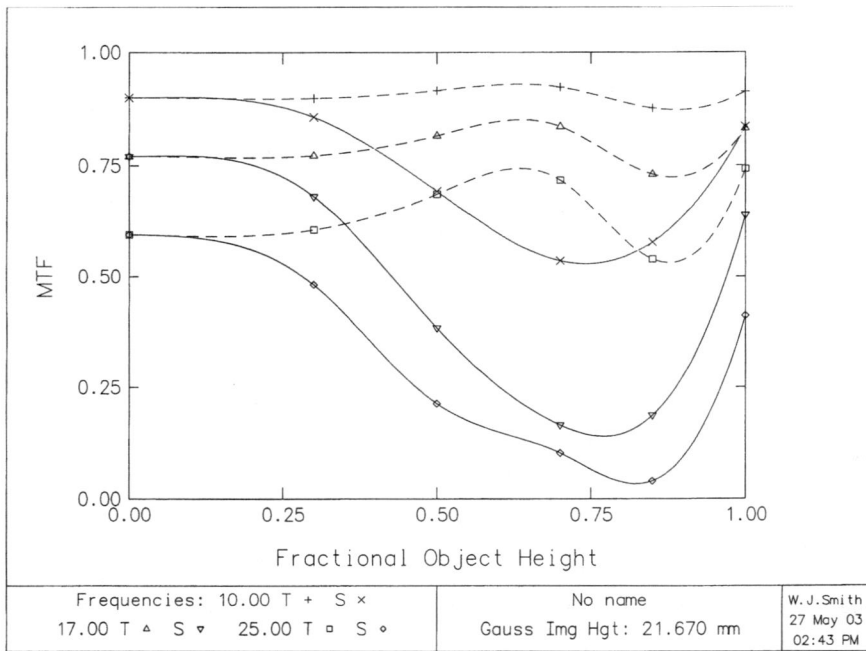

Figure 11.12 The MTF versus field at 10, 17, 25 lpm for the lens of Fig. 11.10, Table 11.3.

This design is modestly, but definitely, better than the triplet designs, which resulted from our efforts in Chap. 8. Possibly if the Cooke and the Dogmar were compared in a more demanding application (i.e., a wider field of view and/or a faster speed) their differences might be more pronounced. The early meniscus design forms from which the Dogmar is descended typically covered larger angular fields than the ±23° we are pursuing here.

At this point we present an example of the type of thing that makes lens design fascinating. Browsing through the literature, we notice that in a few cases there have been liberties taken with the shape of the flint elements. There are designs with identical flints, with symmetrical flints, and even with an equi-concave flint or flints. These are features that are often used to reduce the cost of production.

So we elect to try the identical flints, i.e., $R3 = R6$ and $R4 = R7$. We allow the central airspaces to vary, but with a lower bound of 0.8. We get a design, but it's actually a bit worse. Next we try the symmetrical arrangement with $R7 = -R3$ and $R6 = -R4$. This is no better, and now we discover that we have forgotten to control the *edge thickness* (ET) of the crowns; the thicknesses have gone wild. We get this back under control by heavy weighting of the ET targets in the *merit function* (MF) and freezing the crown thicknesses, but no improvement. Then for no really good reason, we go back to the original conditions, optimize a bit, and find a design, that is much better than the one above. We have accidentaly stumbled upon a different and better local optimum. To the eye it looks very little different but the performance is much better as shown in Figs. 11.13 to 11.15. The lens data are shown in Table 11.4.

After this, we thought it might be profitable to "fiddle around" with the design. This is something designers often do toward the end of the design work. It is a process that is almost impossible to describe because it consists of many (unrecorded) quick changes and reoptimizations; most of them come out badly, a few good. Some are abandoned, some are followed up. It is a more or less random process of trying different approaches to the design operation, those which the designer thinks may be beneficial. It has aptly been described as "following one's nose." Some of these moves are optimizing with the image plane fixed at the paraxial focus, or allowing the image plane to shift; allowing the flint glasses to vary, or not; fixing certain spacings, or not; and manually changing a space or an element shape and holding it for a few cycles and then releasing it to vary. To attack specific aberrations that seem the most troublesome, we may radically adjust the weighting of the targets in the merit function, or add targets to the MF, or indeed switch to a totally different MF. At each step we look at the MTF,

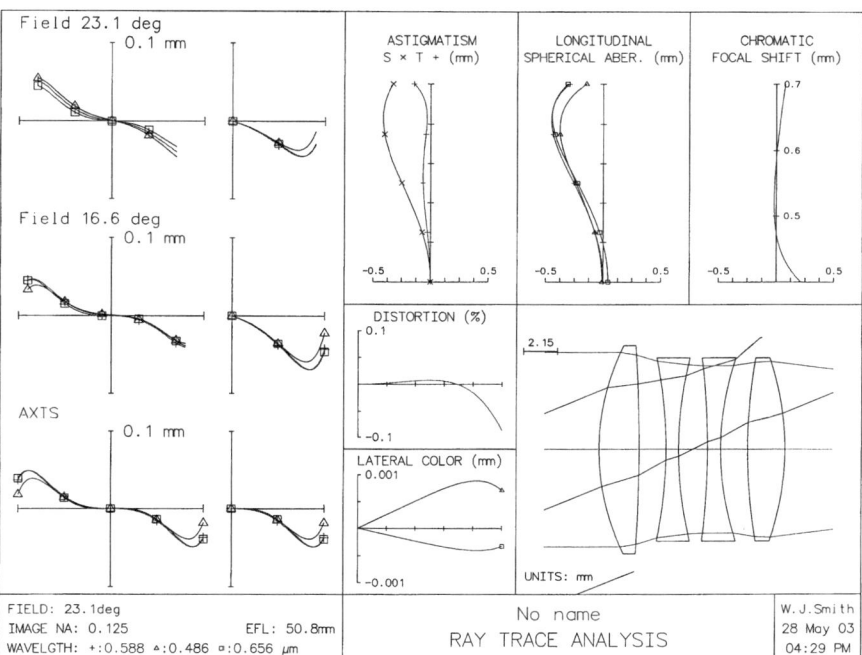

Figure 11.13 The ray aberrations of the Table 11.4 redesign from Fig. 11.10, Table 11.3.

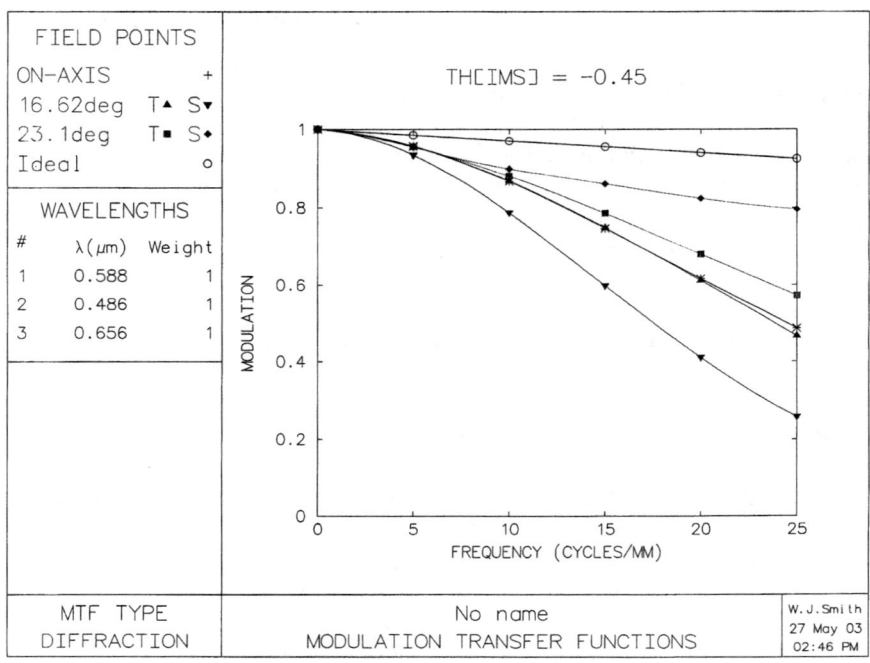

Figure 11.14 The MTF versus frequency for the lens of Fig. 11.13, Table 11.4.

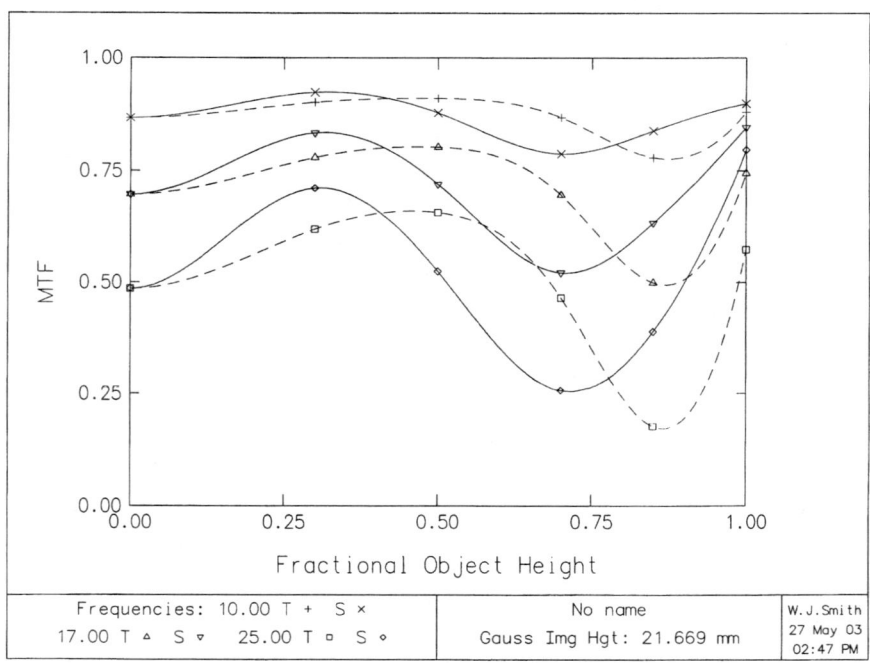

Figure 11.15 The MTF versus field at 10, 17, 25 lpm for the lens of Fig. 11.13, Table 11.4.

TABLE 11.4 The Dogmar Prescription that Resulted from Modifying the Design Symmetry of the Flints and then Allowing them to Vary (See Figs. 11.13 to 11.15)

$R1 = +15.966v$	$T1 = 2.50$	SK16	620603
$R2 = -109.01v$	$T2 = 1.672v$		
$R3 = -31.067v$	$T3 = 0.800$	$567433v$	(near LF8)
$R4 = +26.482v$	$T4 = 0.898v$		
$R5 = $ Stop	$T5 = 0.898p$		
$R6 = -51.115v$	$T6 = 0.800p$	$567433p$	
$R7 = +19.768v$	$T7 = 1.672p$		
$R8 = +37.763v$	$T8 = 2.30$	SK16p	620603
$R9 = -19.364v$	BF = 44.882		

or spot size, or whatever metric we are following, and proceed in the direction that seems likely. This is often effective when the merit function does not correlate well with the metric being used.

Actually, we are meandering around in the n-dimensional design space. The changes we try probably tend to follow our personal bias as to what we think the lens should look like; we tend to keep the design under control. At one point in our wandering we found an indication that very thick crowns might be advantageous; because we were after a thin element Dogmar type lens, we didn't follow this up. The reader may wish to go down this avenue.

In the course of the "fiddle" we again stumble (and "stumble" seems to be an appropriate verb) upon another design form, as follows (see Table 11.5). The performance plots are shown in Figs. 11.16 to 11.18.

This is a quite different form than the previous lens: the big difference is that the element powers are about 20 to 25 percent stronger (and more costly to fabricate); the marginal spherical is corrected versus undercorrected; the optimum defocus is slightly smaller; the Petzval radius is 30 percent longer, i.e., flatter; the sagittal field curvature at

TABLE 11.5 Prescription for Another Dogmar Version that Resulted from Several Cycles of Manually "Fiddling Around" and Re-optimizing (See Figs. 11.16 to 11.18)

$R1 = +14.543v$	$T1 = 2.990v$	SK16	620603
$R2 = -53.918v$	$T2 = 0.768v$		
$R3 = -28.147v$	$T3 = 0.8$	$542469v$	(near LLF2)
$R4 = +18.066v$	$T4 = 1.0$		
$R5 = $ Stop	$T5 = 1.0p$		
$R6 = -28.147p$	$T6 = 0.8p$	$542469p$	
$R7 = +18.066p$	$T7 = 0.821v$		
$R8 = +26.933v$	$T8 = 2.447v$	SK16p	620603
$R9 = -17.932v$	BF = 44.999		

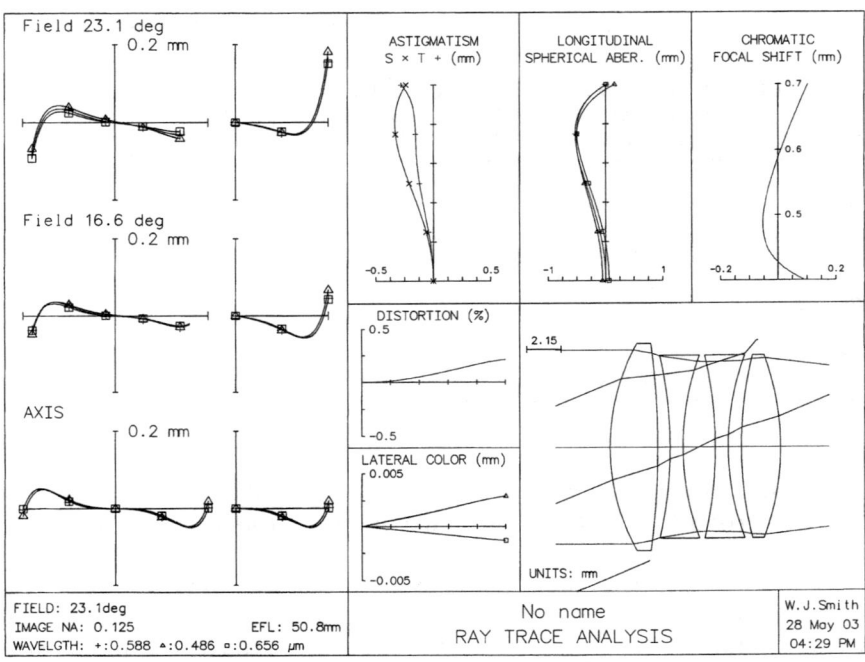

Figure 11.16 The ray aberrations of the final lens of our design story of the Dogmar, Table 11.5.

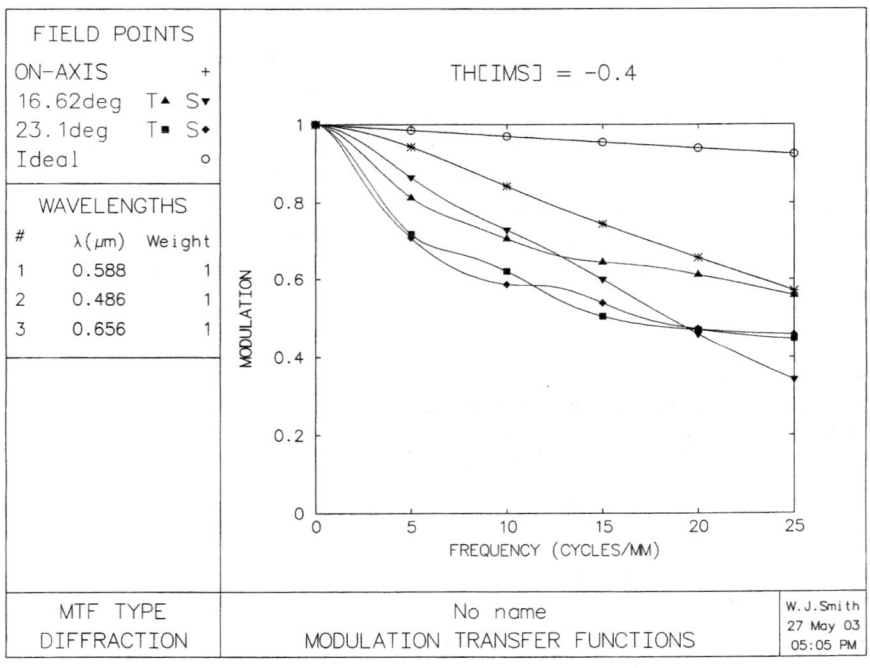

Figure 11.17 The MTF versus frequency for the lens of Fig. 11.16, Table 11.5.

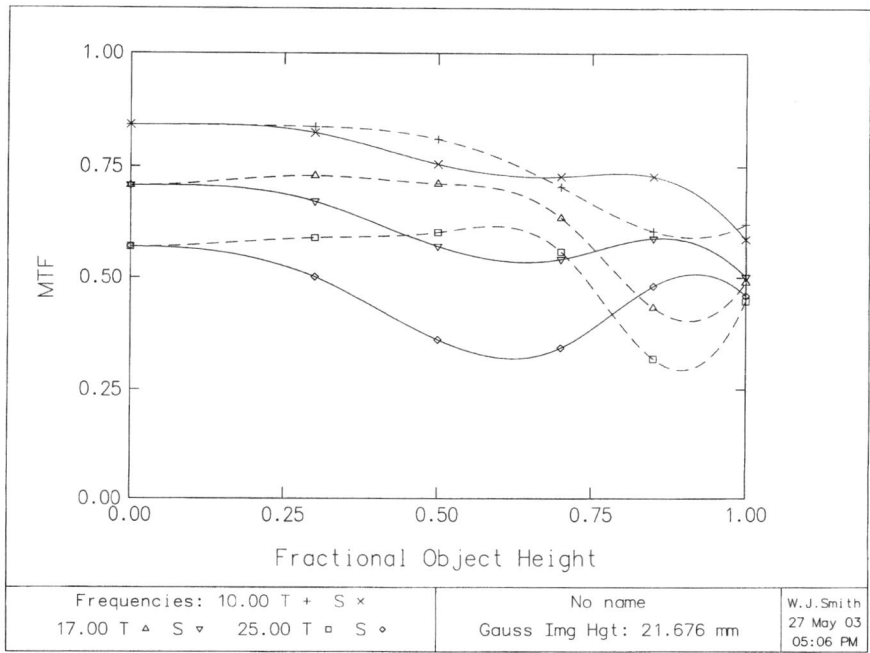

Figure 11.18 The MTF versus field at 10, 17, 25 lpm for the lens of Fig. 11.16, Table 11.5.

the "belly" is smaller; the flints are identical (an economical feature); the vertex length is about eight percent shorter so that the vignetting is better; and the rms spot size is larger. But the performance of the two designs is quite similar, although if we compare the MTF versus field plots, this design shows less variation and a better edge-of-the-field performance.

This "fiddling around" process seems to be innate in the nature of most successful lens designers. Their curiosity impels them to continue to poke and prod the design for as long as possible. But again we must leave our design study incomplete. We have done no systemic study of glasses, thicknesses, symmetries, and the like. There is an obvious trade-off between the flint glass type and the airspaces $T2$ and $T7$, which we have totally negected. But as a wise designer once said, "A lens design is finished only when you have run out of time or money." And we have run out of pages.

Chapter 12

The Biotar or Double-Gauss Lens

12.1 The Basic Six-Element Version

The Biotar or double-Gauss type is a descendant of the double-meniscus anastigmat lenses discussed in Chap. 11. One of the many variants of the double-meniscus form consisted of outer positive singlets and inner cemented meniscus negative doublets, in a symmetrical construction. These lenses had the speed and angular coverage typical of their genre, i.e., good angular coverage at a quite modest aperture; however, a departure from symmetry allowed the speed of the lens to be increased (initially) to $f/2$, and a tremendously useful and powerful design form was born.

This design type is the basis of most normal-focal-length, 35-mm camera lenses, having supplanted the Sonnar and Ernostar types, which evolved at about the same time. It is found in many applications where extremely high performance is required of a lens. It can be made into a moderately wide-angle lens, an enlarger lens, a high-resolution objective, or a lens of extremely high speed. It has been subject to almost every imaginable modification, including splits, compoundings, doublings, inserted components, and even complete doubling or duplication of *all* components.

The basic six-element version is shown in Figs. 12.1 to 12.9, arranged in order of speed, which ranges from $f/1.25$ to $f/8.0$ in these examples. Typically, the positive front element is meniscus in shape, usually of a lanthanum flint glass with an index of about 1.7 and a V-value of about 48. Dense barium crowns and barium flints are also used. The second element tends to be meniscus, although a plano-convex or a mild

Chapter Twelve

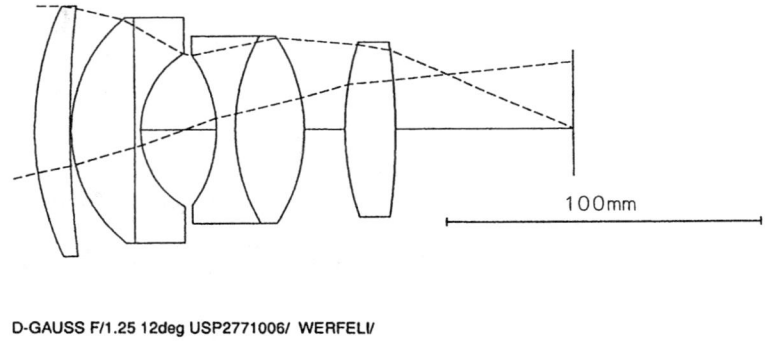

D-GAUSS F/1.25 12deg USP2771006/ WERFELI/

radius	thickness	mat'l	index	V-no	sa
93.320	11.320	LAF3	1.717	48.0	40.0
358.290	0.400	air			40.0
46.320	20.000	BAF9	1.643	48.0	36.0
	2.000	LF2	1.589	40.9	36.0
28.680	14.000	air			24.5
	10.000	air			24.3
-41.320	6.000	SF14	1.762	26.5	24.0
60.800	22.000	LAF2	1.744	44.7	30.0
-55.000	13.000	air			30.0
90.200	16.000	LAF3	1.717	48.0	28.0
-212.580	56.424	air			28.0

EFL = 100.1
BFL = 56.42
NA = -0.3992 (F/1.25)
GIH = 22.03 (HFOV=12.41)
PTZ/F = -3.801
VL = 114.72
OD infinite conjugate

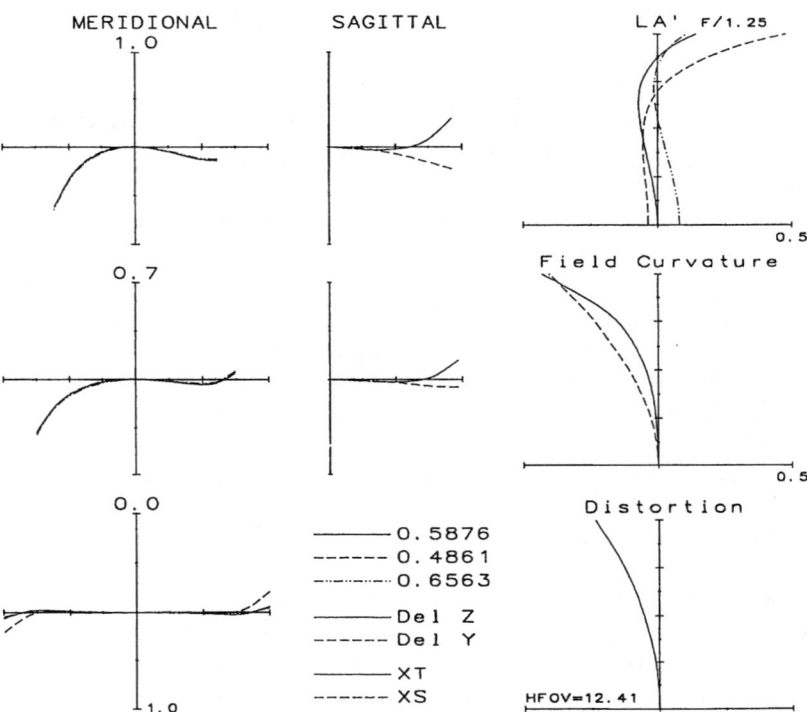

Figure 12.1 $f/1.25 \pm 12°$ double-Gauss lens. Note the large rear airspace.

The Biotar or Double-Gauss Lens   321

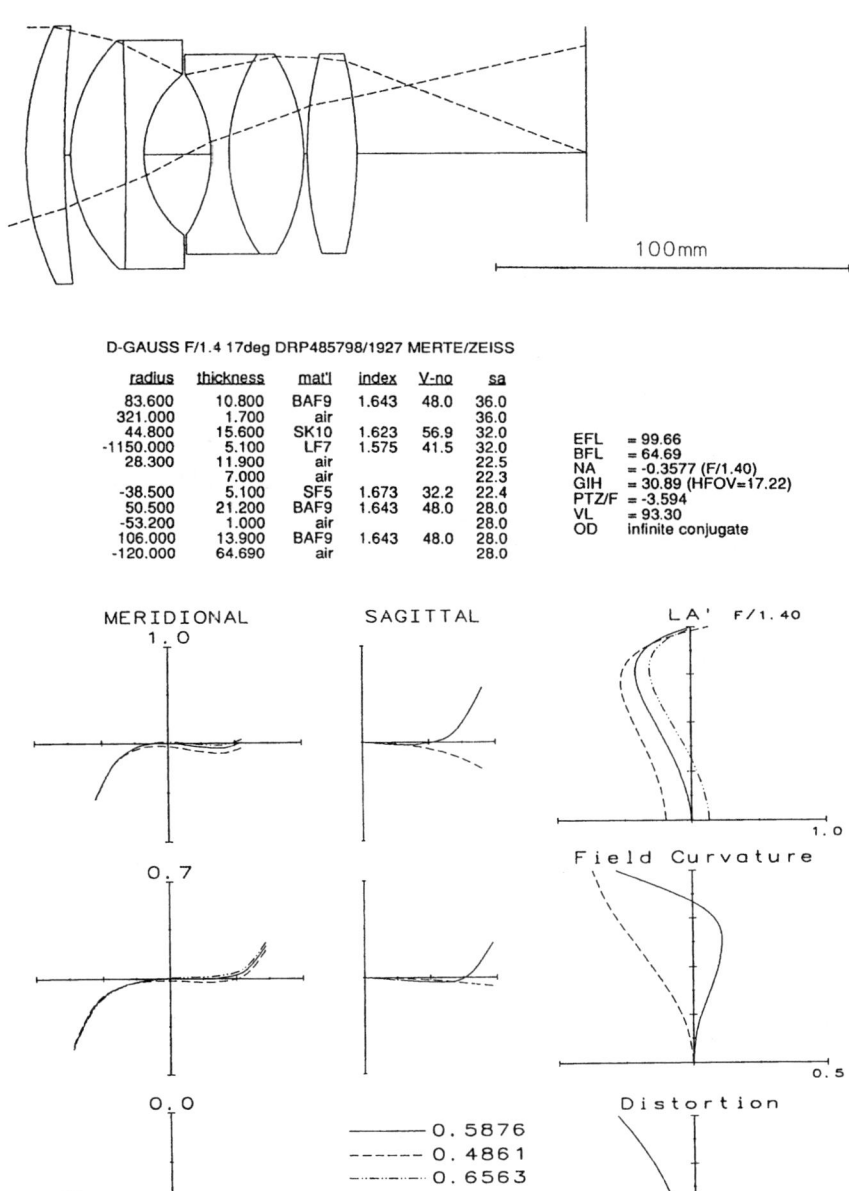

D-GAUSS F/1.4 17deg DRP485798/1927 MERTE/ZEISS

radius	thickness	mat'l	index	V-no	sa
83.600	10.800	BAF9	1.643	48.0	36.0
321.000	1.700	air			36.0
44.800	15.600	SK10	1.623	56.9	32.0
-1150.000	5.100	LF7	1.575	41.5	32.0
28.300	11.900	air			22.5
	7.000	air			22.3
-38.500	5.100	SF5	1.673	32.2	22.4
50.500	21.200	BAF9	1.643	48.0	28.0
-53.200	1.000	air			28.0
106.000	13.900	BAF9	1.643	48.0	28.0
-120.000	64.690	air			28.0

EFL = 99.66
BFL = 64.69
NA = -0.3577 (F/1.40)
GIH = 30.89 (HFOV=17.22)
PTZ/F = -3.594
VL = 93.30
OD   infinite conjugate

Figure 12.2 $f/1.4$ $\pm 17°$ double-Gauss lens.

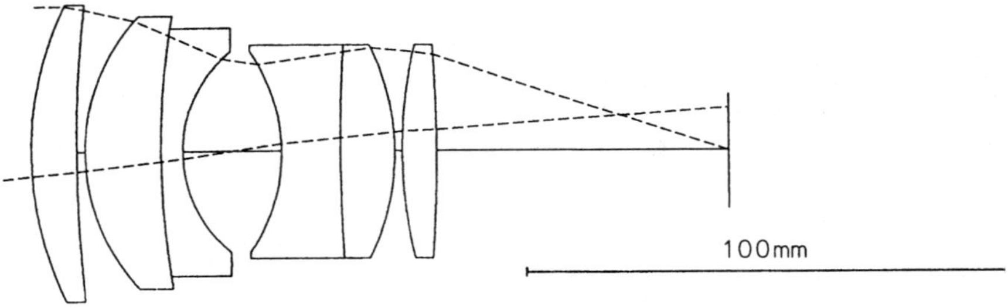

322 Chapter Twelve

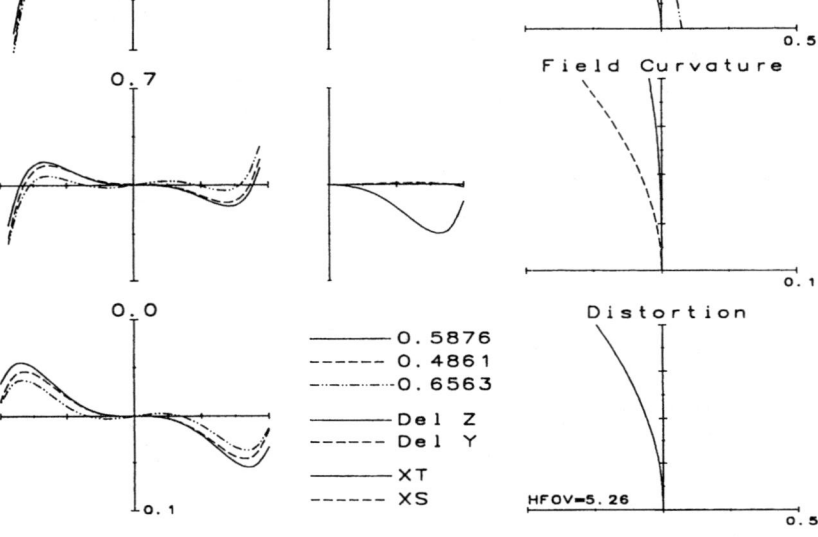

F/1.6 5degHFOV DOUBLE GAUSS

radius	thickness	mat'l	index	V-no	sa
70.525	9.490	SK4	1.613	58.6	31.2
274.170	1.820	air			31.2
41.860	15.860	SK4	1.613	58.6	28.6
166.920	4.550	F5	1.603	38.0	26.0
26.910	10.400	air			21.1
	10.400	air			19.2
-37.245	12.220	SF1	1.717	29.5	21.1
357.500	11.310	LAK16	1.734	51.8	22.4
-50.557	1.560	air			22.4
92.040	7.280	LAK16	1.734	51.8	22.4
-309.712	60.762	air			22.4

EFL = 98.42
BFL = 60.76
NA = -0.3129 (F/1.60)
GIH = 9.05 (HFOV=5.26)
PTZ/F = -4.622
VL = 84.89
OD infinite conjugate

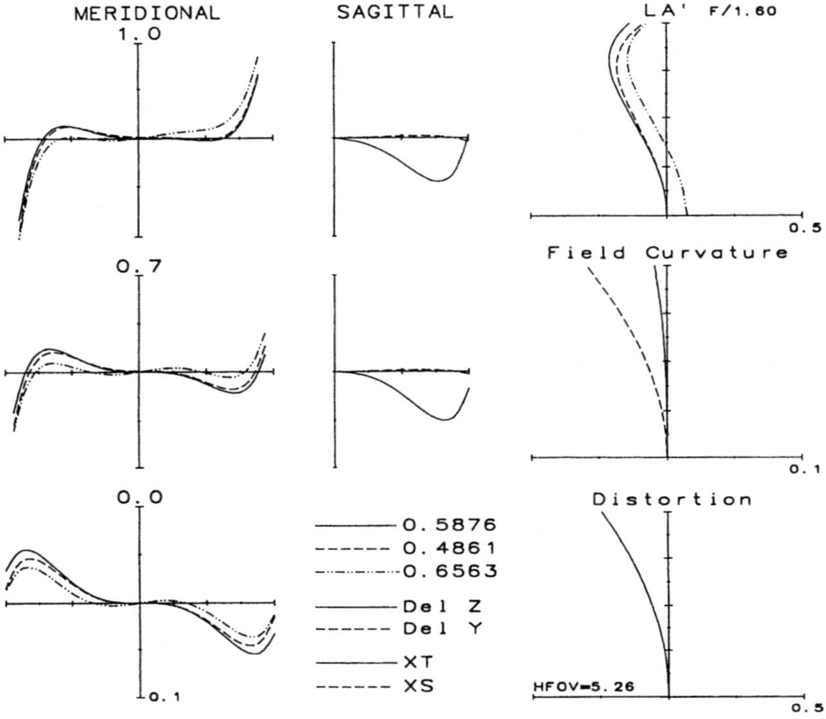

Figure 12.3 $f/1.6$ ±5° double-Gauss, designed as a commercial $f = 2$ in 16-mm projection lens.

The Biotar or Double-Gauss Lens 323

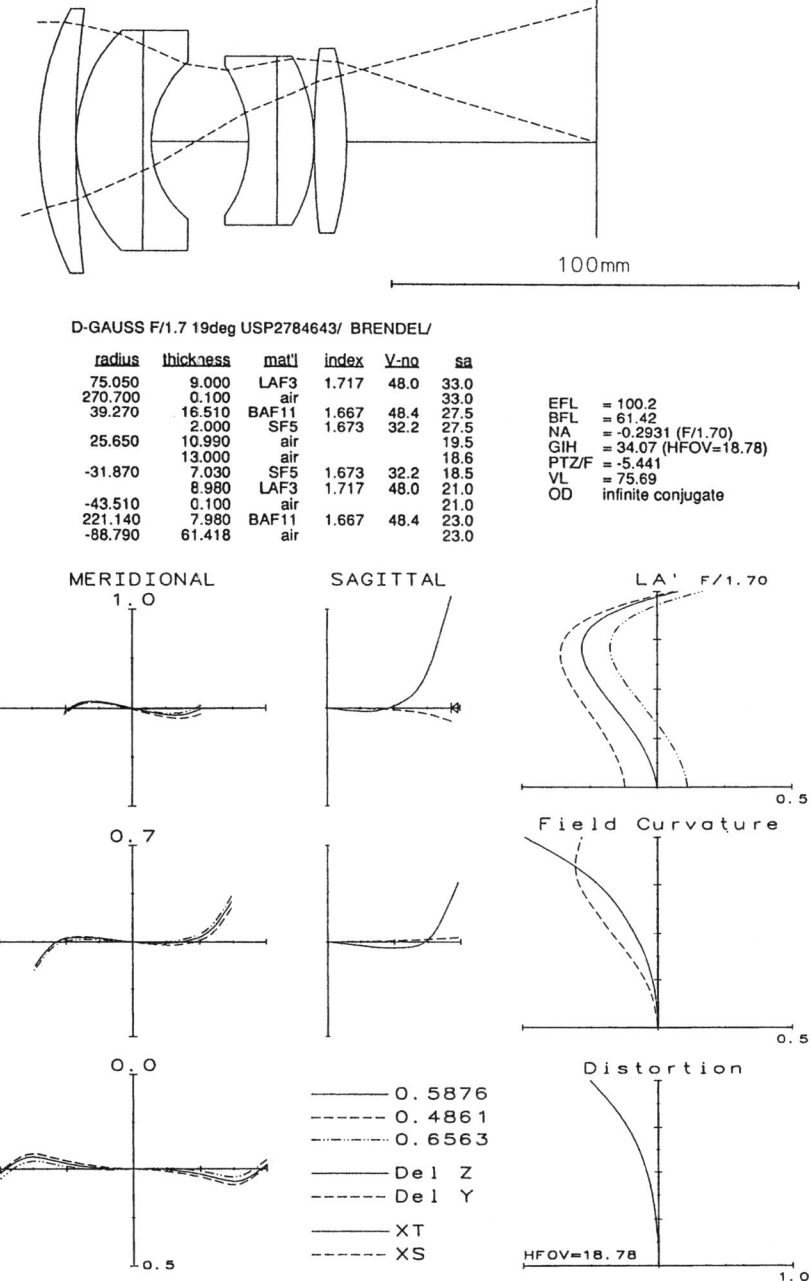

D-GAUSS F/1.7 19deg USP2784643/ BRENDEL/

radius	thickness	mat'l	index	V-no	sa
75.050	9.000	LAF3	1.717	48.0	33.0
270.700	0.100	air			33.0
39.270	16.510	BAF11	1.667	48.4	27.5
	2.000	SF5	1.673	32.2	27.5
25.650	10.990	air			19.5
	13.000	air			18.6
-31.870	7.030	SF5	1.673	32.2	18.5
	8.980	LAF3	1.717	48.0	21.0
-43.510	0.100	air			21.0
221.140	7.980	BAF11	1.667	48.4	23.0
-88.790	61.418	air			23.0

EFL = 100.2
BFL = 61.42
NA = -0.2931 (F/1.70)
GIH = 34.07 (HFOV=18.78)
PTZ/F = -5.441
VL = 75.69
OD infinite conjugate

Figure 12.4 $f/1.7$ ±19° double-Gauss lens.

324 Chapter Twelve

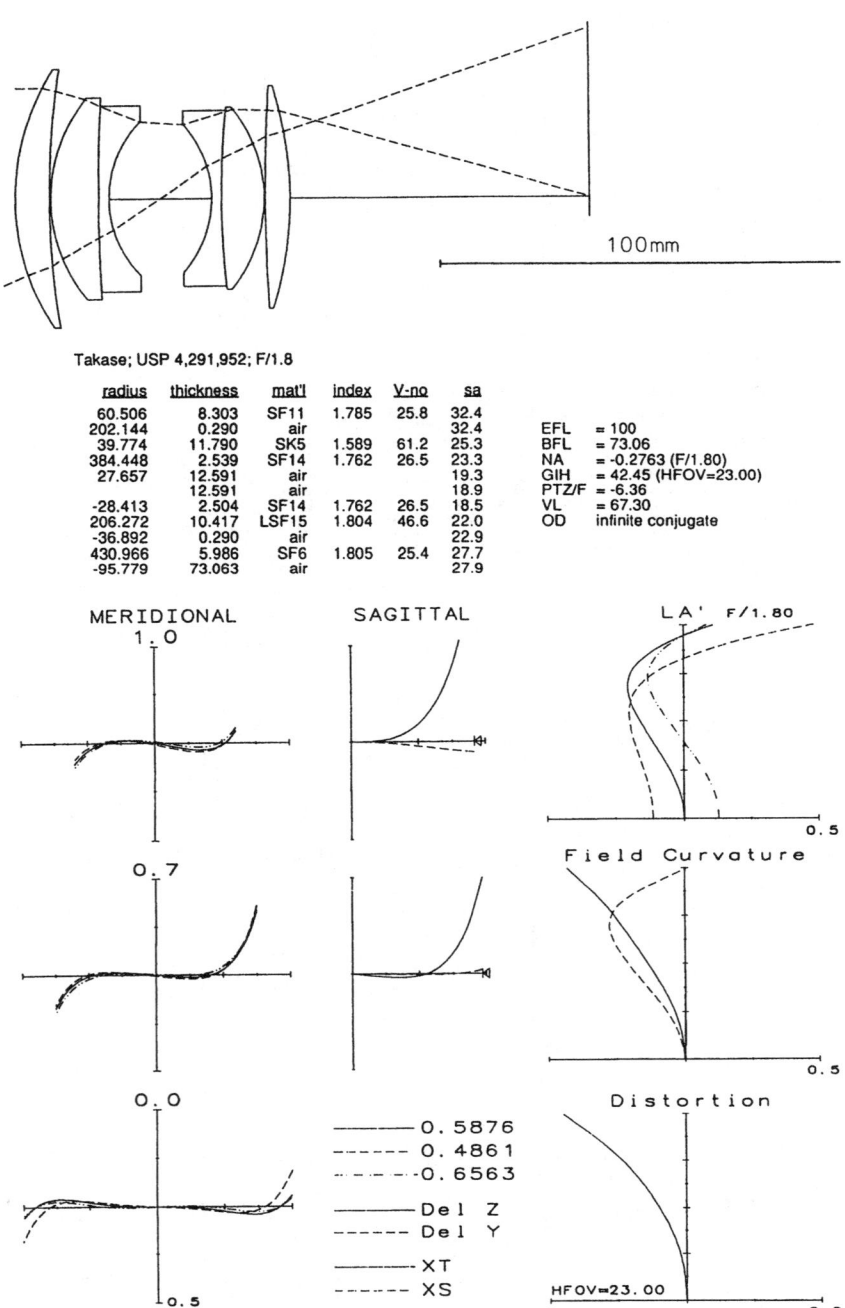

Figure 12.5 $f/1.8$ ±23° double-Gauss camera lens. Note that the outer crowns are actually heavy flints.

The Biotar or Double-Gauss Lens

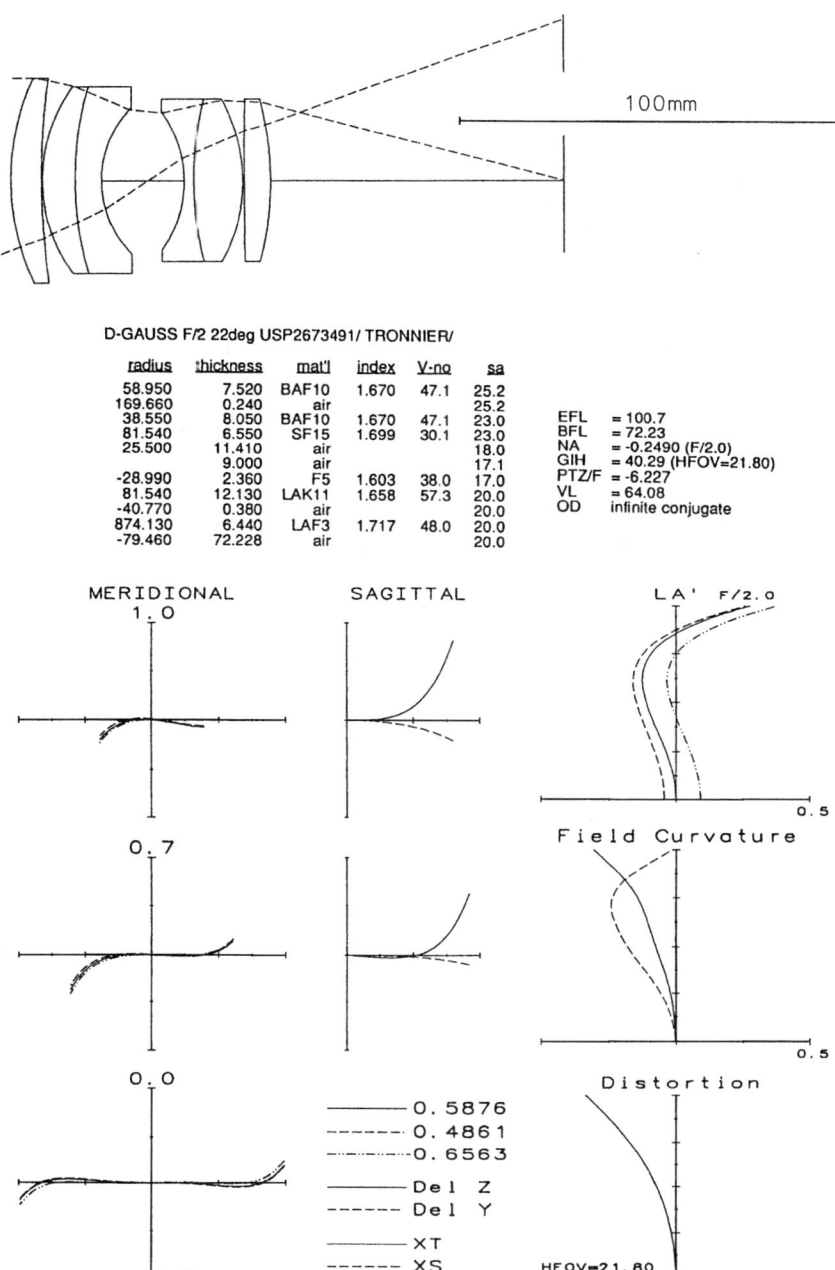

D-GAUSS F/2 22deg USP2673491/ TRONNIER/

radius	thickness	mat'l	index	V-no	sa
58.950	7.520	BAF10	1.670	47.1	25.2
169.660	0.240	air			25.2
38.550	8.050	BAF10	1.670	47.1	23.0
81.540	6.550	SF15	1.699	30.1	23.0
25.500	11.410	air			18.0
	9.000	air			17.1
-28.990	2.360	F5	1.603	38.0	17.0
81.540	12.130	LAK11	1.658	57.3	20.0
-40.770	0.380	air			20.0
874.130	6.440	LAF3	1.717	48.0	20.0
-79.460	72.228	air			20.0

EFL = 100.7
BFL = 72.23
NA = -0.2490 (F/2.0)
GIH = 40.29 (HFOV=21.80)
PTZ/F = -6.227
VL = 64.08
OD infinite conjugate

Figure 12.6 $f/2.0$ ±22° double-Gauss camera lens.

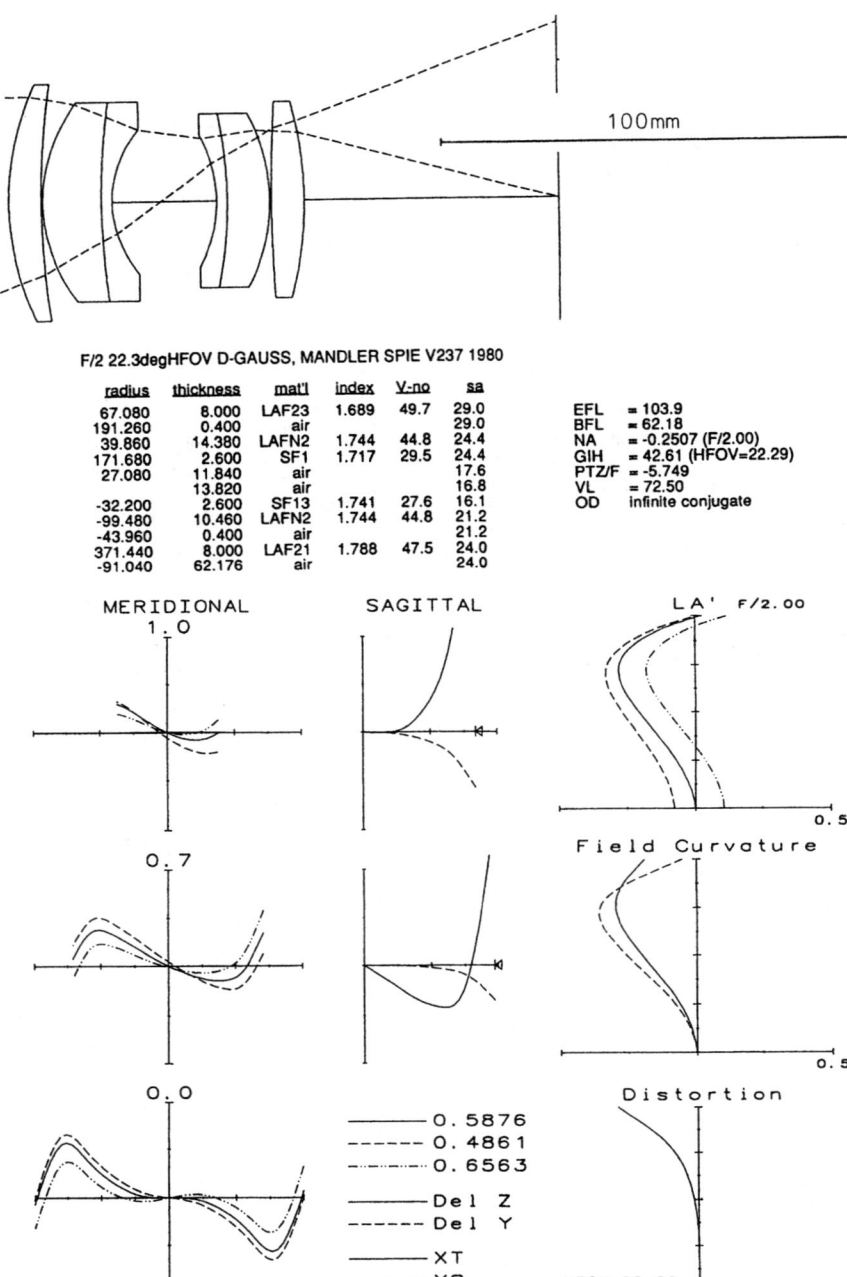

Figure 12.7 $f/2.0$ ±22.3° double-Gauss camera lens.

The Biotar or Double-Gauss Lens

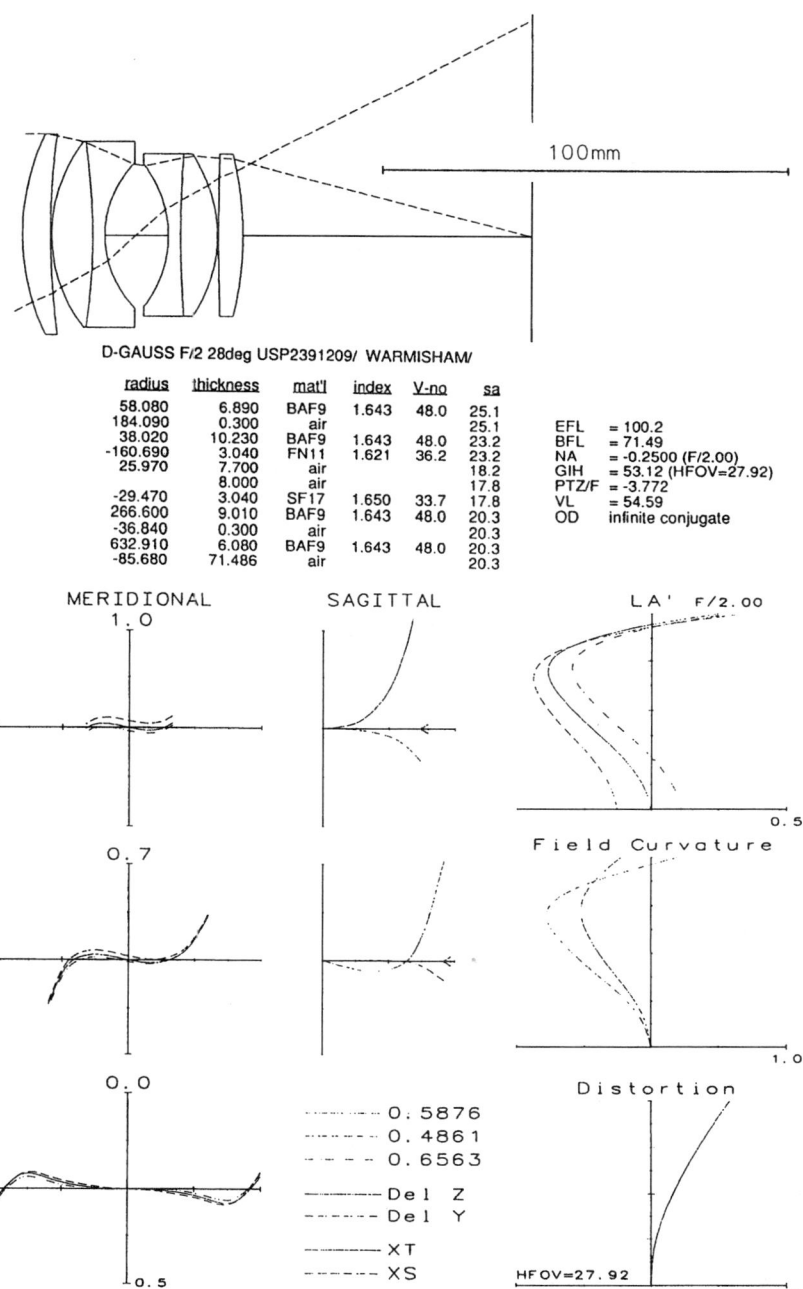

Figure 12.8 $f/2.0$ ±28° double-Gauss lens of shorter vertex length and wider field.

328 Chapter Twelve

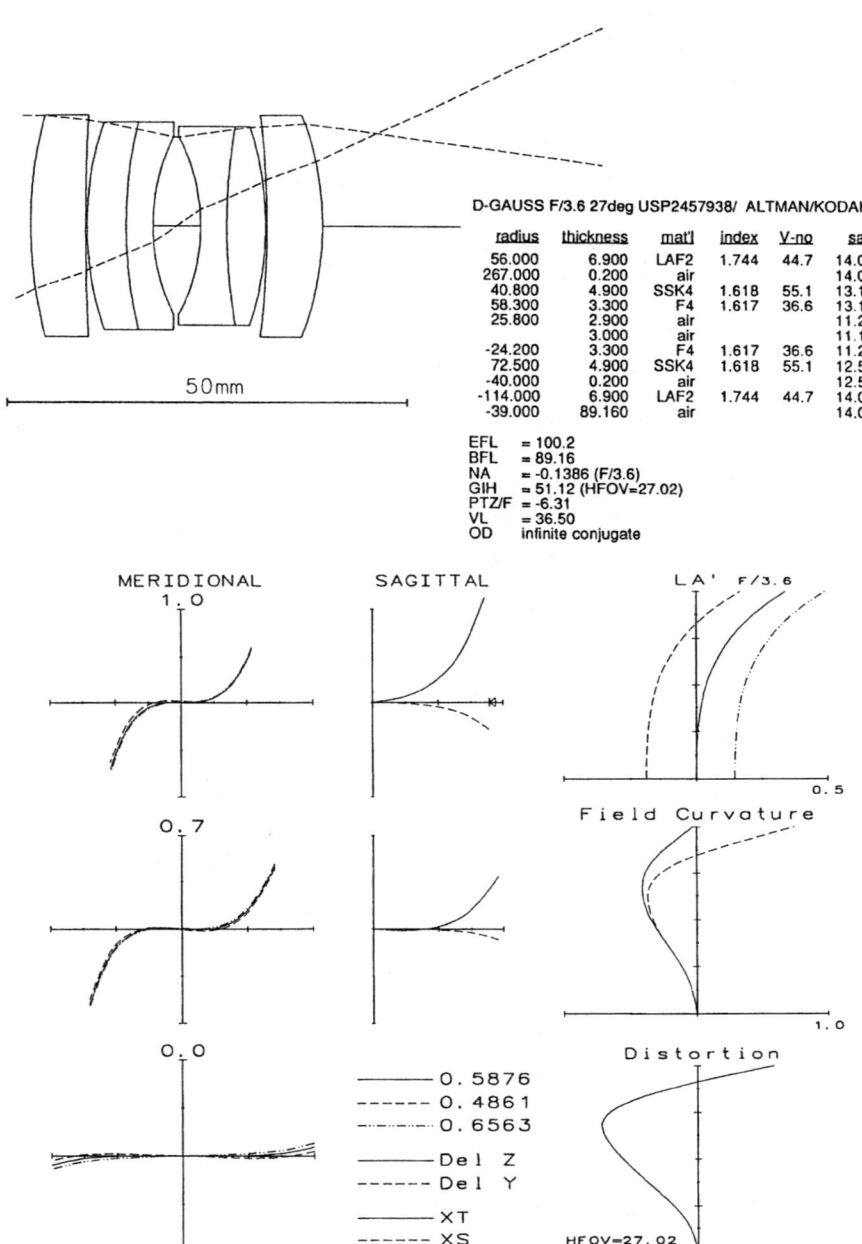

Figure 12.9 f/3.6 ±27° double-Gauss lens; shorter, slower speed, capable of good definition (if recorrected).

biconvex form is not unusual; its index is typically a bit lower and its V-value higher than the first element. The third element usually has an index close to that of the second element, and it is a dense flint from along the glass line. The fourth element is usually biconcave and made of a glass-line flint with a slightly higher index than the third element. The fifth and sixth elements are ordinarily both biconvex and of lanthanum flint glass.

Often the rear positive elements are of higher-index glass than those in front. Since these rear elements are more powerful than those in front, the higher index is more effective there; if cost limits the amount of high-index glass that may be used, the smaller rear elements are the logical (and economical) place to use it. Occasionally, the front element is made of dense flint glass; Fig. 12.5 shows a design in which both outer elements are made of SF-type glass. The dense flint glass is a less expensive choice than a lanthanum glass, and its dispersion can affect the spherochromatic, just as discussed in Sec. 6.4.

Figure 12.3 was designed for use as a commercial 2-in $f/1.6$, 16-mm movie projection lens. Note the ordinary (SK4) glass in the front elements and the economical (i.e., thin) use of the more costly lanthanum glass in the rear elements.

A slower lens is shown in Fig. 12.9 covering about a 54° total field. Figure 12.9 is a relatively standard configuration at $f/3.6$; however, in spite of its slow speed, it is a good example of one of the common problems with the double-Gauss form, i.e., a strongly overcorrecting oblique spherical aberration. This problem can often be reduced by shaping the doublets to a more meniscus form so that the oblique rays pass more normally through them.

Figure 12.7 is one of the ultimate lenses that resulted from an extensive study of $f/2$ normal-focal-length 35-mm camera lenses on which Mandler reported at the 1980 International Lens Design Conference (Ref. 28).

Two unusual varieties of six-element Biotar type lenses are shown in Figs. 12.10 and 12.11. In Fig. 12.10 the inner doublets are split and the front doublet is reversed; note also the reversal of the usual V-value arrangement in elements 2 and 3, with the negative element having the higher V-value rather than the usual lower one. In Fig. 12.11 the inner doublets are reversed and reshaped accordingly.

12.2 Twenty-Eight Things That Every Lens Designer Should Know About the Double-Gauss/Biotar Lens

The following ideas apply to the six-element double Gauss, but many of them can be equally well applied to the many modifications of the double-Gauss (or even, in some cases, to other design forms).

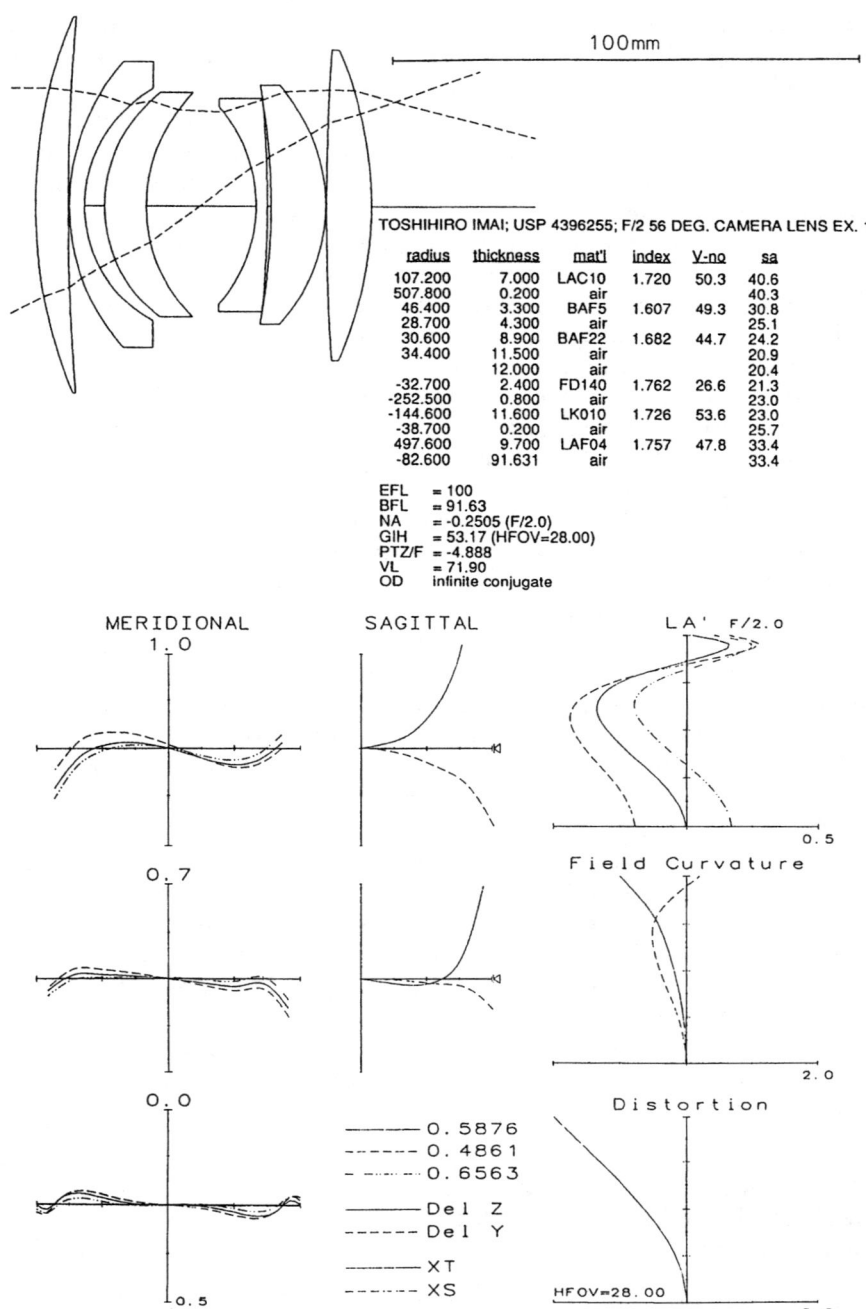

Figure 12.10 f/2.0 ±28° double-Gauss with both inner doublets decemented.

The Biotar or Double-Gauss Lens 331

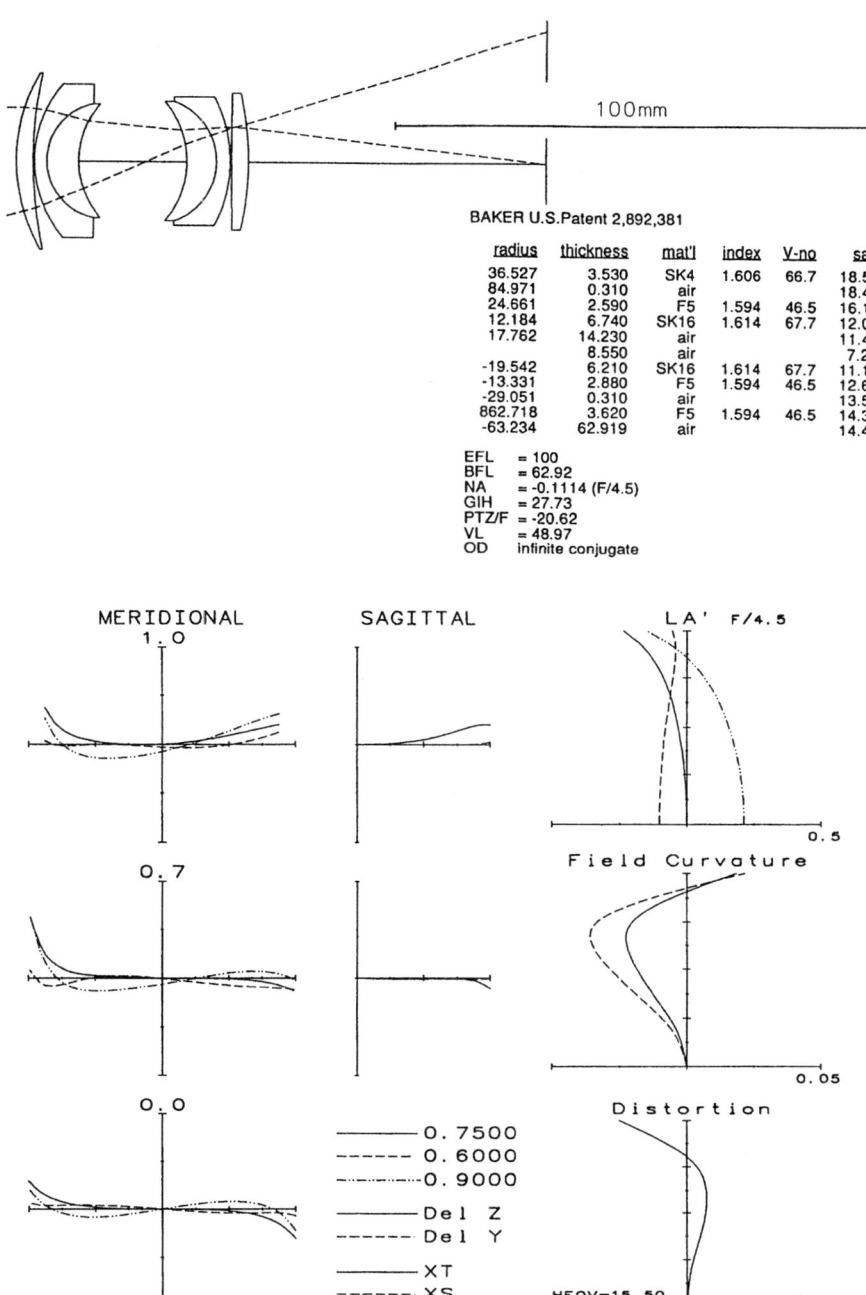

BAKER U.S.Patent 2,892,381

radius	thickness	mat'l	index	V-no	sa
36.527	3.530	SK4	1.606	66.7	18.5
84.971	0.310	air			18.4
24.661	2.590	F5	1.594	46.5	16.1
12.184	6.740	SK16	1.614	67.7	12.0
17.762	14.230	air			11.4
	8.550	air			7.2
-19.542	6.210	SK16	1.614	67.7	11.1
-13.331	2.880	F5	1.594	46.5	12.6
-29.051	0.310	air			13.5
862.718	3.620	F5	1.594	46.5	14.3
-63.234	62.919	air			14.4

EFL = 100
BFL = 62.92
NA = -0.1114 (F/4.5)
GIH = 27.73
PTZ/F = -20.62
VL = 48.97
OD infinite conjugate

Figure 12.11 $f/4.5$ $\pm 16°$ double-Gauss with inner doublets reversed.

1. The origin of the double-Gauss design is not a doubled, flint-to-flint version of the Gauss telescope objective. It is based on the 1896 $f/4$ Planar of Rudolph, which was a symmetrical system with outer singlets and inner meniscus doublets, oriented concave to the stop. Around 1920 Lee thickened the doublets, changed the doublet glasses, abandoned symmetry, reduced the field angle to ±20° to 23° to raise the speed to $f/2$, and created the progenitor of the modern camera objective.
2. High-index glasses are more effective in the rear elements (which are usually the stronger elements) and less effective in the front.
3. In the front doublet, if the crown index is higher than the flint, it can help coma.
4. In the rear doublet, approximately equal indices are OK.
5. High-index crowns help both spherical and Petzval.
6. When the Petzval sum and astigmatism are a problem, try higher-index materials, or try flattening the Petzval field (and let the axial quality degrade a bit).
7. In the final analysis, the field coverage of a design is almost always limited by oblique spherical. Oblique spherical is especially significant in the double-Gauss because of the high speed of most designs.
8. Oblique spherical is reduced by an increased central airspace and less strongly curved surfaces. This will tend to increase the vignetting if the vertex length is also increased.
9. The oblique ray angles in the double-Gauss are moderate except at the inner surfaces of the meniscus doublets.
10. Undercorrected axial spherical can offset overcorrected oblique spherical, but at the expense of a focus shift as the lens is stopped down.
11. The tangential oblique spherical can be vignetted out, but vignetting is much less effective in the sagittal direction. This makes the sagittal oblique spherical important to control.
12. The front and rear airspaces are almost always kept as small as possible to help the angular coverage; however, a large rear airspace has been known to reduce the sagittal oblique spherical.
13. The Petzval sum of the double-Gauss is typically less than that of a comparable Cooke triplet, e.g., its Petzval radius is about minus four to seven times the focal length, occasionally reaching 10 or 20 times the *effective focal length* (EFL).
14. In the double-Gauss the Petzval sum is reduced by the same mechanism as in the Cooke triplet or any of the airspaced anastigmats—by

large ray heights at converging surfaces and low ray heights at the negative, diverging surfaces.

15. Reducing X_S usually brings more LA_z (because you do it by reducing the Petzval, which means higher-element/surface powers, that is, more high-order aberration).

16. The central airspace will affect the astigmatism with only a little effect on the other aberrations.

17. The front airspace in the double-Gauss is shaped like, and works like, the airspace in the Gauss telescope objective, in that the ray height change (or drop) is amplified for the marginal rays by the greatly increased spacing at the margin. The marginal ray height at surface no. 3 is reduced more for blue light than for red light because of the undercorrected chromatic of the front element. Note that the V-value of the front element glass will thereby affect the spherochromatism. (See the discussion of spherochromatism in Chap. 6.) This probably at least partly explains the use of high-dispersion (low V-value) glasses in the front element of many designs.

18. The thick meniscus doublets (as well as the airspaces) affect the high-order aberrations by nonlinearly shrinking or expanding the beam diameter from one end of the doublet or space to the other. The double-Gauss spherical is affected by the doublet thickness just as in the telescope objective (see Secs. 6.4 and 6.5) because there is undercorrected spherical and chromatic aberration within the doublet.

19. Split the outer crowns to reduce the spherical zonal. The front split may restrict the field angle coverage; split the rear crown if the field is a consideration.

20. Break the front doublet cement to help spherical and field curvature. Splitting the rear doublet is not as advantageous.

21. Surface no. 3 is often close to being aplanatic (Sec. 17.2), with little or no spherical, coma, or astigmatism contributions.

22. With almost no exceptions, the double-Gauss has barrel distortion. For most applications 1 to 1.5 percent seems to be acceptable.

23. The double-Gauss is quite sensitive to refocusing for close objects. The image deterioration can be reduced if the rear element is fixed and only the front five elements are moved to focus, or vice versa (Chap. 21).

24. The overall length is what controls vignetting. Of course this is true for all lenses. The length also affects high-order spherical and astigmatism.

25. Approximate symmetry and wrapping the surface(s) around the stop reduce the variation of the angle of incidence (among other things) with obliquity.
26. In fabrication, the thickness of the doublets requires critical control. This requirement is usually met by pairing the crown and flint thickness (which individually are much less critical) to get the correct doublet thickness. Indeed, to some extent balancing the total thickness of both doublets may increase the permissible variation of the individual doublets. The front doublet is the most critical because the marginal ray slope is large there. The central airspace is sometimes adjusted during assembly to control the astigmatism.
27. The large number of variable parameters in the double-Gauss (10 radii + 10 spacings + 10 indices + 10 dispersions = 32) might be expected to make the design task quite easy. It does make possible an excellent design, but the abundance of variables means that there is also an even greater abundance of local optima, most of which are quite bad. Often allowing everything to vary at once is an invitation for disaster. Designers frequently reduce the number of variables at first and then gradually increase them; each designer seems to have a preferred method. As an example, consider the following published approach:

   a. Vary all curvatures.
   b. Add all glass thicknesses as variables (with edge thickness (ET) control).
   c. Add index and dispersion for the glass of the doublets as variables. Fix to real glasses, one at a time, optimizing each time.
   d. Try additional glasses; remove ET control; vary all thicknesses and spaces; reweight the *merit function* (MF); and change rays in the MF; in other words "fiddle around."

28. An often overlooked modification (useful if the six-element double-Gauss is better than needed) is to substitute a thick meniscus singlet for one of the doublets. Other alternatives are the Sonnar and Ernostar, with the three front elements like the double-Gauss and the rear of the lens like the Tessar or triplet.

12.3 The Seven-Element Biotar—Split-Rear Crown

Splitting the rear singlet of the Biotar is an excellent way to improve on the basic configuration. It is a commonly used technique to allow an increased speed, and is often seen in the faster 35-mm camera lenses. Four

The Biotar or Double-Gauss Lens 335

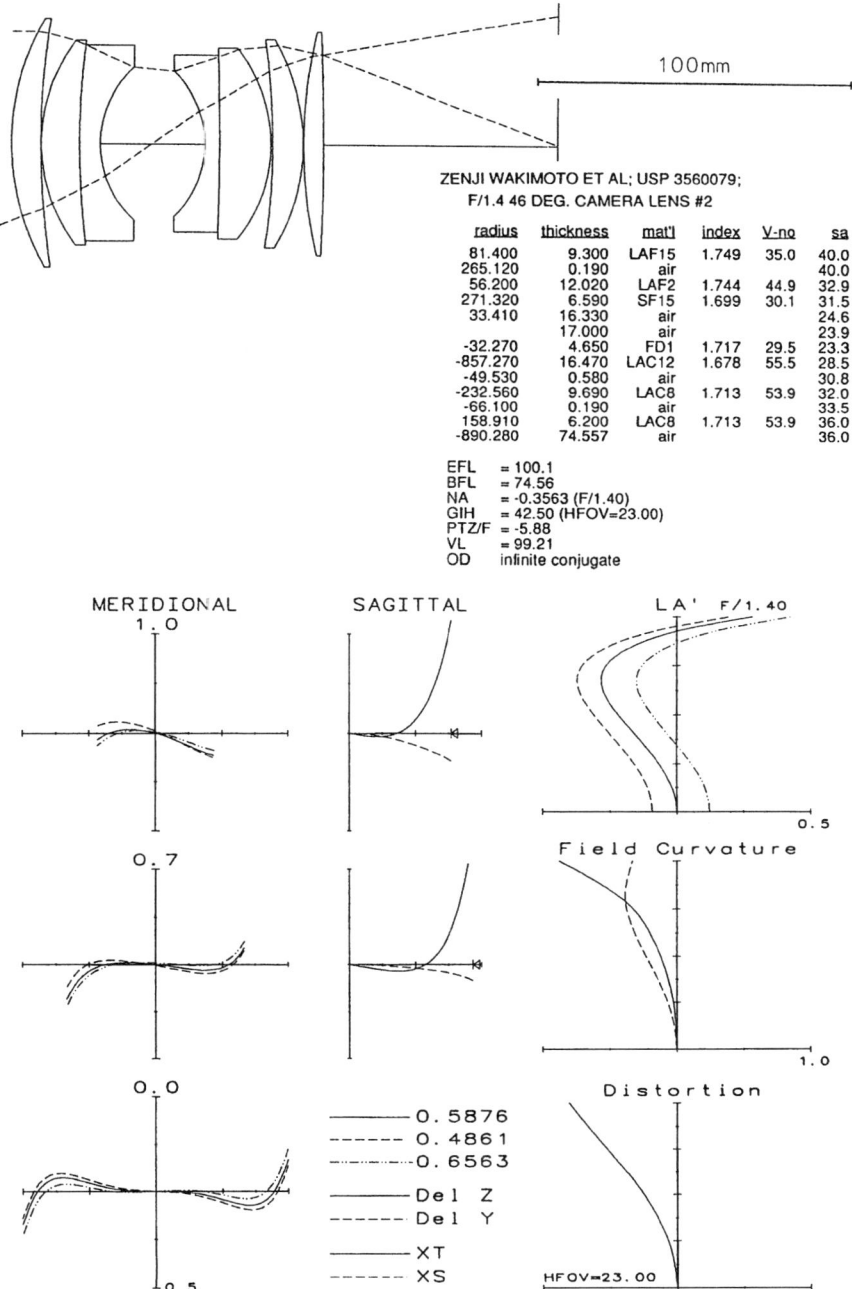

ZENJI WAKIMOTO ET AL; USP 3560079;
F/1.4 46 DEG. CAMERA LENS #2

radius	thickness	mat'l	index	V-no	sa
81.400	9.300	LAF15	1.749	35.0	40.0
265.120	0.190	air			40.0
56.200	12.020	LAF2	1.744	44.9	32.9
271.320	6.590	SF15	1.699	30.1	31.5
33.410	16.330	air			24.6
	17.000	air			23.9
-32.270	4.650	FD1	1.717	29.5	23.3
-857.270	16.470	LAC12	1.678	55.5	28.5
-49.530	0.580	air			30.8
-232.560	9.690	LAC8	1.713	53.9	32.0
-66.100	0.190	air			33.5
158.910	6.200	LAC8	1.713	53.9	36.0
-890.280	74.557	air			36.0

EFL = 100.1
BFL = 74.56
NA = -0.3563 (F/1.40)
GIH = 42.50 (HFOV=23.00)
PTZ/F = -5.88
VL = 99.21
OD infinite conjugate

Figure 12.12 $f/1.4$ ±23° a seven-element, split-rear crown high-speed camera lens.

336 Chapter Twelve

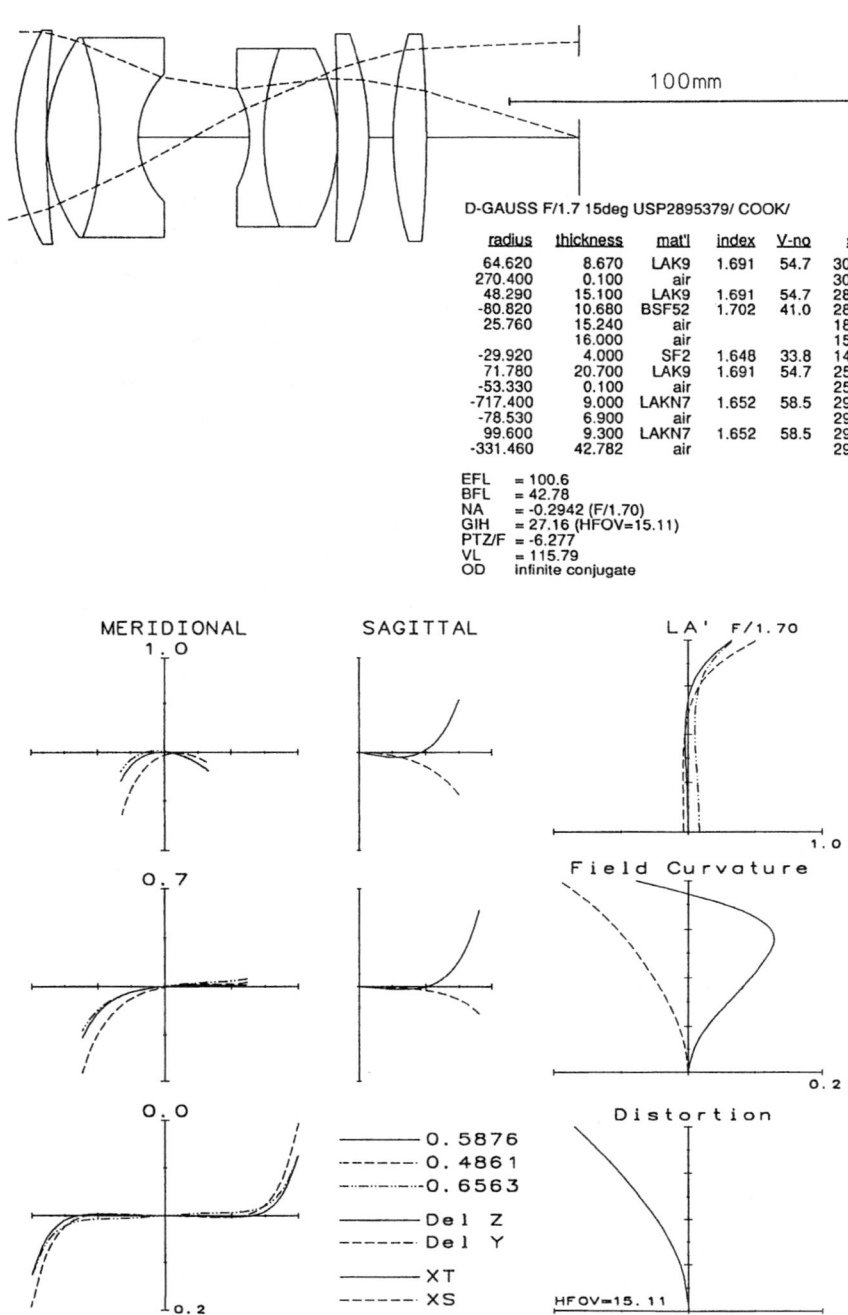

Figure 12.13 $f/1.7$ ±15° a seven-element, split-rear crown with increased rear airspace.

The Biotar or Double-Gauss Lens

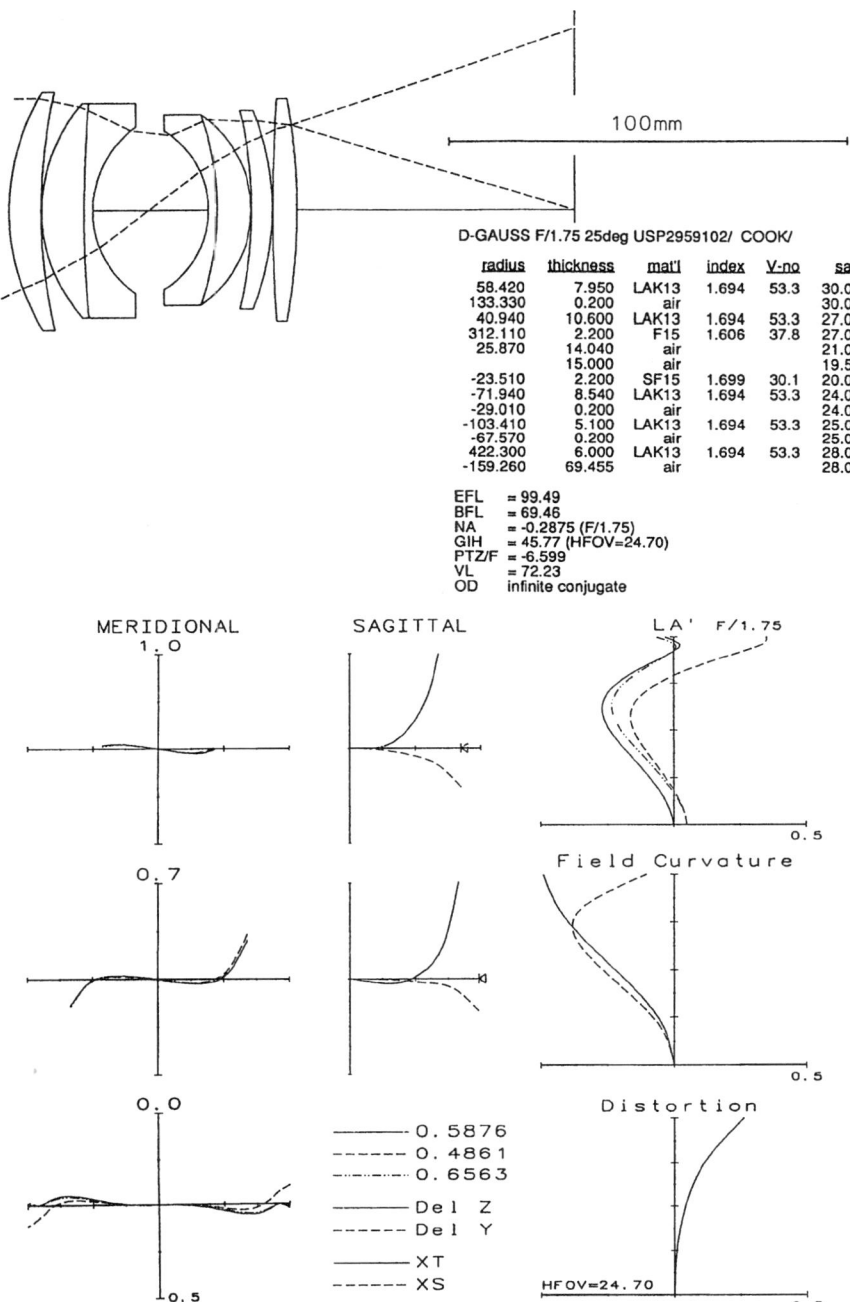

D-GAUSS F/1.75 25deg USP2959102/ COOK/

radius	thickness	mat'l	index	V-no	sa
58.420	7.950	LAK13	1.694	53.3	30.0
133.330	0.200	air			30.0
40.940	10.600	LAK13	1.694	53.3	27.0
312.110	2.200	F15	1.606	37.8	27.0
25.870	14.040	air			21.0
	15.000	air			19.5
-23.510	2.200	SF15	1.699	30.1	20.0
-71.940	8.540	LAK13	1.694	53.3	24.0
-29.010	0.200	air			24.0
-103.410	5.100	LAK13	1.694	53.3	25.0
-67.570	0.200	air			25.0
422.300	6.000	LAK13	1.694	53.3	28.0
-159.260	69.455	air			28.0

EFL = 99.49
BFL = 69.46
NA = -0.2875 (F/1.75)
GIH = 45.77 (HFOV=24.70)
PTZ/F = -6.599
VL = 72.23
OD infinite conjugate

Figure 12.14 $f/1.75$ ±25° a seven-element, split-rear crown double-Gauss.

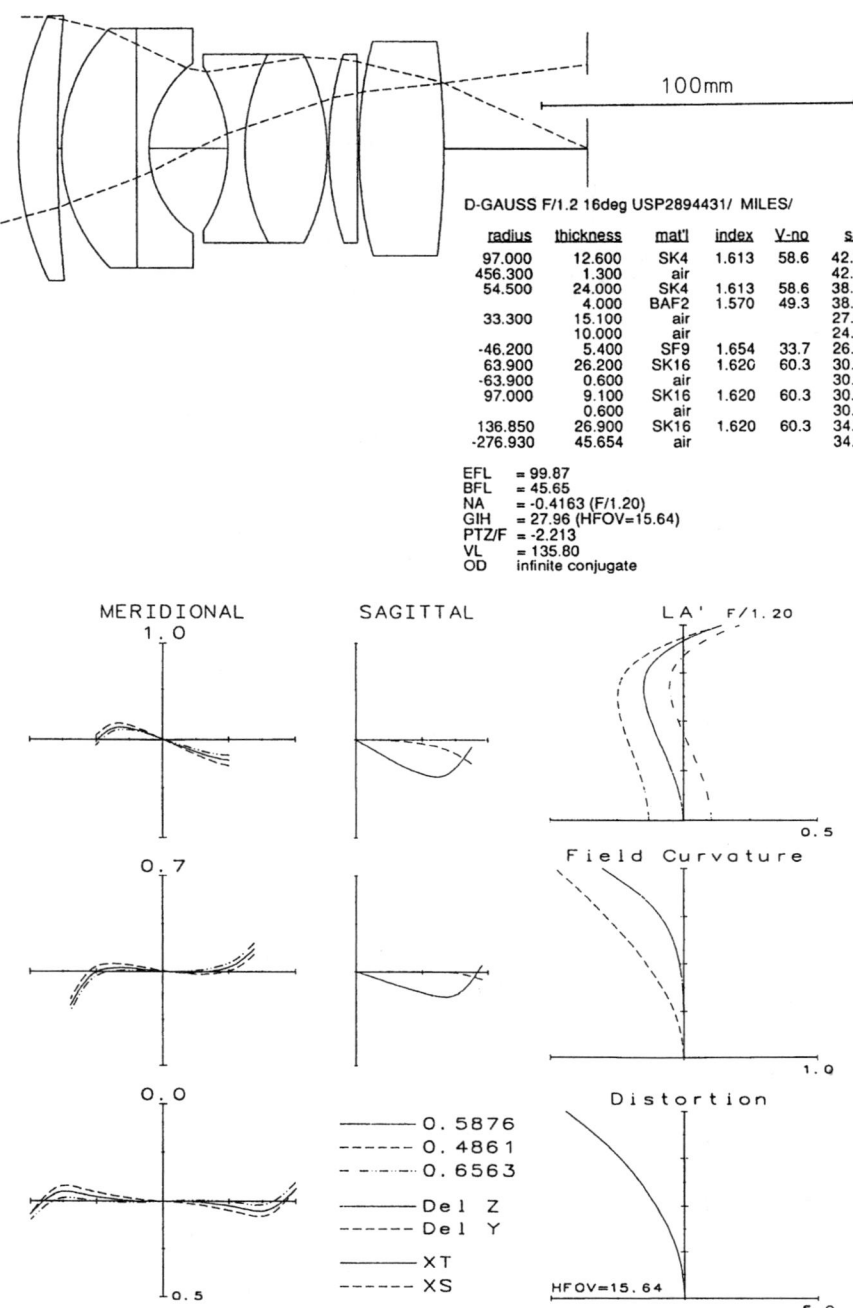

Figure 12.15 f/1.2 ±16° a seven-element, split-rear crown double-Gauss. Compare with Fig. 12.16.

The Biotar or Double-Gauss Lens 339

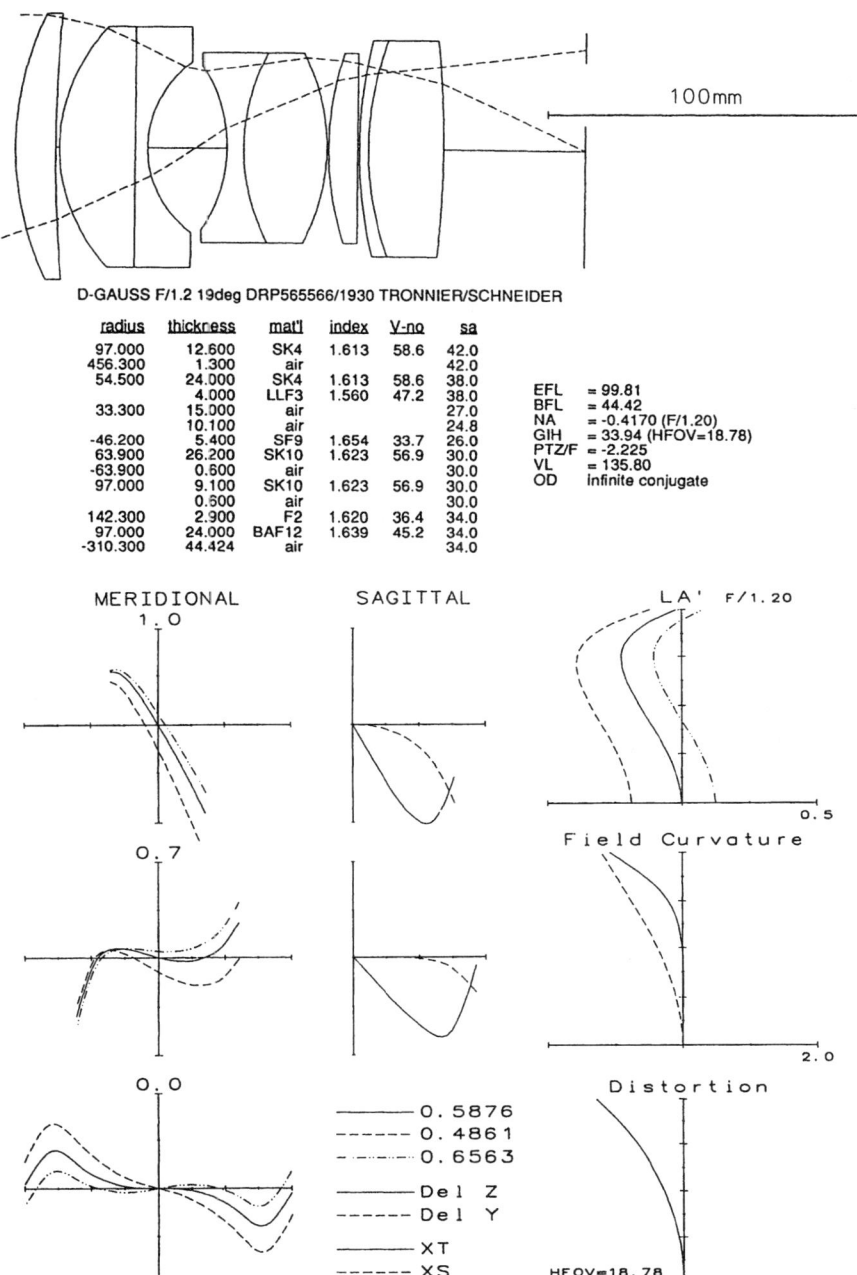

Figure 12.16 $f/1.2$ ±19° an eight-element, split-rear and compounded lens. Compare with Fig. 12.15.

examples of the type are shown in Figs. 12.12 to 12.15, all of relatively high speed. Figure 12.15 is an older design using SK glass; compare it with Figs. 12.12 to 12.14, which utilize higher-index glasses. Most of these designs maintain the two rear elements at a minimal thickness and close spacing. Figure 12.13 uses a larger rear airspace; this can be helpful in controlling the sagittal oblique spherical. Figure 12.15 is an interesting lens in that it is a simplification of an older design (Fig. 12.16) by Tronnier, which has a cemented doublet as the thick rear component. In the original unmodified (i.e., by us) versions, the prescriptions and glasses were identical, with the exception of the last component(s). This was a fairly obvious change since the seventh element of Fig. 12.16 is of very low power.

12.4 The Seven-Element Biotar—Broken Contact Front Doublet

The added freedom gained by breaking the cemented contact in the front doublet has been utilized recently in many camera lenses. Figure 12.17 shows a high-speed design, which is modification of the split-rear-singlet form described in the preceding section. Figure 12.18 is a slower version.

12.5 The Seven-Element Biotar—One Compounded Outer Element

Figures 12.19 and 12.20 show $f/1.4$ lenses of modest angular coverage, one with a compounded front, the other with a compounded rear. Another lens at a speed of $f/2$ is shown in Fig. 12.21. Note that, in all cases (except Fig. 12.19), the doublet is a new achromat as described in Sec. 3.5. Although infrequently encountered, both outer singlets may be compounded. Figure 12.22 is a compounded rear version at an $f/2.5$ speed.

12.6 The Eight-Element Biotar

Three high-speed, eight-element designs are shown in Figs. 12.23 through 12.25. In Fig. 12.23, both outer elements are split and the capability for a reasonable state of correction is achieved using only medium high-index glass. [Note that the undercorrected axial chromatic of this prescription can be corrected by using a glass such as BaSF6 (668:419) in the front element instead of SF5 (673:322).] Two versions of compounding the rear singlet of a seven-element, split-rear singlet type are shown in Figs. 12.16 and 12.24. Note that Fig. 12.16 is the design, which was presented in simplified form in Fig. 12.15. Figure 12.25, while a

The Biotar or Double-Gauss Lens   341

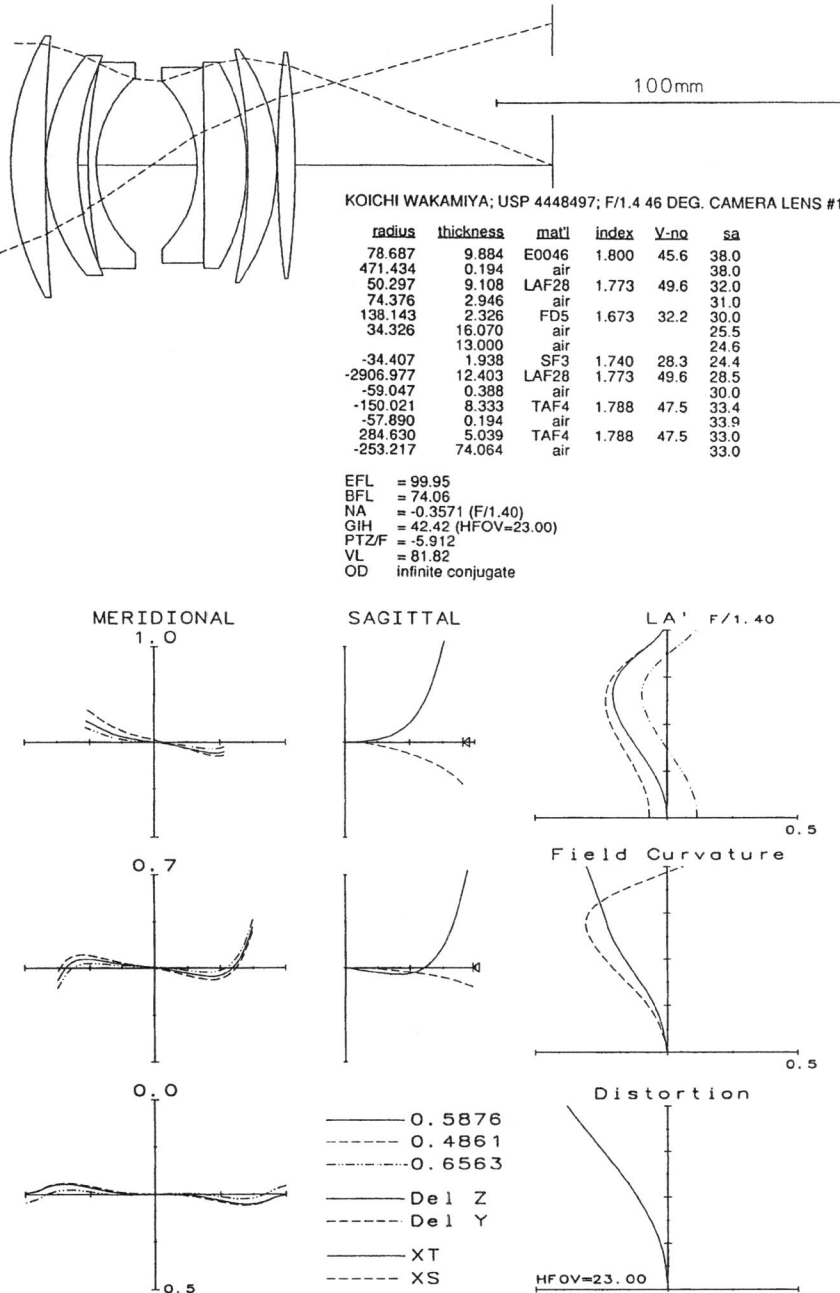

KOICHI WAKAMIYA; USP 4448497; F/1.4 46 DEG. CAMERA LENS #1

radius	thickness	mat'l	index	V-no	sa
78.687	9.884	E0046	1.800	45.6	38.0
471.434	0.194	air			38.0
50.297	9.108	LAF28	1.773	49.6	32.0
74.376	2.946	air			31.0
138.143	2.326	FD5	1.673	32.2	30.0
34.326	16.070	air			25.5
	13.000	air			24.6
-34.407	1.938	SF3	1.740	28.3	24.4
-2906.977	12.403	LAF28	1.773	49.6	28.5
-59.047	0.388	air			30.0
-150.021	8.333	TAF4	1.788	47.5	33.4
-57.890	0.194	air			33.9
284.630	5.039	TAF4	1.788	47.5	33.0
-253.217	74.064	air			33.0

EFL = 99.95
BFL = 74.06
NA = -0.3571 (F/1.40)
GIH = 42.42 (HFOV=23.00)
PTZ/F = -5.912
VL = 81.82
OD infinite conjugate

Figure 12.17 $f/1.4$ ±23° a seven-element, split-rear crown double-Gauss. This form with the front doublet decemented is widely used for high-speed lenses for SLR cameras.

342 Chapter Twelve

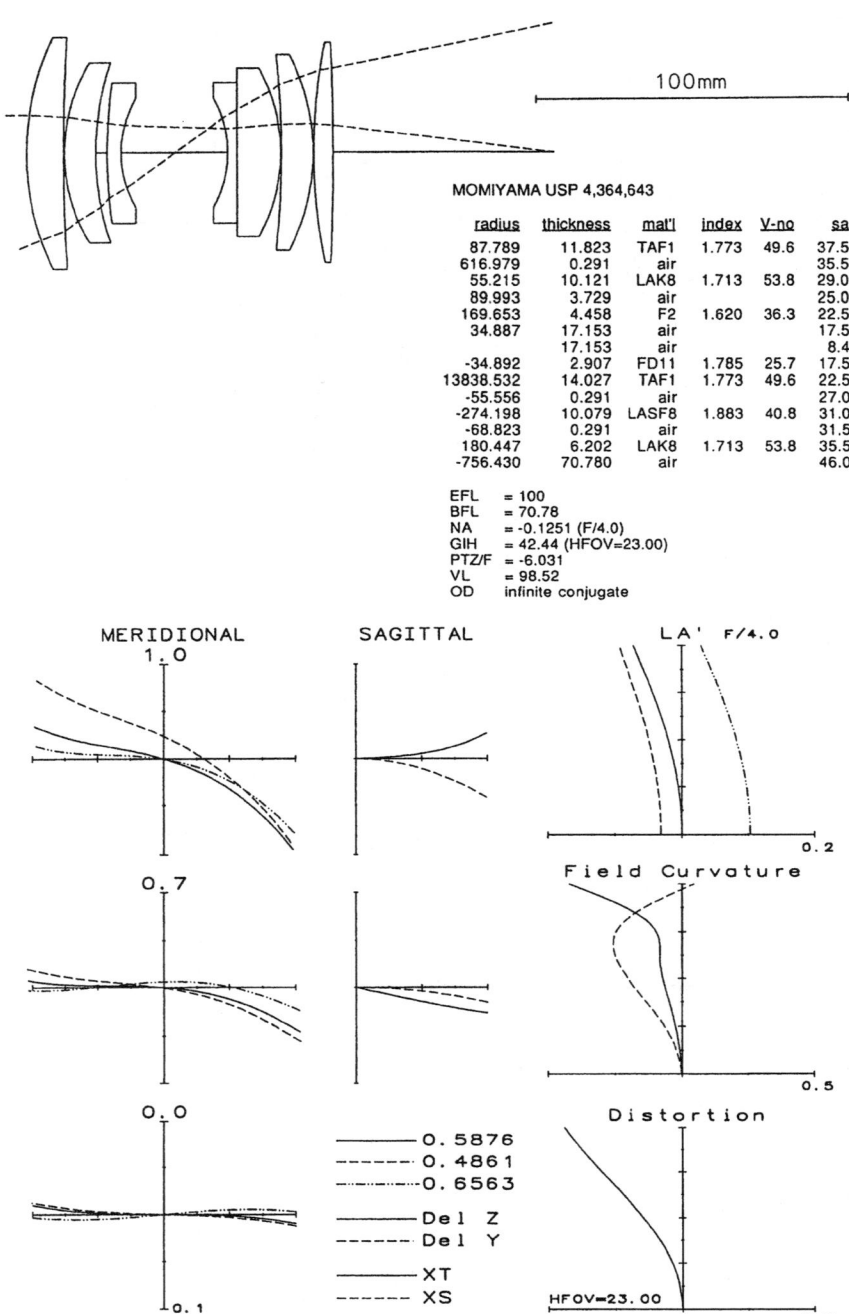

MOMIYAMA USP 4,364,643

radius	thickness	mat'l	index	V-no	sa
87.789	11.823	TAF1	1.773	49.6	37.5
616.979	0.291	air			35.5
55.215	10.121	LAK8	1.713	53.8	29.0
89.993	3.729	air			25.0
169.653	4.458	F2	1.620	36.3	22.5
34.887	17.153	air			17.5
	17.153	air			8.4
-34.892	2.907	FD11	1.785	25.7	17.5
13838.532	14.027	TAF1	1.773	49.6	22.5
-55.556	0.291	air			27.0
-274.198	10.079	LASF8	1.883	40.8	31.0
-68.823	0.291	air			31.5
180.447	6.202	LAK8	1.713	53.8	35.5
-756.430	70.780	air			46.0

EFL = 100
BFL = 70.78
NA   = -0.1251 (F/4.0)
GIH = 42.44 (HFOV=23.00)
PTZ/F = -6.031
VL   = 98.52
OD   infinite conjugate

Figure 12.18 $f/4.0$ ±23° a much slower version of Fig. 12.17.

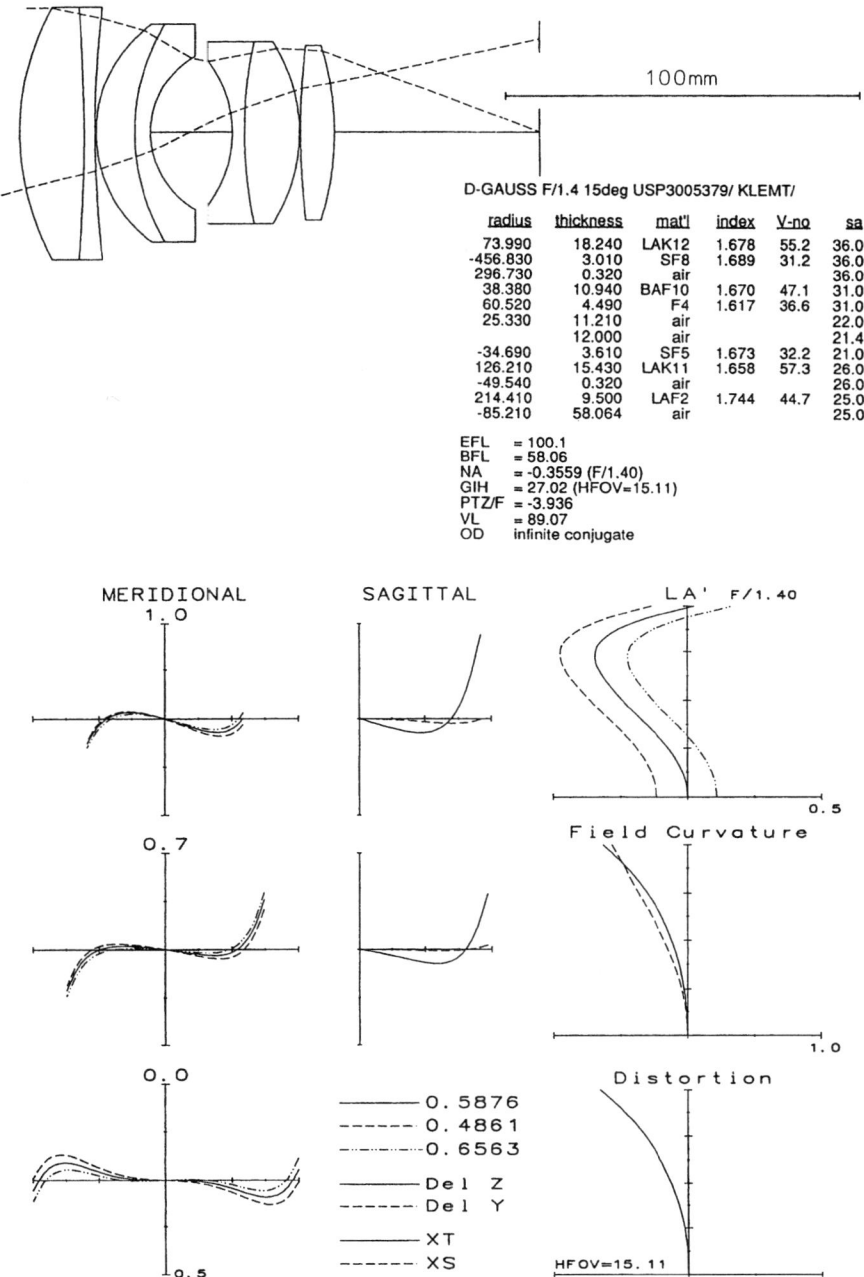

Figure 12.19 $f/1.4$ ±15° a double-Gauss with a compounded front element.

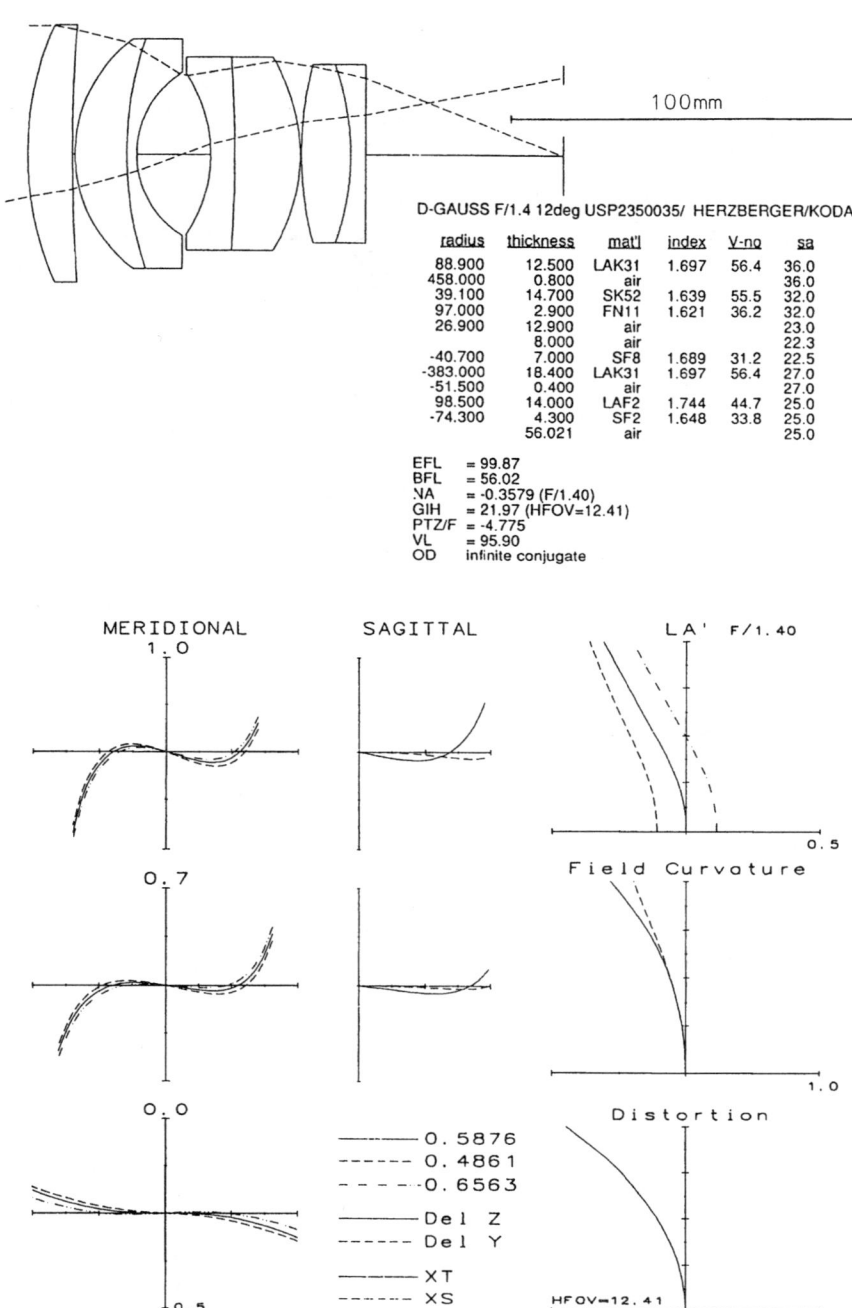

Figure 12.20 $f/1.4 \pm 12°$ a double-Gauss with a compounded rear element.

The Biotar or Double-Gauss Lens   345

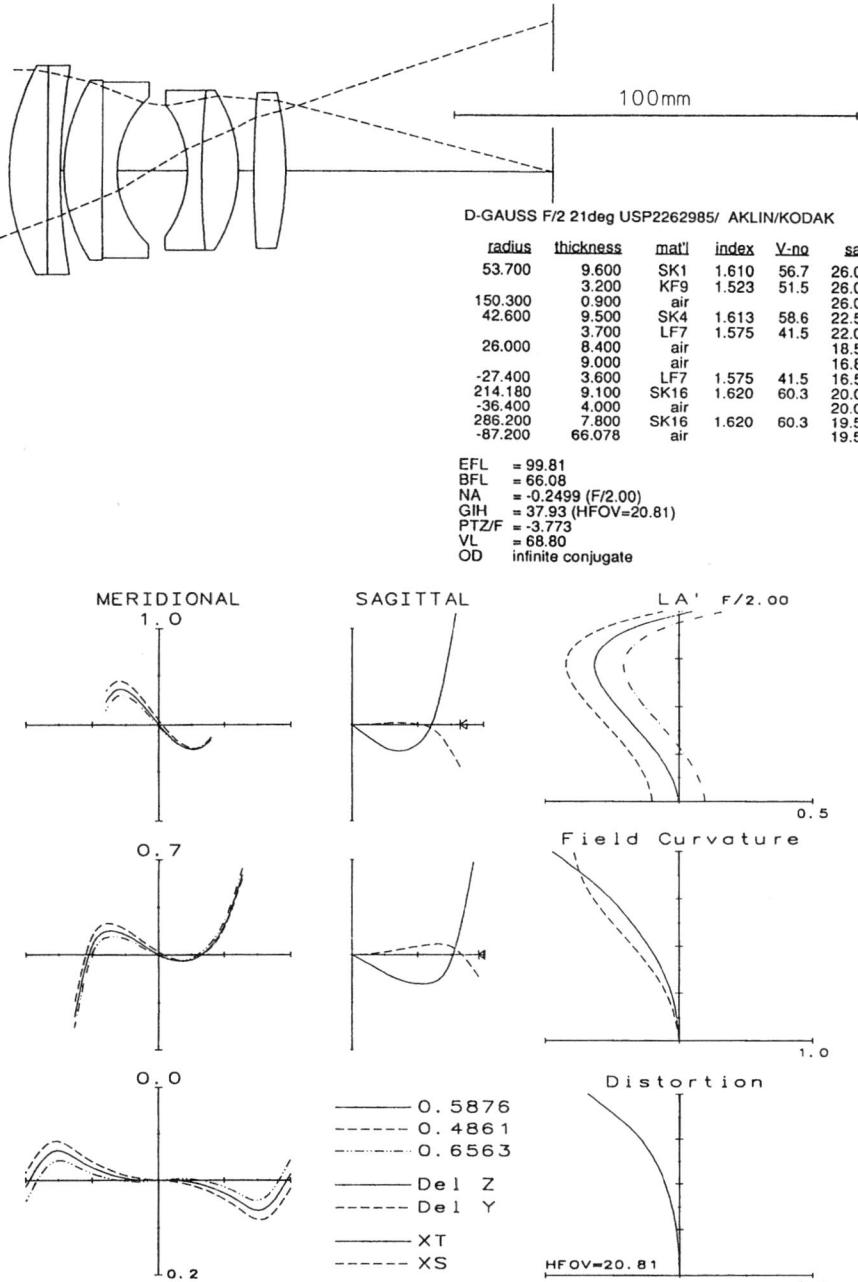

D-GAUSS F/2 21deg USP2262985/ AKLIN/KODAK

radius	thickness	mat'l	index	V-no	sa
53.700	9.600	SK1	1.610	56.7	26.0
	3.200	KF9	1.523	51.5	26.0
150.300	0.900	air			26.0
42.600	9.500	SK4	1.613	58.6	22.5
	3.700	LF7	1.575	41.5	22.0
26.000	8.400	air			18.5
	9.000	air			16.8
-27.400	3.600	LF7	1.575	41.5	16.5
214.180	9.100	SK16	1.620	60.3	20.0
-36.400	4.000	air			20.0
286.200	7.800	SK16	1.620	60.3	19.5
-87.200	66.078	air			19.5

EFL = 99.81
BFL = 66.08
NA = -0.2499 (F/2.00)
GIH = 37.93 (HFOV=20.81)
PTZ/F = -3.773
VL = 68.80
OD infinite conjugate

Figure 12.21 $f/2.0$ ±21° a double-Gauss with a compounded front element.

346   Chapter Twelve

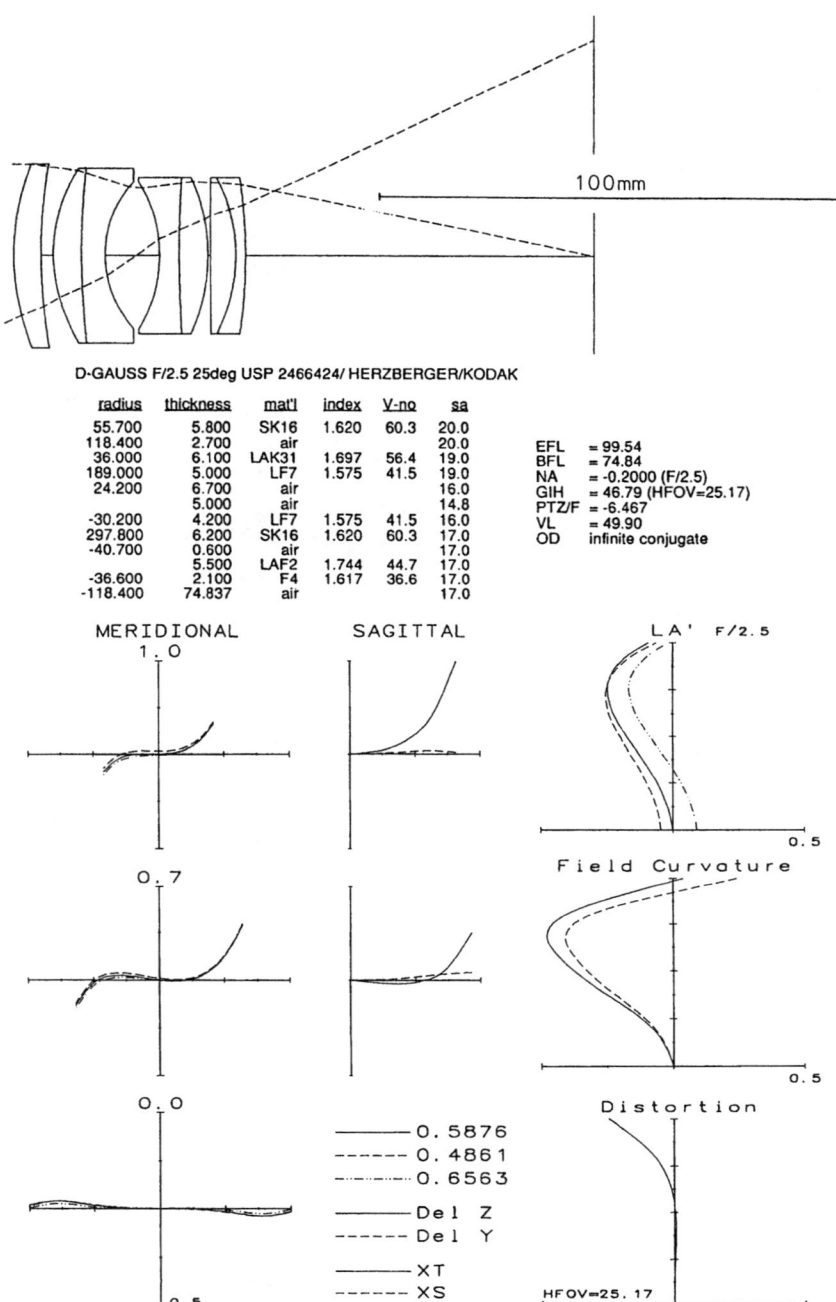

Figure 12.22 $f/2.5$ ±25° a double-Gauss with a compounded rear element.

The Biotar or Double-Gauss Lens 347

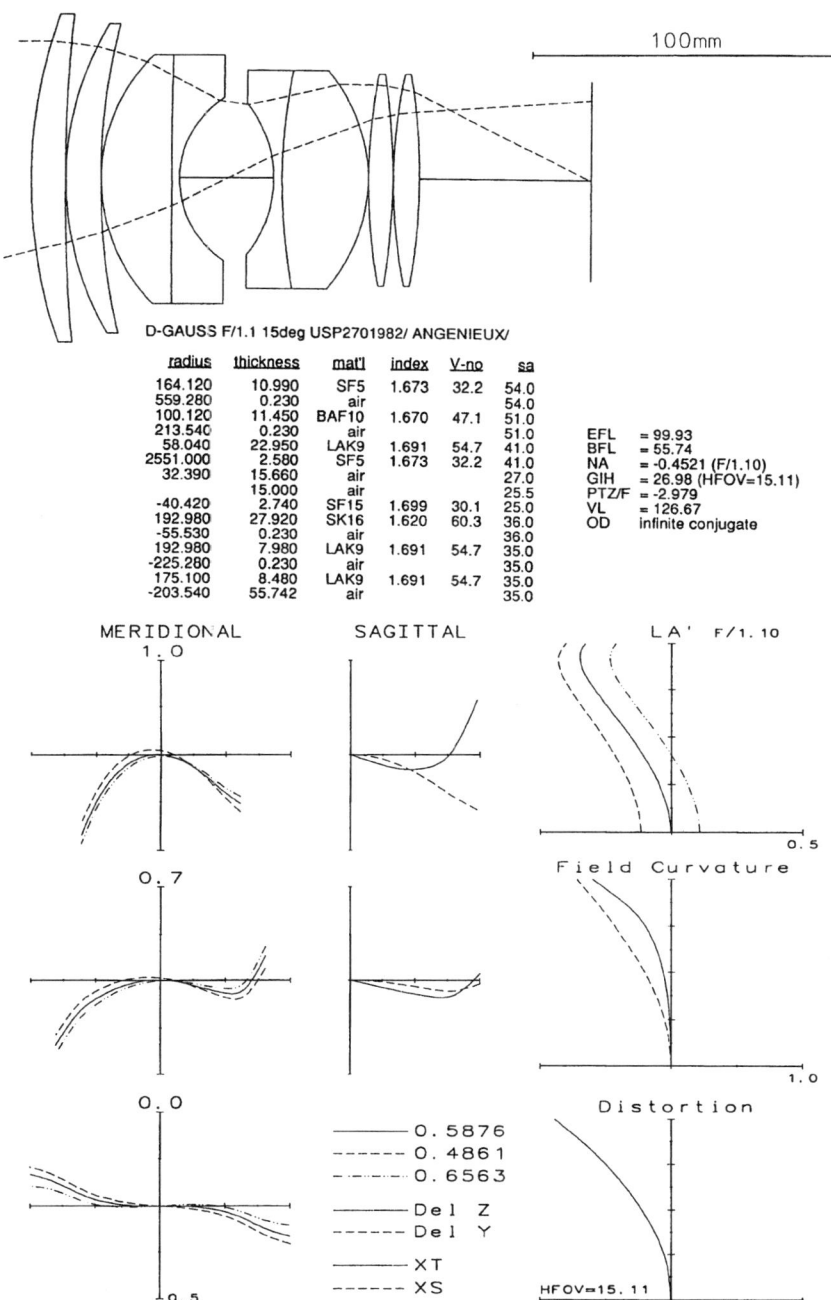

Figure 12.23 $f/1.1$ ±15° a very fast eight-element double-Gauss with both outer elements split.

348 Chapter Twelve

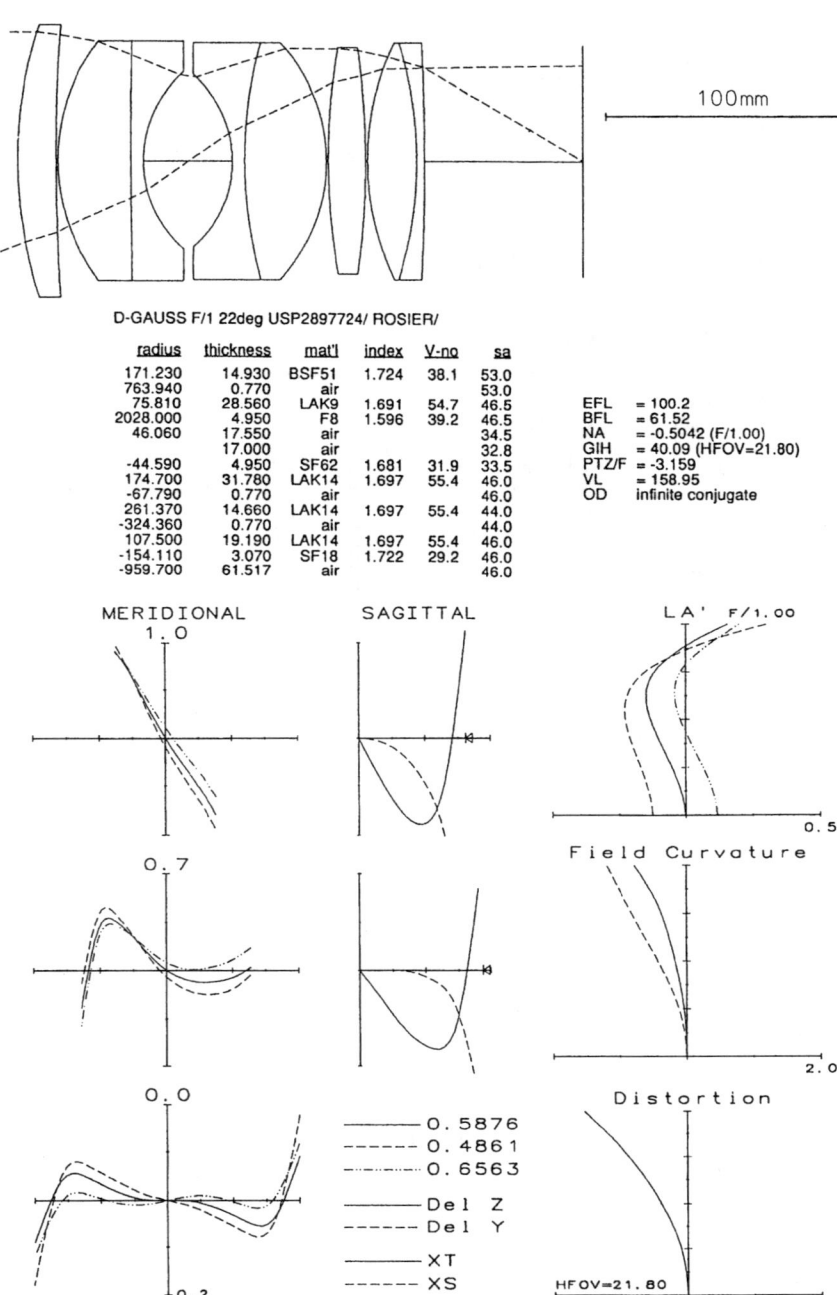

Figure 12.24 $f/1.0$ ±22° an extremely fast eight-element double-Gauss with the last element compounded.

The Biotar or Double-Gauss Lens   349

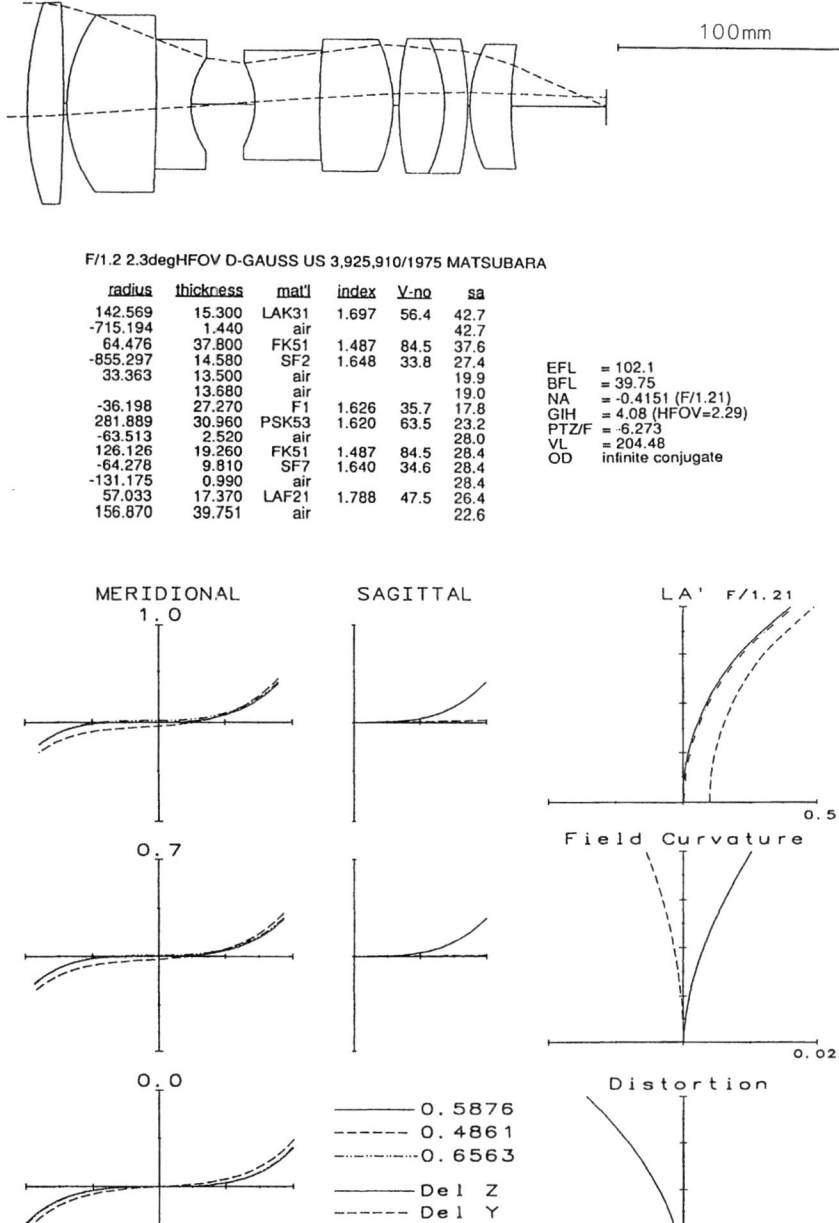

F/1.2 2.3degHFOV D-GAUSS US 3,925,910/1975 MATSUBARA

radius	thickness	mat'l	index	V-no	sa
142.569	15.300	LAK31	1.697	56.4	42.7
-715.194	1.440	air			42.7
64.476	37.800	FK51	1.487	84.5	37.6
-855.297	14.580	SF2	1.648	33.8	27.4
33.363	13.500	air			19.9
	13.680	air			19.0
-36.198	27.270	F1	1.626	35.7	17.8
281.889	30.960	PSK53	1.620	63.5	23.2
-63.513	2.520	air			28.0
126.126	19.260	FK51	1.487	84.5	28.4
-64.278	9.810	SF7	1.640	34.6	28.4
-131.175	0.990	air			28.4
57.033	17.370	LAF21	1.788	47.5	26.4
156.870	39.751	air			22.6

EFL = 102.1
BFL = 39.75
NA = -0.4151 (F/1.21)
GIH = 4.08 (HFOV=2.29)
PTZ/F = -6.273
VL = 204.48
OD infinite conjugate

Figure 12.25 $f/1.2$ $\pm 2.3°$ a fast, but very narrow field eight-element double-Gauss.

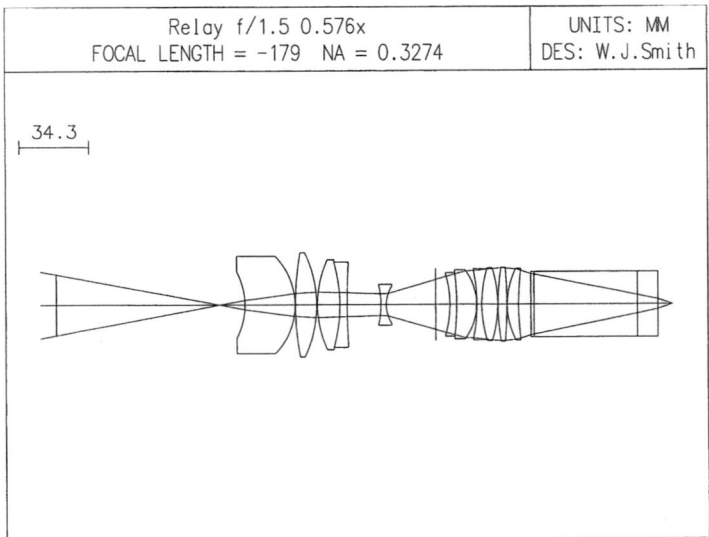

Figure 12.26 $f/1.5$ working f number 0.576x relay, Table 12.1.

high-speed lens, covers only a narrow angular field; it uses unusual partial-dispersion glasses such as FK51 and PSK53 to reduce the secondary spectrum.

12.7 A "Doubled Double-Gauss" Relay

Figure 12.26 is a 0.576x relay with a working f number of $f/1.5$ (NA = 0.33), designed for the purpose of allowing professional motion picture lenses to be used in a digital movie camera. The entrance pupil is 80 mm before the object plane to accommodate the exit pupil of the regular camera lenses, which were used wide open. The actual iris diaphragm is at surface 10. The final space after the lens is filled by a three-color, beam-splitting prism and its filters. The prescription is given in Table 12.1; Figs. 12.27 and 12.28 give the ray aberrations and the *modulation transfer function* (MTF) of the design.

Obviously the need for an external pupil and a working distance large enough to accommodate the RGB color prism were significant factors in the design. The design was begun with a doubled double-Gauss lens. The specified working f/number of $f/1.5$ combined with the 0.576x magnification required a system speed equivalent to a lens with an infinity f number of about $f/0.97$. It was felt that the specified image quality would require a fairly complex lens structure to meet the specs, and this

TABLE 12.1 Prescription for the NA 0.33 (Infinity f/0.97) 0.576 Relay Shown in Figs. 12.26 to 12.28

OBJECT	$T0 = 12.236$			
$R1 = -37.09$	$T1 = 24.43$	SF5	673322	EFL = -170.05
$R2 = -36.47$	$T2 = 0.20$			Mag = -0.576
$R3 = +151.846$	$T3 = 10.30$	LAK9	691547	Obj. Ht. = ± 13.62
$R4 = -63.69$	$T4 = 0.455$			GIH = 7.84
$R5 = +56.74$	$T5 = 10.80$	LAK9	691547	PTZ/F = -2.25
$R6 = -81.95$	$T6 = 3.50$	LAFN7	750349	NA = 0.33 (working f/1.5)
$R7 = +341.894$	$T7 = 16.639$			
$R8 = -47.12$	$T8 = 2.65$	SF8	689312	
$R9 = +18.222$	$T9 = 24.034$			
$R10$ = Stop	$T10 = 7.00$			
$R11 = -56.029$	$T11 = 3.30$	SF4	755276	
$R12 = -136.92$	$T12 = 9.75$	BASF51	724381	
$R13 = -29.82$	$T13 = 0.20$			
$R14 = -84.199$	$T14 = 2.35$	SF6	805254	
$R15 = +63.96$	$T15 = 7.95$	SK51	621603	
$R16 = -63.96$	$T16 = 0.20$			
$R17 = +117.862$	$T17 = 4.30$	LAK9	691547	
$R18 = -228.27$	$T18 = 0.20$			
$R19 = +41.386$	$T19 = 5.50$	LAK9	691547	
$R20 = +146.76$	$T20 = 6.0$			
$R21$ = plano	$T21 = 1.50$	PK50	521697	
$R22$ = plano	$T22 = 50.30$	SK3	609589	
$R23$ = plano	$T23 = 10.00$	BK7	517642	
$R24$ = plano	$T24 = 7.230$			

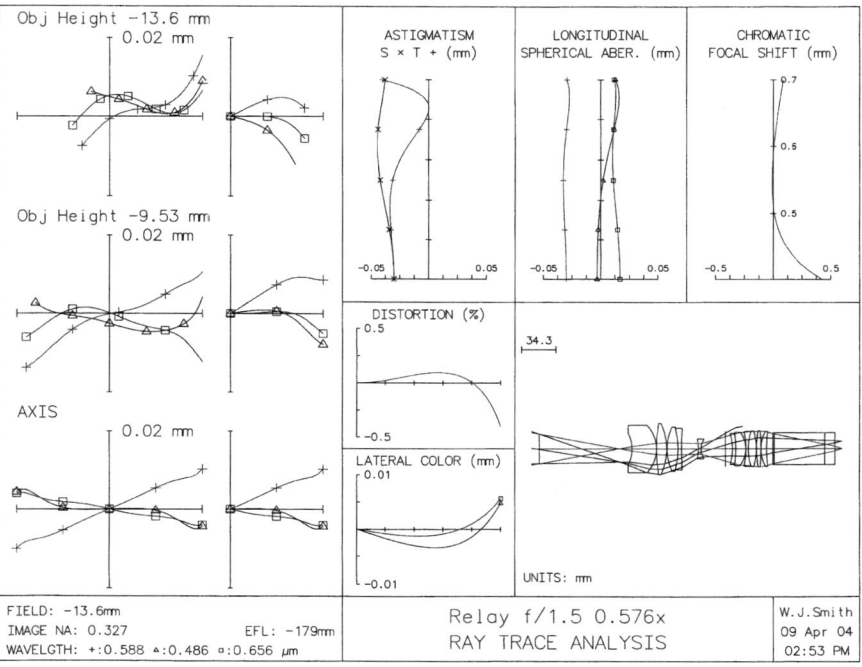

Figure 12.27 Ray aberrations for Fig. 12.26, Table 12.1.

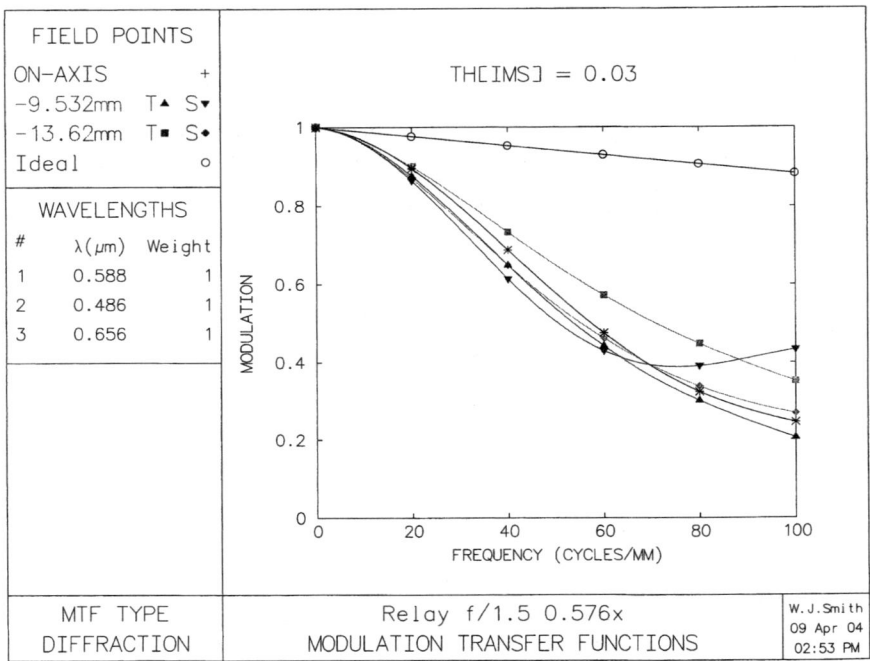

Figure 12.28 MTF vs. frequency for Fig. 12.26, Table 12.1.

form (initially) had 12 elements with which to work. The starting form was two positive singlets, two negative doublets followed by the stop, two more negative doublets, and two positive singlets, all followed by the prism.

The design was done with a fifty-two item merit function, which used calculated aberrations, weighted so as to approximate their wavefront deformations. (This merit function was a precursor of what OSLO describes as its default GENII expert merit function.) As the design progressed, the merit function weightings were adjusted frequently to put pressure on the problem areas as they popped up. The crown element of one of the doublets became weak enough to remove, and the shapes of the components drifted far away from the original double-Gauss configuration. Note that the final EFL is negative; this design is a good example of why, in a finite conjugate situation, the magnification and the total track length should be specified, but never the focal length.

The glasses were allowed to vary; occassionally they were reset to the nearest acceptable catalog glass for a few cycles and then were allowed to vary again. In the end, if an even better image quality had been required, glasses with unusual partial dispersions would have been necessary to further improve the design by reducing the secondary spectrum.

About ten times in the course of the design the program "bottomed out," reaching a local optimum. The escape from these optima was usually affected by making a radical change in one or more dimensions of the prescription and then reoptimizing (hopefully to a better value). The thick meniscus first element has its concave surface close enough to the object plane to act as a field flattener. This allowed the negative meniscus doublets to relax to more benign configurations; they actually became positive doublets. The negative singlet near the stop and the following optics can be regarded as a reverse telephoto, which is providing the long working distance needed to accommodate the thick color prism.

One reason that the design was successful is that enough space and volume were allowed so that a positive field lens, located at the initial image, was not required to control the size of the optics. Instead, the concave $R1$ surface acts as a negative field flattener.

Chapter 13

Telephoto Lenses

13.1 The Basic Telephoto

The arrangement shown in Fig. 13.1, with a positive component followed by a negative component, can produce a compact system with an effective focal length F that is longer than the overall length L of the lens. The ratio of L/F is called the *telephoto ratio*, and a lens for which this ratio is less than unity is classified as a telephoto lens. Typical ratios range from 0.6 to 0.85. The smaller the ratio, the more difficult the lens is to design. Note that many camera lenses sold as telephoto lenses are simply long-focal-length lenses and are not true telephotos.

Many of the comments in Chap. 14 regarding retrofocus or reverse telephoto lenses are equally applicable to the telephoto lens. The usual Petzval problem is with a backward-curving field, just as with the retrofocus, and the same glass choices are appropriate for the telephoto. Since the system is unsymmetrical, each component must be individually achromatized if both axial and lateral color are to be corrected. The aperture stop is usually at the front member or partway toward the rear. Since a telephoto lens usually covers only a relatively small angular field, coma, distortion, and lateral color (which in many other lenses are reduced by an approximate symmetry about the stop) are not as troublesome as they would be with a wider field. Distortion is often corrected by a strong collective interface in the rear (either a cemented surface or a positive power airspace).

The equations below may be used to determine the component powers for the telephoto (or for any two component system).

$$f_a = \frac{D \cdot F}{(F - B)} = \frac{D \cdot F}{(D + F[1 - K])} \tag{13.1}$$

$$f_b = -\frac{D \cdot B}{(F - B - D)} = \frac{D(D + K \cdot F)}{F(1 - K)} \tag{13.2}$$

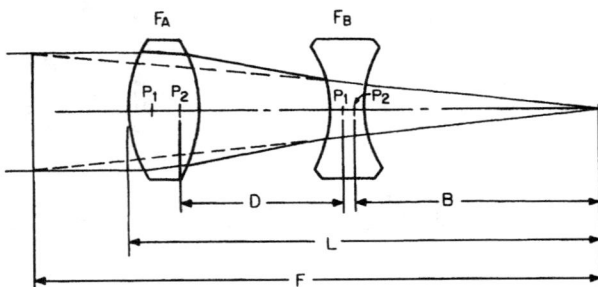

Figure 13.1 The basic power arrangement for a telephoto lens yields a compact lens with an overall length less than its effective focal length.

where f_a, f_b = focal lengths of the two components (see Fig. 13.1)
F = effective focal length of the combination
D = spacing between the components (principal points)
K = telephoto ratio

$$K = \frac{L}{F} = \frac{(D+B)}{F}$$

The minimum power for the second component ϕ_b occurs

$$\text{if } D = 0.5K \cdot F = \left[\frac{L}{2}\right]$$

The Petzval sum is approximately zero if $\phi_b = -\phi_a$ and

$$D = 0.5F(2K - 1 \pm \sqrt{4K - 3})$$

13.2 Close-up or Macro Lenses

The aberration correction of a long-focal-length unsymmetrical lens is usually quite sensitive to a change in object distance, and, for most telephoto lenses, the image quality deteriorates severely when they are focused on nearby objects. Note that this effect varies inversely with the object distance expressed in focal length units; i.e., for a given design type, the image quality may remain acceptable as long as the object distance exceeds some number of focal lengths. Thus, for a given object distance, this effect is more of a problem for a long-focal-length lens than for a short. Since retrofocus lenses tend to have short focal lengths, this problem is somewhat less frequently encountered, in spite of their asymmetry.

Many newer telephoto lenses and the specialized close-focusing lenses (called *macro* lenses) utilize a floating component or separately moving elements to maintain the aberration correction when the lens is focused at a close distance. For many lens designs, the spherical aberration and the astigmatism become undercorrected at close conjugates. Thus a relative motion of the elements to increase the marginal ray height on a negative (or overcorrecting) element/component can be used to stabilize the spherical. The astigmatism can be controlled by a motion that increases the height of the chief ray on a component that contributes overcorrected astigmatism, or reduces the chief ray height on an undercorrecting one.

The design of such a macro system is carried out just like the design of a zoom lens. Two (or more) configurations are set up, one with a long (perhaps infinite) object conjugate distance and the other with a short one. The computer then uses the same lens elements with different spacings for each configuration and optimizes the merit function for both configurations simultaneously.

One could allow all the airspaces to vary, one at a time, examine the results to see which spaces are most effective and degrade the image the least, then repeat the multiconfiguration optimization, possibly with a selected group of spaces, relating their motions with pickup solves and looking for synergies between the optical imagery and mechanical motions. Often anchoring one component and moving the other works out favorably. See Figs. 13.7 to 13.9.

Figures 21.3 to 21.6 and 20.5 show nontelephoto designs with macro features.

Less motion is required to focus if only the front (or only the rear) component is moved. If the exit pupil is not moved, the *numerical aperture* (NA) (f number) can remain the same through the focus adjustment. It would seem apparent that if the aberrations of the components are small, the image quality change produced by a spacing change to focus should also be small.

The aberrations of the front component, especially spherical, spherochromatic, and secondary spectrum, are magnified by the rear component. These must be corrected in the front component; it's very difficult to fix the aberrations of the front with the rear component.

Placing the stop at or near the front avoids the steeply sloped oblique rays at the image, which occur when the stop is at the rear of the lens.

13.3 Telephoto Designs

Figure 13.2 shows a very basic telephoto lens; it consists of just two cemented achromatic doublets, about as simple a construction as possible. Figure 13.2 covers less than 5° at $f/5.6$ and has a telephoto ratio of 0.85; it uses BK7 and F2 glasses and was done as an illustrative design exercise.

Figure 13.3 covers a 30° field at $f/4.5$ with excellent distortion correction, illustrating the benefits derived from the added degrees of freedom gained by splitting the cemented doublets into widely airspaced components. The large telephoto ratio of 0.91 and the modestly high-index glasses are also helpful.

Figures 13.4 illustrates the use of unusual partial dispersion glasses (as described in Chap. 6) to reduce the secondary spectrum. The term *superachromat* implies that at least four wavelengths are brought to a common focus, whereas the term *apochromat* indicates that three wavelengths are corrected. Notice, however, that the spherochromatism and zonal spherical aberration in this lens are much larger than the axial chromatic aberration. These are the aberrations that will determine the limiting performance of this lens; they can be reduced by splitting and spacing elements.

Figures 13.5 and 13.6 each have five elements and illustrate some of the different ways that the inherent capabilities of this configuration can be utilized. In Figs. 13.5 and 13.6, the crown element of the front doublet is split into two elements to reduce the zonal spherical aberration (among others). Figure 13.5 is the result of a classroom exercise that specified a 200-mm $f/5.0$ lens with a telephoto ratio of 0.80 for a 35-mm camera. It uses quite ordinary glasses and achieves an excellent level of performance. Figure 13.6 uses unusual partial dispersion glasses and breaks the contact in the front doublet to achieve what is potentially a high level of correction, although the telephoto ratio is only a modest 0.95.

Figure 13.7 uses seven elements to produce a well-corrected $f/5.6$ ±3° field lens with an extremely short telephoto ratio of 0.66. Notice the overcorrected Petzval field, with $p/f = +2.1$; this backward-curving field is one reason that small telephoto ratios are troublesome. This lens can be focused by shifting the negative doublet.

Figures 13.8 and 13.9 show an internal-focusing telephoto with a modest ratio of 0.92. The front component is fixed and the lens is focused for close-ups by moving the rear component toward the image plane. This could be considered as a sort of macro-style lens. Note the CaF_2 and FK5 glass in the front crowns used to reduce the secondary spectrum (although the spherochromatism certainly could stand some attention).

Telephoto Lenses

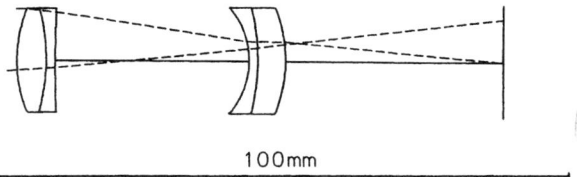

KINGSLAKE TELEPHOTO MODIFIED BY HOPKINS

radius	thickness	mat'l	index	V-no	sa
24.607	5.080	BK7	1.517	64.2	9.2
-36.347	1.600	F2	1.620	36.4	9.2
212.138	12.300	air			9.0
	21.699	air			6.7
-14.123	1.520	BK7	1.517	64.2	9.4
-38.904	4.800	F2	1.620	36.4	9.4
-25.814	37.934	air			9.4

EFL = 101.6
BFL = 37.93
NA = -0.0893 (F/5.6)
GIH = 7.44
PTZ/F = -19.38
VL = 47.00
OD infinite conjugate

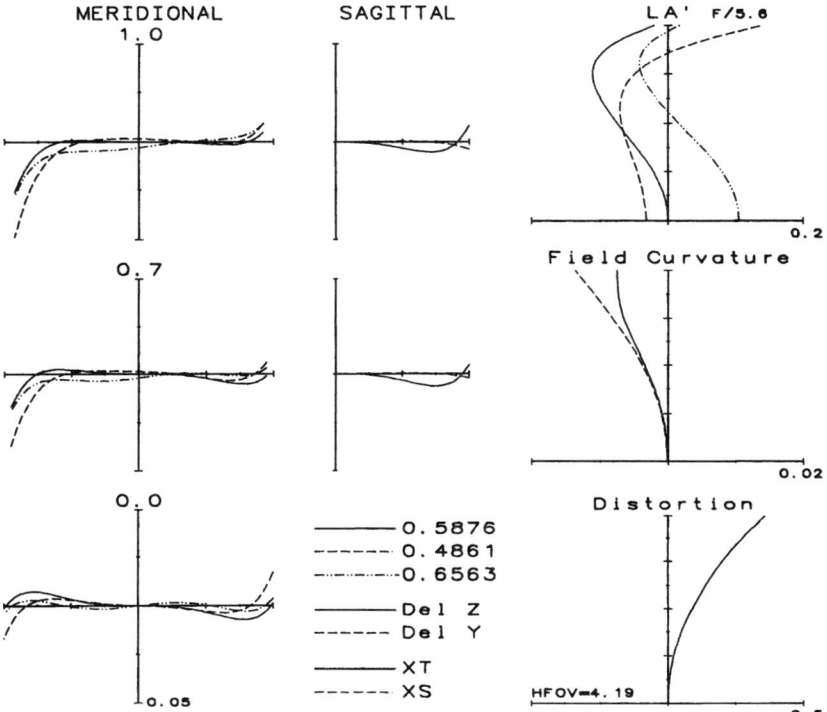

Figure 13.2 f/5.6 ±4.2° simple telephoto lens.

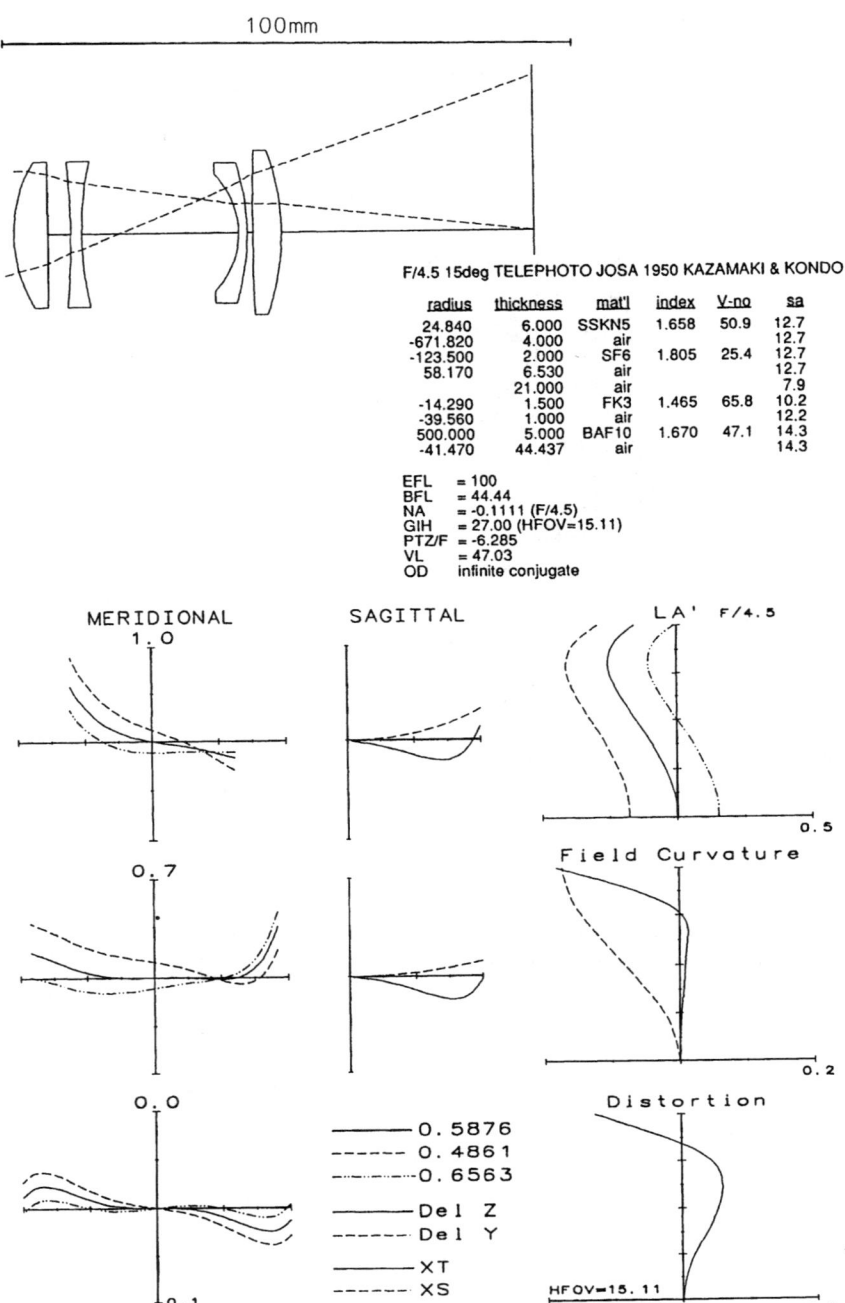

Figure 13.3 f/4.5 ±15° airspaced four-element telephoto lens.

Telephoto Lenses 361

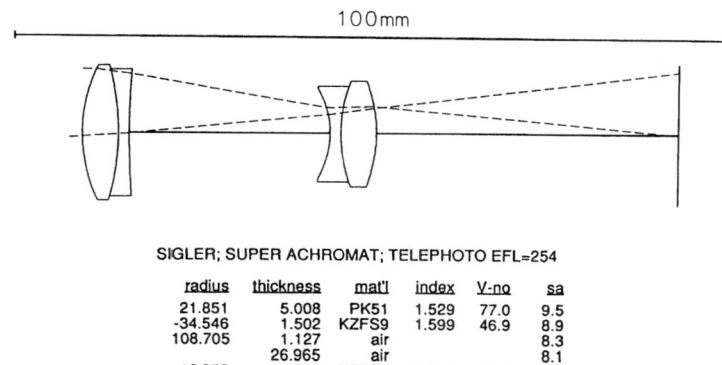

SIGLER; SUPER ACHROMAT; TELEPHOTO EFL=254

radius	thickness	mat'l	index	V-no	sa
21.851	5.008	PK51	1.529	77.0	9.5
-34.546	1.502	KZFS9	1.599	46.9	8.9
108.705	1.127	air			8.3
	26.965	air			8.1
-12.852	1.502	KZFS1	1.613	44.3	6.3
19.813	5.008	BASF5	1.603	42.5	6.7
-20.378	42.174	air			7.4

EFL = 100
BFL = 42.17
NA = -0.0898 (F/5.6)
GIH = 8.75
PTZ/F = -40.38
VL = 41.11
OD infinite conjugate

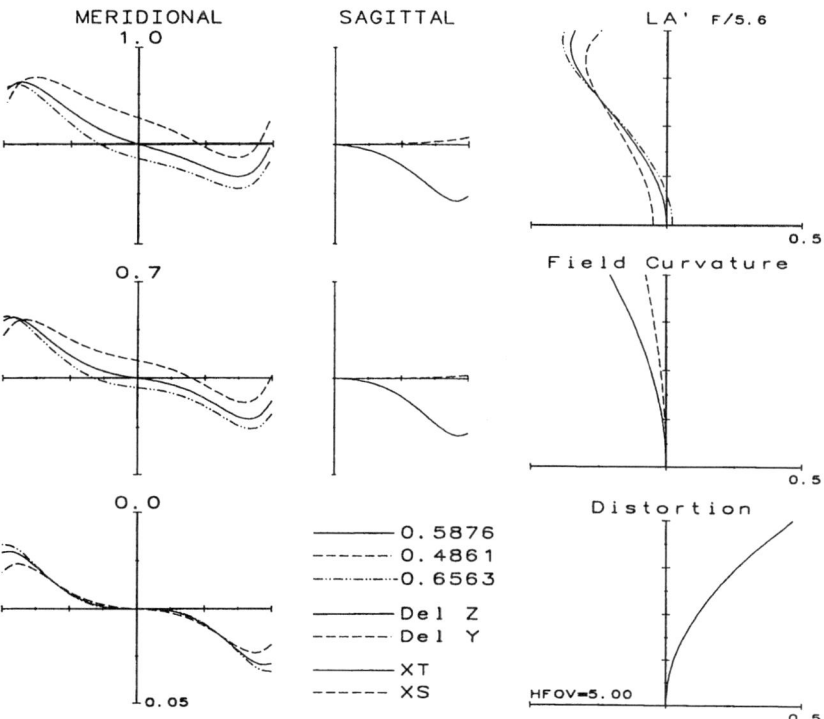

Figure 13.4 f/5.6 ±5° super achromat telephoto lens.

362 Chapter Thirteen

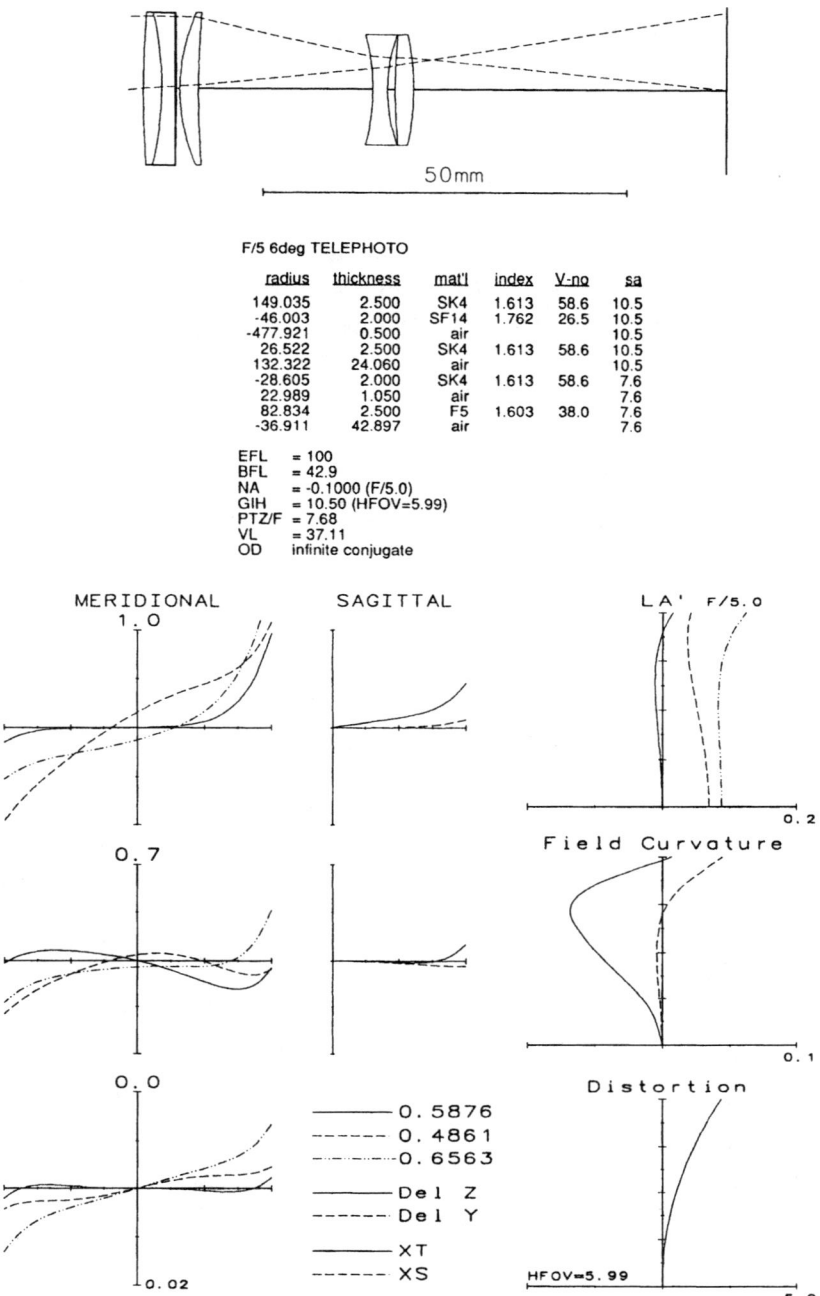

F/5 6deg TELEPHOTO

radius	thickness	mat'l	index	V-no	sa
149.035	2.500	SK4	1.613	58.6	10.5
-46.003	2.000	SF14	1.762	26.5	10.5
-477.921	0.500	air			10.5
26.522	2.500	SK4	1.613	58.6	10.5
132.322	24.060	air			10.5
-28.605	2.000	SK4	1.613	58.6	7.6
22.989	1.050	air			7.6
82.834	2.500	F5	1.603	38.0	7.6
-36.911	42.897	air			7.6

EFL = 100
BFL = 42.9
NA = -0.1000 (F/5.0)
GIH = 10.50 (HFOV=5.99)
PTZ/F = 7.68
VL = 37.11
OD infinite conjugate

Figure 13.5 $f/5.0$ ±6° five-element telephoto lens.

Telephoto Lenses 363

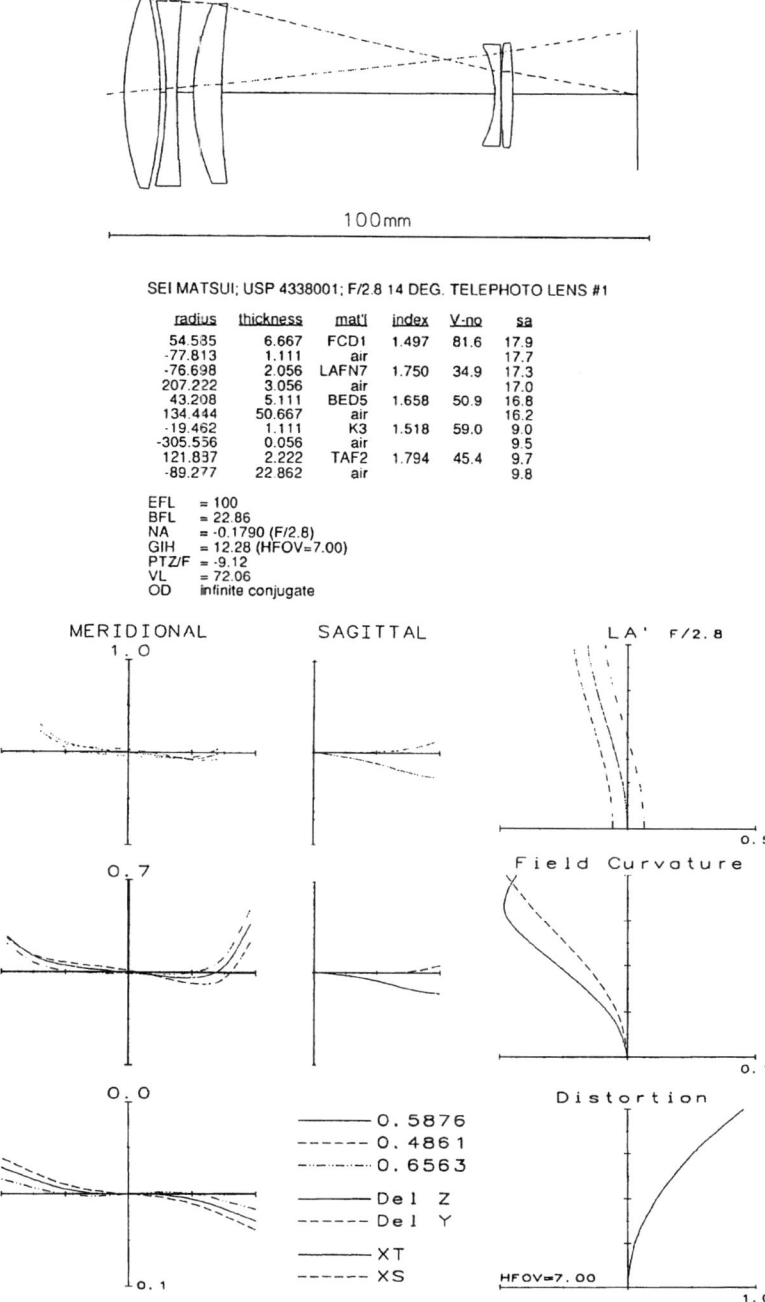

SEI MATSUI; USP 4338001; F/2.8 14 DEG. TELEPHOTO LENS #1

radius	thickness	mat'l	index	V-no	sa
54.535	6.667	FCD1	1.497	81.6	17.9
-77.813	1.111	air			17.7
-76.698	2.056	LAFN7	1.750	34.9	17.3
207.222	3.056	air			17.0
43.208	5.111	BED5	1.658	50.9	16.8
134.444	50.667	air			16.2
-19.462	1.111	K3	1.518	59.0	9.0
-305.556	0.056	air			9.5
121.837	2.222	TAF2	1.794	45.4	9.7
-89.277	22.862	air			9.8

EFL = 100
BFL = 22.86
NA = -0.1790 (F/2.8)
GIH = 12.28 (HFOV=7.00)
PTZ/F = -9.12
VL = 72.06
OD infinite conjugate

Figure 13.6 $f/2.8$ ±7° five-element telephoto lens.

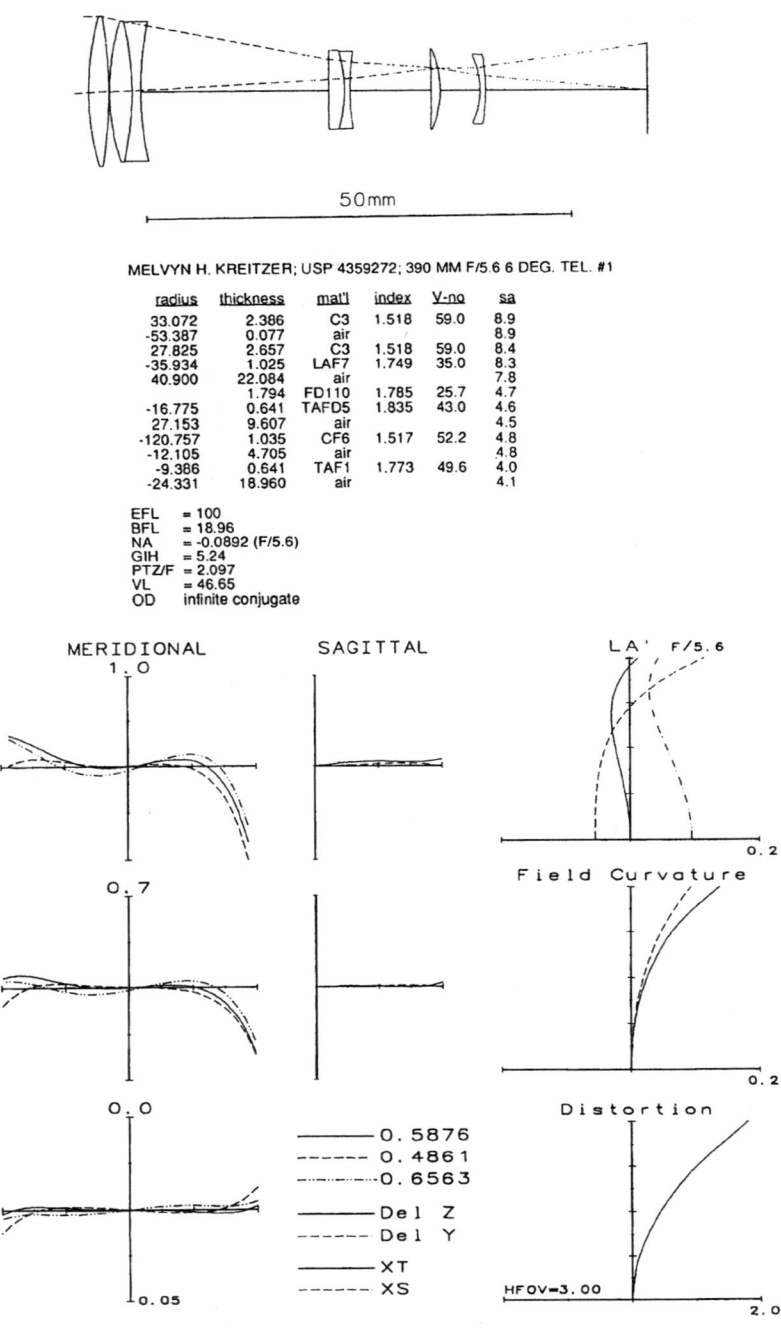

Figure 13.7 f/5.6 ±3° seven-element telephoto lens.

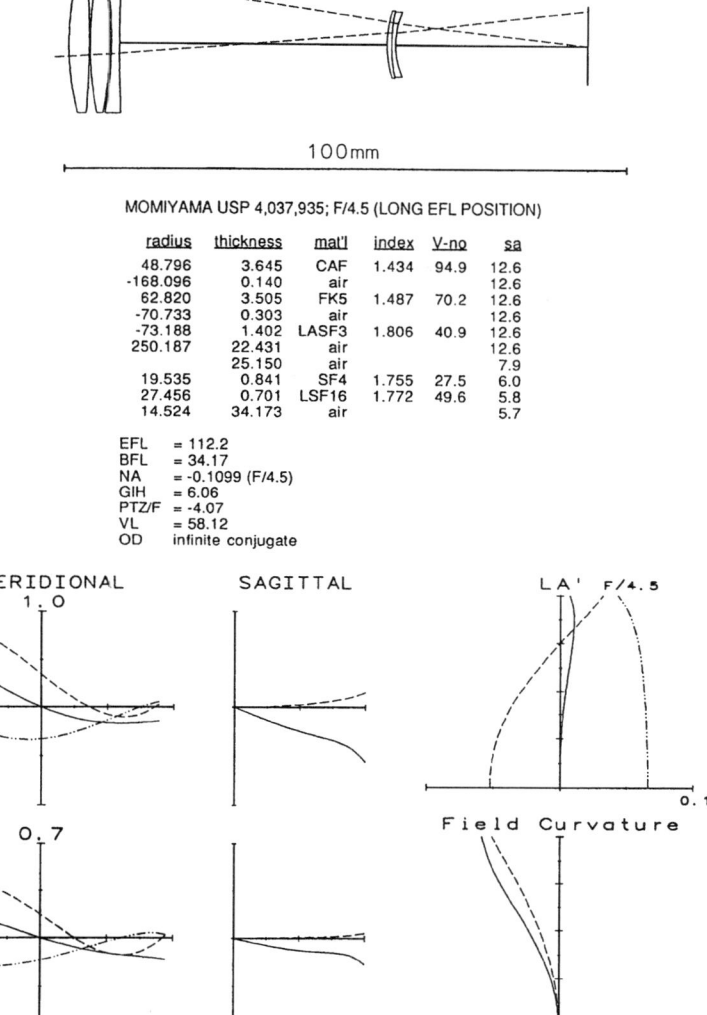

Figure 13.8 f/4.5 ±3.1° five-element internal focusing telephoto, at infinity.

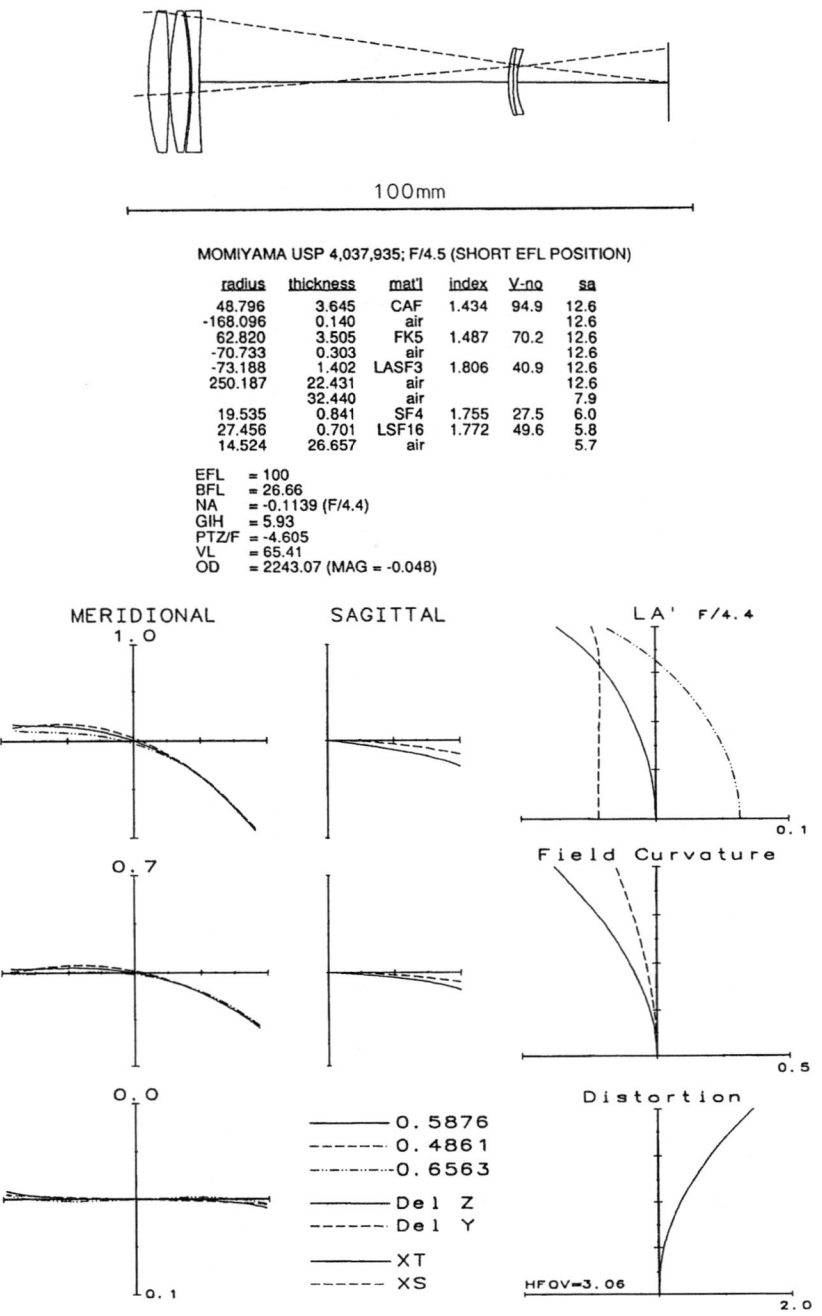

Figure 13.9 f/4.5 ±3.1° five-element internal focusing telephoto, focused at 22 focal lengths.

13.4 Design of a 200-mm *f*/4 Telephoto for a 35-mm Camera "from Scratch"

Our design task is a 200-mm focal length telephoto lens for a 35-mm camera, with a speed of *f*/4, a telephoto ratio of 0.8, and zero vignetting. The telephoto ratio of 0.8 means that the lens length plus its back focus is equal to $0.8 \times 200 = 160$ mm. The 35-mm frame size is 24 mm by 36 mm, with a diagonal of 42.3 mm; thus with a 200-mm focal length, the half-field angle is $\pm 6.2°$.

We start our layout with two thin components; their powers are given by Eqs. (13.1) and (13.2).

$$\phi_a = \frac{(F-B)}{DF}$$

$$\phi_b = \frac{(B+D)-F}{DB}$$

where F is the desired focal length, B is the back focal length, and D is the spacing between the a and b components. We want $F = 200$ mm and $(B + D) = 160$ mm, but to get a solution we must arbitrarily select a value for either B or D. Our selection will determine the Petzval curvature of the lens.

The Petzval sum of published telephoto designs varies wildly. In a literature survey of a dozen designs, the Petzval radius varied from $-40F$ to $+17F$, with the shortest radii at $-4F$ and $+2F$. At this stage we might just as well settle on a zero Petzval sum, which means that the two components of our telephoto will have equal, but oppositely signed, focal lengths; in other words $\phi_a = -\phi_b$. If we now solve for the powers, we find that $f_a = -f_b = (F - B)$ and $D = (F - B)^2/F$.

With the telephoto ratio set at 0.8, and after a bit of algebra, we solve a quadratic equation in B, and find that the component focal lengths are either $\pm 0.276F$ or $\pm 0.724F$, and the spacing $D = 0.524F$ or $0.076F$.

Obviously we want the weaker, longer solution. So for a 200-mm focal length we have

$$f_a = 144.7 \text{ mm}$$

$$D = 104.7 \text{ mm}$$

$$f_b = -144.7 \text{ mm}$$

$$B = 55.3 \text{ mm}$$

Our next step is to select the glass types. The usual choice in most anastigmats (e.g., the Cooke triplet, the Tessar, and the like) is to use high-index, high-V-value crown glasses for the positive elements, and low-V-value (i.e., high-dispersion) flint glasses from along the glass line for the negative elements. This arrangement holds for our positive front component, but is of course reversed for the negative rear component. Because a telephoto lens has absolutely no front-to-back symmetry to correct the lateral color, both components must be independently achromatized to get full color correction.

Again, our survey of published designs indicates that BK7 (517642) and SK16 (620603) are common choices for the positive element of the front component, and SF1 (717295) for the negative element. (Our survey is skewed a bit by designs with low-index crown glasses, which obviously were chosen to reduce secondary spectrum.) There is less unanimity for the rear component glasses. A representative choice might be SK16 combined with an ordinary flint. For simplicity we elect to use SK16 (620603) and SF1 (717295) in both components.

To achromatize we set the element powers using Eqs. (6.1) and (6.2).

$$\phi_i = \frac{V_i}{f(V_i - V_j)} = \frac{60.3}{f(60.3 - 29.5)} = \frac{1.958}{f}$$

$$\phi_j = \frac{V_j}{f(V_j - V_i)} = \frac{29.5}{f(29.5 - 60.3)} = -\frac{0.958}{f}$$

and since the element curvature $c = \phi/(n - 1)$, we get:

$$c_i = \frac{+3.158}{f} \quad \text{and} \quad c_j = \frac{-1.336}{f}$$

where the subscripts i and j indicate the elements of the components, V_i and V_j are the V-values, and f is the component focal length.

A few paragraphs above we arrived at focal lengths of ±144.7 mm for the front and rear components, respectively. Thus the curvatures and the radii (assuming equi-convex and equi-concave shapes) for the four elements are shown in the following table.

$c = +0.021824$ (SK16)	$R1 = +91.6$	$R2 = -91.6$
$c = -0.009233$ (SF1)	$R3 = -216.6$	$R4 = +216.6$
$c = +0.009233$ (SF1)	$R5 = +216.6$	$R6 = -216.6$
$c = -0.021824$ (SK16)	$R7 = -91.6$	$R8 = +91.6$

The Petzval radius is +26,291 and the axial and lateral chromatic are both less than 1.0e − 03, indicating the success of our preliminary layout work. We draw the system on the monitor screen and add the appropriate surface spacings to accomodate the 50-mm beam diameter and the ±6.2 field°. The spacings are, in order: 9.0, 0.1, 4.0, 104.7, 5.0, 0.1, and 3.0, and the back focus is 33.6. The airspace is now adjusted to make the telephoto ratio exactly 0.8; the airspace is now 103.4065 and the back focus is 35.3935.

Before we start optimizing this, we decide it would be interesting to see how well the software would deal with this part of the problem on its own. So we start with the elements as plano-plano plates (usually a rather stupid move) using the glasses and spacings we have chosen, and allow the central airspace and one radius of each of the SK16 (thin) elements to vary. We target the focal length, the telephoto ratio, and the Petzval sum. The optimization blows sky high, yielding a 160-mm airspace, a negligible back focus, a focal length of 1.1e − 16, and very short radii (≈1e − 07). Undaunted, we next try varying one radius in each element, and add both axial and lateral chromatic aberration targets to the merit function. Our merit function now includes: the focal length, the telephoto ratio, and both chromatic aberrations and the Petzval sum. Happily, this optimization is successful and yields the same thin lens system that we derived above. Some days, virtue triumphs!

Now we create a merit function using the first-, third-, and fifth-order aberration coefficients, plus the telephoto ratio (i.e., we target the overall length plus the back focus at 160 mm). We put the aperture stop on surface no. 4, an angle solve of −0.125 on surface no. 8 to hold the focal length at 200 mm and the speed at f/4.0, and vary all the curvatures plus thickness no. 4 (the central airspace). We save this starting lens in case something goes sour when we optimize. It is worth noting here that by begining with a first-, third-, and fifth-order merit function we avoid a problem common in starting almost any design: the optimization may fail due to the trigonometric rays missing a surface or encountering total internal reflection. Using aberration coefficients avoids this problem because no real rays are traced; only paraxial rays (which do not fail) are used for first-, third-, and fifth-order calculations.

We optimize for 10 cycles and adjust the spacings as necessary ($T1 = 12$), and put edge contact solves on surfaces no. 2 and no. 6. After another 20 cycles the process has settled down; the worst aberrations are spherical ($TA_m \approx +1.5$) and field curvature ($X_s \approx X_t \approx +0.4$). Not too bad for a third- and fifth-order effort. But resolution is only a few *line pairs millimeter* (lpm). Again, we cautiously save this result.

Now we switch to the software's default rms-spot-size merit function, opting for the $(D - d)$ δn chromatic correction and also for distortion control at 1 percent. We add 160-mm length control to hold the telephoto ratio at 0.8. We continue to hold the focal length with an angle solve on the last surface. We accept the default ray set, thereby assuming zero vignetting. After about 70 cycles of optimization (i.e., clicking on iterate seven times)

TABLE 13.1 Prescription for the First Optimized Design of the 200-mm, f/4, Telephoto Lens (Using the rms Spot size Merit Function). See Figs. 13.10 and 13.11

R1 = +57.209	T1 = 12.0	SK16	620603
R2 = −168.09	T2 = 0.3072		
R3 = −144.65	T3 = 4.0	SF1	717295
R4 = +242.37	T4 = 71.0845		
R5 = +61.576	T5 = 5.0	SF1	717295
R6 = +127.94	T6 = 5.3880		
R7 = −41.421	T7 = 3.0	SK16	620603
R8 = +5144.5	BFL = 59.22		

we have a design that resolves about 20 lpm as shown in Table 13.1 and Figs. 13.10 and 13.11.

Wave front (rms-OPD) optimization usually produces a better *modulation transfer function* (MTF) than rms-spot-size optimization does, so we switch to the default rms-OPD (*optical path difference*) merit function, opting for three wavelengths (instead of $D - d$ for chromatic) and keeping

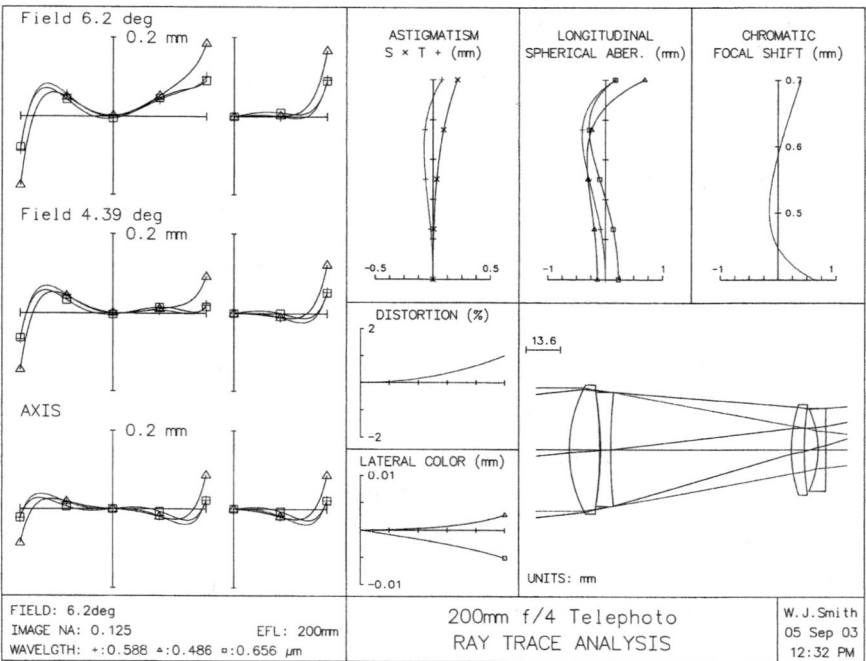

Figure 13.10 200-mm f/4.0 ±6.2° ray aberrations for the first design in the telephoto design story, Table 13.1.

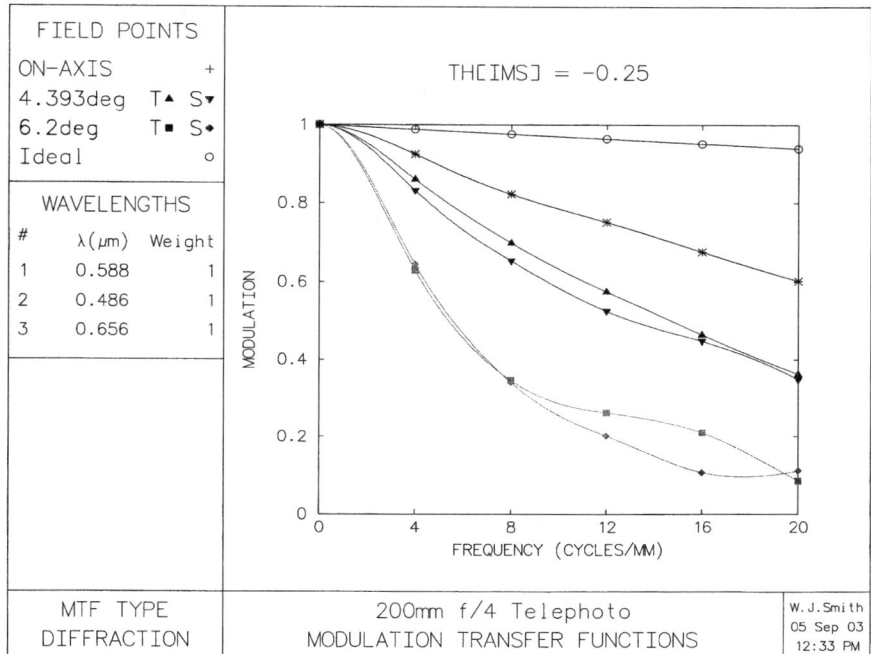

Figure 13.11 MTF for Fig. 13.10, Table 13.1 lens.

the distortion control. The result is an improvement in the MTF at 20 lpm, but the sagittal field is still significantly overcorrected at a bit less than +0.2 mm backward curving (as is the Petzval at about +0.3). The design and performance are shown in Table 13.2 and Figs. 13.12 and 13.13. Note the increased frequency scale in Fig. 13.13 versus Fig. 13.11.

Targets for X_s and X_t, heavily weighted at 100x, are now added to the merit function. This produces a significant improvement in the off-axis MTF, and the resolution is pushed out to about 40 lpm. The design doesn't

TABLE 13.2 The 200-mm f/4 Telephoto Lens After Optimization with an rms OPD Merit Function (See Figs. 13.12 and 13.13)

$R1 = +61.160$	$T1 = 12.0$	SK16	620603
$R2 = -139.53$	$T2 = 0.2418$		
$R3 = -126.27$	$T3 = 4$	SF1	717295
$R4 = +363.67$	$T4 = 74.4963$		
$R5 = +131.91$	$T5 = 5.0$	SF1	717295
$R6 = -425.87$	$T6 = 3.5899$		
$R7 = -42.786$	$T7 = 3.0$	SK16	620603
$R8 = +410.25$	BFL = 57.74		

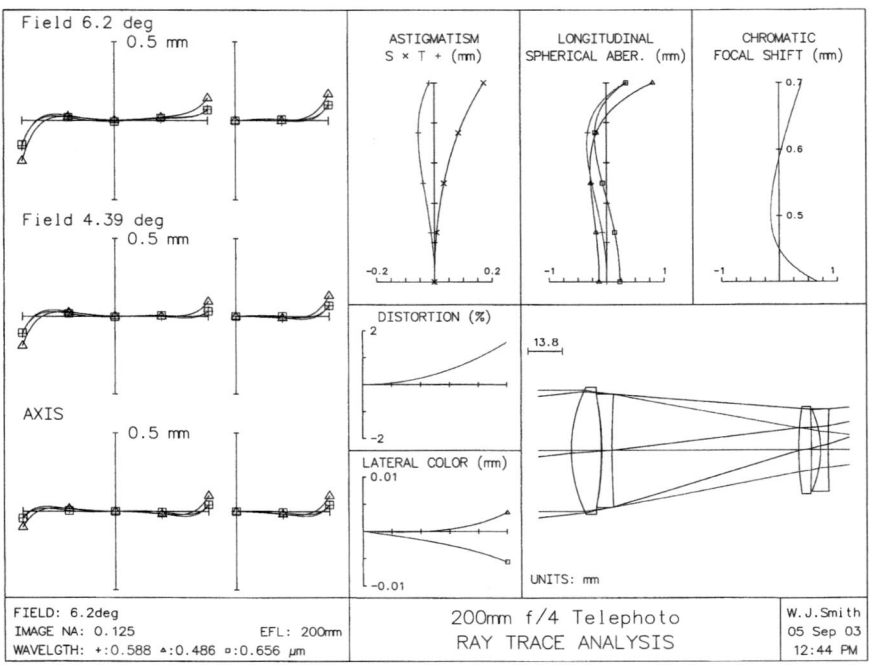

Figure 13.12 Ray aberrations for the second design, Table 13.2.

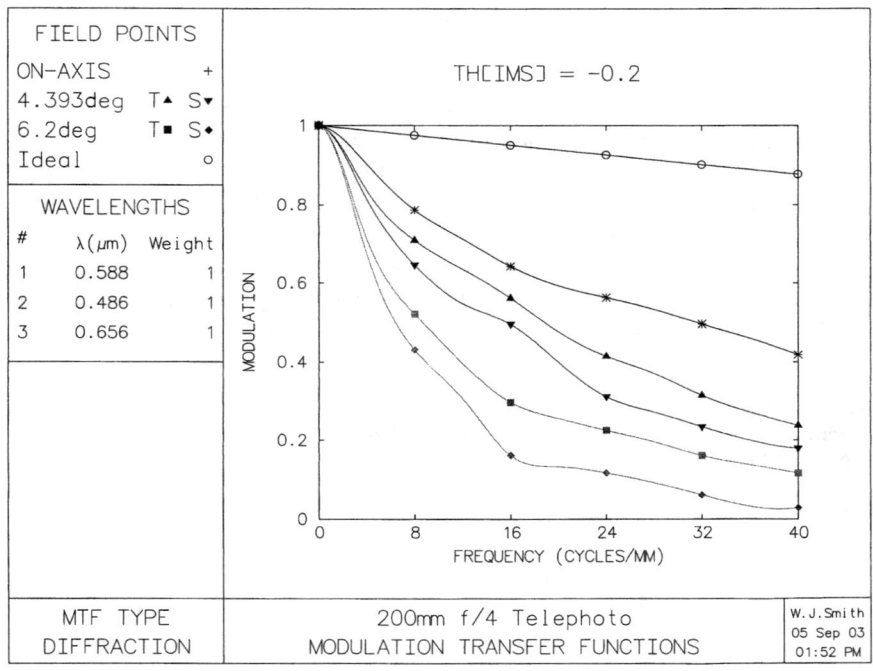

Figure 13.13 MTF for Fig. 13.12, Table 13.2 lens.

TABLE 13.3 The Prescription for the 200-mm f/4 Telephoto Lens Resulting from Increasing the Weight on the X_s and X_t Field Curvature Operands in the Merit Function (See Figs. 13.14 and 13.15)

$R1 = +54.124$	$T1 = 12.0$	SK16	620603
$R2 = -165.23$	$T2 = 0.2841$		
$R3 = -144.02$	$T3 = 4.0$	SF1	717295
$R4 = +219.55$	$T4 = 62.0250$		
$R5 = +191.17$	$T5 = 5.0$	SF1	717295
$R6 = -233.26$	$T6 = 3.6663$		
$R7 = -39.321$	$T7 = 3.0$	SK16	620603
$R8 = -546.36$	BFL = 70.24		

look too bad. There's really nothing wildly outrageous about the design. The front crown is strong; the rear negative element is strong and makes a big air lens between it and the positive element. The two flint glass elements are quite reasonably shaped. The aberrations and MTF are shown in Table 13.3 and Figs. 13.14 and 13.15, and the lens data is in Table 13.3.

Now the question before the house is: "What do we do next?" There are several immediately obvious and simple options:

a. Start over, manually controlling the form of the lens (to a different form if we have one that we feel will be better).

b. Reverse the order of the elements in the doublets.

c. Vary the flint glasses along the glass line.

d. Use higher-index glasses all around.

e. Add more elements, probably by splitting one or both of the two stronger elements.

f. Advance the art by considering a faster lens or a longer focal length. Either option will require significantly better correction than we have achieved to date.

g. The performance can be improved by allowing the telephoto ratio to increase. Even a small increase (say to 0.85 from our targeted 0.80) will allow a noticeable improvement.

13.4.1 Restart

Starting over is always an option in any design project. In the case at hand our system is so simple and minimal that there is little room for change. The big airspace in the rear doublet looks suspicious. But we've hardly scratched the surface, so this option seems a bit premature at this point. Any restart should include a literature study for possible alternate design types.

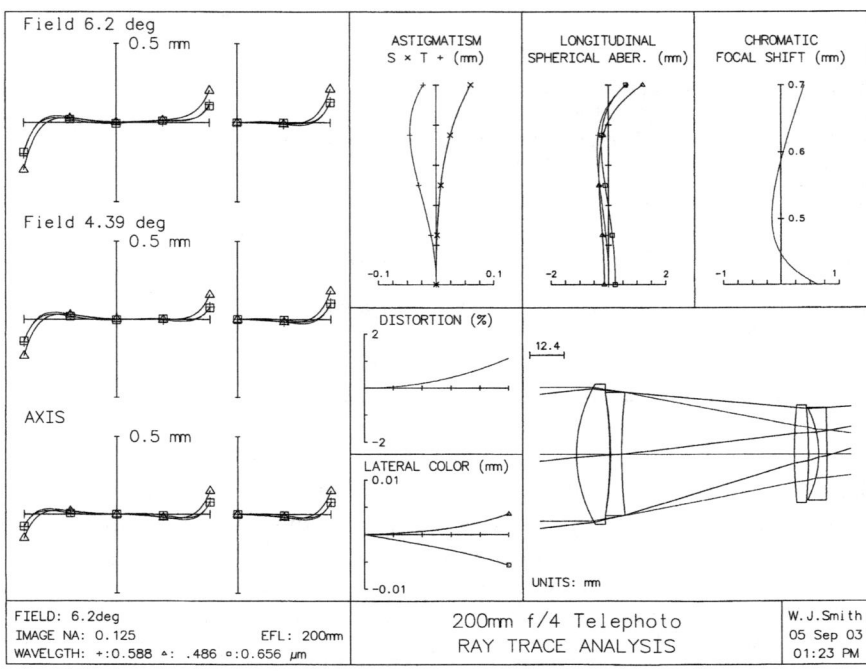

Figure 13.14 Ray aberrations for the third design, Table 13.3.

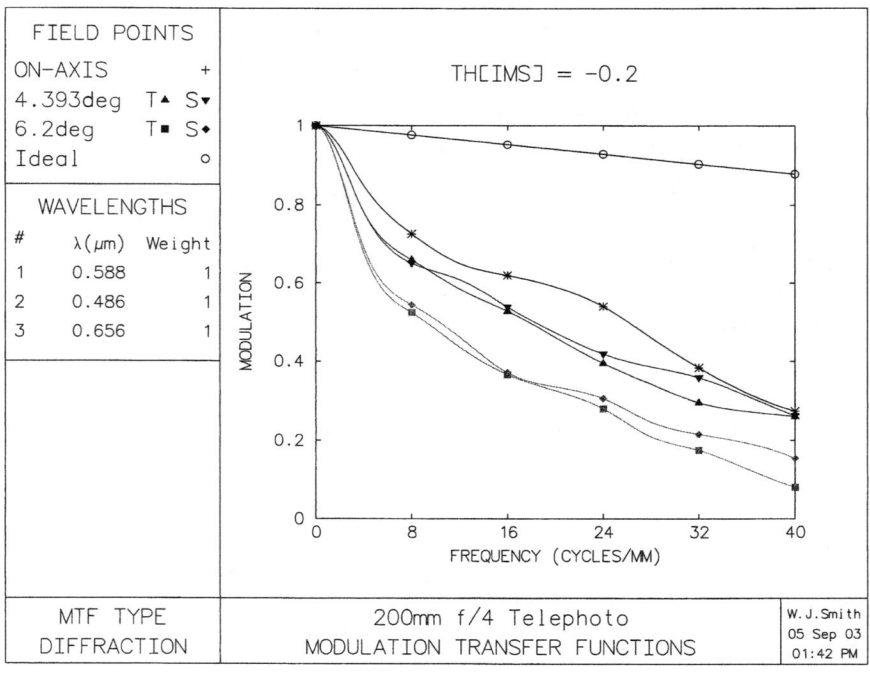

Figure 13.15 MTF for Fig. 13.14, Table 13.3.

13.4.2 Reverse elements

The front doublet is probably best left alone. Reversing it will increase both the diameter of the crown element and the surface curvatures. Since the Fraunhofer and Steinheil doublets (crown-in-front and flint-in-front, respectively, Figs. 6.2 and 6.3) are quite similar in performance, we don't see much to be gained here. But the rear doublet is another story; that wild airspace begs for attention.

So we reverse the rear doublet by simply turning it around, and iterate about a dozen times. The optimized result is quite poor. There is now more high-order spherical and a bigger "belly" of X_t in the zonal field. The rear doublet is still edge contacted with a large space. The MTF is much worse.

We now resort to a lot of more or less random shocking of the system in hopes of escaping from this poor optimum and finding a better one. The things we try include:

a. Arbitrarily and significantly changing several radii at a time, selecting the changes so as to seriously disrupt the design while minimizing the change to the system power.
b. Changing radii (often one at a time), freezing them for several cycles of optimization, then releasing them to vary during the subsequent optimizations.
c. Changing the central airspace, freezing, optimizing, then releasing, and repeating this cycle several times.

After mucking about this way for a bit, the process (aka the poor man's random search and global optimization) pays off, and the system is improved to a performance that is approximately equal to the previous one. The rear doublet, in addition to being reversed, has a completely different shape, without edge contact and with the negative element meniscus shaped and concave toward the front, as shown in Table 13.4 and Figs. 13.16 and 13.17.

TABLE 13.4 The 200-mm f/4 Telephoto Lens Which Resulted from Reversing the Rear Doublet and Introducing Repeated Random Parameter Changes (Attempting to Find a Better Local Optimum for the Merit Function). See Figs. 13.16 and 13.17

$R1 = +56.417$	$T1 = 12.0$	SK16	620603
$R2 = -157.79$	$T2 = 0.3137$		
$R3 = -136.62$	$T3 = 4.0$	SF1	717295
$R4 = +283.85$	$T4 = 77.0819$		
$R5 = -31.511$	$T5 = 3.0$	SK16	620603
$R6 = -111.45$	$T6 = 0.1$		
$R7 = +236.66$	$T7 = 5.0$	SF1	717295
$R8 = -236.70$	BFL = 58.71		

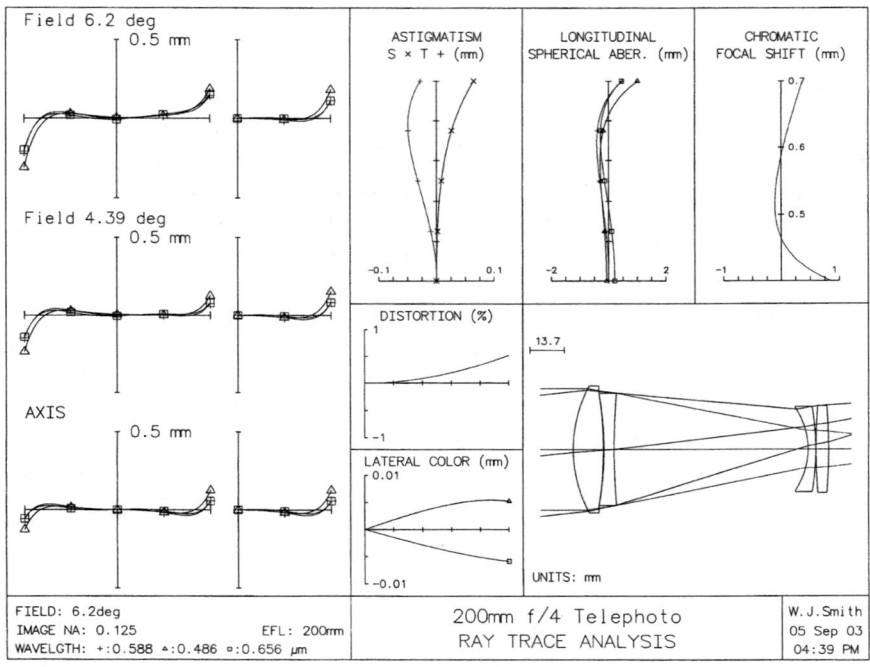

Figure 13.16 Ray aberrations for the fourth design, with reversed rear component, Table 13.4.

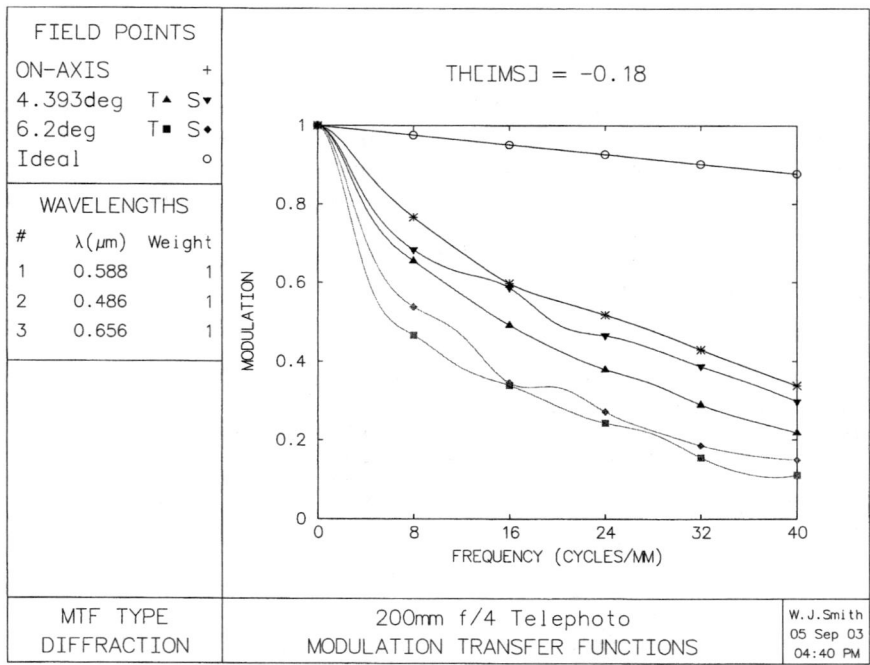

Figure 13.17 MTF for Fig. 13.16, Table 13.4 lens.

TABLE 13.5 The Telephoto Prescription Found When the Rear Doublet is Re-reversed Back to its Original Orientation (See Figs. 13.18 to 13.21)

$R1 = +54.393$	$T1 = 12.0$	SK16	620603
$R2 = -161.99$	$T2 = 0.3235$		
$R3 = -141.24$	$T3 = 4.0$	SF1	717295
$R4 = +225.56$	$T4 = 62.4388$		
$R5 = +213.33$	$T5 = 5.0$	SF1	717295
$R6 = -195.69$	$T6 = 3.4970$		
$R7 = -39.606$	$T7 = 3.0$	SK16	620603
$R8 = -751.54$	BFL = 69.97		

At this point an interesting question is: "Should we re-reverse the rear doublet to see if there is an improvement lurking there?" We give it a try and, not too surprisingly, it requires the same "shock treatment" that the first reversal needed. The resulting design (in Table 13.5 and Figs. 13.18 to 13.21) is almost identical to the unreversed design, but its performance is better than either the reversed or unreversed.

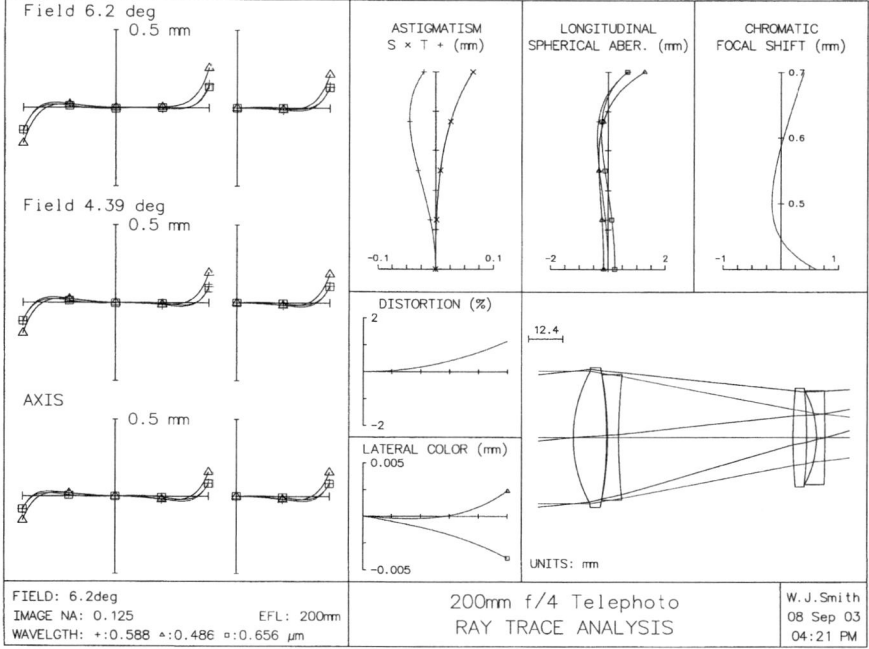

Figure 13.18 Ray aberrations for the fifth design, with re-reversed rear component, Table 13.5.

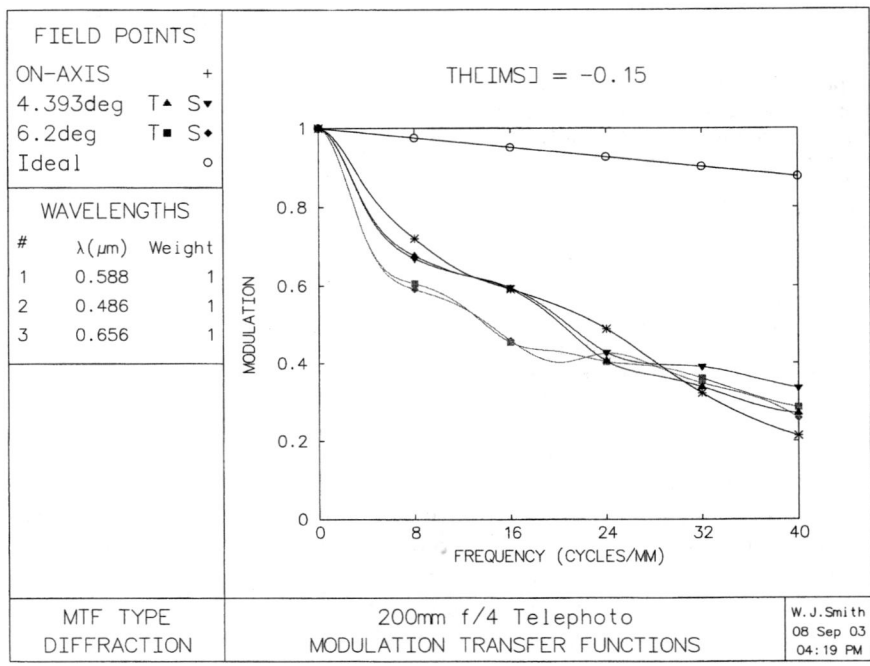

Figure 13.19 MTF vs. frequency for Fig. 13.18, Table 13.5.

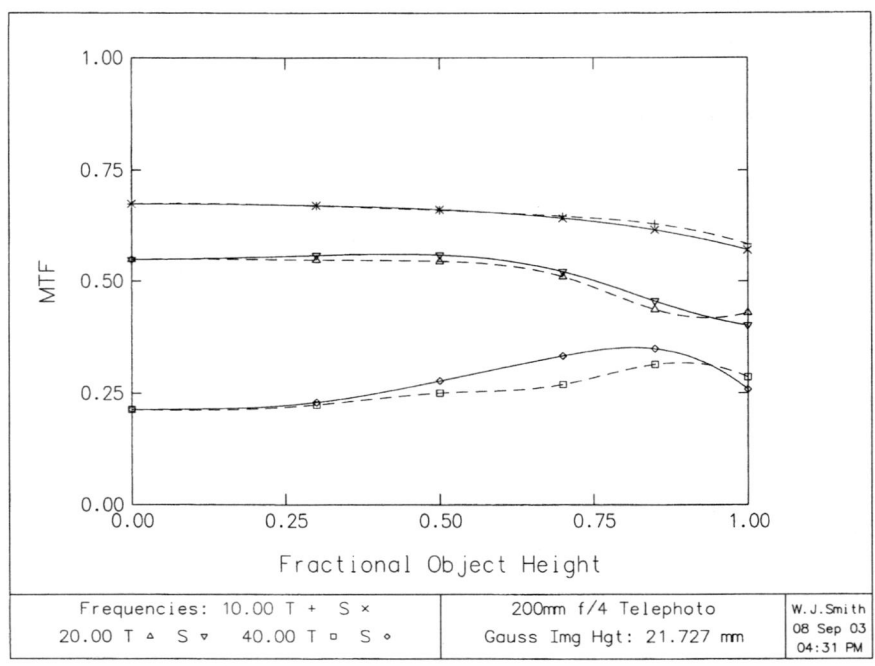

Figure 13.20 MTF vs. field for 10, 20, 40 lpm for Fig. 13.18, Table 13.5.

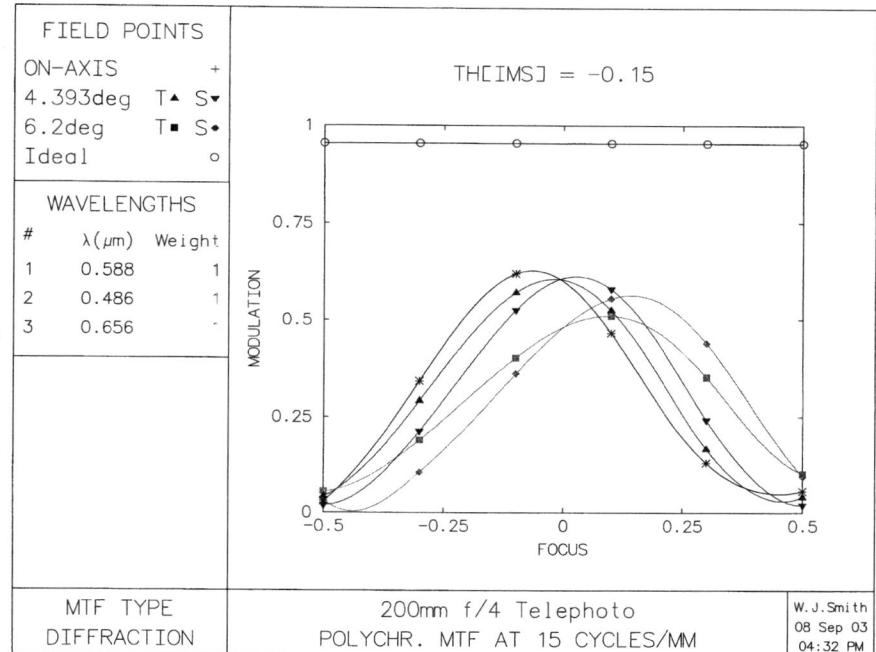

Figure 13.21 Through-focus MTF at 15 lpm for Fig. 13.18, Table 13.5 lens.

13.4.3 Vary the flints

Since the system at this point is completely airspaced, it seems unlikely that allowing the flint glasses to vary will produce much of an improvement. But we give it a try, adding the no. 3 and no. 5 glasses to the variable list. Optimizing, we find that glass no. 3 wants to go to a high index, so we fix it at SF57 (847238) (which is about as far up the glass line as I care to go). Glass no. 5 heads toward a lower index and settles near SF2 (648338). The system is modestly better, but it suffers from a "belly" in the X_t field (as did all of the last few lenses). See Table 13.6 and Figs. 13.22 to 13.25.

TABLE 13.6 The 200-mm f/4 Telephoto Design Found with Variable Flint Glasses (Here Fixed to Real Glasses). See Figs. 13.22 to 13.25

$R1 = +53.446$	$T1 = 12.0$	SK16	620603
$R2 = -573.42$	$T2 = 0.6570$		
$R3 = -276.11$	$T3 = 4.0$	SF57	847238
$R4 = +306.85$	$T4 = 60.1989$		
$R5 = +94.902$	$T5 = 5.0$	SF2	648338
$R6 = -243.76$	$T6 = 3.2307$		
$R7 = -43.528$	$T7 = 3.0$	SK16	620603
$R8 = +190.72$	$BFL = 72.06$		

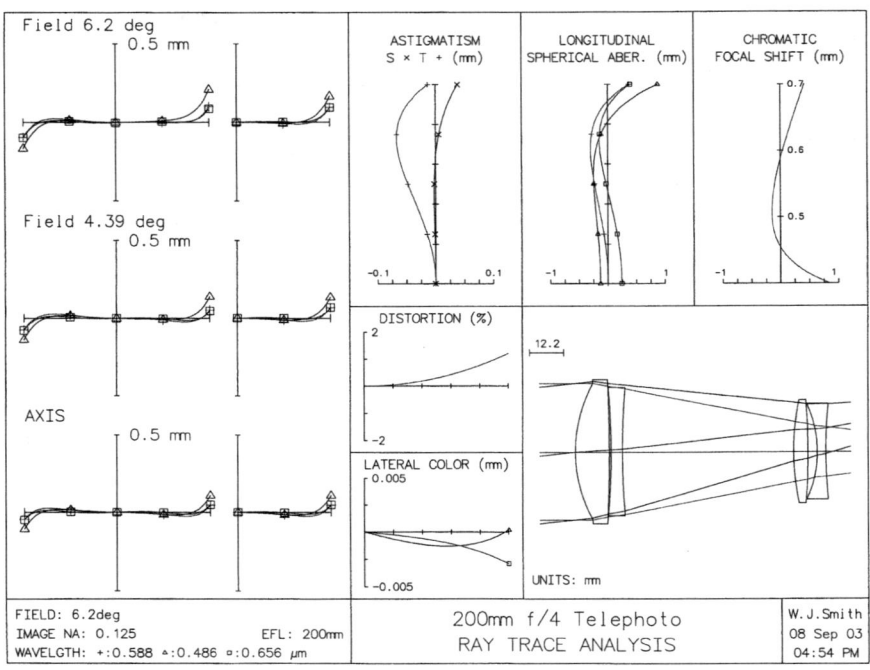

Figure 13.22 Ray aberrations for the sixth design with varied flint glasses, Table 13.6.

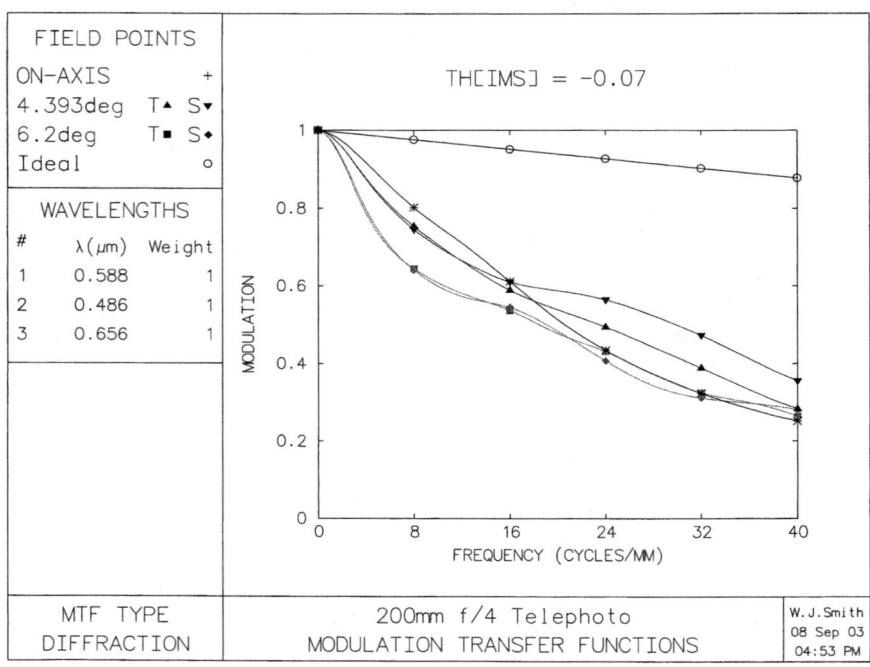

Figure 13.23 MTF vs. frequency for Fig. 13.22, Table 13.6 lens.

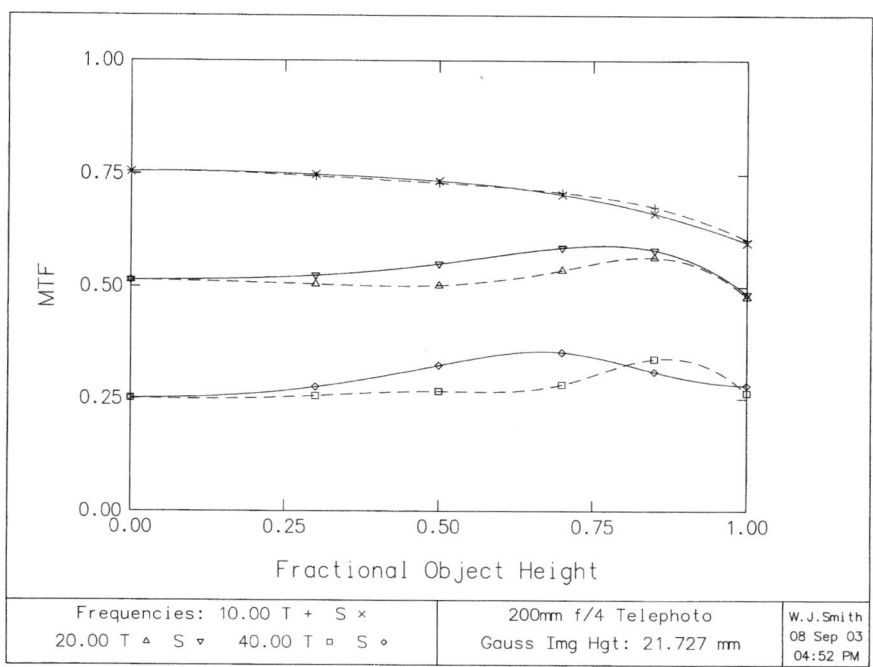

Figure 13.24 MTF versus field for 10, 20, 40 lpm for Fig. 13.22, Table 13.6 lens.

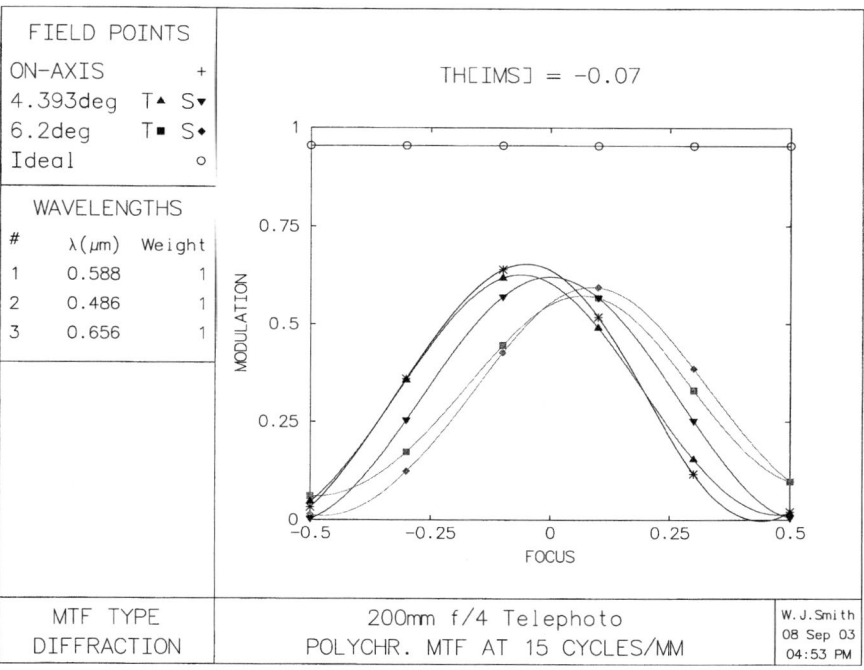

Figure 13.25 Through-focus MTF at 15 lpm for Fig. 13.22, Table 13.6 lens.

382 Chapter Thirteen

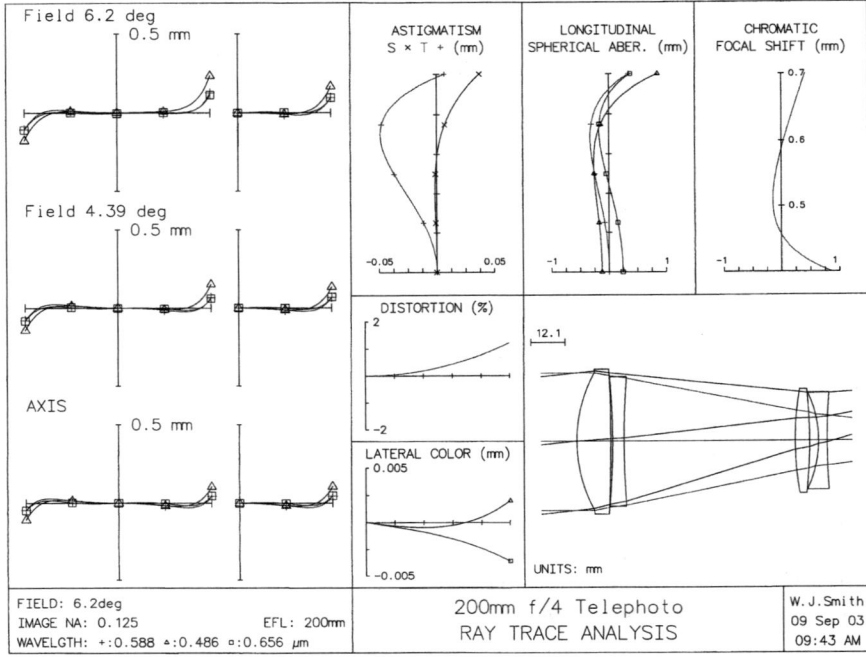

Figure 13.26 Ray aberrations for the seventh design, revised merit function, Table 13.7.

We attack this problem by adding targets for X_s and X_t at the 0.7 field, weighted 100x, the same as the full field curvatures. This reduces the "belly" from $X_t = -0.07$ mm to $X_t = -0.05$ mm, and there is a barely detectable improvement in the MTF as seen in Figs. 13.26 to 13.29. The lens data is in Table 13.7.

TABLE 13.7 The Lens of Table 13.6, Improved by Adding Zonal Field Curvature Targets to the Merit Function (See Figs. 13.26 to 13.29)

$R1 = +53.525$	$T1 = 12.0$	SK16	620603
$R2 = -550.80$	$T2 = 0.6522$		
$R3 = -271.78$	$T3 = 4.0$	SF57	847238
$R4 = +315.65$	$T4 = 60.1097$		
$R5 = +110.88$	$T5 = 5.0$	SF2	648338
$R6 = -185.74$	$T6 = 3.1257$		
$R7 = -42.500$	$T7 = 3.0$	SK16	620603
$R8 = +230.45$	BFL $= 72.27$		

13.4.4 Use higher-index glass

In real life, using higher-index glass almost always necessitates a judgement call. Among the lanthanum glasses it seems that the higher the index, the higher the cost. And another factor is the poor stain characteristics of many of these glasses. Thus the designer is faced with the problem of determining the relative importance of these (and other) factors versus performance. But since this is a tutorial design exercise, we put all this aside and simply select for trial three glasses from along the upper left top of the glass map as shown in the following table:

Name	Six digit code	Rel. cost	Stain factors
N-LAK8	713538	9x	3, 2, 52.3, 1, 3.3
N-LAF21	785475	18x	1, 1, 51.3, 1, 1.3
N-LASF31	881410	52x	1, 0, 2, 1, 1

We substitute these, one at a time, for the SK16 glass in the immediately previous design, and optimize using the same merit function. The N-LAK8

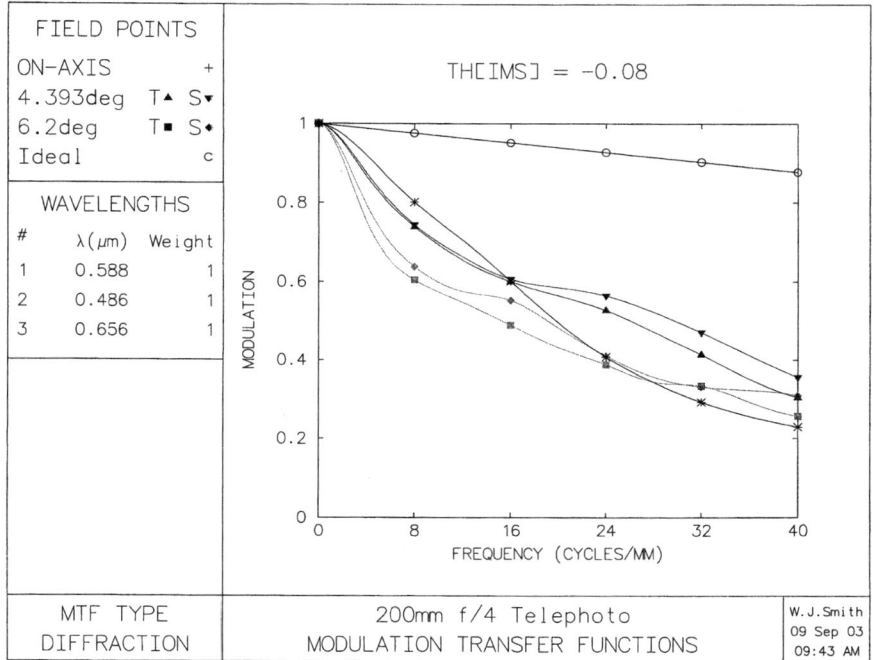

Figure 13.27 MTF versus frequency for Fig. 13.26, Table 13.7 lens.

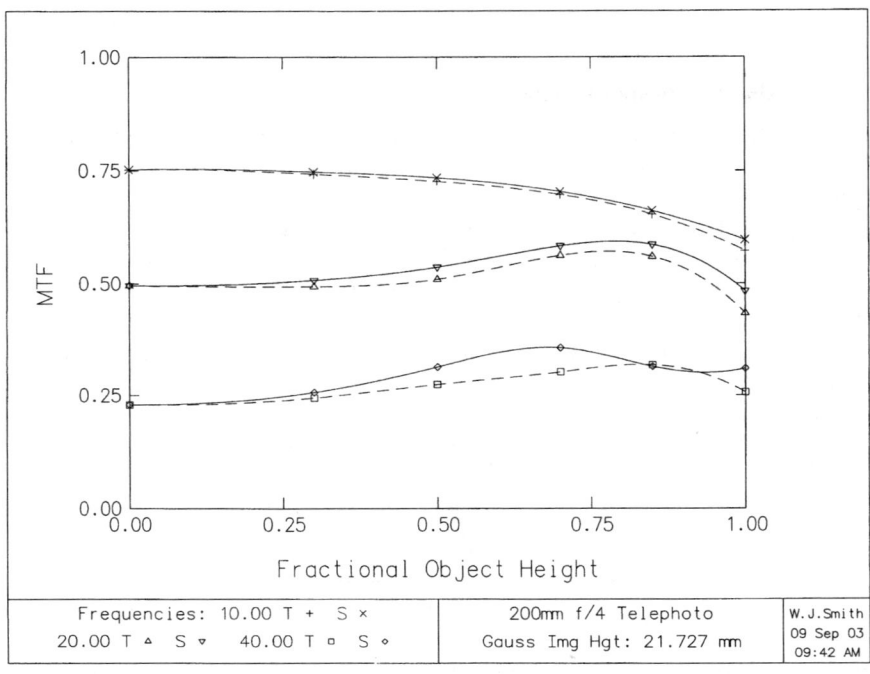

Figure 13.28 MTF versus field for 10, 20, 40 lpm for Fig. 13.26, Table 13.7 lens.

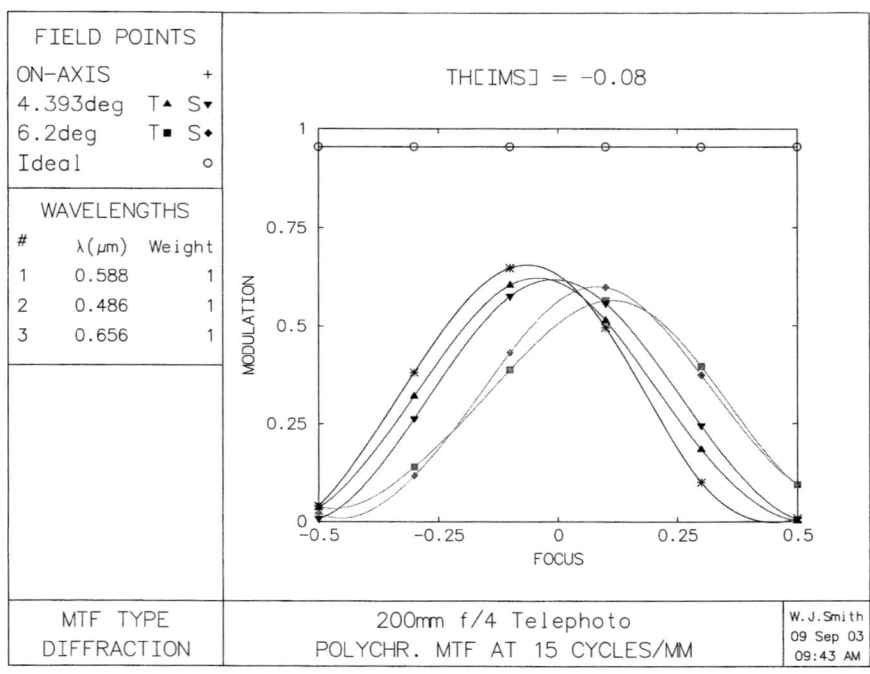

Figure 13.29 Through-focus MTF at 15 lpm for Fig. 13.26, Table 13.7 lens.

and N-LAF21 designs have essentially the same performance as that of the SK16, design, with perhaps a miniscule improvement. The N-LASF31 design is actually a bit worse. On a hunch we try BK7 (517642) for the crowns with a complete lack of success. Finally as a last resort, we decide to let all the glasses vary. The result is a noticeably better design (and the best to date) as shown in Table 13.8 and Figs. 13.30 to 13.33.

Since N-LAK33 is not an inexpensive glass, one might want to try some of the adjacent glasses, such as N-LAK8 (713538), N-LAK10 (720506), or the old LAK16A (734518), which are not as far up on the glass map, but are notably less expensive than N-LAK33.

13.4.5 Add more elements

In adding or splitting elements to improve a design, it is rational to try to reduce the aberration contributions of the elements, which are the worst offenders. These are usually the strongest elements, which have the shortest radii and the largest angles of incidence and refraction. In our case it is the crown elements that are strongest, specifically the positive first element and the negative last element.

One must also bear in mind that in a telephoto the negative rear component magnifies the aberrations of the positive front component, and for this reason a well-corrected front component is mandatory. Since the ray intercept plots of our current design are quite similar for all three field positions (with high-order spherical aberration being the dominant aberration), we suspect that the front component is the culprit, since, located at the stop, it affects all obliquities about equally. A printout of the surface aberration contributions confirms that the first element has the largest third- and fifth-order spherical aberration contributions (although $R7$ has the largest incidence angle).

We decide to split the front crown. Usually the best split is into two elements each with half the power of the original element. Now we must decide which configuration to pursue. With three singlets in the front

TABLE 13.8 The Improved 200-mm f/4 Telephoto Design Found by Allowing All Glasses to Vary (See Figs. 13.30 to 13.33

$R1 = +57.081$	$T1 = 12.0$	SK51	621603
$R2 = -357.34$	$T2 = 0.5986$		
$R3 = -222.03$	$T3 = 4.0$	SF57	847238
$R4 = +489.38$	$T4 = 67.1332$		
$R5 = +188.43$	$T5 = 5.0$	N-SF56	785261
$R6 = -232.11$	$T6 = 3.2668$		
$R7 = -42.832$	$T7 = 3.0$	N-LAK33	754524
$R8 = -458.42$	$BF = 65.06$		

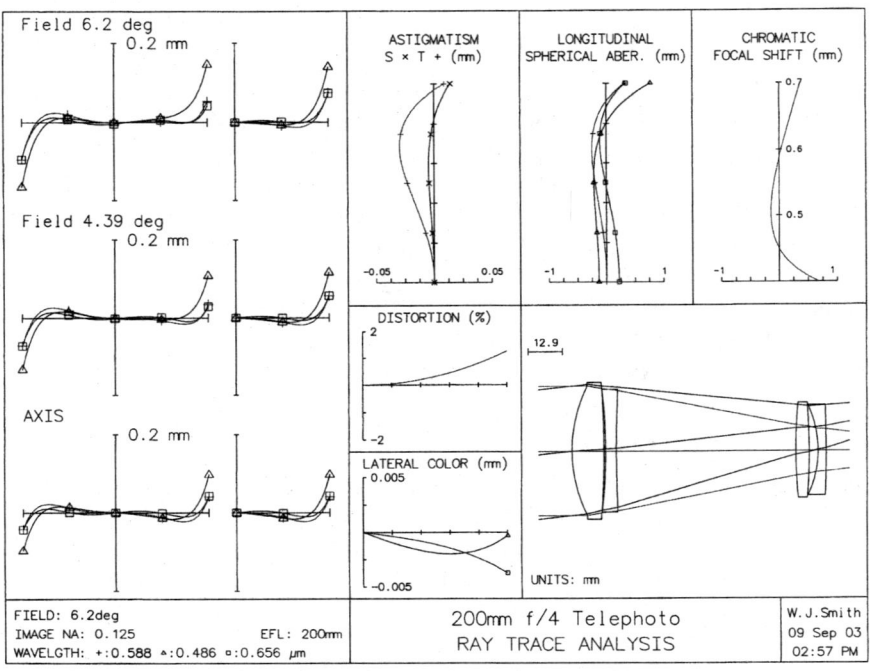

Figure 13.30 Ray aberrations for the eighth design, vary all glass, Table 13.8.

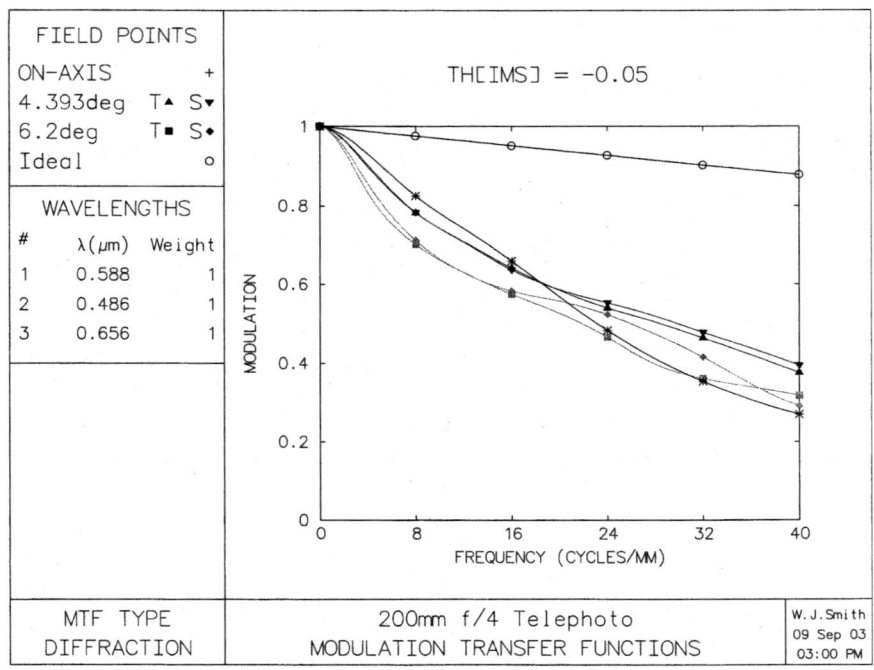

Figure 13.31 MTF vs. frequency for Fig. 13.30, Table 13.8 lens.

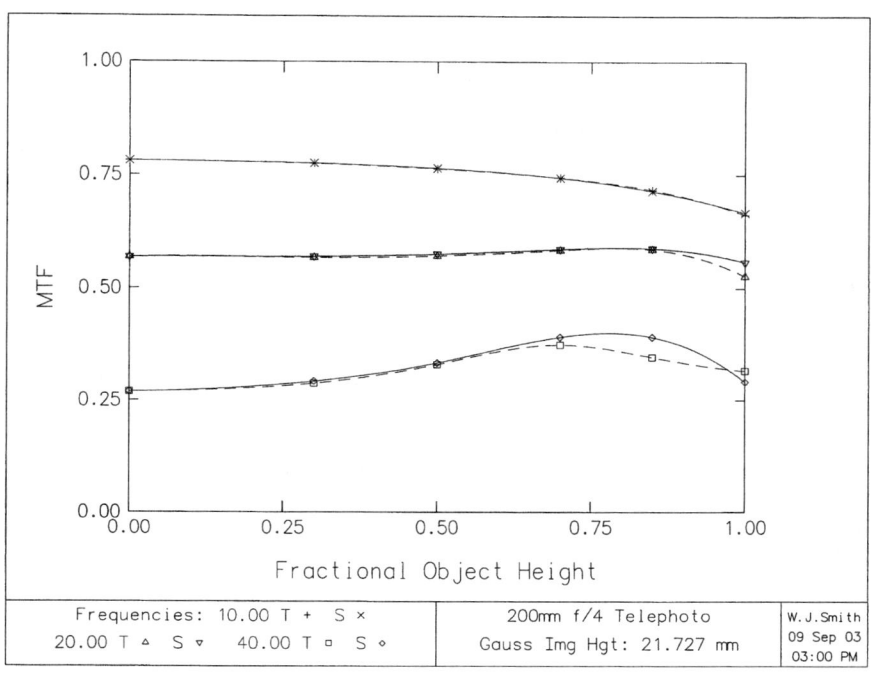

Figure 13.32 MTF vs. field for 10, 20, 40 lpm for Fig. 13.30, Table 13.8 lens.

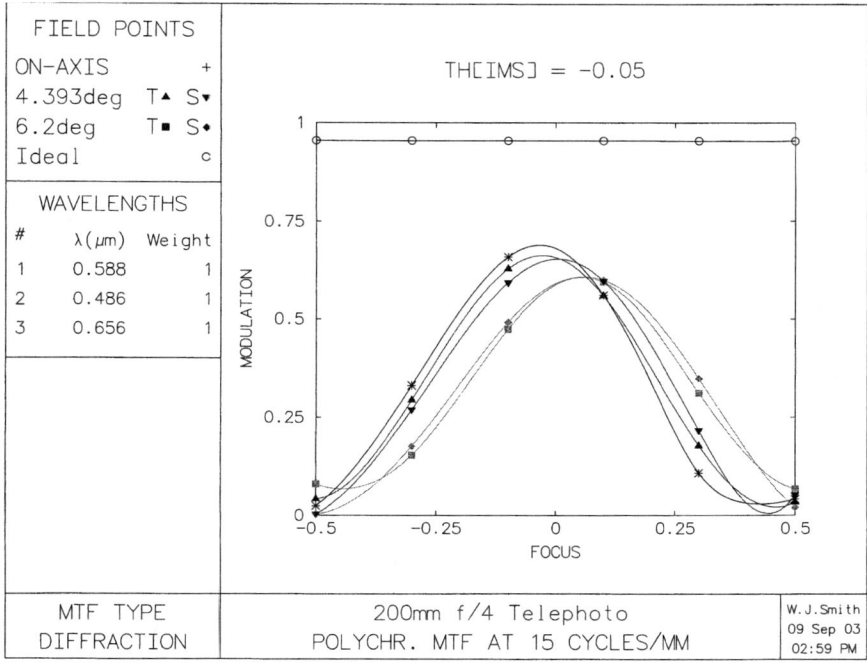

Figure 13.33 Through-focus MTF at 15 lpm for Fig. 13.30, Table 13.8 lens.

component we have three possible configurations: + + −, + − +, or − + +. The last one is an unlikely choice (it's a reverse telephoto), although it may well be a good design form. The first two are more reasonable. There is also the possibility of cementing two of the elements together, i.e., + [+ −], + [− +], or [+ −] +. Although it costs us a variable to do so, cementing has the benefit that sensitivity to misalignment will be less than with three singlets. The third version [+ −] + is frequently used and allows the possibility of making the second positive element a meniscus aplanat (more or less), as do the + + − and + − + arrangements.

We elect to pursue the + + − configuration as the most obvious (and one that increases the *effective* airspace because of its own telephoto arrangement of power). As our starting point we choose the SK16 design of Table 13.7 (from which the latest and best design was derived). We divide the curvature of the first element in two, and make both of the new elements convex-plano. The thickness of the new crowns is made one half that of the old. The old and new front components are shown in the following table.

Starting Front Component Prescription When the Front Crown Element is Split into Two Elements

Old front			New front		
$R1 = +53.525$	$T1 = 12.0$	SK16	$R1 = +96.7$	$T1 = 6.0$	SK16
$R2 = -550.80$	$T2 = 0.652$		$R2 =$ plano	$T2 = 0.2$	
$R3 = -271.78$	$T3 = 4.0$	SF57	$R3 = +96.7$	$T3 = 6.0$	SK16
$R4 = +315.65$	$T4 = 60.11$		$R4 =$ plano	$T4 =$ ET solve	
			$R5 = -271.8$	$T5 = 4.0$	SF57
			$R6 = +315.6$	$T6 = 69.0$	

We use the same merit function, somewhat warily since it has targets that were added to accomodate the specific residual aberrations of the four-element lens. We vary all the curvatures, with an angle solve on the last radius. We vary the central airspace. The optimization goes well, but the performance in only so-so. It worked well before, so we decide to vary all the glasses.

The resulting changes in glass types produced by optimization are interesting. The two front crowns drift lower in index and higher in V-value. We settle on BK10 (498670) for both, although we are reasonably certain that (the more costly and more fragile) FK51, FK52, and FK54 crowns would be better, because of their higher V-values and their unusual partial dispersions. The front flint drops along the glass line to SF5 (673322), and the rear flint wants a high index; we settle for SF57 (847238), again avoiding the more costly and less stable higher-index flints (SF58, SF59). The rear crown (negative element) heads for the LAK31 area of the glass map. We reject this in favor of the less expensive LAK8, with only

TABLE 13.9 The Five-Element 200 mm f/4 Telephoto Design Resulting from Allowing the Glasses to Vary (See Figs. 13.34 to 13.36, 13.38, and 13.39)

$R1 = +54.068$	$T1 = 9.0$	BK10	498670
$R2 = -353.85$	$T2 = 0.20$		
$R3 = +61.316$	$T3 = 6.0$	BK10	498670
$R4 = +152.91$	$T4 = 2.7937$		
$R5 = -424.81$	$T5 = 4.0$	SF5	673322
$R6 = +72.431$	$T6 = 61.6707$		
$R7 = +1013.1$	$T7 = 5.0$	SF57	847238
$R8 = -135.47$	$T8 = 3.3101$		
$R9 = -41.149$	$T9 = 3.0$	LAK8	713538
$R10 = -156.64$	BF = 65.08		

a small loss in performance. The resulting design is in Table 13.9 and Figs. 13.34 to 13.39.

At this point there are still many paths that we could follow. But tutorial discretion indicates that this would be a good point at which to call a halt to our telephoto design exercise. There are still a number of possible continuations that seem promising. Of the seven possible arrangements for the front elements, namely + + −, + − +, − + +, [+ −] +, + [− +], + [+ −], and [− +] +, we have investigated only the first. We have also not conducted any investigation of the effects of the airspaces *within* the components (as in Fig. 13.3)

The ray intercept plots for this lens indicate that an off-axis chromatic variation of coma and/or spherical aberration is probably the most serious remaining problem. This is apparent in the upper rays of the tangential fan plots. Allowing a vignetting of about 25 percent at the upper edge of the off-axis beam and reoptimizing should produce a modest improvement. Since this problem appears to come from the rear component, a likely approach would be to split the last element. Perhaps substituting one of the very high-index lanthanum glasses would help. One could also add a target in the merit function for the difference between the intercept heights of the red-and-blue upper oblique rays to better balance this aberration. An examination of the aberration contributions of the surfaces should be enlightening, as might a check of the incidence angles.

There is about 0.15 mm of secondary spectrum that, at the correction level of the current lens, is not too significant. However, if the lens were improved further, this could become a problem, and the FK glasses mentioned above would be useful to reduce the secondary spectrum. The very high-index flints (SF57, SF58, and SF59) are good in the positive elements of negative doublets; we already have SF57 in the rear component. A KzFS glass in the front flint might also be helpful to reduce the secondary spectrum. But none of these glasses are very easy to love.

At the front component there are two additional possibilities, which might be worth further exploration. In our work with the telescope objective in

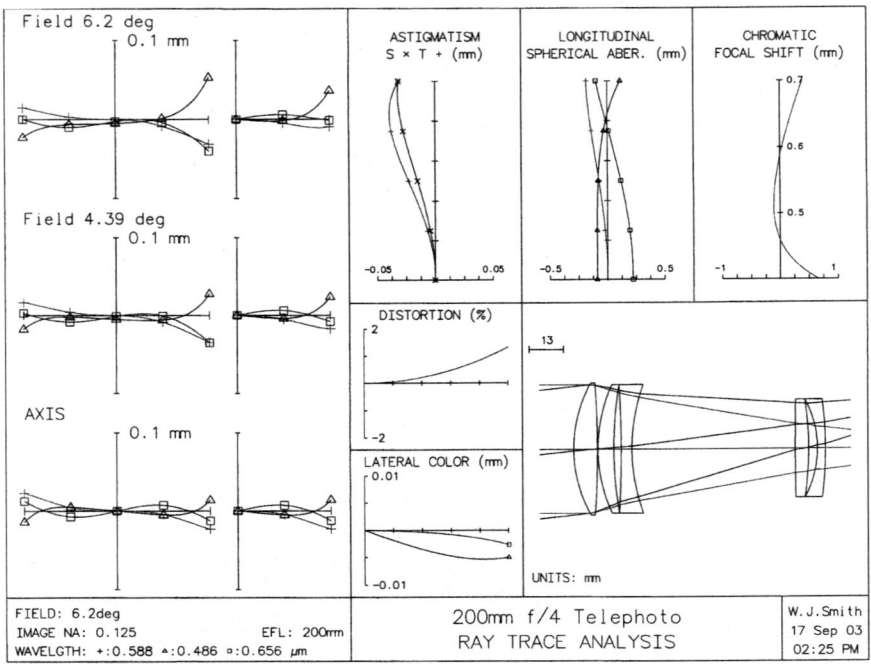

Figure 13.34 Ray aberrations for the ninth design with five elements, Table 13.9.

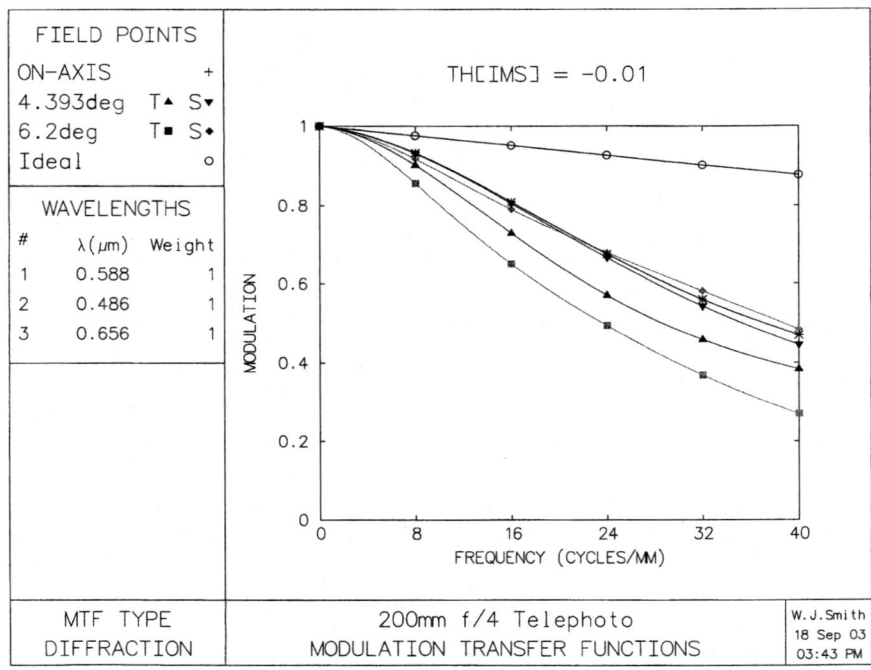

Figure 13.35 MTF vs. frequency for Fig. 13.34, Table 13.9 lens.

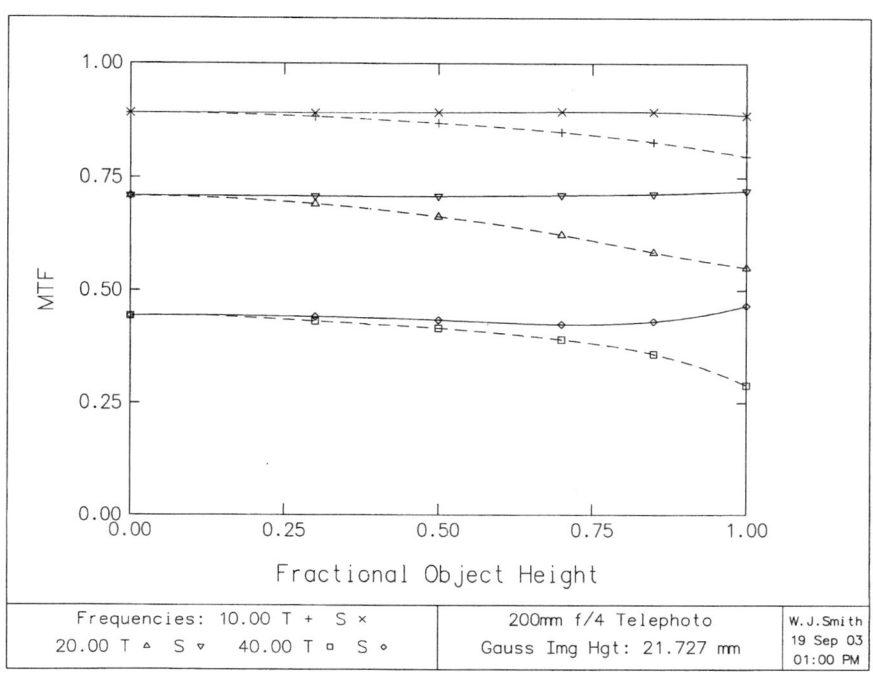

Figure 13.36 MTF vs. field for 10, 20, 40 lpm for Fig. 13.34, Table 13.9 lens.

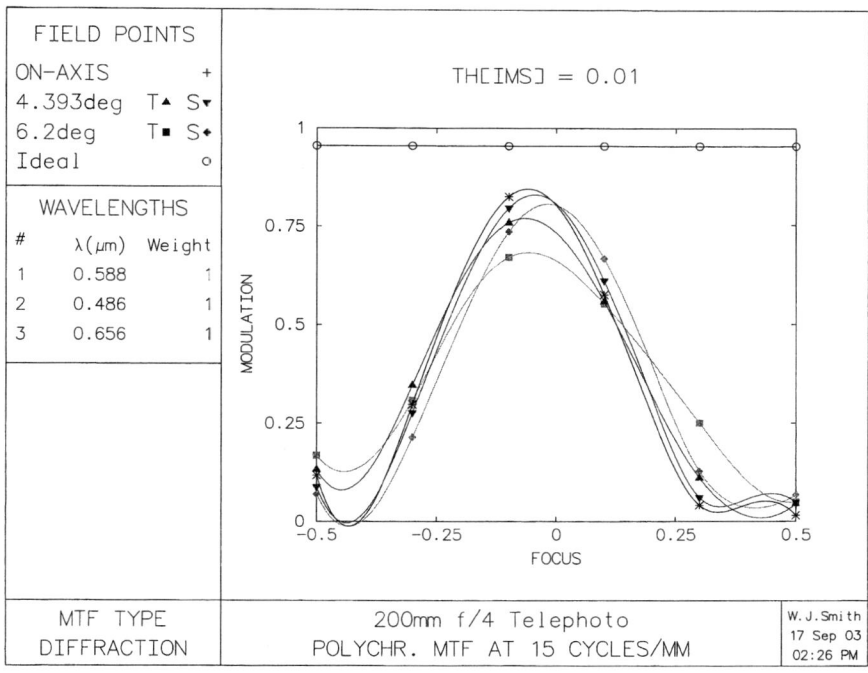

Figure 13.37 Through-focus MTF at 15 lpm for Fig. 13.34, Table 13.9 lens.

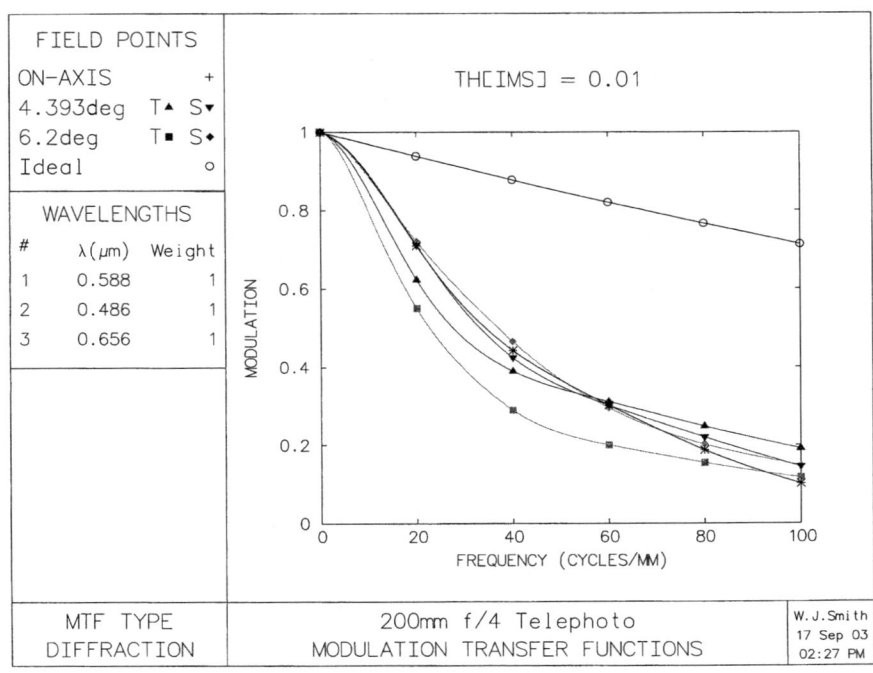

Figure 13.38 MTF vs. frequency, to 100 lpm, for Fig. 13.34, Table 13.9 lens.

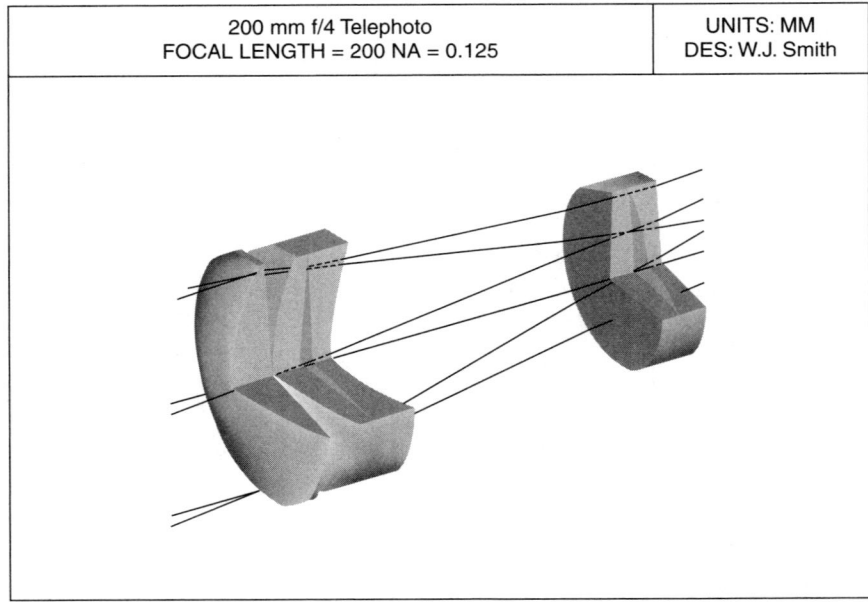

Figure 13.39 Cutaway drawing of Fig. 13.34, Table 13.9 lens.

Chap. 6 we found that increasing the airspace between the crown and flint elements would reduce the spherochromatism by reducing the blue light spherical overcorrection relative to the red. This would reduce the spherochromatism in our current lens. The other possibility would be to make the front component like the front of a double-Gauss, namely a singlet crown followed by a very thick meniscus doublet; this configuration is quite favorable for aberration correction.

Increasing the telephoto ratio from the current 0.8 would certainly allow an improvment of the design.

The distortion correction is quite good in all of our designs, but we note that the distortion is always positive, pushing the 1 percent limit in our merit function. It is a characteristic of the telephoto design form (and also reversed telephoto lenses) that it "wants" to have some distortion. Often a better design results if more distortion is allowed. Allowing an increase to 1.5 or 2 percent distortion could be acceptable for many applications and would probably improve the design.

The aperture stop is located at the front component in the designs above. For a camera lens we must accommodate an iris diaphragm, which means that the aperture stop must be located in the large airspace of the lens. This will either cause some vignetting or require larger diameter front elements. We have omitted this consideration to avoid complicating our design example with mechanical choices, which would be dependent on the specific application or camera in which the lens is to be used. The iris should be located, and the lens so designed, that when the iris is stopped down to a small aperture, the rays that are passed are the best-corrected portion of the oblique beam. Ordinarily this means that the iris should be located at the design aperture stop.

Severely unsymmetrical lenses such as the telephoto (and the retrofocus) tend to be quite sensitive to changes in the object location. Often the aberration change on focusing the lens for a close object can be ameliorated by shifting some elements or components independent of the balance of the lens. The general approach to this would be to analyze the aberration changes and to introduce spacing changes that counteract the changes produced by focusing. The multiconfiguration features of the software can be useful here. We simultaneously optimize two systems with the same elements but with different object distances and independent variable spacings. One approach might be to change one space at a time and look for synergies among the spaces that minimize the aberration changes and are mechanically feasible.

Chapter 14

Reversed Telephoto (Retrofocus and Fish-Eye) Lenses

14.1 The Reversed Telephoto Principle

The reversed telephoto camera lens probably had its origin when the three-film Technicolor movie camera required a long enough back focus to accommodate the color separation prism cube. When a negative lens is placed at the front focal point of a camera lens, the back focus is lengthened, but the effective focal length is unchanged. So the early Technicolor efforts simply put a negative achromat out in front of a standard camera lens to make room for the prism. Eventually the potential of a unified design was recognized, and the reversed telephoto came into its own as a major design type. It is an excellent form, achieving good aberration correction, but it is a big lens. Its size is often many times the focal length; in some extreme cases the length is as much as 10 to 50 times the focal length.

The *single lens reflex* (SLR) 35-mm camera requires a lens with a back focus of 38–40 mm to clear the swinging viewfinder 45° mirror that flips up out of the way when an exposure is made. For short focal length lenses this mandates a reversed telephoto type of construction, and this form has become an important design in its own right. The retrofocus lens can cover a respectable field of view at a relatively high speed. Note that even the standard focal length double-Gauss camera lens must be pushed in this direction (with less positive power in front and more in back) to satisfy this requirement for a long back focus.

The arrangement of a positive component following a negative component, as shown in Fig. 14.1, has a (thin-lens) back focal length that

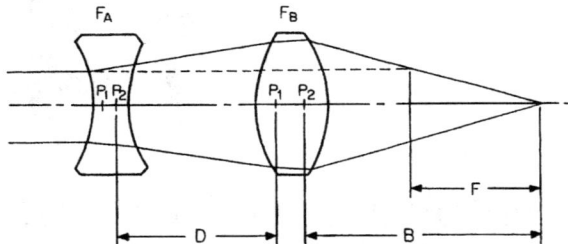

Figure 14.1 The basic power arrangement for a reversed telephoto lens yields a back focal length that is longer than the effective focal length.

is longer than the effective focal length. The necessary power and spacing arrangements can be determined from

$$F_A = \frac{DF}{F-B} \tag{14.1}$$

$$F_B = \frac{-DB}{F-B-D} \tag{14.2}$$

An extremely long back focal length B is possible; however, the Petzval field then has a tendency to become strongly overcorrected (backward-curving). Obviously, the optimum arrangement (from a purely Petzval standpoint) is approximated by the one in which the focal lengths of the two components are nearly equal (and of opposite sign). This will occur when D is chosen so that $D = (F-B)^2/F$; then $F_A = F - B$ and $F_B = B - F$. See also Eqs. (13.1) and (13.2).

However, the retrofocus construction typically departs somewhat from this arrangement, and, as indicated above, the usual problem is a backward-curving Petzval surface. Thus ordinary glasses (low-index crowns and high-index flints) are appropriate for the positive rear component because this combination increases its negative, inward-curving contribution to the Petzval sum. For the negative front component, the crown glasses (which are used for the negative elements) should be high-index glasses and the flint glasses (in the positive elements) should be low-index glasses; this reduces the overcorrected Petzval contribution from this component.

It is apparent that there is absolutely no hope of any semblance of front-to-back symmetry in this type of lens. Thus we can expect the correction of coma, distortion, and lateral color to be difficult. In many other design types, these aberrations are reduced or corrected by an approximately symmetrical arrangement of the elements about the aperture stop. In order to correct both the axial and lateral chromatic of the retrofocus, both components must be individually achromatized. (Although in a system that has only a modestly extended back focus, the front negative component may be weak enough that a high-V-value crown singlet will be acceptable.) The simplest *fully* corrected form is thus a pair of achromatic components. Usually the aperture stop is at the rear component; the natural shape for the front negative achromat is then that of a meniscus, concave toward the stop.

In an unsymmetrical system, as noted above, coma, distortion, and lateral color are difficult to control. Of these, only distortion does not affect the image definition. If it can be tolerated, allowing a few percent distortion will often permit a better level of correction for all the other aberrations. Thus it is usually worthwhile in the course of the optimization process to greatly reduce the weight on distortion in the merit function to see if an overall improvement will result. Some designers actually target the distortion to something like (–) 2 percent (or do without a distortion operand) to take advantage of the natural proclivity of this design form toward negative distortion.

If space is not a problem, a large airspace will reduce the power of the components, especially that of the front component, e.g., see Fig. 14.3. In a wide-angle reversed telephoto it is not uncommon to use an aspheric or conic on the inner surface of a front negative element.

14.2 The Basic Retrofocus Lens

Figure 14.2 is a relatively simple retrofocus design with an airspaced doublet achromatic front and a rear component that is a split-rear-crown triplet form. Designed as a single-frame 35-mm projection lens, it covers a modest 37° field at a speed of $f/3$. Note that the glass types are all rather ordinary. As is true of most retrofocus designs, this lens is quite sensitive to changes in object distance, and the design should be modified if it is to be used at a short conjugate distance. In this particular case, shortening both the second and third airspaces can rebalance the lens reasonably well for use at close object distances.

Figure 14.3 is quite well corrected for high speed ($f/1.2$) and a relatively modest field. Note the use of a significant airspace in the negative front component. The rear components are almost a simple stack of positive power; note the arrangement of the three elements nearest the short conjugate to increase the speed. The negligible back focus distance that accompanies this is very unusual for a retrofocus lens; the design may be intended to be contacted to the film. Note the extreme length, equal to 10.9 times the focal length. The red-blue lateral color is corrected, but there is a quite noticeable amount of secondary spectrum of lateral color apparent.

Figures 14.4 and 14.5 each have seven elements but are totally different designs. The front doublet of Fig. 14.4 has reversed glass types; the usual arrangement of glass in a negative achromatic doublet is for

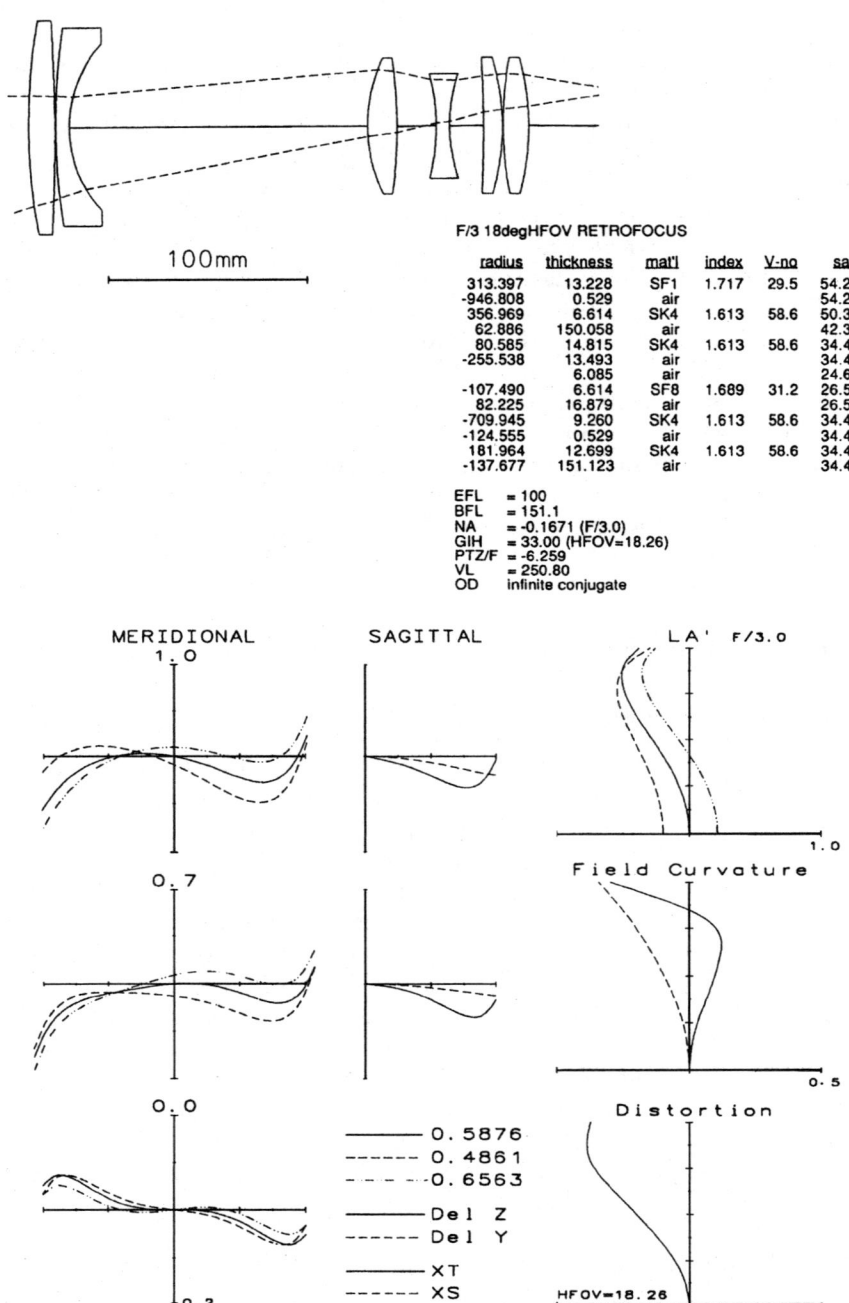

Figure 14.2 $f/3.0$ ±18° a basic, simple six-element retrofocus.

Reversed Telephoto (Retrofocus and Fish-Eye) Lenses

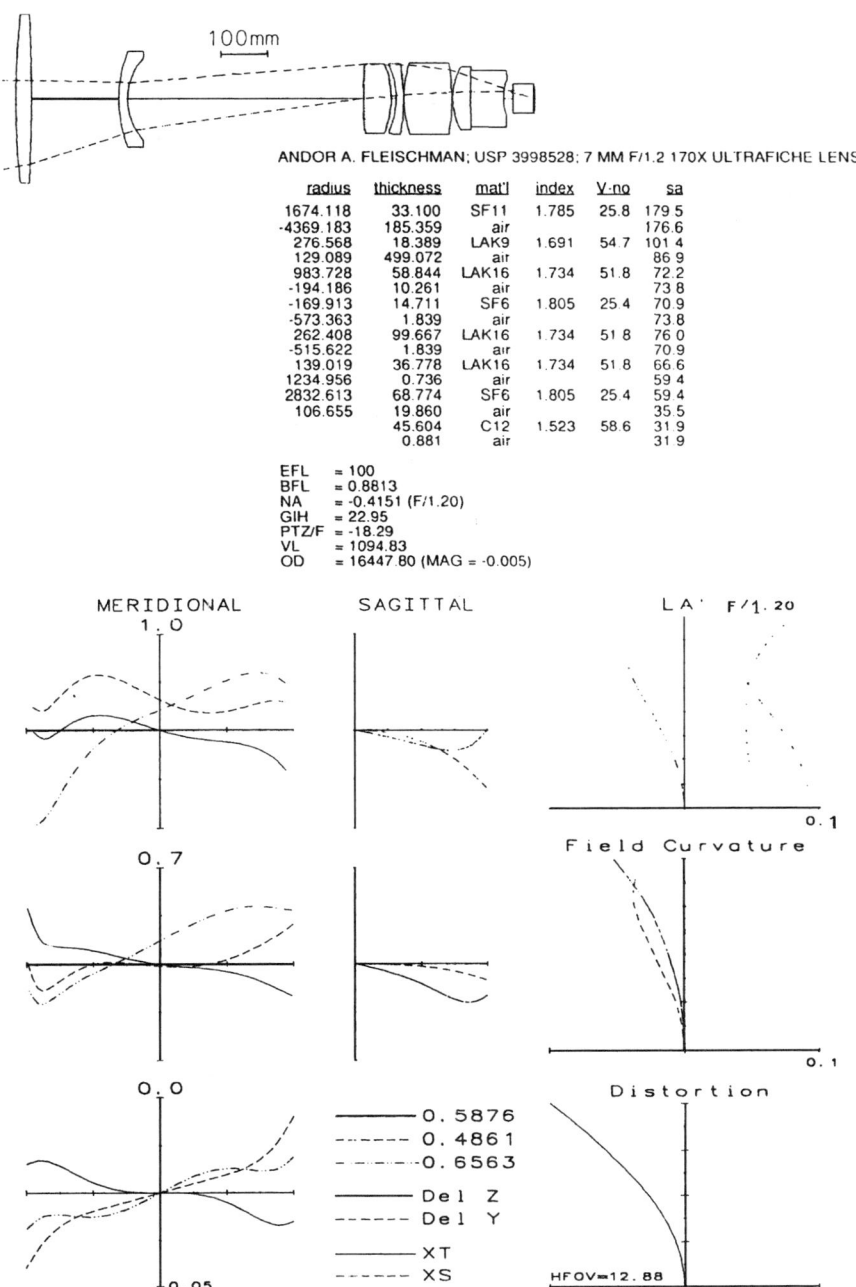

Figure 14.3 $f/1.2$ ±13° a fast, very long high-resolution ultrafiche lens.

400 Chapter Fourteen

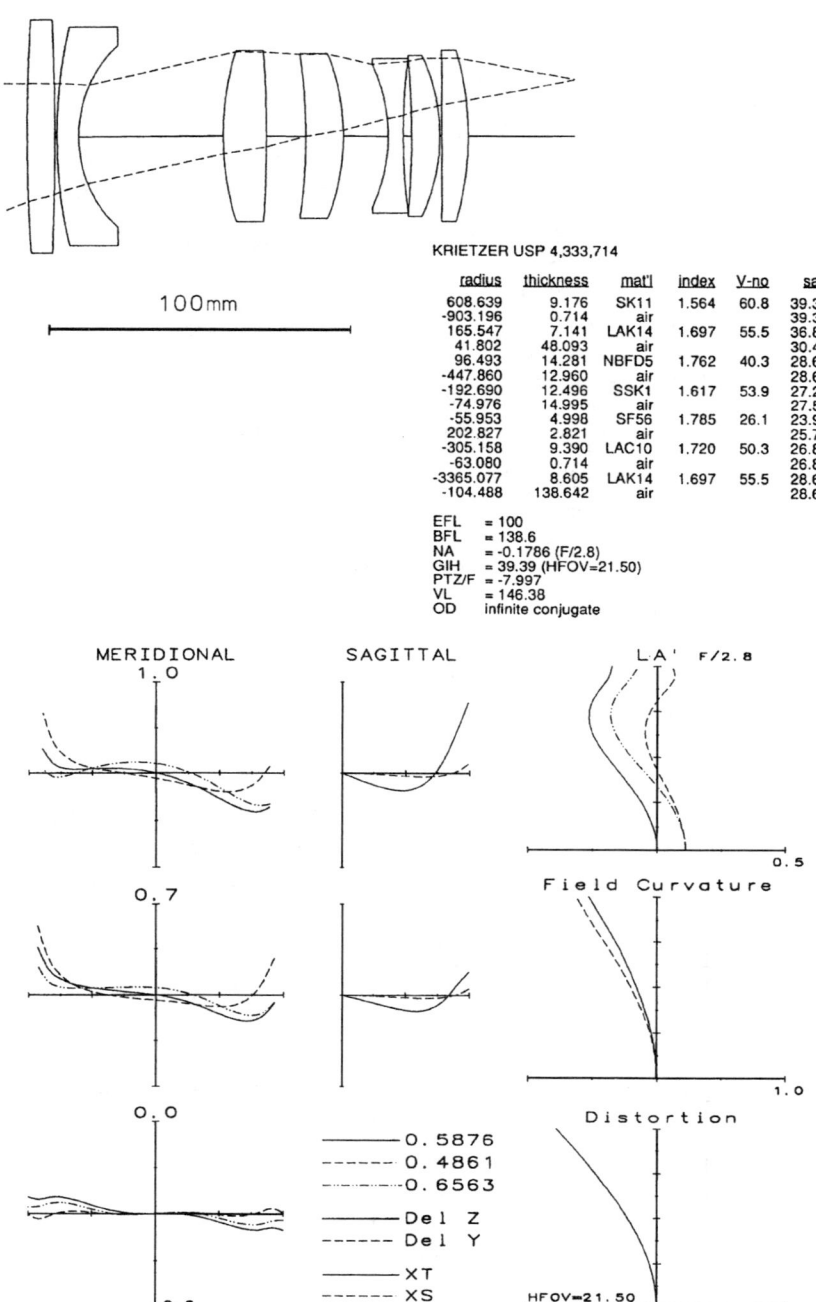

Figure 14.4 $f/2.8$ ±22° a seven-element retrofocus lens.

Reversed Telephoto (Retrofocus and Fish-Eye) Lenses 401

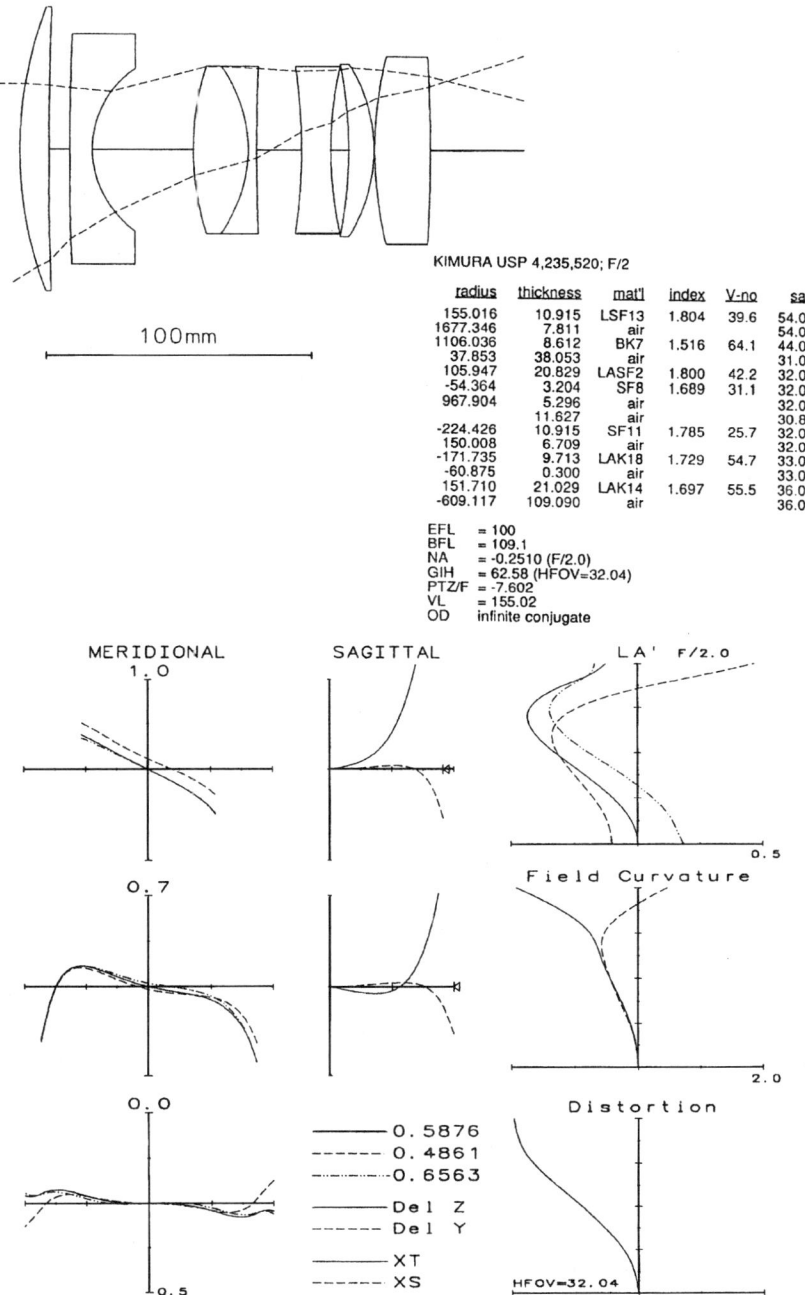

KIMURA USP 4,235,520; F/2

radius	thickness	mat'l	index	V-no	sa
155.016	10.915	LSF13	1.804	39.6	54.0
1677.346	7.811	air			54.0
1106.036	8.612	BK7	1.516	64.1	44.0
37.853	38.053	air			31.0
105.947	20.829	LASF2	1.800	42.2	32.0
-54.364	3.204	SF8	1.689	31.1	32.0
967.904	5.296	air			32.0
	11.627	air			30.8
-224.426	10.915	SF11	1.785	25.7	32.0
150.008	6.709	air			32.0
-171.735	9.713	LAK18	1.729	54.7	33.0
-60.875	0.300	air			33.0
151.710	21.029	LAK14	1.697	55.5	36.0
-609.117	109.090	air			36.0

EFL = 100
BFL = 109.1
NA = -0.2510 (F/2.0)
GIH = 62.58 (HFOV=32.04)
PTZ/F = -7.602
VL = 155.02
OD infinite conjugate

Figure 14.5 $f/2.0 \pm 32°$ a seven-element retrofocus lens.

the negative element to be made of a low-dispersion, high-V-value (crown) glass and the positive element to be made of a high-dispersion, low-V-value (flint) glass. Figure 14.5 covers a wider field (64° versus 42°) at a higher speed ($f/2.0$ versus $f/2.8$); for this reason, its aberration correction is not as good and its back focus is significantly shorter. Note that both designs have fronts that are almost afocal (the first three elements in Fig. 14.4 and the first four in Fig. 14.5).

The rear component of Fig. 14.6 is quite similar to that of Fig. 14.5, but the front negative component of Fig. 14.6 (and also that of Fig. 14.7) is a singlet of crown glass. These simple constructions achieve angular coverages of 64° and 74° at a speed of $f/2.8$ with back focal lengths that exceed their focal lengths by 10 and 20 percent, respectively.

Fields of 80° to 90° are covered by Figs. 14.8 to 14.10. Note the similarity of the construction of the first four elements of Figs. 14.8 to 14.10, although the higher speeds of Figs. 14.9 and 14.10 require a more complex configuration to correct the aberrations over the larger ($f/2.0$ versus $f/2.8$) aperture.

14.3 Fish-Eye, or Extreme Wide-Angle Reversed Telephoto, Lenses

When the reversed telephoto form is carried to extreme, a total field of 180° or more can be covered. In such a system, the front elements are very strongly bent negative meniscus elements, concave toward the image and the aperture stop. These elements have very heavily overcorrected spherical aberration of the pupil; this serves to deviate the high-obliquity principal rays through a large angle, thereby directing them into the stop. In the process, a large amount of barrel distortion is introduced. It is, of course, apparent that if a field of 180° or more is to be imaged on a finite-sized flat surface, such distortion is inevitable. In an extreme wide-angle lens such as this, the focal length ceases to have its usual meaning of $H' = f \tan \theta$, because the large distortion negates this relationship except for small angles. Some lenses are made to follow an $H' = f \theta$ rule; occasionally a lens is designed to $H' = f \sin \theta$, which can result in a more uniform illumination over the field. In any case, the barrel distortion helps to offset the illumination falloff that usually results from the cosine-fourth-power rule.

The origin of the fish-eye lens was probably the Hill sky lens, which was conceived for the purpose of photographing the entire sky from horizon to horizon; this obviously required a field of view of 180° or more. The original sky lens was a strong meniscus negative front lens with one or two elements forming a positive rear component located behind the aperture stop. The name *fish-eye* comes from the fact that the view from underwater into air is 180°, compared to a field of only 97° in the water; this is what a fish sees, looking up.

Reversed Telephoto (Retrofocus and Fish-Eye) Lenses 403

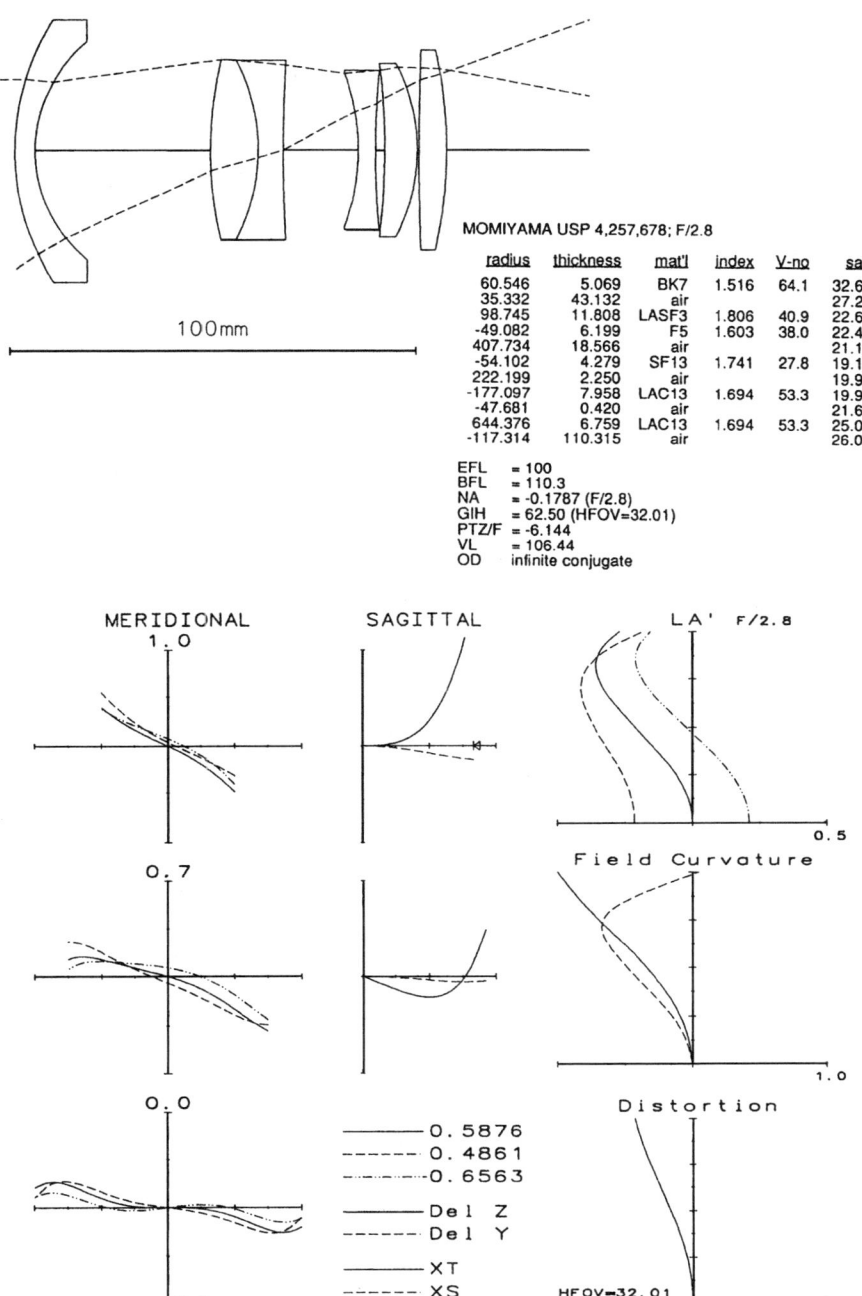

Figure 14.6 $f/2.8$ ±32° a six-element retrofocus lens.

404 Chapter Fourteen

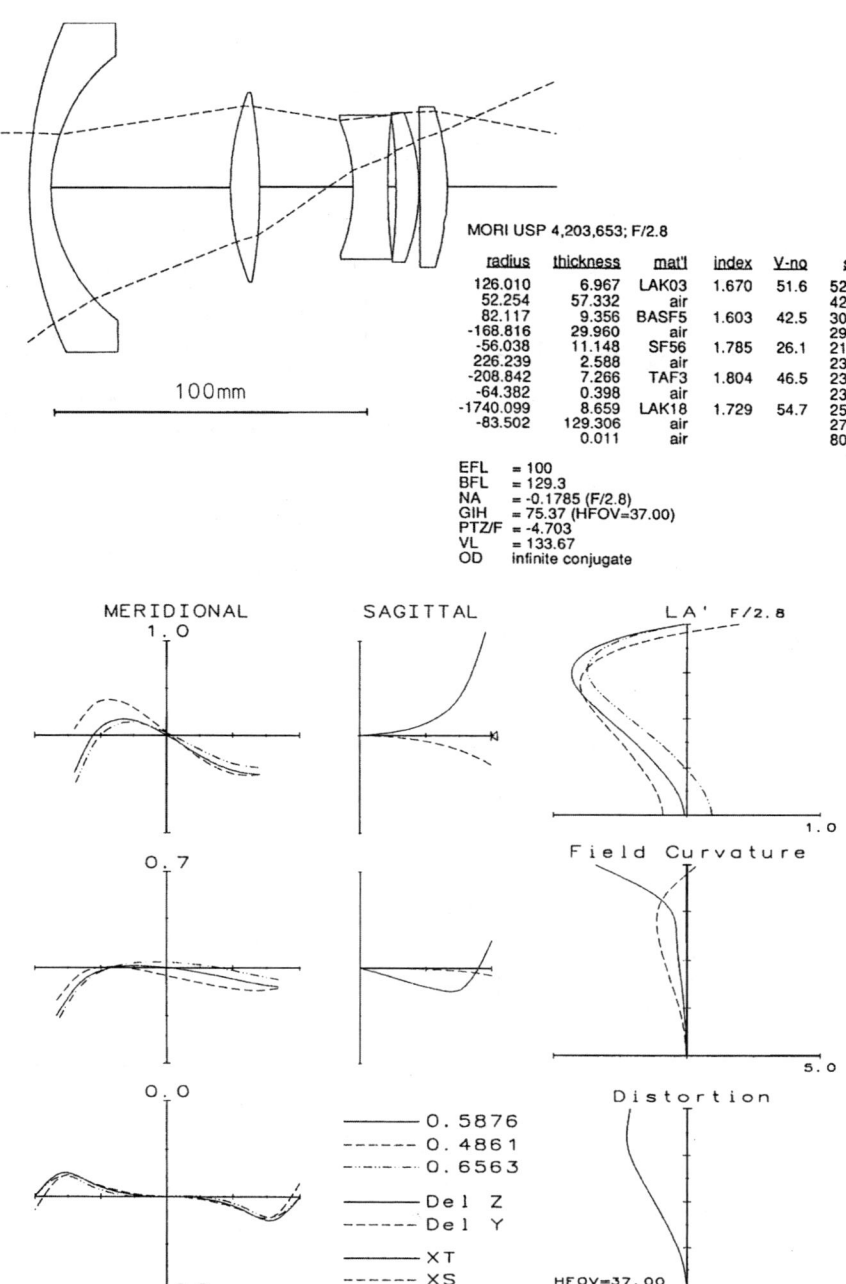

MORI USP 4,203,653; F/2.8

radius	thickness	mat'l	index	V-no	sa
126.010	6.967	LAK03	1.670	51.6	52.4
52.254	57.332	air			42.0
82.117	9.356	BASF5	1.603	42.5	30.1
-168.816	29.960	air			29.8
-56.038	11.148	SF56	1.785	26.1	21.4
226.239	2.588	air			23.0
-208.842	7.266	TAF3	1.804	46.5	23.0
-64.382	0.398	air			23.8
-1740.099	8.659	LAK18	1.729	54.7	25.5
-83.502	129.306	air			27.0
	0.011	air			80.0

EFL = 100
BFL = 129.3
NA = -0.1785 (F/2.8)
GIH = 75.37 (HFOV=37.00)
PTZ/F = -4.703
VL = 133.67
OD infinite conjugate

Figure 14.7 $f/2.8$ ±37° a five-element retrofocus lens.

Reversed Telephoto (Retrofocus and Fish-Eye) Lenses

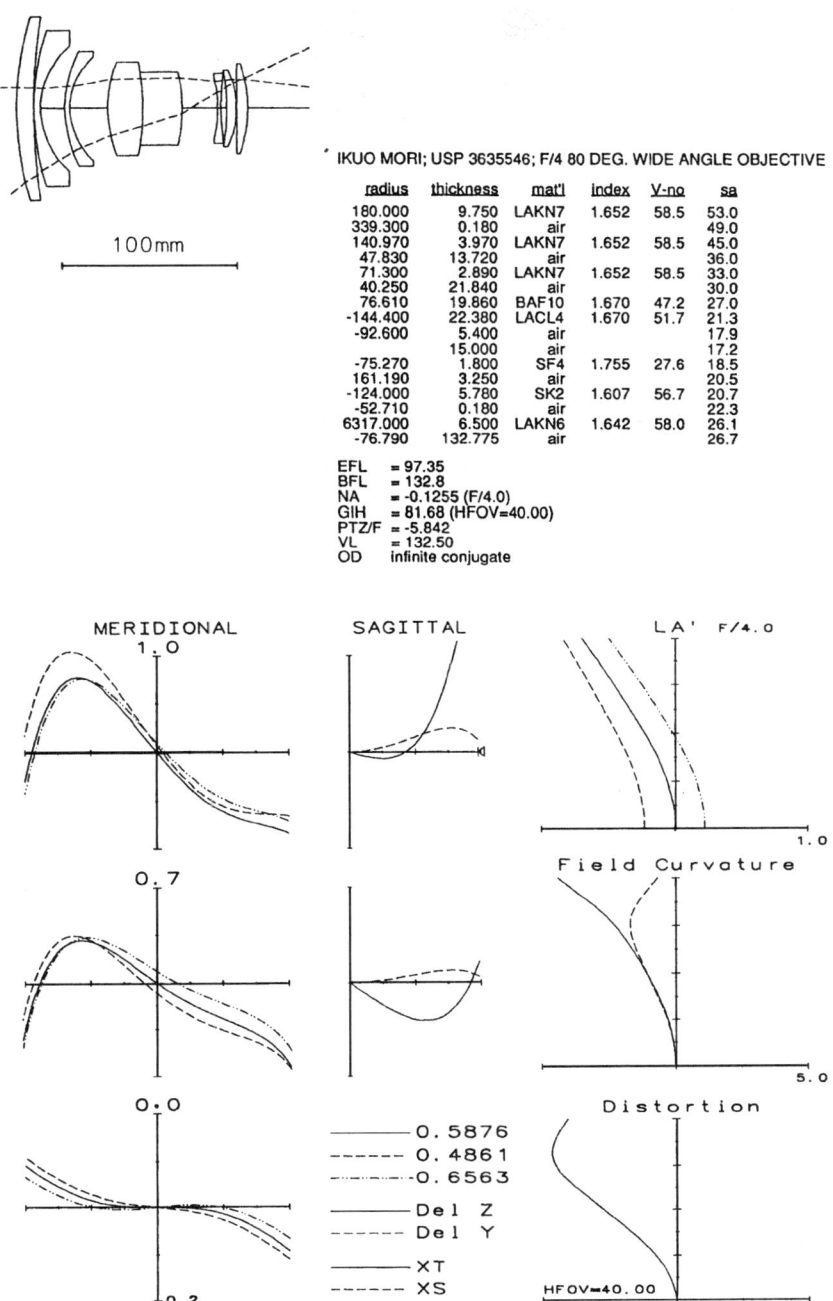

IKUO MORI; USP 3635546; F/4 80 DEG. WIDE ANGLE OBJECTIVE

radius	thickness	mat'l	index	V-no	sa
180.000	9.750	LAKN7	1.652	58.5	53.0
339.300	0.180	air			49.0
140.970	3.970	LAKN7	1.652	58.5	45.0
47.830	13.720	air			36.0
71.300	2.890	LAKN7	1.652	58.5	33.0
40.250	21.840	air			30.0
76.610	19.860	BAF10	1.670	47.2	27.0
-144.400	22.380	LACL4	1.670	51.7	21.3
-92.600	5.400	air			17.9
	15.000	air			17.2
-75.270	1.800	SF4	1.755	27.6	18.5
161.190	3.250	air			20.5
-124.000	5.780	SK2	1.607	56.7	20.7
-52.710	0.180	air			22.3
6317.000	6.500	LAKN6	1.642	58.0	26.1
-76.790	132.775	air			26.7

EFL = 97.35
BFL = 132.8
NA = -0.1255 (F/4.0)
GIH = 81.68 (HFOV=40.00)
PTZ/F = -5.842
VL = 132.50
OD infinite conjugate

Figure 14.8 $f/4.0$ ±40° an eight-element retrofocus lens.

406 Chapter Fourteen

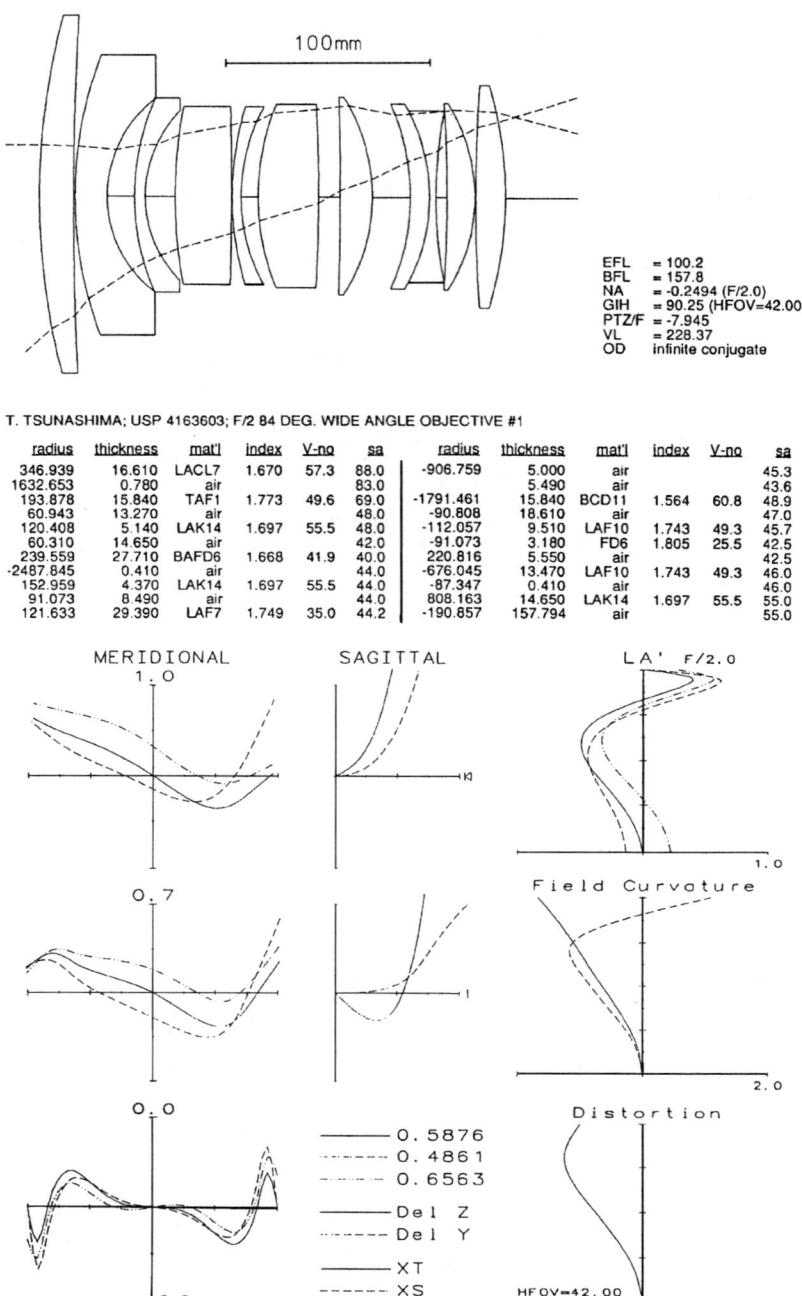

T. TSUNASHIMA; USP 4163603; F/2 84 DEG. WIDE ANGLE OBJECTIVE #1

radius	thickness	mat'l	index	V-no	sa	radius	thickness	mat'l	index	V-no	sa
346.939	16.610	LACL7	1.670	57.3	88.0	-906.759	5.000	air			45.3
1632.653	0.780	air			83.0		5.490	air			43.6
193.878	15.840	TAF1	1.773	49.6	69.0	-1791.461	15.840	BCD11	1.564	60.8	48.9
60.943	13.270	air			48.0	-90.808	18.610	air			47.0
120.408	5.140	LAK14	1.697	55.5	48.0	-112.057	9.510	LAF10	1.743	49.3	45.7
60.310	14.650	air			42.0	-91.073	3.180	FD6	1.805	25.5	42.5
239.559	27.710	BAFD6	1.668	41.9	40.0	220.816	5.550	air			42.5
-2487.845	0.410	air			44.0	-676.045	13.470	LAF10	1.743	49.3	46.0
152.959	4.370	LAK14	1.697	55.5	44.0	-87.347	0.410	air			46.0
91.073	8.490	air			44.0	808.163	14.650	LAK14	1.697	55.5	55.0
121.633	29.390	LAF7	1.749	35.0	44.2	-190.857	157.794	air			55.0

Figure 14.9 $f/2.0$ ±42° an 11-element retrofocus lens.

Reversed Telephoto (Retrofocus and Fish-Eye) Lenses

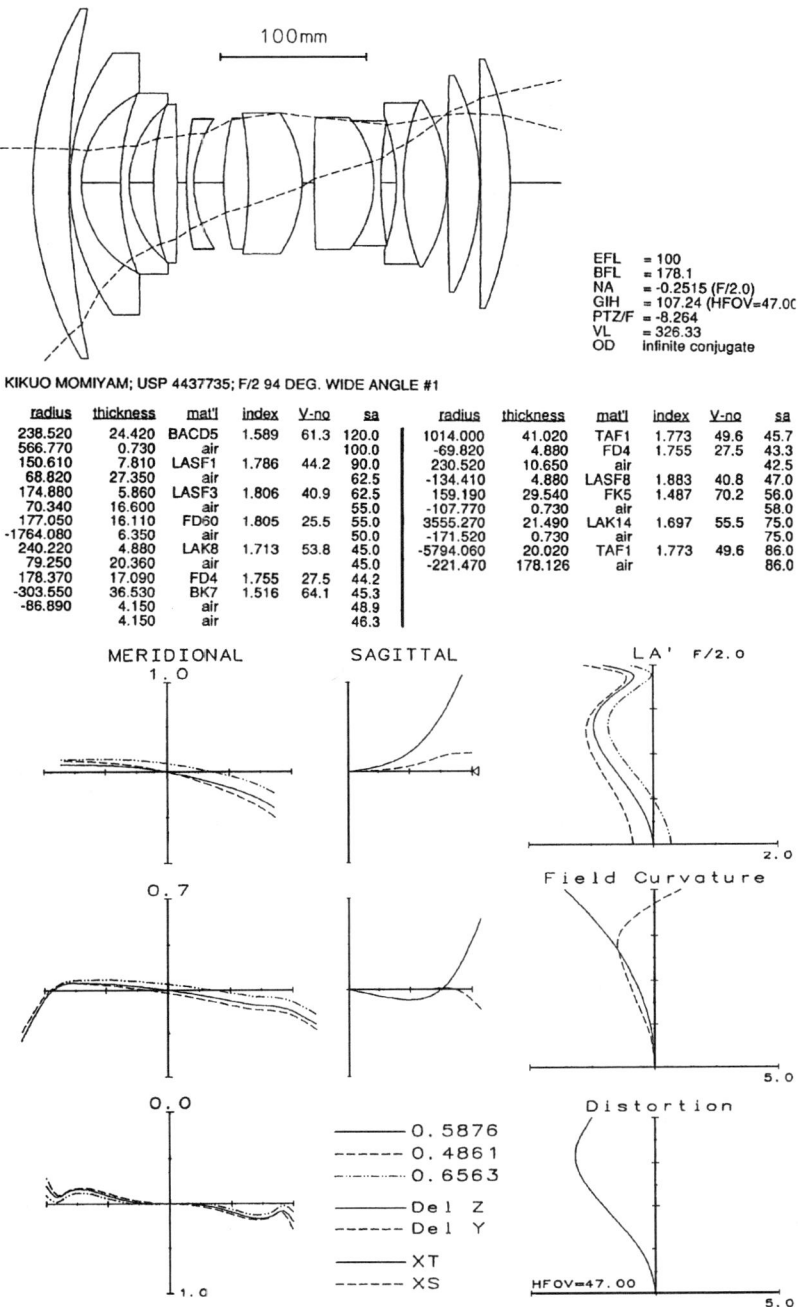

```
EFL   = 100
BFL   = 178.1
NA    = -0.2515 (F/2.0)
GIH   = 107.24 (HFOV=47.00
PTZ/F = -8.264
VL    = 326.33
OD    infinite conjugate
```

KIKUO MOMIYAM; USP 4437735; F/2 94 DEG. WIDE ANGLE #1

radius	thickness	mat'l	index	V-no	sa	radius	thickness	mat'l	index	V-no	sa
238.520	24.420	BACD5	1.589	61.3	120.0	1014.000	41.020	TAF1	1.773	49.6	45.7
566.770	0.730	air			100.0	-69.820	4.880	FD4	1.755	27.5	43.3
150.610	7.810	LASF1	1.786	44.2	90.0	230.520	10.650	air			42.5
68.820	27.350	air			62.5	-134.410	4.880	LASF8	1.883	40.8	47.0
174.880	5.860	LASF3	1.806	40.9	62.5	159.190	29.540	FK5	1.487	70.2	56.0
70.340	16.600	air			55.0	-107.770	0.730	air			58.0
177.050	16.110	FD60	1.805	25.5	55.0	3555.270	21.490	LAK14	1.697	55.5	75.0
-1764.080	6.350	air			50.0	-171.520	0.730	air			75.0
240.220	4.880	LAK8	1.713	53.8	45.0	-5794.060	20.020	TAF1	1.773	49.6	86.0
79.250	20.360	air			45.0	-221.470	178.126	air			86.0
178.370	17.090	FD4	1.755	27.5	44.2						
-303.550	36.530	BK7	1.516	64.1	45.3						
-86.890	4.150	air			48.9						
	4.150	air			46.3						

Figure 14.10 $f/2.0$ $\pm 47°$ a 13-element retrofocus lens.

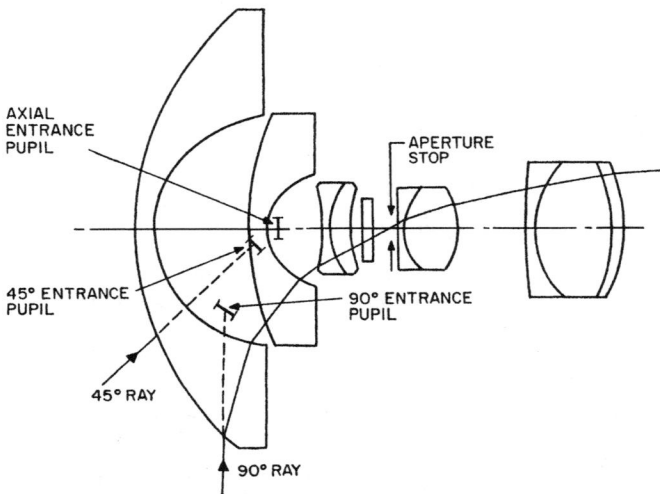

Figure 14.11 In a fish-eye lens, the overcorrected spherical aberration of the principal ray at the front components produces a shift and a rotation of the entrance pupil so that the lens can image a field of 180° or more. (*After R. Kingslake, Optical System Design, Academic Press, N.Y., 1983.*)

In short-focal-length, wide-angle lenses, the longitudinal chromatic and spherochromatic aberration and secondary spectrum are usually not a severe problem, simply because of the short focal length; however, the secondary spectrum of *lateral* color may become serious and chromatic variation of distortion (or higher-order lateral color) can also be troublesome because of the extremely wide angles involved.

In a lens that covers a field of 180° or more, it is obvious that the entrance pupil (the image of the aperture stop) *must* rotate and move if a light beam at 90° to the axis is to enter the lens. Figure 14.11 shows the situation in one such design (Fig. 14.14). It is also fairly common for the entrance pupil to increase in size as the field angle increases because of the aberrations of the front component; this, of course, helps to offset the cosine-fourth effect and thereby produces a more uniform illumination across the field. The large shifts of the position of the entrance pupil can cause problems in the course of the design process if the oblique rays are aimed at the paraxial image of the aperture stop, as is the case in some simple design programs. In Fig. 14.11 it is apparent that, for the large-obliquity ray bundles, the actual entrance pupil is well forward of the paraxial pupil, and, if these rays are aimed at the paraxial image of the stop, they will fail spectacularly. One way of handling this problem that will work with many design programs is as follows: First, use a very long radius parabola or similar surface centered on the lens as the object

Reversed Telephoto (Retrofocus and Fish-Eye) Lenses 409

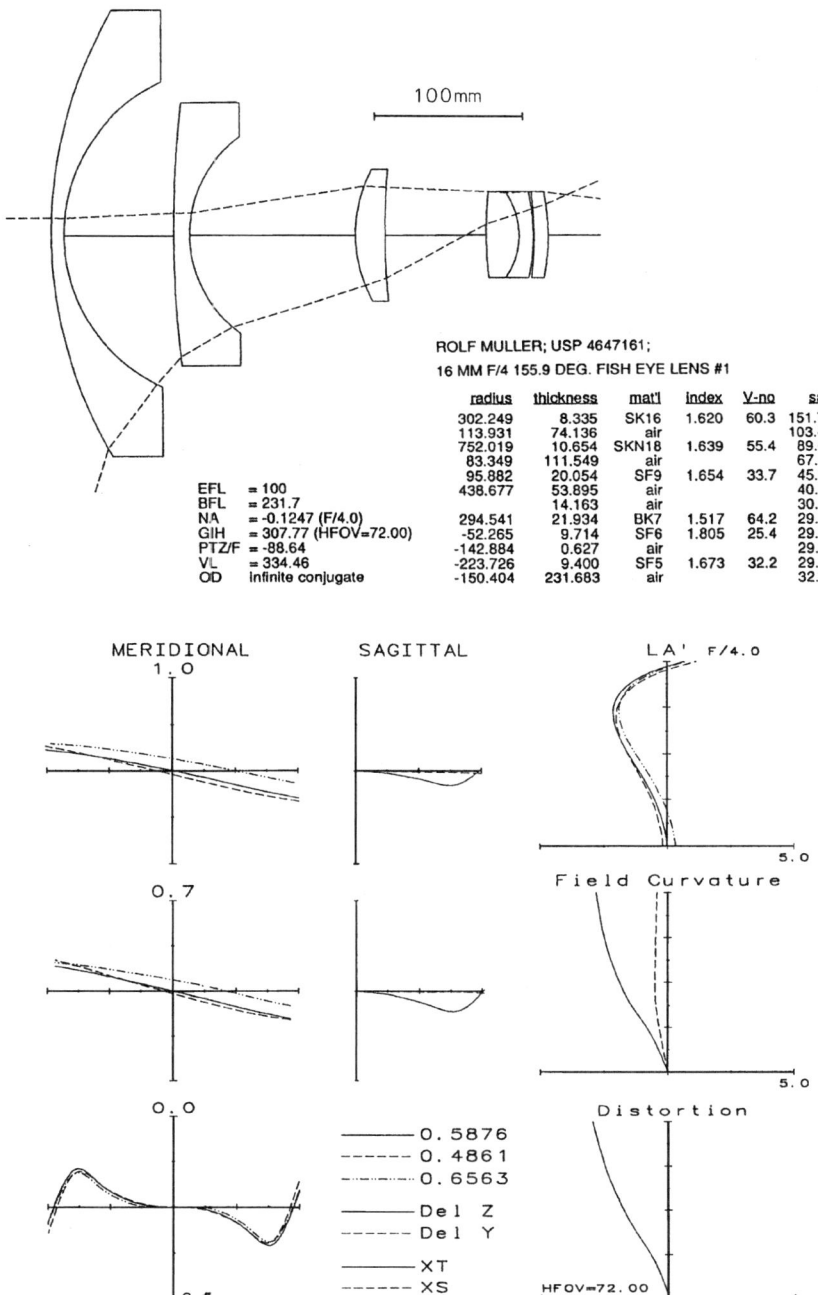

Figure 14.12 $f/4.0$ $\pm 72°$ a fish-eye lens.

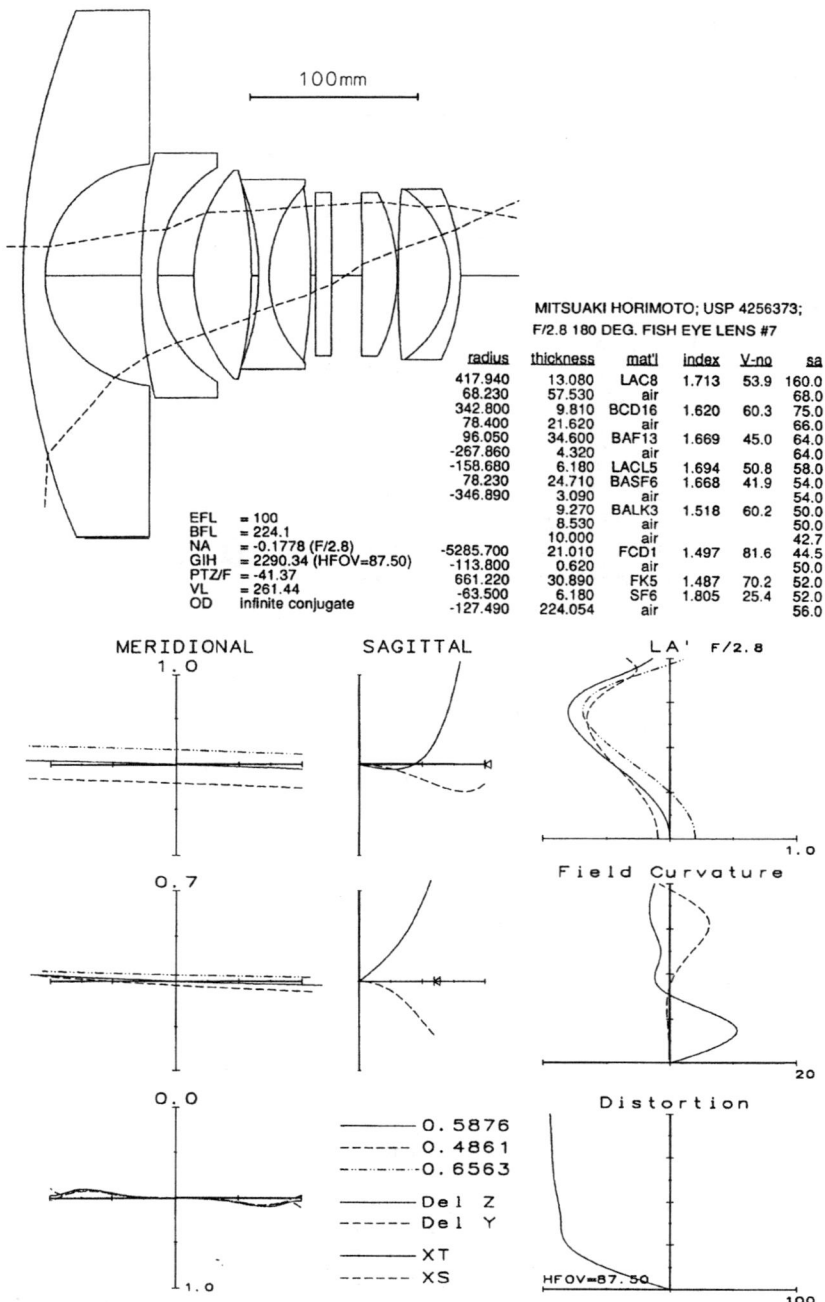

Figure 14.13 f/2.8 ±88° a fish-eye lens.

Reversed Telephoto (Retrofocus and Fish-Eye) Lenses 411

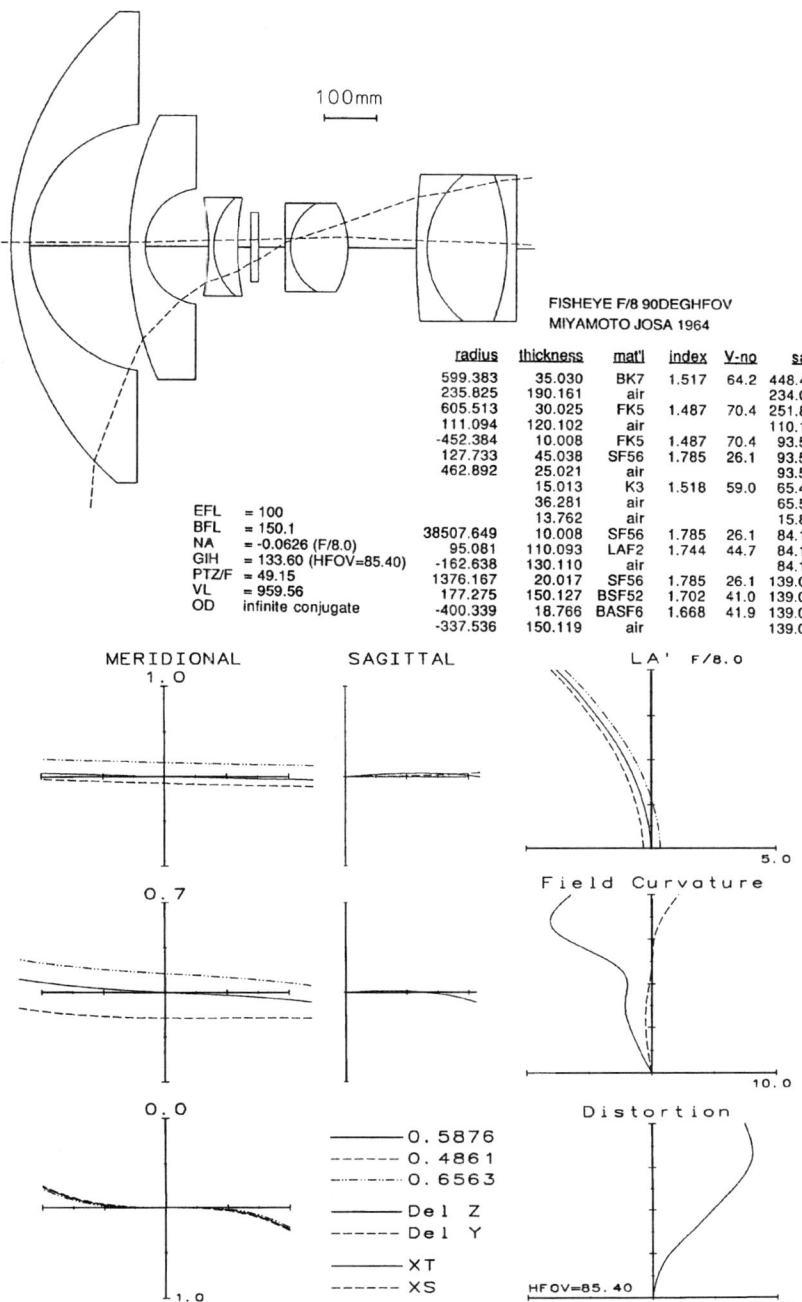

Figure 14.14 $f/8.0$ $\pm 85°$ a fish-eye lens.

412   Chapter Fourteen

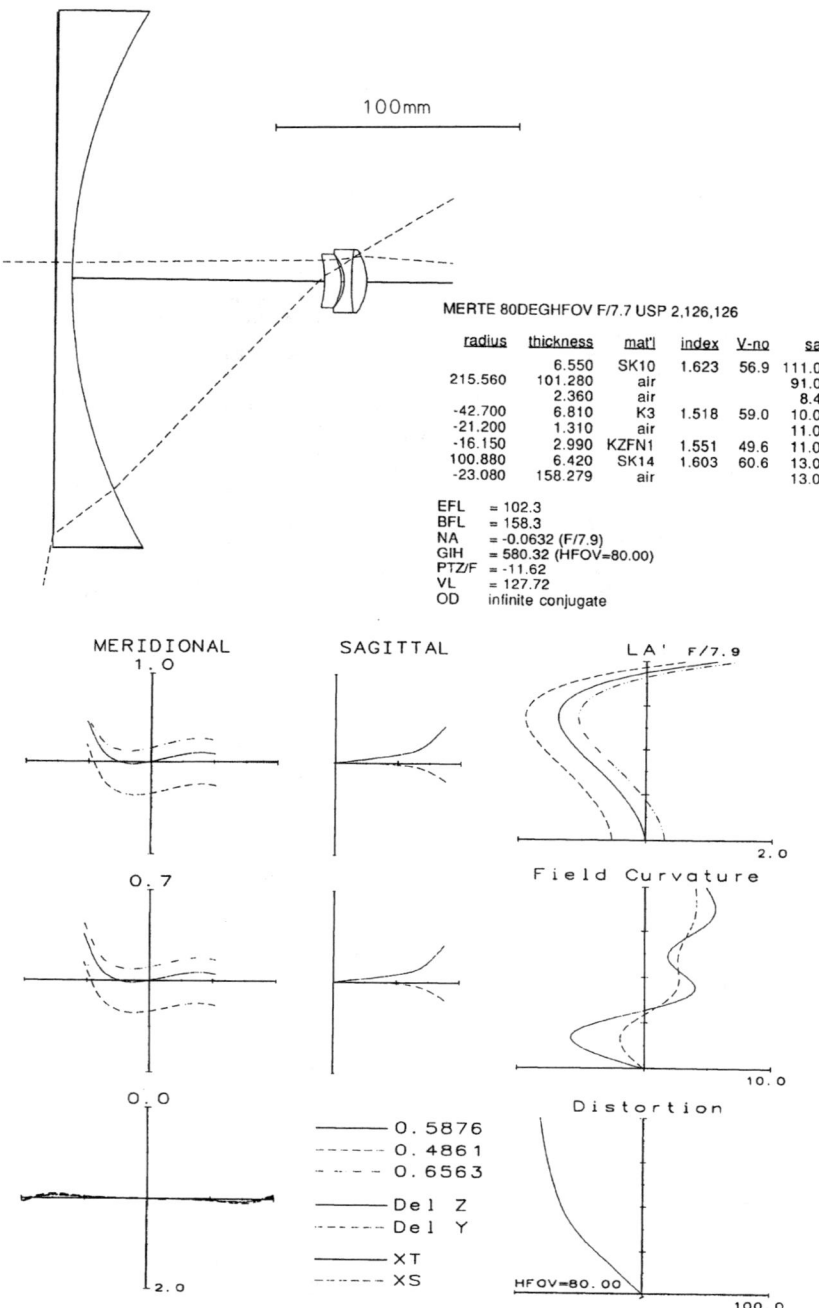

Figure 14.15 $f/8.0\ \pm 80°$ a simple four-element fish-eye lens.

surface (rather than a plane). Obviously a curved object surface is necessary to get a field of 180° or more. Second, set up a separate configuration (just as you would for a zoom lens) for each field point, with each configuration having its own entrance pupil position. Use the height of the principal ray at the aperture stop as an item in the merit function with a target of zero, and allow the entrance pupil position to vary independently in each configuration to satisfy this target. In the early stages of the design process, it may be necessary to adjust the size of the entrance pupil as well, so that the upper- and lower-rim rays pass through the aperture stop at the correct heights. Several programs have this capability built-in and iteratively search for the proper rays to trace.

In many of the early fish-eye designs, the front component consisted of simple negative elements. If this form is used, the lateral color will not be corrected. As can be seen from the examples, a negative achromat or hyperchromat can be used to control the lateral color.

Figures 14.12 to 14.15 show an assortment of extreme retrofocus lenses that cover angular fields large enough to put them into the fish-eye category.

Chapter 15

Wide-Angle Lenses with Negative Outer Elements

This chapter presents several lenses (Figs. 15.1 to 15.5) that derive their ability to cover a wide angular field from a construction that incorporates negative meniscus elements as the outer members. A glance at the path of the principal ray in any of the lens drawings will indicate one purpose of this construction. The outer negative lenses take a wide angular field in object and/or image space and convert it to a much smaller angular field inside the lens, a smaller field that is within the ability of the inner elements to handle. In addition, the wide spacing between the positive center member and the negative outer members provides the spacing between positive and negative power used to flatten the Petzval curvature.

To some extent these negative components can be viewed in the same light as the negative front member of the retrofocus type discussed in Chap. 14; however, the reverse telephoto lens has a long back focal length and absolutely no symmetry. In the lenses of this chapter the possibility of at least a roughly symmetrical construction is apparent, since (with one exception) they all have meniscus negative outer elements on both sides of the lens, and the inner members are of a roughly symmetrical construction as well. The back focal length of a retrofocus lens is typically greater than its focal length; in these lenses the back focal length ranges from as little as 23 to 65 percent of the focal length.

All these lenses tend to have quite good off-axis illumination. This is because of the negative meniscus outer elements whose large pupil aberration tilts the pupil and increases its effective size. This ameliorates the cosine-fourth illumination problem (which assumes that the pupil aberrations and distortion are negligible). The front elements of

416   Chapter Fifteen

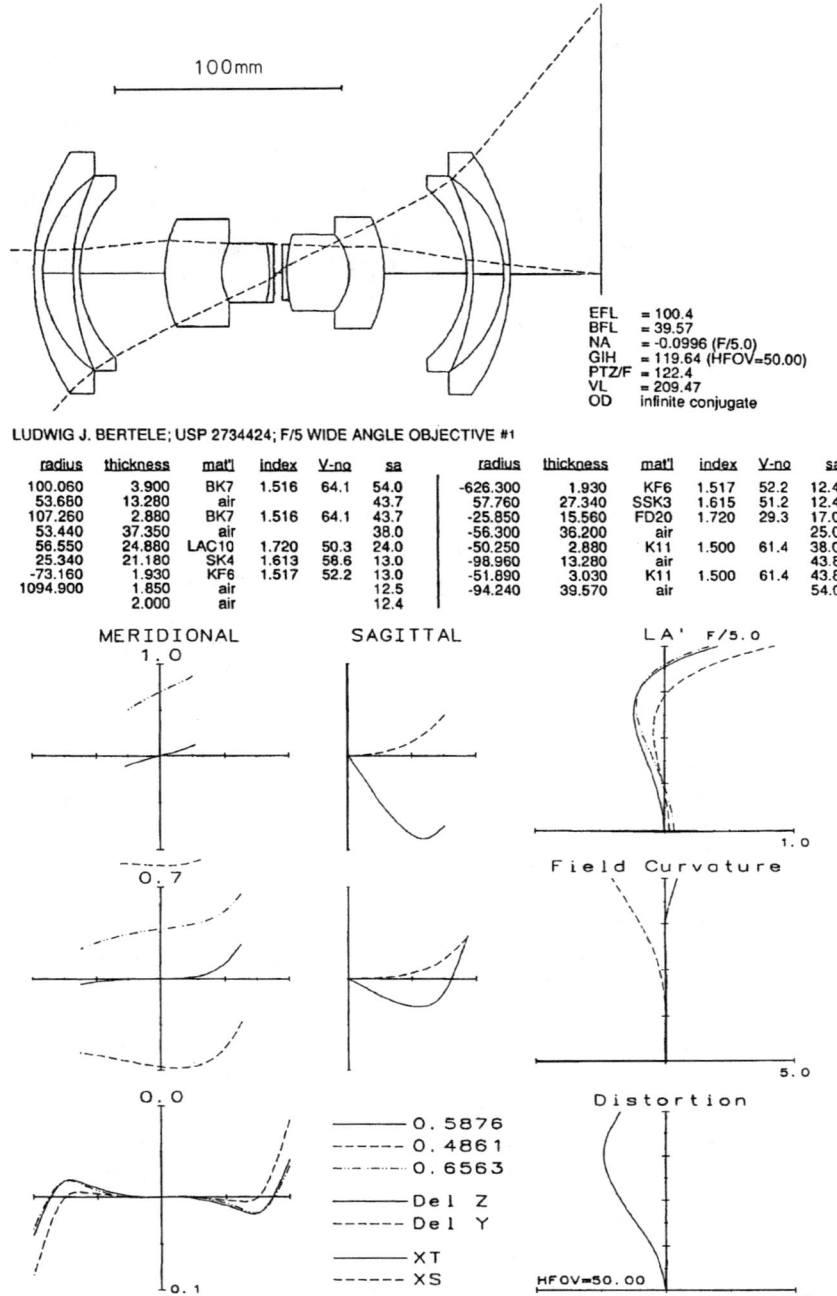

Figure 15.1 $f/5.0$ ±50° ten-element wide-angle objective.

Wide-Angle Lenses with Negative Outer Elements 417

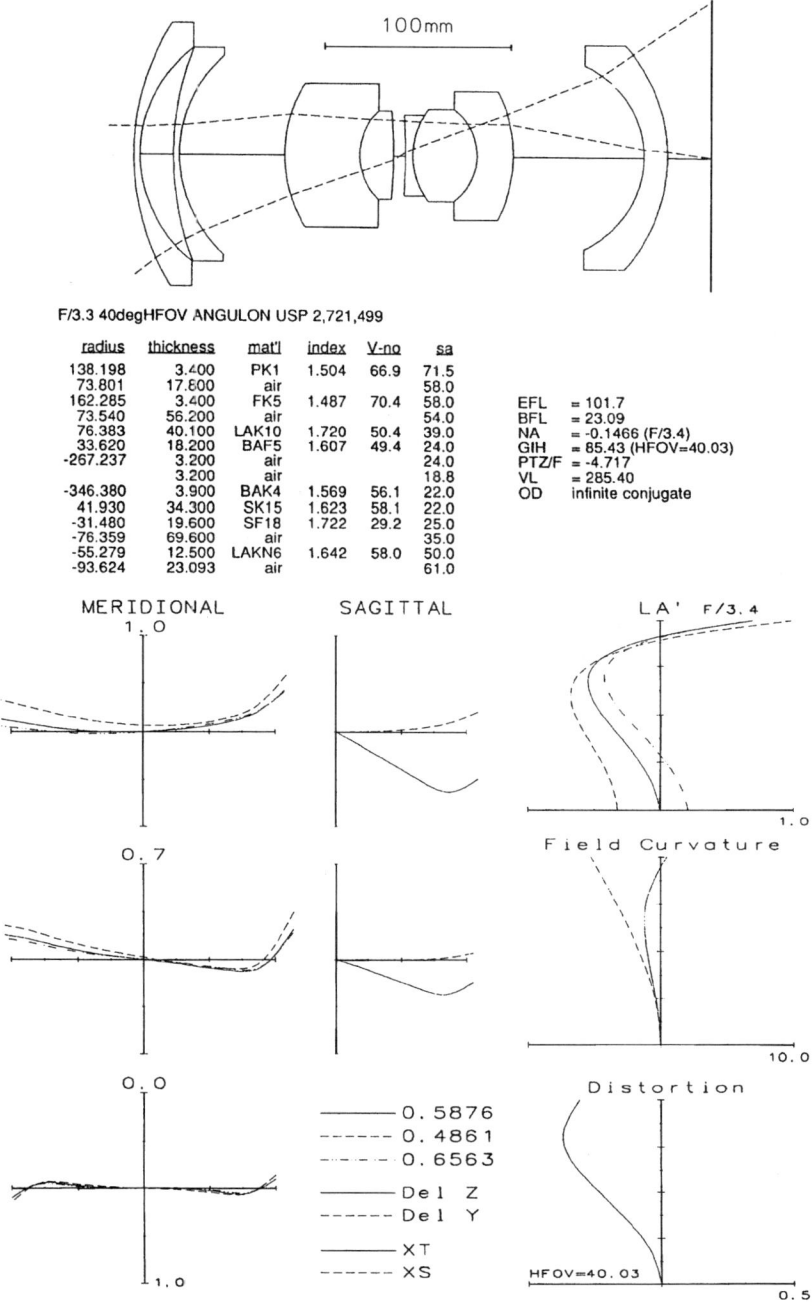

F/3.3 40degHFOV ANGULON USP 2,721,499

radius	thickness	mat'l	index	V-no	sa
138.198	3.400	PK1	1.504	66.9	71.5
73.801	17.800	air			58.0
162.285	3.400	FK5	1.487	70.4	58.0
73.540	56.200	air			54.0
76.383	40.100	LAK10	1.720	50.4	39.0
33.620	18.200	BAF5	1.607	49.4	24.0
-267.237	3.200	air			24.0
	3.200	air			18.8
-346.380	3.900	BAK4	1.569	56.1	22.0
41.930	34.300	SK15	1.623	58.1	22.0
-31.480	19.600	SF18	1.722	29.2	25.0
-76.359	69.600	air			35.0
-55.279	12.500	LAKN6	1.642	58.0	50.0
-93.624	23.093	air			61.0

EFL = 101.7
BFL = 23.09
NA = -0.1466 (F/3.4)
GIH = 85.43 (HFOV=40.03)
PTZ/F = -4.717
VL = 285.40
OD infinite conjugate

Figure 15.2 $f/3.4$ ±40° eight-element wide-angle objective.

Chapter Fifteen

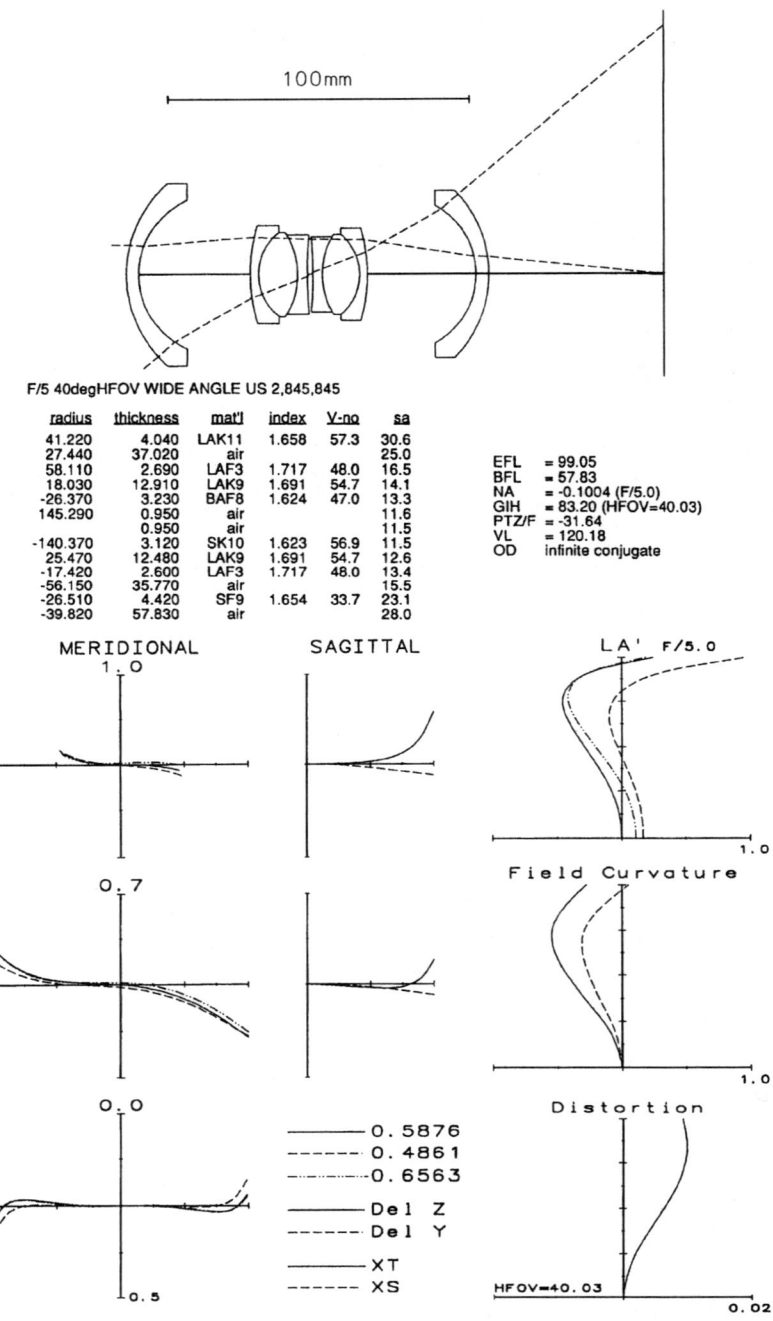

Figure 15.3 $f/5.0 \pm 40°$ eight-element wide-angle objective.

Wide-Angle Lenses with Negative Outer Elements 419

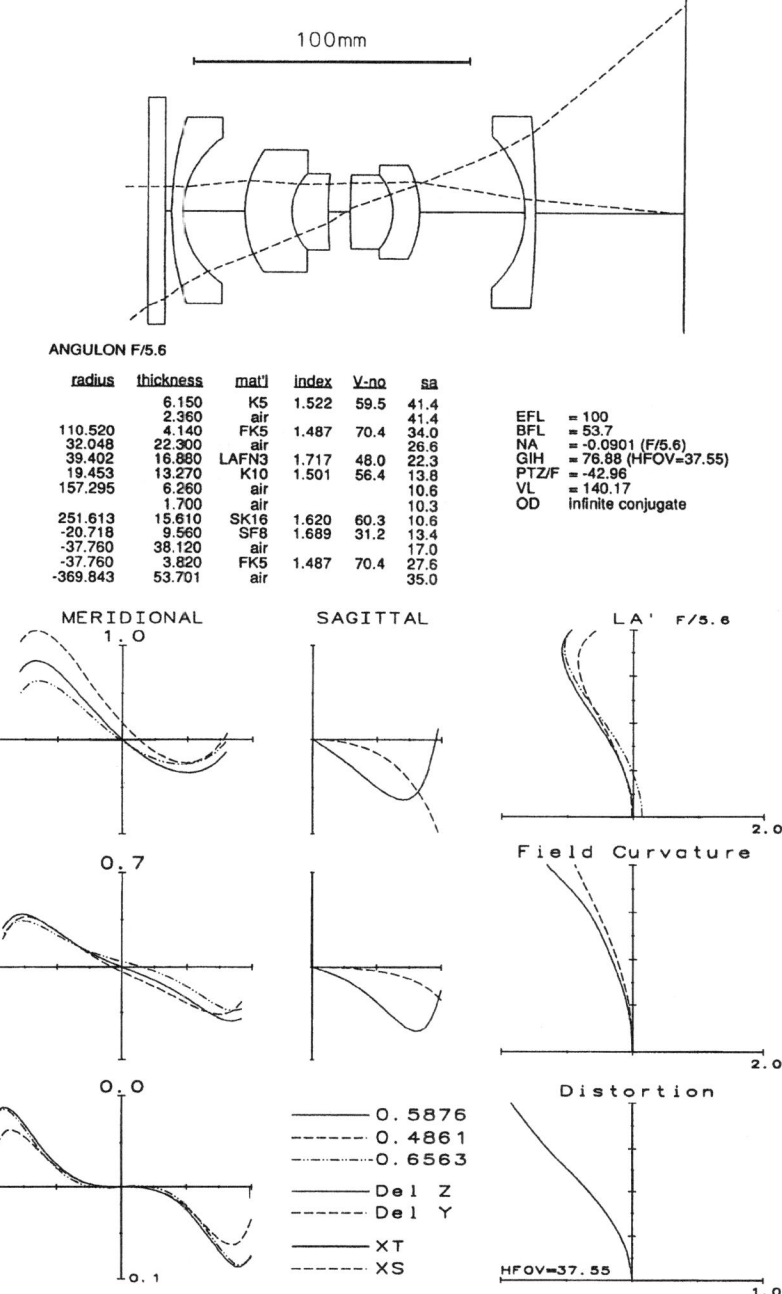

ANGULON F/5.6

radius	thickness	mat'l	index	V-no	sa
	6.150	K5	1.522	59.5	41.4
	2.360	air			41.4
110.520	4.140	FK5	1.487	70.4	34.0
32.048	22.300	air			26.6
39.402	16.880	LAFN3	1.717	48.0	22.3
19.453	13.270	K10	1.501	56.4	13.8
157.295	6.260	air			10.6
	1.700	air			10.3
251.613	15.610	SK16	1.620	60.3	10.6
-20.718	9.560	SF8	1.689	31.2	13.4
-37.760	38.120	air			17.0
-37.760	3.820	FK5	1.487	70.4	27.6
-369.843	53.701	air			35.0

EFL = 100
BFL = 53.7
NA = -0.0901 (F/5.6)
GIH = 76.88 (HFOV=37.55)
PTZ/F = -42.96
VL = 140.17
OD infinite conjugate

Figure 15.4 $f/5.6 \pm 38°$ six-element wide-angle objective.

Chapter Fifteen

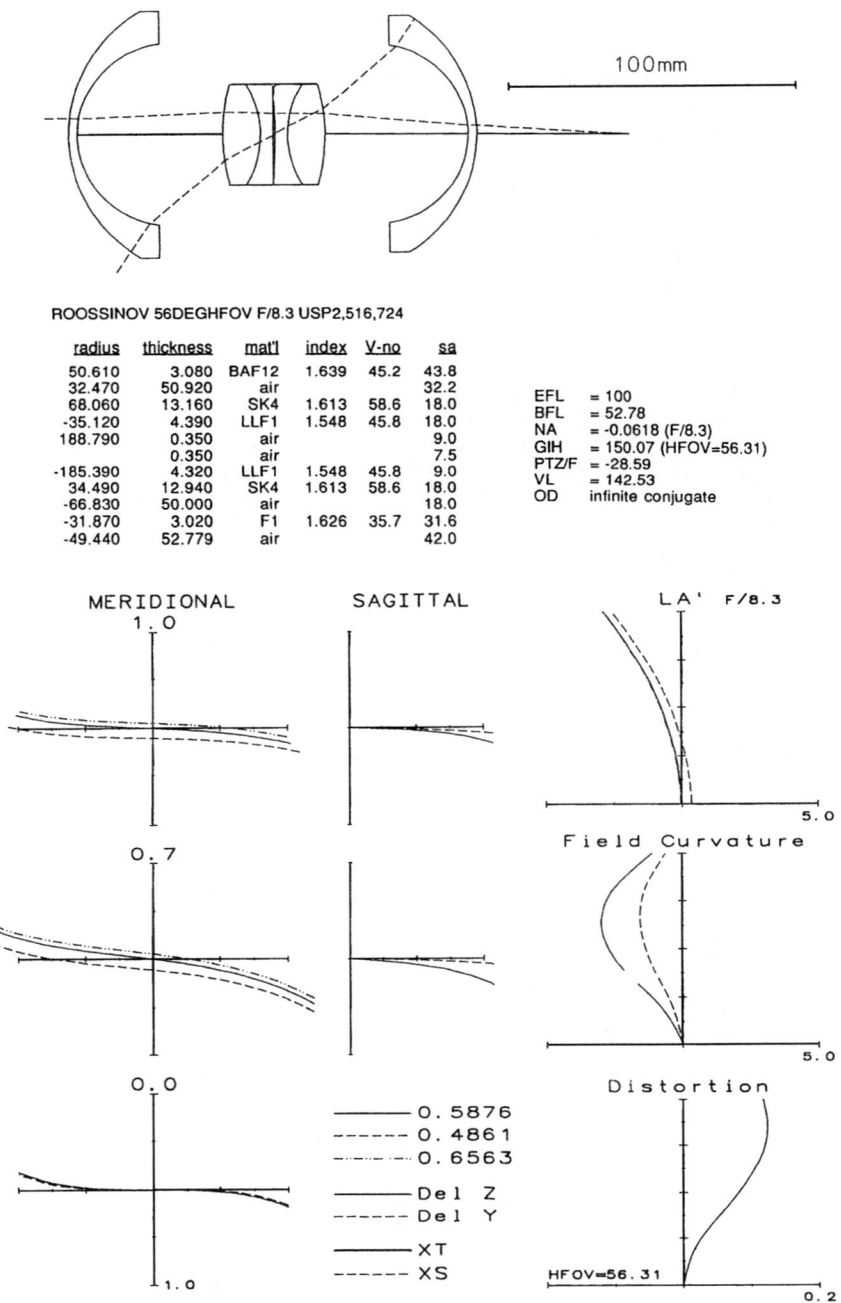

ROOSSINOV 56DEGHFOV F/8.3 USP2,516,724

radius	thickness	mat'l	index	V-no	sa
50.610	3.080	BAF12	1.639	45.2	43.8
32.470	50.920	air			32.2
68.060	13.160	SK4	1.613	58.6	18.0
-35.120	4.390	LLF1	1.548	45.8	18.0
188.790	0.350	air			9.0
	0.350	air			7.5
-185.390	4.320	LLF1	1.548	45.8	9.0
34.490	12.940	SK4	1.613	58.6	18.0
-66.830	50.000	air			18.0
-31.870	3.020	F1	1.626	35.7	31.6
-49.440	52.779	air			42.0

EFL = 100
BFL = 52.78
NA = -0.0618 (F/8.3)
GIH = 150.07 (HFOV=56.31)
PTZ/F = -28.59
VL = 142.53
OD infinite conjugate

Figure 15.5 $f/8.3 \pm 56°$ six-element wide-angle objective.

Wide-Angle Lenses with Negative Outer Elements 421

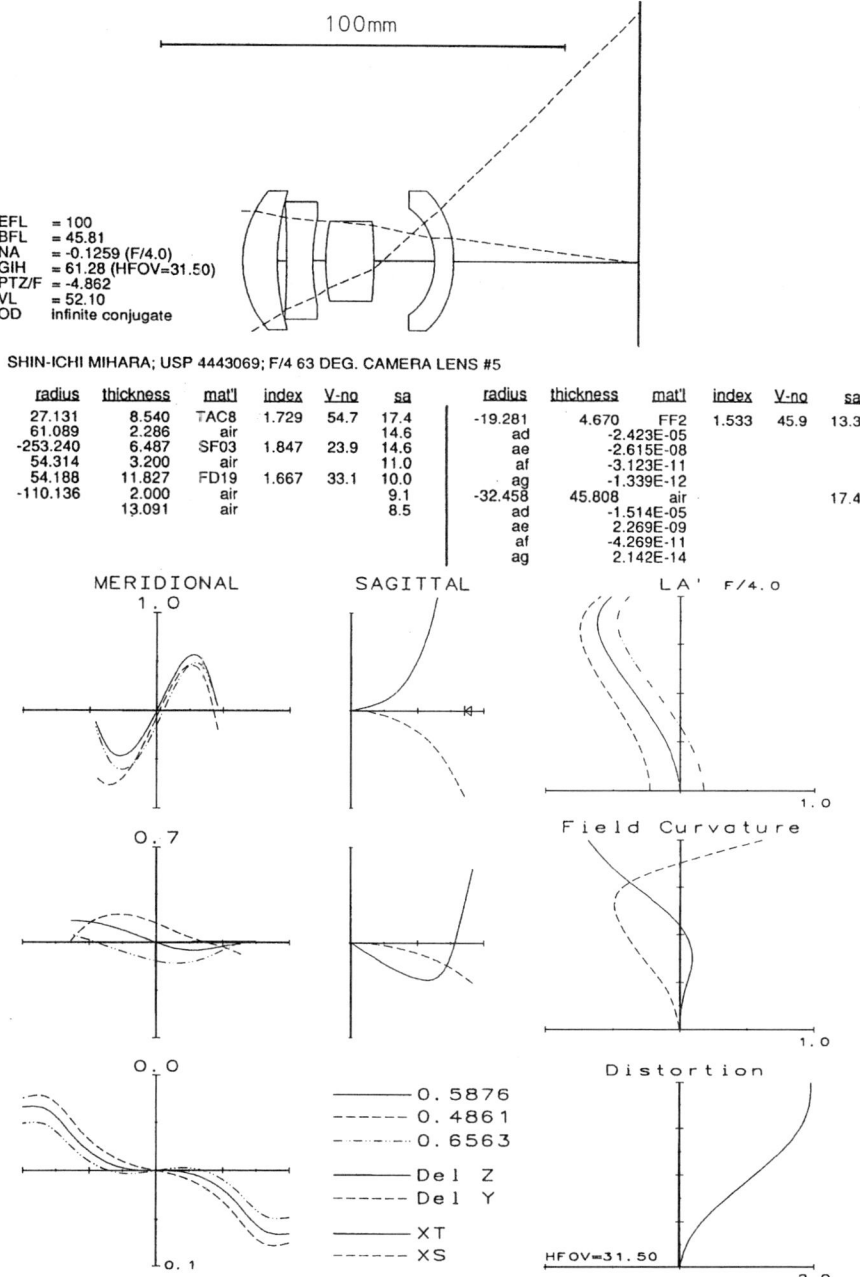

Figure 15.6 $f/4.0 \pm 32°$ modified triplet with aspheric field corrector.

the fish-eye lens have the same effect, as shown in Fig. 14.11. (Note that positive outer elements have the opposite effect on the pupil. See Sec. 11.2.) The pupil aberrations that expand and tilt the pupil are usually coma and spherical aberration.

The "solid glass sphere" central core feature of these designs is said to be good for astigmatism. Note that, with the exception of the Roossinov design in Fig. 15.5, all these designs have a diverging (overcorrecting for spherical) cemented surface in the central component, concave to the stop. Those that also have weak converging surfaces in the central component (to flatten the field), have them oriented convex to the stop. These designs tend to be quite long (ranging from 1.2 to 3 focal lengths) and have a short back focus. These lenses superceded the Topogon and Metrogon as aerial camera lenses because of their better illumination and good distortion correction.

The unusual design of Fig. 15.6 can be regarded as (1) a member of this family with only one outer negative element, (2) a telephoto lens, or (3) a triplet anastigmat with an aspheric field corrector. This is probably an objective for a "point and shoot" camera.

Note that there are also lenses of wide angular coverage included in Chaps. 11 and 14.

Chapter 16

The Petzval Lens; Head-up Display Lenses

16.1 The Petzval Portrait Lens

In 1839 Joseph Petzval, assisted by "two corporals and eight bombadiers skilled in computing" was commissioned by the Archduke Ludwig to create a new lens design. Six months later this group had designed two lenses, one of which was the famed Petzval portrait lens. Ironically, the Petzval lens did not win the contest in which it was entered. The contest winner soon faded into obscurity, while more than 8000 Petzval portrait lenses were sold by 1850. This may have been the first mathematically designed lens.

The basic Petzval lens consists of two positive components, spaced apart so that the astigmatism is controlled to be either zero or slightly positive. Usually the two components are achromats, typically doublets.*

The original Petzval lens (Fig. 16.1) was designed as a portrait lens for the Daguerrotype camera and, at a speed of about $f/3.5$, was (for its day) a very fast lens.

16.2 The Petzval Projection Lens

The more modern version, the Petzval projection lens (Fig. 16.2), is noted for covering a small field at high aperture with excellent image quality.

---

*A significant exception to this is the infrared lens in the form of an airspaced doublet widely spaced from a positive singlet. This can be considered either as a triplet anastigmat or as a Petzval lens. This particular design configuration results from the high index and low dispersion of the infrared materials used—usually silicon and germanium for the mid-infrared, or germanium and zinc sulfide or selenide for the 8- to 12-μm region. See Chap. 19 for infrared lens designs.

424 Chapter Sixteen

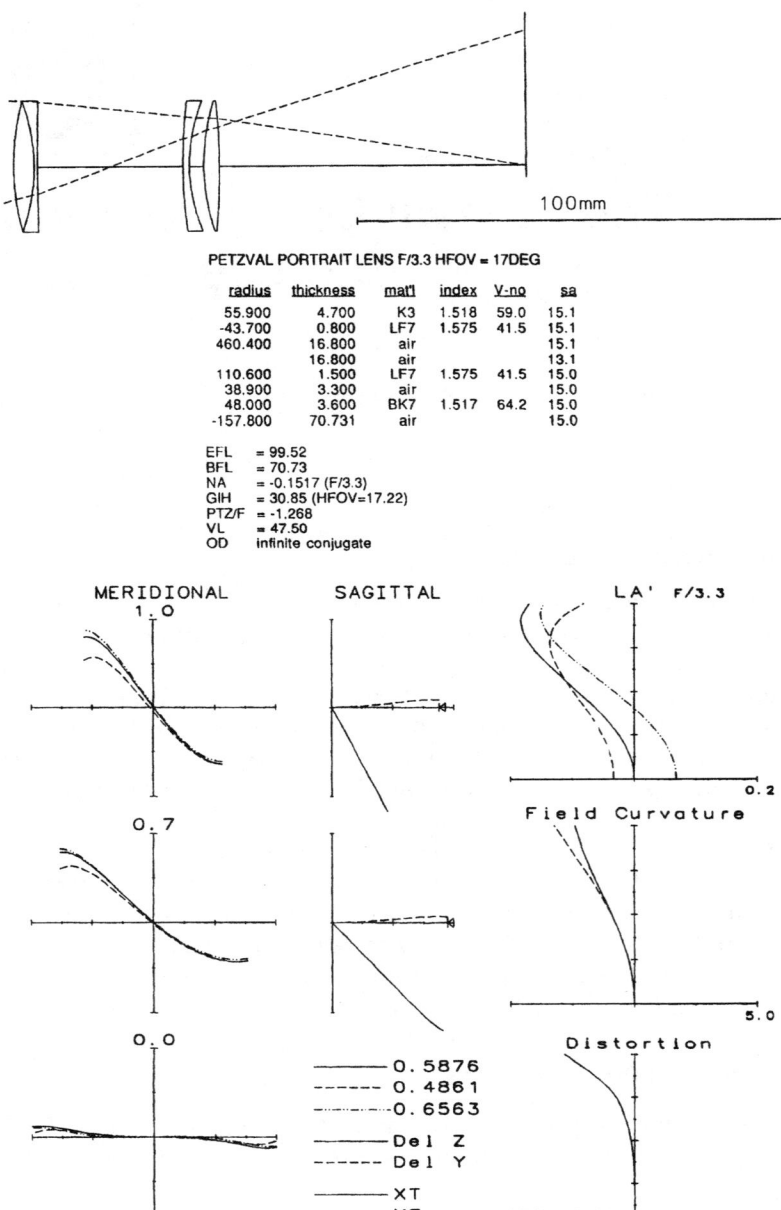

Figure 16.1 The original Petzval portrait lens at $f/3.3$ and $\pm 17°$.

The Petzval Lens; Head-up Display Lenses

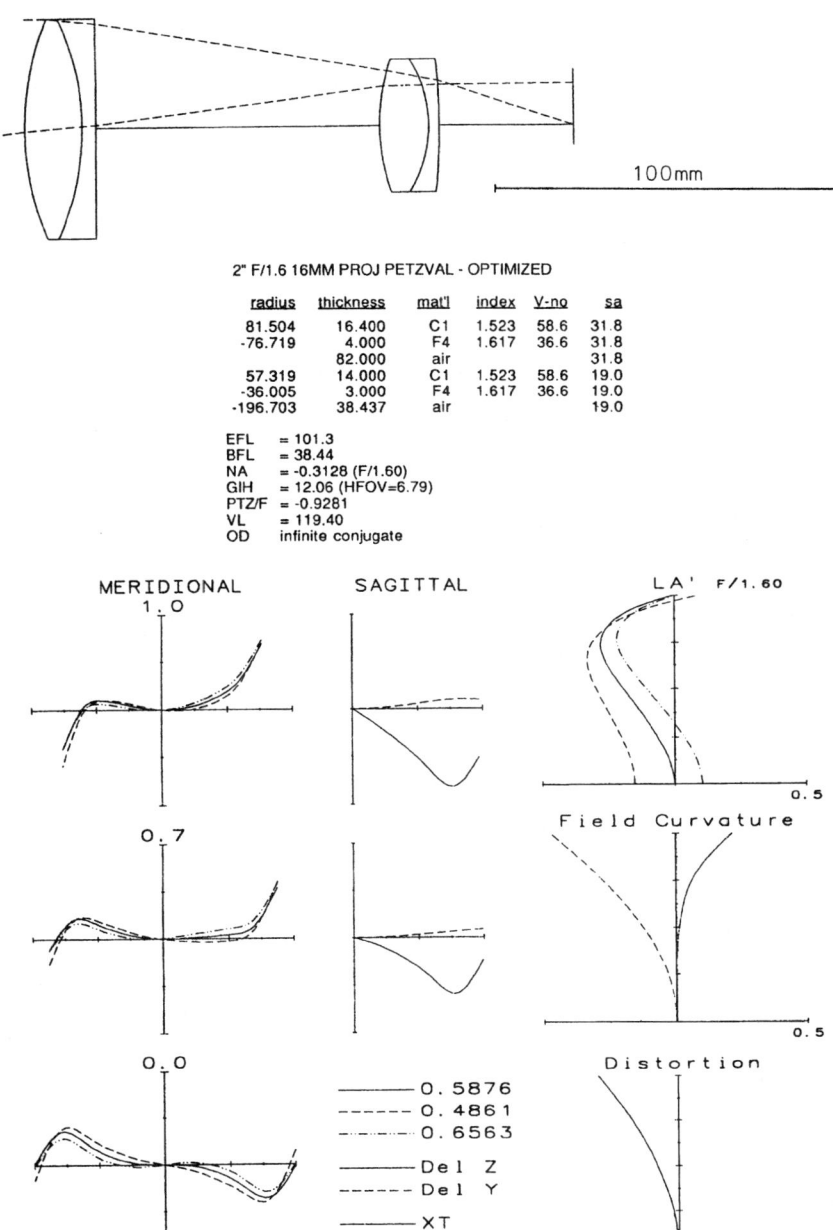

Figure 16.2 The Petzval projection lens is a big modification of the portrait lens, which allows a higher speed over a smaller field, $f/1.6 \pm 7°$.

The zonal spherical and spherochromatism are small, and the secondary spectrum is actually less than that of an achromatic doublet of the same glass. The classic arrangement is two achromats, each bending the axial ray toward the axis by the same amount, so that the work is equally divided—often a good principle to follow. For a system with a focal length of f, the front doublet has a focal length of $2f$, the rear doublet has a focal length of f, and the (thin lens) spacing is equal to f. Then the (thin lens) back focus is equal to $f/2$.

The aperture stop of the Petzval projection lens is located at the front component; stop shift theory tells us that a thin positive lens located at the stop can contribute only negative astigmatism. Therefore the rear component must contribute at least enough overcorrected astigmatism to offset this undercorrection. The astigmatism of the rear lens is, of course, a function of its shape and its distance from the stop, but it is also significantly affected by the cemented surface. If the index break $(n' - n)$ is not large enough, a broken contact at the rear doublet can be utilized to introduce the required overcorrected astigmatism (just as in the portrait lens).

In Fig. 16.3, the airspace between the doublets is made small to increase the working distance (bfl) of the lens. The contact is broken at both doublets to correct the aberrations. The spherical aberration residual is quite a bit bigger than in Fig. 16.2, despite the undercorrecting seventh-order apparent at the margin in the ray intercept plot. The shorter airspace reduces the Petzval field curvature at the same time it increases the bfl and makes the astigmatism correction more difficult.

Higher-index glasses are used in Fig. 16.4. Again, both doublets have been airspaced. The result is an improvement in both the zonal spherical and the field curvature, although the oblique spherical is quite large.

16.3 The Petzval with a Field Flattener

Since the Petzval lens is afflicted with a very large Petzval curvature, a Piazzi-Smyth field flattener is often used to correct the situation. The field flattener is a strong negative lens placed close to the focal plane, where it has little effect on the focal length or on most aberrations, but it corrects the Petzval field curvature. The drawback to this arrangement is the short back focal length, or working distance, that results. Also, the fact that the field flattener is close to the focal plane means that defects or contamination on its surfaces may be apparent in the image. For this reason the field flattener is usually located some distance from the image plane to avoid these problems. When this is done both the field flattener and the basic lens must be made more powerful to maintain the focal length of the lens; this, of course, increases the residual spherical aberration and spherochromatism.

The Petzval Lens; Head-up Display Lenses 427

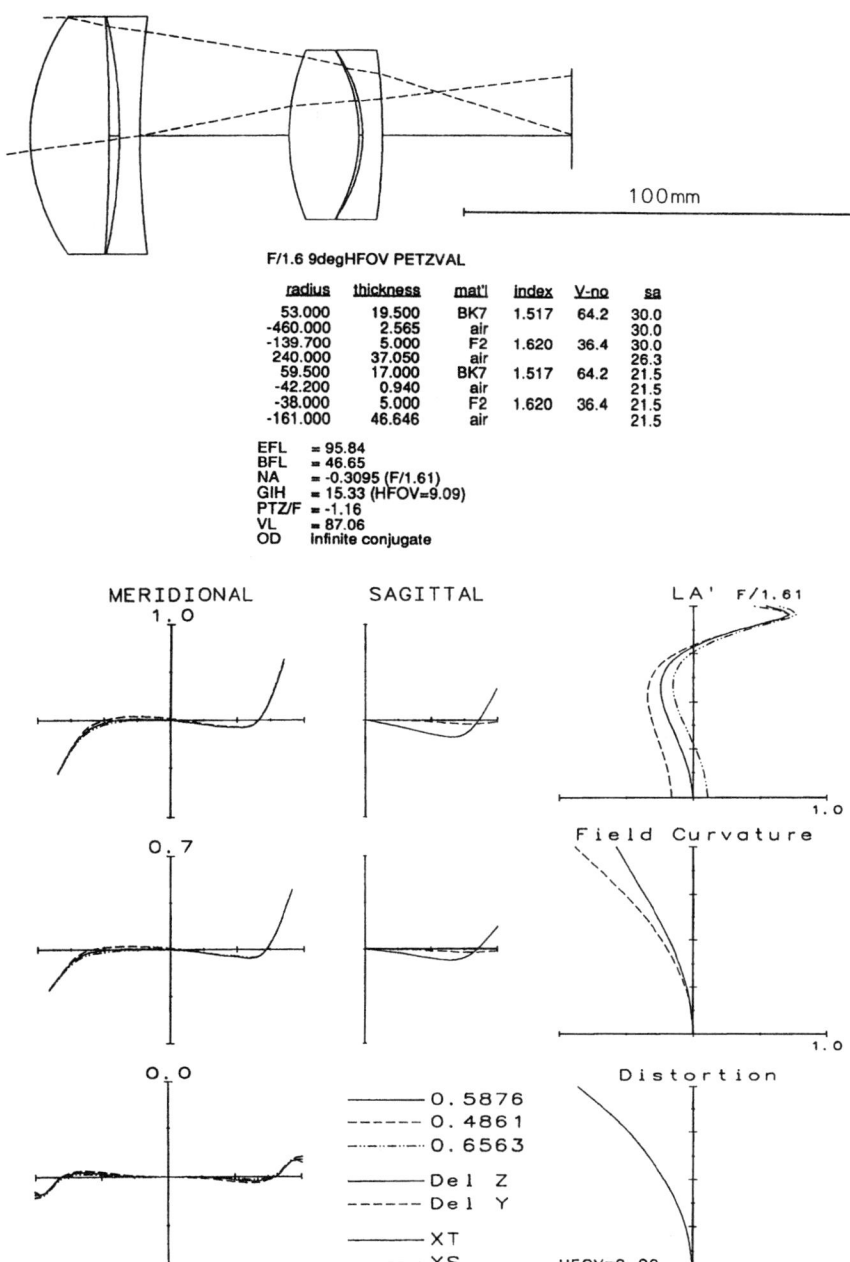

F/1.6 9degHFOV PETZVAL

radius	thickness	mat'l	index	V-no	sa
53.000	19.500	BK7	1.517	64.2	30.0
-460.000	2.565	air			30.0
-139.700	5.000	F2	1.620	36.4	30.0
240.000	37.050	air			26.3
59.500	17.000	BK7	1.517	64.2	21.5
-42.200	0.940	air			21.5
-38.000	5.000	F2	1.620	36.4	21.5
-161.000	46.646	air			21.5

EFL = 95.84
BFL = 46.65
NA = -0.3095 (F/1.61)
GIH = 15.33 (HFOV=9.09)
PTZ/F = -1.16
VL = 87.06
OD Infinite conjugate

Figure 16.3 In this design the Petzval lens has been shortened to achieve a longer working distance, at a sacrifice in performance. $f/1.6 \pm 9°$.

428 Chapter Sixteen

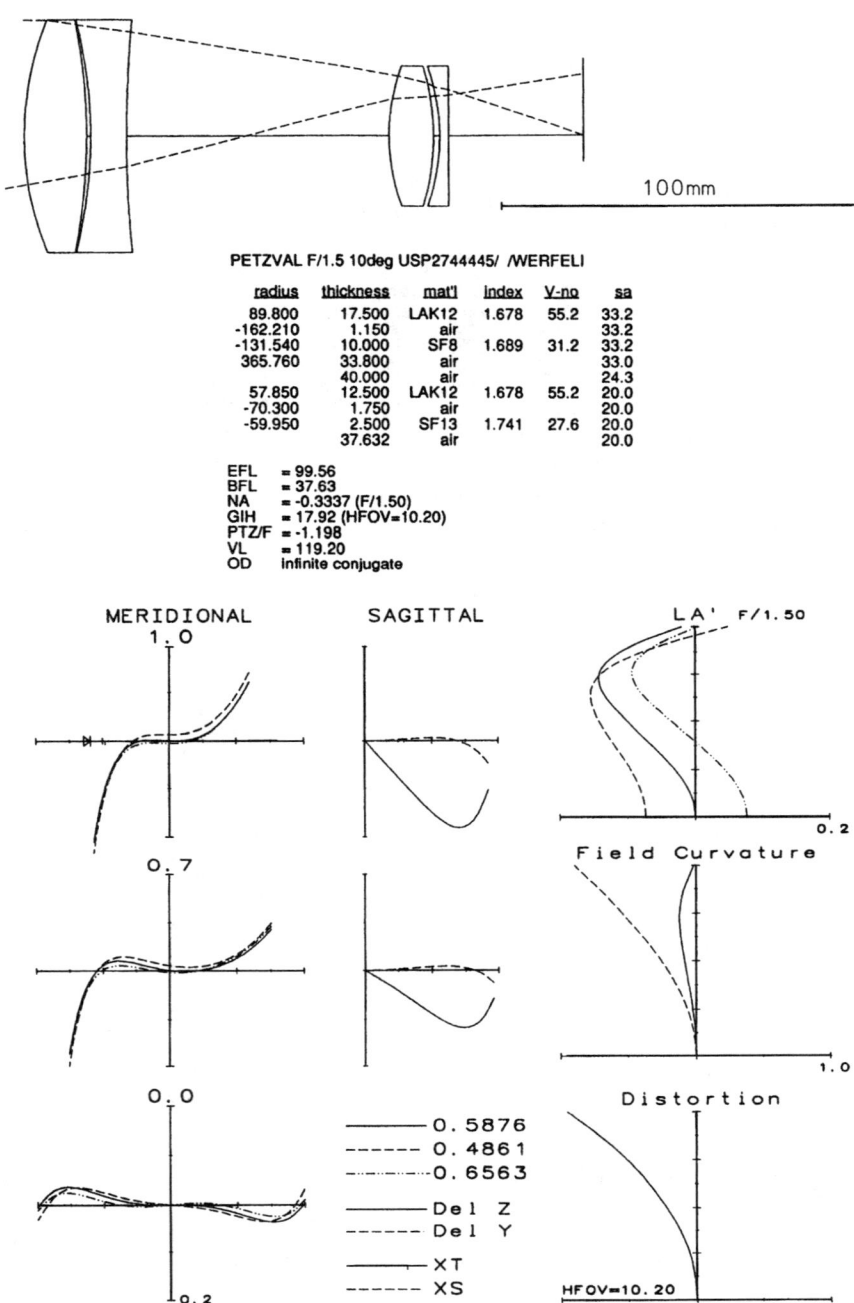

Figure 16.4 An $f/1.5$ Petzval using high-index glass, shown with a field of $\pm 10°$.

The improvement in the field curvature produced by a field flattener is apparent in Fig. 16.5. Note that contact has been broken in the front doublet, probably to correct the spherical; this was not necessary in the rear doublet because the combination of a sizable index break (1.611 to 1.720) at the cemented interface and a reasonably large airspace between the doublets are sufficient to properly control the astigmatism.

In Fig. 16.6 the elements of the rear doublet are spaced so far apart that the flint element acts as a field flattener. Again the front doublet is split to correct the spherical aberration. The spherical zonal is relatively large in this construction. Note that all of the field flatteners in this chapter are made of high-dispersion flint glass. One might expect that the field flattener should be made of a low-index, high-V-value crown glass to reduce its chromatic effects and to increase its overcorrecting Petzval contribution; however, it turns out that with a non-zero back focus a flint glass helps the axial chromatic correction greatly; this benefit is sufficient to make a low-V-value glass the best choice for the field flattener.

The lens of Fig. 16.7 is derived from Fig. 16.6 by splitting each positive element into two parts. The speed is increased to $f/1.4$. The improvement is quite obvious. A limiting aberration in this design is fifth-order coma; this shows up in the off-axis ray intercept plots with overcorrected (third-order) coma in the center of the aperture and undercorrected (fifth-order) at the margin of the aperture. This is often found in lenses where a large edge contact airspace is used to correct the spherical aberration. A redesign that allowed the glasses to vary resulted in Fig. 16.8; the major improvement that this produced was the reduction of this coma and the spherochromatism.

Figures 16.9 and 16.10 are lenses of speed and angular coverage similar to those of Figs. 16.7 and 16.8. Both utilize high-index glasses to achieve a high level of correction. Figure 16.11, shown at a speed of $f/1.25$, uses low-index glasses but has a very short back focal length, which reduces the element powers. All three illustrate the idea of splitting a positive element in two to improve the image quality. Notice that, in Fig. 16.10, in which the rear crown has apparently been split, the front doublet has only about half the power as in Figs. 16.9 and 16.11. Some of the positive power has been shifted from the front doublet to the rear; the front doublet almost functions as a low-power corrector. Another way to view this lens is as a three-doublet Petzval with the third doublet split and widely spaced to flatten the field, as in Fig. 13.6.

16.4 Very High-Speed Petzval Lenses

The Petzval lens is the basis of many extremely fast lenses. With relatively modest modifications, projection lenses with speeds of $f/1.0$ have been made; more extreme modifications have been used to push the speed beyond $f/1.0$.

430   Chapter Sixteen

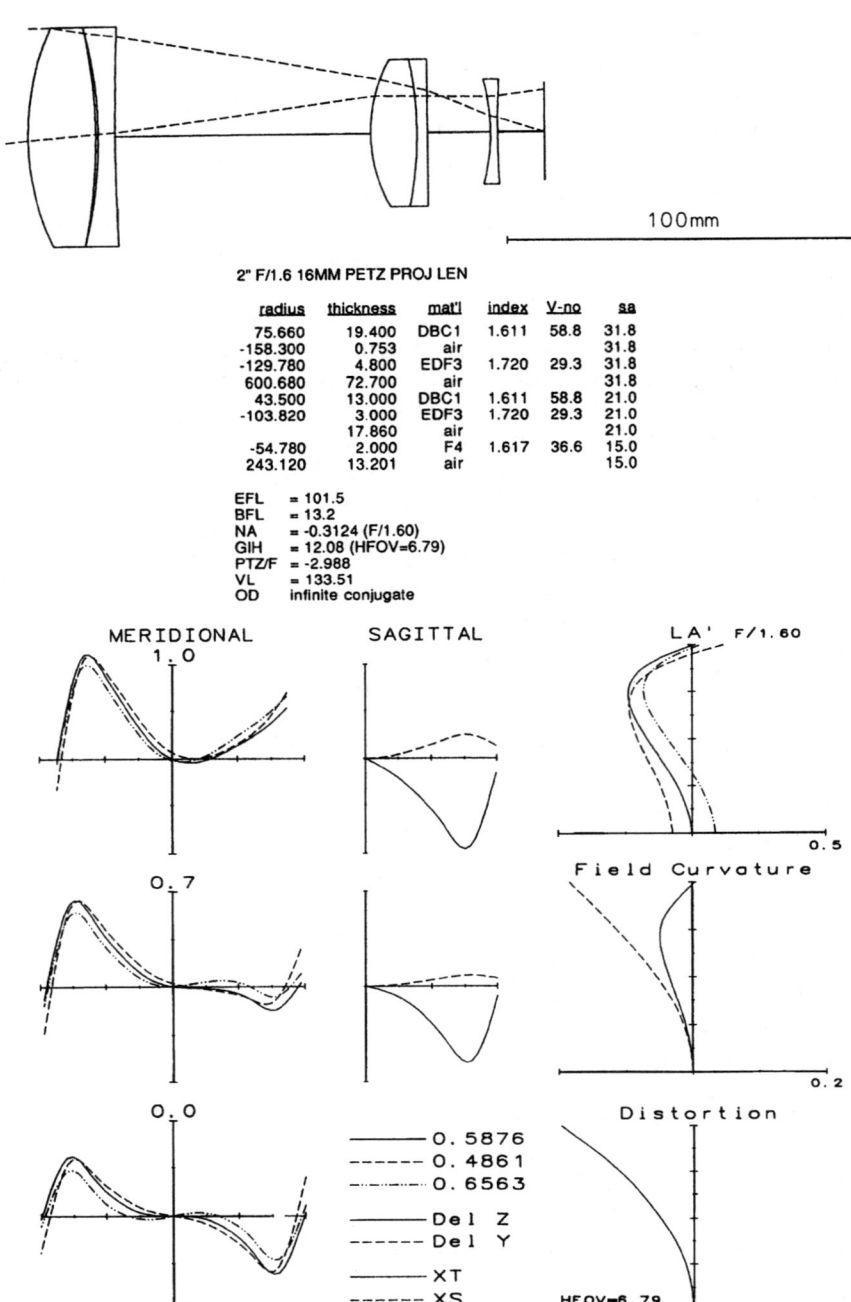

Figure 16.5 A field flattener lens near the focal plane improves the off-axis performance of the f/1.6 Petzval covering a ±7° field.

The Petzval Lens; Head-up Display Lenses 431

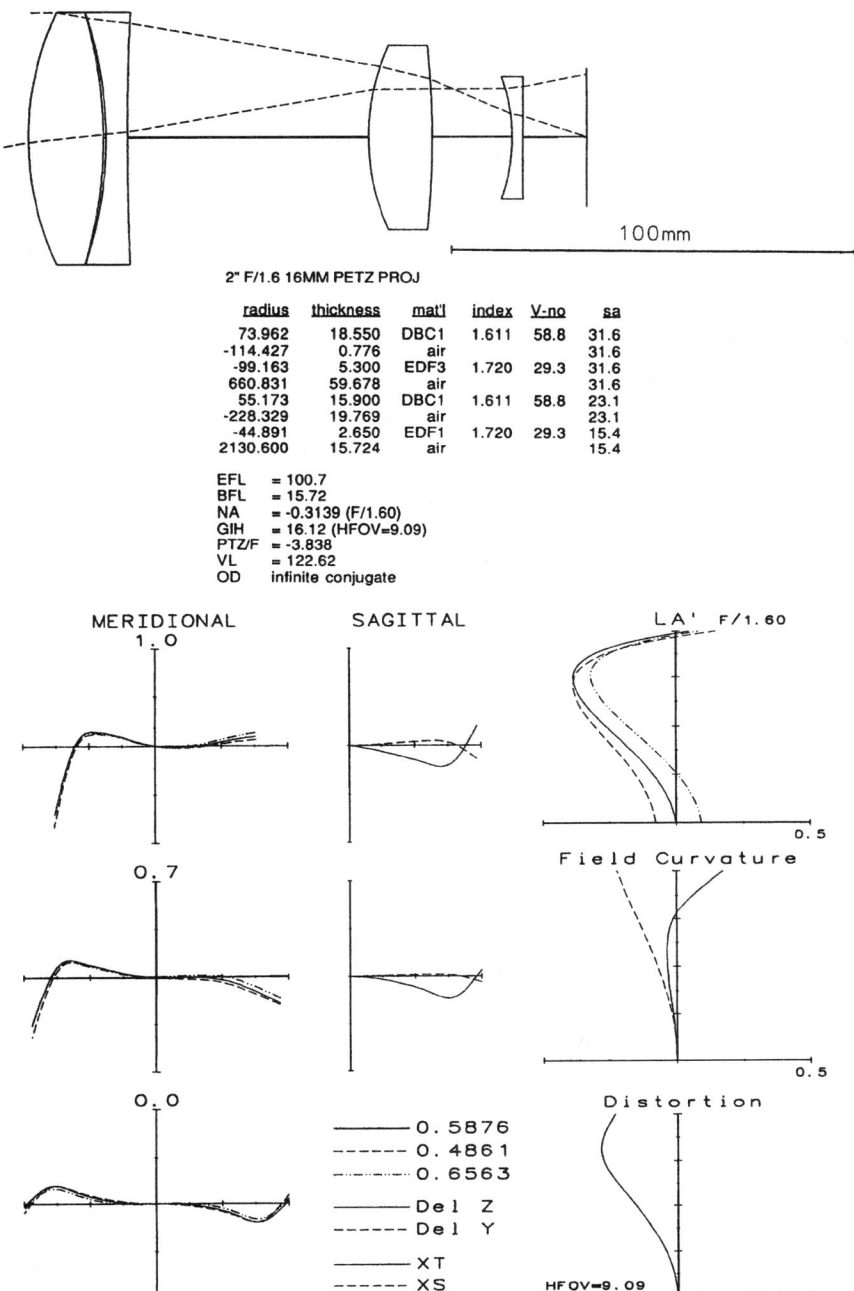

2" F/1.6 16MM PETZ PROJ

radius	thickness	mat'l	index	V-no	sa
73.962	18.550	DBC1	1.611	58.8	31.6
-114.427	0.776	air			31.6
-99.163	5.300	EDF3	1.720	29.3	31.6
660.831	59.678	air			31.6
55.173	15.900	DBC1	1.611	58.8	23.1
-228.329	19.769	air			23.1
-44.891	2.650	EDF1	1.720	29.3	15.4
2130.600	15.724	air			15.4

EFL = 100.7
BFL = 15.72
NA = -0.3139 (F/1.60)
GIH = 16.12 (HFOV=9.09)
PTZ/F = -3.838
VL = 122.62
OD infinite conjugate

Figure 16.6 This four-element Petzval modification splits the rear doublet and spaces the elements so far apart that the flint acts as a field flattener. $f/1.6$ ±9°.

432 Chapter Sixteen

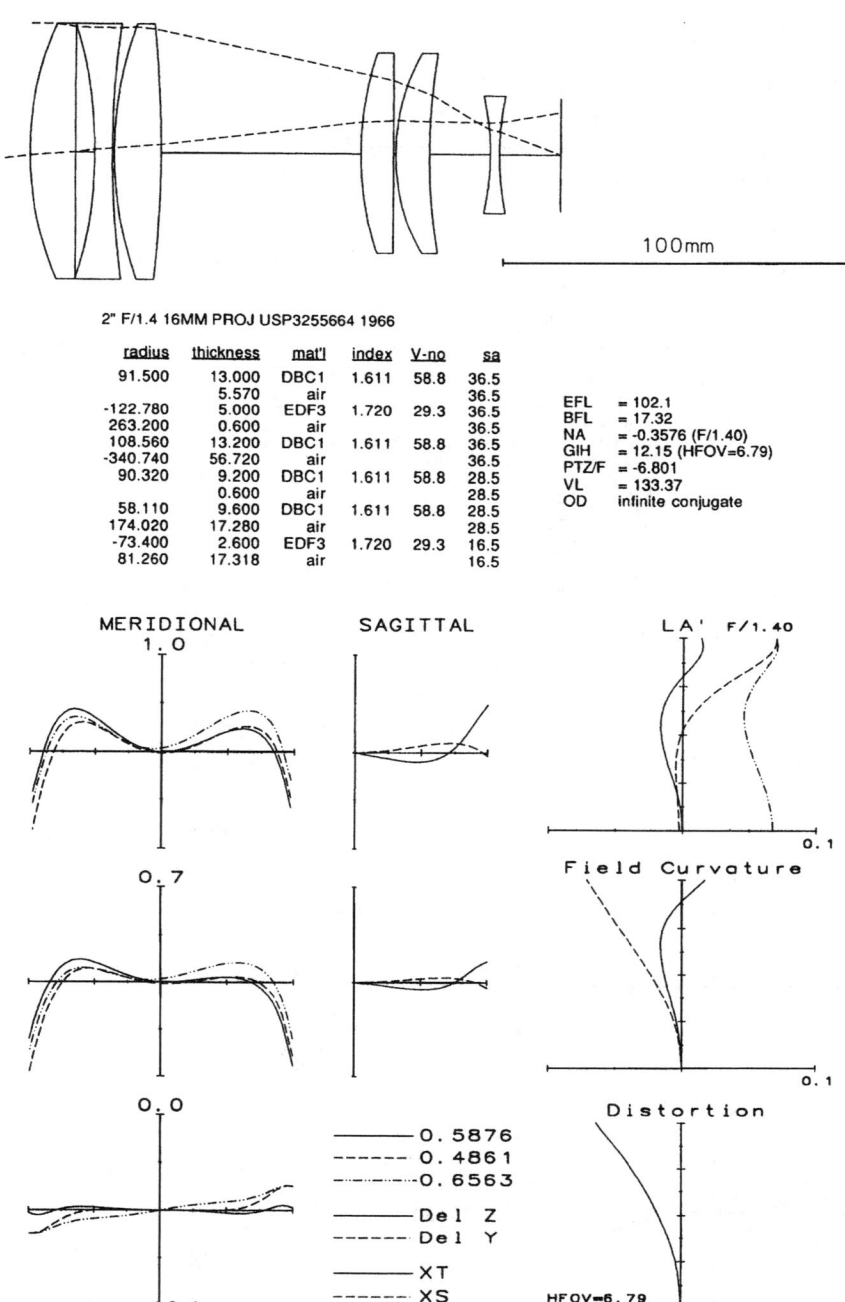

2" F/1.4 16MM PROJ USP3255664 1966

radius	thickness	mat'l	index	V-no	sa
91.500	13.000	DBC1	1.611	58.8	36.5
	5.570	air			36.5
-122.780	5.000	EDF3	1.720	29.3	36.5
263.200	0.600	air			36.5
108.560	13.200	DBC1	1.611	58.8	36.5
-340.740	56.720	air			36.5
90.320	9.200	DBC1	1.611	58.8	28.5
	0.600	air			28.5
58.110	9.600	DBC1	1.611	58.8	28.5
174.020	17.280	air			28.5
-73.400	2.600	EDF3	1.720	29.3	16.5
81.260	17.318	air			16.5

EFL = 102.1
BFL = 17.32
NA = -0.3576 (F/1.40)
GIH = 12.15 (HFOV=6.79)
PTZ/F = -6.801
VL = 133.37
OD infinite conjugate

Figure 16.7 This design was derived from Fig. 16.6 by splitting both crown elements to achieve a speed of $f/1.4$ and an improved performance over a field of $\pm 7°$.

The Petzval Lens; Head-up Display Lenses   433

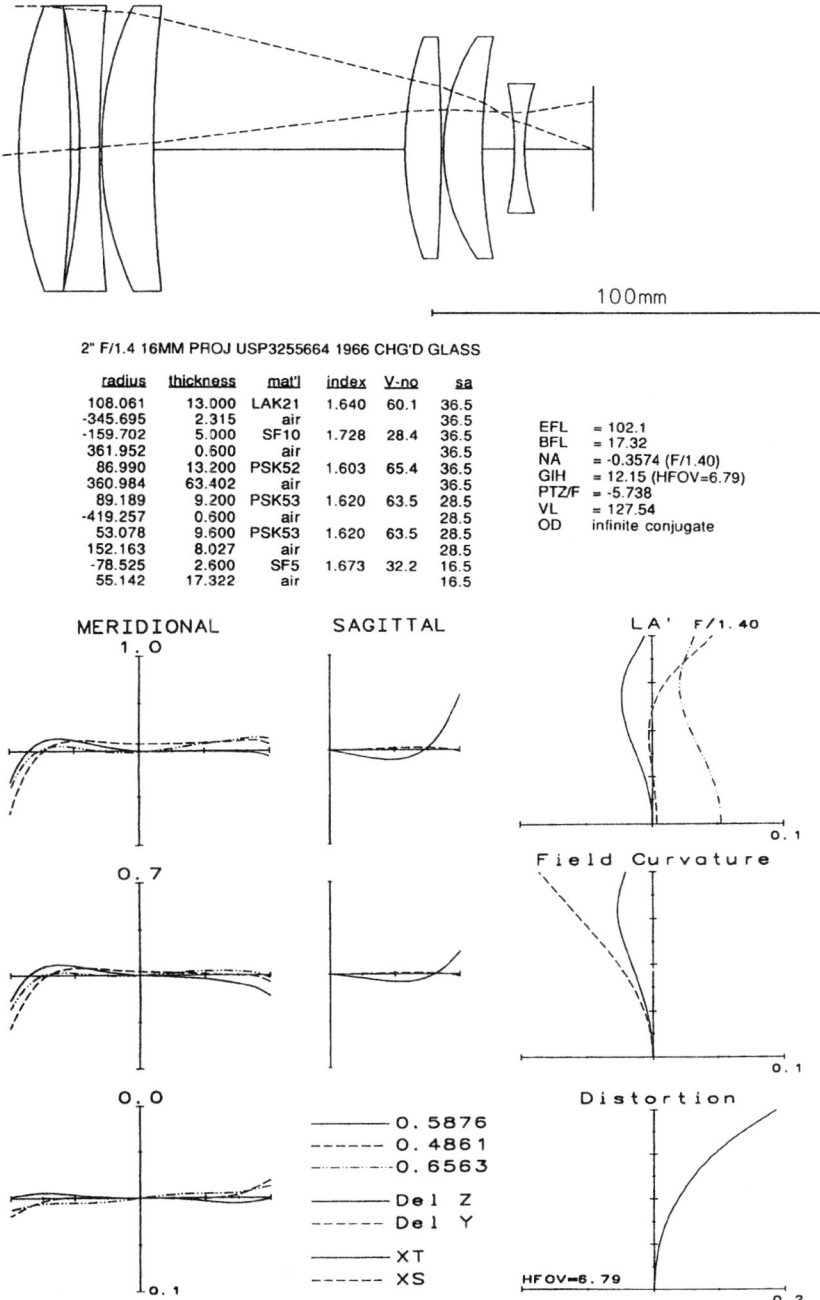

2" F/1.4 16MM PROJ USP3255664 1966 CHG'D GLASS

radius	thickness	mat'l	index	V-no	sa
108.061	13.000	LAK21	1.640	60.1	36.5
-345.695	2.315	air			36.5
-159.702	5.000	SF10	1.728	28.4	36.5
361.952	0.600	air			36.5
86.990	13.200	PSK52	1.603	65.4	36.5
360.984	63.402	air			36.5
89.189	9.200	PSK53	1.620	63.5	28.5
-419.257	0.600	air			28.5
53.078	9.600	PSK53	1.620	63.5	28.5
152.163	8.027	air			28.5
-78.525	2.600	SF5	1.673	32.2	16.5
55.142	17.322	air			16.5

EFL = 102.1
BFL = 17.32
NA = -0.3574 (F/1.40)
GIH = 12.15 (HFOV=6.79)
PTZ/F = -5.738
VL = 127.54
OD infinite conjugate

Figure 16.8 A later redesign of Fig. 16.7 with a more powerful design program (which allowed the glasses to vary) managed to reduce the high-order coma of Fig. 16.7, which had limited the off-axis performance. $f/1.4$ ±7°.

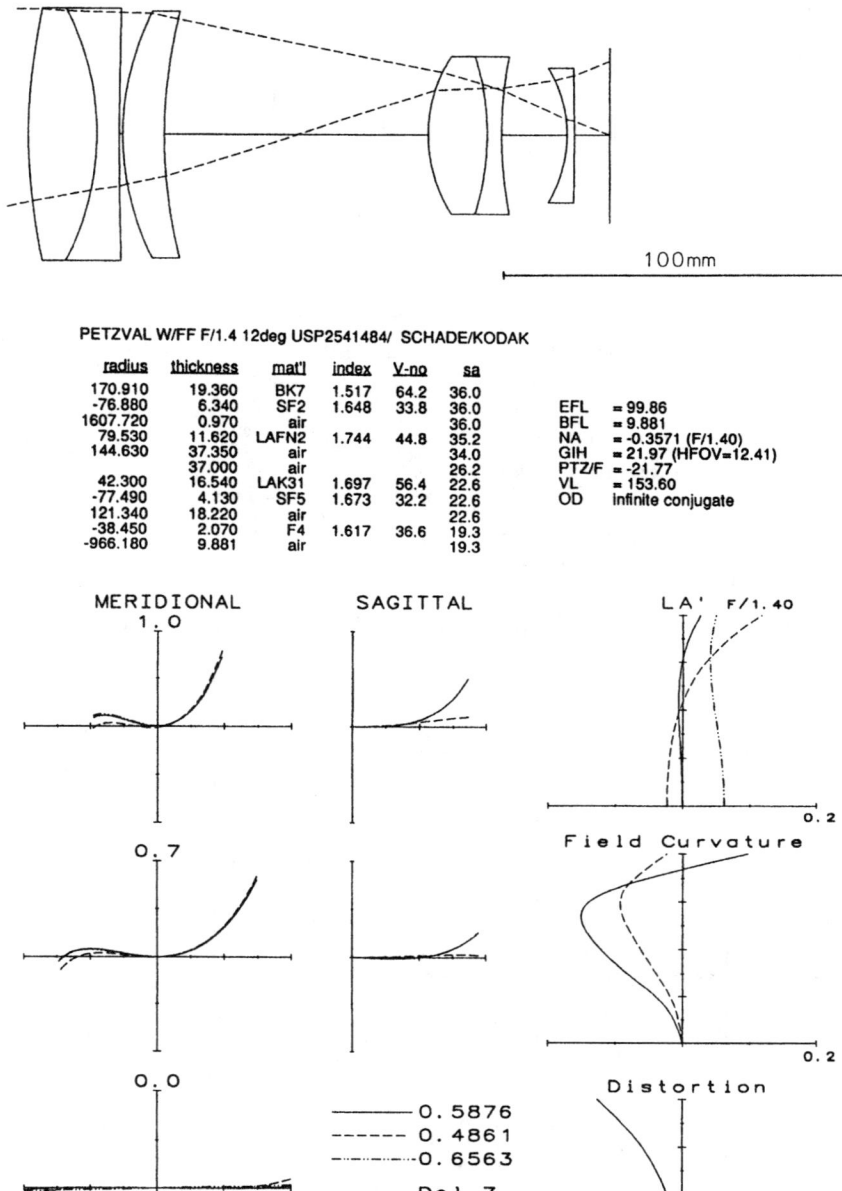

Figure 16.9 A six-element Petzval, which achieved a speed of $f/1.4$ by introducing a meniscus aplanatic element into the airspace near the front doublet. Note the high-index crowns in elements no. 3 and no. 4.

The Petzval Lens; Head-up Display Lenses 435

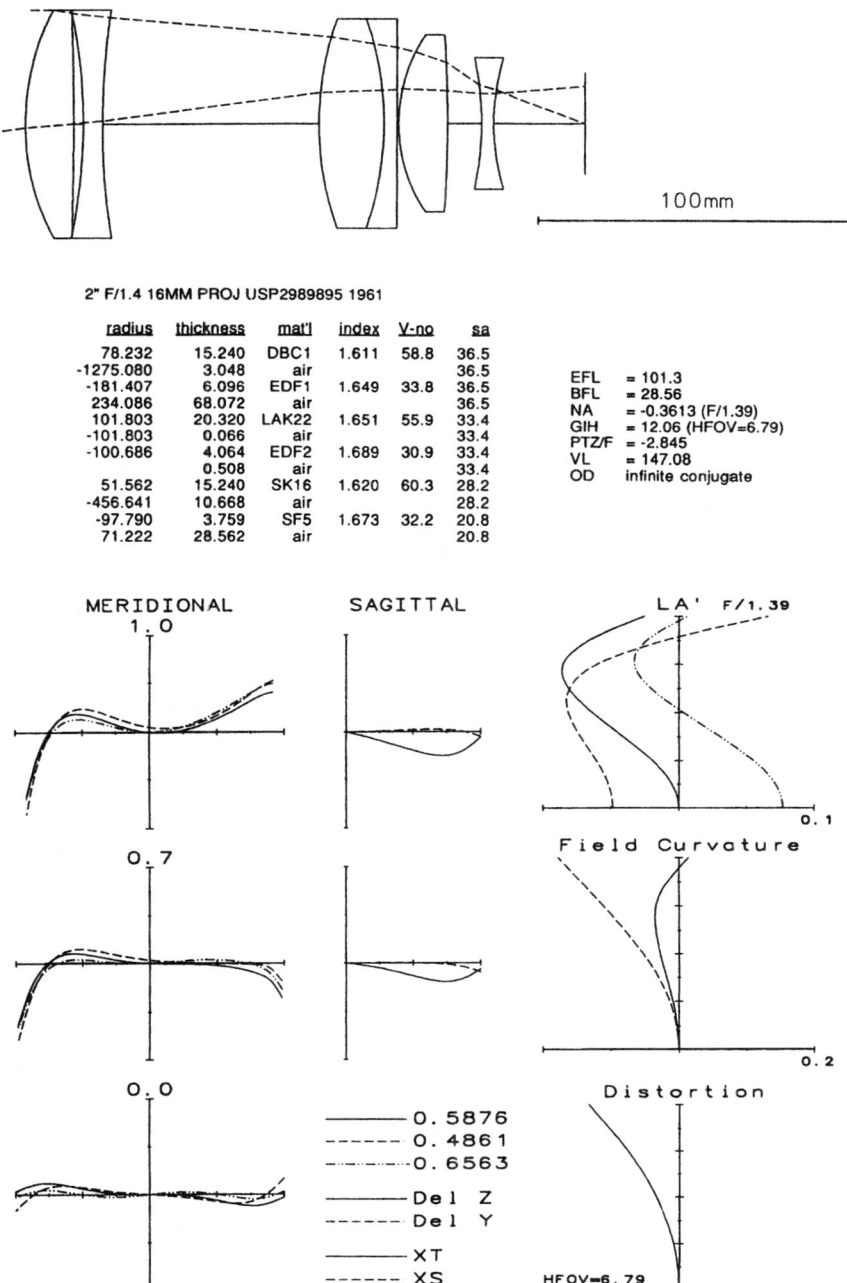

2" F/1.4 16MM PROJ USP2989895 1961

radius	thickness	mat'l	index	V-no	sa
78.232	15.240	DBC1	1.611	58.8	36.5
-1275.080	3.048	air			36.5
-181.407	6.096	EDF1	1.649	33.8	36.5
234.086	68.072	air			36.5
101.803	20.320	LAK22	1.651	55.9	33.4
-101.803	0.066	air			33.4
-100.686	4.064	EDF2	1.689	30.9	33.4
	0.508	air			33.4
51.562	15.240	SK16	1.620	60.3	28.2
-456.641	10.668	air			28.2
-97.790	3.759	SF5	1.673	32.2	20.8
71.222	28.562	air			20.8

EFL = 101.3
BFL = 28.56
NA = -0.3613 (F/1.39)
GIH = 12.06 (HFOV=6.79)
PTZ/F = -2.845
VL = 147.08
OD infinite conjugate

Figure 16.10 This $f/1.4$ modification moved much of the power of the front doublet to the rear, adding a strong singlet after the rear doublet.

Chapter Sixteen

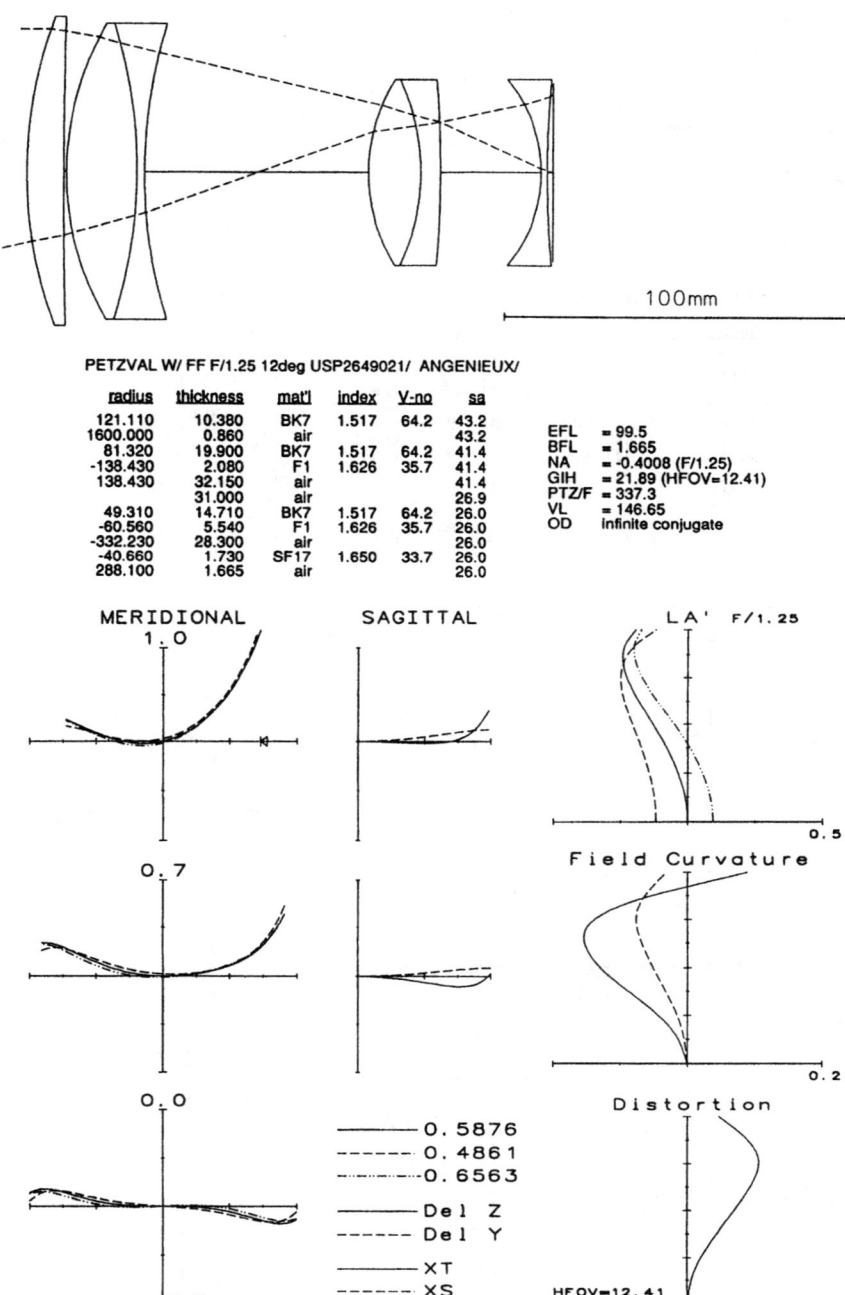

Figure 16.11 Splitting some power from the front crown and reducing the back focus (to an almost totally impractical 1.7% of the focal length) allows quite good correction at a speed of $f/1.25$ over a field of $\pm 12°$.

Doubling up on the rear doublet has been used to increase the speed to $f/1.3$, although this format has not been especially successful. The *R-Biotar* of Fig. 16.12 uses a relatively simple construction to reach a speed of $f/0.9$. Note that the slope of the marginal ray in the figure is increased by about the same amount at each component; this is usually helpful in reducing zonal spherical.

Feder and Hopkins designed a six-element Petzval lens, consisting of two doublets and two singlets, which covered ±15° at a speed of $f/1.0$. It used film that was curved to match the Petzval surface of the lens. The film was pressed against a concave form by gas pressure between the last element and the film.

More recently, Kodak produced an $f/1.0$ 8-mm movie projection lens configured like the R-Biotar of Fig. 16.12 with an added field flattener.

The Petzval lens is also the basis of the classical microscope objective. The standard 10×, NA = 0.25 objective is a Petzval projection lens much like Fig. 16.2, corrected for finite conjugates. The Amici and oil-immersion objectives are Petzval lenses with (more or less) aplanatic elements added to the short conjugate to increase the numerical aperture. See Chap. 17 for microscope objectives.

16.5 Head-up Display (HUD) Lenses, Biocular Lenses, and Head/Helmet-Mounted Display (HMD) Systems

The HUD lens is effectively a collimator that is used to present to the eye an infinitely distant image of a *cathode-ray tube* (CRT), *light emitting diode* (LED), and the like. This image is reflected and superimposed (by a semireflecting beam combiner) on a direct view of the scene, seen through the combiner. The system aperture is made large enough so both eyes can be used (hence, biocular). The large-aperture diameter requires a high-speed lens, typically to the order of $f/1.0$ or faster. Note that *binocular* means an optical system for each eye; *biocular* means a single optical system used simultaneously by both eyes. These systems are almost all based on the Petzval lens, usually with added elements, plus a field flattener. An aspheric surface is often used in the front component to control the spherical aberration; occasionally the concave surface of the field flattener is aspherized to help with the field aberrations. The analysis and evaluation of a biocular system is a bit tricky, since each eye uses only about a 5-mm-diameter part of the image beam. What counts most in this sort of system is the unpleasant "swimming" appearance of the image caused by spherical aberration when the eye is moved, and the parallax between the directions at which the image is seen by the two eyes. This may take the form of

438 Chapter Sixteen

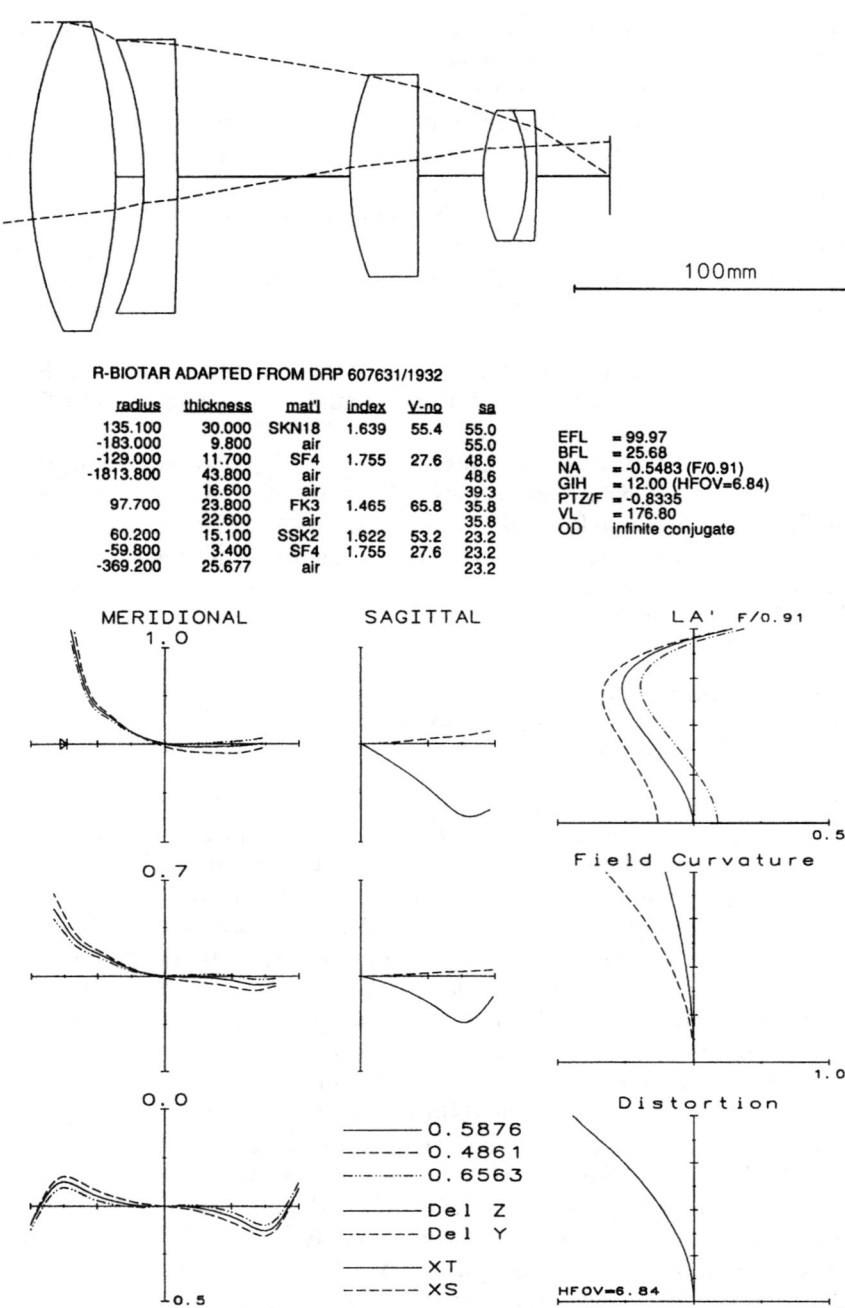

R-BIOTAR ADAPTED FROM DRP 607631/1932

radius	thickness	mat'l	index	V-no	sa
135.100	30.000	SKN18	1.639	55.4	55.0
-183.000	9.800	air			55.0
-129.000	11.700	SF4	1.755	27.6	48.6
-1813.800	43.800	air			48.6
	16.600	air			39.3
97.700	23.800	FK3	1.465	65.8	35.8
	22.600	air			35.8
60.200	15.100	SSK2	1.622	53.2	23.2
-59.800	3.400	SF4	1.755	27.6	23.2
-369.200	25.677	air			23.2

EFL = 99.97
BFL = 25.68
NA = -0.5483 (F/0.91)
GIH = 12.00 (HFOV=6.84)
PTZ/F = -0.8335
VL = 176.80
OD infinite conjugate

Figure 16.12 This extremely high-speed (f/0.9) lens was designed as a lens to photograph low-brightness fluorograph screens. It is a carefully balanced design that controls the high-order aberrations nicely.

convergence (of the eyes), which is quite tolerable, of divergence, or of dipvergence (a vertical difference of direction). HUD designers usually aim for a parallax of about a minute of arc or less over most of the field and aperture.

The field of view of a HUD is equal to the clear aperture of the lens divided by its distance from the eye. Obviously the position of the eye, both laterally and longitudinally, will affect both the size and direction of the field of view. The situation is analagous to viewing a scene through a window that has the size and location of the HUD lens. Each eye sees a slightly different field; the result is that there is a central field seen by both eyes, plus an extended field on either side seen by only one eye.

Aircraft HUDs are usually folded so as to fit into the very limited space available; the Petzval design affords a large central airspace suitable for a folding mirror. In one very compact military HUD the leading lens and the fold mirror are combined into a large, aspheric-surfaced prism. The resolution of an HUD should be one minute or better on axis and about three minutes or better at the edge of the field. The surface quality and accuracy of the HUD optics only need to be about as good as window glass.

The spectral band used in an HUD is usually truncated, often to a very narrow band. The rationale is that the beam combiner coating is very highly reflective in the narrow band of wavelengths emitted by the CRT, LED, or other display device, and has a very high transmission outside this band. Thus efficiencies in both the HUD and in the visible wavelength can be very high, approaching 100 percent for each (and, no, this doesn't mean we have a total efficiency near 200 percent). Filters and polarizers are often included in the HUD optics so that sunlight entering the HUD is prevented from being reflected back to the pilot's eye.

The optics of an HMD are essentially those of a magnifier viewing a small, high-resolution display. Since weight is a significant factor in any device to be mounted on a pilot's head, these magnifiers are usually moulded of plastics, and diffractive surfaces (Sec. 6.10) have been successful in reducing weight. See Ref. 17 and Melzer and Moffitt, *Head Mounted Displays*, 1997, McGraw-Hill.

Chapter 17

Microscope Objectives

17.1 General Considerations

An ordinary microscope objective has an image field with a diameter of about 20 mm; the image distance is about 160 mm. The total field of view is thus about 7° or less. The exit pupil diameter is typically about 8 mm; with the 160-mm image distance, the *numerical aperture* (NA) of the image cone is about 0.025. Therefore, the object-side NA is about $0.025M$, where M is the magnification of the objective.

A microscope objective may be designed to be used with or without a cover plate (or cover slip) over the object. The cover glass is nominally 0.18 mm (0.16 to 0.19 mm) thick with an index of 1.523 ± 0.005 and a V-value of 56 ± 2 (close to Schott K4 or K5 glass). This thickness of glass in the strongly divergent or convergent cone of light at the object will contribute a significant amount of aberration (especially spherical and chromatic) and must always be included in the design calculations.

Most microscope objectives are designed to work at an image distance of about 160 mm. The aberration correction of the objective is significantly affected by the image distance. Often a microscope body tube will have a calibrated tube length adjustment; this adjustment can be used to fine-tune the spherical aberration correction of a less than perfectly made objective, or to compensate for a variation in cover plate thickness from the nominal value.

Some objectives are designed for an infinite image distance. These infinity-corrected objectives are used with what amounts to a telescope to view the image. One advantage of this arrangement is that a tilted-plate beam splitter can be placed in the collimated beam between the objective and the telescope without introducing the astigmatism that would result if the tilted plate were introduced into a convergent image cone.

It is generally much more convenient to design an optical system with the object at the long conjugate; this widespread practice is usually followed in designing and analyzing microscope objectives. The sample designs in this chapter are ray-traced from the long conjugate (image side) to the short.

The mounting shoulder (or flange) to object distance was once standardized at about 36 mm; in newer microscopes it is 45 mm. If there is a rotating nosepiece mount, the object to lens vertex distance must be less than 50 mm to clear the mount as it is rotated.

The Rayleigh criterion for resolution when applied to the microscope is 0.61 (wavelength)/NA; for the Sparrow criterion or the MTF cutoff, the constant is 0.5 instead of 0.61.

The combination of a high NA and a narrow field leads to systems of great length, often many times the objective's focal length. See, for example, Figs. 17.5 to 17.10.

17.2 Classical Objective Design Forms; The Aplanatic Front

The classical low-power microscope objective of Fig. 17.1a is a simple cemented achromatic doublet or triplet, essentially a telescope objective that has been designed for a finite object distance. Occasionally, a more complex arrangement of spaced-apart elements is used to obtain a long focal length in a compact package. The medium-power (Fig. 17.1b) (10×, NA = 0.25) objective is the same design type as an ordinary $f/2$ Petzval projection lens (see Fig. 16.2, for example). Except for the finite-distance long conjugate, the design techniques for these lenses are the same as those outlined in Chaps. 6 and 16, respectively.

There are usually three possible combinations of shapes for which two cemented achromatic doublets are corrected for both coma and spherical aberration. One of these combinations is the Petzval projection lens configuration, where the astigmatism is controlled to reduce the field curvature. Another combination is the so-called divisible or separable Lister objective. In this form, both doublets are independently corrected for spherical and coma. Thus they can be used in combination as a 10× objective, or the small doublet nearest the object can be removed and the large doublet can be used alone as a low-power objective. The very severe drawback to this arrangement is that, since both components are corrected for spherical and coma, they both must contribute large amounts of negative astigmatism (as in Eqs. (24.83) and (24.95)). The result is a badly inward-curving field, astigmatism, and poor off-axis imagery.

The higher-power objectives make use of aplanatic elements located near the object. The function of these elements is to convert a large-numerical-aperture (NA = $n \sin U$), divergent cone of light from an

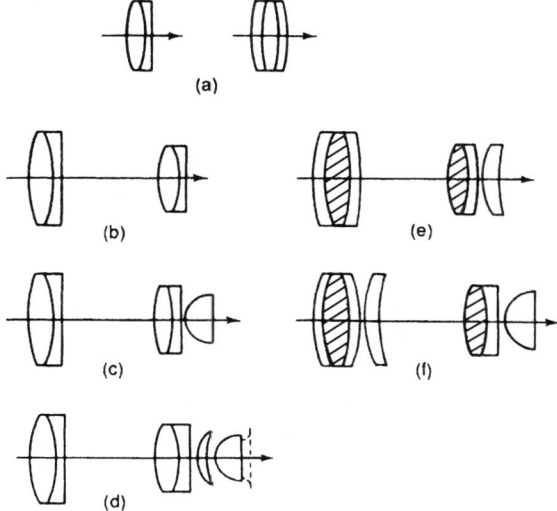

Figure 17.1 Microscope objectives. (*a*) Low-power doublet or triplet. (*b*) Medium-power objective: 10×, NA = 0.25. (*c*) Amici objective: 20×, NA = 0.5 to 40×, NA = 0.8. (*d*) Oil-immersion objective: 100×, NA = 1.2. (*e*) Apochromatic objective: 10×, NA = 0.3 (shading indicates a CaF_2 element). (*f*) Apochromatic objective: 50×, NA = 0.3. Note that the microscope object is to the right and the image is to the left; the arrow indicates the direction of the design raytrace rather than the direction of the light.

object point into a cone with a lower numerical aperture, without introducing any coma or spherical aberration. The reduced numerical aperture is then within the capability of the (Petzval-type) "back" to handle. Figure 17.1 illustrates many of the standard configurations.

17.2.1 Aplanatic surfaces

There are three cases where a simple spherical surface is aplanatic, i.e., free of both coma and spherical aberration. One occurs when the object and image are both located at the surface. The second case is with both the object and image at the center of curvature of the surface. The third case occurs when the object and image distances are related by

$$l' = \frac{R(n'+n)}{n'} = \frac{nl}{n'} \qquad (17.1)$$

where l = object distance
l' = image distance

n = index of object space
n' = index of the image space
R = radius of the surface

Obviously rays from an axial point pass radially through the second case undeviated; in the third case the slope of the ray ($\sin U$) is changed by a factor equal to (n'/n). Note that case one and case three have zero astigmatism contributions, whereas case two contributes overcorrected astigmatism.

The classic full aplanatic front is shown in Fig. 17.2. The object is immersed in a liquid whose index matches that of the front element; note that the higher-than-air immersion index also serves to increase the NA (NA = $n \sin U$) and thus the resolution. Surface R_1 is the third aplanatic case and reduces the ray slope by a factor of the index of the lens. This first element must be hyperhemispheric to meet the requirements of the aplanatic case. Surface R_2 is the second case (with object and image at the center of curvature), and the rays pass through undeviated. Surface R_3 is again the third aplanatic case, so that the meniscus element also reduces the ray slope by a factor equal to its index. Actually, modest departures from the exact aplanatic cases are usually found in real designs; instead of zero spherical contributions, the element shapes are sometimes chosen to contribute some overcorrected spherical aberration and thus reduce the correction load on the Petzval style back.

These classic microscope objectives are usually very well corrected for the on-axis aberrations; however, they suffer from a strongly inward-curving Petzval field curvature. In addition, those with hyperhemispheric or full aplanatic fronts have lateral chromatic that results from their unsymmetrical construction, i.e., the chromatically undercorrected aplanatic front and the overcorrected back. The lateral color can be balanced out by an equal but opposite amount of lateral color designed into a compensating eyepiece, but the Petzval curvature always severely limits the off-axis image quality. The other limiting aberration is secondary spectrum, which can be handled as outlined in Sec. 6.8. Elements of calcium fluoride (or of glass types composed primarily of calcium fluoride) are commonly used in apochromatic microscope objectives as shown in Fig. 17.2 and several other figures.

An unfortunate characteristic of the Amici objective (Fig. 17.1c) is a very short working distance. There is a direct correlation between the working distance and the amount of zonal spherical aberration in this type of objective. This limits the working distance of the higher-NA lenses. Note that occasionally the principle of the aplanatic front is also incorporated near the image in photographic objectives to increase their speed without increasing the aberrations.

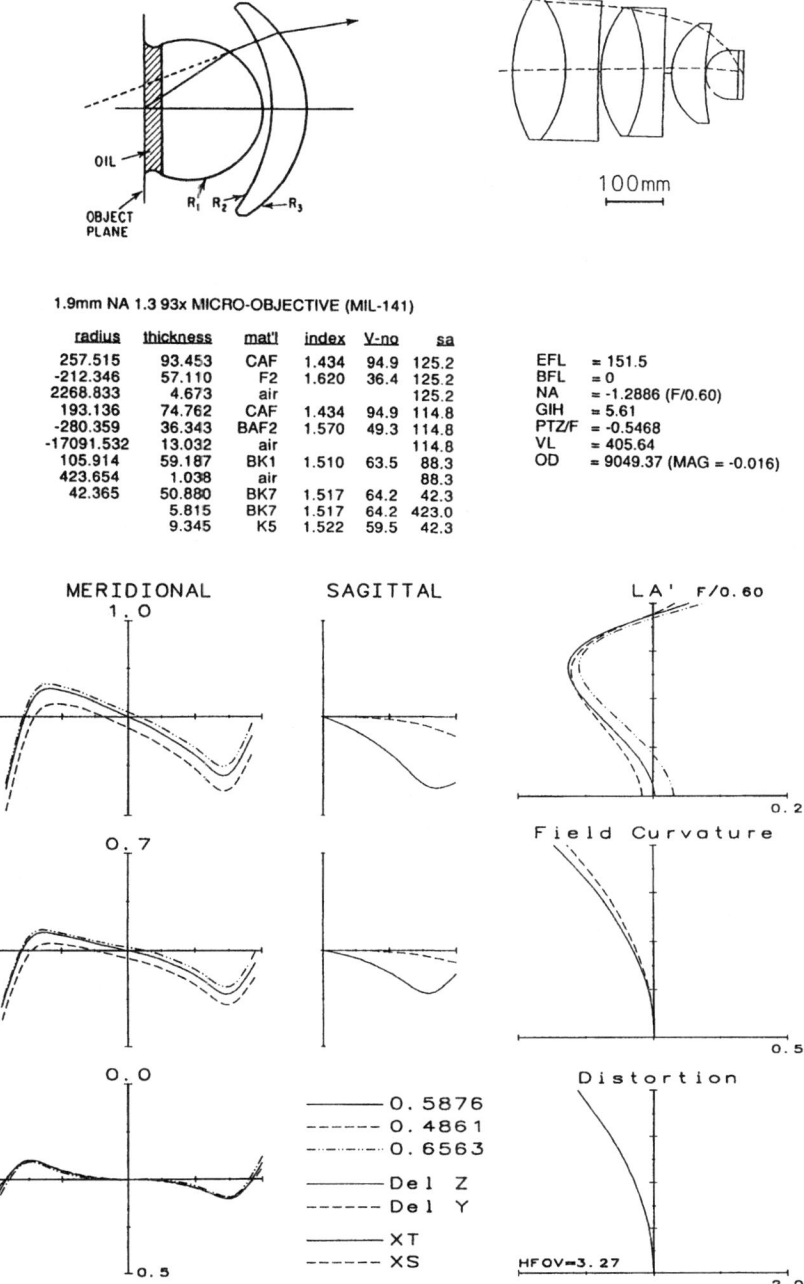

1.9mm NA 1.3 93x MICRO-OBJECTIVE (MIL-141)

radius	thickness	mat'l	index	V-no	sa
257.515	93.453	CAF	1.434	94.9	125.2
-212.346	57.110	F2	1.620	36.4	125.2
2268.833	4.673	air			125.2
193.136	74.762	CAF	1.434	94.9	114.8
-280.359	36.343	BAF2	1.570	49.3	114.8
-17091.532	13.032	air			114.8
105.914	59.187	BK1	1.510	63.5	88.3
423.654	1.038	air			88.3
42.365	50.880	BK7	1.517	64.2	42.3
	5.815	BK7	1.517	64.2	423.0
	9.345	K5	1.522	59.5	42.3

EFL = 151.5
BFL = 0
NA = -1.2886 (F/0.60)
GIH = 5.61
PTZ/F = -0.5468
VL = 405.64
OD = 9049.37 (MAG = -0.016)

Figure 17.2 The full aplanatic front of an oil-immersion objective. The object is immersed in a fluid whose index closely matches that of the hyperhemispheric first element. R_1 is an aplanatic surface of the third kind. The image formed by R_1 is located at the center of curvature of R_2, and R_3 is an aplanatic surface of the same type as R_1. All surfaces are thus aplanatic and the "front" has no coma or spherical aberration.

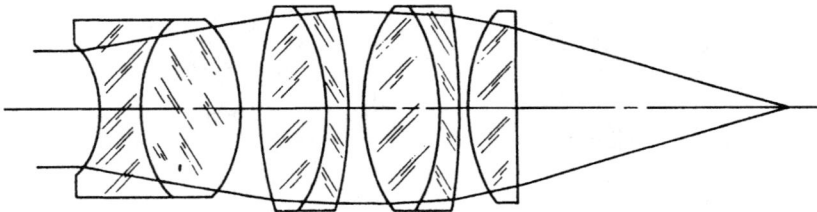

Figure 17.3 A flat-field microscope objective in which the Petzval curvature is corrected by a thick negative achromat placed above the positive components.

17.3 Flat-Field Objectives

The flat-field microscope objective is based on the usual principle of separating positive and negative power to correct the Petzval curvature. The general arrangement is often the same as that found in reverse telephoto (or retrofocus) lenses, as described in Chap. 14. In the retrofocus camera lens, the angular field is usually much larger than in a microscope objective; thus in the retrofocus the general shape of the negative component is almost always a meniscus, concave toward the stop (which is at the rear positive component); however, the narrower field of the microscope objective does not require this shape, and an arrangement with a thick meniscus negative component oriented convex toward the positive component is common. One basic arrangement is shown in Fig. 17.3, with a thick negative achromatic component and a positive component consisting of two cemented doublets and a plano-convex singlet. In some designs, the negative component is a thin negative singlet spaced well away from the positive component. These objectives (like their retrofocus counterparts) have a long working distance; as a result the element diameters tend to be relatively large.

17.4 Reflecting Objectives

The Schwarzschild mirror configuration described in Sec. 18.2 is the basis of the reflecting microscope objective. The most common version is the simple two-spherical-mirror arrangement; increased numerical apertures are possible either by utilizing aspheric surfaces or by incorporating refracting elements as shown in Fig. 17.4. An infinity-corrected two-mirror system can readily be derived using the relationships given in Sec. 18.2; a modest adjustment will convert it to a design corrected for finite conjugates. The central obscuration tends to be large; it can be reduced by refracting elements or by the use of aspheric surfaces.

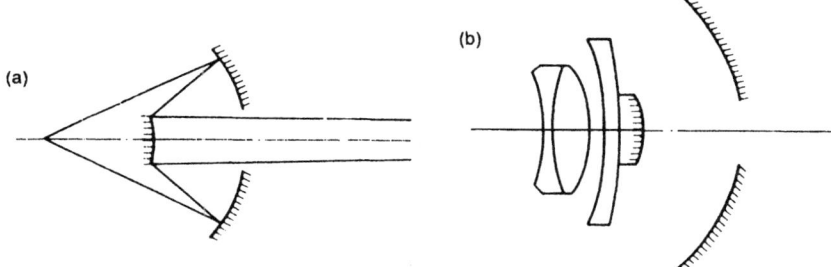

Figure 17.4 Reflecting microscope objectives. (a) The basic reflecting objective consists of simple concentric spherical surfaces in a Schwarzschild arrangement; 30×, NA = 0.5. (b) An ultraviolet modification at 50×, NA = 0.7 attributable to David Grey. The refracting elements are of SiO_2 and CaF_2.

17.5 The Microscope Objective Designs

The flat-field objectives (which usually have the letters *p-l-a-n* in their trade names) reduce the Petzval curvature in part by using an upper* negative component, usually an achromatic doublet, acting a bit like the leading component of a retrofocus lens (Chap. 14). The other techniques found in plano objectives include a strong diverging surface near the focus (i.e., the object), acting like a Piazzi-Smythe field flattener, and occasionally a construction like the inner meniscus components of the double-Gauss lens (Chap. 12).

Figure 17.5 is an infinity-corrected NA 0.57, nine-element objective with an upper portion resembling the inner doublets of the double-Gauss lens. These thick menisci help flatten the field, and affect the higher-order aberrations favorably. The field shown here (±2.1°) is relatively modest. Note the meniscus lower element, which also helps the Petzval while it painlessly increases the NA. The FK51, CaF_2 and KzFS4 elements tend to reduce the secondary spectrum.

Figure 17.6 looks a bit like a modified Petzval lens, but it's really just a cemented doublet followed by two positive elements, the second of

---

* Although the microscope objective designs are shown here at our standardized focal length of 100 mm, they are of course intended for use at much shorter focal lengths, at which their aberrations are correspondingly smaller.

To avoid confusion regarding the parts of an objective, we will identify them by their position in an ordinary vertical microscope as top or upper and bottom or lower. The customary terminology for these are rear and front, respectively, but since microscope objectives are designed from long conjugate to short (the reverse direction of the actual light path), the designer's raytracing front is the microscopist's rear and vice versa.

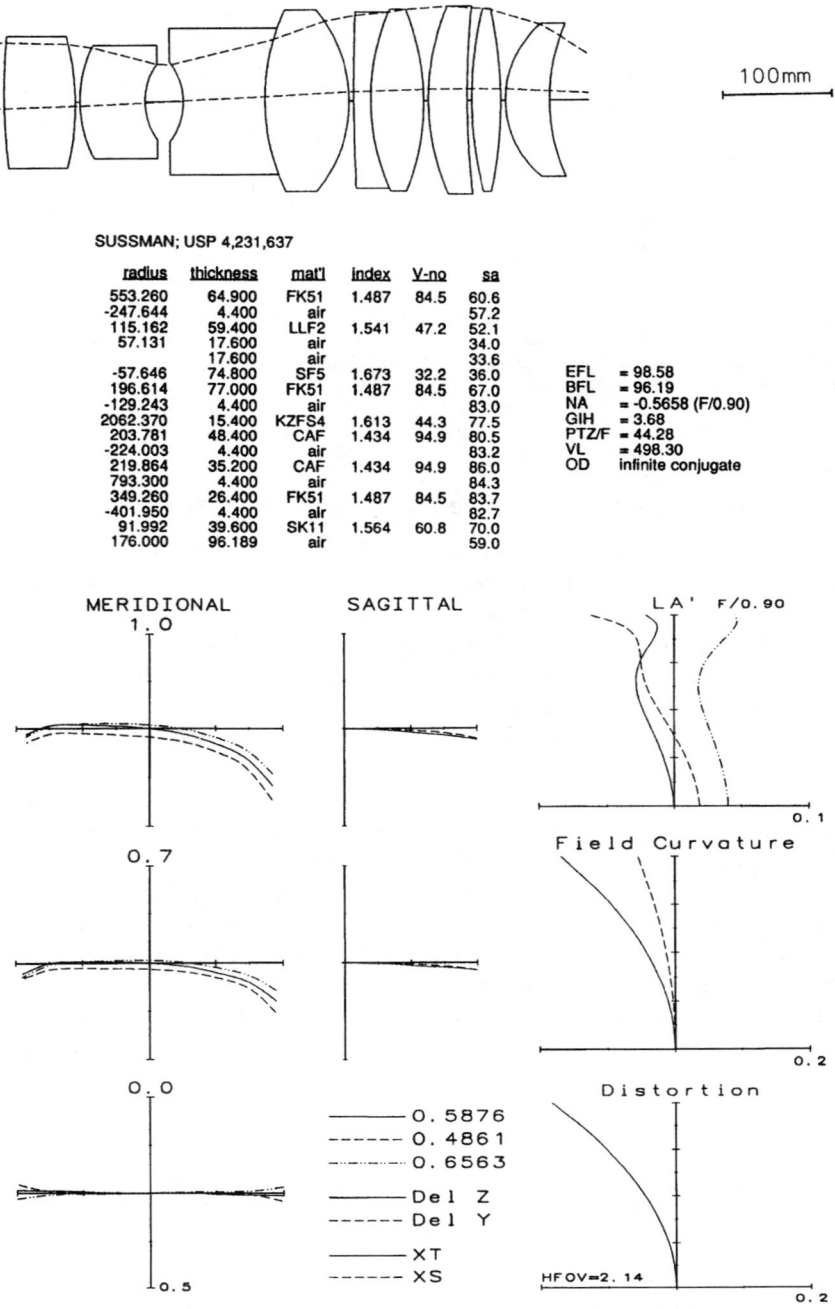

Figure 17.5 An infinity-corrected NA 0.57 ±2.1°, nine-element objective.

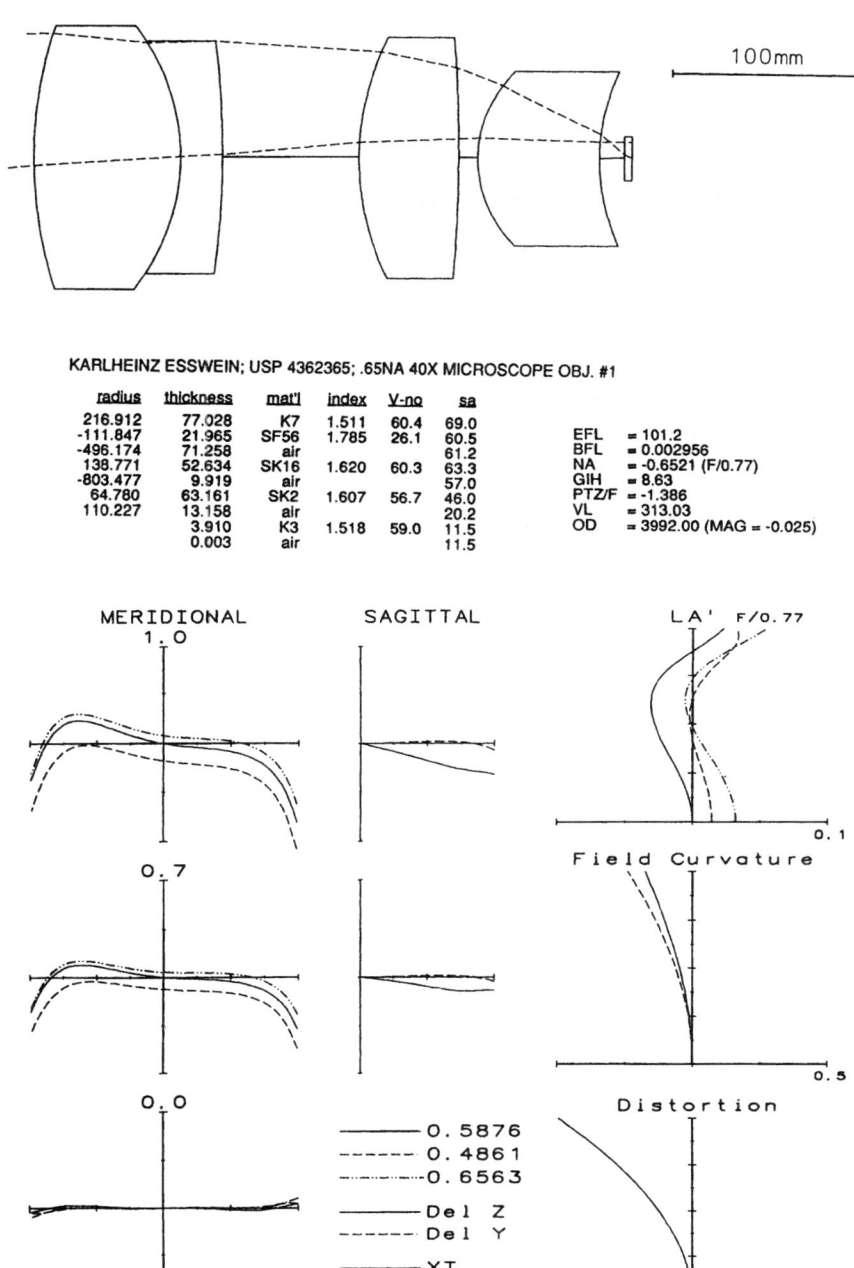

Figure 17.6 A simple four-element NA 0.65 ±4.9°, objective.

450 Chapter Seventeen

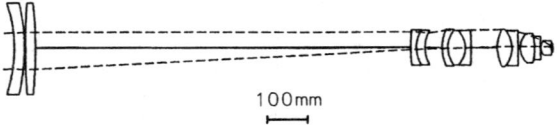

MILTON H. SUSSMAN; USP 4379623; 40X .8NA W/USP 3355234 TELESCOPE

radius	thickness	mat'l	index	V-no	sa
-352.265	24.702	BK1	1.510	63.5	112.8
-504.116	2.685	air			112.8
1729.446	24.702	SSK4	1.618	55.0	112.8
-1136.544	854.287	air			112.8
	69.578	air			0.0
402.742	13.962	KF9	1.523	51.5	44.0
81.612	21.480	SF11	1.785	25.8	43.0
90.085	41.885	air			39.7
117.064	13.962	SF15	1.699	30.1	45.1
72.032	42.959	FK51	1.487	84.5	43.0
-60.175	13.962	BK7	1.517	64.2	43.0
-605.756	63.365	air			43.5
96.819	36.515	FK51	1.487	84.5	46.2
-70.217	13.962	BK7	1.517	64.2	44.0
	6.122	air			40.8
51.551	35.441	FK51	1.487	84.5	36.5
-138.930	12.888	KF9	1.523	51.5	29.0
134.999	1.955	air			22.6
27.043	24.702	LSK02	1.786	50.0	18.6
23.166	3.211	air			8.6
	1.930	K3	1.518	59.0	10.7

EFL = 100
BFL = 0
NA = -0.8021 (F/0.62)
GIH = 2.49
PTZ/F = -2.359
VL = 1324.25
OD = 1876.20 (MAG = -0.025)

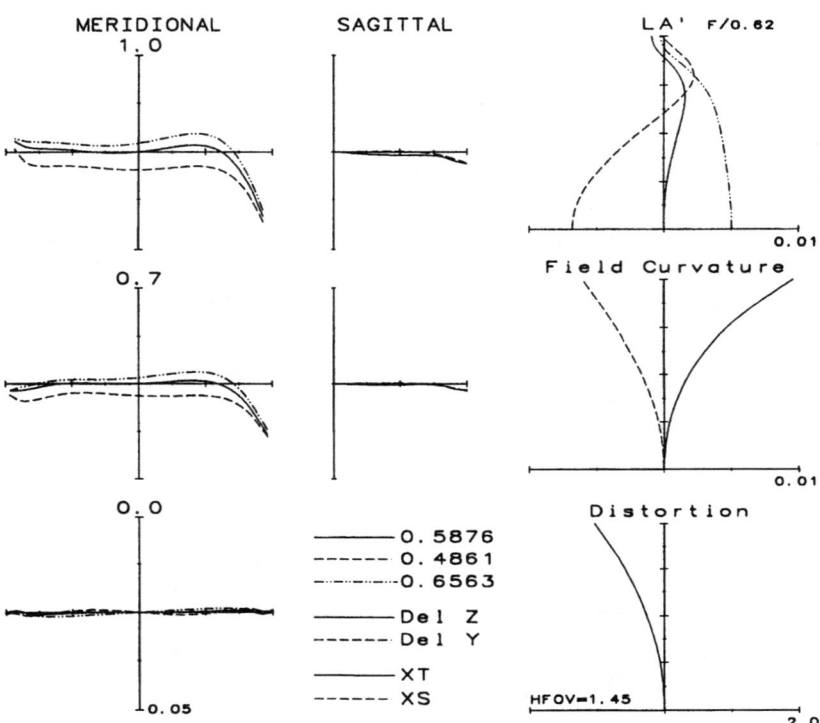

Figure 17.7 An infinity corrected 10-element objective (with the collimating doublet) at NA 0.8 ±1.43° with a meniscus front element.

Microscope Objectives 451

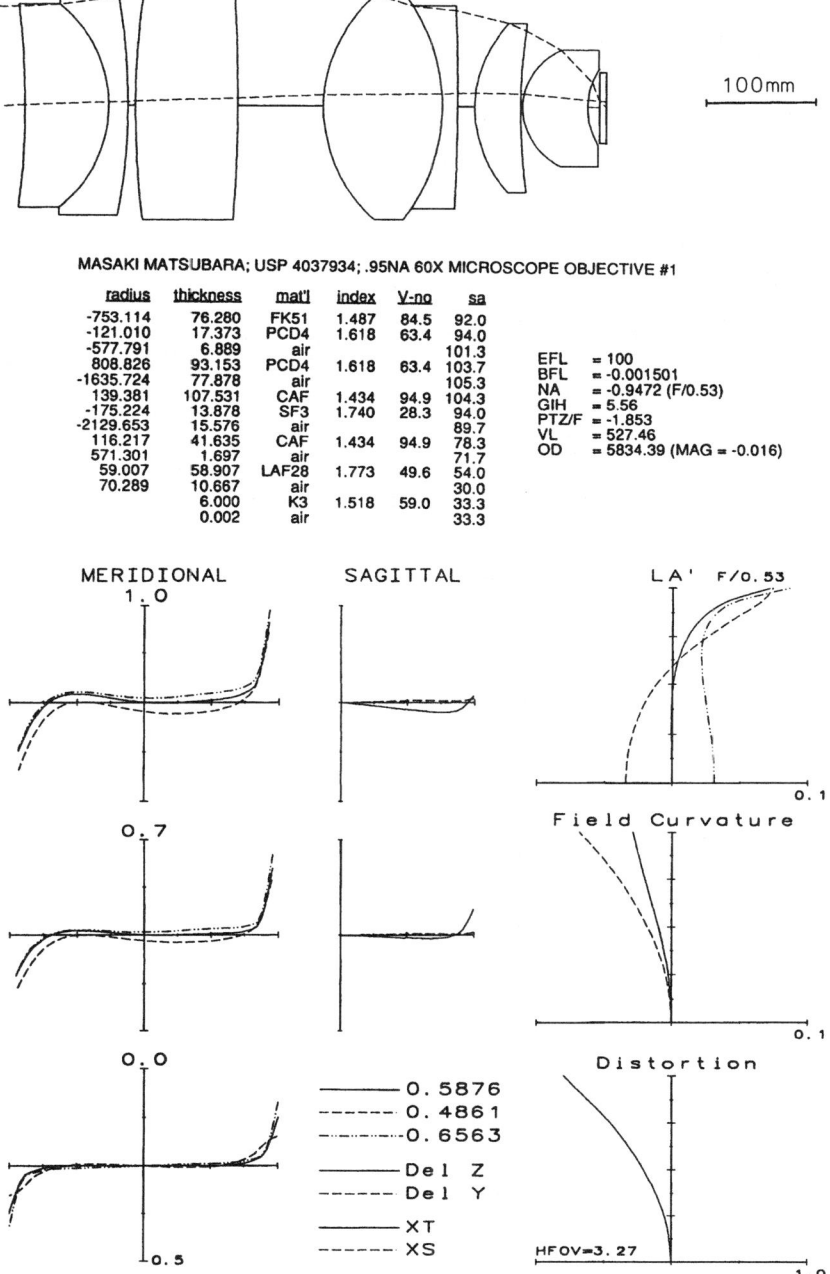

MASAKI MATSUBARA; USP 4037934; .95NA 60X MICROSCOPE OBJECTIVE #1

radius	thickness	mat'l	index	V-no	sa
-753.114	76.280	FK51	1.487	84.5	92.0
-121.010	17.373	PCD4	1.618	63.4	94.0
-577.791	6.889	air			101.3
808.826	93.153	PCD4	1.618	63.4	103.7
-1635.724	77.878	air			105.3
139.381	107.531	CAF	1.434	94.9	104.3
-175.224	13.878	SF3	1.740	28.3	94.0
-2129.653	15.576	air			89.7
116.217	41.635	CAF	1.434	94.9	78.3
571.301	1.697	air			71.7
59.007	58.907	LAF28	1.773	49.6	54.0
70.289	10.667	air			30.0
	6.000	K3	1.518	59.0	33.3
	0.002	air			33.3

EFL = 100
BFL = -0.001501
NA = -0.9472 (F/0.53)
GIH = 5.56
PTZ/F = -1.853
VL = 527.46
OD = 5834.39 (MAG = -0.016)

Figure 17.8 A simple seven-element NA 0.95 ±3.2° objective with a meniscus front element.

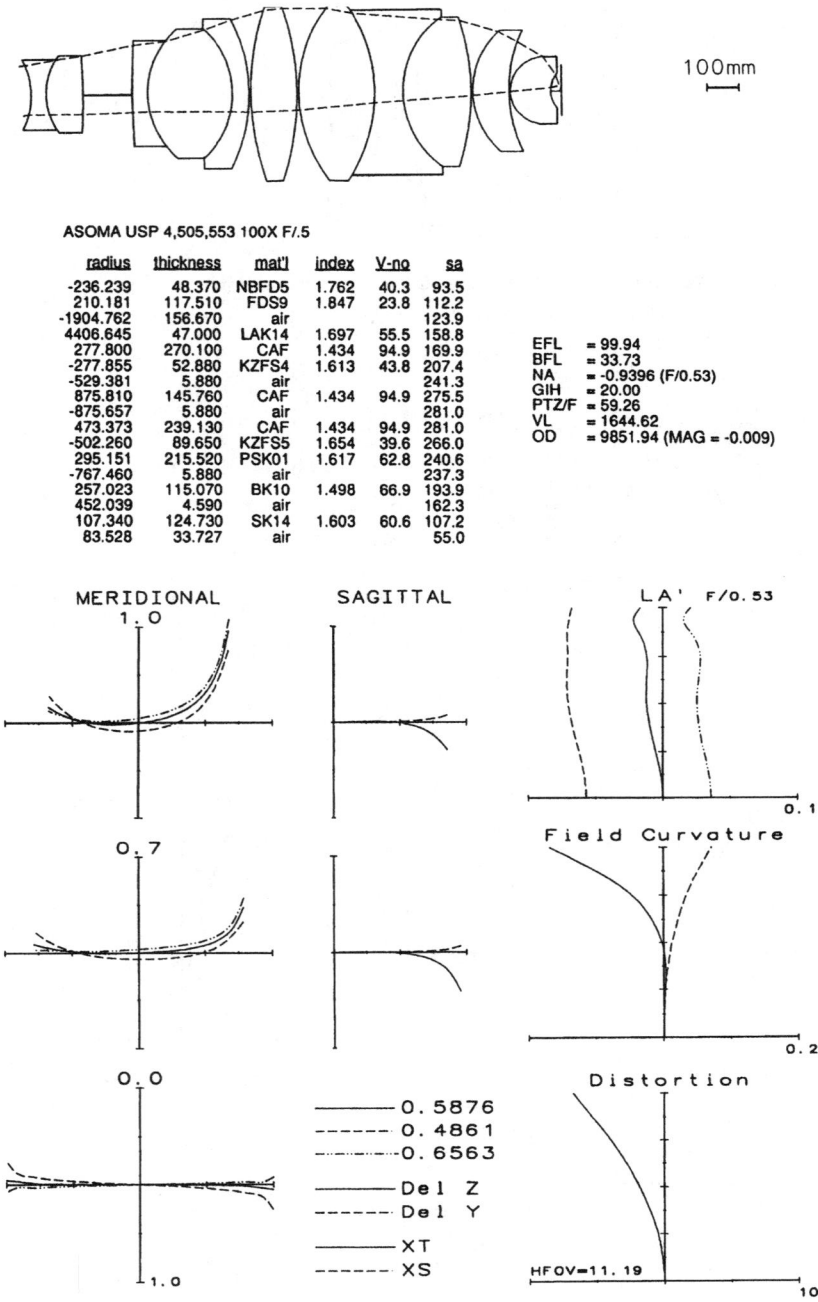

Figure 17.9 An 11-element NA 0.94 ±11.3° objective with a meniscus front and a wide field.

Microscope Objectives

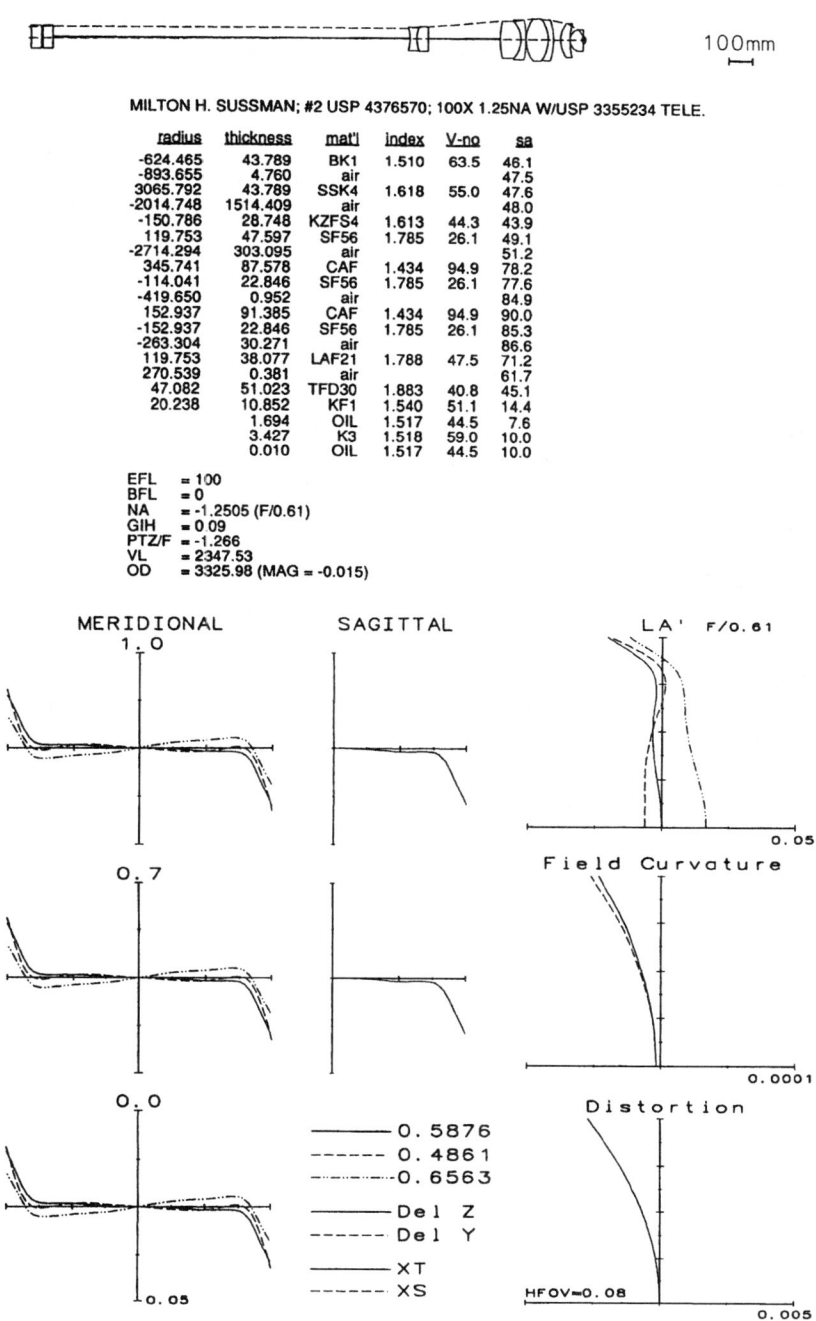

MILTON H. SUSSMAN; #2 USP 4376570; 100X 1.25NA W/USP 3355234 TELE.

radius	thickness	mat'l	index	V-no	sa
-624.465	43.789	BK1	1.510	63.5	46.1
-893.655	4.760	air			47.5
3065.792	43.789	SSK4	1.618	55.0	47.6
-2014.748	1514.409	air			48.0
-150.786	28.748	KZFS4	1.613	44.3	43.9
119.753	47.597	SF56	1.785	26.1	49.1
-2714.294	303.095	air			51.2
345.741	87.578	CAF	1.434	94.9	78.2
-114.041	22.846	SF56	1.785	26.1	77.6
-419.650	0.952	air			84.9
152.937	91.385	CAF	1.434	94.9	90.0
-152.937	22.846	SF56	1.785	26.1	85.3
-263.304	30.271	air			86.6
119.753	38.077	LAF21	1.788	47.5	71.2
270.539	0.381	air			61.7
47.082	51.023	TFD30	1.883	40.8	45.1
20.238	10.852	KF1	1.540	51.1	14.4
	1.694	OIL	1.517	44.5	7.6
	3.427	K3	1.518	59.0	10.0
	0.010	OIL	1.517	44.5	10.0

EFL = 100
BFL = 0
NA = -1.2505 (F/0.61)
GIH = 0.09
PTZ/F = -1.266
VL = 2347.53
OD = 3325.98 (MAG = -0.015)

Figure 17.10 A nine-element NA 1.25 objective with the collimating doublet. Note that the front element is actually a doublet with a buried diverging surface and a flat final surface that permits oil immersion. The ray analysis is shown at a miniscule field of ±0.0005°.

which is a thick meniscus with the concave surface close to the focus. It achieves a high NA of 0.65 and a wider than ordinary field (±5°) with ordinary glasses.

Figure 17.7 has a leading negative achromat followed by a triplet-plus-two-doublets arrangement plus a hemispheric lens with a concave lower surface, all of which tend to flatten the field. It achieves an NA of 0.8 over a relatively narrow field of ±1.4°. Note the FK51 elements to reduce the secondary spectrum.

Figure 17.8 has a weak leading (upper) diverging doublet followed by an almost Petzval-like singlet-doublet combination, followed by a full aplanatic front, the hemispheric element of which has a field-flattening concave lower surface. Again, note the CaF_2 elements for secondary spectrum reduction.

Figure 17.9 again has an upper negative achromat and three CaF_2 elements plus the lower full aplanatic front with the field-flattening lower second surface of the hyperhemisphere. But in between is a horror that includes two cemented triplets. But the NA is a high 0.94 and it covers a relatively wide field of ±11°.

Figure 17.10 also has the upper negative achromat and two CaF_2 elements and one of KzFS4, with a basic back and front quite like the old-fashioned Fig. 17.2. There is a slick stunt in the hemispheric front lens. Instead of a simple concave, field-flattening final surface as in Figs. 17.7 to 17.9, the field-flattening surface is a cemented surface, with the final surface plano, which allows an oil-immersion system. The NA is 1.25, but the raytracing in the figures is for a miniscule ±0.0005° field and gives no clue as to its off-axis performance.

Chapter 18

Mirror and Catadioptric Systems

18.1 The Good and the Bad Points of Mirrors

A mirror element has several advantages over a refracting element. It is completely achromatic, having neither axial nor lateral color, nor chromatic variation of the aberrations (e.g., spherochromatism). A mirror can be used in any spectral region for which a reflecting coating is available (and the wide range of available coatings make this a very broad region), whereas a refractor may be quite severely limited by the transmission characteristics of its materials. Yet a third advantage is the fact that the aberrations of a spherical mirror are inherently smaller than those of a comparable spherical-surfaced lens. For example, the spherical aberration of a mirror is only one-eighth that of an equivalent lens of index 1.5, even with the lens bent for minimum spherical. Another unique feature of mirrors is that the Petzval field curvature is backward-curving for a converging (concave) mirror; this is the reverse of a refracting element.

The drawback to a single mirror is that its image is located in, and thus obscures, the incoming beam of light. Also, in a centered multiple-mirror system, the secondary mirror obscures the incoming beam. The obscuration not only reduces the illumination in the image, but, as indicated in Fig. 18.1, it can drastically reduce the contrast in the image, especially at the lower spatial frequencies. The obscuration can be avoided by decentering the system aperture and/or tilting the mirrors of the system. Figures 18.16 and 18.17 are examples of unobscured systems in which the aperture has been decentered and the axial symmetry of the surfaces has been maintained.

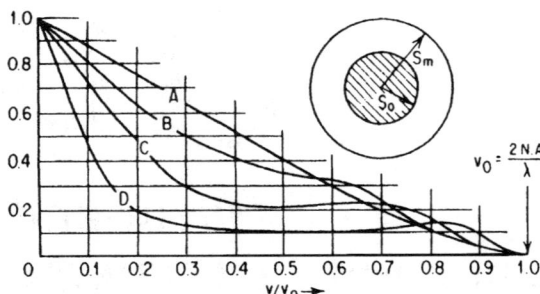

Figure 18.1 The effect of a central obscuration on the MTF of an aberration-free system. (A) $S_o/S_m = 0.00$. (B) $S_o/S_m = 0.25$. (C) $S_o/S_m = 0.50$. (D) $S_o/S_m = 0.75$.

The central obscuration reduces the diameter of the Airy disk by moving light from the central peak to the rings. The result is a slightly better contrast at high spatial frequencies (due to the smaller Airy disk) and a reduced contrast at the lower spatial frequencies (due to the increased energy in the ring structure). An obscuration of 25 to 30 percent of the diameter blocks only 6 to 9 percent of the light and, as indicated in Fig. 18.1, has a relatively modest effect on the contrast. Mirror systems tend to have small fields, which are limited by (a) the central obscuration of the secondary mirror (or the image) and (b) the central perforation of the primary, both of which become more and more troublesome as the field and *numerical aperture* (NA) are increased.

18.2 The Classical Two-Mirror Systems

The third-order aberrations of any two-mirror system can readily be determined from the following equations. Given the focal length F, the back focus B, and the mirror spacing D, one can determine the required mirror curvatures for *any* configuration from Eqs. (18.1) and (18.2). For example, if F is longer than B or D, and B and D are similar, the result is a Cassegrain system, Fig. 18.2a. If the focal length were chosen negative, the result would be a Gregorian objective, Fig. 18.2b. If the focal length is short compared to the back focus, the result can be the Schwarzschild system, Fig. 18.2c.

$$C_1 = \frac{(B-F)}{2DF} \tag{18.1}$$

$$C_2 = \frac{(B+D-F)}{2DB} \tag{18.2}$$

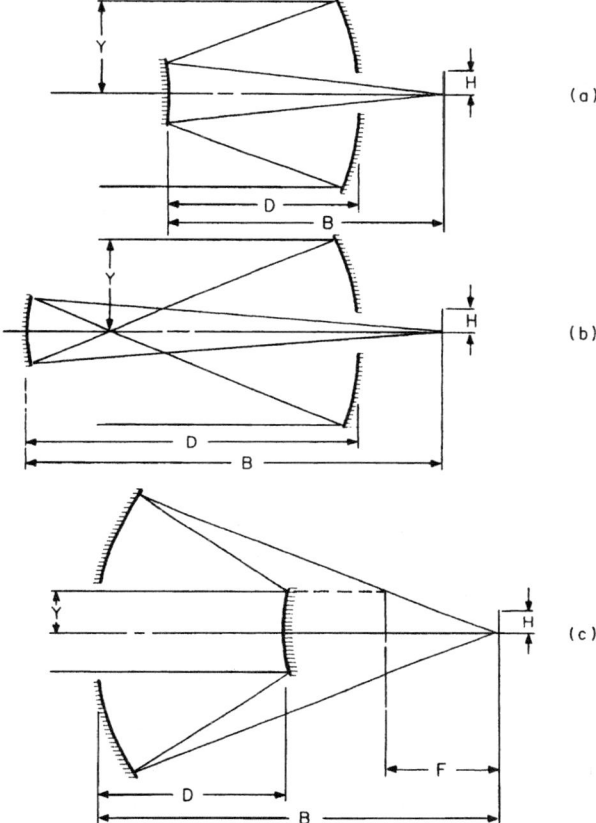

Figure 18.2 Three common two-mirror configurations. (*a*) The Cassegrain arrangement. (*b*) The Gregorian arrangement. (*c*) The Schwarzschild arrangement.

Equations (18.3) to (18.6) give the third-order aberrations of *any* two-mirror system with the object located at infinity and the aperture stop at the primary mirror.

$\Sigma TSC =$

$$\frac{Y^3[F(B-F)^3 + 64D^3F^4K_1 + B(F-D-B)(F+D-B)^2 - 64B^4D^3K_2]}{8D^3F^3} \qquad (18.3)$$

$\Sigma CC =$

$$\frac{HY^2[2F(B-F)^2 + (F-D-B)(F+D-B)(D-F-B) - 64B^3D^3K_2}{8D^2F^3} \qquad (18.4)$$

$$\Sigma\text{TAC} =$$

$$\frac{H^2Y[4BF(B-F)+(F-D-B)(D-F-B)^2-64B^3D^3K_2]}{8BDF^3} \quad (18.5)$$

$$\Sigma\text{TPC} = \frac{H^2Y[DF-(B-F)^2]}{2BDF^2} \quad (18.6)$$

where Y = semiaperture of the system
H = image height
B = distance from mirror no. 2 to image (i.e., the back focal length)
F = system focal length
D = spacing (use positive sign)
ΣTSC = transverse third-order spherical aberration sum
ΣCC = third-order sagittal coma sum
ΣTAC = transverse third-order astigmatism sum
ΣTPC = transverse Petzval curvature sum

and where K_1 and K_2 are the equivalent fourth-order deformation coefficients (see Eq. (5.1)) for the primary and secondary mirrors. For a conic section, K is equal to the conic constant κ divided by eight times the cube of the surface radius. Thus $K = \kappa/8R^3$ and $\kappa = 8KR^3$.

Equations (18.7) to (18.11) describe the case when both mirrors are individually corrected for spherical aberration (and are thus easy to test). This case includes the classical *Cassegrain* (paraboloid primary and hyperboloid secondary) and the classical *Gregorian* (paraboloid primary and ellipsoid secondary). Note well that the third-order coma (per Eq. (18.10)) is exactly the same for any arrangement of the mirrors when each mirror is individually free of spherical aberration.

$$K_1 = \frac{(F-B)^3}{64D^3F^3} \quad (18.7)$$

$$K_2 = \frac{(F-D-B)(F+D-B)^2}{64B^3D^3} \quad (18.8)$$

$$\Sigma\text{TSC} = 0.0 \quad (18.9)$$

$$\Sigma\text{CC} = \frac{HY^2}{4F^2} \quad (18.10)$$

$$\Sigma \text{TAC} = \frac{H^2 Y (D - F)}{2BF^2} \tag{18.11}$$

Equations (18.12) to (18.16) cover the case when the aspherics are chosen so that both the spherical and coma are simultaneously corrected. The *Ritchey-Chretien* design (both mirrors hyperboloidal) falls in this category, as does the Hubble Space Telescope.

$$K_1 = \frac{[2BD^2 - (B-F)^3]}{64 D^3 F^3} \tag{18.12}$$

$$K_2 = \frac{[2F(B-F)^2 + (F-D-B)(F+D-B)(D-F-B)]}{64 B^3 D^3} \tag{18.13}$$

$$\Sigma \text{TSC} = 0.0 \tag{18.14}$$

$$\Sigma \text{CC} = 0.0 \tag{18.15}$$

$$\Sigma \text{TAC} = \frac{H^2 Y (D - 2F)}{4 BF^2} \tag{18.16}$$

Equations (18.17) to (18.21) describe the *Dall-Kirkham* system (with a spherical secondary mirror that is easier to fabricate and test).

$$K_1 = \frac{[F(F-B)^3 - B(F-D-B)(F+D-B)^2]}{64 D^3 F^4} \tag{18.17}$$

$$K_2 = 0.0 \tag{18.18}$$

$$\Sigma \text{TSC} = 0.0 \tag{18.19}$$

$$\Sigma \text{CC} = \frac{HY^2 [2F(B-F)^2 + (F-D-B)(F+D-B)(D-F-B)]}{8 D^2 F^3} \tag{18.20}$$

$$\Sigma \text{TAC} = \frac{H^2 Y [4BF(B-F) + (F-D-B)(D-F-B)^2]}{8 DBF^3} \tag{18.21}$$

While these equations are exact only for the third-order aberrations, they are remarkably accurate for systems of modest aperture, and can provide good starting points even for high-speed designs.

Figures 18.3 to 18.10 are examples of the classical two-mirror forms. So that the reader can compare the various configurations, all of them (except Figs. 18.6 and 18.7) have been designed at a speed of $f/2$ and a total angular field of one degree. This speed is, of course, much, much faster than one would find in an astronomical telescope, but it is not atypical of the speeds used for many other applications, e.g., the infrared, although the Gregorian example *is* rather extreme! The back focus has been set at 30 percent of the focal length to hold the obscuration by the secondary mirror to 30 percent of the aperture diameter. (Note that the back focus determines the obscuration.) The mirror spacing has been set at 20 percent of the focal length so that (in combination with the 30 percent back focus) the final image is behind the primary mirror by 10 percent of the focal length. Obviously these choices significantly affect the aberrations, and other arrangements will differ from these particular examples.

The Cassegrain configuration at $f/2.0$ is sketched in Fig. 18.3; this lens drawing also applies for Figs. 18.4 and 18.5. The prescription and aberration plots for the classical version with a paraboloidal primary and hyperboloidal secondary are given in Fig. 18.3. Note the large overcorrected coma. Figure 18.4 is a conic Ritchey-Chretien, where the conic constants have been modified slightly from the values given by Eqs. (18.12) and (18.13) (which correct the *third-order* spherical and coma) so as to correct the marginal aberrations. The Dall-Kirkham, with a spherical secondary, is shown in Fig. 18.5; while easier to fabricate, its coma is even worse than that of the classic Cassegrain.

Figures 18.6 and 18.7 show Cassegrains at a somewhat more modest speed of $f/5$. The first is a classic Cassegrain, and the second is a general aspheric Ritchey-Chretien, with radii chosen equal to obtain a flat Petzval surface. Note the increased length of this system.

The Gregorian form is infrequently encountered for two primary reasons. With all else equal, it is a longer system than the Cassegrain. In addition, the Cassegrain is preferred because it has a flatter field. The classic parabola-ellipse Gregorian is shown at an extreme speed of $f/2$ in Fig. 18.8. Figure 18.9 (same lens drawing) shows a sort of Ritchey-Chretien-Gregorian combination with the conic constants selected to correct both the spherical and coma.

The Schwarzschild arrangement (Fig. 18.10) suffers from the fact that the concave mirror must be several times as large as the system aperture and its length is more than four times its focal length; however, it can be corrected for spherical, coma, and astigmatism using only spherical surfaces. For an infinite object distance, a mirror spacing of twice the focal length f, a convex radius of $(\sqrt{5}-1)f$, and a concave radius of $(\sqrt{5}+1)f$ will produce a system corrected for third-order spherical aberration. Note that the mirrors have a common center of curvature. In a monocentric

Mirror and Catadioptric Systems 461

F/2 0.5degHFOV CLASSICAL CASSEGRAIN

radius	thickness	mat'l	index	V-no	sa
-57.143	20.000	mirror			26.0
kappa		-1.000			
-24.000	30.000	mirror			9.0
kappa		-3.240			

EFL = 100
BFL = 30
NA = -0.2462 (F/2.0)
GIH = 0.87
PTZ/F = -0.2006
VL = -20.00
OD Infinite conjugate

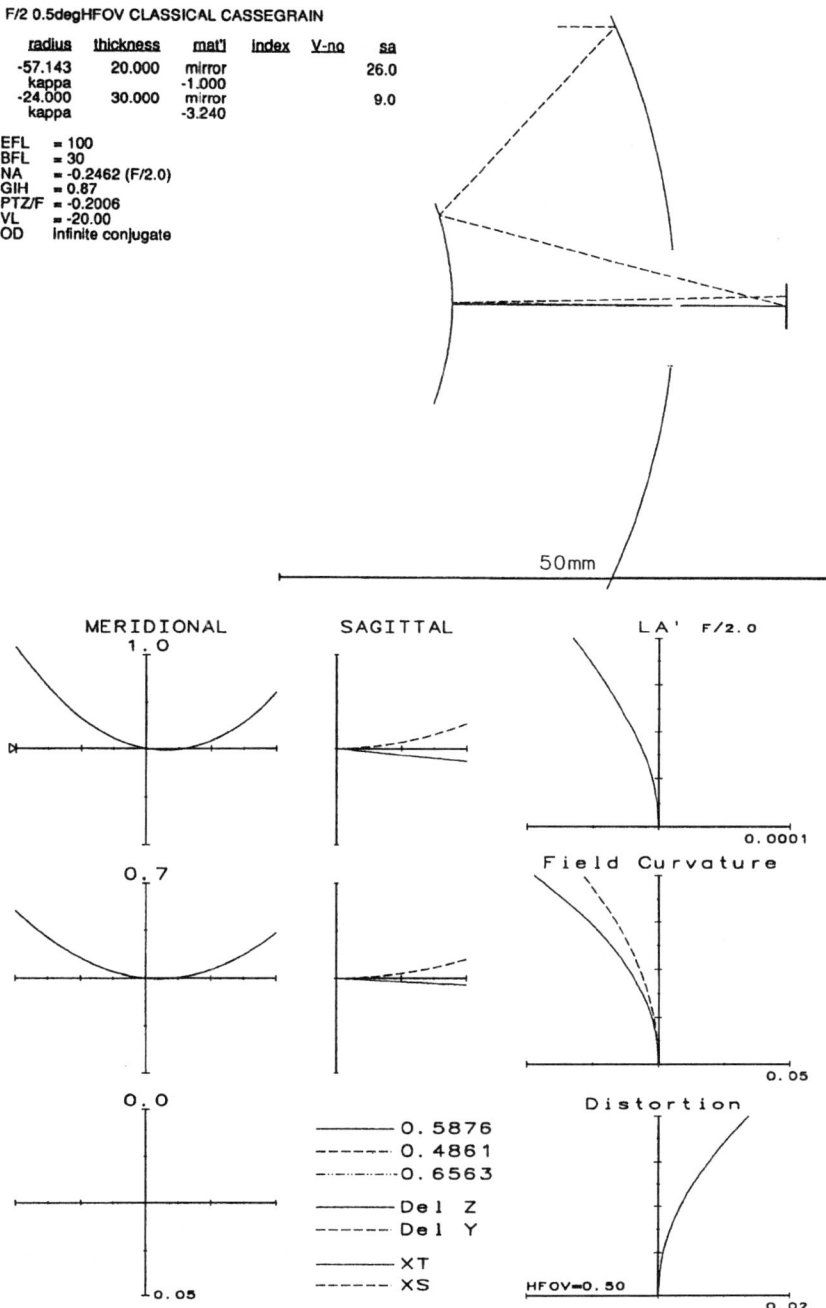

Figure 18.3 $f/2.0$ ±0.5° classic Cassegrain with a paraboloid primary and hyperboloid secondary. Note the coma and inward-curving field.

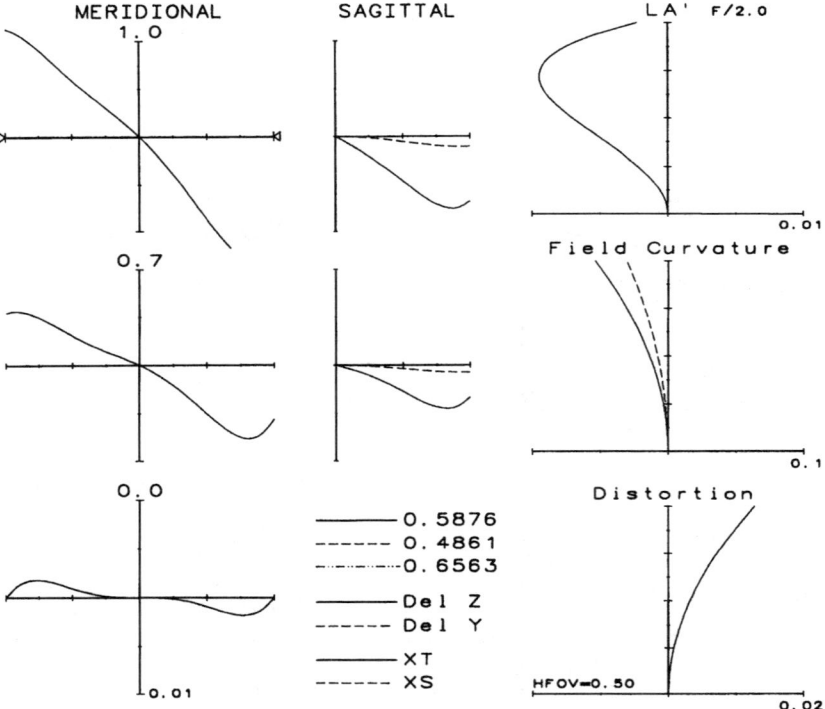

Figure 18.4 f/2.0 ±0.5° Ritchey-Chretien Cassegrain corrected for both spherical and coma, with both mirrors hyperboloids. The zonal aberration could easily be corrected by a general aspheric surface.

system of this type, if the aperture stop is located at the common center, the system is free of coma and astigmatism, and the image surface is a sphere of radius f. There are some other systems of this type that are close to the construction given above but have somewhat differing aberration characteristics; for example, there is a nearby form with very low high-order spherical (which can be found by systematically varying the parameters).

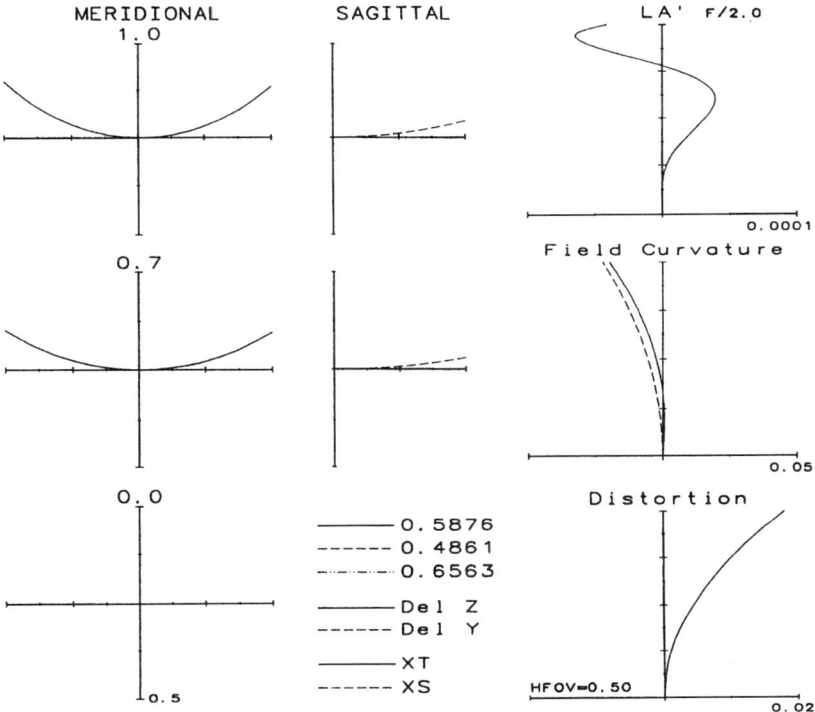

Figure 18.5 $f/2.0$ $\pm 0.5°$ the Dall-Kirkham Cassegrain with a spherical secondary mirror.

An all-spherical-surface Schwarzschild system is shown in Fig. 18.10. The radii have been adjusted (very slightly) from the values that would result from the expressions in the preceding paragraph so that the marginal aberrations are corrected.

The Schwarzschild configuration is often used as a microscope objective, where its relatively large-diameter mirror is not a problem because the short focal length means that the actual size is not inconveniently

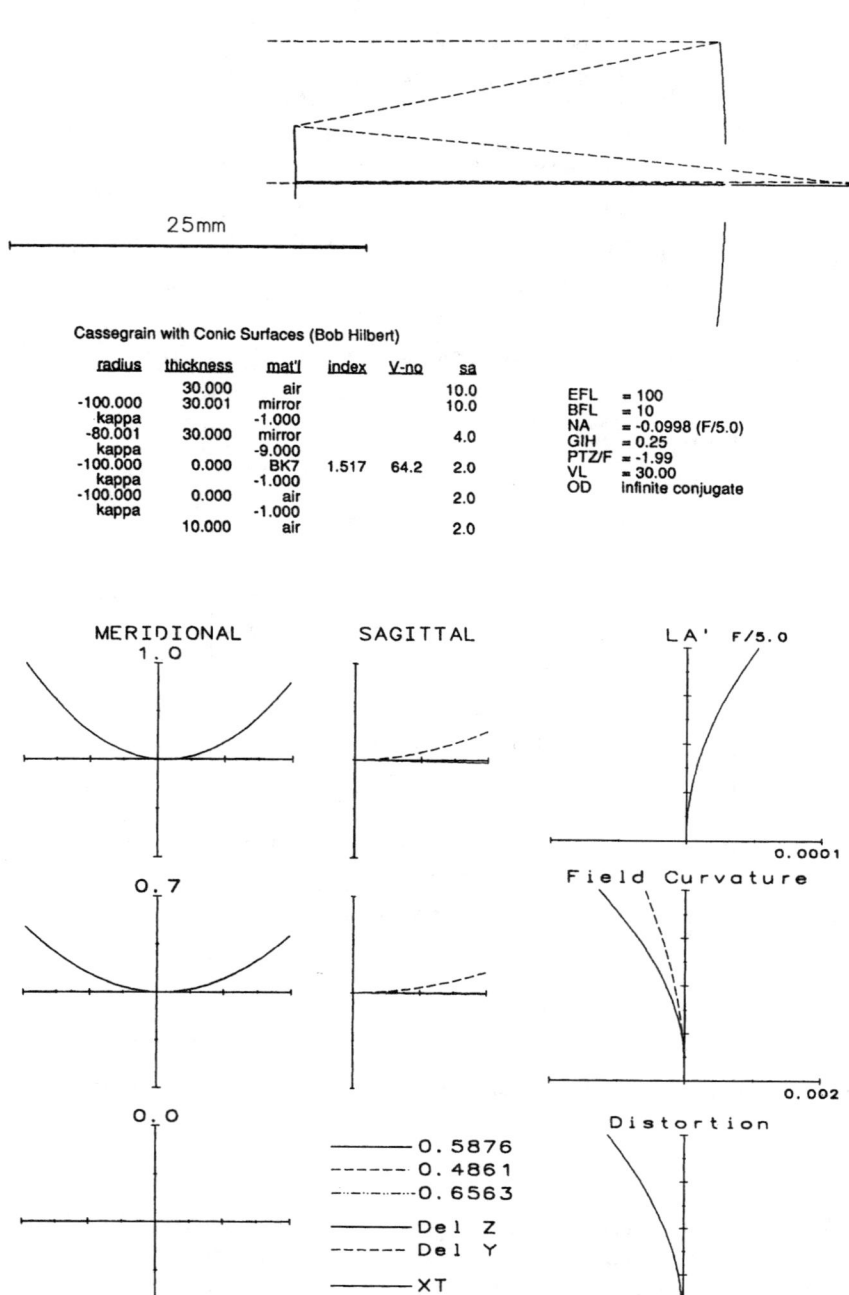

Figure 18.6 $f/5.0$ ±0.14° classic Cassegrain.

Mirror and Catadioptric Systems 465

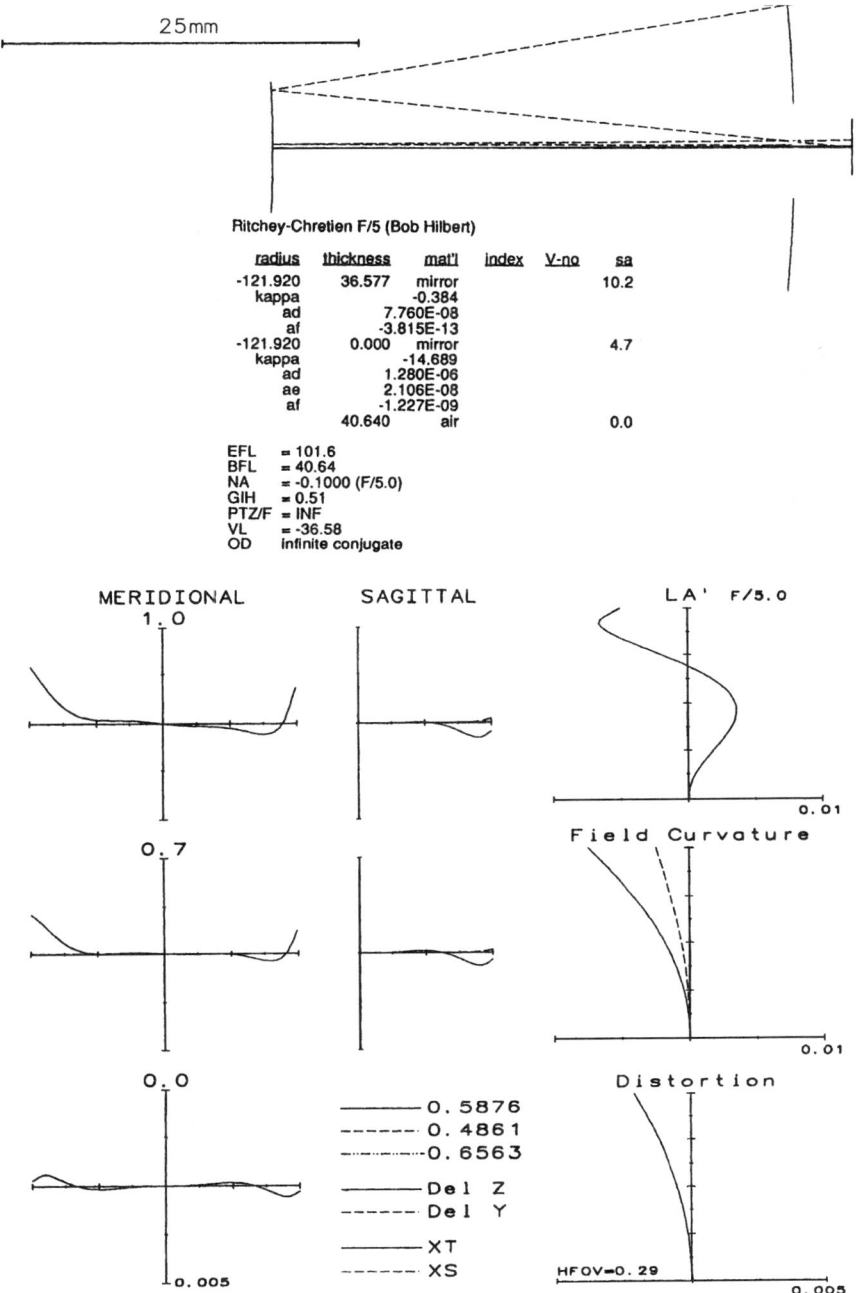

Figure 18.7 $f/5.0$ ±0.29° Ritchey-Chretien Cassegrain.

466 Chapter Eighteen

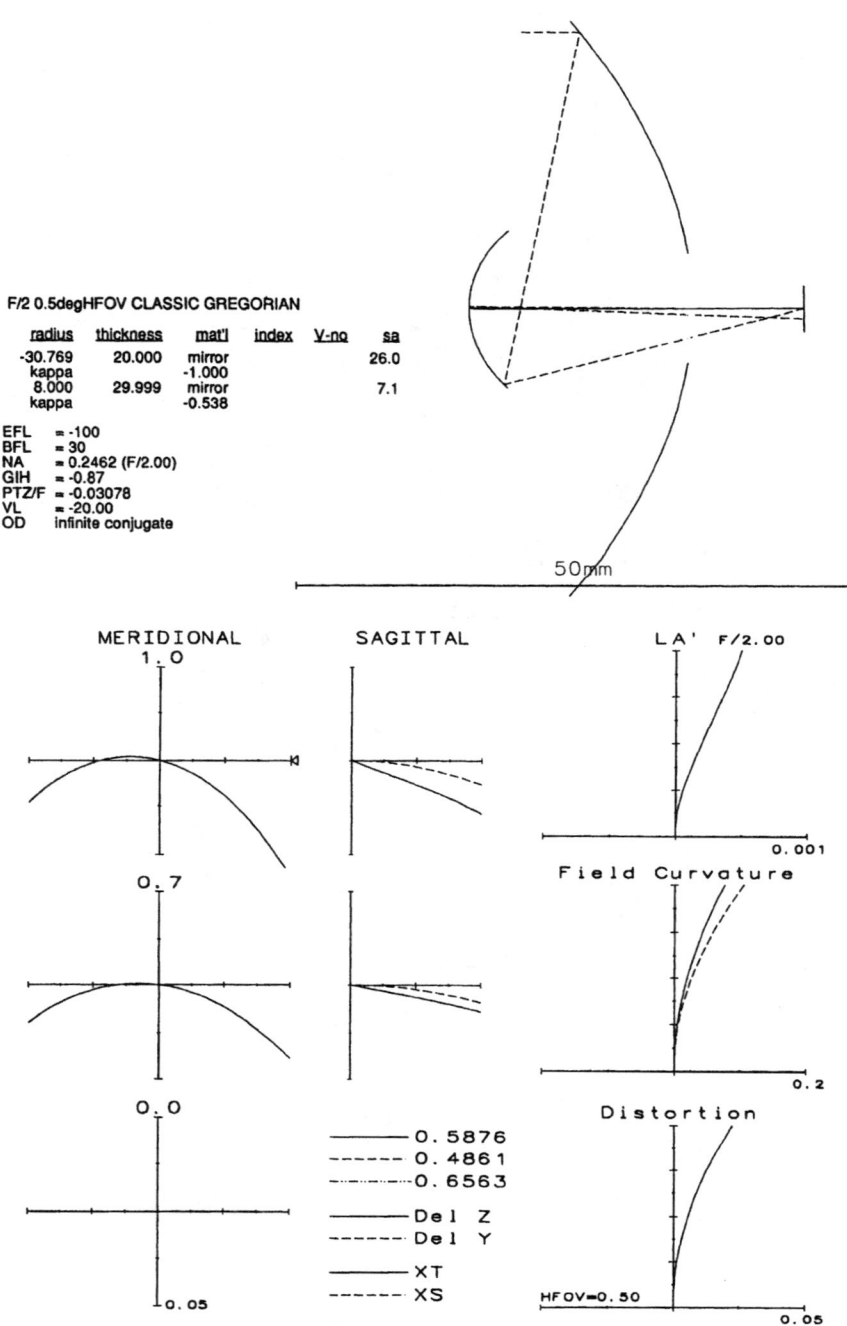

F/2 0.5degHFOV CLASSIC GREGORIAN

radius	thickness	mat'l	index	V-no	sa
-30.769	20.000	mirror			26.0
kappa		-1.000			
8.000	29.999	mirror			7.1
kappa		-0.538			

EFL = -100
BFL = 30
NA = 0.2462 (F/2.00)
GIH = -0.87
PTZ/F = -0.03078
VL = -20.00
OD infinite conjugate

Figure 18.8 $f/2.0$ ±0.5° classic Gregorian.

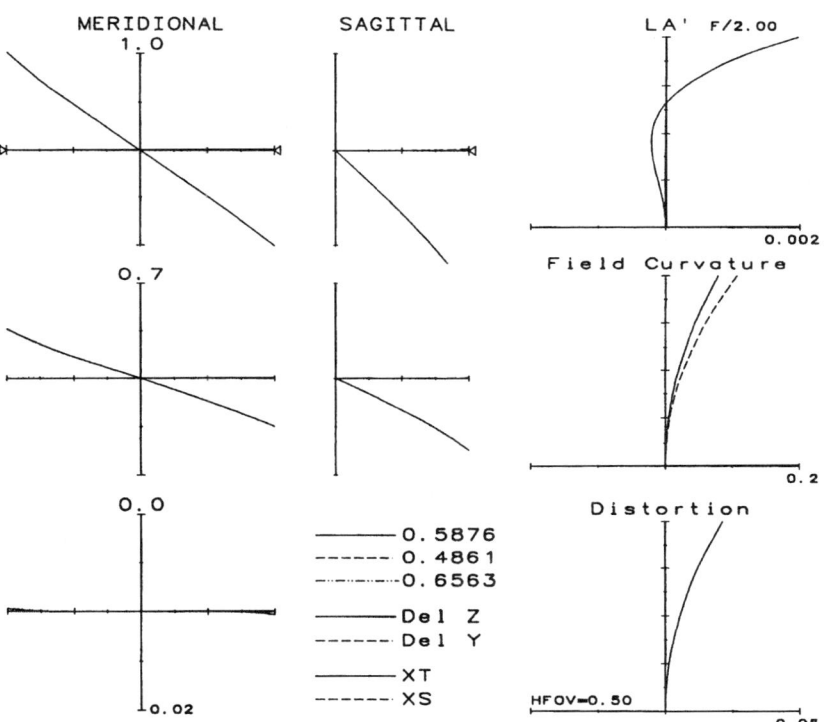

Figure 18.9 $f/2.0 \pm 0.5°$ Gregorian with conic constants adjusted to correct both spherical and coma.

large. The all-mirror construction allows the system to be used in the infrared or in the ultraviolet. Its simple all-spherical construction makes it an economical design to manufacture, and its long working distance is an added convenience for many applications.

The two degrees of freedom represented by the choice of the conic constants in a two-mirror system allow one to correct only two aberrations (assuming that the mirror curvatures have been determined by

468 Chapter Eighteen

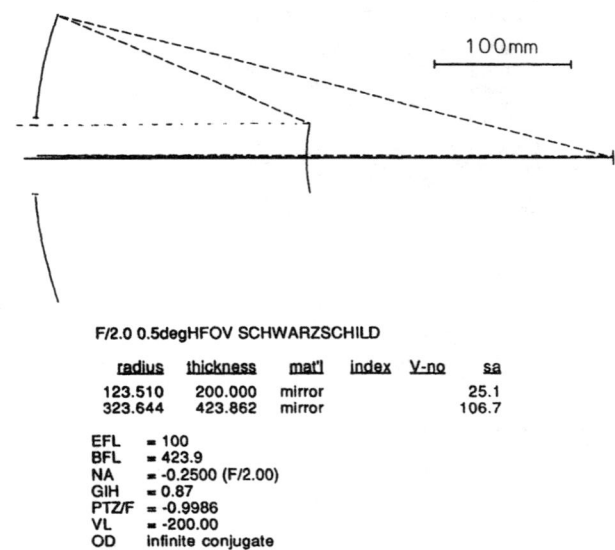
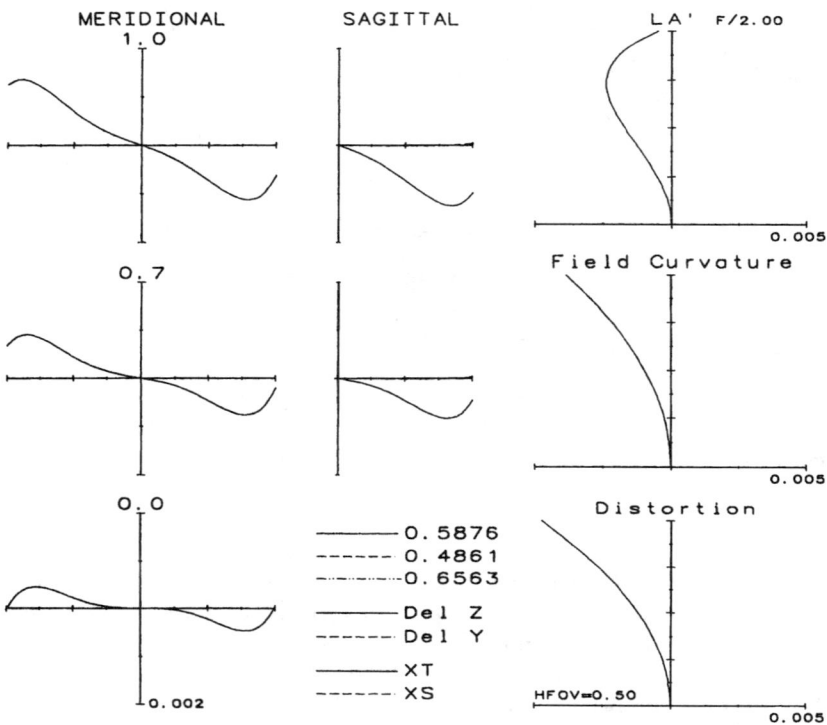

Figure 18.10 $f/2.0$ $\pm 0.5°$ spherical-surfaced Schwarzschild with corrected spherical, coma, and astigmatism.

requirements for f, B, and D). If there is a free choice of mirror curvatures, the Petzval curvature can be corrected in a Cassegrain configuration with identical radii on both primary and secondary, as in Fig. 18.7. Control of astigmatism requires one additional effective degree of freedom, which can be provided by another optical component. There are a few arrangements where the choice of f, B, and D is used to correct the astigmatism.

18.3 Catadioptric Systems

The fundamental philosophy behind most catadioptric systems is to use refracting elements to correct the aberrations of a system of spherical mirrors without introducing any new aberrations. Thus the typical system uses mirrors for most of the focusing power and uses weak refracting elements to provide the aberration correction. The thick meniscus corrector elements of the well-known Bouwers and Maksutov systems are examples of weak refractor elements that are strongly bent to increase their overcorrecting spherical aberration. (See Figs. 18.18 and 18.19.) Where more than one refracting corrector is used, the approach is usually to have them closely spaced and to use approximately equal amounts of positive and negative power. The same glass type is often used for all the elements. The effect of this is that the net chromatic aberration can be made zero; both ordinary chromatic and secondary spectrum can be eliminated. For example, in a two-element zero net power corrector located at the entrance pupil, the positive element is shaped to minimize its spherical aberration contribution and the negative element is shaped to provide enough overcorrected spherical to balance out the undercorrected spherical aberration from the positive element and the spherical mirrors. Both can be bent to simultaneously correct the coma and spherical aberration. Correctors can also be located in the convergent beam in front of the image. In this location, the corrector's function is more likely to be the correction of the field aberrations, coma, astigmatism, and, occasionally, the Petzval curvature. A concentric meniscus centered on the image will reduce the Petzval sum without introducing spherical, coma, or chromatic.

Figure 18.11 incorporates three BK7 refracting correctors near the stop. The axial color is well-corrected, but in this system the spherochromatism is rather large. This system covers a field of 2.5°, is fast ($f/1.5$), and uses no aspheric surfaces. A similar design at $f/1.0$ uses a low power meniscus field lens, concave to the image.

Figures 18.12 and 18.13 show two moderately complex catadioptric designs. They have several features in common. They both incorporate at least one Mangin-type mirror (i.e., with the reflector on the second surface of a lens element). This provides the effect of a lens plus a mirror, in a component that is thus probably at least partially corrected, for about the same cost as the mirror alone. An *ordinary* Mangin mirror is

470   Chapter Eighteen

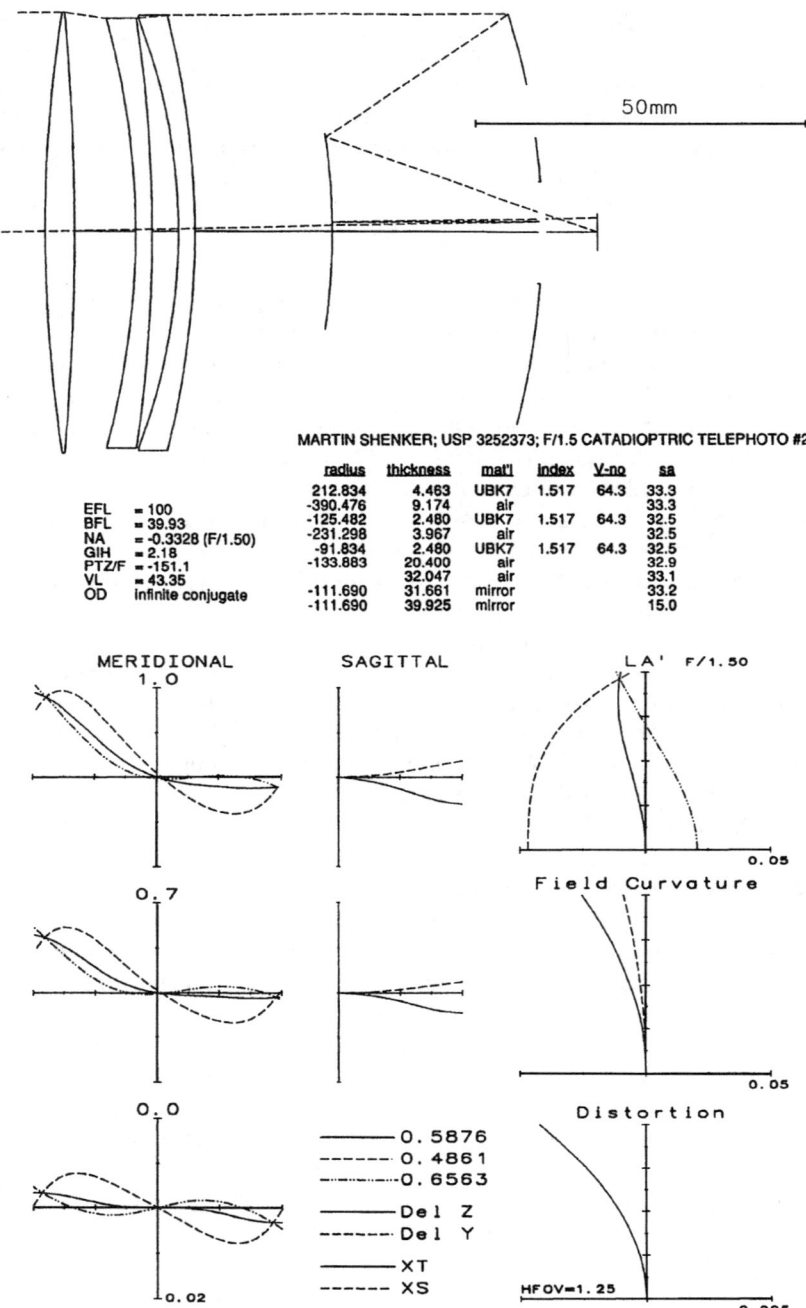

Figure 18.11 $f/1.5$ ±1.25° catadioptric system with all spherical surfaces and a three-element zero power corrector. Note that at the marginal ray the primary and secondary chromatic are zero.

Mirror and Catadioptric Systems   471

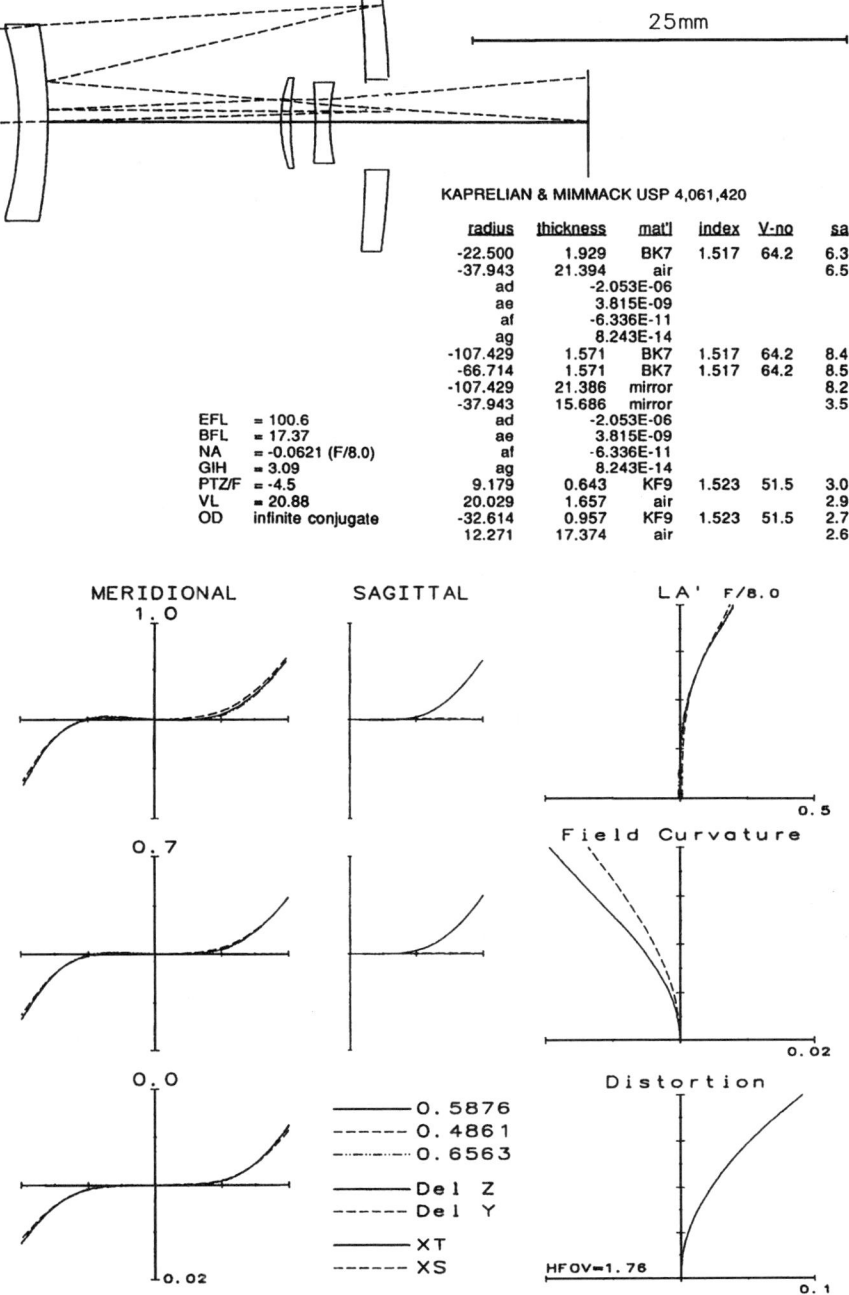

Figure 18.12 $f/8.0$ $\pm 1.8°$ catadioptric with a meniscus corrector window, an aspheric second surface, a Mangin primary, and two field-correcting elements.

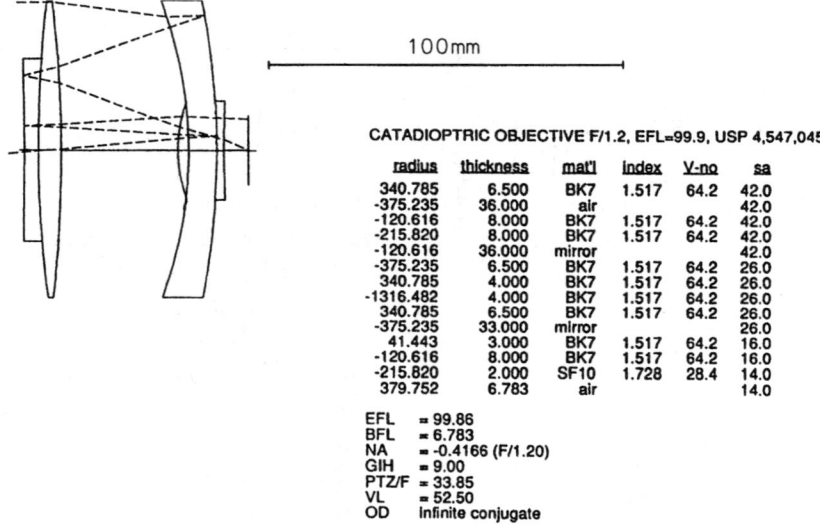

CATADIOPTRIC OBJECTIVE F/1.2, EFL=99.9, USP 4,547,045

radius	thickness	mat'l	index	V-no	sa
340.785	6.500	BK7	1.517	64.2	42.0
-375.235	36.000	air			42.0
-120.616	8.000	BK7	1.517	64.2	42.0
-215.820	8.000	BK7	1.517	64.2	42.0
-120.616	36.000	mirror			42.0
-375.235	6.500	BK7	1.517	64.2	26.0
340.785	4.000	BK7	1.517	64.2	26.0
-1316.482	4.000	BK7	1.517	64.2	26.0
340.785	6.500	BK7	1.517	64.2	26.0
-375.235	33.000	mirror			26.0
41.443	3.000	BK7	1.517	64.2	16.0
-120.616	8.000	BK7	1.517	64.2	16.0
-215.820	2.000	SF10	1.728	28.4	14.0
379.752	6.783	air			14.0

EFL = 99.86
BFL = 6.783
NA   = -0.4166 (F/1.20)
GIH = 9.00
PTZ/F = 33.85
VL   = 52.50
OD   Infinite conjugate

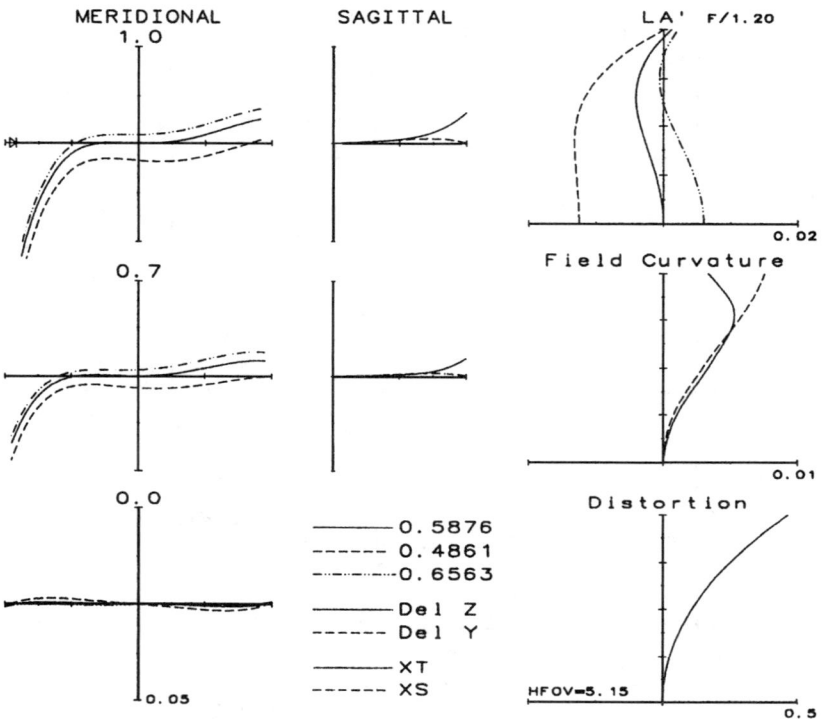

Figure 18.13 $f/1.2$ ±5° catadioptric with Mangin primary and secondary, and a cemented triplet field corrector (which uses the primary as its middle element).

corrected for spherical aberration and also has a reduced coma contribution, but it is afflicted with severe chromatic aberration. The systems shown have a full-aperture element on the object side that can be used as a protecting or sealing window for the system. The secondary mirror is on a surface that does double duty; it is either a surface of a corrector element, or in one case, a Mangin secondary. Both have field correctors in the convergent cone near the image. In one system the primary Mangin is also used in transmission as a refracting corrector in the convergent cone near the image.

Figure 18.14 is what is called a *solid cat*. The volume of the lens is almost completely filled with glass. A modest number of lenses similar to Fig. 18.14 were sold as extremely compact (if weighty) telephoto lenses for 35-mm cameras at a focal length of 800 mm and a speed of $f/11$ (which, taking the obscuration into account, was about the equivalent of an unobstructed $f/16$). Many mirror telephotos are designed along the general lines of Fig. 18.12, with a focal length of 500 to 1000 mm and a speed of about $f/8$ or $f/10$.

It is worth noting that Mangin mirrors have been used with success as a secondary (with a spherical primary) in a Cassegrain configuration, as both primary and secondary in a Cassegrain, and as both primary and secondary in a Schwarzschild arrangement. In some applications, the Mangin can be an inexpensive, uncomplicated substitute for an aspheric surface.

Note that in Fig. 18.13 the leading element is a positive lens, one benefit of which is that by converging the beam it reduces the necessary diameter of the primary mirror, and its undercorrected chromatic tends to offset the overcorrected chromatic of the Mangin primary. Field correctors like those in Fig. 18.12 are typically made of the same glass and, in combination, are approximately of zero power. An infrared catadioptric telescope is shown in Fig. 19.6 of Chap. 19.

18.4 Aspheric Correctors and Schmidt Systems

An aspheric refracting corrector can be used to correct an aberration in a mirror system. Added to a Ritchey-Chretien system, for example, such a corrector can be used in combination with the two aspheric mirrors to correct the astigmatism as well as the spherical and coma. Such a corrector can be located in either the incoming beam or in the converging beam in front of the image. Often an aspheric corrector is used at the entrance pupil with an all-spherical-mirror system (usually in a Cassegrain configuration) to correct the spherical aberration. This *Schmidt-Cassegrain* is a popular form for compact medium-sized commercial telescopes. See the design exercise of Sec. 18.7.

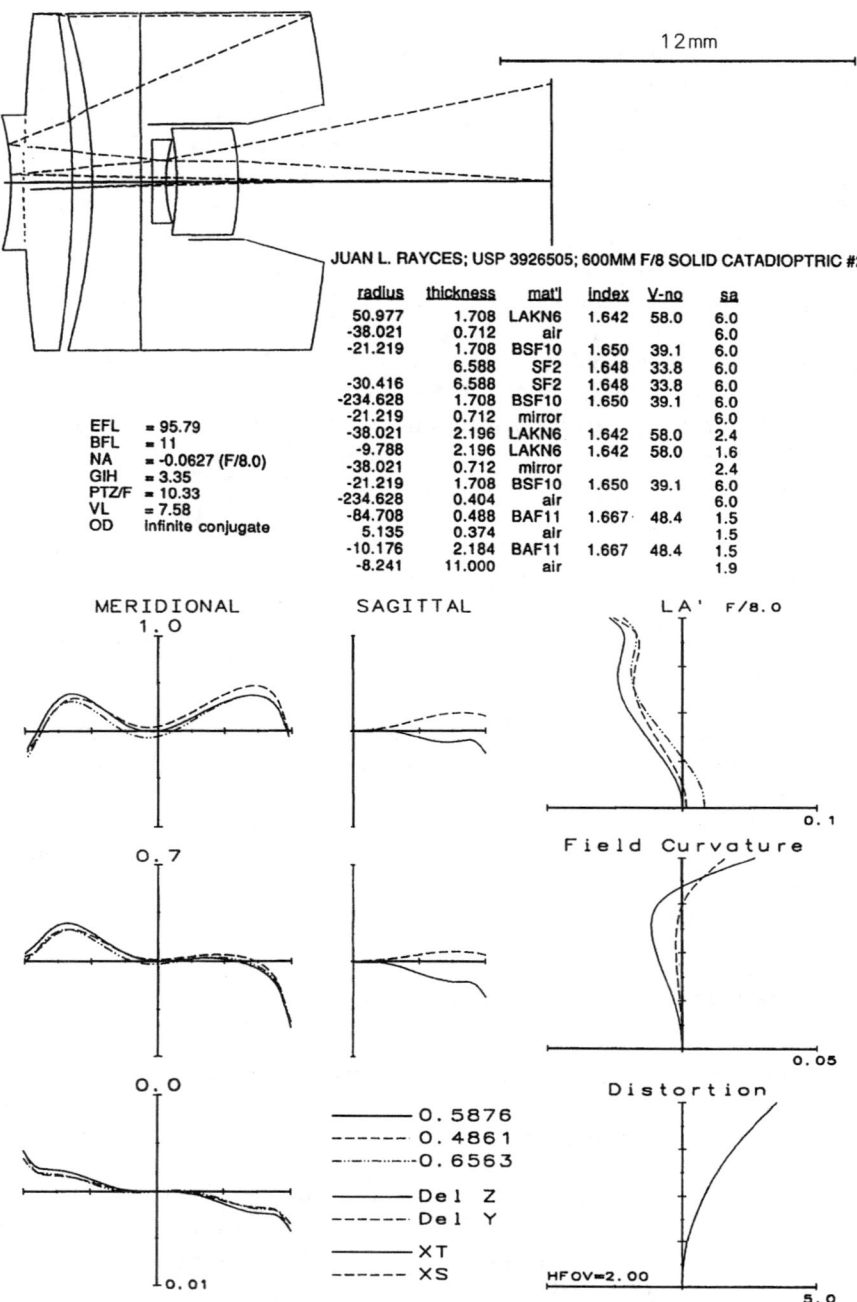

Figure 18.14 *f*/8.0 ±1.9° a solid cat; a catadioptric camera lens that is almost solid glass. Note the balanced fifth-order coma.

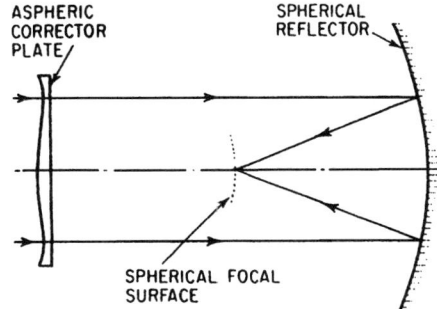

Figure 18.15 The Schmidt system consists of a spherical reflector with an aspheric corrector plate (and the aperture stop) located at its center of curvature. The aspheric surface of the f/1.0 system shown here is greatly exaggerated.

In the *Schmidt system*, as shown in Fig. 18.15, a spherical mirror with the aperture stop at its center of curvature (and thus free of coma and astigmatism) is corrected for spherical aberration by an aspheric corrector located at the stop (where it affects *only* the spherical—see Eqs. (24.100) through (24.106))

If a refracting aspheric corrector is used to correct the spherical aberration, it can exactly and completely correct the spherical at the nominal wavelength; however, at shorter wavelengths its higher index overcorrects the spherical, and at longer wavelengths its lower index leaves the spherical undercorrected. This is ordinary spherochromatism. It can be balanced (but not eliminated) by introducing undercorrected axial chromatic, using some weak positive power to do so. This power is usually combined with the aspheric surface so that the central part of the corrector is convex (converging) and the margin of the aspheric is diverging. The neutral, nondeviating zone is usually located at about 0.866 of the marginal ray height to get the optimum spherochromatic balance. In the case of the classical Schmidt system, the equation for a near-optimum aspheric surface for the corrector plate is

$$x = 0.5Cy^2 + Ky^4 + Ly^6$$

where

$$C = \frac{3}{128(n-1)f(f/\#)^2}$$

$$K = \frac{\left[1 - \dfrac{3}{64(f/\#)^2}\right]^2}{32(1-n)f^3}$$

$$L = \frac{1}{85.8(1-n)f^5}$$

and x = sag of surface
  f = local length (positive if corrector oriented as in Fig. 18.15)
  n = index of corrector material
  $(f/\#)$ = relative aperture of Schmidt system

The 48-in Schmidt camera at the Palomar Observatory has a focal length of 120.9 in, at a speed of $f/2.5$, and covers a 14-by-14-in image area to achieve an angular field of 6° by 6°. Note that while the system aperture is 48 in, the mirror diameter is 72 in, a full 2 ft larger.

The basic Schmidt design can be slightly improved by a little shortening, an aspheric mirror (which reduces the strength of the corrector and its oblique spherical), and slightly undercorrected spherical aberration. A positive field flattener can convert the curved image surface to a plane. The length of the Schmidt at twice its focal length is a significant drawback; the Wright design puts the corrector plate nearly at the focus, cutting the length in half (at the expense of off-axis performance).

18.5 Confocal Paraboloids

If two paraboloid mirrors are arranged with their focal points coincident, the result is an afocal system with rather interesting properties. It is free of spherical, coma, and astigmatism, and is, of course, free of any chromatic aberrations. Both paraboloids may be concave, or one may be concave and the other convex (provided that the convex radius is shorter than the concave). The Petzval curvature is not corrected. For the two-concave-mirror arrangement, the Petzval field is strongly backward-curving; the concave-convex arrangement has a lesser, inward field curvature. Confocal paraboloids are often used as afocal attachments to increase the aperture and focal length, or conversely, to shorten the focal length and increase the field of view.

18.6 Unobscured Systems

Any of the standard mirror systems can be used with a decentered aperture stop to produce a system with an unobscured aperture. The classical Herschel mount is simply a paraboloid with the aperture stop displaced laterally so that the aperture is completely off the optical axis and the image/focal point is outside the beam defined by the aperture. Note that, although this arrangement is often called an *off-axis parabola*, the paraboloid is not used off-axis; the *aperture* is off the axis, but the object and image are both on the optical axis. Since the paraboloid is completely free of spherical aberration for an infinitely distant object, the stop shift equations (specifically Eq. (24.94)) indicate that the large coma of the paraboloid is unchanged by moving the stop off-axis. Note that the astigmatism is no longer axially symmetric.

Two three-mirror systems with axially symmetrical mirrors that are the basis of many of the more complex, tilted or decentered, unobscured systems are illustrated in this section. Figure 18.16 shows a (modified) Baker system that, in its basic form, consists of a confocal paraboloid front followed by a concave spherical reflector. The secondary mirror of the confocal pair is at the aperture stop and is placed at the center of curvature of the tertiary sphere; the secondary is modified by aspheric deformation terms so that it acts as a reflecting Schmidt corrector for the spherical tertiary. (In this design the primary mirror has been modified from its original paraboloid form and the space is less than the radius.) The confocal paraboloid pair is free of spherical, coma, and astigmatism. The tertiary sphere with the stop at its center of curvature is free of coma and astigmatism, and the aspheric deformation of the secondary corrects its spherical aberration. The Petzval curvature can be corrected if the curvature of the convex mirror is made equal to the sum of the concave curvatures. Note that the confocal front may be used in either orientation; with the concave as the primary, the focal length of the system is long, and with the convex as primary, the focal length is short.

The second system (Fig. 18.17) is afocal, with a magnification of 5×. It consists of a classical Cassegrain (with a paraboloid primary and hyperboloid secondary) as the objective, plus a concave paraboloid as the "eyepiece" of the system to collimate the output beam. This arrangement has properties similar to those of confocal paraboloids, being free of spherical, coma, and astigmatism. Again, if the convex curvature is made equal to the sum of the concave curvatures, the Petzval will be corrected. This type of system (i.e., afocal) is often used as a device to reduce a large beam diameter to a smaller, more easily handled beam for the balance of the optical train. Note the external exit pupil, which can be located at the entrance pupil for the rest of the system. These characteristics are often desired to accomodate a scanner, or a cold stop in a Dewar.

18.6.1 Mirror systems with a meniscus shell corrector

Figures 18.18 and 18.19 are examples of systems that incorporate the Gabor-Bouwers-Maksutov principle of a corrector that is a weak, but strongly bent, meniscus element. Each of these designs also incorporates a field flattener near the focal plane—a positive element in one case and a negative in the other. In Fig. 18.18, the Cassegrain-like configuration produces an inward-curving field that requires a negative lens as the

478 Chapter Eighteen

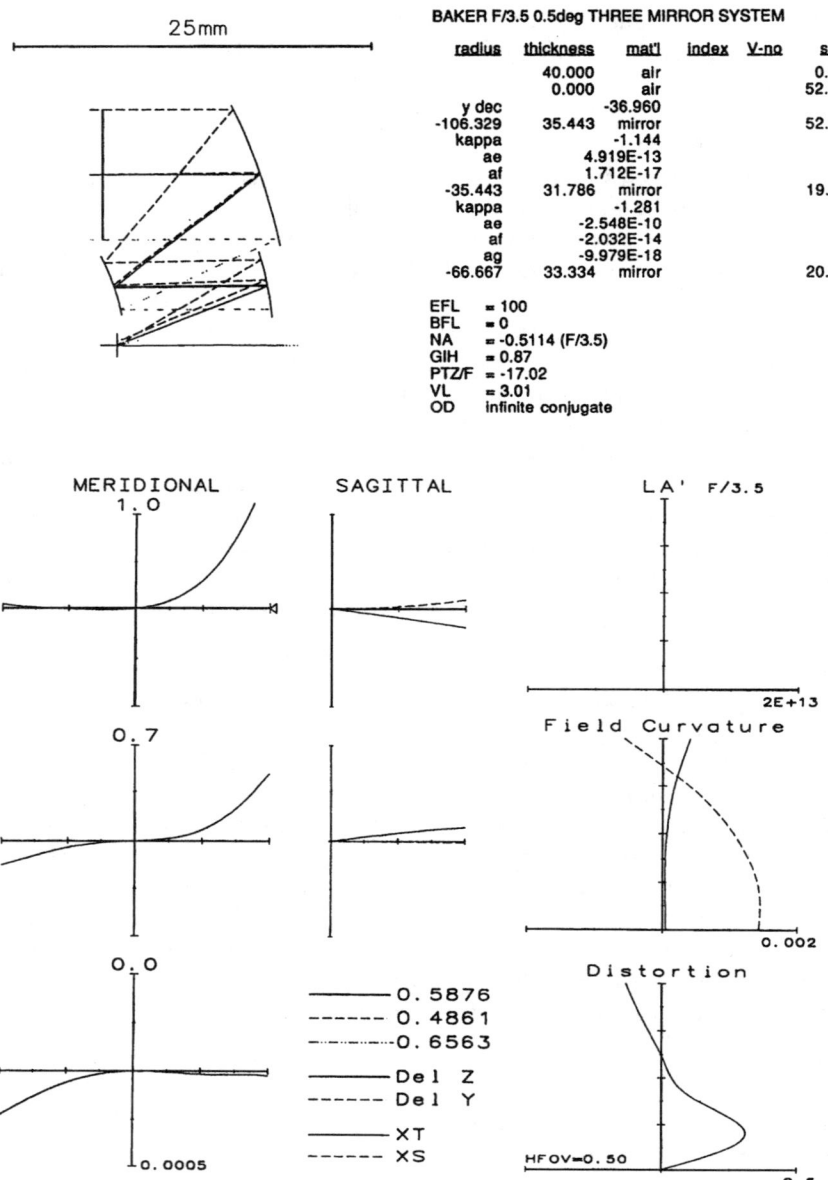

Figure 18.16 $f/3.5$ ±0.5° unobscured aperture triple mirror system. Note that, due to its asymmetry, the field aberrations will be different in each meridian.

Mirror and Catadioptric Systems

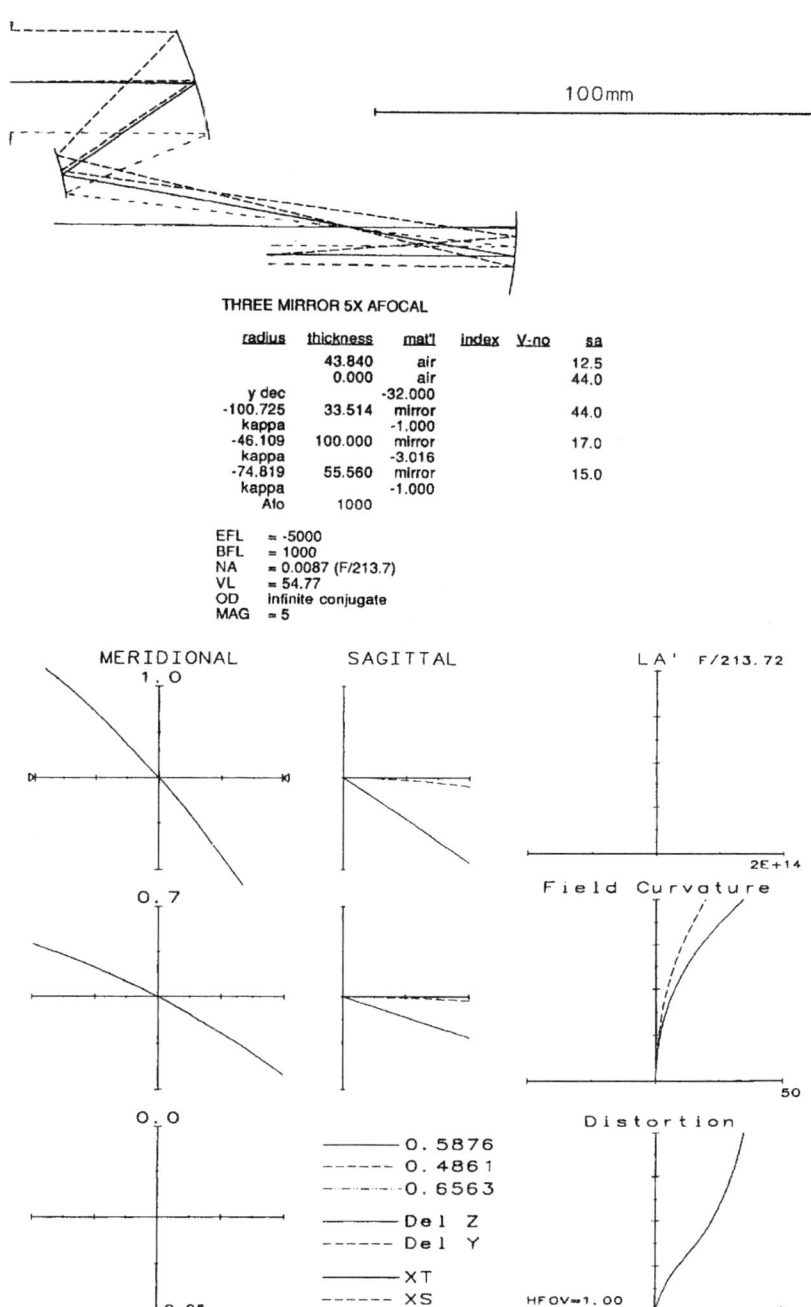

Figure 18.17 Three mirror 5x confocal conic system.

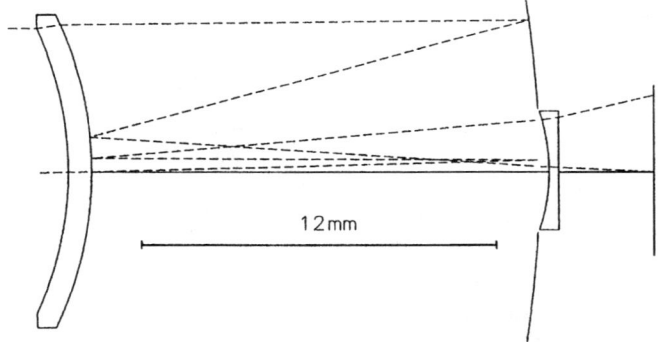

MAKSUTOV CASSEGRAIN F/10 1.5deg HFOV

radius	thickness	mat'l	index	V-no	sa
-11.796	0.831	BK8	1.520	63.7	5.0
-12.528	15.763	air			5.5
-39.711	15.763	mirror			6.0
-12.528	16.096	mirror			2.0
-6.651	0.333	BK8	1.520	63.7	2.0
	3.314	air			2.1

EFL = 99.89
BFL = 3.314
NA = -0.0496 (F/10.0)
GIH = 2.60
PTZ/F = -0.1751
VL = 17.26
OD infinite conjugate

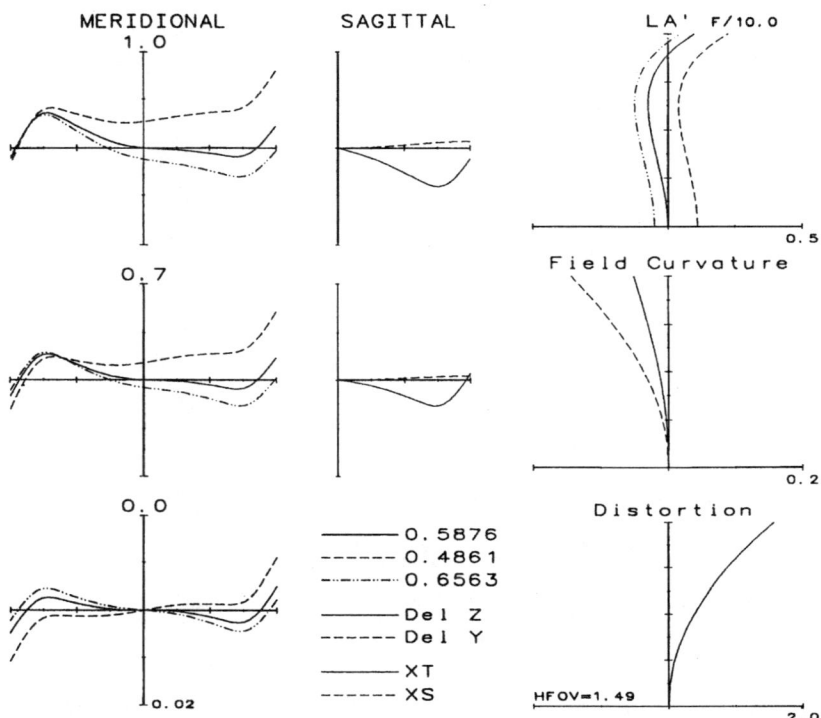

Figure 18.18 $f/10$ ±1.5° Maksutov meniscus corrector Cassegrain with a negative field corrector lens.

Mirror and Catadioptric Systems

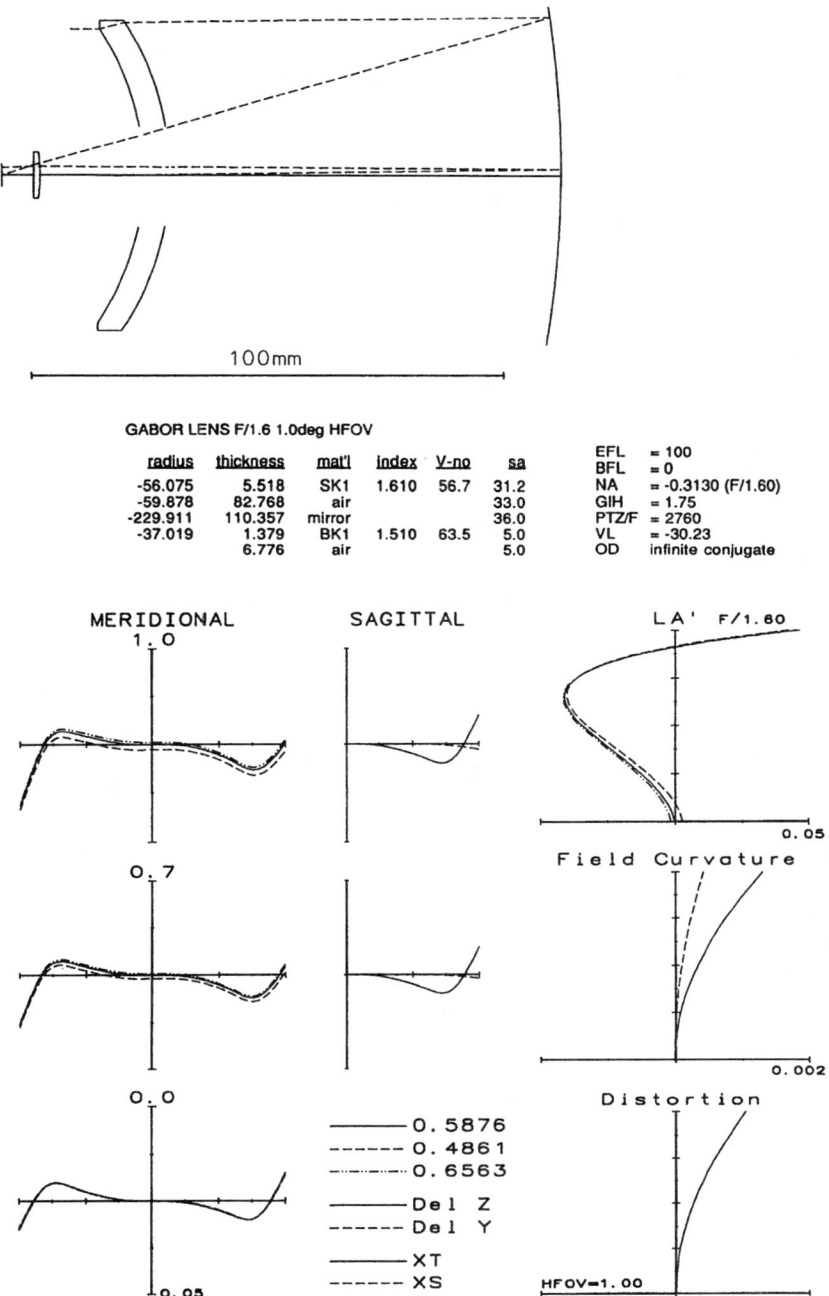

Figure 18.19 $f/1.6 \pm 1.0°$ Gabor meniscus corrector with a positive field corrector lens.

field flattener. On the other hand, the single concave mirror in Fig. 18.19 has a backward-curving field, and a positive field lens is used to flatten it. A drawback to these systems is that the full-aperture meniscus corrector tends to be thick and heavy, and requires a large piece of good quality glass. The *basic* Bouwers system is monocentric, with all radii having a common center located at the aperture stop; the system is thereby free of coma and astigmatism. The Maksutov system corrector has its radii chosen so that the corrector is free of axial chromatic; since it departs slightly from monocentricity, the complete freedom from coma and astigmatism is given up in exchange for the chromatic correction. The meniscus corrector will be achromatic if $t = n^2(R1 - R2)/(n^2 - 1)$. Many designs represent a compromise between these two principles. Note that a thick concentric Bouwers corrector gives a small zonal spherical; a thin Bouwers corrector produces less chromatic and more zonal.

18.7 Design of a Schmidt-Cassegrain "from Scratch"

The *Schmidt-Cassegrain* is just what its name implies: a two-mirror Cassegrain system with an aspheric Schmidt-type corrector plate. At a recent SPIE meeting I happened to notice a big scope on a tracking mount. It appeared to be a Schmidt-Cassegrain and I thought that duplicating it could make an interesting design exercise. It was marked as a 3048-mm focal length, $f/10$. I (very) roughly scaled its size, using the span of my hand as a ruler. My measurements indicated that, with the corrector placed close to the secondary mirror (to get a compact system), the mirror spacing would be about 24 in, with an image location perhaps 2 or 3 in beyond the primary (although this latter value is no more than a guess).

If we plug these approximate dimensions ($F = 3048$ mm, $B = 660$ mm, and $D = 600$ mm) into Eqs. (18.1) and (18.2) for two mirror systems, we find the primary mirror radius should be

$$R = \frac{2DF}{(B-F)} = -1532 \text{ mm}$$

and the secondary radius

$$R = \frac{2DB}{(B+D-F)} = -443 \text{ mm}$$

Assuming a 6-mm thickness for the corrector plate and a 20-mm space between the corrector and the secondary, we optimize with $R4$ as the variable and the back focus ($T4 = 660$) as our target, and our touched-up starting configuration becomes:

The Rough Starting Prescription for the Schmidt-Cassegrain

$R1$ = plano (asph)	$T1 = 6.0$	BK7 517642
$R2$ = plano	$T2 = 20.0$	
$R3$ = plano	$T3 = 600.0$	dummy surface
$R4 = -1531.658v$	$T4 = -600.0$*	mirror
$R5 = -443.953$†	$T5 = 660.0$‡	mirror
$R6$ = image		

*$T4$ is a negative thickness pickup of $T3$.
†$R5$ is an angle solve for $u' = -.05$ ($f/10$) to hold the *effective focal length* (EFL) at 3048 mm.
‡$T5$ is a height solve to put surface no. 6 at the paraxial focus.

1. We use a beam radius of 152.4 mm (6 in) to get $f/10$ at $F = 3048$ (120 in).

2. Ordinary 35-mm film with a format of 24×36 has a diagonal of 43.3 mm. With our 10 ft focal length this subtends a half angle of $0.406°$, so we use $0.4°$ as our half field.

3. We will put the aspheric on surface no. 1, leaving no. 2 plano so that the secondary mirror support can be cemented to it, and there will be no spider needed to support the secondary.

4. We put a 78-mm diameter central obscuration on the dummy surface (no. 3), and the aperture stop on surface no. 1.

Our merit function can be very simple. To start, we will limit the mirrors to spherical surfaces. The basic construction of the system is fixed (to get the F, B, and D that we have assumed). Our only design variable is the corrector plate, and we will vary only its fourth-order deformation coefficient and its curvature. We use these to correct the spherical and to balance the spherochromatism (respectively). So our merit function consists of the height of the marginal axial ray at the focal plane (i.e., the marginal spherical), plus a Conrady $(D-d)\delta n$ sum on the same ray. We accept the d, F, and C default wavelengths.

After optimization, the prescription now has $R1 = +21,229.413$, the fourth-order deformation coefficient AD $= -1.0140531339e - 10$, and $R5 = -439.089777$; the other dimensions are the unchanged. The $(D-d)\delta n$ operand has corrected the axial chromatic at 0.62 zone of the marginal ray height. The spherochromatism is 0.7 mm overcorrected, but it is reasonably balanced between 0.35 mm overcorrected chromatic at the margin and 0.35 mm undercorrected at the axis. The zonal spherical is only 0.06 mm overcorrected, the equivalent of about 0.01 wave *optical path difference* (OPD), hardly worth adding a sixth-order deformation term as a variable to correct it. The on-axis polychromatic *modulation transfer function* (MTF) is within 2 percent of perfect and the Strehl ratio is 0.965—clearly "diffraction limited," on axis, but off-axis is another story.

The Petzval field is not flat; it is inward curving about 0.5 mm at the edge of the field. The astigmatism is a miniscule 0.09 mm. But the big problem is the tangential coma, which is more than 0.07 mm, or about one full wave of OPD. The resolution at the edge of the field is only about 20 line pairs per mm (lpm). Not too bad, but hardly great.

The only degrees of freedom left to us in this simple system are the surface figures. In the ray intercept plots, the coma looks to be essentially of third order, so perhaps a simple conic section aspheric on one of the mirrors will be sufficient to take care of the coma. We add a coma target to the merit function (we can use either an *offence against the sine condition* (OSC) operand or real ray intercepts) and add the conic constant of the primary to our variable list.

This works very well. The field angle is small so we use the OSC operand in the merit function. The coma is now corrected, but the zonal spherical aberration has increased to $LA_z = +0.28$. This can be fixed by adding a sixth-order deformation term to the corrector plate. We add another ray at $Y = 0.707 Y_m$ and add a zonal spherical aberration target to the merit function. As expected, this works well and we achieve a fully corrected spherical aberration. However, a few steps back we goofed; we froze the mirror radii, assuming that the first-order properties were fixed. We overlooked the fact that we allowed the radius of the corrector plate to change (to balance the spherochromatism) and, as a result, the back focus has drifted from the desired 660 mm to 650.6 mm. This is scarcely a major tragedy; we simply add an operand to control $T5$ to 660 mm and allow the radius of the primary mirror ($R4$) to vary so that $T5$ can be controlled. We continue from our last design (rather than starting over from scratch).

The resulting changes are quite modest, the aberrations and performance are shown in Figs. 18.20 to 18.22. The final system is shown in Table 18.1. Realizing that we could aspherize either mirror (or both), we repeat the process, this time putting the conic on the surface of the secondary mirror. The resulting system is given in Table 18.2 and Figs. 18.23 to 18.25.

The two designs are quite similar in performance. If we compare the aspheric primary design to the aspheric secondary design, we find the following differences:

a. In the aspheric primary design the spherochromatism is worse.

b. The field curvature of the aspheric primary design is better (less inward curving) because, although the Petzval sum is the same in both designs, the astigmatism is small and positive compared to larger negative astigmatism in the aspheric secondary design.

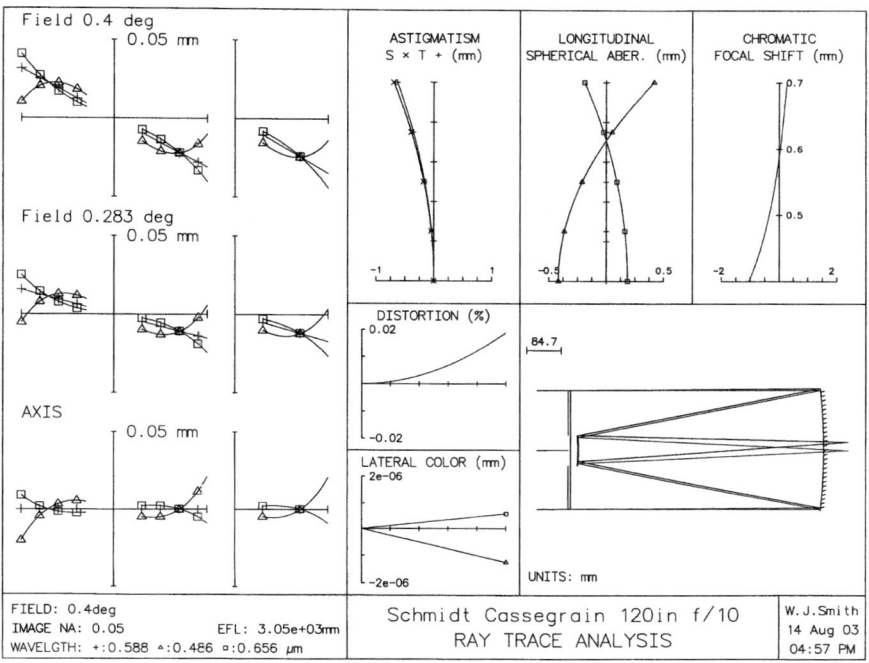

Figure 18.20 Ray aberrations for an $f/10$ ±0.4° 120-in Schmidt-Cassegrain, first design, conic primary, Table 18.1.

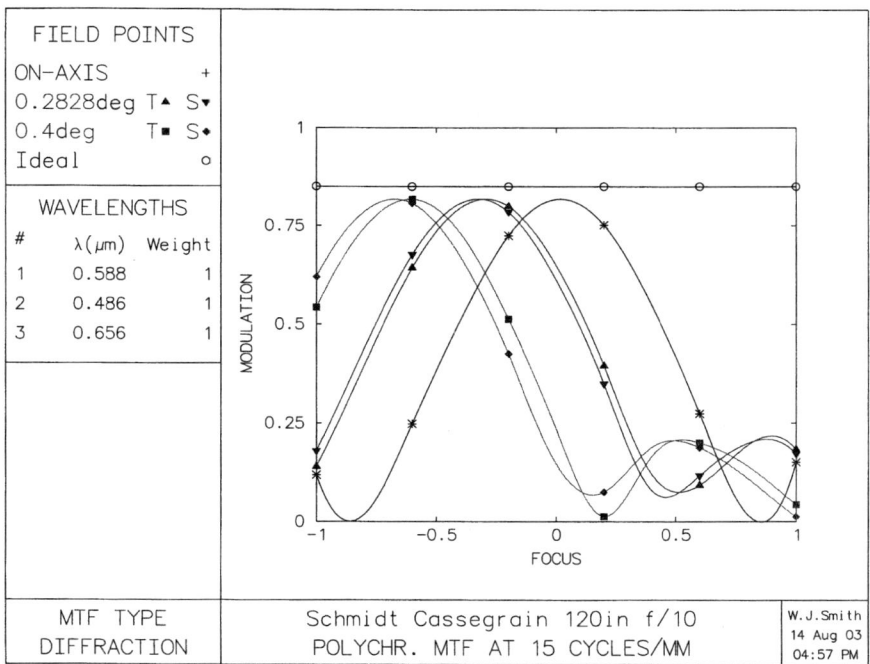

Figure 18.21 Through-focus MTF for Fig. 18.20, Table 18.1.

485

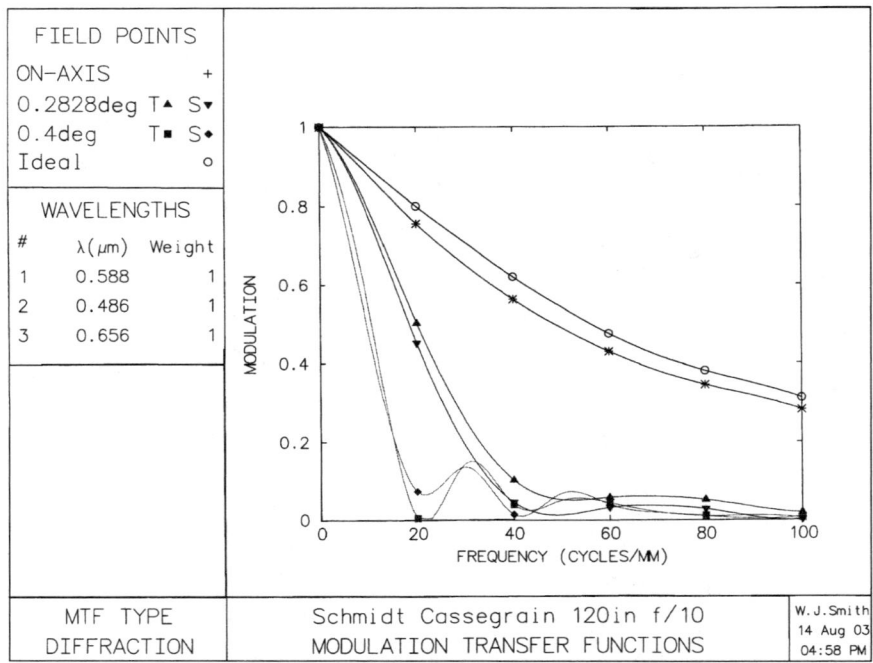

Figure 18.22 Through-frequency MTF for Fig. 18.20, Table 18.1.

c. The Strehl ratio of the conic primary design is worse.

d. The edge-of-field *modulation transfer function* (MTF), encircled energy, and rms spot size of the conic primary are better.

Obviously (c) and (d) result from (a) and (b), respectively. Thus one might prefer the aspheric secondary system for its resolution on axis, or the aspheric primary system for its off-axis performance.

Table 18.1 Prescription for the Basic Schmidt-Cassegrain with a Conic Primary Mirror (See Figs. 18.20 to 18.22)

$R1 = +1.2445e + 05$	$T1 = 6.0$	BK7 (517642)
$AD = -1.7125e - 10$	$AE = -7.4699e - 17$	
$R2 = $ plano	$T2 = 20.0$	
$R3 = $ plano	$T3 = 600.0$	dummy surface
$R4 = -1.5377e + 03$	$T4 = -600.0$	mirror
$CC = 0.544747$		
$R5 = -444.918235$	$T5 = 660.0$	mirror
$R6 = $ image		

Table 18.2 Prescription for the Basic Schmidt-Cassegrain With a Conic Secondary Mirror (See Figs. 18.23 to 18.25)

$R1 = +1.8963e + 05$	$T1 = 6.0$	BK7 (517642)
$AD = -1.1319e - 10$	$AE = -1.4529e - 17$	
$R2 =$ plano	$T2 = 20.0$	
$R3 =$ plano	$T3 = 600.0$	dummy surface
$R4 = -1.5363e + 03$	$T4 = -600.0$	mirror
$R5 = -444.240803$	$T5 = 660.0$	mirror
$CC = -1.099576$		
$R6 =$ image		

We have now fully corrected the coma and spherical aberration, and have balanced the spherochromatism, all the while maintaining the first-order properties of the system to the values we estimated for the original lens at the SPIE exhibit. In the aspheric primary version we have stumbled on an arrangement with very small astigmatism (and thanks may be due to the designer of the original system for this). The Petzval field curvature is now the dominant aberration limiting the performance of our system. On a flat focal surface the resolution is about 150 lpm at

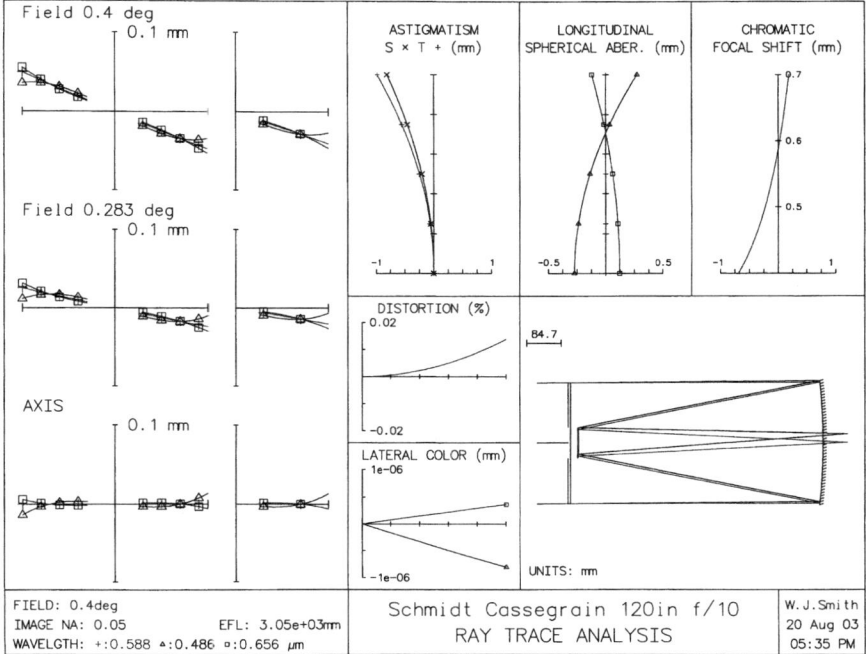

Figure 18.23 Second design of the Schmidt-Cassegrain, conic secondary, Table 18.2.

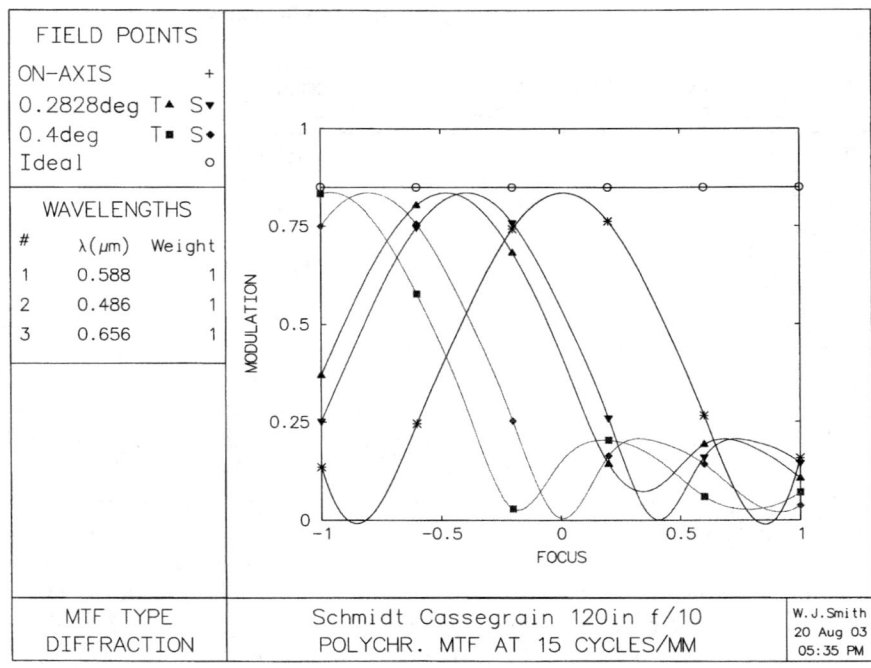

Figure 18.24 Through-focus MTF for Fig. 18.23, Table 18.2.

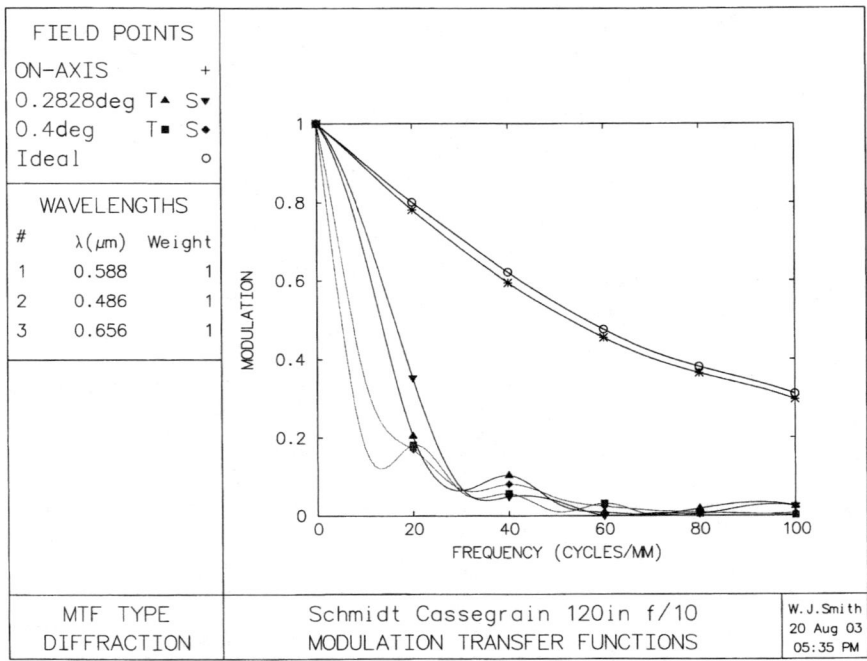

Figure 18.25 Through-frequency MTF for Fig. 18.23, Table 18.2.

the axis, but at ±0.3° off-axis the resolution is about 35 lpm, and it is only about 15 lpm at the corner of our ±0.4° field. If we refocus for the edge of the field, the resolution there improves to about 150 lpm. So, if we allow refocusing, our system can do about 150 lines anywhere in the field. Such refocusing might be acceptable for a visual system, but for photography on a flat film we're still not looking all that great.

Thus it appears that correcting the Petzval curvature ought to be a worthwhile effort. There are two ways that we can go about this: one way is to make both mirror radii identical, so that the Petzval sum contribution of one mirror cancels that of the other, and the second way is to add a negative field-flattener lens near the focal plane.

The first approach requires us to depart from our rather compact initial configuration. A very quick preliminary trial indicates that this approach produces an excellent system, the design resolving about 150 lpm over the full field; however, the system length is about 1200 mm (almost 4 ft). This is 75 percent longer than the 680 mm LOA we want for our project. The added length requires a larger secondary mirror and consequently a larger central obscuration. The data for this "quickie" design is shown in Table 18.3. The aberrations and MTF are shown in Figs. 18.26 to 18.28.

Note that we have accidentally (because we used a curvature pickup) made both mirrors identical conics. If the length of this design were acceptable we might try this design with only one mirror as a conic. Although the Petzval curvature is zero, the through-focus MTF plot and the field curvature plot indicate that there is negative (inward curving) astigmatism, which is reducing the off-axis MTF. The Petzval could easily be adjusted to balance this by reducing the curvature of the secondary mirror relative to the primary. Possibly the astigmatism could be improved by using a different conic constant and curvature for each mirror.

But, since we want to produce an improved compact system, we return to the second alternative and introduce a field flattener. Since the field

TABLE 18.3 Prescription for the Schmidt-Cassegrain System with Identical Mirror Radii in Order to Flatten the Petzval Surface (See Figs. 18.26 to 18.28)

$R1 = +1.0261e + 06$	$T1 = 6.0$	BK7 (517642)
$AD = -2.0912e - 11$	$AE = 2.9505e - 18$	
$R2 = $ plano	$T2 = 20.0$	
$R3 = $ plano	$T3 = 1.1528e + 03$	dummy surface
$R4 = -3.7525e + 03$	$T4 = -1.1528e + 03$	mirror
$CC = 1.711082$		
$R5 = -3.7525e + 03$	$T5 = 1.1728e + 03$	mirror
$CC = 1.711082$		
$R6 = $ image		

490   Chapter Eighteen

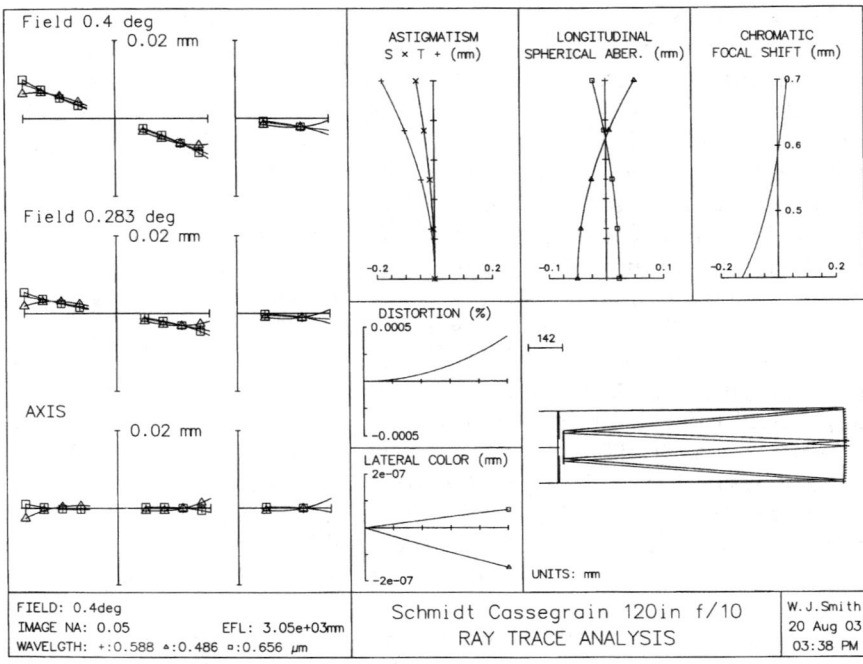

Figure 18.26 Third design of the Schmidt-Cassegrain, equal radius mirrors for flat Petzval field, longer system, Table 18.3.

of our system is inward curving, we need a negative element placed close to the focal plane to flatten the field. The first question is: "How close?" From a lens design standpoint, the closer the better, because (with the exception of Petzval field curvature) the aberrations introduced by an element placed very close to the image are very small; however, any physical imperfections in an element close to the focus will cast a shadow on the film. If we arbitrarily assume a surface quality of 60 – 40 or better for the field-flattener element, the diameter of the maximum bubble, pit or defect allowed by the specification will be 0.4 mm or less. Our system speed is $f/10$, so that the shadow cast by a 0.4-mm defect should be virtually undetectable if we locate the field flattener about 20 mm from the image.

The power of the lens should be such that its Petzval contribution approximately cancels out the field curvature of our design, which is about 0.7 mm at a field height of 21.3 mm. These last numbers indicate that the image surface radius is about –324 mm. (A good estimate; an exact calculation gives –313 mm.) The Petzval radius of a thin singlet is equal to (minus) its focal length times its index. Assuming BK7 glass

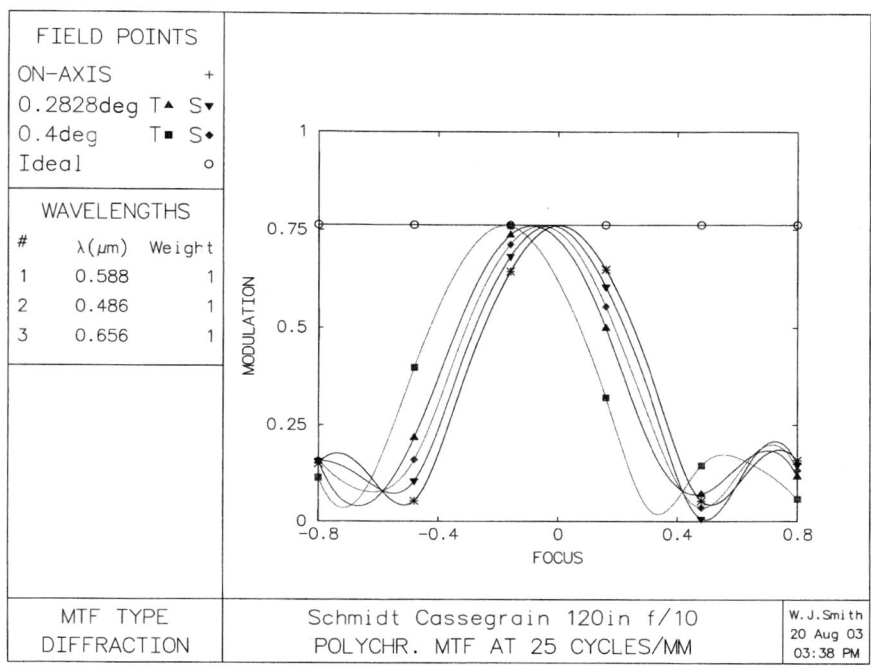

Figure 18.27 Through-focus MTF for Fig. 18.26, Table 18.3.

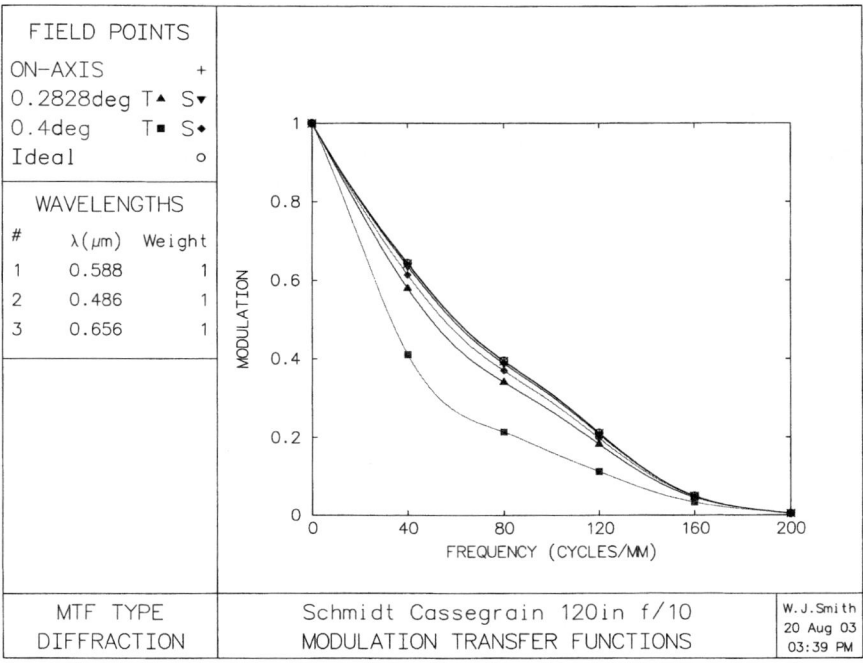

Figure 18.28 Through-frequency MTF for Fig. 18.26, Table 18.3.

with an index of 1.517, we need a focal length of $-324/1.517 = -213.6$. If we use a plano-concave lens its radius will be $213.6 \times 0.517 = 110.4$ mm. (An alternate approach would be to simply insert a lens into the lens data spreadsheet and adjust its radius until the Petzval sum is about zero.) Because the astigmatism is currently slightly overcorrected, we probably won't want to fully correct the Petzval sum, and a starting radius for our field flattener of about 125 mm may be more appropriate.

We are still a long way from home. Inserting the field flattener is obviously going to upset the first-order properties of our system, including focal length and image position. In addition, we must adjust the power of our field flattener to get the best field flatness. We can no longer use the angle solve on the secondary mirror to control the focal length for two reasons. Obviously the secondary mirror is no longer the last surface of the system. Less obvious, but much more significant, is the fact that an angle solve on a surface near the image (e.g., on the field flattener) tends to be quite volatile. This is because any changes to the optics ahead of the angle solve produce relatively large changes in the height of the axial marginal ray upon which the solve is based, and some crazy values may result.

So we must add a merit function target for focal length (3048 mm), a target for the secondary-to-focus distance (660 mm), a target (20 mm) for the back focus, and a target for the field curvature. We add the front radius of the field flattener to our variable list (for the moment maintaining the plano-concave shape). But here we have a chance to do something stupid. If we vary $T5$, and also target both the back focus and the secondary-to-focus distance, we are establishing duplicate targets, which can confound the optimization process. All we need to do is fix $T5 = 635$ mm, set the field-flattener thickness $T6 = 5$ mm, and, if we target the *back focal length* (BFL) = 20 mm, we will get our desired 660-mm value ($635 + 5 + 20 = 660$ = secondary mirror to focus distance) without needing a specific target for it.

We add the field flattener to the previous aspheric primary design, and after optimization we have the following design (see Table 18.4) whose performance is shown in Figs. 18.29 to 18.32. This has worked out quite well, as the figures indicate. The field curvature has effectively been flattened and our merit function targets have been reduced to negligible values; however, the resolution appears to be only about 100 lpm off axis, with the tangential resolution significantly poorer than the sagittal. This resolution is less than we achieved with the equal mirror radius design. We try bending the field flattener, but to no avail. We switch to an rms-spot-size merit function, but are completely unable to improve on the design above. An examination of a spot diagram analysis provides the explanation: we have about 0.025 mm of lateral color as clearly

TABLE 18.4 Prescription of the Schmidt-Cassegrain Design with a Field Flattener Lens (See Figs. 18.29 to 18.32)

$R1 = +1.1285e + 05$	$T1 = 6.0$	BK7 (517642)
$AD = -1.7557e - 10$	$AE = -8.5716e - 17$	
$R2 = $ plano	$T2 = 20.0$	
$R3 = $ plano	$T3 = 600.0$	dummy surface
$R4 = -1.5700e + 03$	$T4 = -600.0$	mirror
$CC = 0.684734$		
$R5 = -504.356093$	$T5 = 635.0$	mirror
$R6 = -153.962865$	$T6 = 5.0$	BK7
$R7 = $ plano	$T7 = 20.0008$	
$R8 = $ image		

indicated in Fig. 18.32. This was introduced by the field flattener, and there is no way to correct it as the system now stands. This is one of the very few times in decades that a three-color spot diagram has ever been of real value to this author (other than for impressing top management).

The obvious answer to this problem is to achromatize the field flattener. Since we have found no gain in bending the field flattener, we elect to stay with the plano-concave shape, and to keep things simple, we

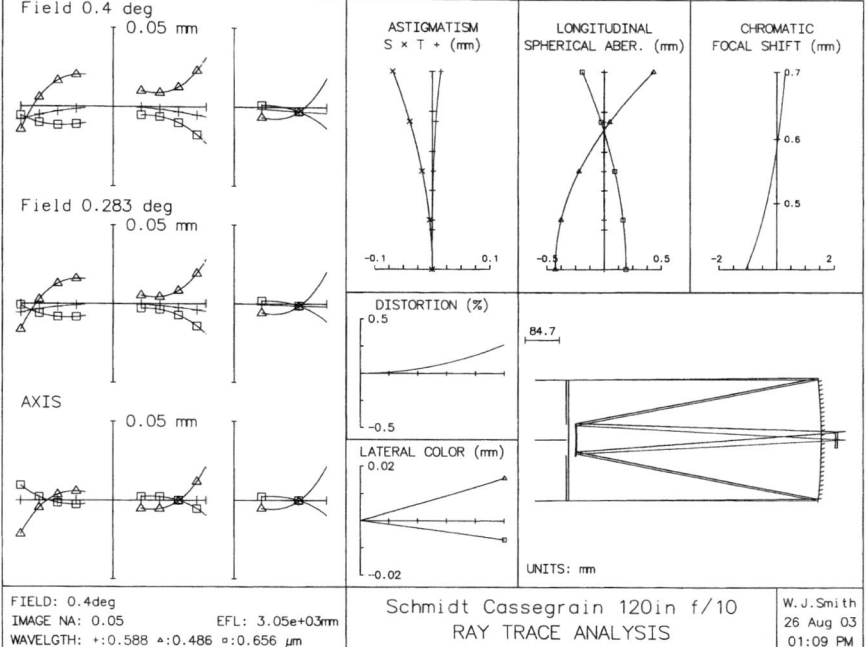

Figure 18.29 Fourth design of the Schmidt-Cassegrain with a field flattener, Table 18.4.

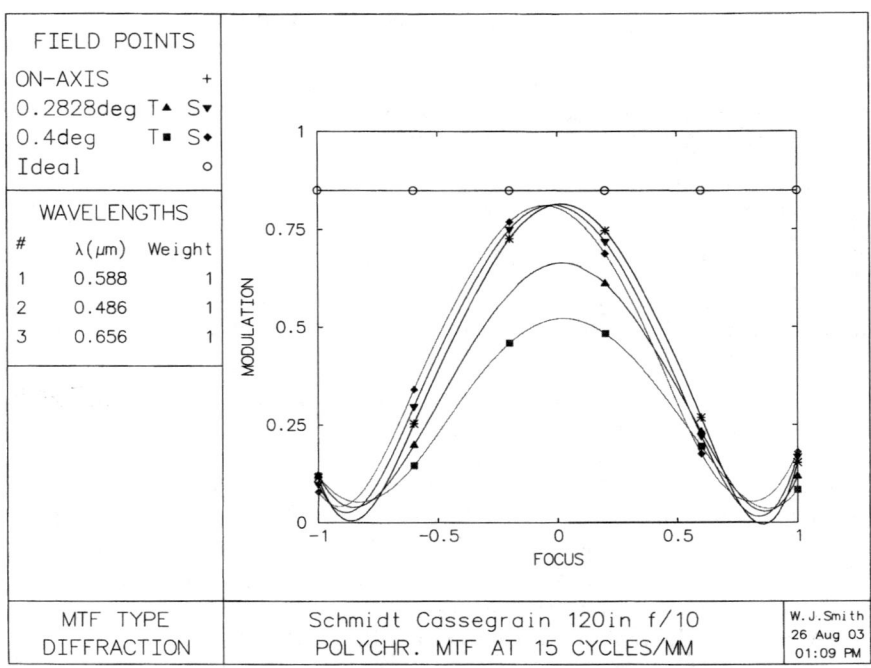

Figure 18.30 Through-focus MTF for Fig. 18.29, Table 18.4.

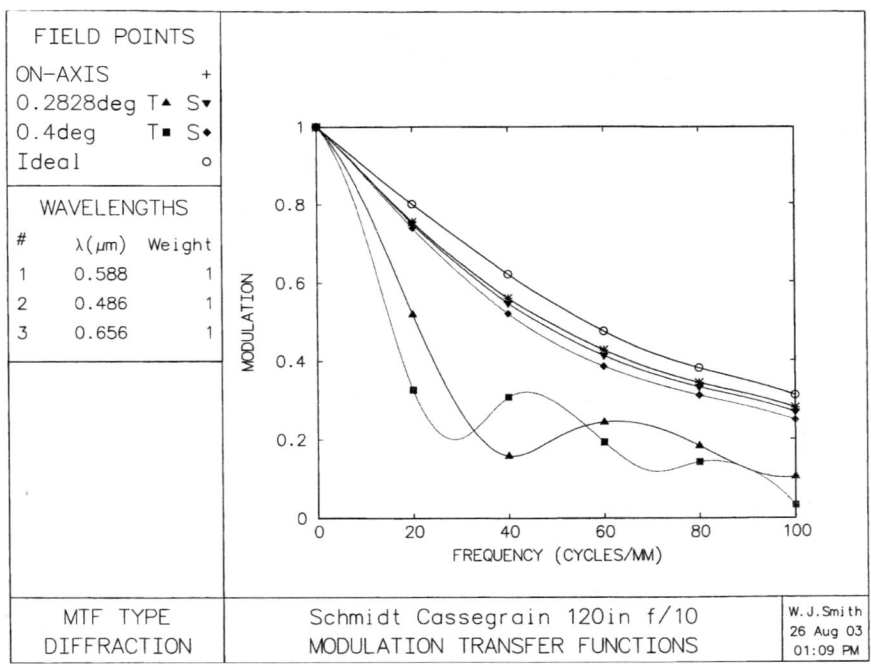

Figure 18.31 Through-frequency MTF for Fig. 18.29, Table 18.4.

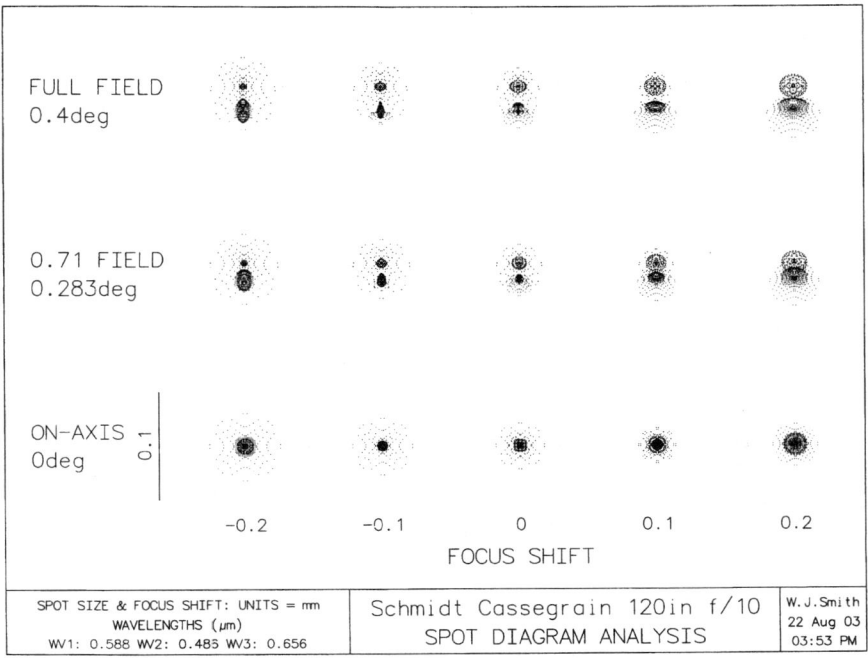

Figure 18.32 Spot diagrams for Fig. 18.28, Table 18.4. Note the separation of the off-axis spots due to a lateral chromatic of about 0.025 mm, which causes the poor off-axis tangential MTF in Figs. 18.30 and 18.31.

select two glasses with the same index for our achromatic field flattener: SK16 (620603) and F2 (620364). To minimize the curvatures we make the diverging SK16 element biconcave and the converging F2 element plano-convex. We optimize with our existing simple merit function, and then switch to the rms spot size default merit function (with *effective focal length* (EFL) and BFL) targets added. The result is quite satisfactory; we get the following design with data given in Table 18.5 and shown in Figs. 18.33 to 18.35.

This is a pretty good design. The resolution is about 150 lpm across the full field. Looking at the ray intercept plot it is apparent that the limiting aberration is the spherochromatism introduced by the aspheric corrector plate, which we balanced by undercorrecting the axial chromatic. Hopefully we can fix this by achromatizing the plate. We do this by adding a second aspheric plate made of flint glass and cementing it to the existing BK7 plate. Just as in an achromatic doublet, the flint will be negative where the crown is positive, so we start very simply by leaving the cementing surface plano and setting the curvature and aspheric

Table 18.5 Prescription for the Schmidt-Cassegrain System with an Achromatic Field Flattener Lens (See Figs. 18.33 to 18.35)

$R1 = +8.9979e + 04$	$T1 = 6.0$	BK7 (517642)
$AD = -1.7120e - 10$	$AE = -5.0372e - 17$	
$R2$ = plano	$T2 = 20.0$	
$R3$ = plano	$T3 = 600.0$	dummy surface
$R4 = -1.5711e + 03$	$T4 = -600.0$	mirror
$CC = 0.644980$		
$R5 = -503.839521$	$T5 = 633.2$	mirror
$R6 = -192.502573$	$T6 = 3.0$	SK16 (620603)
$R7 = +131.543531$	$T7 = 3.8$	F2 (620364)
$R8$ = plano	$T8 = 19.9998$	
$R9$ = image		

deformation coefficients of the flint equal to those of the BK7 plate. We add these to the variable list. Since the plates are large, a flint glass with the same coefficient of thermal expansion as BK7 would seem advisable; we choose SF13 (741276), which has a large dispersion and has the same CTE as BK7.

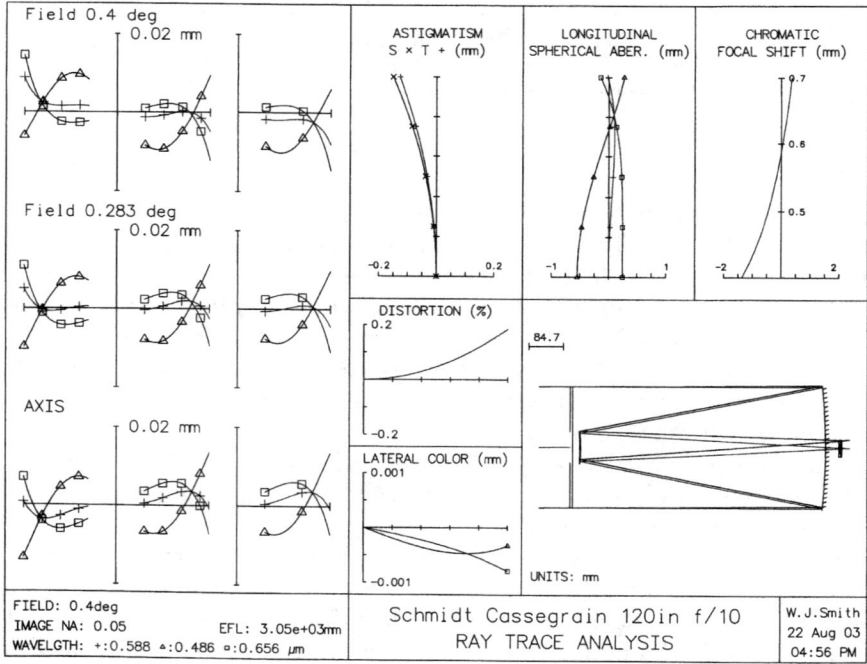

Figure 18.33 Fifth design of the Schmidt-Cassegrain with an achromatic field flattener, Table 18.5.

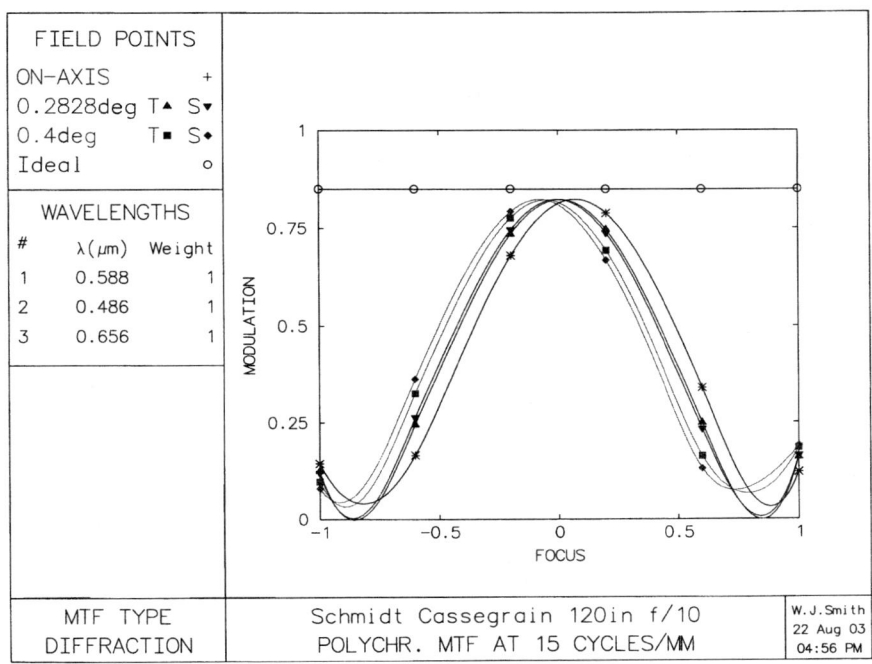

Figure 18.34 Through-focus MTF for Fig. 18.33, Table 18.5.

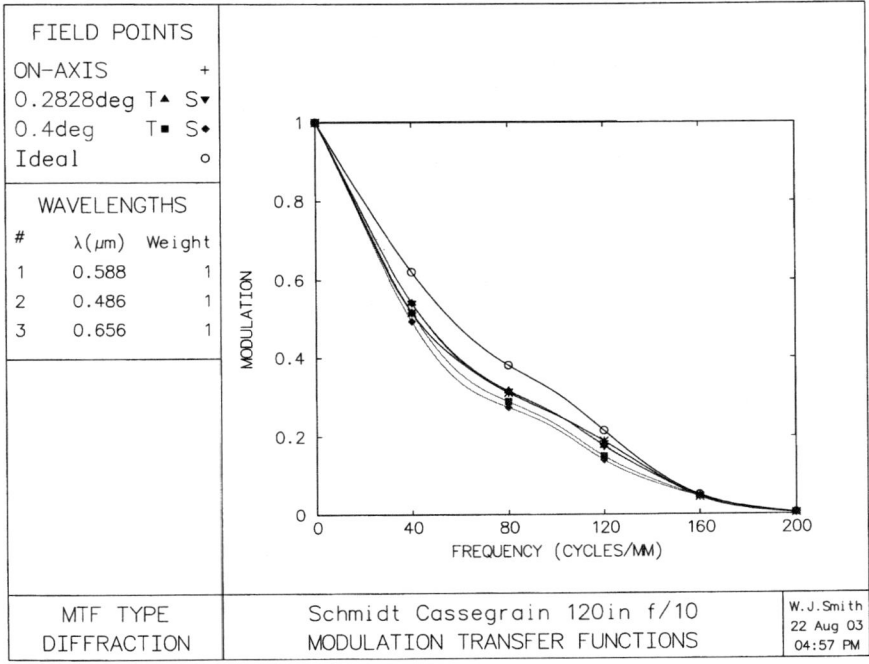

Figure 18.35 Through-frequency MTF for Fig. 18.33, Table 18.5.

This works beautifully; the optimization produces a system that is nearly perfect. We note that the ray intercept plots show a small amount of overcorrected marginal spherical and a slightly inward-curving field. In an attempt to gild the lily, we add targets for LA_m, X_s and X_t to the merit function. This stunt usually works well; this time it totally ruins the design, introducing enough coma to produce a blur spot to the order of a tenth of a millimeter. We abandon this idea and change from the rms-spot-size merit function to the rms OPD. This produces a tiny improvement (there wasn't much room left for improvement) and we decide to stop here. The final design is shown in Table 18.6, with the design plots in Figs. 18.36 to 18.41.

This design is great. (As it should be, with all the bells and whistles that we've put in it.) The MTF plots for all three fields in Fig. 18.38 are almost coincident with the ideal diffraction limited curve. Over the full field, the monochromatic (0.5876 μ) rms OPD shown in Fig. 18.40 is less than 0.026 wave and the rms blur spot is smaller than 0.0016 mm. The MTF plotted against field in Fig. 18.39 is almost perfectly flat at spatial frequencies of 50, 100, and 150 lines (cycles) per mm. The through-focus MTF plot in Fig. 18.37 shows a small inward-curving field, but all things considered, this hardly seems worth any attention. The through-frequency MTF plot in Fig. 18.38 shows only the contrast reduction at low frequencies that one always expects of a system with a central obscuration of the aperture (see Fig. 18.1). The reduction is quite small because the compact basic layout of this system has only 26 percent obscuration by the secondary. For this we are (once again) indebted to the original designer of this lens. We have succeeded in making this design not only "diffraction limited," but close to perfect. There is little point in continuing to try for any additional improvement. But, if there

TABLE 18.6 Prescription for the Schmidt-Cassegrain Objective with an Achromatized Aspheric Corrector Plate and an Achromatic Field Flattener (See Figs. 18.36 to 18.41)

$R1 = +9.7889e + 04$	$T1 = 6.0$	BK7 (517642)
$\quad AD = -3.0399e - 10$	$AE = +8.4940e - 18$	
$R2$ = plano	$T2 = 6.0$	SF13 (741276)
$R3 = +3.4407e + 05$	$T3 = 16.0$	
$\quad AD = -9.2669e - 11$	$AE = +5.6984e - 17$	
$R4$ = plano	$T4 = 600.0$	dummy surface
$R5 = -1.5700e + 03$	$T5 = -600.0$	mirror
$\quad CC = 0.645986$		
$R6 = -508.131867$	$T6 = 633.2$	mirror
$R7 = -176.615022$	$T7 = 3.0$	SK16 (620603)
$R8 = +120.930670$	$T8 = 3.0$	F2 (620364)
$R9$ = plano	$T9 = 20.00$	
$R10$ = image		

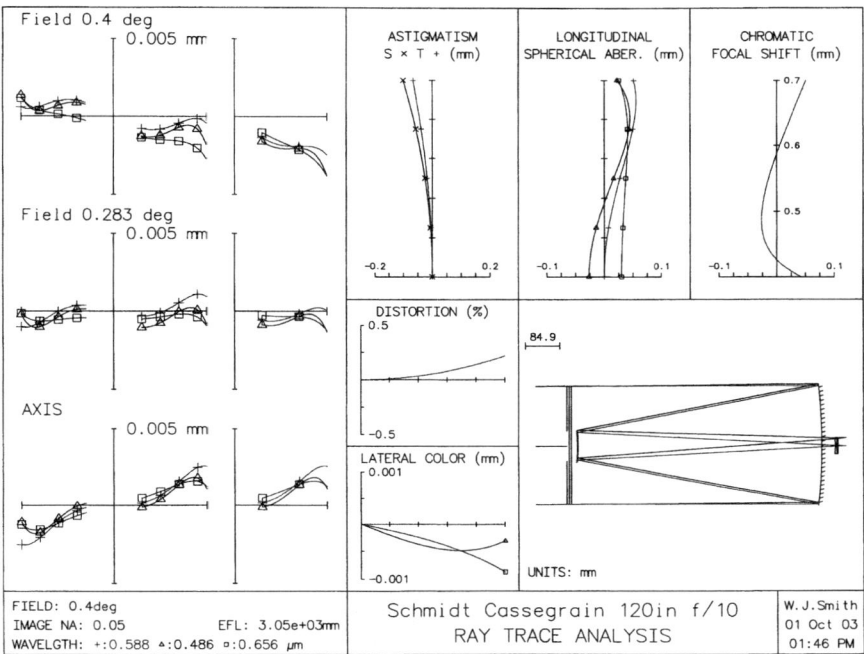

Figure 18.36 Sixth and final design of the Schmidt-Cassegrain with an achromatized Schmidt corrector, Table 18.6.

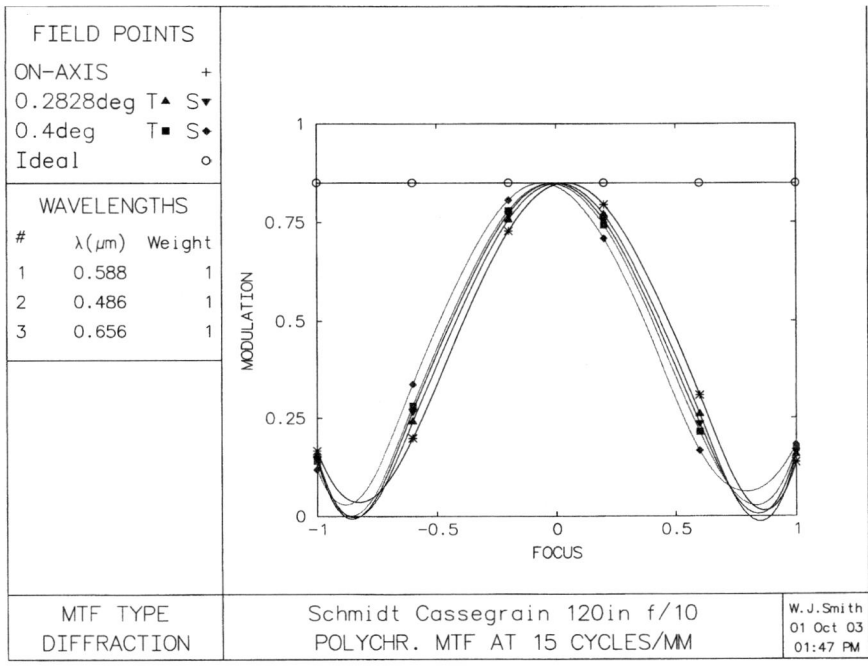

Figure 18.37 Through-focus MTF for Fig. 18.36, Table 18.6.

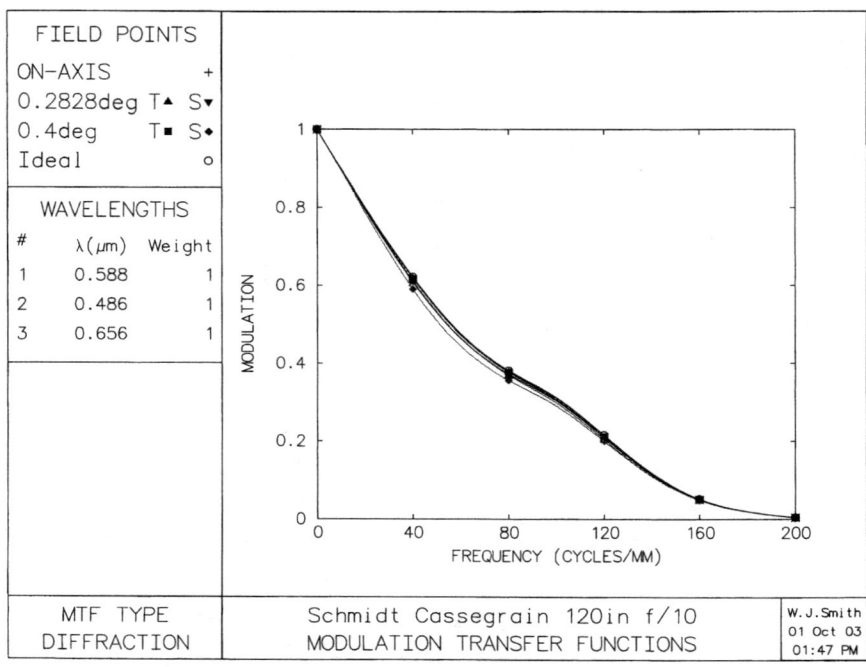

Figure 18.38 Through-frequency MTF for Fig. 18.36, Table 18.6.

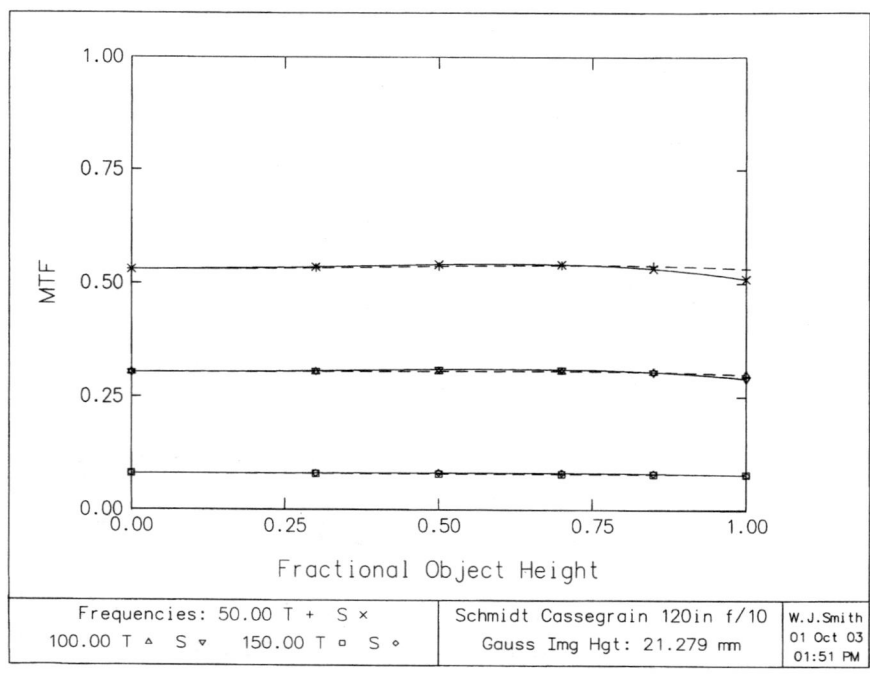

Figure 18.39 Through-field MTF for Fig. 18.36, Table 18.6 at 50, 100, 150 lpm.

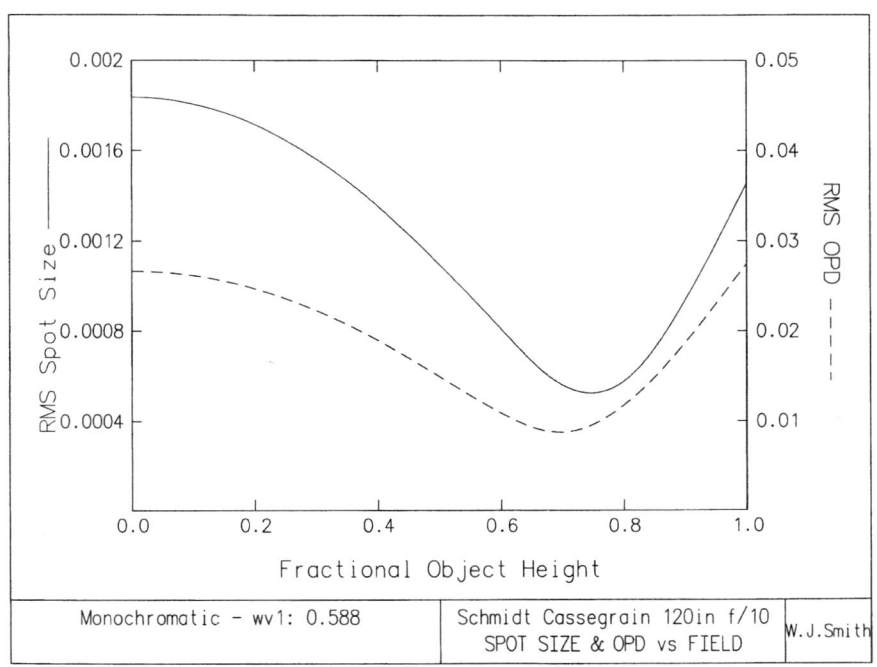

Figure 18.40 Rms spot size and rms OPD vs. field for Fig. 18.36, Table 18.6.

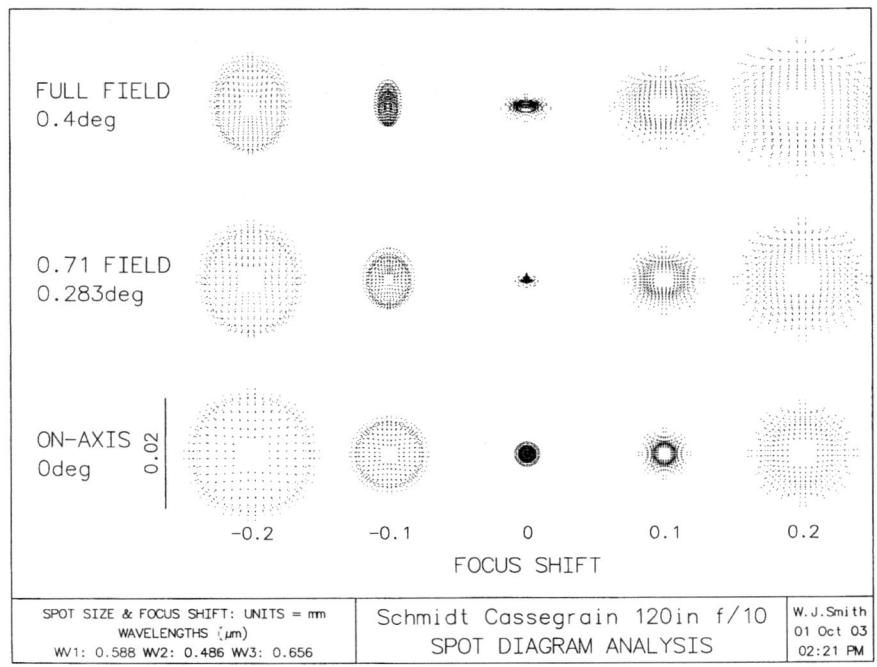

Figure 18.41 Spot diagram for Fig. 18.36, Table 18.6.

were, we might try the following. (In no particular order, here are some items that we might have considered as alternative moves at various stages of the design process):

a. Vary the field flattener glass(es). As usual, we can easily explore this by allowing the flint (F2) to vary along the glass line, taking our clues for further glass changes from the direction in which it varies.
b. Turning the doublet field flattener around.
c. Bending the field flattener. It didn't do much before when the field flattener was a singlet, and probably won't now unless we introduce an index break at the cemented surface.
d. Splitting the achromatic doublet field flattener and bending each element independently.
e. Adding targets for X_s and X_t to the rms merit function will sometimes sharpen up the correction of field curvature. (We tried and abandoned this.)
f. See if aspherizing the secondary is better now. (It was about the same when we tried it before.)
g. Try aspherizing both mirrors.
h. The BFL, at 20 mm, is rather small. A longer BFL might be necessary for mounting considerations and the like. (As, for example, if the system were to be used with an SLR camera, the pivoting mirror would require about a 38- or 40-mm clearance. This would require a significant change in the design emphasis, and the split achromatic field flattener–item (d) above–might then be a good idea).
i. Utilizing fourth-, sixth-, eighth-, and 10th-order surface deformations as the mirror aspherics (instead of conic sections).

Chapter 19

Infrared and Ultraviolet Systems

In terms of lens design techniques, there is little difference between design for the visual region and design for the *infrared* (IR), or design for the *ultraviolet* (UV). The biggest difference is in the optical materials used in each region. Optical quality materials that transmit in the IR are few; those that transmit in the UV are rare.

One other factor crops up in the IR. Everything at a temperature above zero K (–273°C) emits electromagnetic radiation. Ordinarily we think of sources at high temperature that emit well in the visible, and we image objects using the light that is reflected from their surfaces. But even at a low temperature of 300 K (27°C) everything emits radiation; the emission peaks up at 10 μm, right in the middle of the 8- to 12-μm atmospheric window. At 10 μm the world is a light bulb, and if you can detect this radiation, you can see in the dark. Thus, for a device working in the 8- to 12-μm region, the optics and mounts themselves are bright and emit radiation to which the detectors are sensitive.

19.1 Infrared Optics

Infrared optical systems differ from ordinary (visual) systems principally in the materials used for the optics. Out to a wavelength of about 2 μm, ordinary optical glass has a satisfactory transmission. Some IR glasses go out to 4 μm. However, crown and flint glasses reverse their roles at about 1.5 μm; at the longer wavelengths the crown glasses have a greater relative dispersion than do the flints. Thus the wavelength region between 1.0 and 1.5 μm is a transition zone wherein crowns and flints have the same relative dispersion (V-value), and achromatism is not possible with optical glasses. An achromat is possible in this spectral region if (for example) calcium fluoride is used for the positive element.

Many of the other halogen salts (chlorides, iodides, bromides, and fluorides) are excellent transmitters in the IR and have attractive dispersion characteristics, but their physical properties leave a lot to be desired. For this reason their use is almost entirely limited to laboratory-type environments.

The workhorse materials for the IR are silicon, germanium, zinc sulfide, zinc selenide, AMTIR (Ge/As/Se), chalcogenides, and the like. These materials have significantly higher indices than optical glass and the crystals mentioned above. Of course, a higher index means smaller aberrations (see Fig. 3.1, for example), and the very low relative dispersions of silicon and germanium allow reasonably good image quality to be obtained from a singlet element at a relatively high speed. In many IR systems, no attempt is made at achromatism because of this.

Another factor affecting IR system design is that the longer wavelength both limits the resolution of an IR system and simultaneously reduces the impact of the aberrations. As can be seen from the equations in Sec. 24.12, the longer the wavelength, the larger the aberration can be before it produces some given fraction of a wavelength deformation of the wavefront. This in itself allows IR systems to be of comparatively simple construction.

Atmospheric transmission is a factor that greatly influences IR systems. The transmission "windows" of the atmosphere limit the wavelengths that can be used to: (a) the visible and the near IR ($\approx$2.5 µm); (b) the 3- to 5-µm window; and (c) the 8- to 12-µm window. (These wavelengths are not sharply defined, and 3.5- to 4.5-µm and 8- to 14-µm are equally valid.)

19.2 IR Objective Lenses

An achromatic doublet for the 8-to 12-µm window is shown in Fig. 19.1. This is essentially a telescope objective and, despite the fact that high-index materials yield a relatively flat Petzval field, the negative astigmatism present in every thin, stop-in-contact optical system severely limits the off-axis image quality and the useful field of view of these designs. However, the speed of this simple lens is f/1.5; this is made possible by the high indices of refraction. Note that the sign of the secondary spectrum is reversed from that of the usual (glass) lens. A doublet of silicon and germanium makes an excellent achromat, but the transmission of silicon usually limits its usefulness to the shorter wavelengths (e.g., the 3- to 5-µm atmospheric window).

An f/0.75 infrared triplet is shown in Fig. 19.2. This can be regarded as the IR equivalent of the Cooke triplet. The increased length (compared to Fig. 19.1) allows the introduction of some overcorrected astigmatism (by the elements located away from the stop) to offset the negative astigmatism of the components at the stop. Many IR triplets take the form

Infrared and Ultraviolet Systems 505

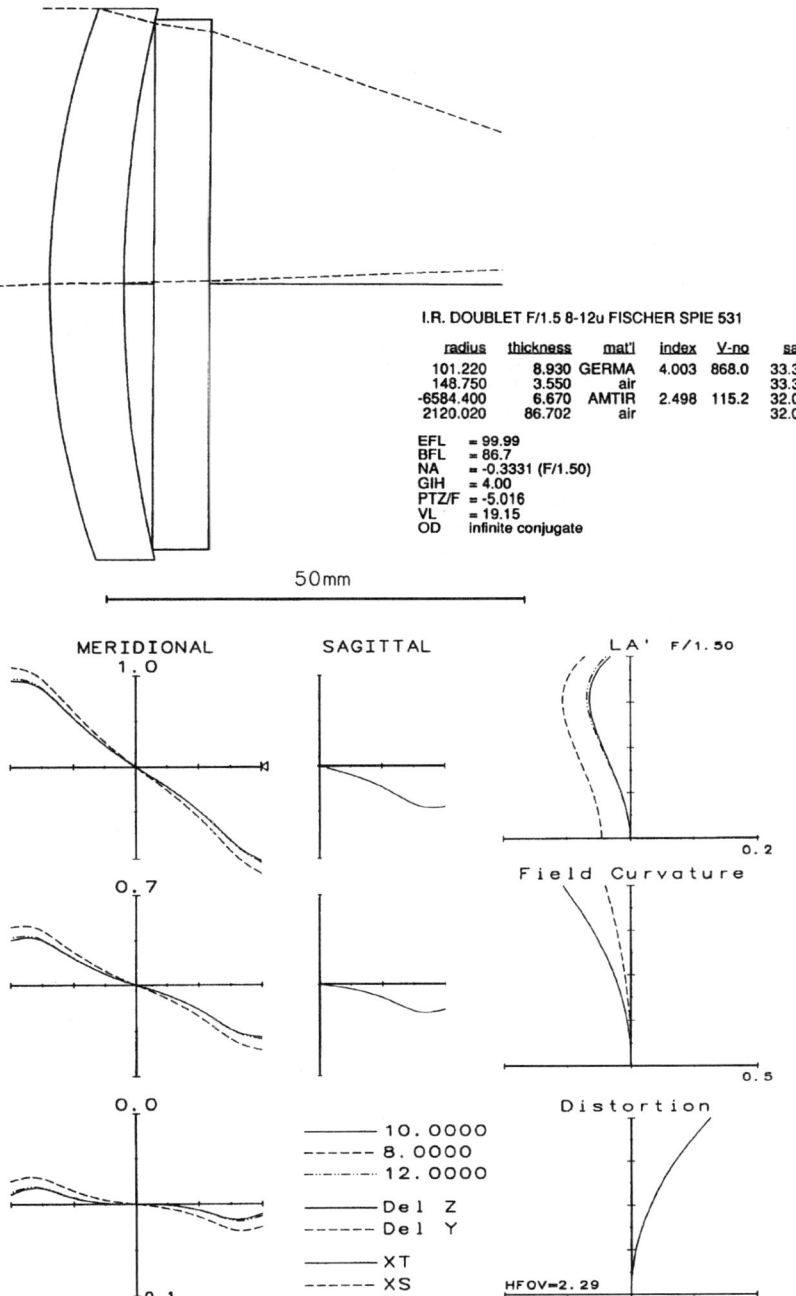

Figure 19.1 $f/1.5$ ±2.3° a simple achromatic doublet for the 8–12 μm region, using germanium and AMTIR.

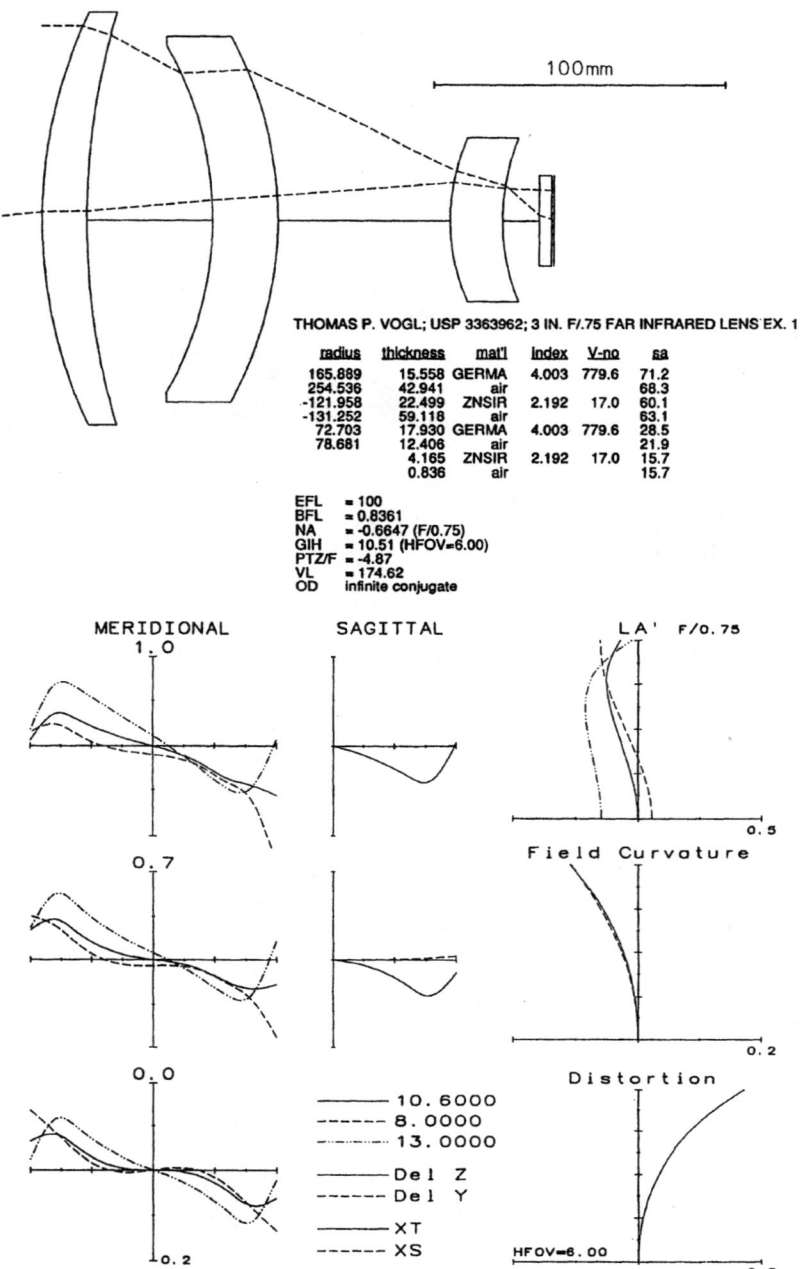

Figure 19.2 $f/0.75$ ±6° a triplet anastigmat of germanium and zinc sulfide for the 8–13 μm region.

of a closely spaced (+ −) doublet followed by a positive singlet spaced well away from the stop; this particular arrangement can also be regarded as a Petzval lens with a rear singlet instead of the usual rear doublet. The high index and low dispersion of silicon and germanium produce many configurations that are quite different from the usual optical glass systems to which they correspond.

An assortment of four-element designs is shown in Figs. 19.3 to 19.5. Figure 19.3 is a highly unusual design. It may be regarded as a reverse telephoto lens with a field flattener, but a more likely view is to realize that the germanium element (no. 3) has almost all the effective positive power and the first two Irtran (ZnS) elements operate as a corrector for its aberrations without introducing much power—a sort of spherical-surfaced Schmidt corrector.

Figure 19.4 is effectively an IR Petzval lens, consisting as it does of two widely spaced achromatic doublets. At $f/1.5$ and over a 6° field, four elements allow a high level of correction. The lens of Fig. 19.5 is a quite sophisticated construction, and, while the level of correction does not match that of the preceding design, the coverage of a 20° field at a speed of $f/0.55$ is no mean feat in itself. (It is often difficult just to find a system at wide field and high speed that will allow all the rays to get through without missing a surface or encountering *total internal reflection* (TIR). See Sec. 2.9 for a technique to handle this problem.)

19.3 IR Telescopes

The infrared telescope usually serves as a collector for a following smaller-diameter system, or as a front for a *forward-looking infrared* (FLIR) scanning system. As a collector, a galilean telescope or set of confocal paraboloids is appropriate. But the purpose of the system as a front for a *scanner* is to reduce the size of the beam to reduce the size and inertia of the scanning mirror. The mirror is placed at the exit pupil of the telescope; the need for an accessible exit pupil obviously requires a positive eyepiece and rules out the galilean configuration for this purpose. Note that the eyepieces in the following designs are very simple, consisting of just two meniscus germanium elements with their convex surfaces facing each other. The telescope aberrations are shown as if there were a perfect 1000-mm focal-length lens following the afocal telescope. Thus the transverse aberrations are shown in units of milliradians.

A "cold stop" is common in systems for the 8- to 12-μm range. This is a refrigerated aperture within an evacuated Dewar. It is located at the exit pupil of the system, and its size matches that of the pupil. The purpose of the cold stop is to prevent the detector from seeing anything except

Chapter Nineteen

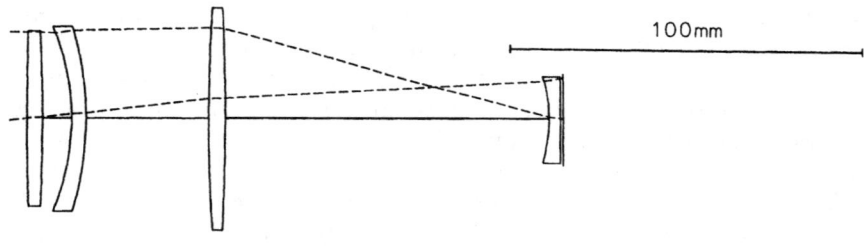

ARTIE D. KIRKPATRICK; USP 3439969; 3 IN. F/2 FAR INFRARED LENS

radius	thickness	mat'l	index	V-no	sa
934.850	4.438	IRTRN	2.185	16.8	25.2
-574.627	8.241	air			25.2
-69.352	4.121	IRTRN	2.185	16.8	26.6
-90.951	34.868	air			26.6
484.959	4.755	GERMA	4.003	779.5	31.9
-636.622	91.612	air			31.9
-42.252	3.169	IRTRN	2.185	16.8	12.6
	0.842	air			12.6

EFL = 100
BFL = 0.8421
NA = -0.2512 (F/2.0)
GIH = 12.28 (HFOV=7.00)
PTZ/F = 0.967
VL = 151.20
OD infinite conjugate

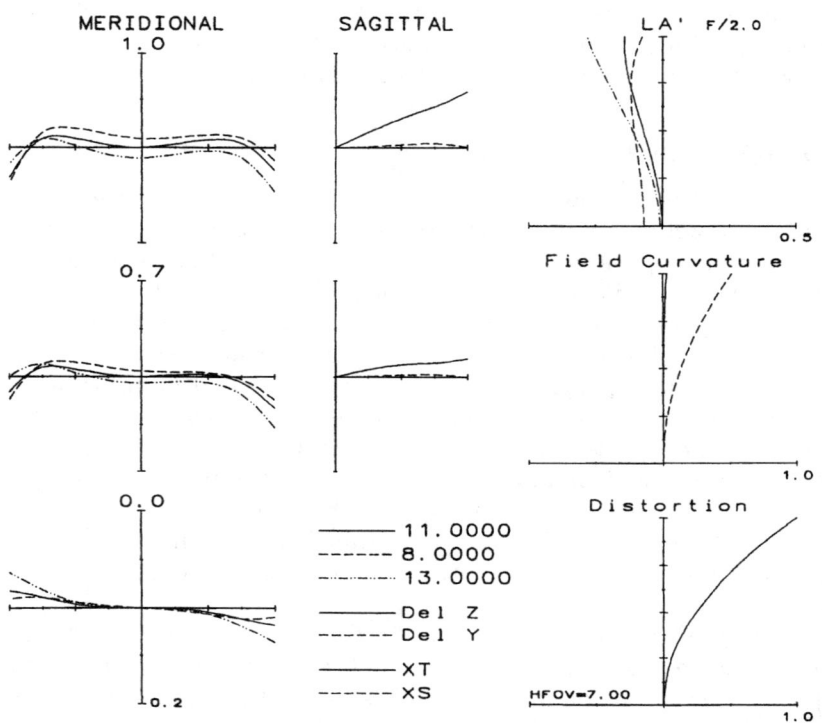

Figure 19.3 $f/2.0$ ±7° an unusual 8–13 µm design of Irtran and germanium, with a leading afocal corrector pair, a germanium singlet that has all the focusing power, and a field flattener.

Infrared and Ultraviolet Systems 509

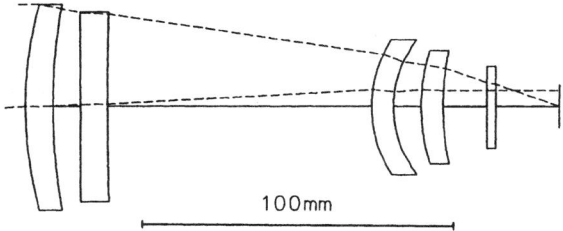

THOMAS P. FJELDSTED; USP 4380363; FOUR ELEMENT F/1.5 IR OBJECTIVE

radius	thickness	mat'l	index	V-no	sa
128.471	8.737	SILCN	3.431	488.9	33.3
186.905	8.737	air			32.1
8457.375	8.737	GERMA	4.029	207.1	30.8
1040.215	85.070	air			30.8
41.030	6.990	GERMA	4.029	207.1	22.0
35.228	9.017	air			22.0
63.397	6.990	SILCN	3.431	488.9	18.5
90.098	13.979	air			18.5
	2.796	SAPIR	1.688	18.9	13.3
	20.477	air			13.3

EFL = 100
BFL = 20.48
NA = -0.3334 (F/1.50)
GIH = 5.24
PTZ/F = -7.231
VL = 151.05
OD infinite conjugate

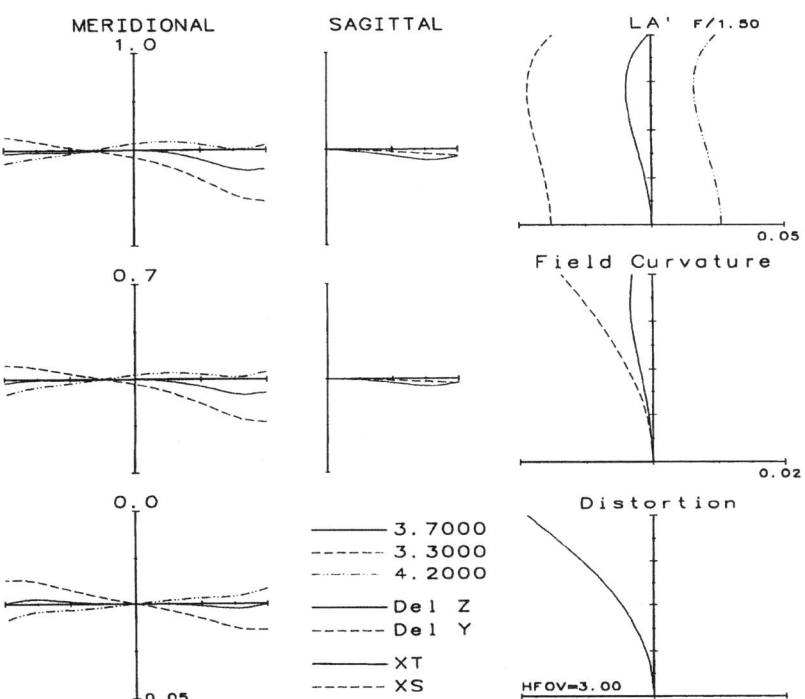

Figure 19.4 $f/1.5$ ±3° an airspaced Petzval of silicon and germanium for the 3.3–4.2 µm region.

510 Chapter Nineteen

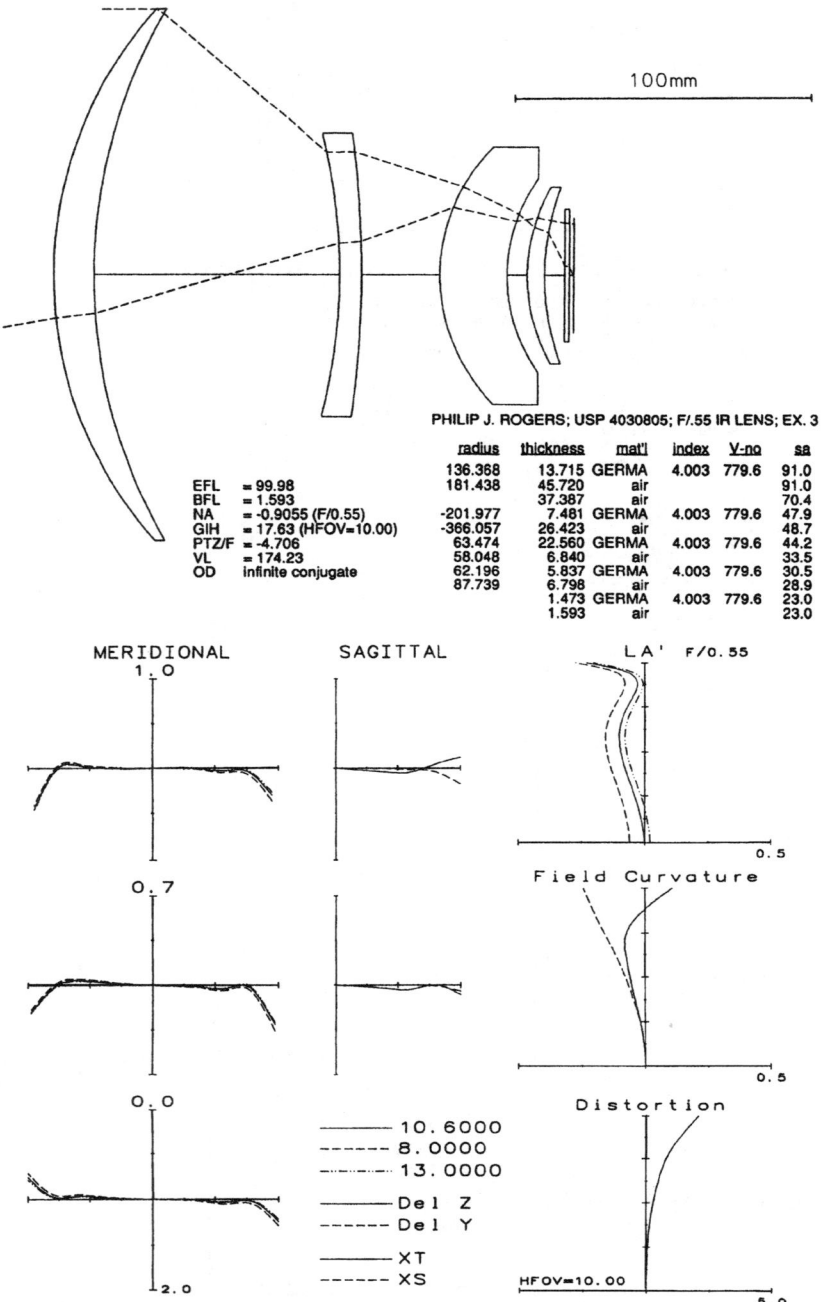

Figure 19.5 $f/0.55$ ±10° an amazingly fast and wide-angle, four-element all-germanium design for the 8–13 μm region.

the optical elements and the image of the object. Radiation from the mounting structures and the like is blocked.

Another problem of long wavelength scanning (e.g., FLIR) systems is called *narcissus*. This is caused by reflections from the optics that focus (or nearly focus) an image of the cold detector back on itself. As the scanning mirror rotates, this image is moved, possibly out of view. The result is a dark spot in the center of the image, sometimes quite distinct. An obvious danger signal is incidence angles near zero on a surface.

The 12× telescope of Fig. 19.6 has a catadioptric objective with a concentric meniscus corrector dome (which allows the balance of the system to be pivoted about the dome's center of curvature if required) and a Mangin-mirror-type secondary. Both reflecting surfaces are aspheric, as is the third surface of the eyepiece.

The simple telescope of Fig. 19.7 manages a magnification of 16× with the standard eyepiece and a sort of telephoto-type construction for the objective.

19.4 Laser Beam Expanders

A laser beam expander usually takes the form of a galilean telescope, with a negative eyepiece and a positive objective, separated by the sum of their focal lengths. The expansion factor is equal to the *magnifying power* (MP) of the telescope, which is simply the ratio of the focal lengths (MP = $-f_o/f_e$). Assuming a sufficiently monochromatic laser, there is no need to achromatize the system. In the visible and near infrared the heavy flint glasses can be used; as always, their high index of refraction is very beneficial and their high dispersion is immaterial. The usual infrared materials have even higher indices and produce an even better design.

The expansion of the beam not only increases the beam diameter, but also reduces the beam divergence by the same factor. Interestingly enough, a 5× beam expander (for example) reduces the beam divergence to one-fifth of its original value on a geometric optics basis, and also on a diffraction basis, since it increases the beam waist diameter by five times.

For a low-power beam expander a design of two spherical-surfaced single elements is quite adequate. The spherical aberration is easily corrected by shaping the elements, but one should keep an eye on the coma, since in real life alignment is seldom perfect enough to justify the assumption that the beam expander's field is zero. Bear in mind that the spherical aberration is a quadratic function of the shape, so there will be more than one solution for spherical correction.

As the expansion factor (the magnifying power) increases, focal length and the spherical aberration of the objective lens increase enough that it becomes difficult for the negative eyepiece to correct it. Here one can either split the objective element in two, which will reduce its spherical

512 Chapter Nineteen

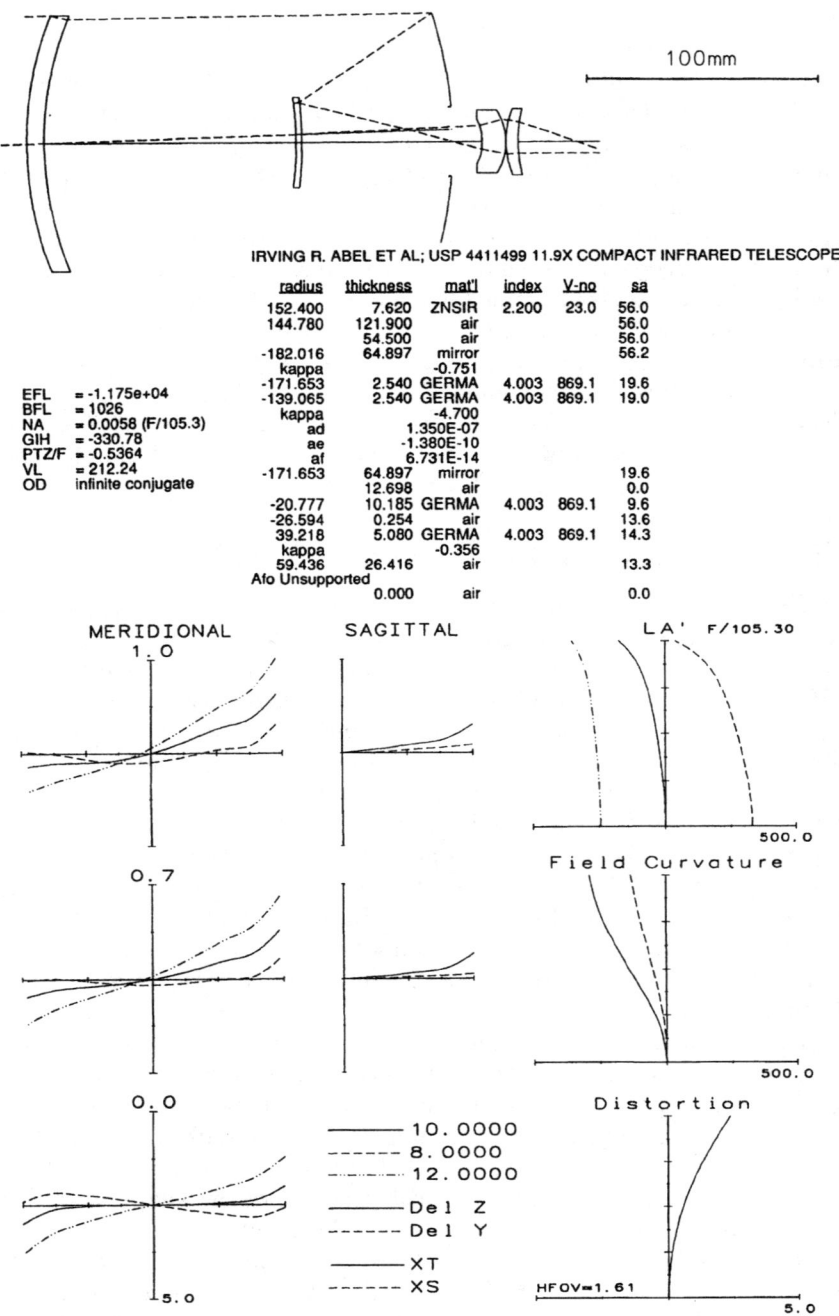

Figure 19.6 A catadioptric 12×, 8–12 μm telescope. Note that the eyepiece is simply two meniscus germanium elements. Their concave surfaces and thicknesses have a favorable effect on the Petzval sum.

Infrared and Ultraviolet Systems 513

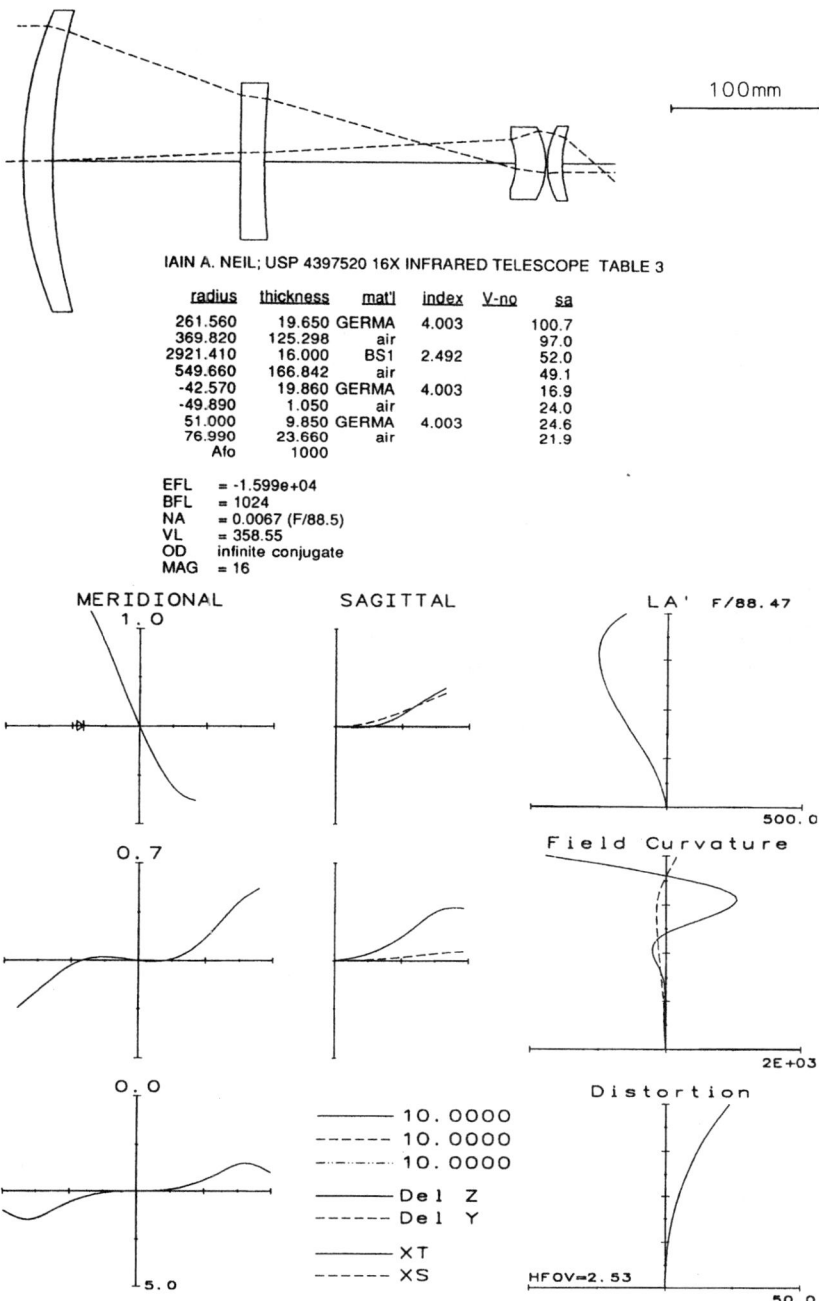

Figure 19.7 A refracting 16×, four-element, all-spherical telescope at 10 µm.

by a factor of five or more depending on its index, or one can convert the objective to a doublet. If a cemented doublet is chosen for stability, the index break at the cemented surface can be an added degree of freedom. If the objective is simply split in two, it too has an added degree of freedom. If the doublet is airspaced, a design like the collimator/focusing doublet of Fig. 22.5 can be utilized to advantage.

19.5 Ultraviolet Systems

The short wavelengths of the UV have the reverse effect that the long wavelengths of the IR have. Fabrication tolerances are proportionately tighter. The Rayleigh quarter-wave criterion at 10 µm is 33 times larger than it is at 0.3 µm. This obviously means that the aberrations must be better corrected in the UV than in the visual, and *much* better corrected than in the IR.

The UV transmission of optical materials is a really significant problem. There are a few ordinary optical glasses that are usable below 400 nm, as well as a few special UV glasses. Many of the optical crystals (e.g., the halogen salts) transmit well in the UV, but their physical characteristics limit them to friendly environments. The two outstanding UV materials are fused quartz (SiO_2) and fluorite (CaF_2). Fused quartz is usable to a bit below 250 nm; fluorite can be used to below 150 nm. Note that some coating materials absorb in the UV, so be sure that the coating transmission is high; it's not enough just to get a low reflection. Ideally, reflection plus transmission should total 100 percent.

19.6 Microlithographic Lenses

The lenses used to produce integrated circuits and computer chips must have extremely high image quality and resolution to produce images in which the details and line widths are only small fractions of a micron (micrometer). The resolution (or detail size) limit of the imaging process is given by

$$\frac{K \cdot \text{wavelength}}{\text{NA}}$$

where NA is the numerical aperture (NA = $n \cdot \sin U$) and K is a constant whose value depends on factors such as the resist, development, and exposure. Thus it is apparent that a short wavelength and a large NA are desirable when high resolution is required.

The wavelengths currently used in microlithography are deep in the UV, and there are only a few optical materials that have sufficient transmission in the UV to be useful. Fused quartz (silica, SiO_2) and, for the

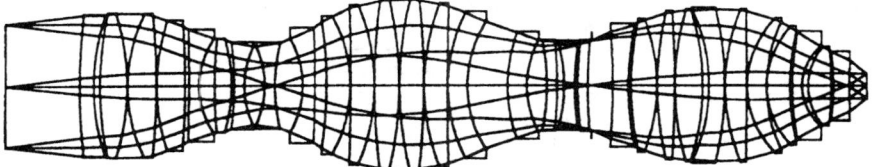

Figure 19.8 A 30-element microlithographic lens.

shortest wavelengths, calcium fluoride (CaF_2) are materials that are useful at the wavelengths of interest. Achromatization is possible with these two materials, but the short wavelength systems typically use only one material and rely on monochromatic illumination to eliminate chromatic aberration. Fused quartz can be used for wavelengths down to less than a quarter micron (μm), and calcium fluoride is used for even shorter wavelengths, to about 150 nm. Immersion systems (akin to oil immersion microscope objectives) have recently been discussed as a way to increase the NA (and the resolution) of the system; there has been promising work at 197 nm.

Two figures show the type of lens system that has evolved for this purpose. Figure 19.8 shows a 30-element design and Fig. 19.9 is a

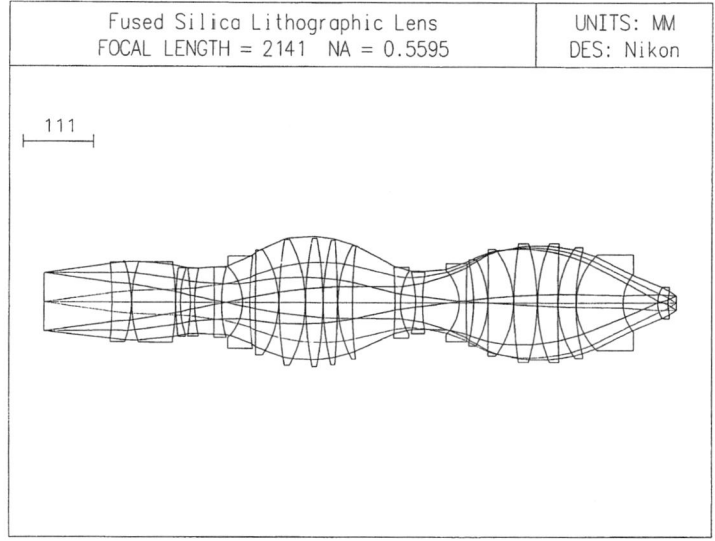

Figure 19.9 A 20-element, afocal, doubly telecentric, fused quartz microlithographic lens. NA = 0.56, 0.25×, for use at 248 nm.

TABLE 19.1 **Prescription for the 20-Element Microlithographic Lens Shown in Figs. 19.10 to 19.13**

	Radius	Space	Semi-diam.	Mat'l
Object	—	107.954	46.80	
1	−617.8800	30.375	61.30	SiO$_2$
2	−207.0893	0.934	64.20	
3	+201.9739	68.636	64.75	SiO$_2$
4	−416.6217	0.865	59.60	
5	+460.0439	7.061	55.25	SiO$_2$
6	+179.6999	15.608	55.25	
7	−373.0162	6.952	54.90	SiO$_2$
8	+249.4960	30.983	54.35	
9	−2591.2	11.541	55.90	SiO$_2$
10	+229.2357	33.165	56.85	
11	−82.3025	11.524	57.45	SiO$_2$
12	+569.8191	9.159	74.85	
13	+5523.6	36.703	79.45	SiO$_2$
14	−156.8200	0.889	85.05	
15	+610.3354	41.168	100.20	SiO$_2$
16	−221.8862	0.883	101.90	
17	+528.5938	26.903	104.20	SiO$_2$
18	−570.2004	0.883	104.05	
19	+423.5775	21.883	101.00	SiO$_2$
20	−1396.3	0.883	100.00	
21	+203.9075	22.715	91.85	SiO$_2$
22	+835.4548	67.972	89.70	
23	−735.8990	8.386	57.50	SiO$_2$
24	+104.6386	23.616	50.55	
25	−184.6683	11.034	49.95	SiO$_2$
26	+288.7053	58.171	46.10	
27	−74.5663	11.343	51.85	SiO$_2$
28	+2319.0	11.371	63.05	
29	−283.4504	22.211	64.75	SiO$_2$
30	−142.5176	1.323	69.90	Stop
31	−5670.5	39.484	81.85	SiO$_2$
32	−146.6908	0.883	86.45	
33	+654.7531	37.168	94.75	SiO$_2$
34	−347.7071	0.883	96.35	
35	+254.9142	31.600	96.45	SiO$_2$
36	+2133.2	0.883	94.50	
37	+164.8042	27.885	89.95	SiO$_2$
38	+349.3775	0.884	86.00	
39	+108.9816	73.045	77.70	SiO$_2$
40	+75.6698	54.069	46.50	
41	+46.2841	16.956	25.70	SiO$_2$
42	+99.3161	13.1684	19.85	
43	Image		11.70	

21-element fused quartz (SiO$_2$) design taken from the OSLO lens library (Nikon, USP 5,805,344, 1998). It works at a wavelength of 0.248 µm (KrF, Eximer), the NA is 0.56 (f/0.9), the magnification is 0.25×, and the object and image sizes are 93.6 mm and 23.4 mm,

Infrared and Ultraviolet Systems

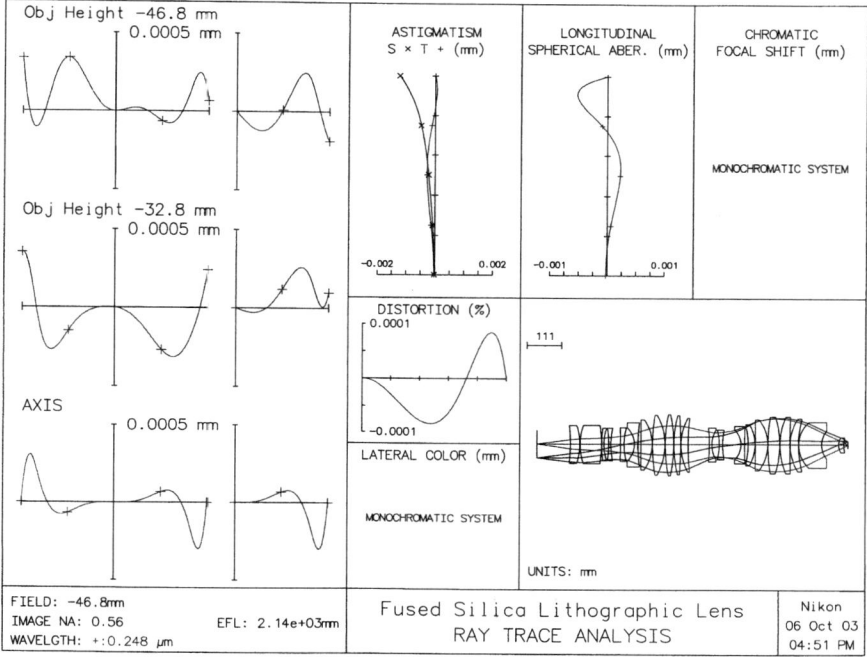

Figure 19.10 The ray aberrations of the lens of Fig. 19.9, Table 19.1.

respectively. It is nearly afocal and doubly telecentric. (Telecentricity is essential to maintain exact image size if the system is slightly defocused.) The overall length is 879 mm. Its Petzval radius is −2.07e + 05, which indicates a Petzval field curvature (sag) of only 0.3 µm at an image height of 11.7 mm.

The prescription for this lens is given in Table 19.1; the index of SiO_2 is 1.5084 at 248 nm. The ray intercept plots and aberrations are shown in Fig. 19.10. The (incoherent) through-frequency and through-focus (at 1000 lpm) *modulation transfer function* (MTF) are shown in Figs. 19.11 and 19.12, respectively. The MTF across the field for frequencies of 1000, 2000, and 3000 lpm are shown in Fig. 19.13.

These two lenses nicely exemplify several of the design techniques we have previously discussed. The concept of splitting elements into several lower-powered elements is very obvious in these designs. Many of the elements are bent to an approximation of the aplanatic shape so that their aberration contributions are minimized.

The marginal ray path shows the large ray heights at positive elements and the small ray heights at the negative elements, which allows the correction of the Petzval sum. This can yield a system where the sum of the

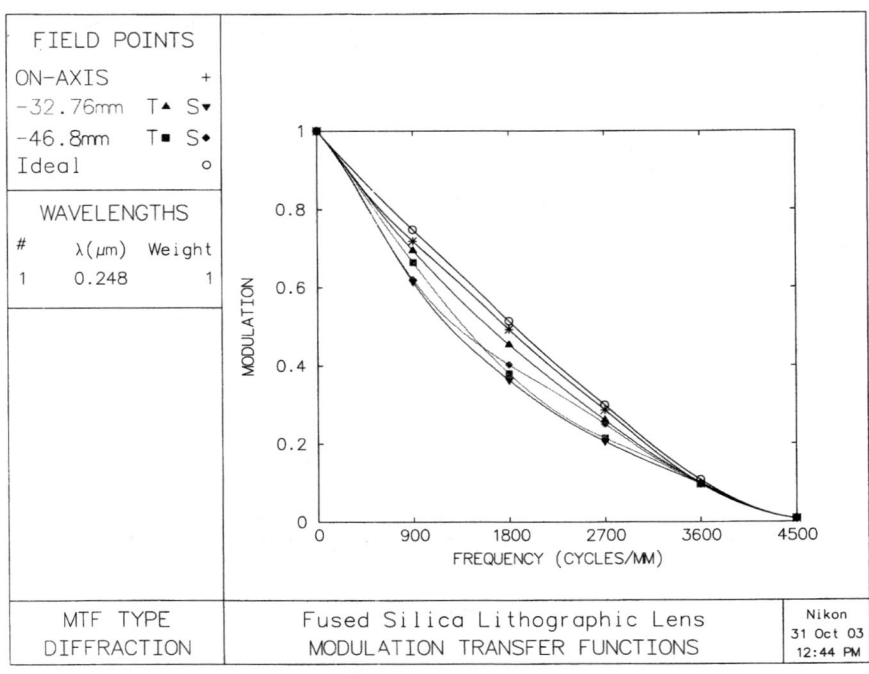

Figure 19.11 The through-frequency MTF of the lens of Fig. 19.9, Table 19.1.

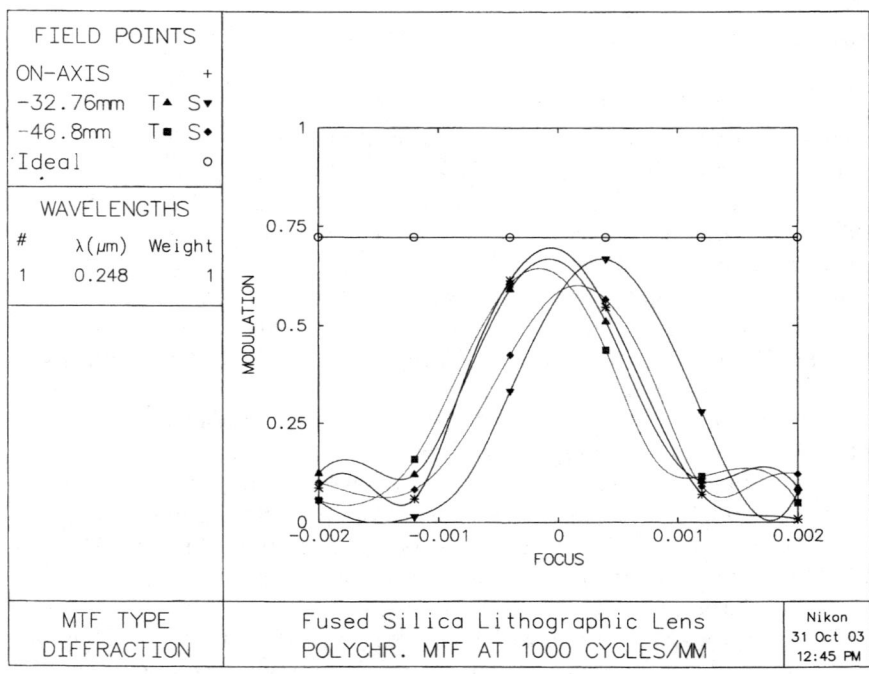

Figure 19.12 The through-focus MTF of the lens of Fig 19.9, Table 19.1.

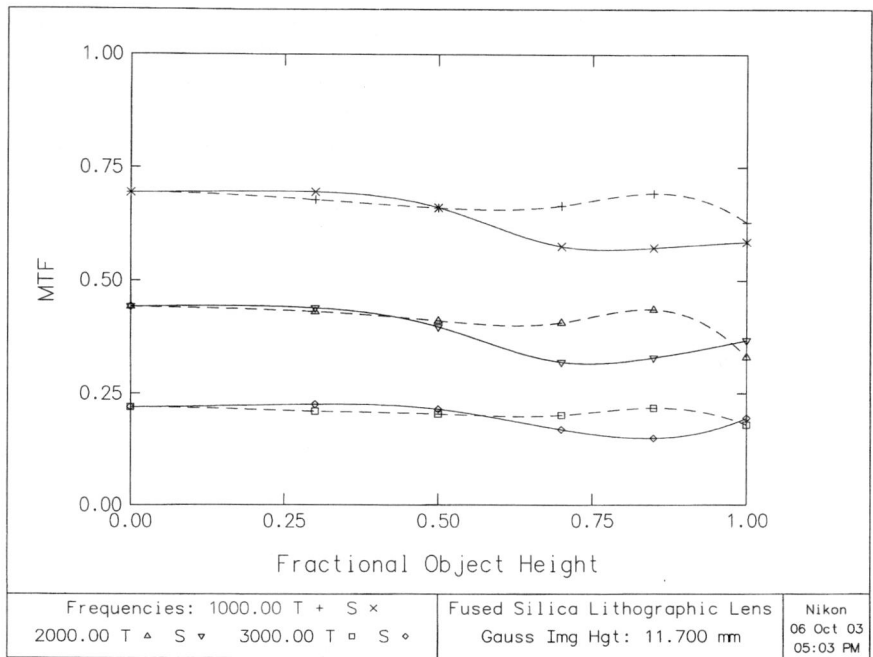

Figure 19.13 The through-field MTF at 1000, 2000 and 3000 lpm for the lens of Fig. 19.9, Table 19.1.

surface refracting powers, $\Sigma(n' - n)/R$, is approximately zero (producing a zero Petzval sum when all the elements have the same index), while the "image forming power," $\Sigma y(n' - n)/R$, is a useful, nonzero power. Note that the element diameters in these systems alternately bulge and shrink, and that in the bulges we find positive elements and in the waists we find negative elements. Of course, this basic idea is found in every anastigmat lens; however, in these lenses there are more bulges and waists than are usually found in other lenses.

In Fig. 19.9 the outer elements appear to be there primarily for the purpose of imaging the entrance and exit pupils at infinity to achieve a telecentric system. This is less obvious in Fig. 19.8, but it is there nonetheless; it just takes a bit of looking to spot it.

The long object-to-image distance means that these lenses are large; diameters to the order of a foot are not uncommon. But the long track length also means that the powers of the elements are low, the field angles are small, and so are the angles of incidence and refraction.

A thick, nearly concentric, meniscus element in Fig. 19.9, located in the convergent beam near the image reduces the Petzval sum and can also be used to correct the field aberrations. Meniscus elements near the object and/or the image have also been used to produce telecentricity. Spacing and thickness have effects on the higher-order aberrations here, just as they do in systems exemplified by such disparate lenses as the airspaced doublet telescope objective in Chap. 6 and the double-Gauss camera lens in Chap. 12. Note, however, that, given the material and transmission problems, thick elements often are less than desirable.

These lenses must approach perfection in both design and fabrication. They are extremely difficult to manufacture, requiring tolerances to wavelength orders of magnitude. Assembly is often an interferometric process. The element diameters are large and the quality required for the glass approaches (or exceeds) the limit of what is readily available. The cost of such a lens easily reaches millions.

Chapter 20

Zoom Lenses

20.1 Zoom Lenses

Other than a single component working near unit magnification, the simplest zoom lens consists of two components with a variable airspace between them to change the focal length, plus, of course, a provision to shift the entire system to keep it in focus. To get the maximum change of focal length, one component is usually made positive and the other negative. Thus the two possible arrangements are similar to either a telephoto or a retrofocus, depending on whether the positive or negative component is on the object side of the lens. The telephoto arrangement is obviously more suited to longer focal lengths, and the retrofocus arrangement to shorter focal lengths and wider angular coverage. Figures 20.1 and 20.2 are systems of the latter type, with the negative component facing the object. Figure 20.1 has a modest zoom ratio (long efl/short efl) of only 1.46 but covers a wide field angle of 83° at its short-focal-length setting. It maintains a speed of $f/3.5$ throughout the zoom. There are nine elements, and the eighth surface is a general aspheric.

Figure 20.2 has a larger zoom ratio of 1.9 and covers a somewhat smaller angle of 61° at the short focus. Its f number changes from $f/3.5$ to $f/4.5$ as the lens is zoomed from short to long focal length. In a modern camera with automatic exposure control this change in lens speed is not a serious problem. But without automatic exposure control, the iris must be located in a fixed position relative to the image plane so that the f number and the image illumination remain constant throughout the zoom. Note that, in this lens, the designer inserted a fixed, weak meniscus lens behind the two moving components. Its weak power (only 9 percent of the midrange power) and fixed location lead one to wonder if it could be removed from the design. Both designs have a back focal length long enough for a single lens reflex camera.

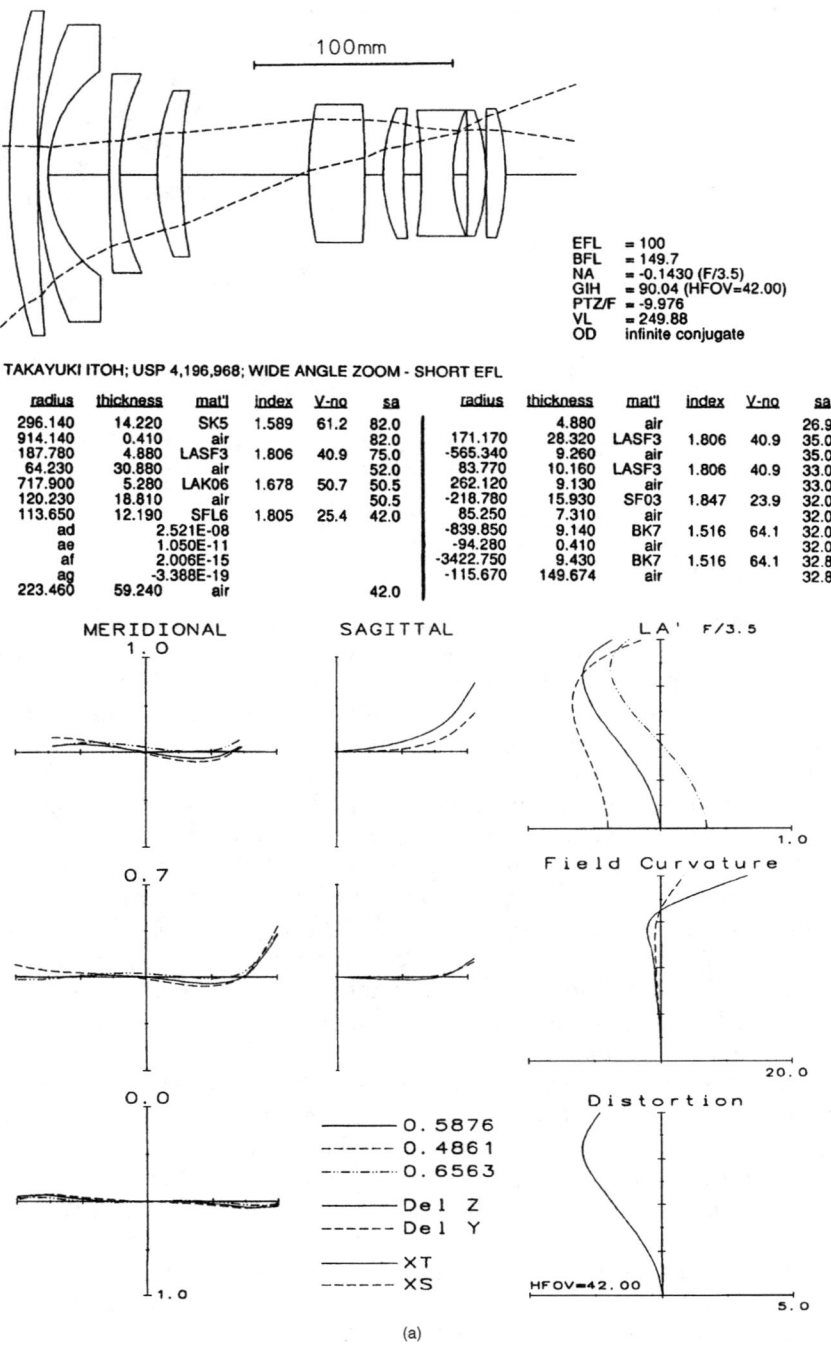

Figure 20.1 A two-component, nine-element, retrofocus type zoom, $f/3.5$, $\pm 42°$ to $\pm 31°$, with a zoom ratio of 1.46x.

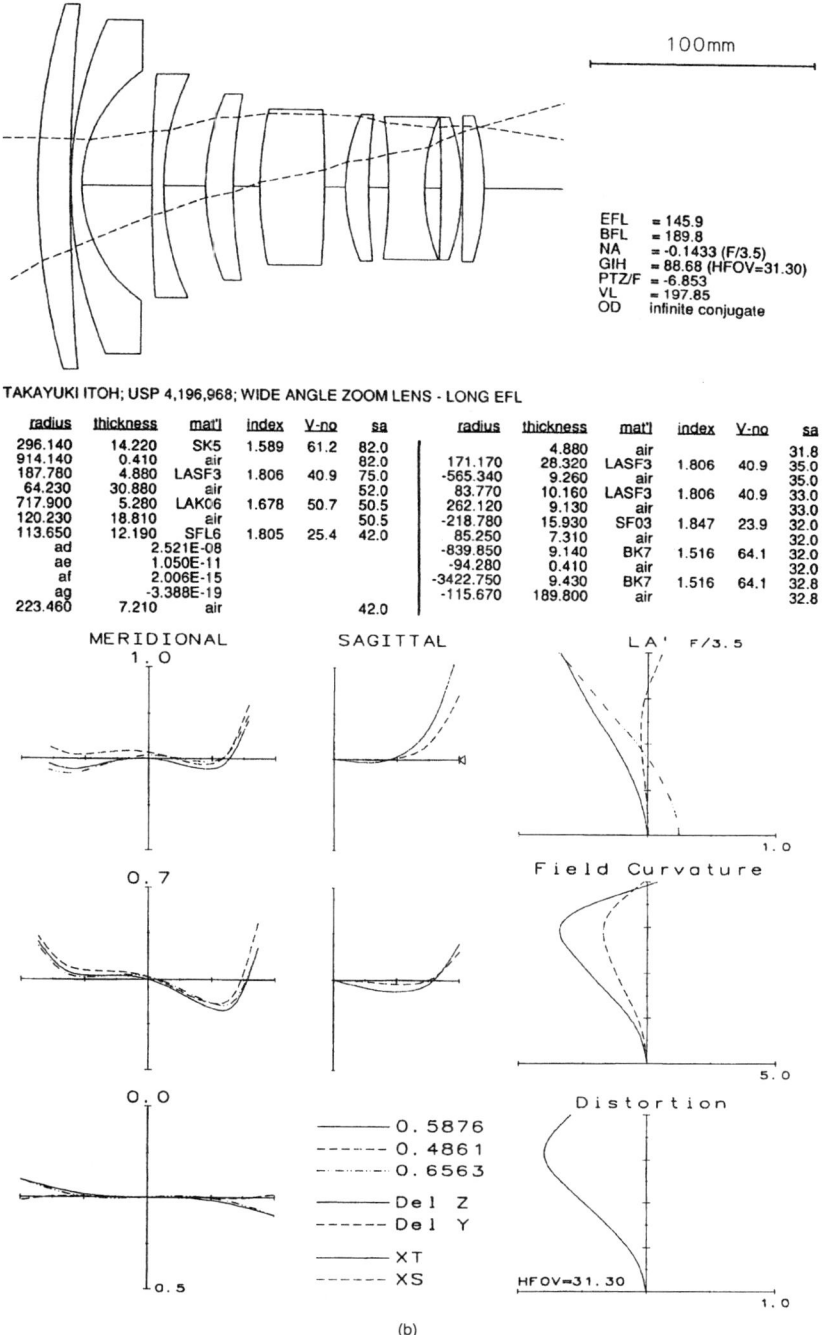

TAKAYUKI ITOH; USP 4,196,968; WIDE ANGLE ZOOM LENS - LONG EFL

radius	thickness	mat'l	index	V-no	sa	radius	thickness	mat'l	index	V-no	sa
296.140	14.220	SK5	1.589	61.2	82.0		4.880	air			31.8
914.140	0.410	air			82.0	171.170	28.320	LASF3	1.806	40.9	35.0
187.780	4.880	LASF3	1.806	40.9	75.0	-565.340	9.260	air			35.0
64.230	30.880	air			52.0	83.770	10.160	LASF3	1.806	40.9	33.0
717.900	5.280	LAKC6	1.678	50.7	50.5	262.120	9.130	air			33.0
120.230	18.810	air			50.5	-218.780	15.930	SF03	1.847	23.9	32.0
113.650	12.190	SFL6	1.805	25.4	42.0	85.250	7.310	air			32.0
ad	2.521E-08					-839.850	9.140	BK7	1.516	64.1	32.0
ae	1.050E-11					-94.280	0.410	air			32.0
af	2.006E-15					-3422.750	9.430	BK7	1.516	64.1	32.8
ag	-3.388E-19					-115.670	189.800	air			32.8
223.460	7.210	air			42.0						

EFL = 145.9
BFL = 189.8
NA = -0.1433 (F/3.5)
GIH = 88.68 (HFOV=31.30)
PTZ/F = -6.853
VL = 197.85
OD infinite conjugate

(b)

Figure 20.1 (*Continued*)

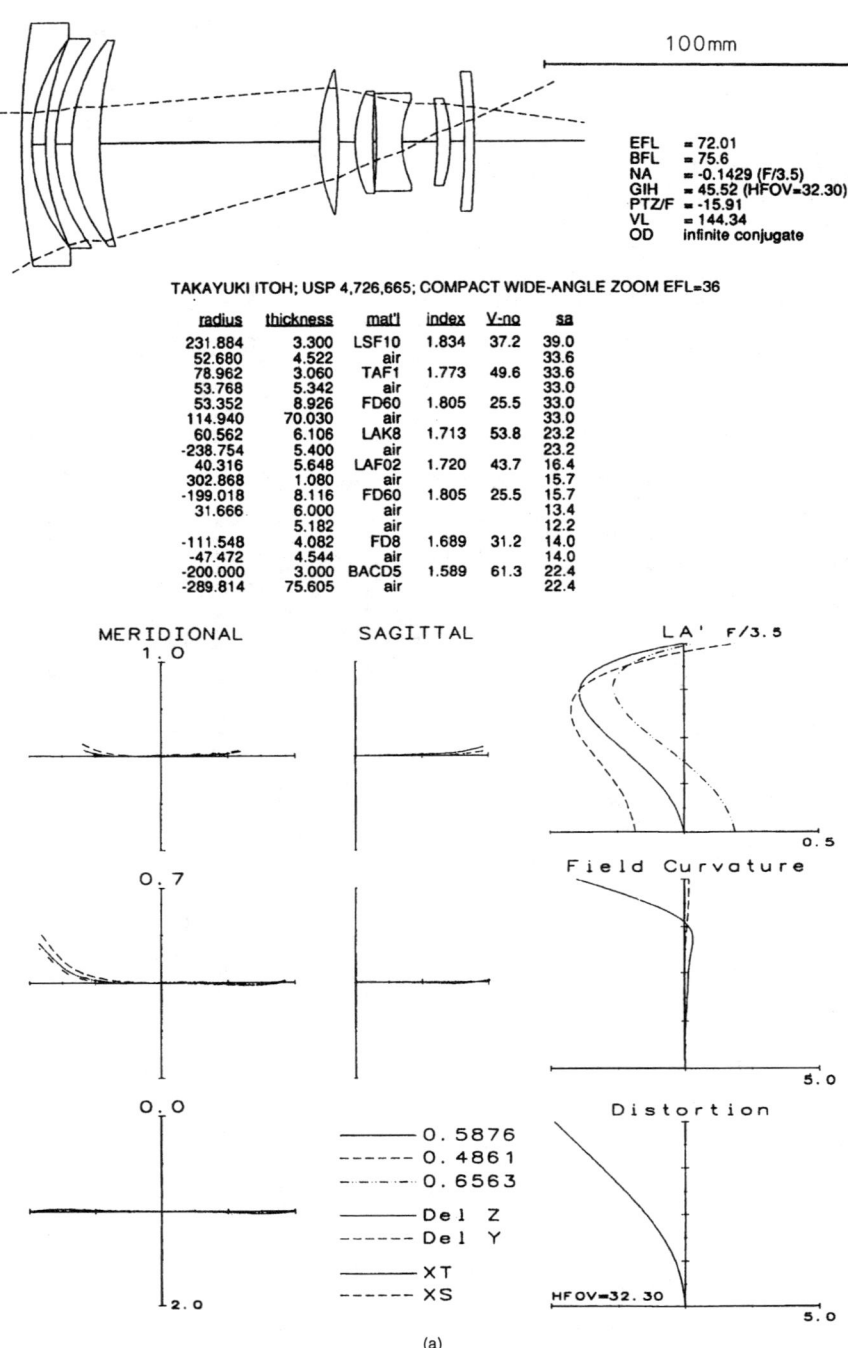

Figure 20.2 A two-component, eight-element, retrofocus type zoom, $f/3.5$ to $f/4.5$, ±32° to ±17°, with a zoom ratio of 1.89x.

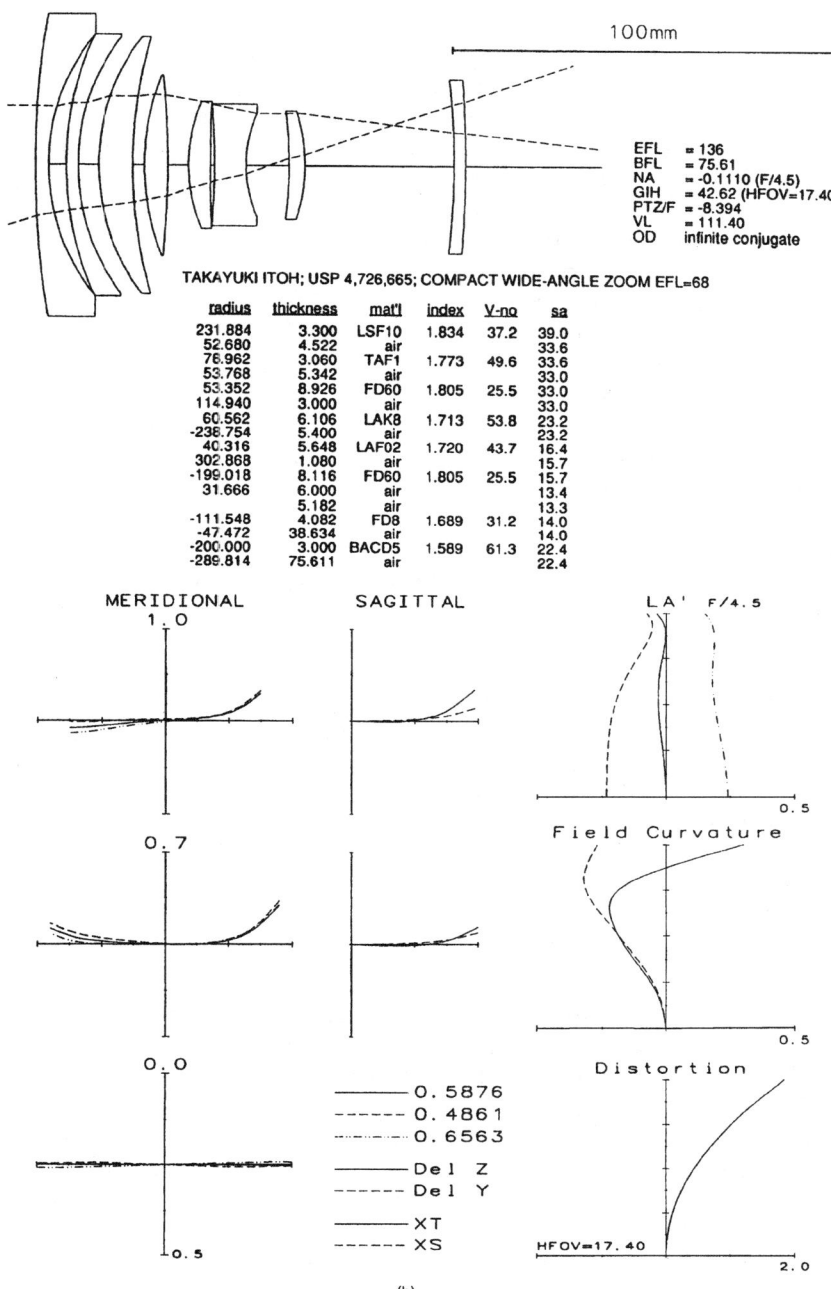

TAKAYUKI ITOH; USP 4,726,665; COMPACT WIDE-ANGLE ZOOM EFL=68

radius	thickness	mat'l	index	V-no	sa
231.884	3.300	LSF10	1.834	37.2	39.0
52.680	4.522	air			33.6
76.962	3.060	TAF1	1.773	49.6	33.6
53.768	5.342	air			33.0
53.352	8.926	FD60	1.805	25.5	33.0
114.940	3.000	air			33.0
60.562	6.106	LAK8	1.713	53.8	23.2
-236.754	5.400	air			23.2
40.316	5.648	LAF02	1.720	43.7	16.4
302.868	1.080	air			15.7
-199.018	8.116	FD60	1.805	25.5	15.7
31.666	6.000	air			13.4
	5.182	air			13.3
-111.548	4.082	FD8	1.689	31.2	14.0
-47.472	38.634	air			14.0
-200.000	3.000	BACD5	1.589	61.3	22.4
-289.814	75.611	air			22.4

EFL = 136
BFL = 75.61
NA = -0.1110 (F/4.5)
GIH = 42.62 (HFOV=17.40)
PTZ/F = -8.394
VL = 111.40
OD infinite conjugate

(b)

Figure 20.2 (*Continued*)

The reverse telephoto arrangement provides a zoom lens with a relatively long back focus. This fact has made this form popular for use on single lens reflex cameras, where the back focus must be long enough (i.e., about 38 to 40 mm) to clear the swinging mirror of the SLR. These design types are usually used for the short focal length, wider angle lenses that need the long working distance of the retrofocus type.

Figure 20.3 is a 12-element, 2.5-zoom-ratio system that works at $f/2.8$ and covers about 30° at its short focal length. It is an almost classic example of a positive-negative-positive (+ − +) afocal zoom unit followed by a four-element focusing lens. The three-element inner negative component moves linearly along the axis to zoom the focal length, and the following positive doublet component moves in a nonlinear path to compensate for the image shift the zooming component produces. Many lenses with large zoom ratios utilize this + − + configuration to produce the zoom and to compensate for the focus shift.

Figure 20.4 covers a wider field (72°) and has a somewhat larger zoom ratio (2.9), but a significantly slower speed ($f/4.0$ to $f/5.5$) than Fig. 20.3. It also has a considerably more complex set of component motions. Each of the four components has a different motion. Note that the component powers are arranged − + − +; in Fig. 20.3 the zoom part of the system was + − +.

Figure 20.5 is more complex still in that it incorporates a macro capability. The system has a zoom ratio of 3.4 and covers a field of 72° at a speed of $f/4$ to $f/5.3$, like Fig. 20.4. Its zoom components are also arranged − + − +, followed by a fixed rear doublet. Note that the small third component is also fixed. The macro version is shown for the short focal length in Fig. 20.5(b).

20.2 Zoom Lenses for Point and Shoot Cameras

Tables 20.1 and 20.2 and Figs. 20.6 and 20.7 are examples of zoom lenses for 35-mm point and shoot cameras. Lenses for these cameras are hallmarked by two features: they are extremely compact, and have a very short back focus in the wide angle setting.

The first example (Table 20.1 and Fig. 20.6) has a 2.6x zoom range, 11 elements, and three moving components, and it works at speeds of $f/4$ to $f/8.2$. There are many modifications of this design style in today's compact cameras. Many have a telephoto ratio of one or less at the long focal length setting.

Note the short back focus at the wide angle setting for this lens and the next (Fig. 20.7 and Table 20.2). This indicates that the design is for a point and shoot camera where an extremely short back focal length is acceptable.

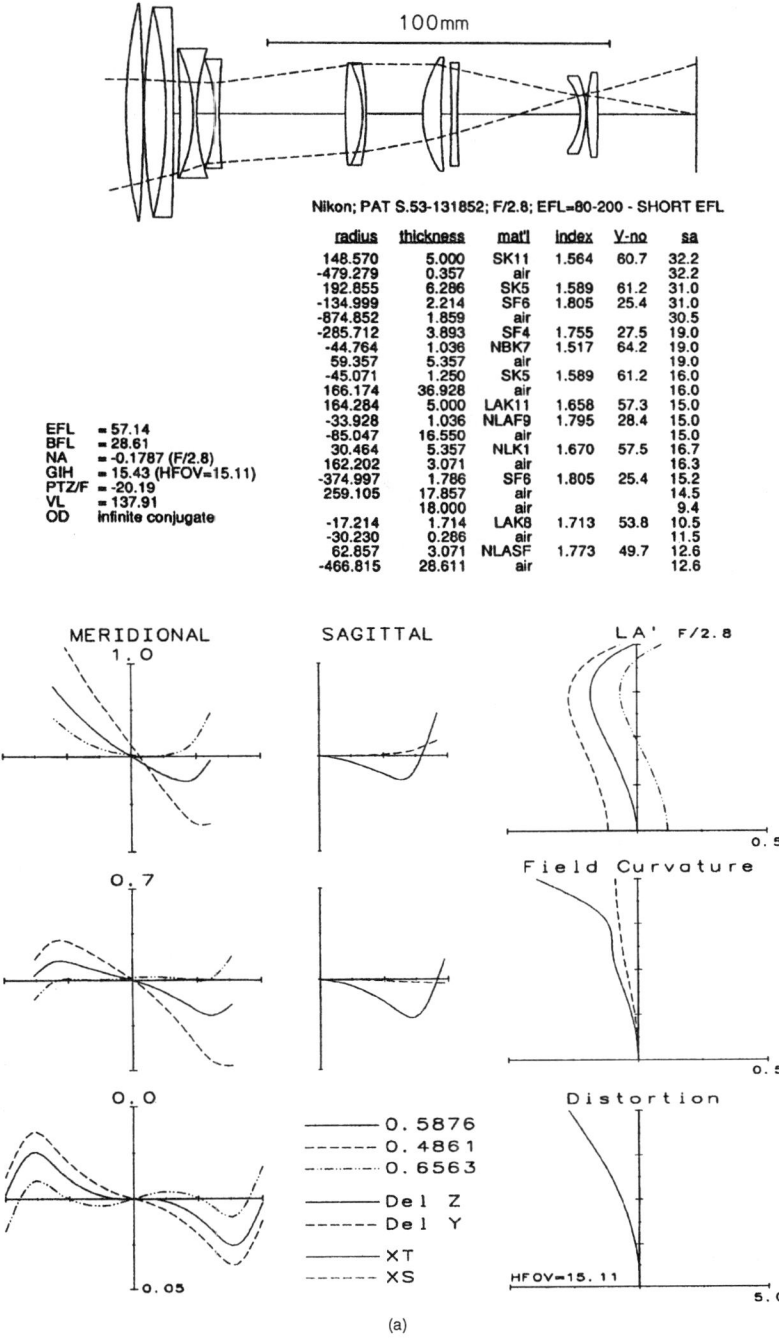

Figure 20.3 A three-component (+ − +), eight-element afocal zoom kernel followed by a fixed, four-element focusing lens, $f/2.8$, ±15° to ±6°, with a zoom ratio of 2.5x.

528 Chapter Twenty

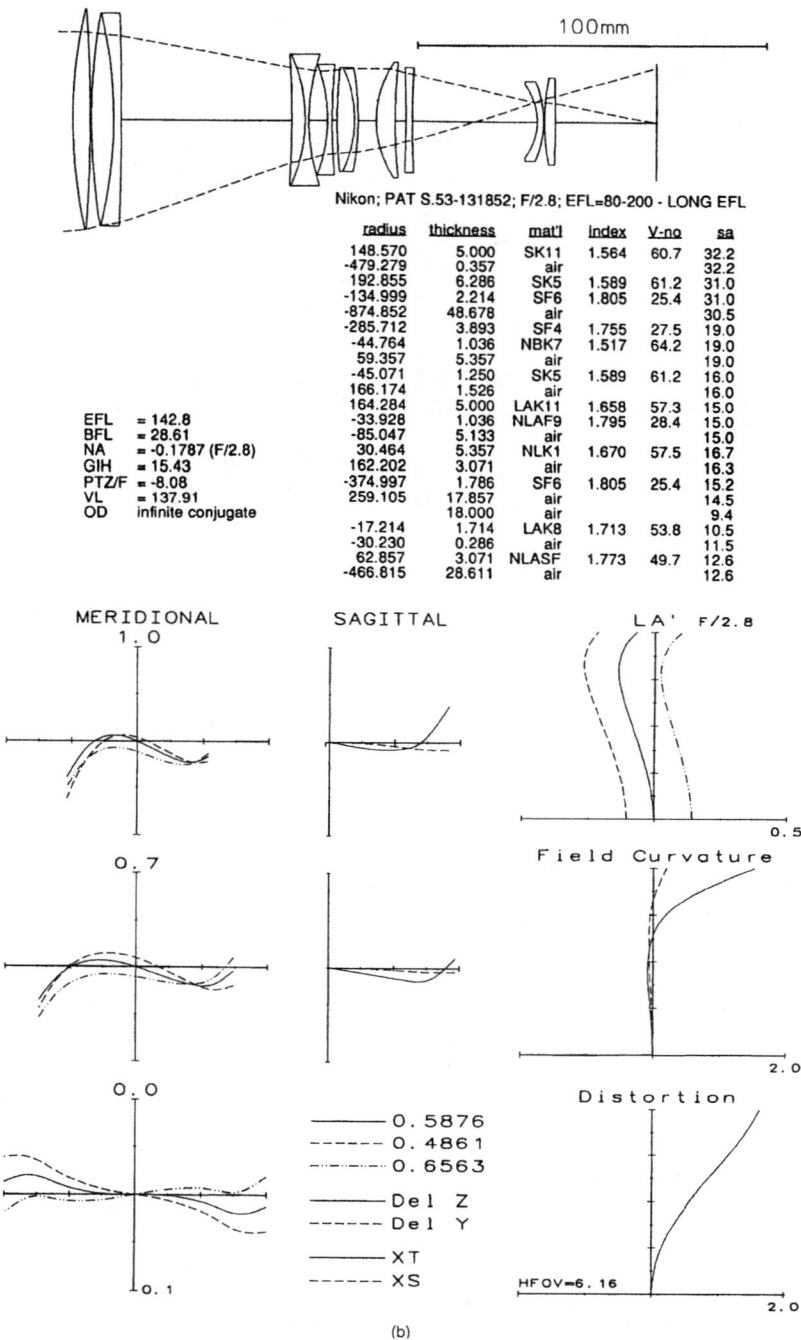

Nikon; PAT S.53-131852; F/2.8; EFL=80-200 - LONG EFL

radius	thickness	mat'l	index	V-no	sa
148.570	5.000	SK11	1.564	60.7	32.2
-479.279	0.357	air			32.2
192.855	6.286	SK5	1.589	61.2	31.0
-134.999	2.214	SF6	1.805	25.4	31.0
-874.852	48.678	air			30.5
-285.712	3.893	SF4	1.755	27.5	19.0
-44.764	1.036	NBK7	1.517	64.2	19.0
59.357	5.357	air			19.0
-45.071	1.250	SK5	1.589	61.2	16.0
166.174	1.526	air			16.0
164.284	5.000	LAK11	1.658	57.3	15.0
-33.928	1.036	NLAF9	1.795	28.4	15.0
-85.047	5.133	air			15.0
30.464	5.357	NLK1	1.670	57.5	16.7
162.202	3.071	air			16.3
-374.997	1.786	SF6	1.805	25.4	15.2
259.105	17.857	air			14.5
	18.000	air			9.4
-17.214	1.714	LAK8	1.713	53.8	10.5
-30.230	0.286	air			11.5
62.857	3.071	NLASF	1.773	49.7	12.6
-466.815	28.611	air			12.6

EFL = 142.8
BFL = 28.61
NA = -0.1787 (F/2.8)
GIH = 15.43
PTZ/F = -8.08
VL = 137.91
OD   infinite conjugate

——— 0.5876
------ 0.4861
·······-- 0.6563

——— Del Z
------ Del Y

——— XT
------ XS

(b)

Figure 20.3 (*Continued*)

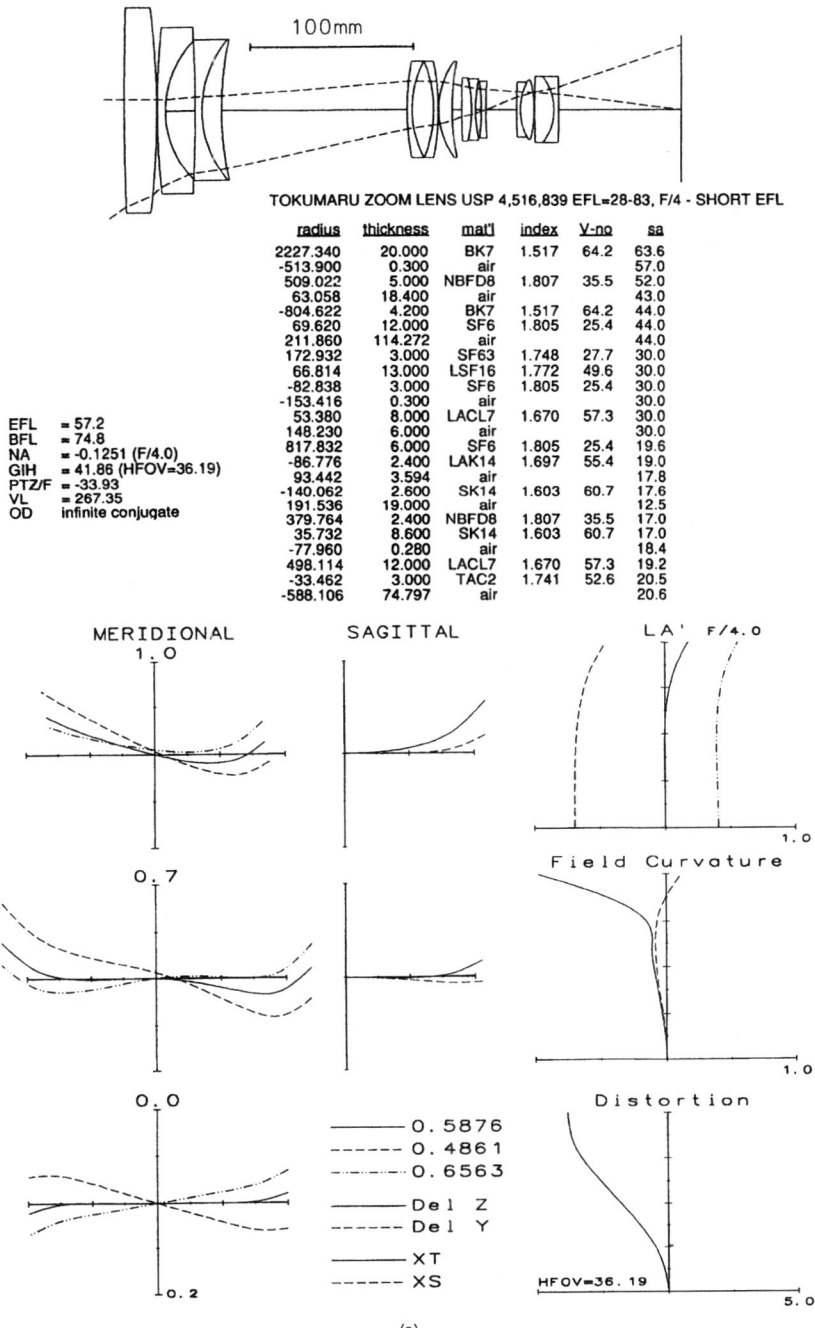

Figure 20.4 A four-component (− + − +) zoom, with four independent motions, f/4.0 to f/5.5, ±36° to ±14°, with a zoom ratio of 2.9x.

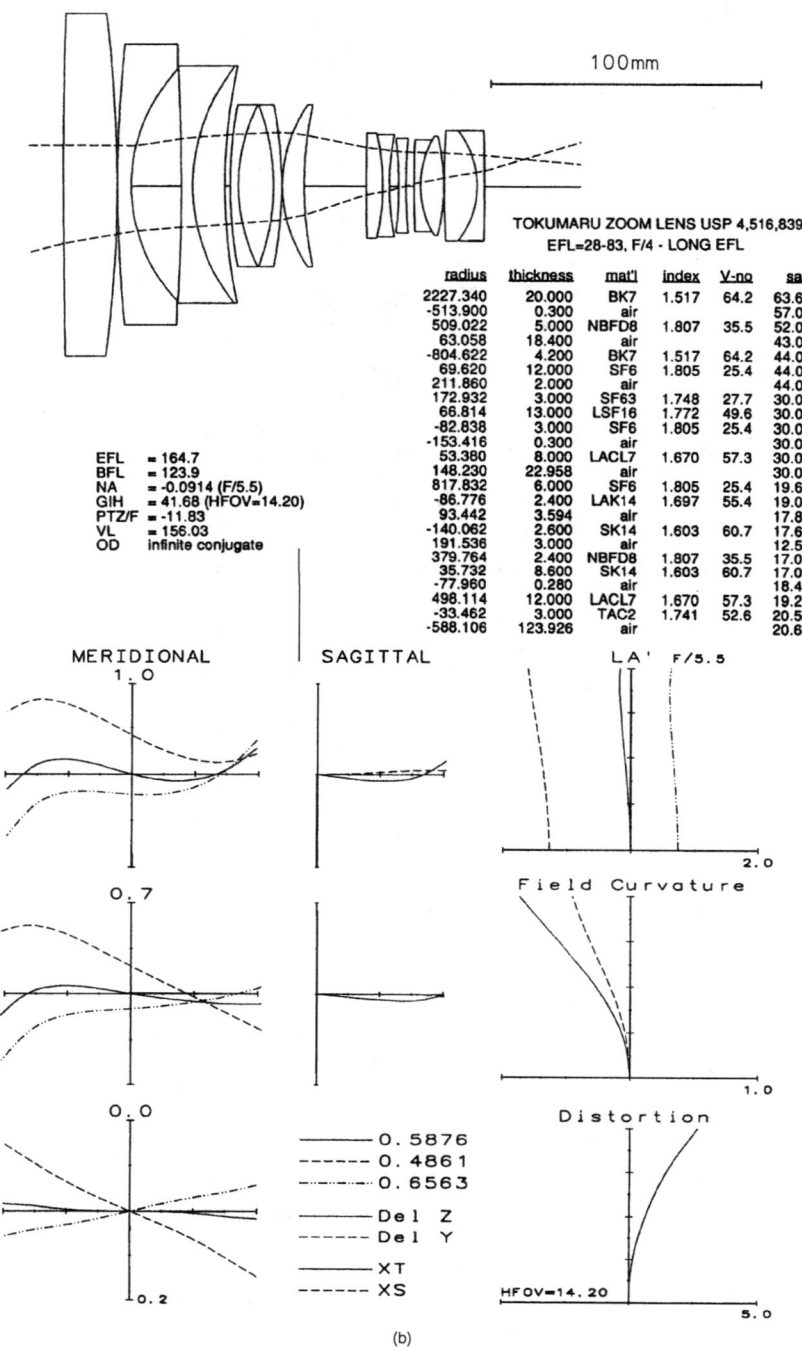

Figure 20.4 (*Continued*)

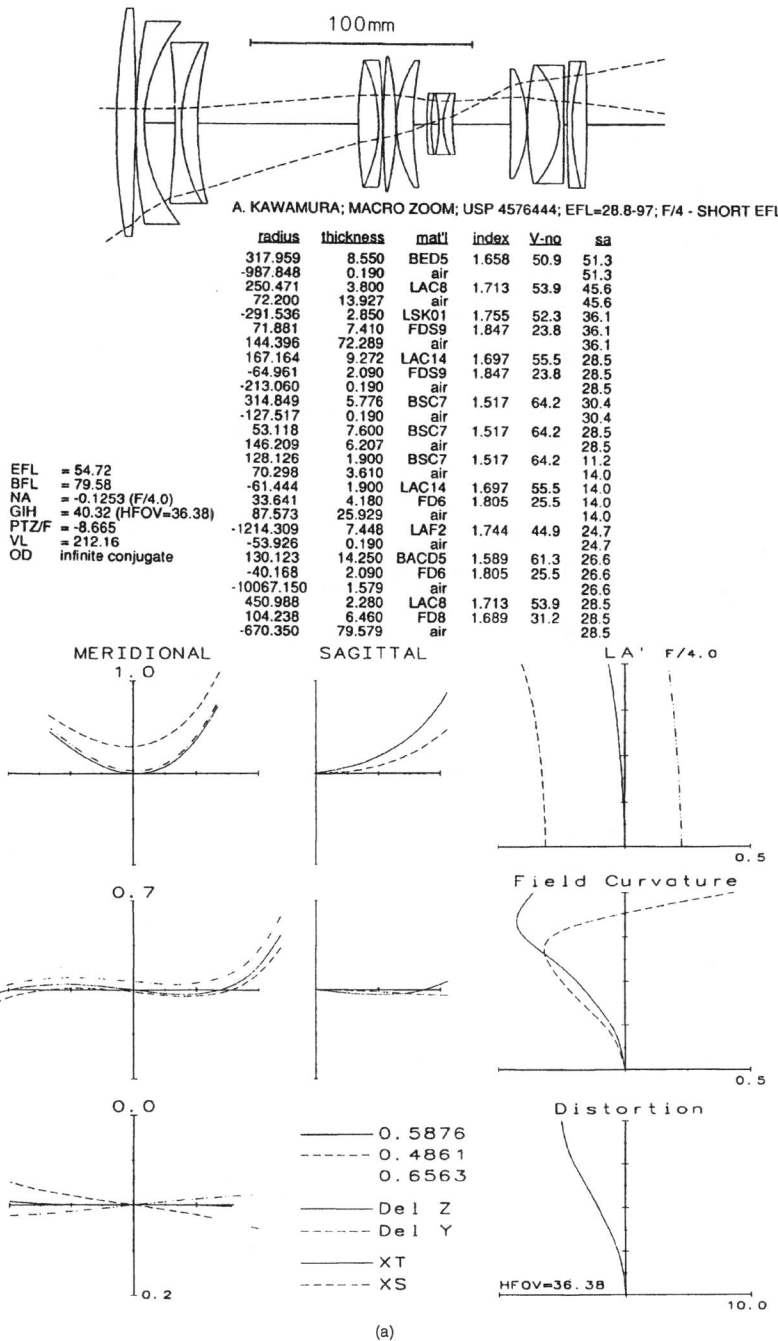

Figure 20.5 A five-component (− + − + +) zoom, the design of which incorporates a macro focus at 0.18x magnification, f/4 to f/5.3, ±36° to ±12°, with a zoom ratio of 3.4x.

532 Chapter Twenty

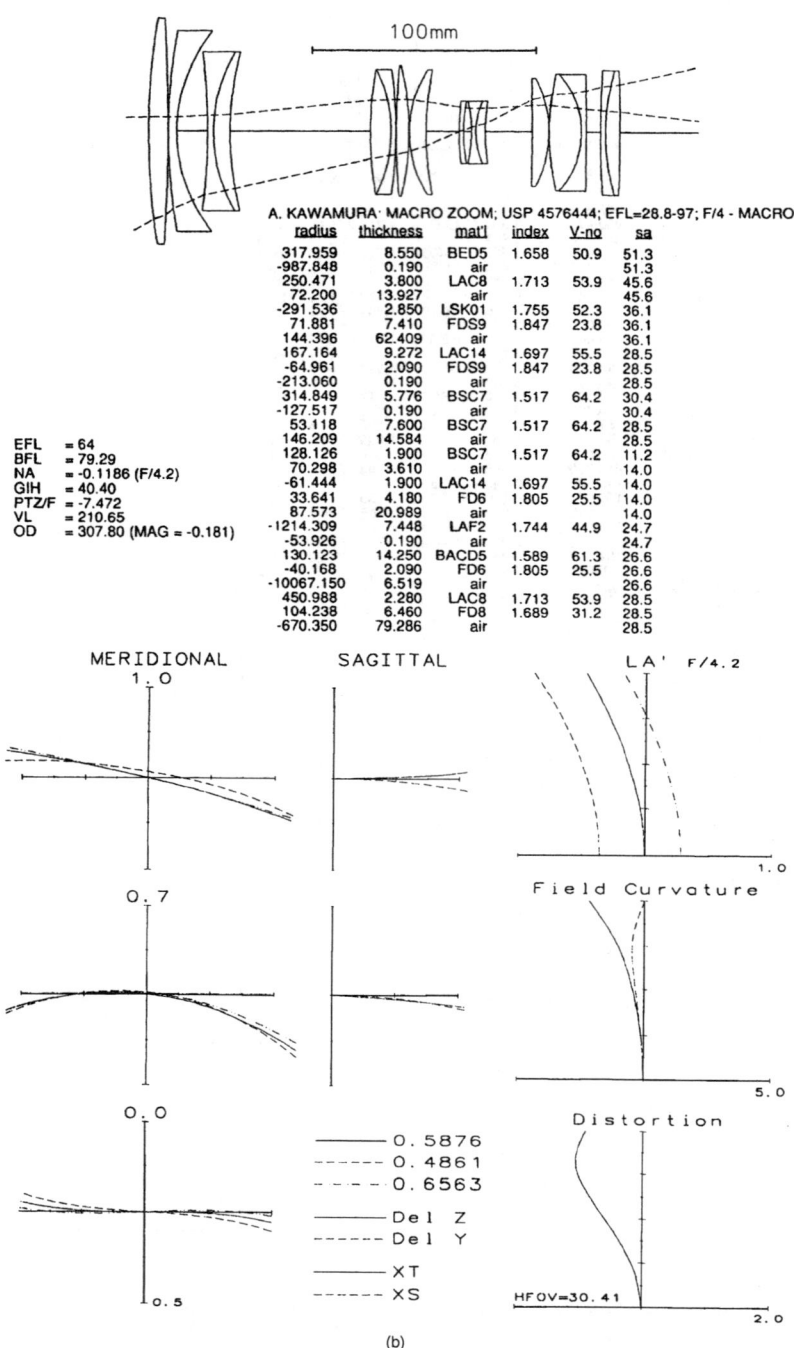

Figure 20.5 (*Continued*)

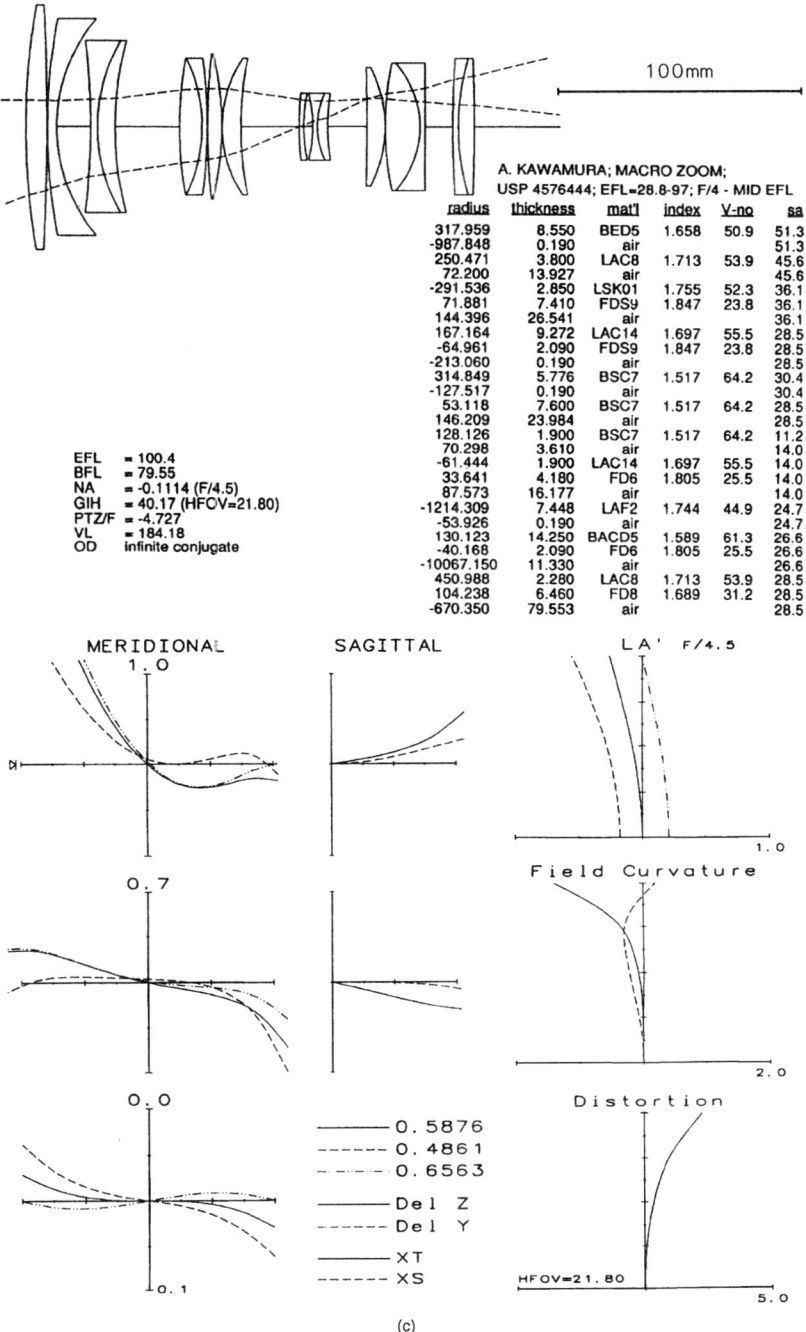

Figure 20.5 (*Continued*)

534 Chapter Twenty

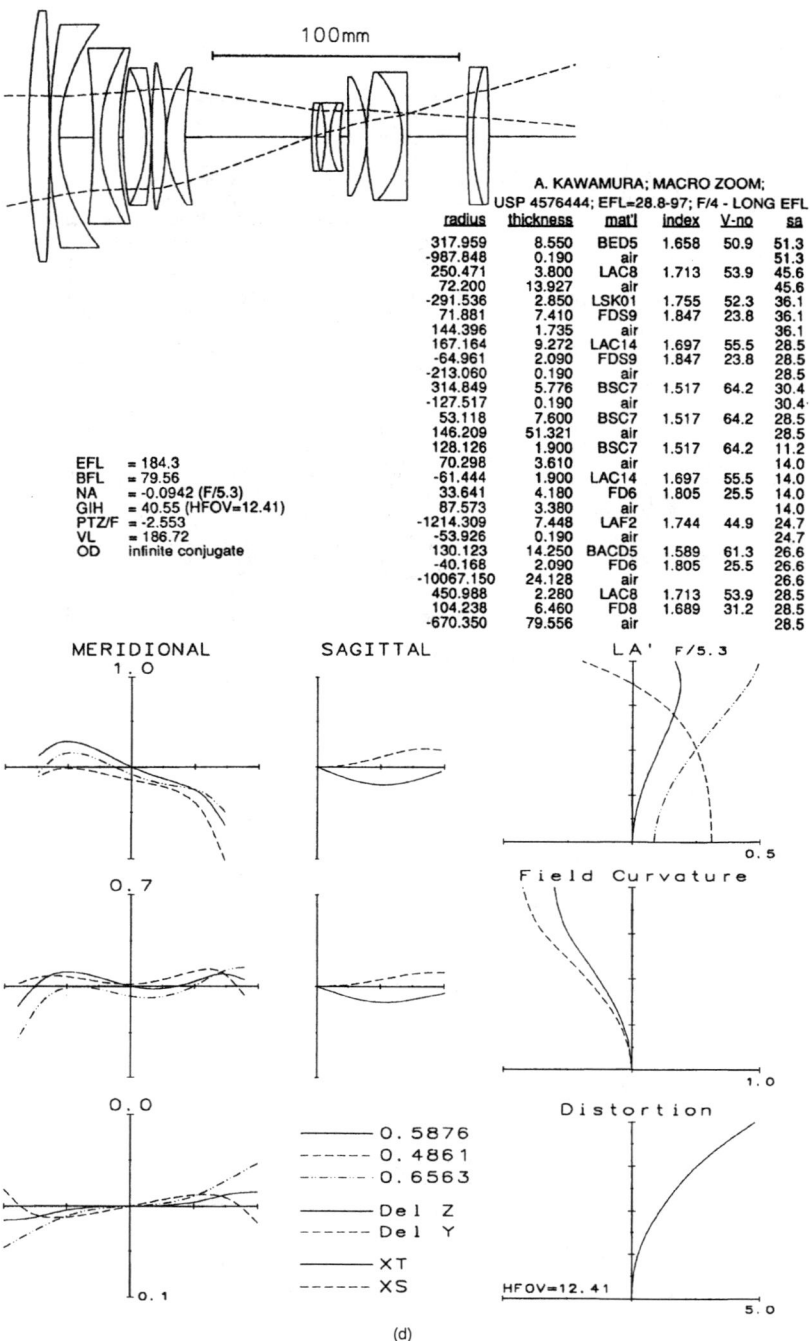

Figure 20.5 (*Continued*)

TABLE 20.1 Prescription of a Zoom Lens for a Point and Shoot Camera (See Figs. 20.6 (a,b,c))

$R1 = -81.208$	$T1 = 1.50$	834372	EFL = 39.00, 69.98, 102.04
$R2 = +36.830$	$T2 = 0.50$		f number = 4.0, 6.5, 8.2
$R3 = +33.534$	$T3 = 5.52$	516641	FOV = ±28.7°, ±16.8°, ±11.8°
$R4 = -38.625$	$T4 = 0.10$		Zoom range = 2.62x
$R5 = +30.331$	$T5 = 3.18$	620603	
$R6 = +250.834$	$T6 = 3.20, 8.93, 12.82$		
$R7 = -25.364$	$T7 = 1.20$	835427	
$R8 = +20.031$	$T8 = 3.24$	805254	
$R9 = -79.760$	$T9 = 7.16$		
$R10 = +69.379$	$T10 = 5.67$	518650	
$R11 = -11.998$	$T11 = 1.35$	805254	
$R12 = -22.095$	$T12 = 0.10$		
$R13 = +145.924$	$T13 = 2.00$	589612	
$R14 = -36.500$	$T14 = 1.00$		
$R15 = $ stop	$T15 = 12.82, 5.56, 2.07$		
$R16 = -101.608$	$T16 = 3.04$	805254	
$R17 = -23.884$	$T17 = 1.99$		
$R18 = -24.348$	$T18 = 1.30$	834372	
$R19 = +341.863$	$T19 = 3.84$		
$R20 = -15.639$	$T20 = 1.40$	772496	
$R21 = -62.435$	BFL = 8.90, 29.74, 49.44		

SOURCE: U.S. Patent #4,978,204, Ito

TABLE 20.2 Prescription of a Zoom Lens for a Point and Shoot Camera (See Figs. 20.7 (a,b,c))

$R1 = -106.485$	$T1 = 1.286$	804466	EFL = 28.50, 100.06
$R2 = +28.800$	$T2 = 3.463$	805254	f number = 3.6, 8.2
$R3 = +186.366$	$T3 = 0.376$		GIH = 21.4, 21.3 (i.e., 35-mm film)
$R4 = +245.442$	$T4 = 1.088$	603380	Field = ±36.9°, ±12.0°
$R5 = +33.193$	$T5 = 8.142, 0.970$		Zoom Ratio = 3.5x
$R6 = +33.045$	$T6 = 0.989$	847238	
$R7 = +19.461$	$T7 = 2.671$	487702	
$R8 = -174.316$	$T8 = 0.791, 7.984$		
$R9 = +29.434$	$T9 = 2.770$	569632	
$R10 = -53.069$	$T10 = 2.572, 20.984$		
$R11 = -19.332$	$T11 = 0.861$	648338	
$R12 = -92.565$	$T12 = 0.989$		
$R13 = $ stop	$T13 = 0.989$		
$R14 = -56.453$	$T14 = 0.772$	487702	
$R15 = +36.151$	$T15 = 2.078$	847238	
$R16 = -45.036$	$T16 = 4.798$		
$R17 = +34.281$	$T17 = 1.088$	847238	
$R18 = +12.684$	$T18 = 5.442$	583594	
$R19 = -23.003$	$T19 = 12.080, 0.782$		
	AD = 1.1875e − 05 AE = −2.7428e − 08 AF = −2.8021e − 10		
$R20 = -33.163$	$T20 = 2.968$	847238	
$R21 = -17.749$	$T21 = 0.148$		
$R22 = -20.569$	$T22 = 1.286$	835427	
$R23 = -169.349$	$T23 = 4.343$		
$R24 = -17.720$	$T24 = 1.484$	785442	
$R25 = -72.540$	$T25 = 5.180, 47.910$		
$R26 = $ image			

SOURCE: From the OSLO Betensky Library. U.S. Patent #5,691,851.
NOTE: Rounded to three figures after the decimal point

536 Chapter Twenty

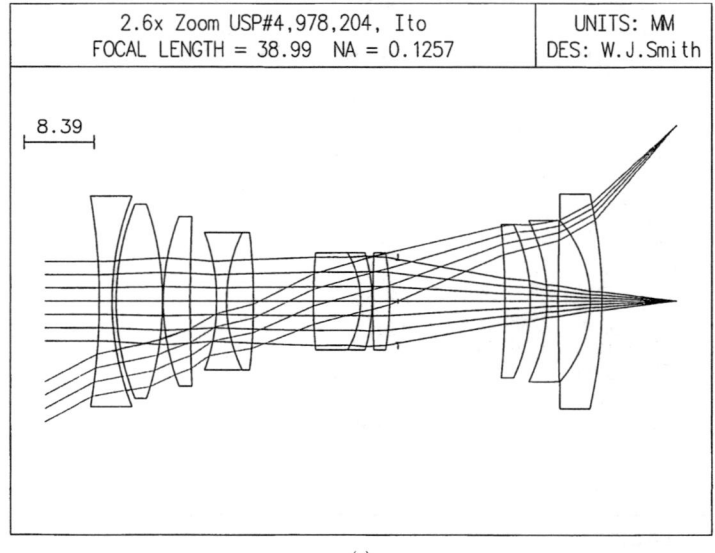

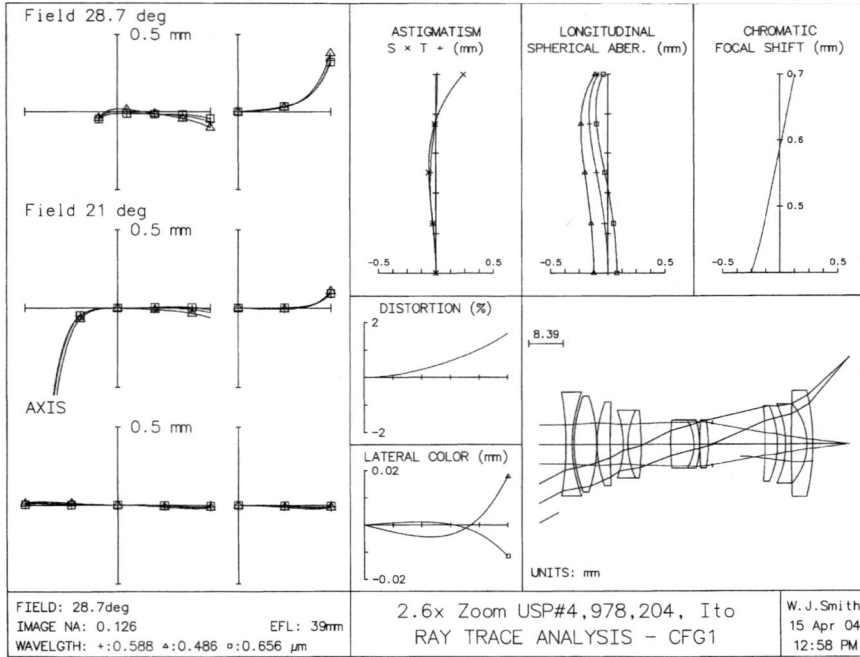

Figure 20.6 An 11-element zoom suitable for a 35-mm point and shoot camera with a 2.6x zoom ratio, ±29° to ±12°, f/4.0 to f/8.2, Table 20.1.

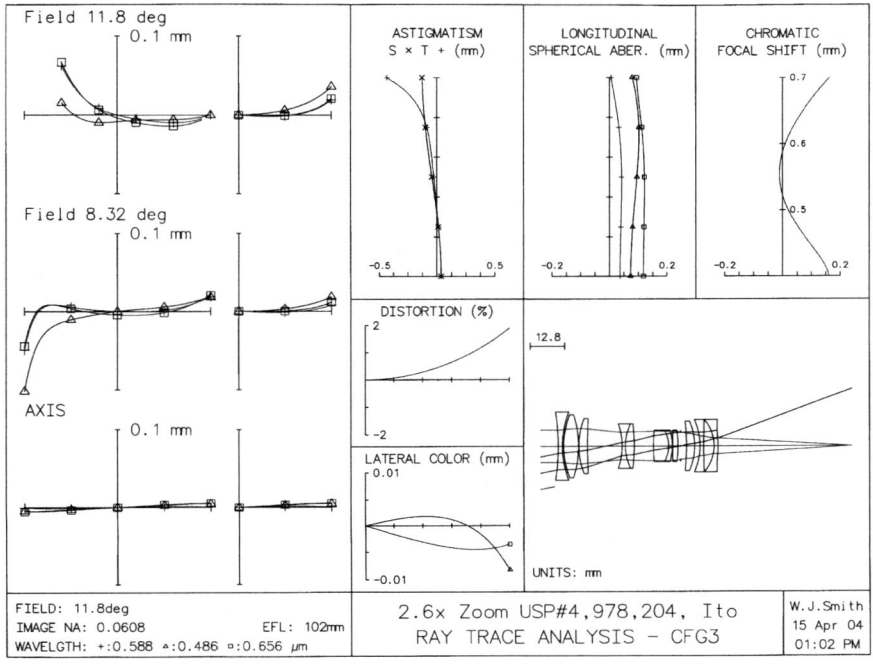

(c)

Figure 20.6 (*Continued*)

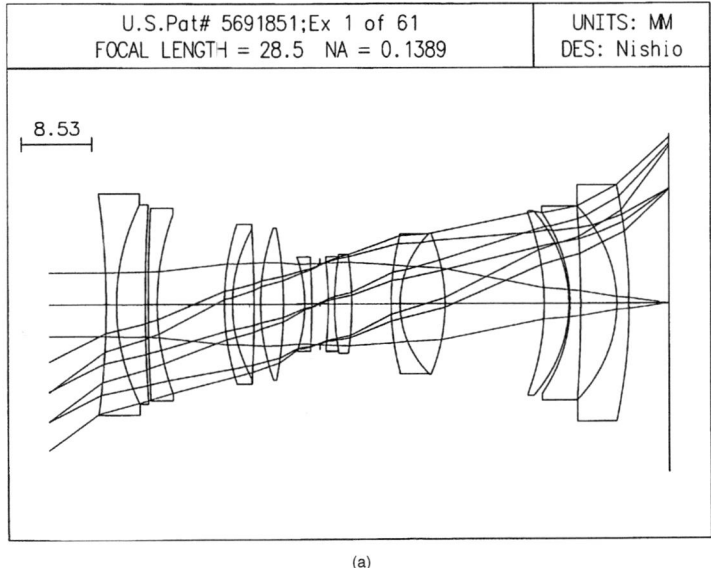

(a)

Figure 20.7 A 14-element, one aspheric, zoom suitable for a point and shoot 35-mm camera, with a 3.5x zoom ratio, ±37° to ±12°, f/3.6 to f/8.2, Table 20.2.

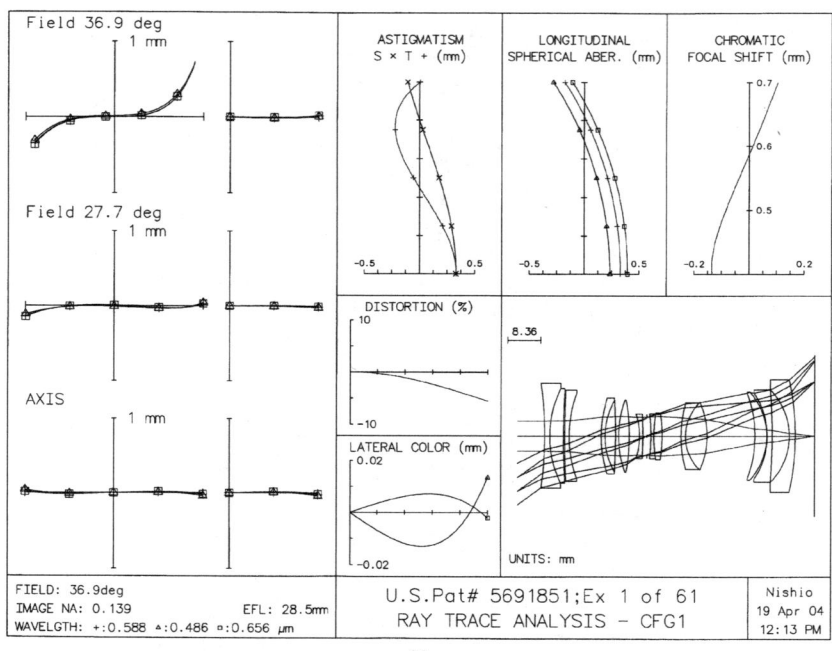
(b)

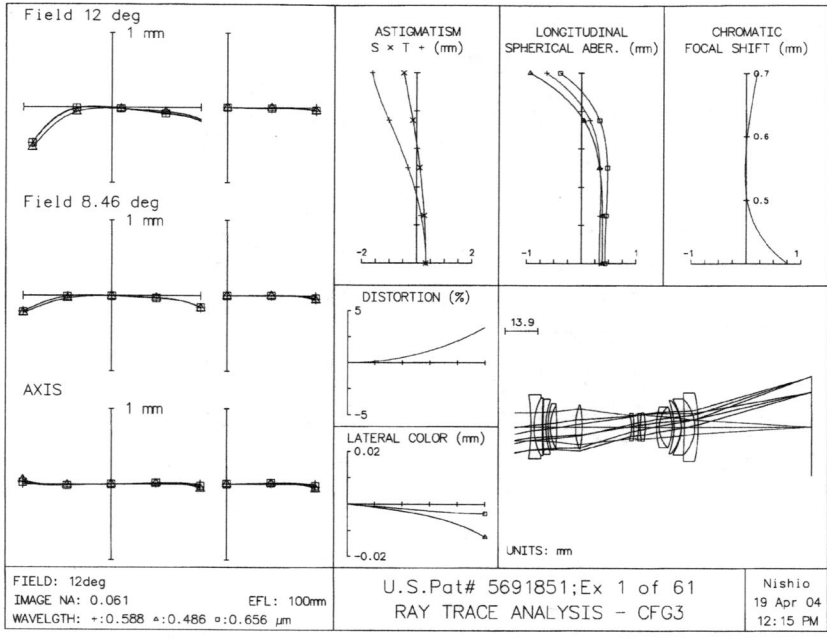
(c)

Figure 20.7 (*Continued*)

The second example (Table 20.2 and Fig. 20.7) has a wider, 3.6x zoom range, 14 elements, one aspheric surface, and five moving components. It works at speeds of $f/3.6$ to $f/8.2$.

20.3 A 20x Video Zoom Lens

The lens of Table 20.3 and Fig. 20.8 is an extreme example of a video zoom lens, from the OSLO demo library. It has a zoom ratio of 19.5x; the focal length varies from 12 mm to 234 mm. Its length is almost the same as the long focal length. The speed is $f/1.6$ at wide angle and drops to $f/2.8$ toward the long focal length end of the zoom. The image size is

TABLE 20.3 Prescription for the 20X Zoom Lens Shown in Figs. 20.8 (a,b,c,d, and e)

$R1 = +151.327$	$T1 = 2.80$	TIH23	755263	EFL = 12.002, 54.006, 234.033
$R2 = 86.000$	$T2 = 14.4$	PHM52	618633	f number = 1.60, 1.60, 2.80
$R3 = -704.000$	$T3 = 0.20$			GIH = 4.13, 4.25, 4.20
$R4 = +93.500$	$T4 = 8.00$	FPL51	497815	Field = ±19.0°, ±4.5°, ±1.03°
$R5 = +249.207$	$T5 = 2.00,$			Zoom Ratio = 19.5x
	62.15, 83.22			
$R6 = +275.000$	$T6 = 1.90$	PHM52	618633	
$R7 = +50.165$	$T7 = 4.80$			
$R8 = -5.4025e3$	$T8 = 6.20$	TIH53	847238	
$R9 = -86.000$	$T9 = 1.90$	LAM55	762401	
$R10 = -160.000$	$T10 = 1.90$			
$R11 = -64.907$	$T11 = 1.60$	LAL18	729547	
$R12 = +96.200$	$T12 = 62.32,$			
	4.84, 21.55			
$R13 = -69.120$	$T13 = 1.600$	BSM28	618498	
$R14 = +17.380$	$T14 = 2.900$	TIH6	805254	
$R15 = +35.200$	$T15 = 42.89,$			
	40.23, 2.44			
$R16 =$ plano	$T16 = 0.500$	BSL7	516641	
$R17 =$ plano	$T17 = 1.000$			
$R18 =$ stop	$T18 = 3.000$			
$R19 = +121.500$	$T19 = 3.500$	LAH65	804466	
$R20 = -121.500$	$T20 = 0.100$			
$R21 = +98.849$	$T21 = 3.500$	LAL14	697555	
$R22 = -2.1000e3$	$T22 = 0.100$			
$R23 = +54.344$	$T23 = 7.000$	FSL5	487702	
$R24 = -54.344$	$T24 = 1.400$	TIH6	805254	
$R25 = +167.164$	$T25 = 0.100$			
$R26 = +26.041$	$T26 = 4.700$	FPL51	497815	
$R27 = +57.091$	$T27 = 13.200$			
$R28 = +21.500$	$T28 = 2.400$	LAM7	749353	
$R29 = +13.900$	$T29 = 10.400$			
$R30 = +27.682$	$T30 = 3.700$	BSL7	516641	
$R31 = -84.000$	$T31 = 17.130$			
$R32 =$ plano	$T32 = 3.500$	BSL7	516641	
$R33 =$ plano				

540   Chapter Twenty

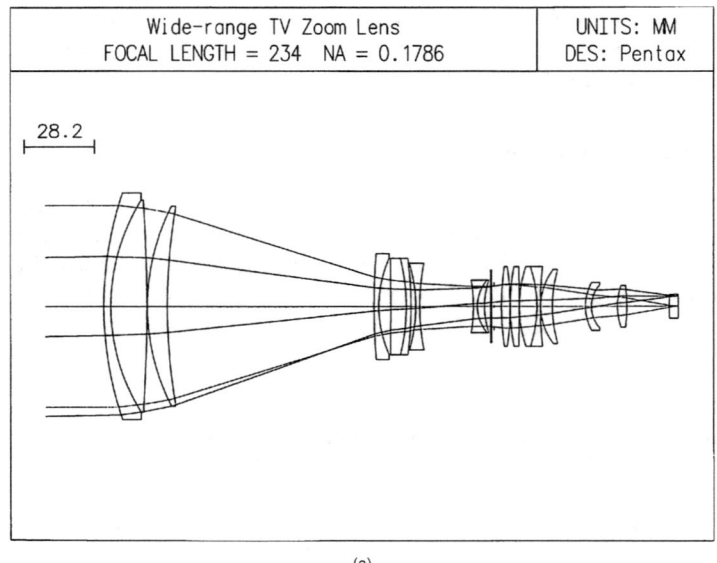

(a)

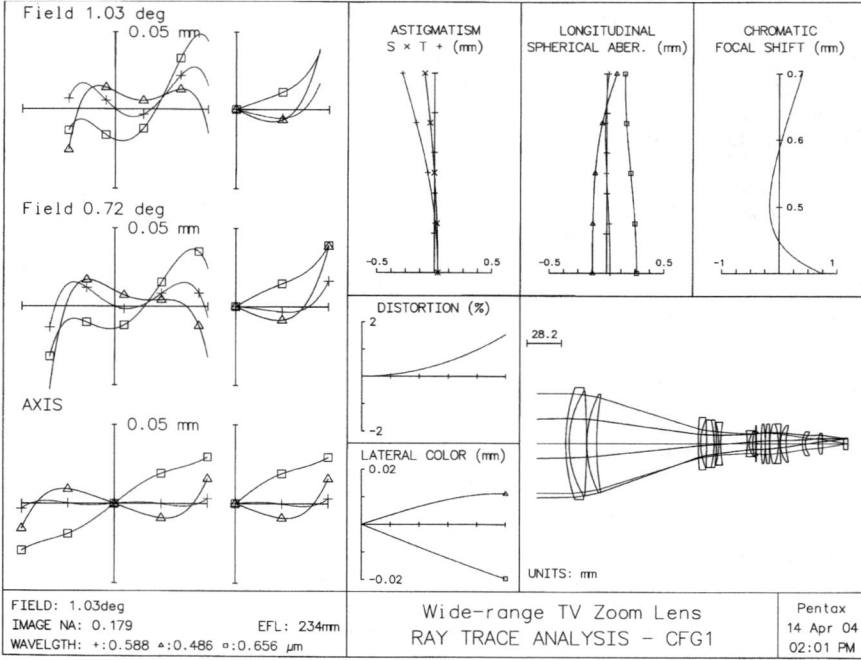

(b)

Figure 20.8 A 16-element (plus two plano elements) 19.5x zoom for a video camera, ±19° to ±1°, f/1.6 to f/2.8, Table 20.3.

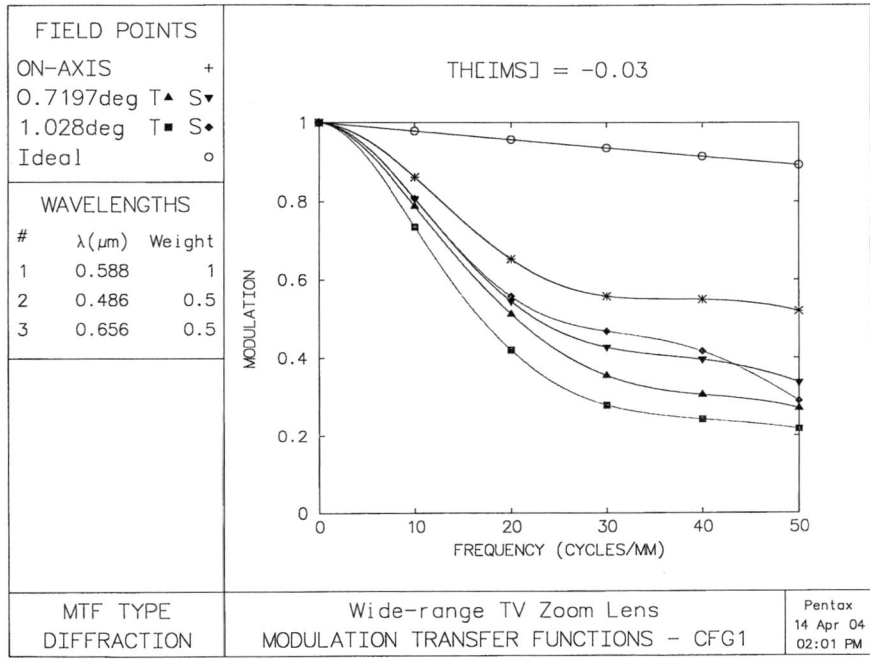

(c)

Figure 20.8 (*Continued*)

±4.2 mm, suitable for a video sensor. It uses 16 elements in four components, three of which move. It maintains a quite respectable image quality through the zoom.

20.4 A Zoom Scanner Lens

Figure 20.9 shows a zoom scanner lens. It uses a fixed input scan angle of ±14° and a constant HeNe laser beam diameter; as a result, its speed and the scan line length vary throughout the zoom. The external stop is located at the scan mirror. Here again we see an almost classic arrangement of + − +, with the negative component producing the zoom and the third (positive) component moving to maintain the focus. The first and last elements are fixed and the two inner components move. (The alert reader may have noticed that data transmittal errors have resulted in incorrect spacings in Fig. 20.9.)

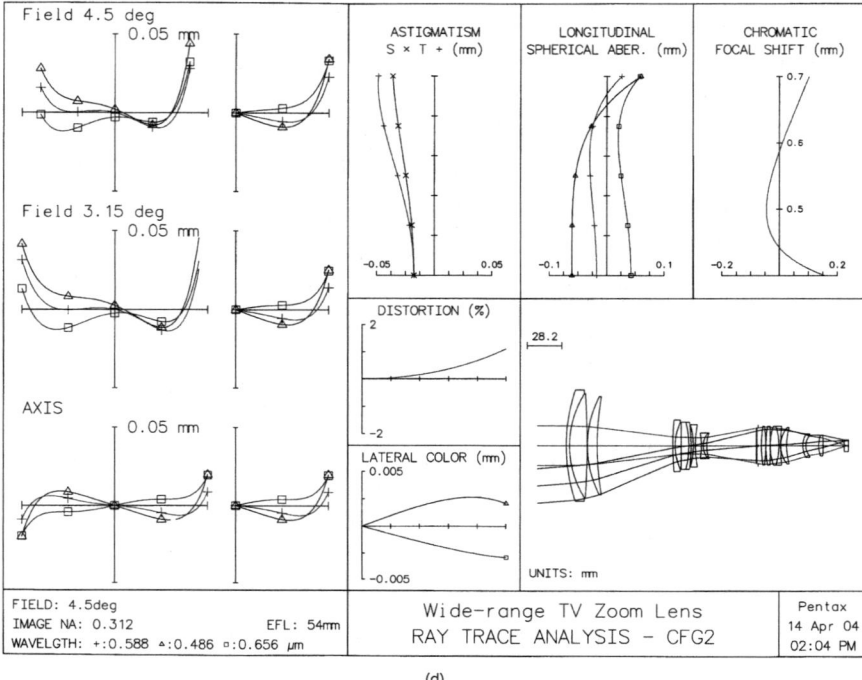

Figure 20.8 (*Continued*)

20.5 A Possible Zoom Lens Design Procedure

The multiconfiguration option in lens design programs is what makes the design of modern zoom lenses practical. This feature allows several configurations with most dimensions (i.e., radii, thickness, and material) the same in all configurations, but allows the spacings (or whatever) to be independently varied.

20.5.1 The modification route

There are two main ways to begin a zoom lens design. One can locate an existing design and stretch and/or improve it to meet the requirements at hand. To do this one will probably need to add more "horsepower" to the design. To simply improve, one can add elements and/or use higher index and/or abnormal dispersion glasses, and/or introduce aspheric surfaces. To stretch the design by increasing the zoom range, the elements will become stronger and/or the system will become longer, likely both.

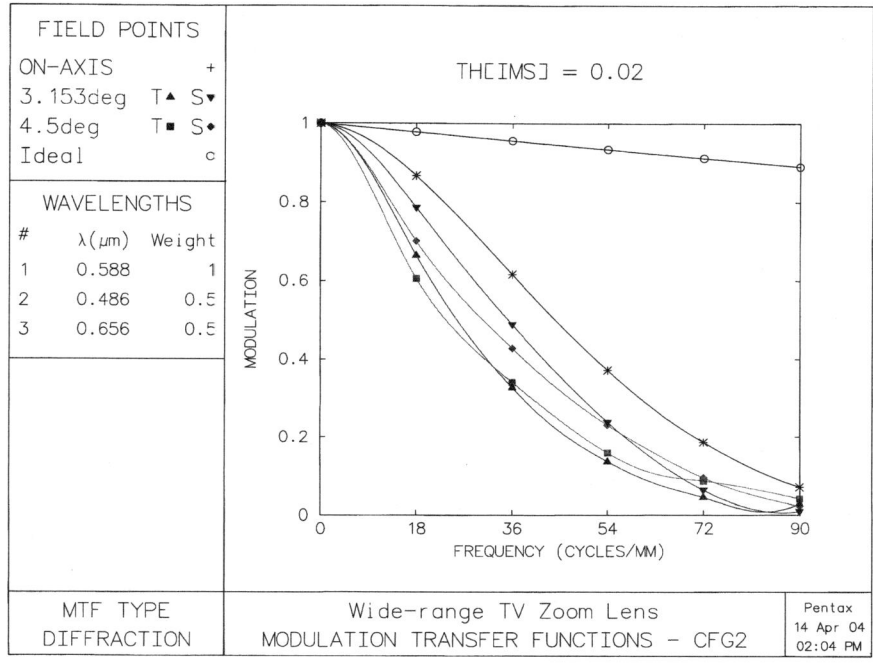

Figure 20.8 (*Continued*)

The zoom lens area is a minefield of patents, some of them hotly contested. If the starting design one locates is from a patent, the claims to be avoided are obvious (even if their language may not be). But many patents have broad and poorly written claims, and a complete patent search is apt to be both trying and costly.

20.5.2 Design from scratch

The alternative method is to start from scratch. (Note that this does not make the patent problem go away.) One simply defines the number, sign (positive or negative), and the order of the components, sets up the conditions (spatial limits), motion limits, and the like and lets the computer optimize to find the powers, spacings, and motions that are necessary to achieve the desired focal lengths and working distances. Again the multiconfiguration feature is used, keeping the same components for all configurations and allowing different spacings. A worthwhile addition to the merit function (which is basically a paraxial one with

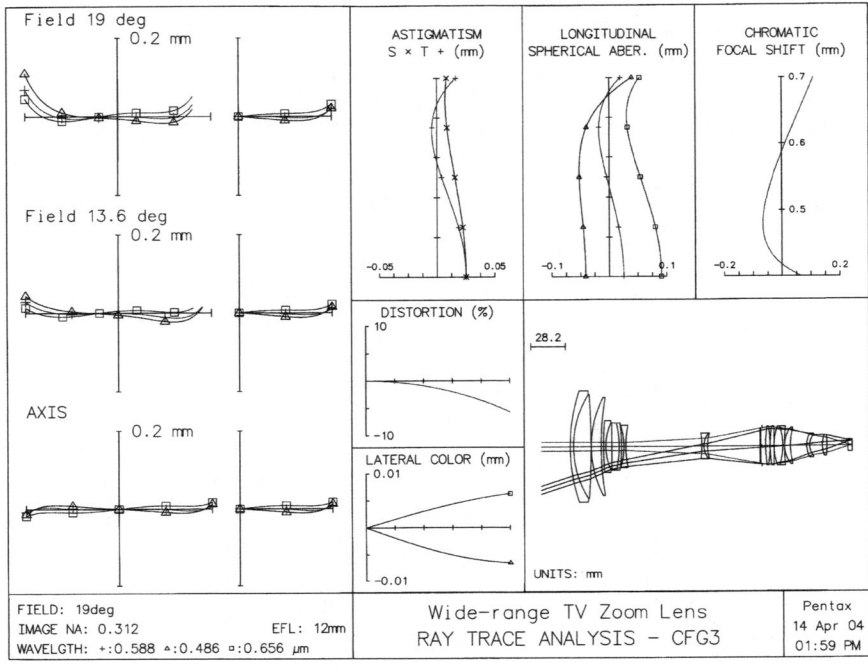

Figure 20.8 (*Continued*)

operands for the focal lengths and motion limits) is to minimize the sum of the absolute values of the powers, $\Sigma |\phi|$.

The odds are that the early runs will be rife with crashes where the components run into each other; it is not sufficient to carry out this process with only the long and short focal length setups. One or more intermediate focal lengths must be included, and generous spacing clearances must be established as boundary conditions. Each component can be represented by a plano-convex or plano-concave singlet to start; the curvatures and the spaces are the optimization variables. Add thicknesses to simulate reasonably realistic components. It is possible, or even likely, that this process will produce ridiculous elements; raytracing these presents no problem at this level because we are working with paraxial rays, and paraxial raytraces cannot fail. But at this point it is the better part of valor to split the too-strong components, even if only to keep the system drawings looking reasonable.

The next step is to aspherize the singlets and then optimize (and modify) until the design looks promising. Note that from the beginning

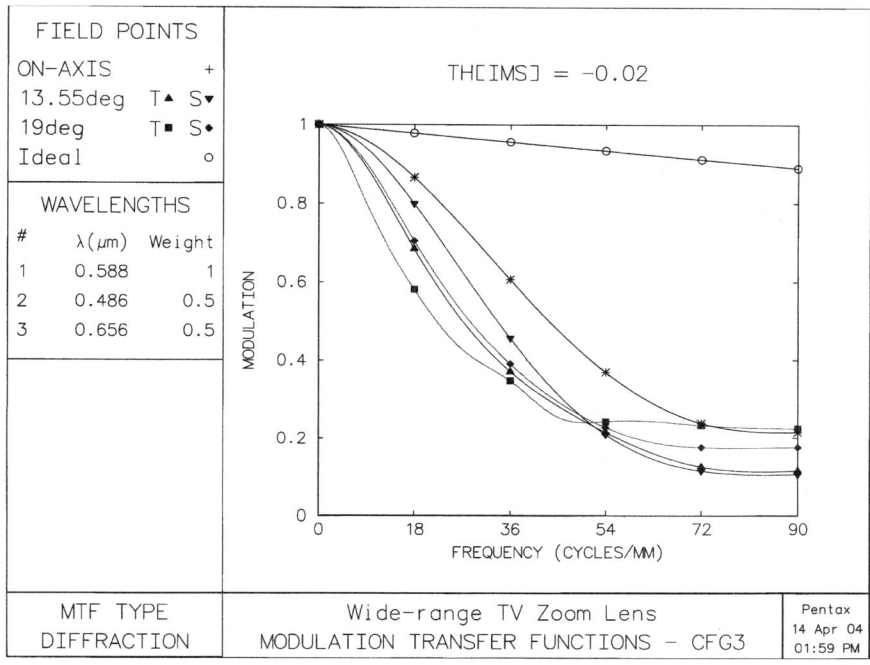

(g)

Figure 20.8 (*Continued*)

through this stage we use only the central wavelength, i.e., the system is not achromatized yet.

The final stage is to convert the aspheric singlets into achromatic doublets (or triplets), splitting strong elements as necessary. One secret of zoom lens design is that the aberrations of the separate components should never be allowed to become large. In a two-component zoom, for example, both components must be well corrected, and to keep the aberrations low we typically find four to seven elements in each component. In a three- (or more) component zoom, the zooming aberration change in one component may be balanced by a suitable change in another.

At least two motions are necessary to change the focal length and compensate for the focus shift. If there are more than two motions, the extra(s) can be used to maintain the image quality. Focusing is usually done by moving an element or component located before the first zooming component; however, with automatic focusing, this is now a less significant item. In some designs the focus motion(s) also behave as in a macro lens to maintain image quality at close focusing distances.

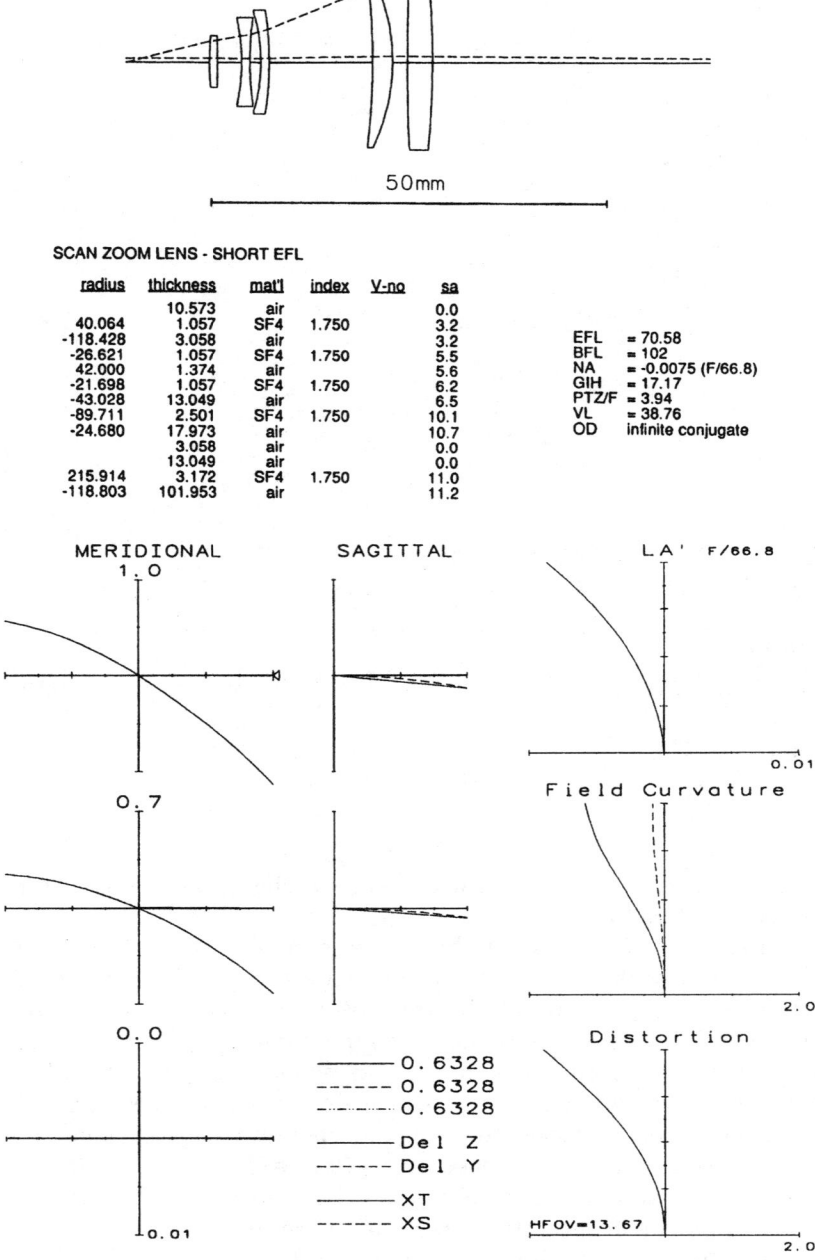

Figure 20.9 A zoom scanner lens (+ − +) $f/67$ to $f/134$, ±13.7°, with a zoom ratio of 2x. Note that the scan angle of the mirror (at the stop) is constant, and the width of the scan is zoomed.

Zoom Lenses 547

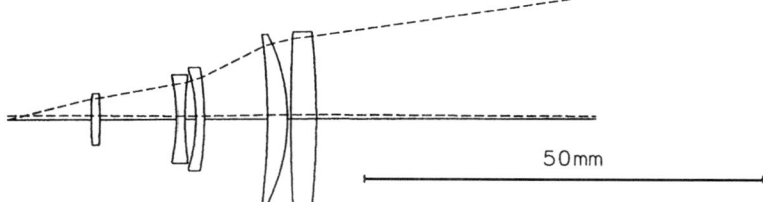

SCAN ZOOM LENS - MID EFL

radius	thickness	mat'l	index	V-no	sa
	10.573	air			0.0
40.064	1.057	SF4	1.750		3.2
-118.428	9.649	air			3.2
-26.621	1.057	SF4	1.750		5.5
42.000	1.374	air			5.6
-21.698	1.057	SF4	1.750		6.2
-43.028	7.900	air			6.5
-89.711	2.501	SF4	1.750		10.1
-24.680	17.973	air			10.7
	9.649	air			0.0
	7.900	air			0.0
215.914	3.172	SF4	1.750		11.0
-118.803	98.954	air			11.2

EFL = 100
BFL = 98.95
NA = -0.0053 (F/94.6)
GIH = 24.33
PTZ/F = 2.779
VL = 38.76
OD infinite conjugate

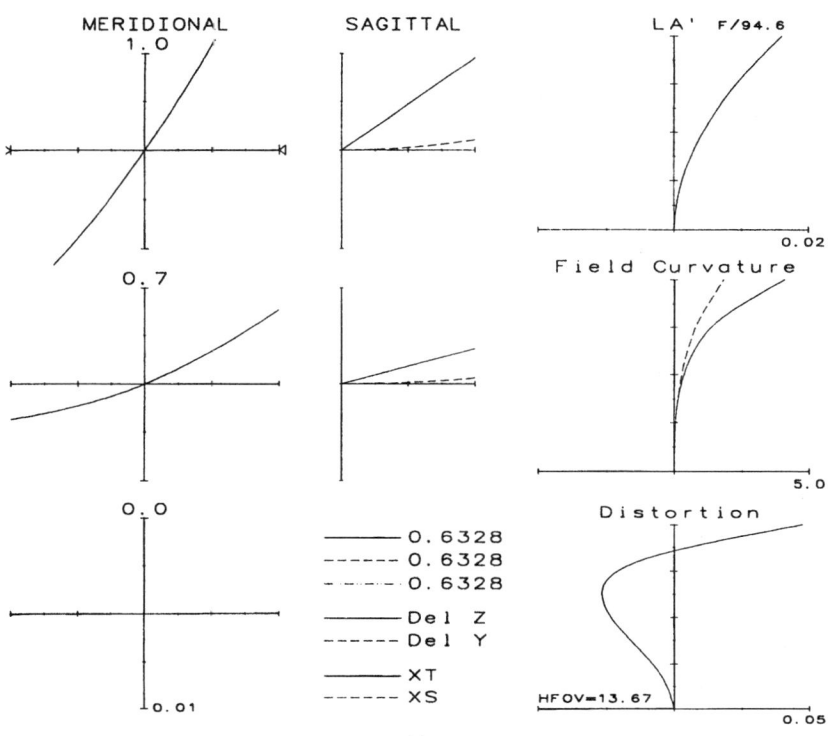

(b)

Figure 20.9 (*Continued*)

548   Chapter Twenty

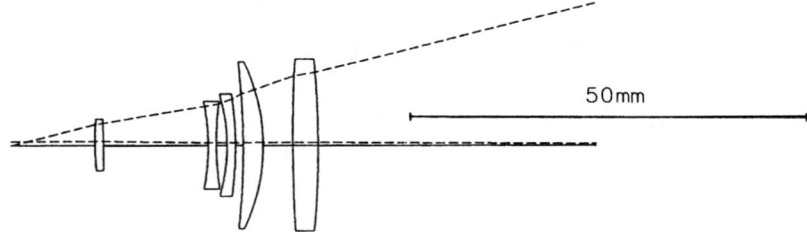

SCAN ZOOM LENS - LONG EFL

radius	thickness	mat'l	index	V-no	sa
	10.573	air			0.0
40.064	1.057	SF4	1.750		3.2
-118.428	13.278	air			3.2
-26.621	1.057	SF4	1.750		5.5
42.000	1.374	air			5.6
-21.698	1.057	SF4	1.750		6.2
-43.028	0.972	air			6.5
-89.711	2.501	SF4	1.750		10.1
-24.680	17.973	air			10.7
	13.278	air			0.0
	0.972	air			0.0
215.914	3.172	SF4	1.750		11.0
-118.803	101.910	air			11.2

EFL = 141.6
BFL = 101.9
NA = -0.0037 (F/133.9)
GIH = 34.44
PTZ/F = 1.963
VL = 38.76
OD infinite conjugate

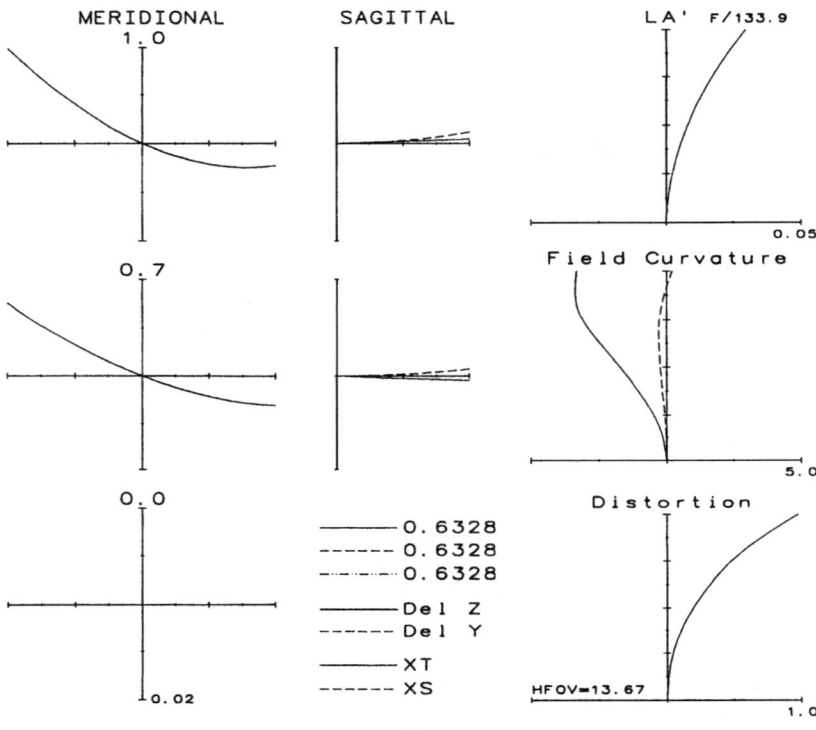

(c)

Figure 20.9   (*Continued*)

Interference between the glass components is usually the first problem that must be fixed when thin lenses are converted to thick ones. It is necessary to test the clearances for at least three (or more) configurations to be certain that there is no interference throughout the entire zoom. Distortion is a common problem; it is often balanced so that it's barrel at one end of the zoom and pincushion at the other. Linkage of alternate components is often useful in balancing the aberrations, and it also reduces the fabrication cost of the mechanism by eliminating a cam that would otherwise be needed to provide a separate motion.

In a zoom camera lens that must be focused for various object distances, the focusing cannot be conveniently accomplished by shifting the entire lens, because the amount of focus shift needed for a given object distance will vary with the square of the focal length. For this reason, zoom camera lenses are focused by a spacing change between the front elements so that the object position for the zoom kernel of the lens is a constant one regardless of the actual object position.

Figures 22.8 and 22.9 in Chap. 22 show a zooming lens for use as a laser collimator or focusing lens.

Chapter 21

Projection TV Lenses and Macro Lenses

21.1 Projection TV Lenses

The lenses used to project large-screen television must have a high speed (usually $f/1$ or faster) to project a bright enough image, and must also cover a relatively wide field to produce a large picture in a short throw. However, they do not need to be fully chromatically corrected, since the three color images are independently projected and each lens covers only a limited spectral band. The resolution requirements are relatively modest, in keeping with the low resolution of television, but, of course, a high *modulation transfer function* (MTF) is needed to maintain the image quality at this low resolution. Plastic elements are used for low cost and light weight, and (primarily) because they allow for the economical inclusion of aspheric surfaces, which are absolutely essential to the design. The resultant high tooling and start-up costs are readily amortized over the large production runs.

The basic concept behind most of these lenses begins with a positive element and a field flattener, plus some aspheric surfaces. The two sample designs shown here are fairly typical. Figure 21.1 is a basic three-element version with a weak, aspheric-surfaced leading element and an aspheric field flattener, both made of acrylic (which is the crown of the plastic materials). The primary function of the leading element is to act as a corrector for the rest of the system. The strong positive element is made of glass and has spherical surfaces, since a large aspheric glass element would be prohibitively expensive. A problem encountered with plastic optics is that they shift focus with temperature because of their high thermal expansion coefficient and their large negative change

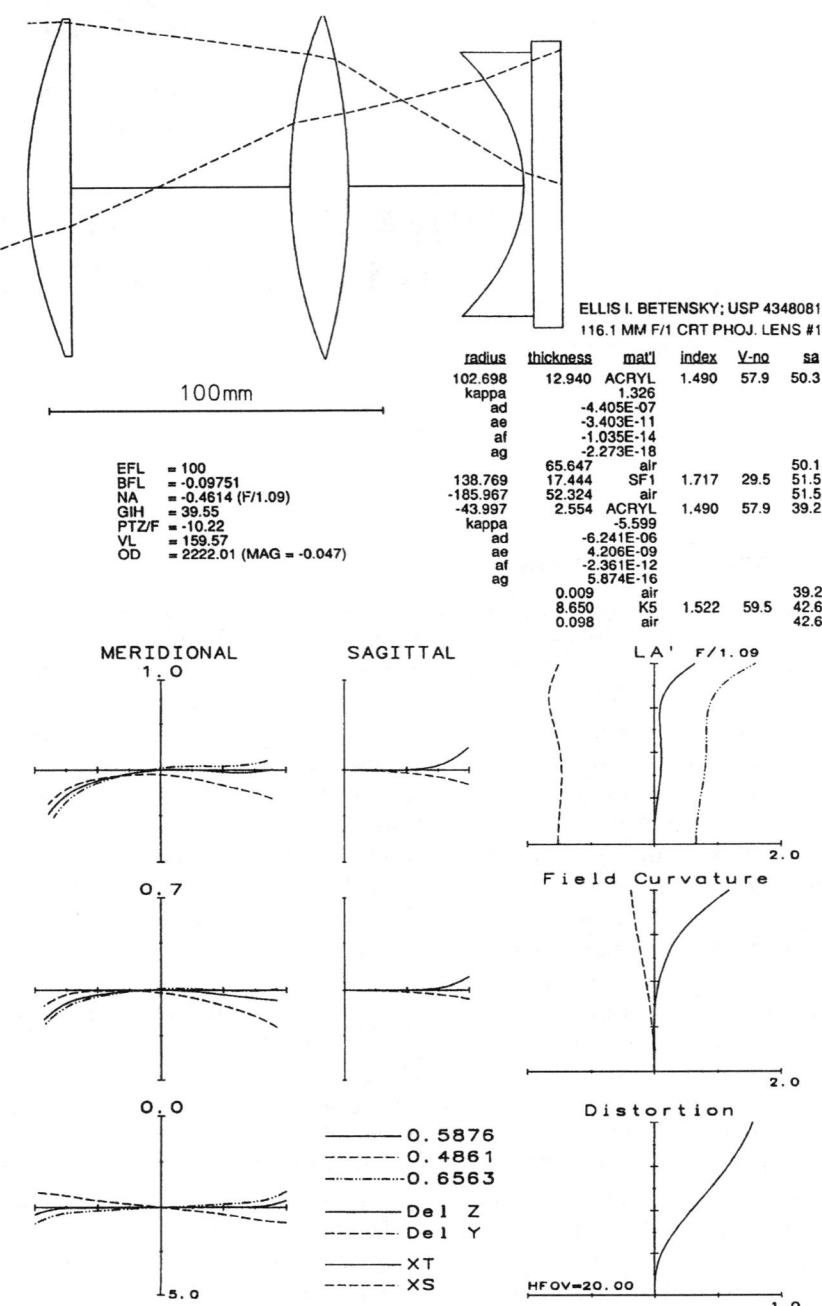

Figure 21.1 $f/1.1$ ±22° three-element projection TV lens with a spherical-surfaced glass element (for thermal stability) and two aspheric acrylic elements.

of index with temperature (dN/dT), both of which work to increase the focal length as the temperature rises. Thus, putting most of the lens power into a much more stable glass element, as in this design, tends to correct this problem. Figure 21.2 uses four acrylic elements, each with one aspheric surface.

Note that both Figs. 21.1 and 21.2 show severe undercorrected axial chromatic aberration but very little lateral chromatic. The axial chromatic is tolerable because the projection TV uses three lenses and three cathode ray tubes (CRTs), one for each color, and the lenses can be individually focused. The lateral color must be corrected because the three projected images must not show any color fringing (which would result if there were chromatic differences of image size).

21.2 Macro Lenses

The aberration correction of a lens changes with object distance. It is impossible to correct all the aberrations of a lens for more than one object distance. Some design forms are more sensitive in this regard than others. High-speed lenses are usually more subject to this effect, and, for a given object distance, the effect is obviously more pronounced for longer focal-length lenses. Pupil spherical is a big factor in producing the changes to the aberrations that result from conjugate changes. The two sample lenses in this section are the double-Gauss or Biotar form, which is a very high-quality lens, but one that is relatively sensitive to changes in conjugate distances. The designs have been stabilized as discussed in Sec. 13.2 by changing airspaces as the lens is focused on nearby objects.

There are several ways of approaching the macro problem. One can aim for mechanical simplicity by "anchoring" a part of a system and shifting the rest to change the focus. Figures 21.3 and 21.4 show an example of this approach. The front five elements are fixed with respect to the film plane and the last element is moved away from the film to focus at a distance of 40 focal lengths. Of course the last airspace must be large enough to accomodate this motion. Another approach to using this same lens as a macro is to refocus by shifting the front three elements.

While considering these examples, bear in mind that the important thing in a macro lens is the stability of the aberrations as the focus is changed. Once this has been achieved, presumably the aberrations can be brought to the desired state of correction by moderate adjustments to the design. The retrofocus lens of Fig. 14.1 could be considered a macro lens if it were mounted so as to reduce the second and third airspaces when the long conjugate is shortened. The telephoto lens of Figs. 13.8 and 13.9 is an example of a lens that is focused by moving its rear

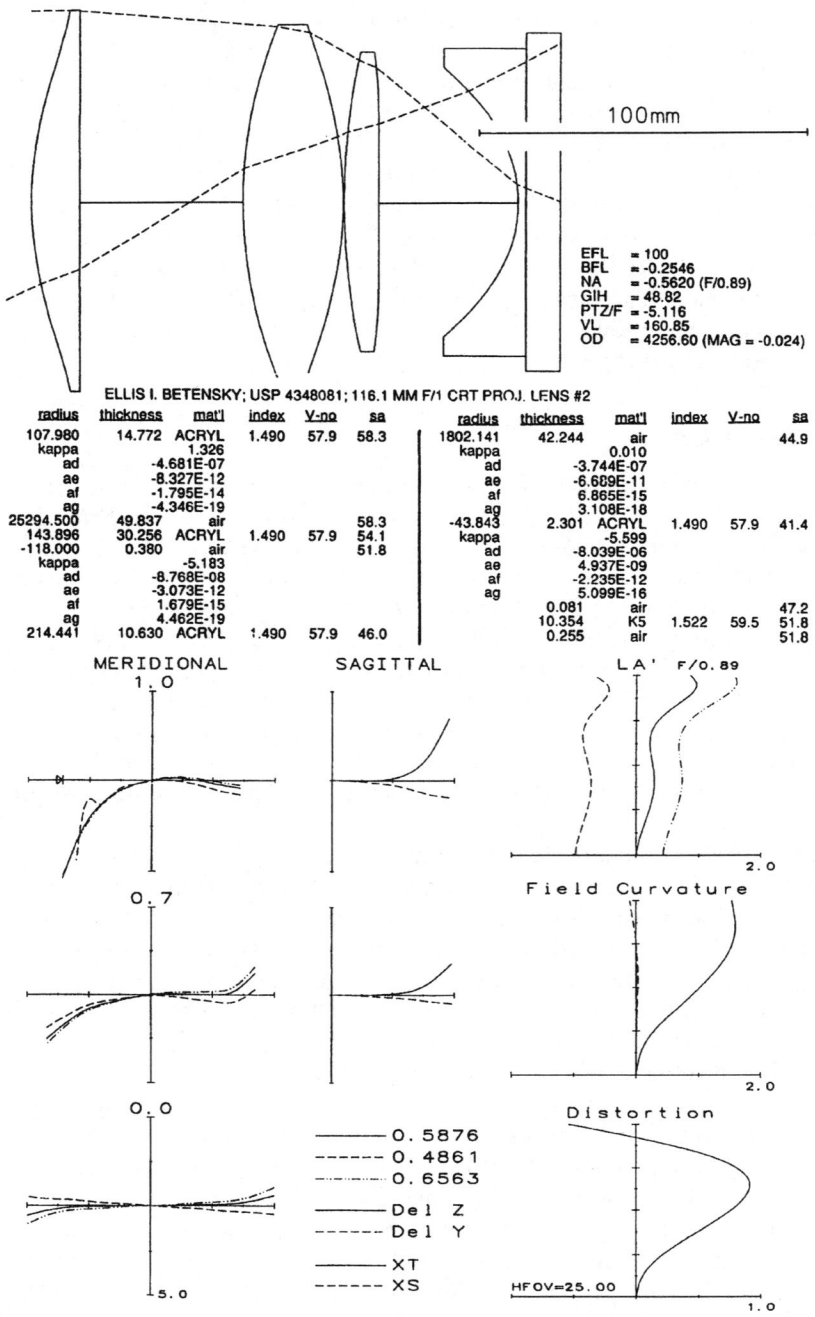

Figure 21.2 $f/0.9 \pm 26°$ four-element, all-acrylic, all-aspheric TV projection lens.

Projection TV Lenses and Macro Lenses 555

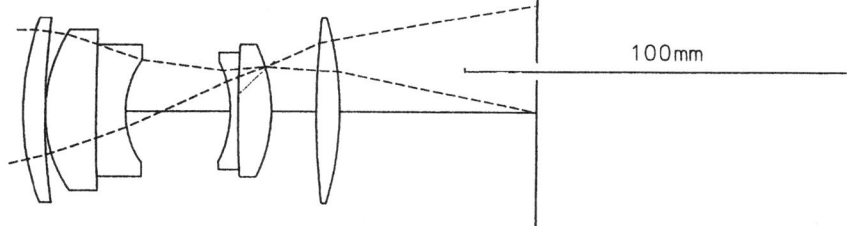

IKUO MORI; USP 4390252; F/2.5 30 DEG. TELEPHOTO LENS #1 INF.

radius	thickness	mat'l	index	V-no	sa
65.731	5.500	TAC6	1.755	52.3	23.1
166.522	0.130	air			22.2
37.250	12.500	LACL7	1.670	57.3	20.1
575.000	7.250	F3	1.613	37.0	16.6
23.130	9.000	air			12.9
	16.750	air			0.0
-28.630	1.880	LAF11	1.757	31.7	12.5
312.500	8.380	LAF20	1.744	44.9	14.7
-38.880	11.000	air			16.6
218.151	5.630	TAF2	1.794	45.4	23.0
-84.847	48.274	air			23.2

EFL = 99.99
BFL = 48.27
NA = -0.2003 (F/2.5)
GIH = 26.79 (HFOV=15.00)
PTZ/F = -7.04
VL = 78.02
OD infinite conjugate

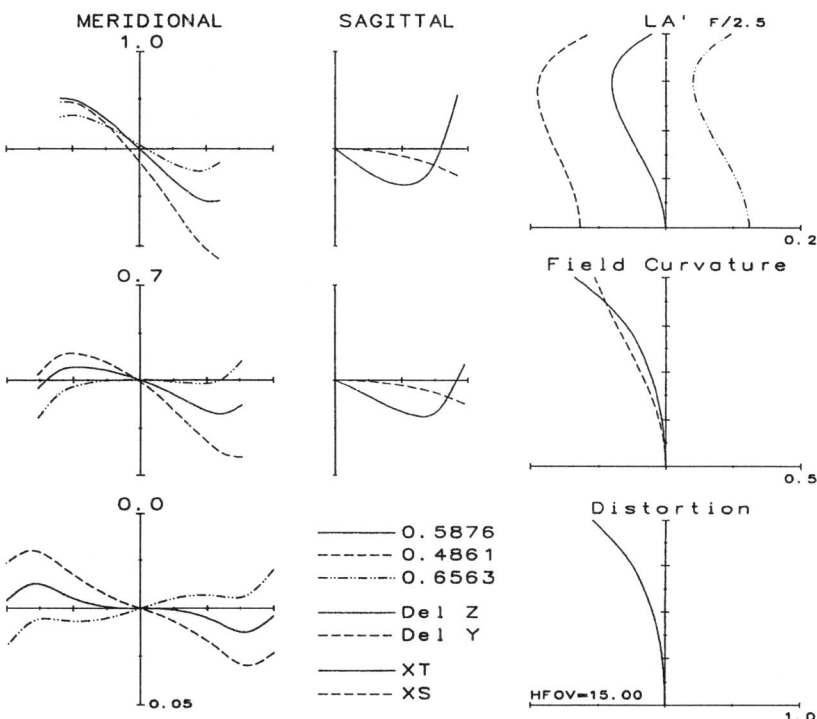

Figure 21.3 $f/2.5$ ±15° macro lens at infinity focus.

Chapter Twenty-One

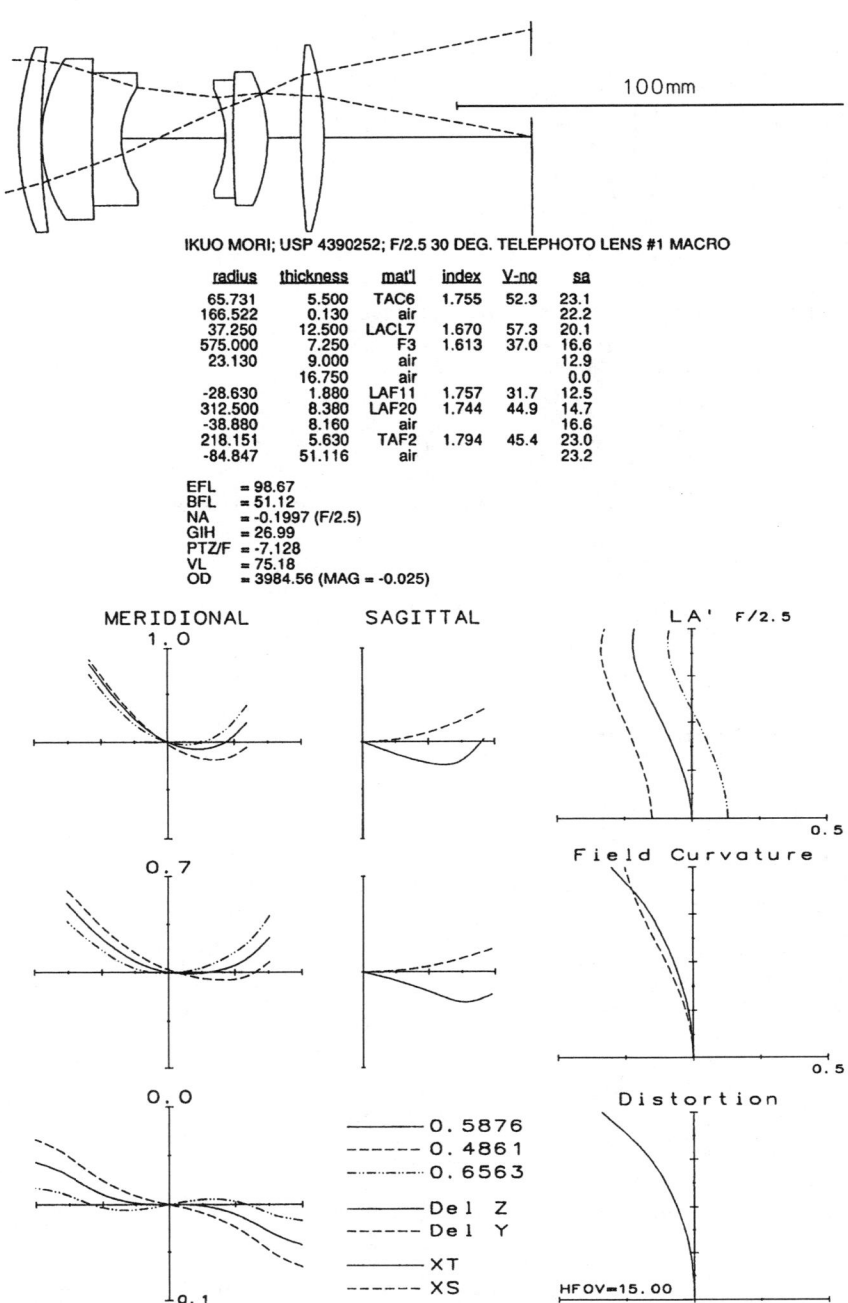

IKUO MORI; USP 4390252; F/2.5 30 DEG. TELEPHOTO LENS #1 MACRO

radius	thickness	mat'l	index	V-no	sa
65.731	5.500	TAC6	1.755	52.3	23.1
166.522	0.130	air			22.2
37.250	12.500	LACL7	1.670	57.3	20.1
575.000	7.250	F3	1.613	37.0	16.6
23.130	9.000	air			12.9
	16.750	air			0.0
-28.630	1.880	LAF11	1.757	31.7	12.5
312.500	8.380	LAF20	1.744	44.9	14.7
-38.880	8.160	air			16.6
218.151	5.630	TAF2	1.794	45.4	23.0
-84.847	51.116	air			23.2

EFL = 98.67
BFL = 51.12
NA = -0.1997 (F/2.5)
GIH = 26.99
PTZ/F = -7.128
VL = 75.18
OD = 3984.56 (MAG = -0.025)

Figure 21.4 f/2.5 ±15° macro lens at 1:40 magnification, focused by moving the front three components toward the image.

Projection TV Lenses and Macro Lenses

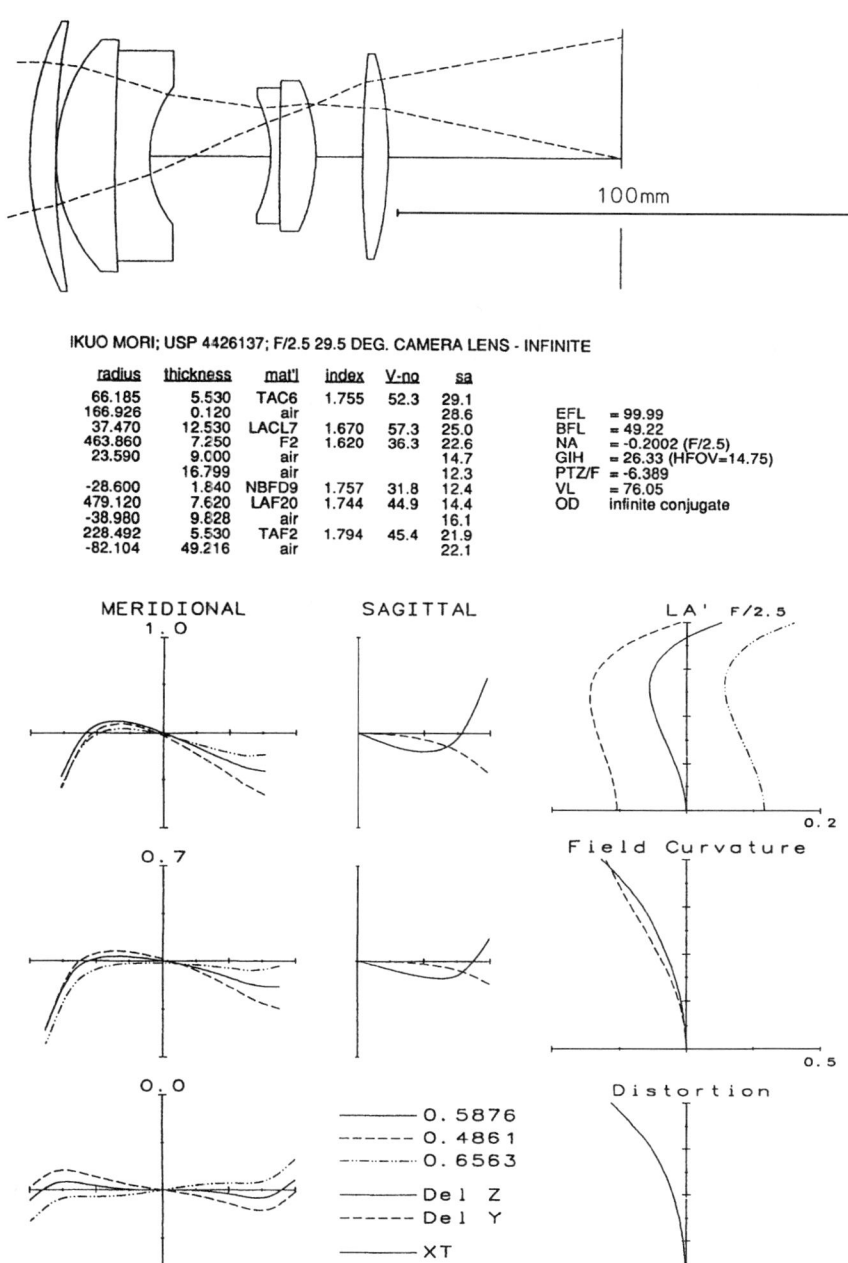

IKUO MORI; USP 4426137; F/2.5 29.5 DEG. CAMERA LENS - INFINITE

radius	thickness	mat'l	index	V-no	sa
66.185	5.530	TAC6	1.755	52.3	29.1
166.926	0.120	air			28.6
37.470	12.530	LACL7	1.670	57.3	25.0
463.860	7.250	F2	1.620	36.3	22.6
23.590	9.000	air			14.7
	16.799	air			12.3
-28.600	1.840	NBFD9	1.757	31.8	12.4
479.120	7.620	LAF20	1.744	44.9	14.4
-38.980	9.828	air			16.1
228.492	5.530	TAF2	1.794	45.4	21.9
-82.104	49.216	air			22.1

EFL = 99.99
BFL = 49.22
NA = -0.2002 (F/2.5)
GIH = 26.33 (HFOV=14.75)
PTZ/F = -6.389
VL = 76.05
OD infinite conjugate

Figure 21.5 $f/2.5 \pm 15°$ macro lens at infinity focus.

558 Chapter Twenty-One

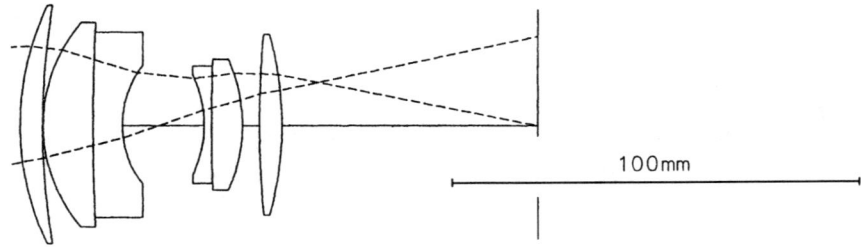

IKUO MORI; USP 4426137; F/2.5 29.5 DEG. CAMERA LENS - MACRO

radius	thickness	mat'l	index	V-no	sa
66.185	5.530	TAC6	1.755	52.3	29.1
166.926	0.120	air			28.6
37.470	12.530	LACL7	1.670	57.3	25.0
463.860	7.250	F2	1.620	36.3	22.6
23.590	9.000	air			14.7
	11.176	air			12.3
-28.600	1.840	NBFD9	1.757	31.8	12.4
479.120	7.620	LAF20	1.744	44.9	14.4
-38.980	4.205	air			16.1
228.492	5.530	TAF2	1.794	45.4	21.9
-82.104	62.819	air			22.1

EFL = 93.94
BFL = 62.82
NA = -0.1913 (F/2.6)
GIH = 21.64
PTZ/F = -6.78
VL = 64.80
OD = 872.55 (MAG = -0.114)

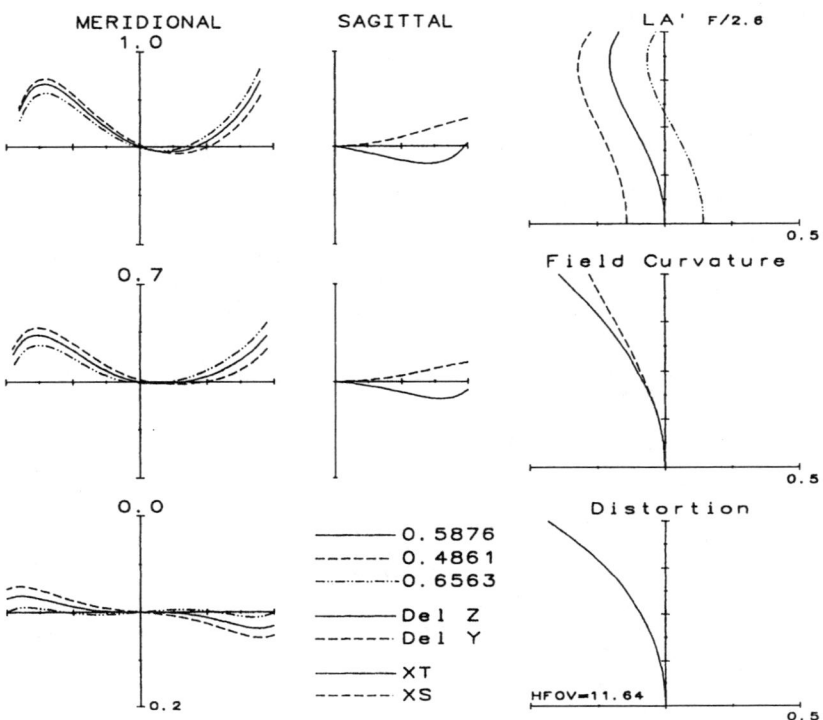

Figure 21.6 $f/2.6$ ±13° macro lens at 1:9 magnification, focused by shifting elements (1, 2, 3), (4, 5), and 6 independently.

doublet (i.e., internal focusing). In Fig. 13.7 the lens may be focused by moving the central doublet.

In Figs. 21.5 and 21.6 three sections of the lens (the first three elements, the rear doublet, and the rear element) are independently moved to maintain the aberration correction and to focus the lens to an object distance of only nine focal lengths. This sort of complexity can raise the costs of the mount.

Figure 20.5 in Chap. 20 shows a zoom lens with macro capability.

Chapter 22

Scanner/*f-θ*, Laser Disk, and Collimator Lenses

22.1 Monochromatic Systems

In a system that is truly monochromatic, the designer is no longer constrained by the need to achromatize the lens system. Thus high-index flint glasses can be used in positive elements and low-index crown glasses can be used in negative elements. This is obviously beneficial as regards the Petzval curvature and obviates any need to use the expensive lanthanum glasses for the high-index elements. The resulting lens is, of course, a hyperchromat and is suitable only for use with very monochromatic light sources.

22.2 Scanner Lenses

The scanner lens operates with an oscillating mirror that scans the point image across the field. To minimize the size of the scanning mirror, the pupil of the system is located at the mirror. An ordinary distortion-free lens has an image height (distance from the axis) that follows the rule: $h = f \tan \theta$. When the image is scanned across the field by a mirror rotating with a constant angular velocity, its linear velocity changes; the exposure produced will vary with the velocity. In order to achieve a uniform exposure across the field, distortion is deliberately introduced so that the image position relationship becomes $h = f\theta$. The fractional distortion works out to be equal to $(\theta/\tan \theta) - 1.0$. Note that all of the designs in this section have a negative distortion of this type.

The simplest scanner lens is a single meniscus lens, similar to the meniscus landscape lens. A two-element lens with the negative element facing the scan mirror is the next step of complexity. Figure 22.1 is a

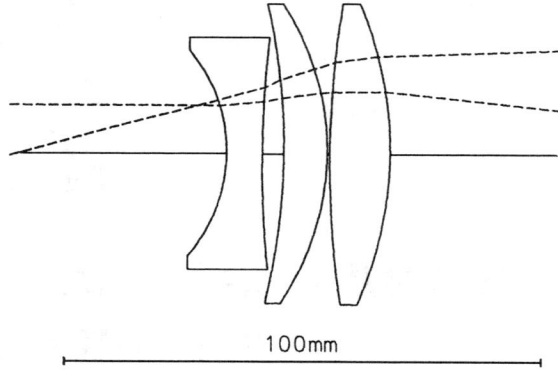

HOPKINS LASER DIODE SCAN LENS F/5, EFL=55, H'=14.31, FOV=30

radius	thickness	mat'l	index	V-no	sa
	43.550	air			10.0
-33.679	7.349	BK7	1.511		21.7
227.078	4.536	air			24.6
-137.219	9.073	SF11	1.765		27.6
-57.486	0.544	air			31.7
207.716	12.702	SF11	1.765		31.9
-80.622	129.740	air			33.8

EFL = 100
BFL = 129.7
NA = -0.1002 (F/5.0)
GIH = 26.79 (HFOV=15.00)
PTZ/F = -31.85
VL = 77.76
OD = infinite conjugate

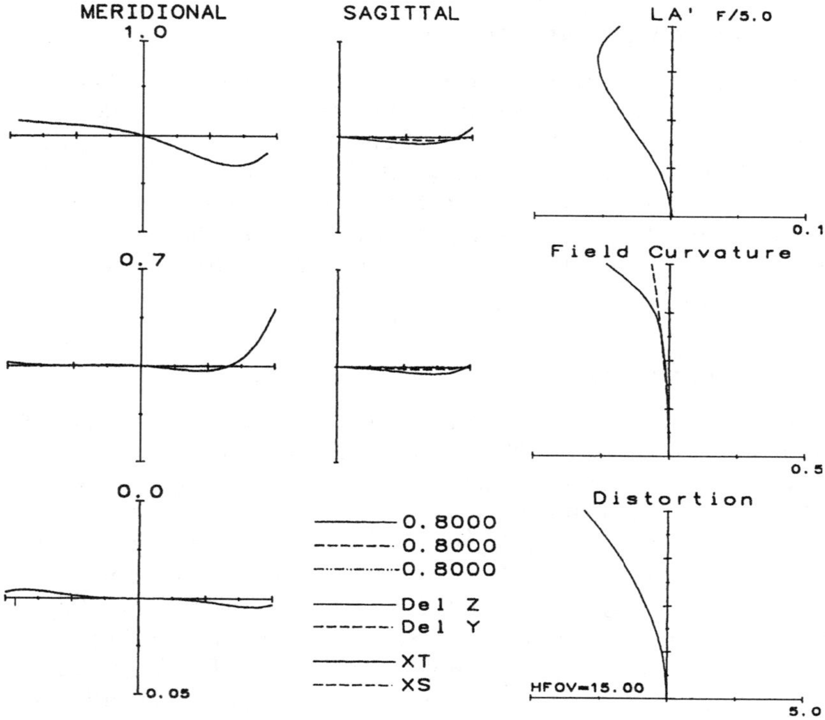

Figure 22.1 $f/5.0$ ±15° a three-element scanner lens. Note the use of a low-index crown ($n = 1.511$) for the negative element and the use of a high-index flint ($n = 1.765$) in the positive elements. This helps to reduce the Petzval field curvature. The lens is of course a hyperchromat.

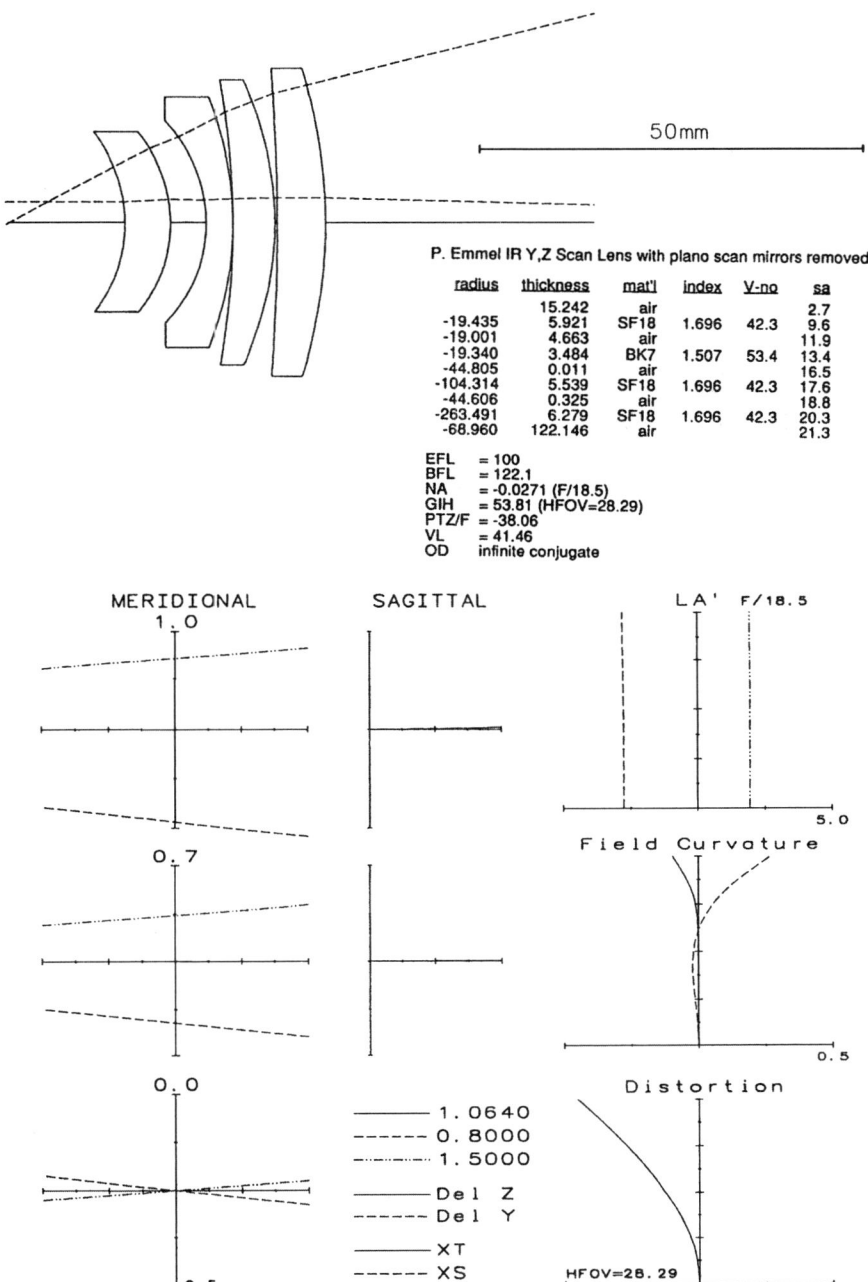

Figure 22.2 $f/18 \pm 28°$ a four-element scanner lens. Note the glass choice.

564 Chapter Twenty-Two

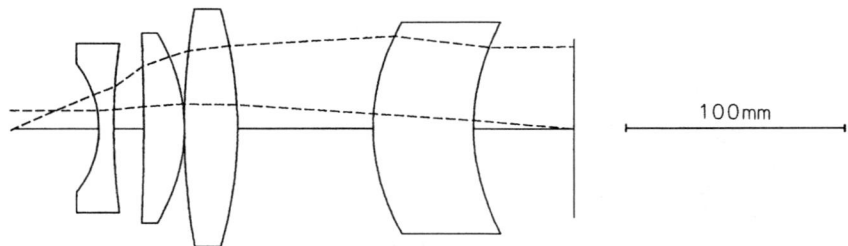

HOPKINS DIODE; F/6.1, Y'=12.5, FOV=45.4, TELECENTRIC IMAGE SPACE

radius	thickness	mat'l	index	V-no	sa
	39.301	air			8.2
-48.237	6.729	BK7	1.511		29.9
283.340	13.710	air			39.1
-1155.949	18.411	SF6	1.784		40.5
-84.616	0.003	air			43.9
322.746	24.553	SF6	1.784		54.5
-192.547	61.960	air			54.5
96.625	45.631	BK7	1.511		46.0
100.698	45.660	air			48.6

EFL = 100
BFL = 45.66
NA = -0.0818 (F/6.1)
GIH = 41.89 (HFOV=22.73)
PTZ/F = -25.06
VL = 210.30
OD infinite conjugate

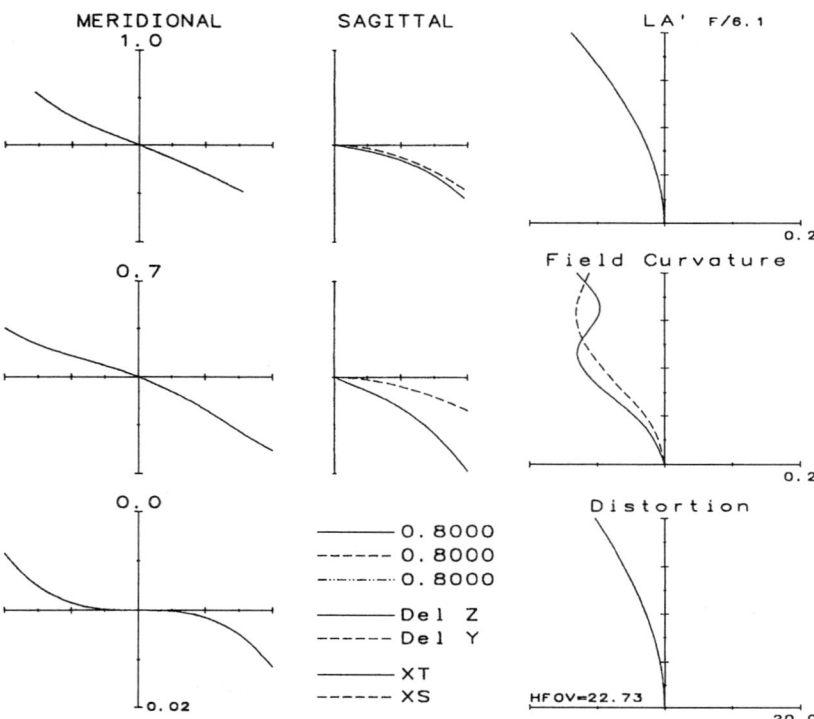

Figure 22.3 $f/6.0$ ±23° a telecentric scanner lens.

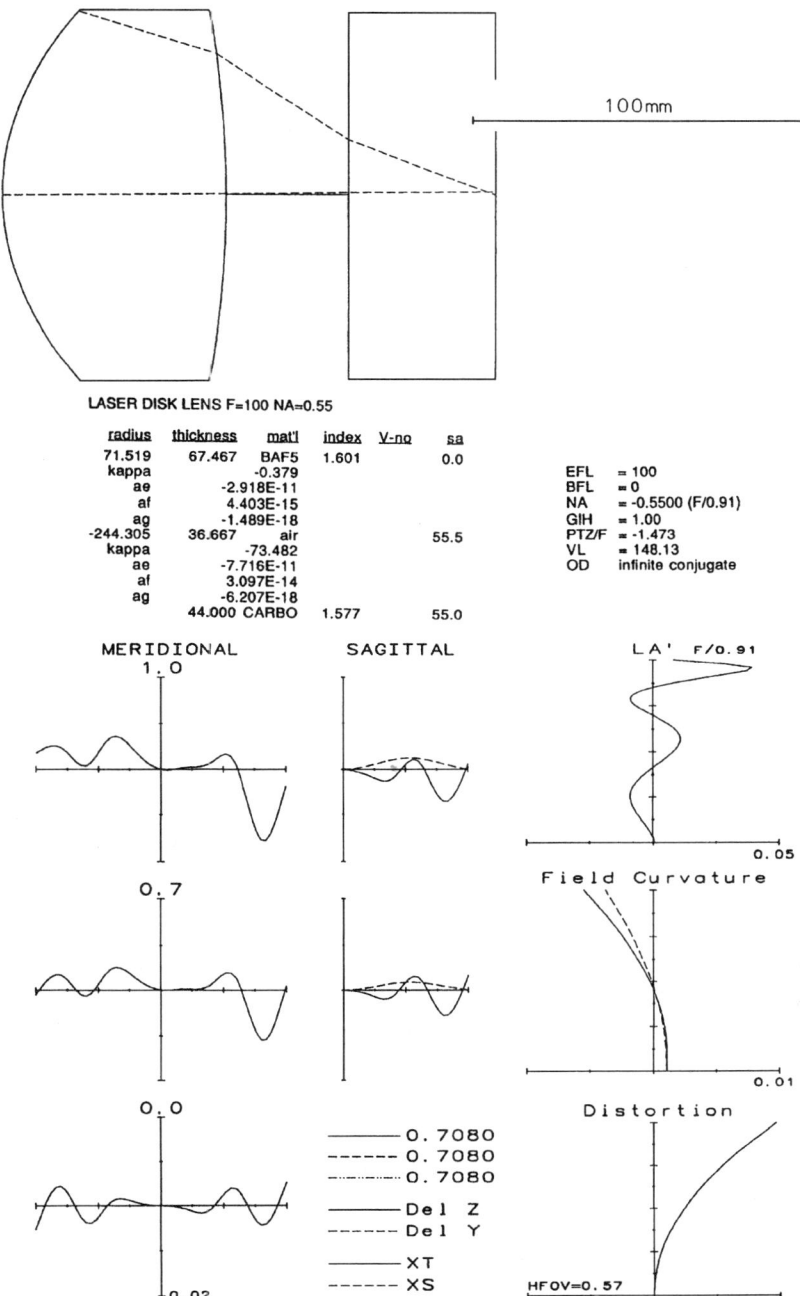

Figure 22.4 $f/0.9$ ±0.6° a double-aspheric laser disk lens. Note that the large thickness allows some correction of astigmatism.

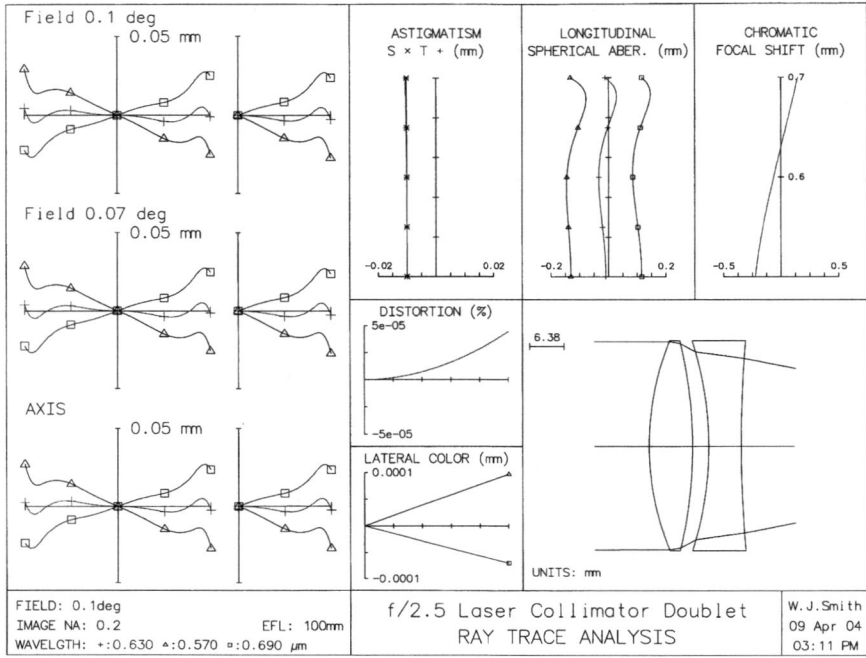

Figure 22.5 $f/2.5$ ±0.1° a doublet laser focusing (or collimating) lens, corrected for a band of wavelengths, i.e., with very low spherochromatism, achieved by the airspace and glass choice.

simple three-element lens, in a − + + configuration, with the negative element made of low-index BK7 glass (517-642) and the positive elements made of SF11 (785258), an inexpensive, stable, high-index flint. This very logical basic configuration (− + +) is not only nearly ubiquitous but quite versatile; it has even been used for long-wavelength (10.6-µm) scanners (with suitable materials). Figure 22.2 covers a wider angle at a slower speed. It uses four elements, adding a weakly positive meniscus element between the stop and the lens. The positive elements are SF18 (722293) and the negative element is BK7 (517642).

TABLE 22.1 Prescription for a Laser Focusing (or Collimator) Doublet Which Maintains its Spherical Aberration Correction Over a Range of Wavelengths

$R1$ = Stop	$T1$ = −3.0			EFL = 100.0000	
$R2$ = +55.381	$T2$ = 8.0	LaSFN30	(803464)	BFL = 82.547	
$R3$ = −89.402	$T3$ = 2.978			NA = 0.2 ($f/2.5$)	
$R4$ = −68.677	$T4$ = 6.0	SF6	(805254)	GIH = 0.1745 (0.1°)	
$R4$ = +243.105				PTZ/F = −2.128	

The last scanner lens (Fig. 22.3) is a simple example of a telecentric system in which the exit pupil is located at infinity, so that the principal ray of the imaging cone is always normal to the focal plane as the image is scanned across the field. As can be expected, telecentricity not only tends to require a complex design, but also requires that the lens aperture be larger than the image field.

Note that Fig. 20.6 shows a scanner lens with zooming capability.

22.3 Laser Disk, Focusing, and Collimator Lenses

Figure 22.4 shows a typical molded glass laser disk lens. Both surfaces are conics with general aspheric deformations. The lens thickness is important to the design in that it allows for some correction of the astigmatism. At the speed of $f/0.9$ of this example, it is of course absolutely necessary that the plastic cladding on the disk be included in the design. The actual focal length of this type of lens is to the order of 5 mm, at which focal length the design wavefront aberration is a tiny fraction of a wavelength. This type of lens is often molded in plastic as well as glass, although there is a problem with the thermal focus shift when the lens is plastic.

Figure 22.5 is an airspaced doublet, whose correction is based on the same principles as outlined in Chap. 6 except that, as a monochromatic system, both elements can be made from high-index flint glasses LASFN 30, 803464 and SF6, 805-254. Note that if the configuration is chosen so that the spherochromatic aberration is well-corrected, then the lens can be used for several different wavelengths, although it will require refocusing for each wavelength. The prescription is given in Table 22.1.

Figures 22.6 and 22.7 are examples of spherical-surfaced laser disk lenses. Note that, in each case, the final positive element is spaced well away from the aperture stop to introduce some overcorrected astigmatism. In Fig. 22.7, the designer has completely flattened the tangential field and almost completely eliminated the spherochromatism, whereas in Fig. 22.6, because the automatic focus control of the tracking system could tolerate a curved field, the elimination of astigmatism was the higher priority.

Figures 22.8 and 22.9 show a 2x zooming version of an $f/1.67$ to $f/3.3$ laser collimating (or focusing) lens at a field of $\pm 0.9°$, designed for use at 442 nm. This design could be improved by substituting a higher-index glass for the BK7, but one would want to be certain that the substituted glass had a high transmission at the wavelength of interest.

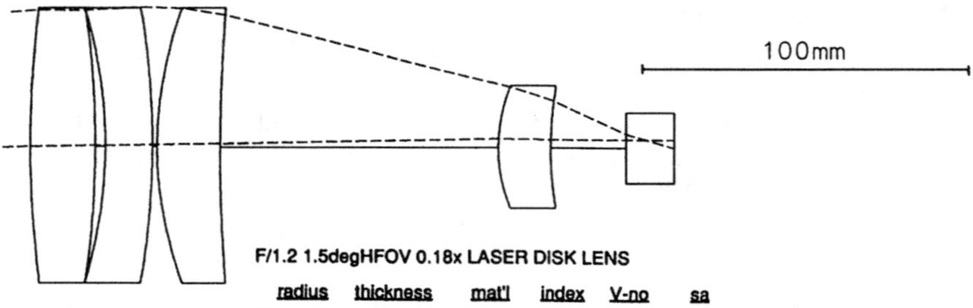

568 Chapter Twenty-Two

F/1.2 1.5degHFOV 0.18x LASER DISK LENS

radius	thickness	mat'l	index	V-no	sa
362.650	20.160	SF11	1.764		42.8
-281.130	3.134	air			42.8
-142.820	14.400	SF11	1.764		42.8
-247.400	1.440	air			42.8
125.430	19.080	SF11	1.764		42.8
569.900	85.250	air			40.7
50.978	16.130	SF11	1.764		19.1
112.790	22.920	air			15.1
	14.630	ACRY	1.486		11.0
	0.003	air			11.0

EFL = 100
BFL = -0.002722
NA = -0.4216 (F/1.18)
GIH = 2.51
PTZ/F = -1.134
VL = 197.14
OD = 551.10 (MAG = -0.179)

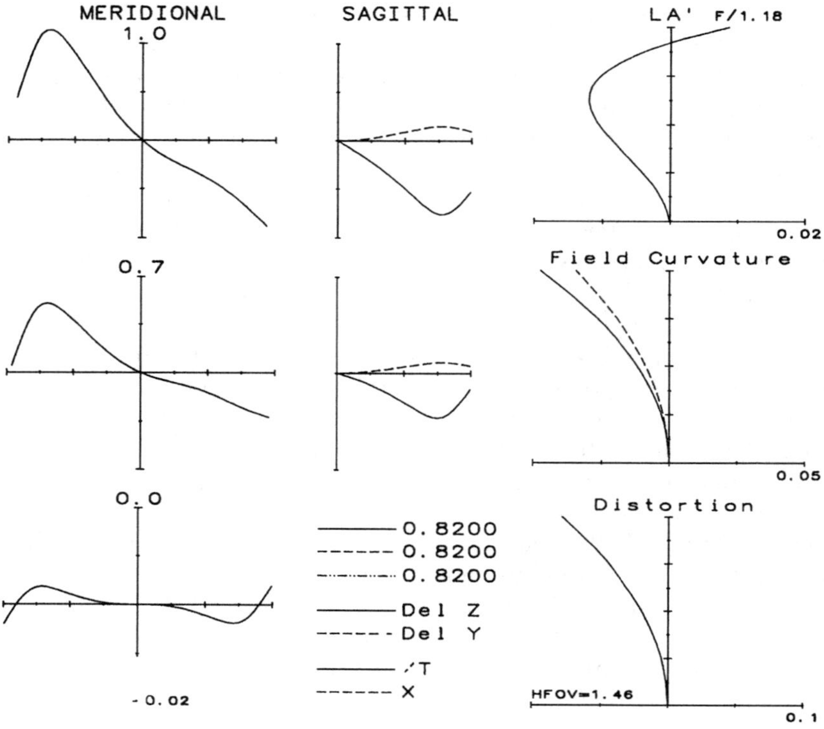

Figure 22.6 $f/1.2$ ±1.4° a four-element laser disk lens with low astigmatism.

Scanner/f-θ, Laser Disk, and Collimator Lenses   569

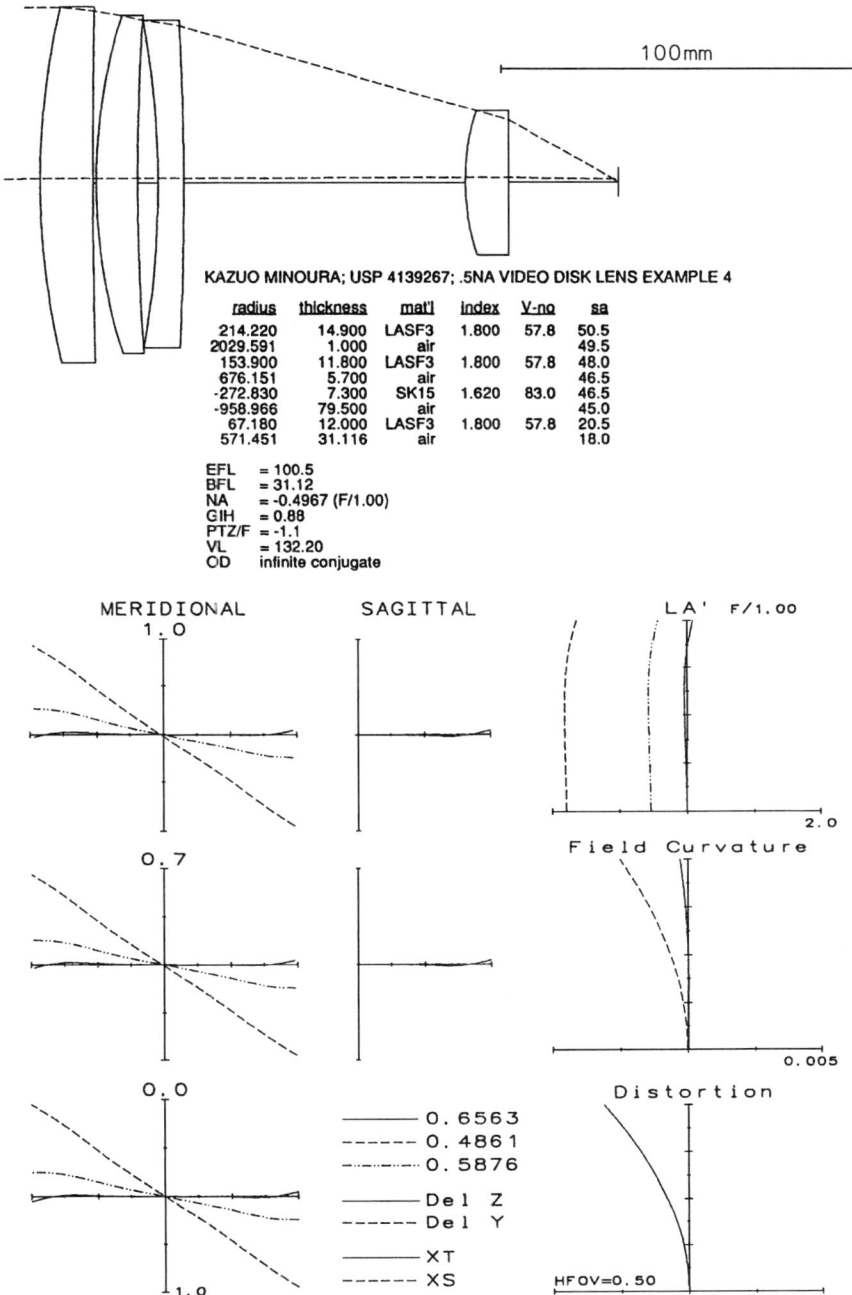

Figure 22.7 f/1.0 ±0.5° a four-element laser disk lens.

570 Chapter Twenty-Two

LASER ZOOM COLLIMATOR FEED LENS; EFL=4-8 MM - SHORT EFL

radius	thickness	mat'l	index	V-no	sa
-566.418	31.802	BK7	1.526		79.5
866.479	768.000	air			79.5
148.977	31.802	BK7	1.526		67.1
539.639	3.235	air			64.4
140.065	39.752	BK7	1.526		62.3
-1167.605	1.590	air			56.0
100.733	39.752	BK7	1.526		48.8
154.535	15.320	air			33.7
-135.259	15.901	BK7	1.526		27.9
88.765	74.032	air			21.7

EFL = 100
BFL = 74.03
NA = -0.2912 (F/1.67)
GIH = 1.58
PTZ/F = 5.19
VL = 947.15
OD infinite conjugate

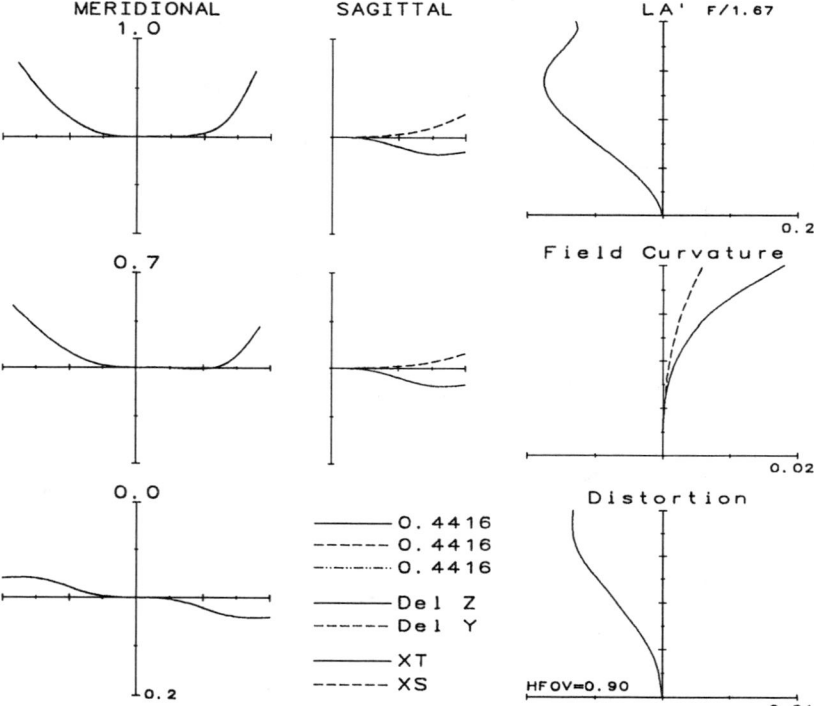

Figure 22.8 $f/1.7$ ±0.9° a laser zoom collimator lens at short focal length.

Scanner/f-θ, Laser Disk, and Collimator Lenses

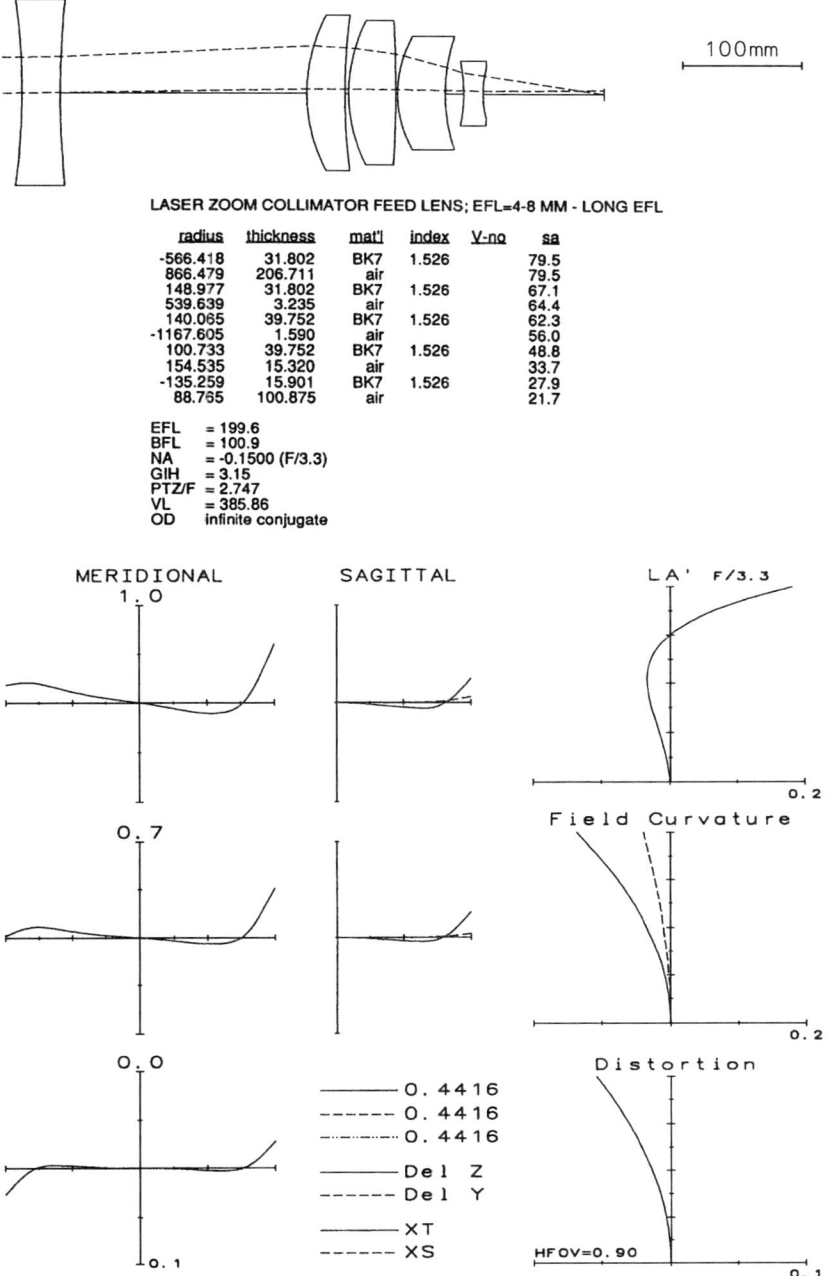

LASER ZOOM COLLIMATOR FEED LENS; EFL=4-8 MM - LONG EFL

radius	thickness	mat'l	index	V-no	sa
-566.418	31.802	BK7	1.526		79.5
866.479	206.711	air			79.5
148.977	31.802	BK7	1.526		67.1
539.639	3.235	air			64.4
140.065	39.752	BK7	1.526		62.3
-1167.605	1.590	air			56.0
100.733	39.752	BK7	1.526		48.8
154.535	15.320	air			33.7
-135.259	15.901	BK7	1.526		27.9
88.765	100.875	air			21.7

EFL = 199.6
BFL = 100.9
NA = -0.1500 (F/3.3)
GIH = 3.15
PTZ/F = 2.747
VL = 385.86
OD infinite conjugate

Figure 22.9 $f/3.3$ ±0.9° a laser zoom collimator lens at long focal length.

Chapter 23
Tolerance Budgeting

23.1 The Tolerance Budget

A lens design is not truly complete until the designer has defined a set of tolerances that will assure that when fabricated, the lens will perform as desired. Establishing a tolerance budget is almost as much an art as is lens design itself. The budget should take into account not only the characteristics of the optical system (and its application) but also the capabilities of the shop that is to build it, as well as the effect of the tolerances on the cost and delivery schedule.

Lens designers often assume that the specifications given to them are reasonable, possible, and necessary. Sometimes they are not. It is often very wise to question the specifications even before the optical design is begun, to be quite certain that the specifications are indeed reasonable, possible, and necessary, and to be sure that they are not arbitrarily and capriciously overspecified "just to be safe." The list of possible specifications in Chap. 1 makes a good checklist to be sure that nothing vital has been overlooked, and also to be sure that there isn't a better way of specifying some desired characteristic.

The dimensions and characteristics for which tolerances should be defined may include the following:

Surface geometry
- Radius value
- Departure from nominal shape
- Aspheric deformation

Surface finish
- Quality (scratch and dig)
- Roughness, scatter, and the like

Surface separation
- Element thickness
- Spacing

Index of refraction
- At the central wavelength
- Total dispersion (V-value)
- Partial dispersion (P-value)
- Homogeneity
- dn/dt

Alignment
- Surface tilt
- Element tilt and /or decentration
- Component tilt and/or decentration
- Prism or mirror angles and alignment

Transmission
- Optical material
- Filters
- Coatings

Physical characteristics
- Thermal effects (α and dn/dt)
- Stability
- Durability

The nominal surface geometry is specified by a spherical radius of curvature (in most cases), by a conic section, or as a general (power series) aspheric (e.g., Eq. (5.1)). For a spherical surface, the specification is often given with respect to a specific test plate. The value of the radius is indicated by the departure of the lens surface from the test plate, given in units of interference fringes, or Newton's rings (at a specified wavelength). Each fringe indicates a change in the space between lens and test plate of one-half wave. The accuracy of the geometry of the surface is specified as a departure from the best-fit perfect sphere, and is often referred to as *asphericity* or *irregularity*.

Because it is difficult to detect a small irregularity if there is a large number of fringes visible in a test, it is common practice to require that there be no more than four or five times as many fringes present as the number of fringes that the irregularity specification allows. This requirement is equally valid when the irregularity is tested on any other type of interferometer. The same general approach is used to specify conics or general aspherics, i.e., a tolerance on the value of the basic surface curvature and a separate tolerance on the departure from the nominal geometry.

The effects produced by an error in the value of a radius varies as a function of ray heights, index break, angle of incidence, and the like at that surface. For a given surface the significance of the error is not a function of δR, or even $\delta R/R$, but is a function of the change in surface curvature, $\delta c = \delta(1/R) = \delta R/R^2$. Thus, with all else equal, the size of a radius tolerance should be proportional to R^2. For example, the change in focal length produced by a change in a 1.0-in radius of $\delta R = \pm 0.001$ in is equivalent to a change of $\delta R = 0.10$ in in a 10.0-in radius, or a change of $\delta R = \pm 10.0$ in in a 100-in radius, or a change of $\delta R = \pm 1000$ in in a 1000-in radius. This last is not quite as ridiculous as it seems at first glance, because the curvature change we are considering corresponds to $cv = 1/(1.0 \pm 0.001) \approx 0.999$ or 1.001, i.e., a curvature change of ± 0.001. And of course the curvature of a 1000-in radius is just exactly 0.001. Note also that the number of Newton's rings is a direct function of the change in curvature (see Eqs. (23.1) and (23.2)). A specified tolerance of radius $R \pm \delta R$ (e.g., $R = 5 \pm 0.005$ in) is very bad; a tolerance of $R + x\% \cdot R$ (e.g., $R = 5$ in ± 0.1 percent of 5 in) is better, but still bad; the size of the tolerances should be proportioned to the square of the radius.

The difference in surface radius corresponding to N fringes departure from a test plate is given by

$$\Delta R = N\lambda \left(\frac{2R}{d}\right)^2 = N\lambda \left(\frac{R}{y}\right)^2 \qquad (23.1)$$

The change in curvature is given by

$$\Delta C = \frac{4N\lambda}{d^2} = \frac{N\lambda}{y^2} \qquad (23.2)$$

where R = nominal radius
λ = test wavelength
d = diameter
y = semidiameter over which the N fringes are observed

The effect of a surface irregularity on the shape of the system wavefront can be calculated from

$$\text{OPD} = 0.5(n' - n) \text{ (number of fringes)} \qquad (23.3)$$

where OPD is the wavefront deformation in wavelengths, $(n' - n)$ is the index change across the surface, and (number of fringes) is the height of the surface bump or irregularity in interference fringes. Note that, in most cases, irregularity takes the form of an astigmatic or toric surface, and that a compromise focus usually reduces its effect by a factor of 2. In an assembly, even a random orientation of the toric axes can be expected to effectively cancel out much of the astigmatism. However, axially symmetrical irregularities of the type referred to by opticians as a *gull-wing* or *hair-pin* surface (after the shape of the test-plate fringes) are always directly additive in a worst-case sense, and can be a serious problem when the optics have been fabricated "one up."

For most optical systems, scratches, digs, pits, bubbles, inclusions, and the like are purely cosmetic defects with little or no effect on the function of the system. Under such circumstances, their specification can best be regarded as an agreement between buyer and vendor as to an acceptable level of workmanship; this level usually depends on whether or not the optic can be seen. The functional effect is primarily scattering. If one expresses the area of the defect as a fraction of the area of the light beam cross section, this will give the percentage of the light intercepted by the defect; most of this is scattered, some may be absorbed. The scattered light is probably distributed into 4π steradians; one can easily determine the resultant level of relative illumination in the image plane that is caused by the defect. It is usually totally negligible. If the defect is on a surface close to an image plane, the defect may then be visible; its visibility (contrast) and significance can be used to determine an acceptable size for such a defect.

There are, of course, exceptions to this. In laser systems, particularly those with high power levels, scattering resulting from surface roughness may be significant, and even small defects may become damage centers that can lead to the degradation or destruction of the optic. Systems that are sensitive to low levels of scattered light, especially those where the defects may be brightly illuminated, constitute another class of exceptions.

Figure 23.1 is a table of typical tolerances that can be used as a rough guide to tolerance sizes commonly considered appropriate. While these values are in fact fairly typical of ordinary optical shop practice, one should remember that there are many special cases that cannot be covered in a summary tabulation of this type.

A thin airspace that is introducing (or correcting) higher-order aberrations is usually quite critical, both as to its thickness and its wedge.

	Surface Quality	Diameter, mm	Deviation (concentricity), min	Thickness, mm	Radius	Regularity (asphericity)	Linear Dimension, mm	Angles
Low cost	120-80	± 0.2	> 10	± 0.5	Gage	Gage	± 0.5	Degrees
Commercial	80-50	± 0.07	3–10	± 0.25	10 Fr	3 Fr	± 0.25	± 15'
Precision	60-40	± 0.02	1–3	± 0.1	5 Fr	1 Fr	± 0.1	± 5'–10'
Extraprecise	60-40	± 0.01	< 1	± 0.05	1 Fr	⅕ Fr	As req'd.	Seconds
Plastic	80-50		1	± 0.02	10 Fr	5 Fr	0.02	minutes

Figure 23.1 Tabulation of typical optical fabrication tolerances.

Often the angles of incidence and refraction are quite large in this situation and small errors can produce huge problems. A thin metal spacer is difficult to fabricate with sufficient accuracy, simply due to its thinness. Edge contact is of course preferable, but usually doesn't produce the spacing that is needed. Trueing and cementing the elements in precision cells is probably the best answer to this problem.

Tilt and decenter (along with surface irregularity) are usually among the most severe fabrication errors. Our usual panoply of Seidel aberrations is based on the assumption of axial symmetry. Tilt, decenter, and astigmatism obviously destroy axial symmetry. They introduce a whole new array of second-order aberrations, such as a transverse image shift, axial coma and astigmatism, independently tilted sagittal and tangential field curves, tangential (i.e., not radial) distortion, and axial lateral color. Of these, the tilted fields are usually the worst.

For the *classic* or *normal* optical production methods, the cost of an element varies roughly as the following expressions:

$$\text{Quantity cost factor} = 1.07 + 2.26 \, Q^{-0.42} \tag{23.4}$$

$$\text{Quality cost factor} = \frac{1.5}{\sqrt[20]{S \cdot D \cdot P \cdot R \cdot T}} \tag{23.5}$$

where Q = quantity of pieces to be made
S = scratch number
D = dig number
P = surface power tolerance in fringes
R = surface regularity tolerance in fringes
T = thickness tolerance in millimeters

Cost also varies with the number of elements that can be blocked on a tool for the grinding and polishing operations.

$$\text{Number per tool} \approx \frac{0.75 D^2}{d^2} - \frac{1}{2} \tag{23.6}$$

where D is the tool diameter and d is the element diameter.

The glass catalog tolerance on index is ±0.001 for most glass types; tolerances of ±0.0005 or ±0.0002 can be had by selection, at a modest increase in cost (and sometimes delivery time). V-value tolerances are typically about ±0.8 percent; again, tighter tolerances can be obtained. The catalog tolerance on index homogeneity is usually given as 0.0001. It is easy to calculate that, if the index varied this much within an element, the resultant wavefront deformation ($\Delta n \times$ thickness) would be overwhelming. This variation is, however, that which may occur within a *melt*, not a blank. The index variation within an element is one or two orders of magnitude less than this, but, in high-quality systems, even this can be important.

Unfortunately, index tolerances and the like for other (i.e., nonglass) optical materials are not well established. There is a tendency to regard the optical characteristics of crystals and other materials as if they were exact constants, despite the fact that significant variations often show up in measurements made on different material samples. This is an area that could profit from further investigation.

The effects of a temperature change can be considered along with the tolerance effects. A *thermal soak* (where the temperature change is uniform and without gradients) produces mostly first-order changes, such as changes in focal length or image position. Aberration changes are usually relatively small. In a thermal analysis, remember that the spacing changes are caused by changes in the length of the spacers, whose length is not the same as the axial airspace. What counts is the edge thickness of the airspace.

If you can get your merit function to reflect the lens performance reasonably well, the merit function can be very useful in tolerancing. (Obviously, to do this the merit function should have only image quality operands.) Strip out the focal length, working distance, magnification, edge thicknesses, and the like; remember that we're tolerancing, not optimizing. Then determine the relationship between the merit function and the performance, and establish an acceptable percentage degradation of the merit function. Again, this is justifiable on the basis that the tolerance changes are small enough to be considered linear. Then the performance specification can be equated to some level of the merit function, or to some percentage increase in the merit function. This obviously depends not only on the construction of the merit function, but, to a lesser extent, on the type of lens at hand. Along the same lines, if each dimension is individually changed by a tolerance sized amount, then the change that each tolerance produces in the value of the merit function is a good measure of the sensitivity of that dimension.

23.2 Additive Tolerances

The variation of curvatures, spacings, and index are simple and straightforward; we can expect that in many cases the change in a characteristic produced by a dimension may be offset (at least partially) by a change in another dimension. Decentration, surface tilt, astigmatic surfaces, and the like are simple two-dimensional errors; they have plus or minus magnitudes as well as an angular orientation. This means that their effects can also randomly cancel out, just as the other tolerance effects do. However the axially symmetric asphericity known as a *gull-wing* or *hair-pin* surface (so-called by opticians because of the shape of the test plate fringes when the plate is tilted a bit) is different. This type of surface error results from ordinary "one up" fabrication, and does not randomly cancel out. The reason is that this is always a surface that is high at both the center and edge of the surface and low at the zone. The surface is shaped like a very weak Schmidt corrector plate, and its effect is to produce overcorrected spherical aberration like a Schmidt corrector. But the gull wing is *always* directly additive regardless of whether it is on a convex or concave surface; there is no random cancellation as with an astigmatic surface, and it must be toleranced much more stringently than other surface irregularities.

In analyzing an optical system to determine the size of the tolerances to be applied to specific dimensions, one can readily calculate the partials of the system characteristics with respect to the dimensions under consideration. Thus one obtains the value of the partial derivative of the focal length (for example) with respect to each thickness, spacing, curvature, and index; likewise for the other characteristics, which may include back focus, magnification, and field coverage, as well as the aberrations or wavefront deformations. Then each dimensional tolerance, multiplied by the appropriate derivative, indicates the contribution of that tolerance to the variation of the characteristic. (In performing a tolerance analysis we can assume that the effects produced by a tolerance-sized dimensional change are a linear function of the size of the change. We are justified in this assumption because the tolerance sizes are so relatively small that any nonlinearities are negligible.)

Now if it were necessary to be *absolutely* certain that (for example) the focal length did not vary more than a certain amount, one would be forced to establish the parameter tolerances so that the sum of the absolute values of the derivative × tolerance products did not exceed the allowable variance. Although this worst-case approach is occasionally necessary, one can frequently allow much larger tolerances by taking advantage of the laws of probability and statistical combination.

As a simple example, let us consider a stack of disks, each 0.1 in thick. We will assume that each disk is made to a tolerance of ±0.005 in and that the probability of the thickness of the disk being any given value between 0.095 and 0.105 in is the same as the probability of its being any other value in this range. This situation is represented by the rectangular (i.e., uniform) frequency distribution curve of Fig. 23.2a. (This is not a very realistic assumption, but one chosen for simplicity and convenience.) Thus, for example, there is a 1 in 10 chance that any given disk will have a thickness between 0.095 and 0.096 in. Now if we stack two disks, we know that it is *possible* for their combined thicknesses to range from 0.190 to 0.210 in. However, the *probability* of the combination having either of these extreme thickness values is quite low. In a frequency

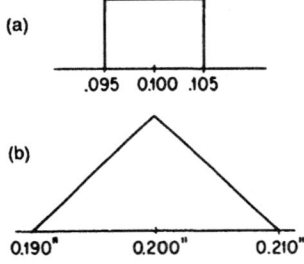
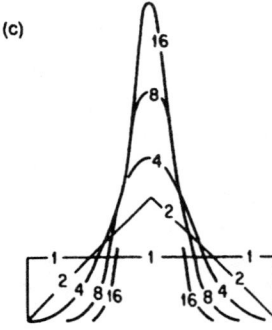

Figure 23.2 How additive tolerances combine in a random assembly. (*a*) Uniform probability distribution in a dimension of a single piece. (*b*) Resulting frequency distribution when two such pieces are combined. (*c*) Normalized curves for assemblies of 1, 2, 4, 8, and 16 pieces.

distribution curve such as those shown in Fig. 23.2, the area under the curve between two abscissa values represents the (relative) number of pieces that will fall between the two abscissa values. Thus the probability of a characteristic falling between two values is the area under the curve between the two abscissas divided by the total area under the curve.

Since the probability of either of the disks having a thickness between 0.095 and 0.096 in is 1 in 10, if we randomly select two disks, the probability of *both* falling in this range is 1 in 100. Thus the probability of a pair of disks having a thickness between 0.190 and 0.192 in is 1 in 100; similarly for a combined thickness of 0.208 to 0.210 in. The probability of a combined thickness of 0.190 to 0.191 in (or 0.209 to 0.210 in) is much less: 1 in 400.

The frequency distribution curve representing this situation is shown in Fig. 23.2*b* as a triangular distribution. Figure 23.2*c* shows frequency distribution curves for 1-, 2-, 4-, 8-, and 16-element assemblies. These curves have been normalized so that the area under each is the same and the extreme variations have been equalized. The important point here is that the probability of an assembly taking on an extreme value is tremendously reduced when the number of elements making up the assembly is increased. For example, in a stack of 16 disks with a nominal

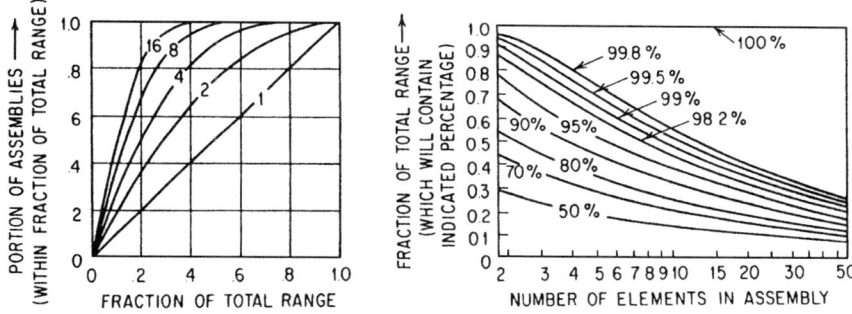

Figure 23.3 Probability distributions of additive tolerances in multiple assemblies, assuming a uniform distribution for a single part.

total thickness of 1.6 in and a possible variation in thickness of ±0.080 in, the probability of a random stack having a thickness less than 1.568 in or more than 1.632 in (i.e., ±0.032 in) is less than 1 in 100.

The importance of this in setting tolerances is immediately apparent. In the stacked-disks example, if the range of thicknesses represented by 1.568 to 1.632 in for 16 disks was the greatest variation that could be tolerated, we could be absolutely sure of meeting this requirement only by tolerancing each individual disk at ±0.002 in. However, if we were willing to accept a rejection rate of 1 percent in large-scale production, we could set the thickness tolerance at ±0.005 in. If the cost of the pieces made to the tighter tolerance exceeded the cost of the pieces made to the looser tolerance by as little as 1 percent (plus one sixteen-hundredth of the assembly, processing, and final inspection costs), the looser tolerance would result in a less costly product.

This peaking-up characteristic of multiple assemblies can also be represented by the two plots shown in Fig. 23.3. The graph on the left shows the percentage of assemblies that fall within a given central fraction of the total possible tolerance range as a function of that fraction. The number of elements per assembly is indicated on each curve. These curves were derived from Fig. 23.2c. The graph on the right in Fig. 23.3 is simply another way of presenting the same data. If one were interested in an assembly of 10 elements, the intersection of the abscissa corresponding to 10 and the appropriate curve would indicate that all but 0.2 percent (using the 99.8 percent curve) of the assemblies would fall within 0.55 of the total tolerance range represented by the absolute sum of all 10 tolerances, and that more than one-half the assemblies (using the 50 percent curve) would fall within 0.15 of the total possible range.

The preceding discussion has been based upon the simplified assumptions that (1) each individual piece had a rectangular frequency distribution, and (2) each tolerance was equal in effect. This is very rarely true in practice. The frequency distribution will, of course, depend on the techniques and controls used in fabricating the part, and the tolerance sizes may represent the partial derivative tolerance products from such diverse sources as tolerances on index, thickness, spacing, and curvature. Note, however, that in Fig. 23.2c the progression of curves may be started at any point. If, for example, the production methods produce a triangular distribution (such as that shown for an assembly of two elements), then the curve marked 4 (for four elements) will be the frequency distribution for two elements (of triangular distribution) and so on.

Note also that as more and more elements are included in the assembly, the curve becomes a closer and closer approximation to the normal distribution curve that is so useful in statistical analysis (except that the tolerance-type curves do not go to infinity as do normal curves). One useful property of the normal curve for an additive assembly is that its "peakedness" is proportional to the square root of the number of elements in assembly. Thus if 90 percent of the individual pieces are expected to fall within some given range, then for an assembly of 16 elements, 90 percent would be expected to fall within $\sqrt{1/16}$, or one-quarter of the total range. A brief examination will indicate that even the rectangular distribution assumed for Figs. 23.2 and 23.3 tends to follow this rule when there are more than a few elements in the assembly.

A rule frequently used to establish tolerances may be represented as follows:

$$T = \sqrt{\sum_{i=1}^{n} t_i^2} \qquad (23.7)$$

This is frequently referred to as the RSS rule, shorthand for the square Root of the Sum of the Squares. What the RSS rule means is this: If some percentage (say 99 percent, for example) of the parts have tolerances that produce effects less than t_i (and that vary according to a normal, or gaussian, distribution), then the same percentage (i.e., 99 percent in our example) of the assemblies will show a total tolerance effect less than T. For *normal* tolerance distributions, the central limit theorem reduces to the RSS rule.

While this section may seem to be a far cry from optical design, consider that a simple Cooke triplet has the following dimensions that affect its focal length and aberrations: six curvatures, three thicknesses, two spacings, three indices, and three V-values. These total

14 for monochromatic characteristics and 17 for chromatic aberrations. Such a system is eminently qualified for statistical treatment. (To these we can add tilts, decenters, surface asphericities, index in homogeneity, and the like.)

Note well that the validity of this approach does not depend on a large production quantity; it depends on a random combination of a certain number of tolerance effects.

There are two obvious features of the RSS rule that are well worth noting. One is the square root effect: If you have n tolerance effects of a size $\pm x$, then the RSS rule says that a random combination will produce an effect equal to $\pm x$ times the square root of n. For example, given 16 tolerance effects of ±1 mm, we should expect a variation of only ±4 mm, not ±16 mm. The other feature is that the larger effects dominate the combination. As an example, consider a case with nine tolerances of ±1 mm and one tolerance of ±10 mm. If we use the RSS rule on this, we get an expected variation equal to the square root of 109, or ±10.44 mm. Compare this with the fact that the single ±10-mm tolerance has an RSS of ±10 mm. The addition of the nine ±1-mm tolerances changes the expected variation by only 4.4 percent.

23.3 Establishing the Tolerance Budget

The tolerance budget analysis must be separately done for the image quality and other types of characteristics (i.e., the focal length, image location, magnification, distortion, collimation, aiming, and the like). One possible way to establish a tolerance budget using this principle is as follows:

1. Calculate the partial derivatives of the aberrations (and other characteristics to be controlled) with respect to the fabrication tolerances (radius, asphericity, thickness and spacing, index, homogeneity, surface tilt, and the like). Express the aberrations as OPDs (wavefront deformation) where appropriate.
2. Select a preliminary tolerance budget. Figure 23.1 can be used as a rough guide to appropriate tolerance values.
3. Multiply the individual tolerances by the partial derivatives calculated in step 1.
4. Compute RSS from Eq. (23.7) for all the aberrations for each tolerance *individually*. This will indicate the relative sensitivity of each tolerance.
5. Compute RSS for all of the effects calculated in step 4. (Note that the RSS values calculated in step 4 can be RSS'd together in this step.)

6. Combine the RSS tolerance effects found in step 5 with the OPD of the nominal design to determine the expected performance of the toleranced fabricated design (OPD$_{fab}$). This combining is done by RSS:

$$\text{OPD}_{fab} = (\text{OPD}^2_{des} + \text{OPD}^2_{tol})^{1/2} \qquad (23.8)$$

where OPD$_{des}$ is the OPD of the nominal design and OPD$_{tol}$ is the OPD caused by the tolerances. The nominal design OPD$_{des}$ can be determined directly in some cases, or deduced from a calculated MTF or Strehl Ratio, using the relationships described in Chap. 4.

The *acceptable* OPD$_{tol}$ can be solved for from Eq. (23.8) above:

$$\text{OPD}_{tol} = (\text{OPD}^2_{spec} - \text{OPD}^2_{des})^{1/2} \qquad (23.9)$$

where OPD$_{spec}$ is the maximum OPD allowed by the system performance specification.

For example, if the design OPD$_{des}$ is a quarter wave and the specified performance requires an OPD$_{spec}$ of less than a half wave, then the permissible OPD$_{tol}$ from tolerance effects would be given by:

$$\text{OPD}_{tol} = (0.5^2 - 0.25^2)^{1/2} = 0.433 \text{ waves}$$

7. Adjust the tolerance budget so that the result of step 6 is equal to the required performance. Since the larger effects dominate the RSS, if you are tightening the tolerances (as is quite likely on the first go-round), you should tighten the most sensitive ones as determined in step 4 (and possibly loosen the least sensitive). Note that there is no economic gain if you loosen tolerances beyond the level at which costs or prices cease to go down. Conversely, you should be sure that the tolerances are not tightened beyond a level at which fabrication becomes impossible—since cost rises asymptotically toward infinity as this level is approached.

8. After one or two adjustments (steps 2 through 7), the tolerance budget should converge to one that is reasonable economically and will yield an acceptable product.

If the tolerances necessary to get acceptable performance are too tight to be fabricated economically, there are several ways that are commonly used to ease the situation (as described at greater length in Sec. 2.9).

In evaluating tolerance effects on image quality, refocus is almost always allowed to compensate for any longitudinal image shift. If refocus is not allowed, the fabrication tolerances become extremely (almost impossibly) tight.

1. A *test plate fit* is a redesign of the system using the measured values of the radii of existing test plates. This eliminates the radius tolerance (except for the variations due to the test glass fit in the shop and any error in the measurement of the radius).
2. A *melt fit* can effectively eliminate the effects of index and dispersion variation. Again, this is a redesign, using the measured index of the actual piece of glass to be used, instead of the catalog values.
3. A *thickness fit* uses the measured thicknesses of the actual fabricated elements; this amounts to an adjustment of the airspaces during the assembly process.
4. A *reverse adjustment*. Measure the aberrations, wave front deformation, and the like of the assembled lens system. Calculate the fabrication errors that would produce this performance by optimizing the design to have aberrations equal to the measured aberrations. Adjust the lens to eliminate the defects by introducing dimensional changes opposite to those calculated by the optimization.

Mounting design can change the effect of element thickness variations. When an element is mounted against one surface, the airspace on the other side is changed by the opposite amount of a thickness change. If the airspace has the same sign effect (i.e., partial differential) of the aberrations, at least some cancellation will occur. Allowing the element to flow into the airspace on one side of an element may be more favorable than allowing it to flow into the other.

The redesigns called for in the fitting operations above, while hardly trivial, are not major undertakings when an automatic lens design program is used.

While the above may tend to induce a desirable relaxation in tolerances, one or two words of caution are in order. As previously mentioned, the index of refraction distribution within a melt or lot of glass may or may not be centered about the nominal value. When it is centered about a non-nominal value, the preceding analysis is valid only with respect to the central value, not the nominal value. Further, in some optical shops there is a tendency to make lens elements to the high side of the thickness tolerance; this allows scratched surfaces to be reprocessed and will, of course, upset the theoretical probabilities. Another tendency is for polishers to try for a hollow test glass fit, i.e., one in which there is a convex air lens between the test plate and the work. This is done because a block of lenses that is polished "over" (i.e., so that there is center contact between the work and the test glass) is difficult to bring back. That is, in ordinary polishing it is difficult to change the radius to a shorter one for a concave surface or a longer one for a convex surface. Surprisingly, these non-normal distributions have very little effect on Eq. (23.7) (if there are enough elements in the assembly). Actually, Eq. (23.7) seems to be quite conservative in actual practice, by a factor of about two.

Thus the situation is seen to be a complex one, but, nonetheless, one in which a little careful thought in relaxing tolerances to the greatest allowable extent can pay handsome dividends. For those who wish to avoid the labor of a detailed analysis, the use of Eq. (23.7), or even the assumption that the tolerance buildup will not exceed one-half or one-third of the maximum possible variation, are fairly safe procedures in assemblies of more than a few elements. Above all, when cost is important, one should try to establish tolerances that are readily held by normal shop practices.

Chapter 24

Formulary

This section is a condensed collection of formulas that are regularly used in optical design. It is assumed that the reader is generally familiar with the subject and that no detailed explanations are necessary. Thus this formulary is more in the style of a handbook than that of a textbook, since its purpose is that of a reference listing rather than a tutorial. However, each section is appropriately labeled, defined, and illustrated so that one generally familiar with the subject can readily apply the formulas. If a more extensive explanation is desired, the reader is referred to Smith.[23]

24.1 Sign Conventions, Symbols, and Definitions

A primed symbol refers to a quantity after refraction (or reflection) by a surface or by a lens, or to a quantity associated with the image. The subscript k is used as a generalization for the last surface or lens of the system. Lowercase symbols are used for paraxial quantities and uppercase for trigonometrically exact quantities. The data of a chief, or principal, ray is signified by the subscript p, e.g., u_p, y_p (in some works a bar over the symbol is used). Letter or numerical subscripts are used to identify surfaces or lenses in a sequence.

1. *Heights.* A height above the axis or above a reference point is positive; below is negative.

    h, H Object or image height; the intersection of a ray with the object or image surface

    y, Y Ray height; the height at which a ray intersects a surface or element

2. *Distances.* A distance to the right of a reference point is positive, to the left is negative. The reference point may be at a surface, a lens, a principal point, or a focal point. In ray tracing, the reference point moves sequentially from surface to surface (or from lens to lens).

    s Axial distance from a principal point
    l Axial distance from a surface
    x Axial distance from a focal point
    d, t Axial spacing between surfaces, elements, or principal planes

3. *Angles.* An angle or ray slope is positive if the ray is rotated clockwise to reach the axis or surface normal.

    u, U Ray slope angle
    i, I Angle of incidence or refraction

4. *Radius of curvature.* A radius of curvature is positive if the center of curvature is to the right of the surface.

    r Radius of curvature
    c Curvature, reciprocal of the radius, equal to $1/r$

5. *Focal length and power.* The focal length and power are positive if the lens or surface bends the ray toward the axis, i.e., if the lens converges light.

    f Effective focal length; the distance from the second principal point to the second focal point
    ϕ Power; for a lens, the reciprocal of the focal length; for a surface, $\phi = (n' - n)/r$

6. *Index of refraction.* The index is positive if light rays travel from left to right, negative if they travel right to left (as after a reflection).

    n Index of refraction, equal to the velocity of light in a vacuum (in practice, the velocity in air is usually used) divided by the velocity in the medium

24.2 The Cardinal Points (Fig. 24.1)

First/second focal point: The focus of rays emanating from an axial object point an infinite distance to the right/left

First/second principal point: The axial point corresponding to the locus at which the first/second focal-point-defining ray appears to be bent

First/second nodal points: A pair of axial points such that an oblique ray directed toward the first nodal point appears to emerge from the second, making the same angle to the optical axis. For a system in air, they coincide with the principal points

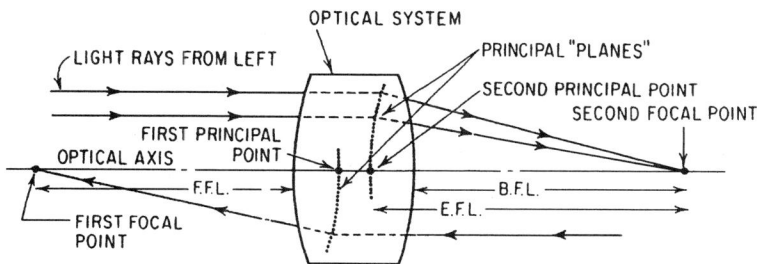

Figure 24.1 The locations of the focal points and principal points of an optical system.

efl = f: Effective focal length: the distance from the second principal point to the second focal point

bfl: Back focal length: the distance from the vertex of the last surface to the second focal point

ffl: Front focal length: the distance from the vertex of the first surface to the first focal point

To determine efl and bfl, trace a paraxial ray with $u_1 = 0$; then efl = $f = -y_1/u_k'$ and bfl = $-y_k/u_k'$.

For a single element:

$$1/f = \phi = (n-1)[c_1 - c_2 + t(n-1)c_1 c_2/n] \quad (24.1)$$

$$\text{bfl} = f[1 - t(n-1)c_1/n] \quad (24.2)$$

Saggital height:

$$S = R - (R^2 - Y^2)^{1/2} \quad (24.3)$$

$$R = (Y^2 + S^2)/2S \quad (24.4)$$

$$Y = (2RS - S^2)^{1/2} \quad (24.5)$$

where S = sagittal height
R = surface radius
Y = height

Diameter at which an element is sharp-edged:

$$D = 2\left[-T_2(2R_2 + T_2)\right]^{1/2} \tag{24.6}$$

where
$$T_2 = \frac{2R_1T - T^2}{2(R_1 - R_2 - T)} \tag{24.7}$$

24.3 Image Equations (Fig. 24.2)

Newton's equation:

$$x' = \frac{-f^2}{x} \tag{24.8}$$

Gauss equation:

$$\frac{1}{s'} = \frac{1}{s} + \frac{1}{f} \tag{24.9}$$

$$s' = \frac{sf}{(s+f)} \tag{24.10}$$

$$s = \frac{s'f}{(f-s')} \tag{24.11}$$

$$f = \frac{ss'}{(s-s')} \tag{24.12}$$

Lateral magnification:

$$m = \frac{h'}{h} = \frac{s'}{s} = \frac{-x'}{f} = \frac{f}{x} \tag{24.13}$$

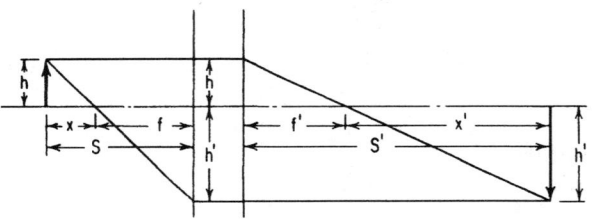

Figure 24.2 The dimensions used in the imaging equations.

$$x' = -mf \tag{24.14}$$

$$x = \frac{f}{m} \tag{24.15}$$

$$s' = f(1-m) \tag{24.16}$$

$$s = \frac{f(1-m)}{m} \tag{24.17}$$

$$\text{Object to image distance} \equiv T \equiv s' - s \tag{24.18}$$

$$T = \frac{-f(m-1)^2}{m} \tag{24.19}$$

$$f = \frac{-Tm}{(m-1)^2} \tag{24.20}$$

$$s = 0.5\,[-T \pm (f^2 - 4fT)^{1/2}] \tag{24.21}$$

$$f/\# = \frac{1}{2(u-u')} \tag{24.22}$$

$$f/\# = \frac{1}{2(m-1)\text{NA}'} \tag{24.23}$$

$$f/\# = \frac{m}{2(m-1)\text{NA}} \tag{24.24}$$

$$u' = mu = \frac{1}{2(f/\#)(m-1)} \tag{24.25}$$

$$u = \frac{u'}{m} = \frac{m}{2(f/\#)(m-1)} \tag{24.26}$$

$$\text{NA} = \frac{m}{2(m-1)(f/\#)} \tag{24.27}$$

$$\text{NA}' = \text{NA}/m = \frac{1}{2(m-1)(f/\#)} \tag{24.28}$$

Longitudinal magnification:

$$m = \frac{s_2' - s_1'}{s_2 - s_1} = \frac{s_1'}{s_1} \cdot \frac{s_2'}{s_2} \tag{24.29}$$

$$= m_1 \times m_2 \quad \text{(approaches } m^2 \text{ as } s_2 \text{ approaches } s_1\text{)}$$

24.4 Paraxial Ray Tracing (Surface by Surface) (Figs. 24.3 and 24.4)

$$u = \frac{-y}{l} \quad l' = \frac{-y}{u'} \quad y = -lu = -l'u' \tag{24.30}$$

$$n'u' = nu - \frac{y(n'-n)}{r} = nu - y(n'-n)c \tag{24.31}$$

$$y_{j+1} = y_j + du'_j \tag{24.32}$$

$$\frac{n'}{l'} = \frac{n}{l} + \frac{n'-n}{r} \tag{24.33}$$

$$i = cy + u \tag{24.34}$$

$$i' = \frac{ni}{n'} \tag{24.35}$$

$$u' = u - i + i' \tag{24.36}$$

$$m = \frac{h'}{h} = \frac{nu}{n'u'} = \frac{\text{object } nu}{\text{image } nu} \tag{24.37}$$

The (a) *angle solves* and (b) *height solves* available in lens design programs: (a) solve for a surface curvature (or radius), which will produce

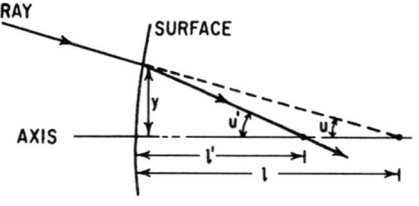

Figure 24.3 The relationship $y = -lu = -l'u'$ for paraxial rays.

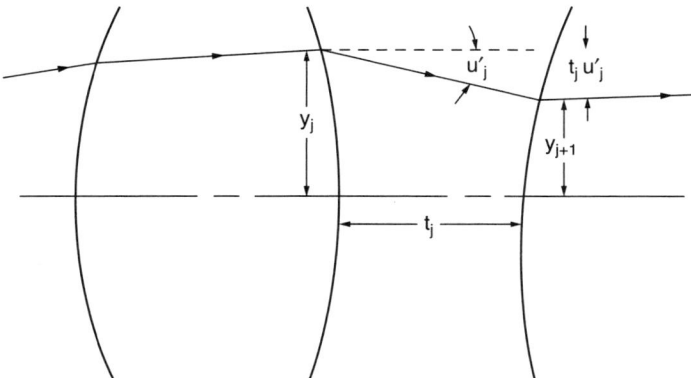

Figure 24.4 The transfer of a paraxial ray from surface to surface by $y_{j+1} = y_j + tu_j'$.

a desired ray slope following the surface, and (b) solve for the following space to produce a desired ray height on the next surface. They may be applied to either the axial ray or the principal ray.

The angle solve:

$$c = \frac{(nu - n'u')}{y(n' - n)} \qquad (24.38)$$

where u' is the desired ray slope.

The height solve:

$$t = \frac{(y_{j+1} - y_j)}{u_j'} \qquad (24.39)$$

where y_{j+1} is the desired ray height.

To get the data of a new (third) ray from the data of two previously traced rays, given the data of the three rays at some (any) location in the system, evaluate:

$$A = \frac{(y_3 u_1 - u_3 y_1)}{(u_1 y_2 - y_1 u_2)} \qquad (24.40)$$

$$B = \frac{(u_3 y_2 - y_3 u_2)}{(u_1 y_2 - y_1 u_2)} \qquad (24.41)$$

Then, anywhere in the system where the data of rays no. 1 and no. 2 are known, we can determine (without additional raytracing) the data for ray no. 3 from:

$$y_3 = Ay_1 + By_2 \qquad (24.42)$$

$$u_3 = Au_1 + Bu_2 \qquad (24.43)$$

Note that rays no. 1 and no. 2 are often the axial ray and the principal ray. This technique is often used to determine the focal length without tracing a specific ray for the purpose, and to locate the pupil(s), given the location of the aperture stop.

24.5 Invariants

$$\text{Inv} = y_p nu - y n u_p = y_p n'u' - y n'u'_p \qquad (24.44)$$

At the image $y = 0$ and $\text{Inv} = hnu = h'n'u'$ $\qquad (24.45)$

At the pupil $y_p = 0$ and $\text{Inv} = y n u_p = y'n'u'_p$ $\qquad (24.46)$

The stop shift coefficient Q is invariant throughout the system

$$Q = \frac{y_{p1} - y_{p2}}{y} \qquad (24.47)$$

24.6 Paraxial Ray Tracing (Component by Component) (Fig. 24.5)

$$u = \frac{-y}{s} \qquad u' = \frac{-y}{s'} \qquad (24.48)$$

$$u' = u - y\phi \qquad (24.49)$$

$$y_{j+1} = y_j + du'_j \qquad (24.50)$$

$$m = \frac{h'}{h} = \frac{u_1}{u_k'} \qquad (24.51)$$

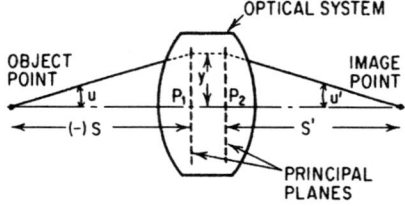

Figure 24.5 The principal planes are *planes of unit magnification*, so that a ray appears to leave the second principal plane at the same height (y) that it appears to strike the first principal plane. Note that $y = -su = -s'u'$.

24.7 Two-Component Relationships

- Given: f_a, f_b, and d (or ϕ_a and ϕ_b) (Fig. 24.6)

$$\phi_{ab} = \phi_a + \phi_b - d\phi_a\phi_b \tag{24.52}$$

$$f_{ab} = \frac{f_a f_b}{f_a + f_b - d} \tag{24.53}$$

$$B = \frac{f_{ab}(f_a - d)}{f_a} \tag{24.54}$$

- Given: f_{ab}, d, and B (Fig. 24.6)

$$f_a = \frac{df_{ab}}{f_{ab} - B} \tag{24.55}$$

$$f_b = \frac{-dB}{f_{ab} - B - d} \tag{24.56}$$

- Given: s, s', d, and m = h'/h = u/u' (Fig. 24.7)

$$\phi_a = \frac{ms - md - s'}{msd} \tag{24.57}$$

$$\phi_b = \frac{d - ms + s'}{ds'} \tag{24.58}$$

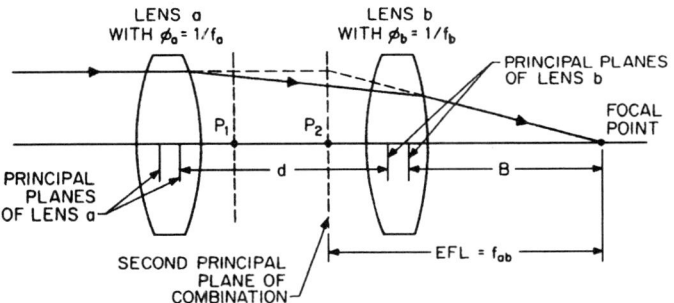

Figure 24.6 Two separated components, object at infinity, showing the symbols used in Sec. 24.7.

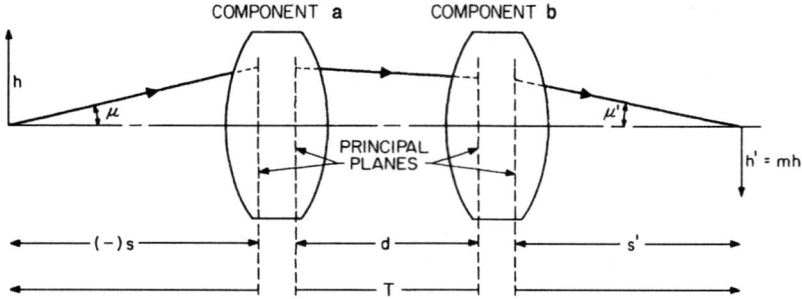

Figure 24.7 Two separated components, object at a finite distance, showing the symbols used in Sec. 24.7.

- Given ϕ_a, ϕ_b, T, and $m = h'/h = u/u'$ (Fig. 24.7)

    Solve this quadratic for d:

    $$0 = d^2 - dT + T(f_a + f_b) + \frac{(m-1)^2 f_a f_b}{m} \tag{24.59}$$

then

$$s = \frac{(m-1)d + T}{(m-1) - md\phi_a} \tag{24.60}$$

$$s' = T + s - d \tag{24.61}$$

24.8 Third-Order Aberrations—Surface Contributions

Given: the paraxial ray data from an axial marginal ray (y, u, i) and a principal ray (y_p, u_p, i_p), calculated for each surface using Eqs. (24.31), (24.32), and (24.34); h is the image height (i.e., y_p at the paraxial focus).

$$B = \frac{n(n'-n)y(u'+i)}{2n'hn_k'u_k'} \tag{24.62}$$

$$B_p = \frac{n(n'-n)y_p(u'_p + i_p)}{2n'hn_k'u_k'} \tag{24.63}$$

The following are the contributions of the surface to the transverse aberrations. The aberrations at the image are the sum of all the surface contributions:

Spherical:
$$TSC = Bi^2 h \tag{24.64}$$

Sagittal coma:
$$CC = Bii_p h \tag{24.65}$$

Astigmatism:
$$TAC = Bi_p^2 h \tag{24.66}$$

Petzval:
$$TPC = \frac{(n'-n)ch^2 n_k' u_k'}{2nn'} \tag{24.67}$$

Distortion:
$$DC = h\left[B_p ii_p + \frac{u_p'^2 - u_p^2}{2}\right] \tag{24.68}$$

Axial chromatic:
$$TAchC = \frac{-yi(dn - dn'n/n')}{n_k' u_k'} \tag{24.69}$$

Lateral chromatic:
$$TchC = \frac{-yi_p(dn - dn'n/n')}{n_k' u_k'} \tag{24.70}$$

where
$$dn = n_F - n_C = \frac{n-1}{V}$$

Conversion from transverse to longitudinal aberration:
$$\text{Longitudinal} = \frac{\text{transverse}}{-u_k'} \tag{24.71}$$

$$PC = \frac{-TPC}{u_k'} \tag{24.72}$$

$$AC = \frac{-TAC}{u_k'} \tag{24.73}$$

$$x_s = \Sigma PC + \Sigma AC \tag{24.74}$$

$$x_T = \Sigma PC + 3\Sigma AC \tag{24.75}$$

$$\text{Petzval radius } \rho = \frac{h^2}{2\Sigma PC} \tag{24.76}$$

$$SC = \frac{-TSC}{u_k'} \tag{24.77}$$

$$\text{Axial chrom.} = -\frac{TA_{ch}C}{u_k'} \tag{24.78}$$

Conversion to Seidel coefficients

$$S = \text{transverse } (-2n_k'u_k') \tag{24.79}$$

e.g., $SI = -TSC\,(2n_k'\,u_k') = 2TSC \cdot NA$

24.9 Third-Order Aberrations—Thin Lens Contributions; The G-Sum Equations

Given: the paraxial ray data (y, u, y_p, u_p) for an axial marginal ray and a principal ray (calculated for each element using Eqs. (24.49) and (24.50), the element power ϕ, the element curvature $c = \phi/(n-1)$, the curvature of the left surface c_1, $v = u/y$, and n = the index.

$$\begin{aligned}
G_1 &= \frac{n^2(n-1)}{2} & G_2 &= \frac{(2n+1)(n-1)}{2} \\
G_3 &= \frac{(3n+1)(n-1)}{2} & G_4 &= \frac{(n+2)(n-1)}{2n} \\
G_5 &= \frac{2(n+1)(n-1)}{n} & G_6 &= \frac{(3n+2)(n-1)}{2n} \\
G_7 &= \frac{(2n+1)(n-1)}{2n} & G_8 &= \frac{n(n-1)}{2}
\end{aligned} \tag{24.80}$$

With the stop at the element, the thin lens contributions are given by the following equations. Use the stop shift equations from Sec. 24.10 with $Q = y_p/y$ to determine the contributions with the stop displaced from the lens.

Spherical:

$$\text{TSC} = \left(\frac{y^4}{u_k'}\right)(G_1 c^3 - G_2 c^2 c_1 - G_3 c^2 v + G_4 cc_1^2 + G_5 cc_1 v + G_6 cv^2) \quad (24.81a)$$

$$= \left(\frac{y^4}{u_k'}\right)(G_1 c^3 + G_2 c^2 c_2 + G_3 c^2 v' + G_4 cc_2^2 + G_5 cc_2 v' + G_6 cv'^2) \quad (24.81b)$$

Sagittal coma:

$$CC = -hy^2(0.25 G_5 cc_1 + G_7 cv - G_8 c^2) \quad (24.82a)$$

$$= -hy^2(0.25 G_5 cc_2 + G_7 cv' + G_8 c^2) \quad (24.82b)$$

Astigmatism:

$$\text{TAC} = \frac{h^2 \phi\, u_k'}{2} \quad (24.83)$$

Petzval:

$$\text{TPC} = \frac{h^2 \phi\, u_k'}{2n} \quad (24.84)$$

$$\rho = -\Sigma \frac{n}{\phi} \quad (24.85)$$

Distortion:

$$\text{DC} = 0 \quad (24.86)$$

Axial chromatic:

$$\text{TAchC} = \frac{y^2 \phi}{V u_k'} \quad (24.87)$$

where

$$V = \frac{n_d - 1}{n_F - n_C}$$

Lateral chromatic:

$$\text{TchC} = 0 \tag{24.88}$$

Secondary chromatic:

$$\text{TSchC} = \frac{y^2 \phi P}{V u_k'} \tag{24.89}$$

where

$$P = \frac{n_d - n_C}{n_F - n_C} \tag{24.90}$$

24.10 Stop Shift Equations

The stop shift coefficient Q is defined by

$$Q = \frac{(y^*_p - y_p)}{y} \tag{24.91}$$

(The asterisk [*] indicates data with the stop in a new, shifted position.) Since Q is an invariant, it may be determined at any surface or element of a system, because it is everywhere the same. For this reason the following expressions may be applied to the third-order contributions of a single surface, or a group of surfaces, or a complete lens. Note that because the *thin lens* equations of Sec. 24.9 apply with the stop in contact, they assume y_p is zero. Thus when the stop shift equations are used to determine the thin lens aberrations, Q reduces to $Q = (y^*_p - 0)/y$, or, when one has traced the principal ray through the center of the actual aperture stop to

$$Q = \frac{y_p}{y} \tag{24.92}$$

$$\text{TSC}^* = \text{TSC} \tag{24.93}$$

$$\text{CC}^* = \text{CC} + Q \cdot \text{TSC} \tag{24.94}$$

$$\text{TAC}^* = \text{TAC} + 2Q \cdot \text{CC} + Q^2 \text{TSC} \tag{24.95}$$

$$\text{TPC}^* = \text{TPC} \tag{24.96}$$

$$\text{DC}^* = \text{DC} + Q(\text{TPC} + 3\text{TAC}) + 3Q^2 \text{CC} + Q^3 \text{TSC} \tag{24.97}$$

$$\text{TAchC}^* = \text{TAchC} \tag{24.98}$$

$$\text{TchC}^* = \text{TchC} + Q \cdot \text{TAchC} \qquad (24.99)$$

Note that these equations are exact in the third order.

24.11 Third-Order Aberrations— Contributions from Aspheric Surfaces

Determine the contributions from the spherical surface or thin lens, then add the following to determine the total contribution from the aspheric surface or thin lens.

$K =$ equivalent fourth-order deformation coefficient (AD in Eq. (5.1))

$$= \frac{\text{conic constant } \kappa}{8r^3} \quad \text{(for pure conic sections)}$$

$$= \frac{\kappa}{8r^3} \text{ plus the fourth-order deformation coefficient} \\ \text{(for conics with aspheric deformations)}$$

$$W = \frac{4K(n'-n)}{h\, n_k' u_k'} \quad (h\, n_k' u_k' = \text{the invariant})$$

$$\text{TSC}_a = Wy^4 h \qquad (24.100)$$

$$\text{CC}_a = Wy^3 y_p h \qquad (24.101)$$

$$\text{TAC}_a = Wy^2 y_p^2 h \qquad (24.102)$$

$$\text{TPC}_a = 0 \qquad (24.103)$$

$$\text{DC}_a = Wy y_p^3 h \qquad (24.104)$$

$$\text{TAchC}_a = 0 \qquad (24.105)$$

$$\text{TchC}_a = 0 \qquad (24.106)$$

Note that if the aspheric surface is at the aperture stop, then y_p is zero and only TSC_a is nonzero.

24.12 Conversion of Aberrations to Wavefront Deformation (Optical Path Difference)

The following expressions convert aberrations to peak-to-peak or peak-to-valley wavefront deformations, when the reference point is chosen to minimize the *optical path difference* (OPD). Note that for low-order aberrations, rms OPD and peak-to-valley (P-V) OPD are related by:

$$\text{rms OPD} = \frac{\text{P-V OPD}}{3.5} \qquad (24.107)$$

For random errors:

$$\text{rms OPD} \approx \frac{\text{P-V OPD}}{5} \qquad (24.108)$$

and the Strehl ratio is approximated by:

$$\text{Strehl ratio} = e^{-(2\pi w)^2} \qquad (24.109)$$

where w = rms OPD. In the following, NA = numerical aperture = $n_k' \sin u_k'$ and λ = wavelength.

Longitudinal defocus, field curvature:

$$\text{OPD} = \frac{(\text{defocus}) \cdot \text{NA}^2}{2\lambda n_k'} \text{ waves} \qquad (24.110)$$

Transverse third-order spherical (at best focus, $\delta = LA_m/2$):

$$\text{OPD} = \frac{\text{TA}_m \cdot \text{NA}}{16\lambda} \text{ waves} \qquad (24.111)$$

Transverse zonal spherical ($\text{TA}_m = 0$) (at best focus, $\delta = 0.75\ LA_z$):

$$\text{OPD} = \frac{\text{TA}_z \cdot \text{NA}}{16.8\lambda} \text{ waves} \qquad (24.112)$$

Tangential coma:

$$\text{OPD} = \frac{\text{coma}_t \cdot \text{NA}}{6\lambda} \text{ waves} \qquad (24.113)$$

Transverse axial chromatic:†

$$\text{OPD} = \frac{\text{TAch} \cdot \text{NA}}{4\lambda} \text{ waves} \qquad (24.114)$$

Lateral chromatic:†

$$\text{OPD} = \frac{\text{TchA} \cdot \text{NA}}{2\lambda} \text{ waves} \qquad (24.115)$$

For angular aberrations (as in afocal systems):

Marginal spherical:

$$\text{OPD} = \frac{AA \cdot n \cdot D}{32\lambda} \text{ waves} \qquad (24.116)$$

Zonal spherical:

$$\text{OPD} = \frac{AA_z \cdot n \cdot D}{33.6\lambda} \text{ waves} \qquad (24.117)$$

Tangential coma:

$$\text{OPD} = \frac{AA \cdot n \cdot D}{12\lambda} \text{ waves} \qquad (24.118)$$

where D is the beam diameter and "AA" represents the appropriate angular aberration.

---

†These chromatic expressions yield the OPD for the extreme wavelengths of the spectral band. For values correlating well with the other relationships given in this section as to the effect on image quality (e.g., MTF), divide the chromatic OPD from these expressions by 2.5 for ordinary chromatic, and by 4.5 for secondary spectrum.

Glossary of Optical Design Terms

The following is a glossary of terms common to (and peculiar to) the field of lens design. A new lens designer might well profit from a thorough perusal of this list, since the jargon of lens design may differ from that of physics or even basic optics.

Abbe V-number The reciprocal relative dispersion of an optical material. For visual work $V = (n_d - 1)/(n_F - n_C)$, where d, F, and C indicate the Fraunhofer wavelengths—0.5876, 0.4861, and 0.6563 µm. Often called *V-value* or *v-value*.

Abbe sine condition A condition where magnification is constant across the aperture and coma near the axis is zero. Paraxial magnification ($m = u/u'$) and trigonometric magnification ($M = \sin U/\sin U'$) should be equal, so that $(u' \cdot \sin U)/(u \cdot \sin U') = 1.0$. A departure from this equality is called the *offense against the sine condition*, or OSC. Near the axis OSC = (Sagittal Coma)/h'.

aberration A defect of the image whereby all rays from a point source do not converge to a point image at the desired location. An aberrated wavefront departs from a perfect sphere centered on the desired image point. The primary aberrations are spherical, coma, astigmatism, distortion, axial chromatic, and lateral color. Relative to the paraxial image, an aberration may be measured as a transverse displacement, a longitudinal displacement, an angular deviation, or a wavefront deformation.

achromat An optical system free of primary chromatic aberration. Usually defined as a system where two different wavelengths come to a focus at the same location. Usually accomplished by the use of materials of differing V-values.

afocal system An optical system which forms an image of an infinitely distant object at infinity, i.e., a system where both input and output beams are collimated. A telescope is an afocal system. An afocal system may also form a real image of an object at a finite distance.

air path, equivalent The distance in air, which is equivalent to a distance in a medium. Equal to $\Sigma(D/n)$.

airy disk The central bright patch of the diffraction pattern, which is formed as the image of a point source. The disk size is defined by the diameter of the first dark ring of the pattern, equal to $1.21 \cdot$ wavelength/NA. Usually implies a perfect or near perfect lens with a circular aperture.

Glossary of Optical Design Terms

anamorphic   A system having a different magnification or focal length in each of two mutually perpendicular meridians. It may be composed of prisms or cylindrically surfaced lenses.

anastigmat   Strictly, without astigmatism. The term is usually applied to a lens system where an attempt has been made to flatten the field and reduce the astigmatism. An anastigmat often has a *node* with zero astigmatism at some field angle.

angle of incidence, refraction, reflection   The angle between an incident, refracted, or reflected ray and the normal to a surface at the point where the ray intersects the surface.

aperture stop   That feature of an optical system which most severely limits the diameter of the axial beam and which can pass through the system. The feature is usually the clear aperture of a lens element or a mechanical aperture, such as the iris diaphragm in a camera lens. The *chief* or *principal* ray crosses the axis at, and passes through the center of, the aperture stop. For many compound optical systems (e.g., a telescope) the aperture stop is located at the objective lens. Note that for off-axis points the beam size may be limited (vignetted) by more than one physical feature of the system.

aplanat(ic)   A lens (or surface) which is free of spherical aberration and coma.

apochromat   A lens in which three colors are brought to a common focus. Usually requires the use of materials with unusual partial dispersions. Pronounced APO-chro-mat, *not* a-POCH-ro-mat.

apodization   Where the transmission of a system is varied over the aperture in order to modify the diffraction pattern. Originally to eliminate the rings (feet = pod) around the *Airy* disk in the diffraction pattern. Similar effects may be produced by varying the beam intensity distribution, e.g., as in a gaussian beam.

apparent field of view   The angular field of view as seen by the eye through a telescope. The apparent field equals the *real* object field times the magnification of the telescope.

aspheric surface   A surface which departs from a spherical shape. The conic section surfaces (paraboloid, ellipsoid, and hyperboloid) are aspherics, as are more general aspheric surfaces. Aspherics are used to correct aberrations.

astigmatism   An aberration which causes a fan of rays in one meridian to focus at a different location than a fan in the orthogonal meridian. May be caused by a toroidal surface or by off-axis imagery. Primary astigmatism varies as the square of the field angle.

axial chromatic   An aberration which causes different wavelengths to be focused at different distances from the lens.

axial ray   A ray from the axial intercept of the object to the edge or margin of the entrance pupil. Marginal ray.

axis, optical   The common axis of rotational symmetry for an optical system. For a spherical surfaced element, the line connecting the centers of curvature of the surfaces.

back focal length or BFL   The distance from the vertex of the last surface of a system to the second focal point.

baffle   Opaque shielding to reduce or eliminate stray light.

beam expander   An afocal system (usually Galilean) used to increase the beam diameter and reduce the beam divergence, usually of a laser beam.

bending   The process of changing the shape of an element while maintaining its power constant.

binary optics   Diffractive optics where the ideal smooth surface is approximated by a stepped or staircase surface.

buried surface   A cemented interface separating two materials with very nearly identical refractive indices but with different dispersions. Used (now rarely) by some designers during the late stages of a design process to correct chromatic aberration without affecting the other aberrations.

cardinal points   The first and second focal points and the principal points; the Gauss points; the nodal points are often considered as cardinal points as well.

catadioptric and catoptric   Optical systems consisting of only mirrors (catoptric), or of mirrors and refracting surfaces (catadioptric). A purely refracting system is dioptric.

CDM (*chromatic difference of magnification*)   See **chromatic aberration** and **lateral chromatic**.

chief ray or principal ray   There are several definitions, depending on the use to which the concept is put:

- The oblique ray which passes through the center of the aperture stop of an optical system
- The central ray of a (vignetted) oblique beam
- The oblique ray aimed toward center of the entrance pupil

chromatic aberration   An aberration resulting from the dispersion of the materials used in an optical system. See **axial chromatic** and **lateral chromatic (color)**.

clear aperture   The diameter of the transmitting (or, if a mirror, the reflecting) portion of a surface, lens, or system.

coherent   Light equivalent to that from a true point source (spatial coherence) and/or monochromatic light (temporal coherence).

cold stop   A cooled aperture within a vacuum dewar in an infrared system, which is (ideally) the aperture stop or a pupil of the system. Its purpose is to prevent the detector from seeing anything but the optics and the imaged scene, especially the (warm) interior of the system.

collimated light   A beam of light wherein all the rays originating at a point are parallel to each other, and wherein all wavefronts are plane. Light from a point source at infinity is collimated. A collimated beam from a source which is not a true point expands as it travels; its divergence angle equals the source size divided by the focal length of the optical system.

coma An off-axis aberration where annular zones of the aperture have different magnifications. The resulting image of a point looks like a comet. Axial coma is produced by tilted surfaces. See also **Abbe sine condition** and **OSC**.

component One or more lens elements (usually in a group) which are treated as a unit.

concave surface A hollow curved surface, i.e., one which is lower in the center, *sunken*. The inner surface of a hollow sphere.

condenser The lens or component in a projection system which collects the light from the source and directs it through the aperture of the projection lens. A form of field lens. In Koehler illumination the condenser images the source into the projection lens pupil in order to produce uniform illumination on the screen.

conjugate Object and image are conjugates, as are the distances associated with them.

Conrady $(D - d)\delta n$ A technique used to control or correct chromatic aberration by making chromatic variation of the optical path of the marginal ray ($\Sigma D \delta n$) equal to that of the axial or principal ray ($\Sigma d \delta n$), where δn is the dispersion.

converging lens or surface One which bends rays toward the optical axis. A positive lens or surface.

convex surface A surface which is higher at the center than at the edge, outward curving, and bulging. The outer surface of a sphere.

coordinate system The optical axis is the z coordinate, the y coordinate is vertical, and x is the coordinate normal to the meridional plane. A right-handed system. The coordinate origin is usually located at the vertex of the surface.

cosine fourth The illumination in the image plane of a nominal optical system varies as the fourth power of the cosine of the angle of obliquity of the chief ray. Assumes no distortion of the image or the pupil.

critical angle See **TIR (total internal reflection)**.

crown An optical shop term for a convex shaped element.

crown glass A low dispersion glass. A glass with a V-value of more than 50 (for index > 1.6) or more than 55 (for index < 1.6). Named for the irregular lump in the center of the sheet when window glass was blown.

curvature The reciprocal of the surface radius. ($c = 1/r$) A measure of the departure from a flat surface.

$(D - d)\delta n$ See **Conrady $(D - d)\delta n$**.

damped least squares A modification of the least squares (which see) solution technique where the squares of the weighted parameter changes are added to the merit function (as a penalty, to prevent nonlinearities from producing extreme or impossible solutions.)

depth of focus (field) The longitudinal shift of the image sensor (e.g., film) (or of the object) which produces an acceptable image degradation. Depth of focus and depth of field are related by the longitudinal magnification. In photography the criterion is often based on the size of the defocus blur spot compared with the smallest perceptible or the smallest recordable blur. Another criterion is based on the Rayleigh quarter wave limit for OPD; this allows a depth of focus equal to $2 \cdot$ wavelength $\cdot (f/\#)^2$, or $0.5 \cdot$ wavelength/NA^2.

dialyte A lens consisting of two separate positive and negative elements or components.

diffraction The cause of the spreading or divergence of a wavefront which occurs when it encounters an obstruction such as an aperture or an opaque edge. The *airy disk* and the associated rings are caused by the diffraction resulting from the aperture of the optical system.

diffraction limited Strictly, when the system performance is limited solely by diffraction. Often applied to a system with a *strehl ratio* of 0.8 or more, or an OPD of one quarter wave or less.

diffractive optics Optics where the effects are produced by diffraction (as opposed to refraction). A diffractive lens surface is a Fresnel surface modulo 2π, i.e., with a step height optical path difference of one wavelength [$H = $ *wavelength*$/(n' - n)$].

diopter A measure of the power of a lens or surface equal to the reciprocal of the focal length in meters (or $(n' - n)/r$ for a surface). Distances (as reciprocals) may also be expressed in diopters.

dispersion The change of index with wavelength. For visual work it is usually taken as the index difference between the red and blue Fraunhofer hydrogen lines C (656.3 nm) and F (486.1 nm), thus $(n_F - n_C)$. This is the total or principal dispersion. See also *partial dispersion*.

distortion An aberration in which the magnification varies over the field of view. It is called pincushion or positive if the magnification increases toward the edge of the field, and barrel or negative if it decreases. Distortion reverses sign if object and image are interchanged.

diverging lens or surface One which bends light rays away from the optical axis. A negative lens or surface. A lens with a negative focal length.

doublet lens Either (*a*) a closely spaced or cemented pair of elements, one positive and one negative, or (*b*) two separated components with the stop between them.

effective focal length See **focal length**.

element A lens which is a single piece of glass (or a mirror).

empty magnification Magnification (in a telescope or microscope) which is larger than the magnification at which the diffraction limited resolution of the instrument matches the resolution of the eye. An increase in magnification which does not increase the information content of the image, although the increased magnification may make the image easier to recognize, locate or process, and also reduce the strain on the user. For a visual telescope, empty magnification begins when the magnifying power exceeds MP = 11 · D where D is the diameter of the objective aperture in inches. For a microscope it begins when the magnifying power exceeds MP = 225 · NA.

entrance pupil   The image of the aperture stop as seen from object space. All light rays passing through the system must enter through the entrance pupil. The principal/chief ray passes through the center of the pupil.

entrance window   The image of the field stop in object space.

erector lens   A lens which relays the image and reinverts an inverted image to produce a final erect image.

etendue or throughput   The product of the light beam area and the solid angle of the beam. It is constant through the system (which demonstrates the conservation of brightness/luminance/radiance). Related to the square of the Lagrange invariant.

exit pupil   The image of the aperture stop as seen from image space. All light rays passing through the optical system must emerge through the exit pupil. In a visual system the eye must be placed at the exit pupil to see the full field of view.

exit window   The image of the field stop in image space.

eye box   The area or volume within which the eye may be placed and specifications must be met. Usually specified for a system without an (imaged) exit pupil.

eye relief   The relief or clearance distance between the exit pupil (which is the usual location for the eye) and the last surface of a visual optical system such as a telescope or microscope. Sometimes a specified eye location, as in *eye box*.

field lens   A lens placed at or near an internal image of an optical device (e.g., telescope, periscope, or endoscope) in order to converge the oblique beams so that they pass through the clear apertures of the following components (e.g., the eye lens of a telescope). The lens is usually positive with the aim of widening the field of view. Occasionaly a negative field lens is used to flatten the Petzval field or increase the eye relief; this requires a larger diameter eye lens.

field curvature   The departure of the image surface from a plane, when the image is formed as a curved surface due to astigmatism and/or Petzval aberrations.

field flattener   A lens (usually negative) placed close to the image to flatten the Petzval curvature without greatly affecting the image size or the other aberrations. In some mirror systems it is a positive lens.

field of view   That part of the object which is included in the final image. May be expressed as an angle or as a linear dimension. Abbreviated *FOV*. The *real* FOV of a telescope is the angular field on the object side; the *apparent* field is the corresponding field on the image side.

field stop   An aperture, usually located at an image, which limits and defines the field of view.

fifth-order aberration   See **third-order aberration**.

first order   A term applied to paraxial calculations and characteristics.

fish-eye lens   A lens with a field of view of 180° or more. A reversed telephoto type with a large amount of barrel distortion.

flint   An optical shop term for a concave shaped element.

flint glass   An optical glass with a V-value less than 50 (for index > 1.6), or less than 55 (for index < 1.6). Named for the broken flints added to the melt in making fine glass for tableware.

f number   The speed or relative aperture of a lens system. The ratio of the effective focal length to the diameter of the entrance pupil. A measure of the illuminating capability of a lens. Usually written f/n where n is the f number, e.g., $f/6.3$, or 1:6.3, or f:6.3. Sometimes abbreviated as $f/\#$ or $f/$no. For an aplanatic system with the object at infinity, the f number equals 0.5/NA. See also **working f number**.

focal length   The effective (or equivalent) focal length is the distance from the second principal point to the second focal point. Often abbreviated *EFL* or simply f. The limiting value of $f = h'/\tan\theta$ as h' and θ approach zero, where h' is the image height and θ is the angle subtended by an infinitely distant object.

focal point   The image of an infinitely distant axial point source object. The *second* or back focal point is the image of a point which is to the left of the lens, and the *first* or front focal point is the image of a point to the right.

front focal length   The distance from the vertex of the first surface of a lens system to the first focal point, often abbreviated *FFL*.

fraunhofer lines   A series of dark lines in the solar spectrum, e.g., C, d, and F lines.

fresnel surface   A surface wherein annular zones are stepped and the surface of each zone has a curvature and slope corresponding to that of an ordinary surface. Used in condensing, signaling, and illuminating systems. Allows crude imagery with a thinner element. A diffractive lens is one with a Fresnel surface where the step height is $1/(n-1)$ (or about two) wavelengths.

fringes   Dark bands caused by interference, as seen in an interferometer or with a test plate.

f theta lens   A lens with barrel distortion whose image height is given by $h' = f \cdot \theta$, rather than $h' = f \cdot \tan\theta$.

galilean telescope   A telescope with a positive objective lens and a negative eye lens. It produces an erect image and covers a small field of view; usually of low power. Opera glass or field glass.

Gauss points   See **cardinal points**.

gaussian   Having to do with paraxial or first order.

gaussian beam A beam whose intensity cross section is described by a gaussian, e.g., $I(r) = I_0 \exp(-2r^2 w^2)$, where r is the radial distance and w is the value of r at which $I = I_0/e^2$. Often an idealized laser beam. See apodization.

glare stop An aperture placed at a pupil (usually internal) to block stray light.

gradient index An inhomogenous optical material in which the index varies. In a radial gradient, the index varies according to the distance from an axis. In an axial gradient, the index varies with the longitudinal position along the axis.

H-tan U plot A ray intercept plot where the intercept height is plotted against the tangent of the ray slope angle.

incoherent Light from an extended (nonpoint) source and/or nonmonochromatic light.

index See **refractive index**.

invariant In general, an expression which has the same value everywhere in an optical system. The Lagrange or optical invariant is the product of the index, the image (or object) height, and the half convergence (or divergence) angle of the axial beam. Often expressed as $hnu = h'n'u'$ or $m = h'/h = nu/n'u'$. At a general surface the invariant is given by $(y_p nu - y n u_p) = (y_p n'u' - y n' u'_p)$. *Throughput* or *etendue* is the three dimensional version of the invariant, and can be regarded as the product of the area of the object (or image) times the solid angle of collection (or illumination); also the pupil area times the solid angle of the field. Another paraxial invariant (used in stop shift theory) is $\delta y_p / y$, where δy_p is the change in principal ray height at a surface produced by the shift of the stop and y is the axial ray height at the surface.

lateral chromatic (color) The variation of magnification or image height with wavelength. The *chromatic difference of magnification*, CDM = $\delta h'/h'$.

least squares An optimization method which minimizes the sum of the squares of the operands in the merit function, based on the assumption that the operands are linearly related to the variable parameters. It will locate the nearest optimum in parameter space. See damped least squares.

line spread function The distribution of illumination in the width of the image of a line.

longitudinal Having to do with distance measured along, or parallel to, the optical axis.

macro lens A lens corrected, or adjustable, for use with near-by objects.

magnification, angular Telescope magnification. The angle subtended by the image divided by the angle subtended by the object. Usually the ratio of the tangents of the half angles of the fields.

magnification: lateral, linear, or transverse The ratio of image height to object height, measured normal to the axis.

magnification: longitudinal The ratio of the longitudinal motion of the image to that of the object, or the longitudinal length of the image to that of the object.

For small motions or lengths it is equal (in the limit) to the square of the lateral magnification, and is thus always positive, so that the image and object always move in the same direction.

magnification: microscopic   In a microscope or magnifier, the ratio of the angle subtended by the image to that subtended by the object, when the object is viewed at a conventional distance of 10 in. (250 mm), which is assumed to be the distance of most distinct vision. Thus if the image is at infinity, MP = 10 in./f = 250 mm/f, where f is the focal length of the magnifier or microscope.

marginal ray   The ray (usually from an axial object point) through the edge or margin of the lens aperture. An axial ray.

member   In a photographic lens the *front member* consists of those elements before the aperture stop or iris diaphragm, and the *rear member* consists of those following the stop.

meridional plane   Any plane which includes the optical axis—the *tangential plane*.

meridional ray   A ray which lies in the *meridional plane*. In an axially symmetrical system a meridional ray can never leave the meridional plane.

merit function   A collection of weighted terms (or targets, or *operands*) which are squared and summed. The terms may represent aberrations, physical dimensions, magnifications, image locations, or any characteristics which can be calculated; very often it is their departure from a desired value. Since each entry represents an undesired characteristic, the merit function might better be called a defect function or error function. The purpose of the merit function is to represent the worth of an optical system with a single number; thus the smaller the number, the better the system.

microscope, compound   A two component system where the objective lens forms a magnified image of a small object which is further magnified by the eyepiece.

microscope, simple   A magnifying glass.

modulation   The contrast of an object or image whose luminance or illuminance varies sinusoidally, defined as

$$M = (\max - \min)/(\max + \min)$$

where max and min are the maximum and minimum levels of luminance or illuminance.

MTF (*modulation transfer function*)   The ratio of the image modulation or contrast to that of the object, expressed as a function of the spatial frequency, where the object modulation is a sinusoidal variation of brightness or radiance. The real part of the complex *optical transfer function*, in which the imaginary part is the *phase transfer function*. Formerly known as sine wave response, frequency response, and contrast transfer function.

NA See **numerical aperture**.

narcissus The image of a cold detector formed by reflection from a surface of the optical system, which when (nearly) in focus, produces a dark central spot in the image.

nodal points Two axial points such that an oblique paraxial ray aimed toward the first nodal point emerges from the lens parallel to its original direction, and appears to come from the second nodal point. If the system is immersed in the same medium on both sides (e.g., air), the nodal points and the principal points are coincident.

node In an anastigmat, the image height at which the astigmatism is zero, i.e., where the sagittal and tangential fields cross.

null lens An optical system designed to convert the nominal design wavefront of a system under test into a spherical wavefront.

numerical aperture A measure of the convergence or divergence of a light beam, often abbreviated NA. $NA = n \cdot \sin U$, where n is the index and U is the half angle of convergence or divergence. For an object at infinity $NA = 0.5/f$ number, or f number $= 0.5/NA$.

objective lens In a camera, telescope, microscope, or other optical system, the lens which is closest to the object.

oblique beam/ray A beam or ray originating at an off axis object point.

offence against the sine condition (OSC) See **Abbe sine condition**.

operand The name commonly used for an entry or target in the merit function. It usually defines a system characteristic which is desired to be minimized or controlled.

optical axis The common axis of symmetry of a lens or optical system. See **axis, optical**.

optical path (length) The index times the path distance along a ray. $OP = \Sigma (n \cdot D)$. Related to the transit time of light through a system.

optical path difference (OPD) Wavefront aberration. The departure of the actual wavefront from an ideal spherical wavefront. The difference between the optical paths, $\Sigma(n \cdot D)$, of two rays measured from their origin at the same point to their intersection with a reference sphere centered on the ideal image point.

optical transfer function (OTF) The complex function of spatial frequency used to describe the imagery of an optical system. It consists of the real part, the MTF or modulation transfer function, and the imaginary part, the PTF or phase transfer function.

optimum, global The best possible design form for an optical system, i.e., that with the smallest possible merit function. In practice, it is unknowable for even a modestly complex system.

optimum, local   A solution to the design problem which represents a minimum value of the merit function, such that a small parameter change in any direction produces an increase in the merit function. An optical system may have many local optima.

OSC   The *offence against the sine condition*. See **Abbe sine condition**.

parameter   See **Variable**.

paraxial   A region where all angles are treated as infinitesimals, so that $\alpha = \sin \alpha = \tan \alpha$ and the equations for raytracing are simple linear (i.e., nontrigonometric) expressions. "A thin threadlike region about the optical axis." Paraxial equations describe the imagery of perfect, aberration-free optical systems. Also called *first order* or *gaussian*.

paraxial ray   A ray traced according to the refraction law $ni = n'i'$. A ray whose height and angles are infinitesimals, i.e., approaching zero. The linearity of the equations allows the raytrace to use large, finite angles and heights.

partial dispersion   The difference in refractive index for two wavelengths, expressed as a fraction of the total dispersion, e.g., $P_{Fd} = (n_F - n_d)/(n_F - n_C)$.

periscope   An optical system used to image a relatively wide field of view through a long, narrow space. It usually consists of an objective, followed by alternating field and relay lenses, and is terminated by an eyepiece or camera. A medical endoscope is a miniature periscope. A coherent fiber bundle or a radial index gradient rod may substitute for the field lens-relay lens system. Alternate definition: A rhomboid prism or a pair of parallel mirrors used to displace, but not deviate, the line of sight, e.g., a child's toy periscope.

Petzval sum   A measure of the basic field curvature of a lens, equal to $\Sigma(n' - n)/nn'r$. ρ is the radius of the Petzval surface at the optical axis, and equals (minus) the reciprocal of the Petzval sum. For a thin element $\rho = -n/\phi = -nf$.

point spread function   The distribution of illumination in the image of a point.

power   The power of a lens is the reciprocal of its effective focal length. The power of a surface is equal to $(n' - n)/r$. If the dimensions are in meters, the unit of power is the *diopter*. A positive power, converging, lens or surface bends rays toward the axis; a negative, diverging, lens or surface bends rays away from the axis. *Power* may also refer to the magnifying power of a telescope, microscope, or magnifier.

principal plane   The hypothetical surface in a lens at which it appears that an incoming paraxial ray, parallel to the optical axis, is bent. If the incoming and the emerging rays are extended until they intersect, the intersection point is on the principal plane. The *second* principal plane is defined by rays from the left, and the *first* by rays from the right. For real trigonometric (i.e., not paraxial) rays, the surface is curved, approximating a sphere centered on the object or image. The surface is a plane only in the infinitesimal paraxial region. See also cardinal points and focal length.

principal point   The point at which the principal plane (or surface) intersects the optical axis.

principal ray   See chief ray. (Infrequently, a ray directed toward the first principal point.)

pupil   Any image of the aperture stop.

ray, xxx ...   See axial ray, chief ray, marginal ray, meridional ray, oblique ray, paraxial ray, principal ray, rim ray, sagittal ray, trigonometric ray, exact ray, tangential ray, zonal ray.

ray intercept plot   A plot of the intercept locations of a fan of rays against the relative position of the ray in the lens aperture or pupil. Usually plotted for a fan of tangential or sagittal rays. The plotted location is usually the difference between the intersection location of the ray and a reference ray, such as the axis or the principal ray. For tangential rays the y coordinate is plotted; for sagittal rays, the x coordinate. The relative pupil position of the ray is sometimes given as the tangent of the ray slope, hence the term "H-tan U plot."

Rayleigh criterion: image quality   If the OPD or wave front deformation is less than one-quarter of a wavelength, the image is considered to be *sensibly perfect*. A system which meets this criterion is commonly (but incorrectly) referred to as *diffraction limited*.

Rayleigh criterion: resolution   Two points are assumed to be resolvable by a perfect optical system if their images are separated by 0.61(wavelength)/NA, or if the object points have an angular separation of 1.22(wavelength)/D radians (or 5.5/D seconds of arc, where D is the entrance pupil diameter in inches and visual wavelengths are assumed.) The Sparrow and the Dawes resolution criteria are approximately 20 percent smaller.

real field of view   The field of view in object space.

real image   An image which can be formed on a screen, as opposed to a virtual image, which is located inside the optical system and cannot be directly accessed.

refractive index   The ratio of the velocity of light in vacuum (or commonly, in air) to its velocity in the medium being characterized ($n = c/v$). Related to the bending of a light ray at a surface, as in Snell's law: $n \sin I = n' \sin I'$.

relative aperture   The ratio of the aperture of a lens to its focal length. May be given as a ratio, 1:6.3 or as a fraction, $f/6.3$. In this example 6.3 is called the f number. With an object at infinity, f number = 0.5/NA.

relay lens   A lens used to transfer an image longitudinally from one location to another. It reimages and erects a real internal image, as in a periscope or terrestrial telescope.

resolution   The ability to separate or distinguish individual points or lines (or arrays of points or lines) in an image.

reversed telephoto: retrofocus   A lens system whose back focus is longer than its effective focal length. It consists of a negative front component followed by a positive rear component.

rim ray   A ray through the edge/rim of the aperture. The *upper* and *lower* rim rays are the meridional rays through the top and bottom of the aperture.

rms   The acronym for *r*oot *m*ean *s*quare, the square root of the average of the squares of a set of numbers. The rms spot size and the rms OPD are often used as operands in a merit function.

RSS   The acronym for the square *r*oot of the *s*um of the *s*quares. Used to assess the combined effect of multiple random fabrication tolerances.

sagittal plane   A plane normal to the meridional or tangential plane. It is usually defined by the principal ray in object space, and is thus not a simple continuous plane through the system.

sagittal ray   A ray in the sagittal plane, usually as defined in object space.

secondary spectrum   The residual chromatic aberration when the primary chromatic aberration has been corrected. For example, if red and blue light have been brought to a common focus, the distance from that focus to the yellow-green focus is the secondary spectrum.

Seidel aberrations   The third-order aberrations, which are spherical, coma, astigmatism, Petzval, and distortion. Also called primary aberrations.

semicoherent   Light from a very small but finite sized source and/or light with a small spectral bandwidth. The illumination produced when the image of the light source only partially fills the pupil of the lens, as in microscopy or microlithography.

simulated annealing   A controlled ramdom search design process which allows the lens design to temporarily become worse in order to find a new optimum.

Snell's law   The change in the direction of a ray crossing the boundary between two media is governed by Snell's law, which is:

$$n \sin I = n' \sin I'$$

where n and n' are the refractive indices of the two media, and I and I' are the angles of incidence and refraction (the angle between the ray and the surface normal).

solve   A capability of a design program which algebraicly solves for a construction parameter (such as a curvature or spacing) which will produce a desired paraxial ray slope or intersection height. The ray is usually the axial marginal ray or the principal/chief ray. An *angle solve* or a *height solve*.

speed   See *f* number.

spherical aberration   The difference between the focus location of rays through the center of a lens aperture (i.e., paraxial rays) and those through the margin (or other parts) of the aperture.

spherochromatism   The variation of spherical aberration with wavelength.

spot diagram   A plot of the intersection points of rays from an object point in the image surface, where each intersection point is represented by a spot. If the rays are uniformly distributed in the aperture, the spot diagram is a representation of the illumination distribution in the image, when diffraction effects are neglected. Usually several hundred rays are plotted, often in several colors, to make a single diagram.

stigmatic   Perfect imagery; all rays from an object point pass through the same image point.

stop   See **aperture stop** and **field stop**.

Strehl ratio/definition   The ratio of the peak illuminance in the point spread function of an aberrated lens to the peak illuminance for a perfect lens. A Strehl ratio (SR) of 80 percent (the Marechal criterion) is often regarded as the equivalent of the Rayleigh quarter-wave criterion. SR is most useful in comparing well corrected systems.

symbols   Many symbols are defined in Secs. 24.1 and 24.2 of Chap. 24.

tangential plane   The plane containing the optical axis—a *meridional plane*. *Tangential rays* lie in the tangential plane.

telecentric   A system with the *entrance* and/or *exit pupil* located at infinity, so that the associated principal ray is parallel with the axis. Used to prevent a change in image size when the system is slightly defocused, as in microlithography, or to optimally illuminate an LCD.

telephoto lens   A lens whose length from the first surface to its focal point is shorter than its effective focal length. The ratio of the two is called the telephoto ratio, which for a true telephoto lens is less than one. The lens often consists of a positive front component followed by a negative rear component. A term sometimes incorrectly applied to an ordinary lens of long focal length.

telescope   An *afocal system* which produces a magnified image of a distant object. A Keplerian or astronomical telescope comprises two positive components and has an inverted image. A Galilean or Dutch telescope has a positive objective and a negative eyepiece and produces an erect image. A terrestrial telescope includes an erecting lens component.

thin lens   A concept which is useful in preliminary system layout. It assumes that optical components have zero axial thickness, and therefore the principal points and the lens are all coincident.

third order   Aberrations which (in transverse measure) vary as the combined exponents of the aperture (y) and field (h) equaling three, e.g., y^3, y^2h, yh^2, and h^3. The five Seidel aberrations are: spherical, coma, astigmatism, Petzval, and distortion. In fifth-order aberrations the exponents add up to five, and there are two additional aberrations—elliptical coma and oblique spherical aberration.

throughput or etendue   See **invariant**. The invariant product of pupil (or beam) area and the solid angle field of view.

TIR (*total internal reflection*) In passing from a higher index to a lower, a ray is totally reflected back into the higher index medium if the angle of incidence exceeds the critical angle, equal to the arcsin (n'/n).

T-number An equivalent f number which includes the effects of transmission.

$$T\text{-number} = \frac{f\,\text{number}}{\sqrt{\text{transmission}}}$$

track length; total track (TT) The object to image distance.

transverse Measured in a direction normal to the optical axis.

trigonometric ray Also called an *exact ray*. A ray whose path is traced according to Snell's law (which see), as opposed to a paraxial ray.

variable A construction element, such as surface curvature, surface spacing, material index or dispersion, surface asphericity, and the like, which may be varied for the purpose of improving a system. A variable parameter.

vertex length The axial distance from the first optical surface of a system to the last.

vignetting The mechanical clipping or obscuration of the edges of oblique beams by the apertures of elements spaced away from the stop. It reduces the off-axis image illumination (in addition to the cosine-fourth reduction). Often introduced to reduce manufacturing cost and/or to block aberrated portions of the beam. In visual or photographic systems as much as 50% vignetting is not uncommon.

virtual image An image formed within the optical system and which therefor cannot be accessed or focused on a screen, as opposed to a *real* image which can.

visible light Light to which the human eye is sensitive. Usually considered to include wavelengths from 380 nm to 780 nm.

V-value See **Abbe V-number**.

wavefront A surface wherein all points have the same optical path distance (Σnd) from the object point, i.e., where the light has the same phase.

wavefront aberration See optical path difference. The departure of the wavefront from a perfectly spherical shape.

working *f* number Describing the convergence of the imaging cone when the object is at a finite distance; equal to 0.5/NA. This is not an established convention; caution is advised. As opposed to the conventional "infinity" f number.

zonal aberration Aberration of rays in the midzones (e.g., 0.7) of the aperture or field, usually when aberrations of the central and marginal zones are corrected.

zonal ray Usually the ray whose height in the aperture is 0.707 of the height of the marginal ray.

References

1. Conrady, A. E. *Applied Optics and Optical Design,* Part 1, (1929) 1957, Dover, New York.
2. Conrady, A. E. and R. Kingslake. *Applied Optics and Optical Design,* Part 2, 1957, Dover, New York.
3. Cox, A. *A System of Optical Design,* 1964, Focal, New York.
4. Fischer and Tadic. *Optical System Design,* 2000, McGraw-Hill, New York.
5. Geary, J. *Introduction to Lens Design,* 2002, Willmann-Bell, New York.
6. Hartmann and Smith (eds.). Infrared optical design and fabrication, 1991, *Crit. Rev. CR38,* SPIE, Bellingham, WA.
7. Hopkins, R. in *Handbook of Optical Design, MIL-HDBK-141,* 1962, U.S. Govt. Prtg. Office. (Republished 1987 by Sinclair Optics, Fairport, NY.)
8. Kidger, M. *Fundamental Optical Design,* 2002, SPIE, Bellingham, WA.
9. Kidger, M. *Intermediate Optical Design,* 2004, SPIE, Bellingham, WA.
10. Kingslake, R. (ed.) *Applied Optics and Optical Engineering,* a series with several editors, 1965 to 1992, Academic, New York. Especially of interest:
    Vol. III, 1965, Ch. 1 Lens Design, R. Kingslake.
    Vol. III, 1965, Ch. 3 Photographic Objectives, G. Cook.
    Vol. III, 1965, Ch. 9 Eyepieces & Magnifiers, S. Rosin.
    Vol. VIII, 1985, Ch. 1 Photographic Lenses, E. Betensky.
    Vol. VIII, 1985, Ch. 3 Aspheric Surfaces, R. Shannon.
    Vol. VIII, 1985, Ch. 4 Automated Lens Design, W. Peck.
    Vol. VIII, 1985, Ch. 6 The Calculation of Image Quality, W. Wetherell.
    Vol. X, 1987, Ch. 3 Afocal Lenses, W. Wetherell.
11. Kingslake, R. *Lens Design Fundamentals,* 1978, Academic, New York.
12. Kingslake, R. *Optical System Design,* 1983, Academic, New York.
13. Kingslake, R. *A History of the Photographic Lens,* 1989, Academic, New York.
14. Kingslake, R. *Optics in Photography,* 1992, SPIE, Bellingham, WA.
15. Korsch, D. *Reflective Optics,* 1992, Academic, New York.
16. Laikin, M. *Lens Design,* 2001, Dekker, New York.
17. Mouroulis, P. *Visual Instrumentation,* 1999, McGraw-Hill, New York.
18. Opt. Soc. Am. In Bass, M. (ed.), *Handbook of Optics,* 1995, McGraw-Hill, New York.
    Vol. I, part 9: *Optical Design Techniques,* Chs. 32–39.
    Vol. II, part 1: *Optical Elements,* Chs. 1–14.
    Vol. II, part 2: Ch. 16, Camera Lenses, Betensky et al. Ch. 17, Microscopes, Inoue et al. Ch. 18, Reflective and Catadioptric Objectives, L. Jones.
    Vol. II, part 4: Ch. 34, Polymeric Optics, J. Lytle.
19. Riedl, M. *Optical Design Fundamentals for Infrared Systems,* 1995, v. TT20, SPIE, Bellingham, WA.
20. Rutten and van Venrooj. *Telescope Optics,* 1988, Willmann-Bell, New York.
21. Schroeder, D. *Astronomical Optics,* 1987, Academic, New York.
22. Shannon, R. *The art and Science of Optical Design,* 1997, Cambridge, Cambridge, UK.
23. Smith, W. *Modern Optical Engineering: The Design of Optical Systems,* 3rd ed., 2000, McGraw-Hill, New York.
24. Smith, W. *Practical Optical System Layout,* 1997, McGraw-Hill, New York.
25. Smith, W. (ed.) Lens design, 1992, *Crit. Rev., v. CR41,* SPIE, Bellingham, WA.
26. Welford, W. *Aberrations of the Symmetrical Optical System,* 1974, Academic, New York.
27. Wolfe, W. (ed.) *Optical Engineer's Desk Reference,* 2003, OSA, Washington, DC and SPIE, Bellingham, WA.
28. Proceedings of The International Lens Design Conference and The International Optical Design Conference, 1980, 1985, 1990, 1994, 1998, 2002, SPIE, Bellingham, WA.

Index

A
AA (angular aberration), 95
Aberration(s):
  angular, 95
  balancing, 60–66
  chromatic, 63–64
  conversion to wavefront deformation, 601–603
  Cooke triplet anastigmats, 203
  fifth-order, 15, 101
  higher-order, 101
  induced, 96, 124
  intrinsic, 96
  relationship between, 5–6
  seidel, 13
  seventh-order, 101
  source of, 4–5
  third-order, 13
  wavefront, 95, 101–102
  zonal, 16, 123–124
  [see also Spherical aberration(s); Third-order aberrations]
Aberration plots, 90–92
Achromat, 125
  new, 266, 272, 275
  old, 272, 275
Achromatic diffractive doublets, 149–150
Achromatic diffractive singlets, 148–149
Additive tolerances, 578–583
Aerial image modulation (AIM) curve, 82
Airspaces, 33, 576–577
Amici objective, 444
Anastigmats (see Cooke triplet anastigmats; Double-meniscus anastigmats)
Angle solves, 7, 14, 16–17
Angles, symbols for, 588
Angular aberration (AA), 95

Aperture coefficient of coma (S2T), 16
Aperture stop, 21–22
Apochromat, 125, 358
Apochromatic triplet ($f/7$), 133–145
  and characteristics of apochromatic triplets, 134–135
  design process for, 135–145
  partial dispersion in, 133
Aspheric correctors, 473, 475
Aspheric surfaces (surface asphericity):
  Cooke triplet anastigmats, 237, 239–246
  design improvement using, 68–69
  and third-order aberrations, 601
  variables for, 35–36
Asphericity (irregularity), 574
Aspherics, spherical aberration of conic, 5
Astigmatism, 6
  balancing, 64–66
  in eyepieces, 153
  transverse astigmatism contribution, 109
Automatic lens design, xiv, 11
Aviar lens (see Dogmar lens)

B
Back focal length (BFL), 18
Betensky, Ellis, 215, 222, 299
Betensky plastic aspheric triplet camera lens, 215, 222–223
BFL (back focal length), 18
Binary surface, 145
Binocular, eyepiece for 6×30, 160–176
  complexity of designing, 160
  DLS optimization for, 163–165
  with equi-convex crowns, 164–168
  inverted Kellner design, 160–161
  material factors affecting, 174, 176

Binocular, eyepiece for 6 × 30 (*Cont.*):
  MFT plot for, 161, 163
  with plano surfaces, 169–174
  Plössl eyepiece vs., 170, 171, 174–175
  rough starting layout for, 161–162
Binocular lenses, 437
Biocular lenses, 437
Biotar (double-Gauss) lens, 319–353
  "doubled double-Gauss" relay, 350–353
  and double-meniscus form, 319
  eight-element version of, 340, 347–350
  origins of, 332
  seven-element (broken contact front doublet) version of, 340–342
  seven-element (one compounded outer element) version of, 340, 343–346
  seven-element (split-rear crown) version of, 334–340
  six-element version of, 319–331
  tips for working with, 332–334
Books on lens design, 1–2
Bouwers system, 469, 477, 482
Brueke style magnifiers, 155

C

Caldwell, Brian, xiv
Camera obscura, 21
Cardinal points, 588–590
Cassegrain systems, 456–458, 460, 461, 464, 477
  Dall-Kirkham system, 459, 460
  Petzval curvature correction in, 469
  Ritchey-Chretien design, 459, 460, 462, 465, 473
Cathode-ray tube (CRT), 437, 553
CCD (charge-coupled device) cameras, 247
CDM (chromatic difference of magnification), 96
Celor lens (*see* Dogmar lens)
Cemented doublets:
  design of telescope objective ($f/7$), 115–122
  separating, 55
Charge-coupled device (CCD) cameras, 247
Chromatic aberration, 63–64
Chromatic difference of magnification (CDM), 96
Close-up (macro) lenses, 356–357
Coma:
  aperture coefficient of, 16
  for apochromatic triplets, 139, 140
  balancing, 66
  in eyepieces, 151
  tangential, 6
  of telescope objective, 114
  of zero field size systems, 18
Compensating eyepiece, 154
Compounding (of elements), 55–57
Computers, xiii–xiv
Confocal paraboloids, 476
Conic aspherics, spherical aberration of, 5
Conrady $D - d$, 16, 114–115
Constraints, 31
Convertible lenses, 302
Cooke triplet anastigmats (Cooke triplets), 20–21, 33, 201–246
  aspherizing surfaces of, 237, 239–246
  Betensky plastic aspheric triplet camera lens, 215, 222–223
  camera lens anastigmat design ($f/4$), 223–235
  field-aperture combinations in, 210–221
  glass choice for, 205, 236–239
  improvements to, 234, 236
  increasing element thickness in, 244, 246
  index of positive elements in, 209, 210
  inward-curving Petzval field for, 210
  rare earth glasses with, 236–239
  with small field, 208–210
  solutions for primary aberrations in, 201–205
  and Tessar, 259
  undercorrected axial chromatic in, 210
  vertex length, control of, 206–207
Cracked crown triplet, 54
Crown element, 125–128
Crowns, equi-convex, 164–168
CRT (*see* Cathode-ray tube)
Curvatures, variables for, 33

D

Dagor lens, 304
  about, 302
  split, 305–308
Dall-Kirkham system, 459, 460
Damped least-squares (DLS) optimization, 20, 49
  for 6 × 30 binocular eyepiece, 163
  about, 11–13
Defect (error) function, xiv
Defined tolerances, 573–574
Defocusing, 34

Index

"Design of a Triplet Anastigmat of the Taylor Type" (R. E. Stephens), 203
Diffraction MTF, 96–97
Diffractive lens, manufacturability of, 146–147
Diffractive surface, 145
Distance, symbols for, 588
Distortion, 6
　in eyepieces, 151
　and spherical aberration of pupil, 152
DLS optimization (*see* Damped least-squares optimization)
Dogmar (Celor, Aviar) lens, 272, 305, 307–309
　in camera lens anastigmat design, 305, 307–317
"Doubled double-Gauss" relay, 350–353
Double-Gauss lens, 110, 272, 305
　(*See also* Biotar lens)
Double-meniscus anastigmats, 297–317
　Betensky camera lens design, 299, 301
　and Biotar lens, 319
　convertible lenses, 302
　Dagor lens, 302, 304–308
　"descendents" of, 302
　Dogmar lens, 305, 307–309
　Hypergon lens, 297–299
　Metrogon lens, 299
　Protar lens, 302, 303
　split Dagor lens, 305–308
　Topogon lens, 299, 300
　(*See also* Biotar lens)
Doublets, 55–57
　achromatic diffractive, 149–150
　cemented, 115–122
　of high index glass, 121
　of low index glass, 122
　secondary spectrum of, 125, 129–131
　separating cemented, 55
Drawings, lens, 89–90

E

Edmund RKE eyepiece, 160, 164
Effective focal length (EFL), 15, 18, 224, 332
Enlarger lens, 266, 272
Equi-convex crowns, 164–168
Erfle eyepiece, 178
　with field flattener, 196
　five-element, 187–189, 191–193, 200
　six-element, 187, 189–191, 195, 197, 198, 200

Ernostar lens, 253, 257–258, 319
Error (defect) function, xiv
Euryplan lens, 305
Evaluation, 71–84
　fabrication considerations in, 83–84
　of geometric blur spot size vs. certain aberrations, 80–82
　and interpretation of MTF, 82–83
　of OPD vs. measures of performance, 71–80
　preliminary, 71
Evaluation plots, 103–108
Exit pupil:
　of eyepieces, 151
　raytracing from, 154–155
Eye relief, 151
Eyepiece(s), 151–200
　astigmatism in, 153
　compensating, 154
　exit pupil of, 151
　field curvature in, 153, 154
　five-element, 187–189
　four-element, 176–186
　Huygens, 155, 158
　Kellner, 155, 158, 159
　and magnifiers, 155–157
　and prisms, 153
　problems with, 151–152
　Ramsden, 155, 158–159
　range of characteristics of, 152–153
　six- and seven-element, 200
　for 6×30 binocular, 160–176
　very high-index, 187, 190–299

F

Field curvature:
　in eyepieces, 153, 154
　Petzval, 5, 44, 153
　sagittal, 6
　tangential, 6
Fifth-order aberrations:
　about, 101
　merit functions based on, 15
First-order layout, making a, 43–44
Fish-eye lens design, 402, 408–413
Flint glasses, 52
FLIR scanning system (*see* Forward-looking infrared scanning system)
Focal length:
　angle solves to control, 17
　back, 18
　effective (*see* Effective focal length)
　symbol for, 588

Focal power, symbol for, 588
Forward-looking infrared (FLIR)
  scanning system, 507, 511
Fraunhofer lens, 110, 112, 117, 118
Freezing (of variables), 20–21
Fresnel surface, diffractive surface as, 145
Front meniscus, 21–27

G
Gabor-Bouwers-Maksutov principle, 477
Gauss lens, 110, 111
Gauss space, 110
Gaussian image height (GIH), 91
Generalized simulated annealing, 30–31
GENII, xv, 86, 352
Geometric MTF, 96–97
GIH (Gaussian image height), 91
Glass, 51–52
  for Cooke triplet anastigmats, 205, 236–239
  for telephoto lenses, 383, 385
Glass map, 31–33
Global optimization, 30
Graphic representations (plots), 8
Gregorian mirror configuration, 456–458, 460, 466–467
Grey, David, 14

H
Head/helmet-mounted display (HMD) lenses, 439
Head-up display (HUD) lenses, 437, 439
Height, symbols for, 587
Height solves, 7, 14, 17
Hektor lens, 57, 272, 274
Heliar lens, 266, 268–272
  camera lens design example, 285–295
Higher-order aberrations, 101, 124
Hill sky lens, 402
A History of the Photographic Lens (Rudolf Kingslake), 6
HMD (head/helmet-mounted display) lenses, 439
Hopkins, R. E., 203
Hubble Space Telescope, 459
HUD lenses (*see* Head-up display lenses)
Huygens eyepieces, 155, 158
Hypergon anastigmats, 27
Hypergon lens, 297–299

I
Image equations, 590–592
Improvement techniques, 47–69
  with aspheric surfaces, 68–69
  for astigmatism, 64–66
  and balancing of aberrations, 60–66
  cemented doublet, separation of, 55
  for chromatic aberration, 63–64
  for coma, 66
  compounding of elements, 55–57
  Cooke triplet anastigmats, 234, 236
  and elimination of weak elements, 60
  general tips, 49–50
  glass changes, 51–52
  and Merté surfaces, 57–58
  for Petzval field curvature, 64–66
  for spherical aberration, 61–63
  for spherochromatism, 63–64
  splitting of elements, 52–55
  summary of, 47–49
  and symmetrical principle, 67–68
  vignetting, 58–59
Index of refraction, symbol for, 588
Induced aberrations, 96, 124
Infrared systems, 503–511
  objective lenses, 504–510
  optics of, 503–504
  in telescopes, 507, 511–513
Intercept plots, ray, 98–102
Internal focusing eyepiece, 194, 200
Intrinsic aberrations, 96
Inverted Kellner design, 160–161
Iris, 295
Irregularity (asphericity), 574

K
Kellner eyepiece, 155, 158–161
Kingslake, Rudolf, 6, 63, 259
Kinoform, 145
Konig USP eyepiece, 160

L
LA (longitudinal measure), 95
Lambda Research Corp., xiv
Landscape lens, 21–27
  "descendants" of, 27
  front vs. rear meniscus forms of, 21–27
  optical system of, 21
Lanthanum (rare earth) glasses, 236–239
Laser beam expanders, 511, 514
Laser disk lens, 565–571
Least-squares optimization, damped, 11–13, 20

Index

LED (light-emitting diode), 429
Lee, 332
Lens design:
 automatic, xiv, 11
 commonly asked questions about, 4–7
 existence of multiple optima in, 22
 hypothetical/generic process of, 45–46
 resources on, 1–2
 and "starting over," 373
Lens drawings, 89–90
Lens prescriptions, 87–89
LensVIEW, xiv
Light-emitting diode (LED), 429
Limiting resolution, 82
Lister objective, 442
Local minima, 19–21
Local optimum, xiv
Longitudinal measure (LA), 95

M

Macro lenses, 553, 555–559
Macro (close-up) lenses, 356–357
Magnifiers, 156–157
 about, 155
 very high-index, 187, 190–299
Magnifying power (MP), 511
Maksutov system, 469, 477, 482
Mandler, 329
Mangin mirror, 469, 473
Marechal criterion, 72–73
Materials:
 characteristics of, 31–33
 and lens fabrication, 83–84
Melt fits, 38–39
Melt sheet, 38, 39
Meniscus, front vs. rear, 21–27
 (*See also* Double-meniscus anastigmats)
Merit function(s), xiv, 13–19, 110, 114–115
 based on third- and fifth-order aberrations, 15
 default, 110
 hypothetical two-variable, 19–20
 minimum ray-sets for, 15
 and MTF, 18–19
 as n-dimensional space, 14
 operands of typical, 15–16
 and optimization, 11
 sampling by, 15
 third-order, 16
 and tolerance budget, 578
 types of, 28–29
Merté surfaces, 57–58, 272

Metrogon lens, 299
Microlithographic lenses, 514–520
Microscope objectives, 441–454
 aplanatic elements/surfaces in, 442–445
 classical design forms, 442–443
 designs for, 447–454
 flat-field objectives, 446
 general characteristics of, 441–442
 reflecting objectives, 446–447
Microscopes, long-eye-relief eyepiece for, 176–177
Minima, local, 19–21
Mirror systems, 455–502
 advantages of, 455
 aspheric refracting correctors for, 473
 Catadioptric systems, 469–474
 classical two-mirror systems, 456–469
 confocal paraboloids, 476
 Dall-Kirkham system, 459, 460
 disadvantages of, 455–456
 Gregorian configuration, 456–458, 460, 466–467
 Ritchey-Chretien design, 459, 460, 462, 465
 Schmidt systems, 473, 475–476
 Schmidt-Cassegrain design example, 482–502
 Schwarzschild system, 456, 457, 460, 462, 463, 467–468
 unobscured systems, 476–482
 (*See also* Cassegrain systems)
Modulation transfer function (MTF), 16, 50, 65, 71, 74–80
 for 6 × 30 binocular eyepiece, 161
 for Betensky triplet, 222
 for cemented doublet telescope objective, 117, 118
 diffraction MTF, 96–97
 geometric MTF, 96–97
 interpretation of, 82–83
 and merit function, 18–19
Monocentric systems, 67, 68
Monochromatic systems, 561
MP (magnifying power), 511
MTF (*See* Modulation transfer function)
Multiconfiguration feature (of lens design programs), 12–13

N

NA (*see* Numerical aperture)
Nagler eyepiece, 199, 200
Narcissus, 511

Natural stop position, 21
New achromat, 266, 272, 275
Node, 64
Numerical aperture (NA), 153, 441, 456
Nuquist frequency, 82

O
Objectives (see Microscope objectives; Telescope objectives)
Off-axis parabola, 476
Offence against the sine condition (OSC), 16
Old achromat, 272, 275
OP (optical path), 96
OPD (see Optical path difference)
Operands (targets), 8, 13
 of typical aberration merit function, 15–16
 weighting of, 14–15
Optical path (OP), 96
Optical path difference (OPD), 15, 16, 96
 derivation of, 101–102
 performance measures vs., 71–80
 and wavefront deformation, 601–603
Optimization:
 damped least-squares, 11–13
 global, 30
 and merit function, 11
 variables for, 31–33
Optimum, local, xiv
Orthometar lens, 305
Orthoscopic eyepieces:
 classical, 178, 179
 as magnifiers, 155
OSC (offence against the sine condition), 16
OSLO, xiv, xv, 86, 352, 516, 539

P
Palomar Observatory, 476
Parabola, off-axis, 476
Paraboloids:
 confocal, 476
 spherical aberration of, 5
Paraxial ray tracing:
 component by component, 594
 surface by surface, 592–594
Peak-to-valley optical path difference (P-V OPD), 73
Pentac lens, 266
Periscopic anastigmats, 27
Periscopic lens, 297n

Petzval field curvature, 5, 44
 for 6 × 30 binocular eyepiece, 160
 in all anastigmats, 201
 balancing, 64–66
 Cooke triplet anastigmats, 210
 in eyepieces, 153
 of singlets, 55–56
 split triplets, 253
Petzval lens, 423–438
 with field flattener, 426, 429–436
 portrait lens, 423, 424
 projection lens, 423, 425–428, 442
 very high-speed, 429, 437, 438
Phase function, 96
Piazzi-Smyth field flattener, 426
Pick up, 7, 17
Planar lens, 332
Plano surfaces:
 for 6 × 30 binocular eyepiece, 169–173
 conic constants applied to, 36
Plössl eyepiece, 170, 171, 174–175, 178–185
Plots, 8
 aberration, 90–92
Point and shoot cameras, zoom lenses for, 526, 535–539
Point spread function (PSF), 71
Portrait lens, 266
Precision melt sheet, 39
Prescriptions, lens, 87–89
Primary aberrations (see Third-order aberrations)
Primed symbols, 587
Prisms, 153
Programs, lens design, xiii–xiv
 damped least-squares optimization in, 11–13
 features of, 7–8
 multiconfiguration feature of, 12–13
 random search programs, 30
 spectral weighting by, 40–41
Projection TV lenses, 551–554
Protar lens, 302, 303
PSF (point spread function), 71
Pupil, spherical aberration of, 152
Pupil function, 96
P-V OPD (peak-to-valley optical path difference), 73

Q
Quadratic equations, 203, 224
Questions, commonly asked, 4–7

R

Radius of curvature, symbols for, 588
Ramsden eyepieces, 155, 158–159
Random search programs, 30
Rare earth (lanthanum) glasses, 236–239
Ray failure, 36–37
Ray intercept plots, 98–102
Rear meniscus, 21–27
Redesign, estimating effects of modest, 92–96
Resolution, limiting, 82
Retrofocus lens design, 397–407, 521
Reverse aberration fits, 39
Reversed telephoto lenses, 395–413
 basic characteristics of, 395–397
 fish-eye design, 402, 408–413
 origins of, 395
 retrofocus design, 397–407
Ritchey-Chretien systems, 459, 460, 462, 465, 473
Rms OPD, 72
RSS rule, 582–584

S

S2T (aperture coefficient of coma), 16
Sagittal field curvature, 6
Sagittal height, 589
Sampling (by merit function), 15
Scaling, 96–98
Scanner lenses, 562–564
 about, 561, 566, 567
 zoom lenses, 541, 546–548
Schmidt systems, 473, 475–476
Schmidt-Cassegrain design, 473, 482–502
Schott Glass Technologies, Inc., 86, 88
Schwarzschild mirror configuration, 446, 456, 457, 460, 462, 463, 467–468
Secondary spectrum, 125, 129–133
 of doublet, 125, 129
 of triplets, 131–132
Seidel aberrations, 13
 (*See also* Third-order aberrations)
Seventh-order aberrations, 101
Short flints, 133
Single lens reflex (SLR) camera, 395
Singlets, 55–56, 148–149
SLR (single lens reflex) camera, 395
Snell's law, 4–5, 57
Software (*see* Programs, lens design)
Solid cat, 473, 474
Solves (lens design programs), 7–8, 16–17
Sonnar lens, 253–256, 319
Specifications, 2–4

Spectral weighting, 40–41
Speed, increasing the, 36
Spherical aberration(s):
 balancing, 61–63
 of conic aspherics, 5
 and distortion, 152
 of paraboloids, 5
 of spherical surfaces, 5
 third-order, 101
 zonal, 123–124
Spherochromatism:
 balancing, 123
 improvement techniques for, 63–64
 in telescope objective, 118, 123, 139, 140
Split triplets, 247–258
Split-front triplet, 54–55
Stagnation, 29
Steinheil lens, 110, 113
Stephens, R. E., 203
Stop (*see* Aperture stop)
Stop shift equations, 600–601
Strehl ratio (Strehl definition), 71–74
Superapochromat, 358
Sweatt, W. C., 147
Sweatt model, 147–148
Symmetrical eyepieces, 178–185
 as magnifiers, 155
 Plössl, 170, 171, 174–175
Symmetrical principle, 67–68

T

TA (transverse measure), 95
TAC (transverse astigmatism contribution), 109
Tangential coma, 6
Tangential field curvature, 6
Targets (*see* Operands)
Telephoto lens(es), 355–393, 521
 adding more elements to, 385–393
 close-up (macro) lenses, 356–357
 designs for, 358–366
 general characteristics of, 355–356
 reversed, 395–413
 reversing elements in, 375–379
 simple, 358, 359
 200-mm $f/4$ telephoto example, 367–393
 using higher-index glass in, 383, 385
 varying the flints in, 379–384

630   Index

Telephoto ratio, 355
Telescope objectives, 109–150
  achromatic diffractive doublet, 149–150
  achromatic diffractive singlet, 148–149
  apochromatic triplet ($f/7$), 133–145
  cemented doublet telescope objective ($f/7$), 115–122
  coma of, 114
  and Conrady $D - d$ target, 114–115
  and diffraction efficiency, 145–146
  estimating manufacturability of, 146–147
  focal length of, 114
  induced aberrations in, 124
  merit function for, 110, 114–115
  secondary spectrum contribution in, 125, 129–133
  spherochromatism in, 118, 123
  and Sweatt model, 147–148
  thin airspaced doublet design for, 109–113
  three-element objectives, 125–128
  zonal spherical aberration in, 123–124
  [*see also* Eyepiece(s)]
Telescopes, infrared, 507, 511–513
Tessar lens, 259–267, 302
  camera lens design example, 275–285
  doublets in, 56–57
Test plate fits, 37–38
Thermal soak, 578
Thickness, variables for, 34–35
Thickness fits, 39
Thin airspaced doublet design (telescope objectives), 109–113
"Third and Fifth Order Analysis of the Triplet" (R. E. Hopkins), 203
Third-order aberrations (TOAs), 6–7, 13
  aspheric surfaces, contributions from, 601
  Cooke triplet anastigmats, 201–205
  merit functions based on, 15
  surface contributions, 596–598
  thin lens contributions, 598–600
Third-order merit functions, 16
Third-order spherical aberrations, 101
Three-element objectives, 125–128
Tilt/decenter, 577
TIR (*see* Total internal reflection)
TOAs (*see* Third-order aberrations)
Tolerance budgets, 573–586
  additive tolerances, 578–583
  and airspaces, 576–577
  defined tolerances in, 573–574

  establishment of, 583–586
  and merit function, 578
  and surface irregularities, 574–576
  and tilt/decenter, 577
  and V-value tolerances, 577–578
Topogon anastigmats, 27
Topogon lens, 299, 300
Total internal reflection (TIR), 36, 47
Transverse astigmatism contribution (TAC), 109
Transverse measure (TA), 95
Triplets:
  apochromatic, 133–145
  cracked crown, 54
  Dogmar lens, 272
  enlarger lens, 266, 272
  in $f/4$ camera lens design example, 275–295
  Hektor lens, 272, 274
  Heliar lens, 266, 268–272
  Pentac lens, 266
  portrait lens, 266
  secondary spectrum of, 131–132
  split, 247–258
  split-front, 54–55
  Tessar lens, 259–267
  (*See also* Cooke triplet anastigmats)
TV projection lenses, 551–554
Two-component relationships, 595–596

U

Ultraviolet (UV) systems, 503, 514–520
  aberration correction in, 514
  microlithographic lenses, 514–520
Unobscured mirror systems, 476–482
  Herschel mount, 476
  with meniscus shell corrector, 477, 480–482
  off-axis parabola in, 476–477
UV systems (*see* Ultraviolet systems)

V

Variable parameters (variables), 7
  bounds on, 31
  for curvatures, 33
  for defocusing, 34
  freezing of, 20–21
  material characteristics as, 31–33
  for surface asphericity, 35–36
  for thickness, 34–35
Vertex length (Cooke triplet anastigmats), 206–207

Very high-index eyepiece/magnifier, 187, 190–299
Vignetting, 58–59
V-value tolerances, 577–578

W
W. A. Express lens, 305
Wave (wavefront) function, 96
Wavefront aberration, 95, 101–102
Wavefront deformation, aberrations and, 601–603
Weak elements, elimination of, 60
Weighting, spectral, 40–41
Wide-angle lenses (with negative outer elements), 415–422

Z
Zonal aberration(s):
   as merit function operand, 16
   spherical, 123–124
Zoom lenses, 521–549
   design procedures, 542–549
   multiconfiguration with, 12–13
   for point and shoot cameras, 526, 535–539
   scanner lens, 541, 546–548
   telephoto vs. retrofocus arrangements for, 521
   video zoom lens, 539–545
Zoom scanner lens, 541, 546–548

ABOUT THE AUTHOR

WARREN J. SMITH is Chief Scientist at Kaiser Electro-Optics as well as an independent consultant. He is the author of three prior books on lens design, including the first edition of this one, and the classics, *Modern Optical Engineering* and *Practical Optical System Layout*. He lives in Carlsbad, California.